T0393028

FRONTIERS IN CIVIL AND HYDRAULIC ENGINEERING, VOLUME 1

Frontiers in Civil and Hydraulic Engineering focuses on the research of architecture and hydraulic engineering in civil engineering. The proceedings feature the most cutting-edge research directions and achievements related to civil and hydraulic engineering. Subjects in the proceedings including:

- Engineering Structure
- Intelligent Building
- Structural Seismic Resistance
- Monitoring and Testing
- Hydraulic Engineering
- Engineering Facility

The works of this proceedings can promote development of civil and hydraulic engineering, resource sharing, flexibility and high efficiency. Thereby, promote scientific information interchange between scholars from the top universities, research centers and high-tech enterprises working all around the world.

PROCEEDINGS OF THE 8TH INTERNATIONAL CONFERENCE ON ARCHITECTURAL, CIVIL AND HYDRAULIC ENGINEERING (ICACHE 2022), GUANGZHOU, CHINA, 12–14 AUGUST 2022

Frontiers in Civil and Hydraulic Engineering

Volume 1

Edited by

Mohamed A. Ismail
Department of Civil Engineering, Miami College of Henan University, China

Hazem Samih Mohamed
School of Civil Engineering and Geomatics, Southwest Petroleum University, China

CRC Press
Taylor & Francis Group
Boca Raton London New York Leiden

CRC Press is an imprint of the
Taylor & Francis Group, an **informa** business

A BALKEMA BOOK

First published 2023
by CRC Press/Balkema
4 Park Square, Milton Park, Abingdon, Oxon, OX14 4RN
e-mail: enquiries@taylorandfrancis.com
www.routledge.com – www.taylorandfrancis.com

CRC Press/Balkema is an imprint of the Taylor & Francis Group, an informa business

Library of Congress Cataloging-in-Publication Data
A catalog record has been requested for this book

SET
ISBN: 978-1-032-47153-2 (hbk)
ISBN: 978-1-032-47154-9 (pbk)

Volume 1
ISBN: 978-1-032-38247-0 (hbk)
ISBN: 978-1-032-38257-9 (pbk)
ISBN: 978-1-003-34420-9 (ebk)

DOI: 10.1201/9781003344209

Volume 2
ISBN: 978-1-032-47155-6 (hbk)
ISBN: 978-1-032-47161-7 (pbk)
ISBN: 978-1-003-38483-0 (ebk)

DOI: 10.1201/9781003384830

Typeset in Times New Roman
by MPS Limited, Chennai, India

Table of contents

VOLUME 1

Civil engineering structure and geological performance analysis

Numerical simulation and technical optimization of construction engineering

Preface

Hosted by Guangzhou Association of Young Technologists & Scientists and Guangdong-Hongkong-Macao Great Bay Area (Guangdong) Talent Hub, the 2022 8th International Conference on Architectural, Civil and Hydraulic Engineering (ICACHE 2022) was successfully held on August 12–14, 2022 in Guangzhou, China. ICACHE 2022 is an annual conference and has been held consecutively in the past seven years, providing a platform for the presentation of technological advances and research results in related fields.

We had the honor of having invited Prof. Mohamed A. Ismail from Miami College of Henan University, Canada to serve as our Committee Chair. The conference was composed of keynote speeches and oral presentations, attracting 200 leading researchers, engineers and scientists from all over the world. Firstly, keynote speakers are each allocated 30–45 minutes to hold their speeches. Then in the next part, oral presentations, the excellent papers selected are presented by their authors one by one.

During the conference, three distinguished professors invited address their keynote speeches. Among them, Assoc. Prof. Hazem Samih Mohamed from Southwest Petroleum University, Egypt performed a thought-provoking speech on the Rehabilitation of Corroded Tubular Joints with CFRP Laminates. One of the significant challenges confronting the researchers is extending and enhancing the bearing capacity and service life of tubular structures. Therefore, this presentation focused on the rehabilitation of corroded offshore tubular joints by utilizing Carbon Fiber Reinforcement Polymers CFRP laminates as strengthening materials, corrosion inhibition techniques, and a fatigue life extension approach. Moreover, Assoc. Prof. Mianheng Lai from Guangzhou University, China addressed a speech on the title High-Performance Steel Slag Concrete Material and Structure. To solve the problem of concrete's unsoundness, steel slag concrete-filled-steel-tube (SSCFST) column was proposed, in which the concrete's expansion would activate larger confining stress, and thereby enhance the overall behavior of the column. And a theoretical load-strain model was developed to predict the mechanical behavior of traditional CFST and SSCFST columns. Their brilliant speeches had triggered heated discussion in the conference. And every participant praised this conference for disseminating useful and insightful knowledge.

The proceedings of ICACHE 2022 are a compilation of the accepted papers and represent an interesting outcome of the conference. These papers feature but are not limited to the following topics: Engineering Structure, Structural Seismic Resistance, Monitoring and Testing, Intelligent Building, Smart City, etc. All the papers have been checked through rigorous review and processes to meet the requirements of publication.

We would like to express our sincere gratitude to all the keynote speakers, peer reviewers, and all the participants who supported and contributed to ICACHE 2022. Particularly, our special thanks go to the CRC Press / Balkema – Taylor & Francis Group, for all the efforts of its colleague in publishing this paper volume. We firmly believe that ICACHE 2022 had turned out to be a forum for excellent discussions that enable new ideas to come about, promoting collaborative research.

Committee of ICACHE 2022

Committee members

Committee Chair
Prof. Mohamed A. Ismail, *Department of Civil Engineering, Miami College of Henan University, Canada*

Program Committee
Prof. Tetsuya Hiraishi, *Kyoto University, Japan*
A. Prof. Hazem Samih Mohame, *Southwest Petroleum University, Egypt*
A. Prof. Mohammad Arif Kamal, *Aligarh Muslim University, India*
A. Prof. Aeslina Abdul Kadir, *Universiti Tun Hussein Onn Malaysia, Malaysia*
Ph. D. Dayang Zulaika Binti Abang Hasbollah, *Universiti Teknologi Malaysia, Malaysia*
Asst. Prof. Hamza Soualhi, *University of Laghouat, Algeria*
Senior Lecturer Mohammadreza Vafaei, *Universiti Teknologi Malaysia, Malaysia*
Senior Lecturer Au Yong Cheong Peng, *University of Malaya, Malaysia*
Senior Lecturer Nor Hasanah Binti Abdul Shukor Lim, *Universiti Teknologi Malaysia UTM, Malaysia*
Senior Lecturer Libriati Zardasti, *Universiti Teknologi Malaysia, Malaysia*

Technical Committee
Prof. Dr. Mohammad Bin Ismail, *Universiti Teknologi Malaysia, Malaysia*
Prof. Ir. Dr. Hj. Ramli Nazir, *Universiti Teknologi Malaysia, Malaysia*
Prof. Dr. Muhd Zaimi Bin Abd Majid, *Universiti Teknologi Malaysia, Malaysia*
Prof. Lu, Jane Wei-Zhen, *City University of Hong Kong, Hong Kong, China*
Prof. Mingqiao Zhu, *Hunan University of Science and Technology, China*
Prof. QingXin Ren, *Shenyang Jianzhu University, China*
Prof. Bing Li, *Shenyang Jianzhu University, China*
Prof. Jianhui Yang, *Henan Polytechnic University, China*
Prof. Changfeng Yuan, *Qingdao University of Technology, School of Civil Engineering, China*
A. Prof. Bon-Gang Hwang, *National University of Singapore, Singapore*
A. Prof. Zhu Yuan, *School of Architecture, southeast University, China*
A. Prof. Chaofeng Zeng, *Hunan University of Science and Technology, China*
A. Prof. Weijun Cen, *Hohai University, China*
Asst. Professor Dr. Shah Kwok Wei, *National University of Singapore, Singapore*
Dr. Shaoyun Pu, *Southeast University, China*
Dr. Zhongzheng Lyu, *Dalian University of Technology, China*
Dr. Mohd Rosli Mohd Hasan, *Universiti Sains Malaysia, Malaysia*
Dr. Kim Hung Mo, *University of Malaya, Malaysia*
Dr. Yuen Choon Wah, *University of Malaya, Malaysia*
Dr. Huzaifa Bin Hashim, *University of Malaya, Malaysia*
Dr. Suhana Koting, *University of Malaya, Malaysia*
Dr. Sharifah Akmam Syed Zakaria, *Universiti Sains Malaysia, Malaysia*
Dr. Xian Zhang, *School of Architecture of Southeast University, China*
Dr. Zhiming Chao, *University of Warwick, UK*
Dr. Jun Xie, *School of Architecture and Art Central South University, China*
Dr. Derek Ma, *University of Warwick, England*
Dr. Ning Xu, *Department of Science and Technology Development Shanghai Ershiye Construction CO., LTD., China*
Dr. Hongchao Shi, *Chengdu Technological University, China*
Dr. Li He, *Wuhan University of Science and Technology, China*

*Civil engineering structure and
geological performance analysis*

*Frontiers in Civil and Hydraulic Engineering – Mohamed A. Ismail and
Hazem Samih Mohamed (Eds)*
© 2023 The Authors, ISBN 978-1-032-38247-0

Water fracture interaction behavior of cemented sand and gravel (CSG) material: Experimental and numerical study

Jiaojiao Chen*
Department of Roadway Engineering, Nanjing Vocational Institute of Transport Technology, Nanjing, China

Xin Cai*
College of Water Conservancy and Hydropower Engineering, Hohai University, Nanjing, China

ABSTRACT: As a new material for the dam, cemented sand and gravel (CSG) material has become quite popular in recent times, and it is necessary to study the water fracture interaction behavior which is hard to find in the literature. This paper presents an experimental and computational study. A wedge splitting test with water pressure was designed and carried out. The test results show that as water pressure increases, the fracture process zone becomes shorter, and the failure mode tends to be more brittle. The experimental tests were then simulated by the Multiphysics Lattice Discrete Particle Model (M-LDPM), which was properly calibrated to simulate CSG materials. The simulations demonstrated that M-LDPM is well capable of capturing the CSG water fracture interaction behavior and can be a viable tool for engineers to use in practical applications. Furthermore, the size effect on water fracture properties of CSG was discussed. Simple functional relationships between peak load and water pressure for different size specimens were presented. The simulation data fit well with Bažant's size effect law, indicating that CSG material has the characteristics of quasi-brittle material.

1 INTRODUCTION

In the 1990s, Raphael (1992) presented a new dam material, cement-enriched rockfill, whose properties are between concrete and rockfill material at the "Concrete rapid construction conference". Then this emerging material, cemented sand and gravel (CSG) receives widespread attention (Ohtomo et al. 2005). CSG material is made up of sand and gravel, which can be got from the construction site without screening and washing, and mixed with a small number of cementitious materials (cement, fly ash, etc.) and water. Compared with normal concrete, it is characterized by lower cement content, higher water-to-cement ratio, and bigger aggregate size. As a result, the hydration heat can be reduced and the construction speed can be accelerated. With further research on its properties and construction technology, CSG material gradually becomes attractive in the dam construction field after roller compacted concrete.

Usually, the CSG dam operates under water pressure. Original cracks may appear in the structure for many reasons, such as temperature distribution, differential settlement, and chemical action. Water pressure may accelerate the crack propagation and finally lead to structural deterioration. Therefore, studies on the water fracture interaction of CSG material are significant. The most common methods are experimental studies and numerical studies. Experimental studies include laboratory experiments and in-situ experiments. Through laboratory experiments, Jaworski et al. (1981) and Yi et al. (2011) tried to investigate the influencing factors of hydraulic fracture. These include material properties, test velocity, the value of water pressure, and the width of the crack. Bruhwiler and Saouma (1995b) described the water pressure distribution in the fracture

---

*Corresponding Authors: chen1992@njitt.edu.cn and xcai@hhu.edu.cn

DOI 10.1201/9781003344209-1

process zone. This distribution is influenced by the crack opening rate (Slowik & Saouma 2000). The experiment by Mao et al. (2017) indicated that the main failure mode of hydraulic fracture is a tensile one. In order to overcome the shortcomings of the small-size specimen in the laboratory that cannot exactly reflect the real situation, many in-situ experiments were carried out. Compared with experimental studies, numerical studies can be more economical with high efficiency. Many modeling methods, including the distinct element method (DEM), extended finite element method (XFEM), and phase field method, have been developed in past decades.

Despite the large number of available studies involving water fracture interaction in the literature, this research has not been applied to the study of CSG material. In this study, wedge splitting tests with different water pressures are conducted to explore the water fracture interaction behavior of CSG material. Then a new mesoscale method, Multiphysics Lattice Discrete Particle Model (M-LDPM), is introduced to simulate the experiments. The water fracture properties of CSG material and size effect are studied.

2 WEDGE SPLITTING TESTS

The materials under study are cement (P.C.32.5), water, and aggregate. Cement contents is 100 kg/m³. The water-to-cement ratio is 1.0, and the aggregate content is 2130 kg/m³. The maximum particle size is 40 mm. The aggregate consisted of sand (20%) and coarse aggregate (80%); 3% of the coarse aggregate had particle size below 5 mm, 20% from 5 mm to 10 mm, 35% from 10 mm to 20 mm, 42% from 20 mm to 40 mm.

The experiment is inspired by the work of Bruhwiler and Saouma (1995a, 1995b) in which the classic wedge splitting tests were supplemented by a hydraulic loading device and a series of pressure transducers placed along the fracture ligament to study the interaction between water pressure and the fracturing behavior of materials. Figure 1(a) shows the experiment apparatus. The dimension of CSG specimens is 500 mm × 500 mm × 200 mm with an initial notch of 200 mm in length. Three pressure transducers are placed 50 mm, 100 mm, and 150 mm away from the notch tip, marked as transducer 1 (T1), transducer 2 (T2), and transducer 3 (T3) (Figure 1(b)).

The cycle-loading method (Figure 1(c)), which is firstly used by Bruhwiler and Saouma (1995a), is adopted. A loading cycle includes four stages: water pressure is set to zero at first. Meanwhile, crack mouth opening displacement (CMOD) increases linearly to a value. Then, the CMOD remains constant while water pressure is gradually applied to a specific value σ_w. After the water pressure reaches σ_w, it is locked, and the value of CMOD increases. After a while, water pressure is reduced to zero while CMOD is maintained. A loading cycle is finished and then repeated until the specimen is broken. In order to keep a constant CMOD, the applied force increases as the water pressure decrease and vice versa. The horizontal force applied by the wedge-splitting device and CMOD is recorded during the whole simulation. Water pressure σ_w is set to be 0.02 MPa, 0.05 MPa, 0.1 MPa, 0.15 Mpa, and 0.2 MPa respectively. Reference experiment without water pressure ($\sigma_w = 0$ MPa) is also carried out. For each water pressure, four parallel experiments are conducted.

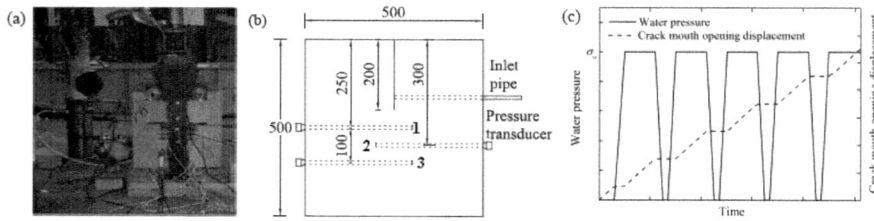

Figure 1. Wedge splitting tests (a) Test setup; (b) Diagram of experiment apparatus in mm; (c) Loading method.

The broken specimen is shown in Figure 2(a). Figure 2(b) shows the force versus CMOD curves for specimens under different water pressure. It reveals that the maximum value of applied force

4

(the peak load) and corresponding CMOD decrease as water pressure increases. What's more, the drop slope of the curves after the peak load becomes sharper. This can be well explained by the theory of fracture process zone (FPZ). As the water pressure increases, the size of FPZ becomes shorter, and it is easier for water to be injected into the crack tip. As a result, a steeper decline is exhibited after the peak load. But the initial slopes of the curves for different water pressure are almost the same. This means that the fracture property is independent of applied water pressure at the beginning of the simulation. That is, no FPZ exists in the specimen before the crack is generated.

Figure 2(c) shows the curves of applied water pressure (the solid line) and that recorded by the three pressure transducers ($\sigma_w = 0.02$ MPa). These curves can be divided into three stages: In the first stage, there is a proportion relation between the four curves. And when the applied water pressure is kept constant, the four curves are parallel. The value decreases progressively from transducers 1 to transducers 3 in the meanwhile. The variation trend of each transducer is consistent with the applied water pressure. This means that the permeability coefficient is the same in the whole specimen and no crack occurs in this stage. At some point, the three curves which represent the three pressure transducers increase nonlinearly. The rise for transducers 1 is the maximum. This indicates that cracks begin to widen and the permeability coefficient is changing in the crack zone. The closer to the crack, the bigger the permeability coefficient is. But the crack tip does not reach transducer 1, so its recording value is still smaller than the applied pressure. Then the crack expands rapidly to the third stage. The following three curves rise vertically, and suddenly all curves overlap together. This suggests that the crack pass through all three transducers and the specimen is broken completely.

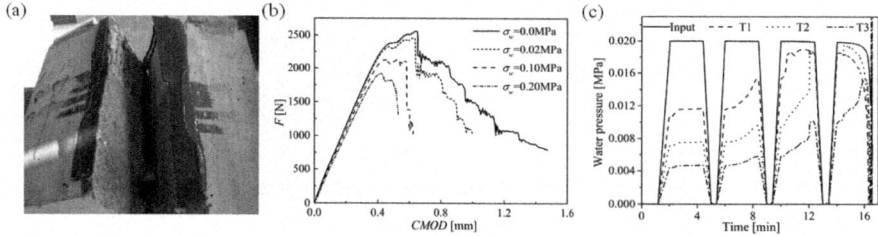

Figure 2. Experimental result (a) Broken specimen; (b) Force versus CMOD curves for different water pressures; (c) The recorded water pressure ($\sigma_w = 0.02$ MPa).

3 MULTIPHYSICS LATTICE DISCRETE PARTICLE MODEL

Multiphysics Lattice Discrete Particle Model is based on Lattice Discrete Particle Model (LDPM), which was proposed and developed by Cusatis et al. (2011a, 2011b), and has shown unprecedented capabilities to simulate the mechanical response of concrete and other quasi-brittle materials under various loading conditions (Ashari et al. 2017, Smith & Cusatis 2017).

3.1 *Construction of mesostructure*

M-LDPM is formulated in a discrete poromechanics setting by adopting two coupled dual lattices simulating mechanical and transport behaviors, respectively. It constructs the geometrical representation of the mesostructure through the following steps: (1) Particle generation. The coarse aggregate pieces, whose shapes are assumed to be spherical, are introduced into a certain volume by a try-and-reject random procedure. Aggregate diameters are determined by sampling an aggregate size distribution function consistent with a Fuller sieve curve: $F(d) = (d/d_a)^{n_F}$, where d is the particle diameter, d_a is the maximum aggregate size, and n_F is the Fuller coefficient. The minimum aggregate size used in the simulation is d_0. Then, zero-radius aggregate pieces (nodes) are placed over the external surfaces to facilitate the application of boundary conditions. Details of the procedure can be found in Cusatis et al. (2011a). (2) Definition of the granular lattice system.

A constrained Delaunay tetrahedralization connects the given particle centers and fills the space without gaps. For 2D illustration, the tetrahedralization corresponds to a triangular mesh. For each particle, combining the relevant tessellation portions from all Delaunay tetrahedral connected to the same node, one obtains a corresponding polyhedral cell (Figure 3(a)) which encloses the spherical aggregate. A 3D domain tessellation, anchored to the Delaunay tetrahedralization but not coinciding with the classical Voronoi tessellation, subdivides the domain into a system of polyhedral cells. Figure 3(b) depicts two adjacent polyhedral cells interacting through shared triangular facets. The triangular facets are assumed to be the potential material failure locations. Each facet has one vertex inside the tetrahedron (tet-point), one on each edge of the tetrahedron (edge-point), and one on the face of the tetrahedron (face-point). The tessellation procedure is described in detail in Cusatis et al.(2011a). And it is formulated in a way to minimize the intersection between each facet and the grains. This is done so that, as often verified in practice, cracks are simulated to occur at the grain interface and in the embedding matrix as opposed to cutting through the grains. (3) Definition of the dual lattice system. The linkage between two contiguous tet-points forms a flow lattice element (FLE) (Figure 3(c)). When all the FLEs connect to each other, a lattice network for water flow is generated (Figure 3(d)). Then, this model can simulate the fluid-solid coupling.

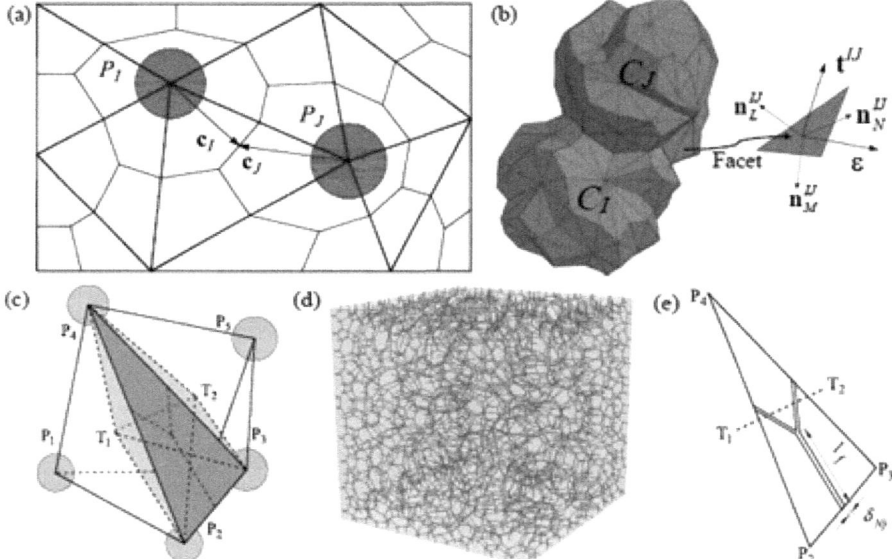

Figure 3. Mesostructure of M-LDPM (a) Mesostructure tessellation; (b) LDPM polyhedral cells for two adjacent aggregate particles; (c) A FLE and its associated control volumes; (d) FLE network (e) The cracked triangle face and the illustration of normal crack opening.

3.2 *Constitutive equations for mechanical behavior*

As the rigid body kinematics to each pair of adjacent cells I and J are applied, the relative displacement vector at the facet interface allows for defining appropriate measures of strain (Cusatis et al. 2017; Lale et al. 2017)

$$\varepsilon_\beta = l^{-1} \left(\boldsymbol{u}^J + \boldsymbol{\theta}^J \times \boldsymbol{c}^J - \boldsymbol{u}^I - \boldsymbol{\theta}^I \times \boldsymbol{c}^I \right) \cdot \boldsymbol{n}_\beta^{IJ} \qquad (1)$$

where $\boldsymbol{u}^I$, $\boldsymbol{u}^J$ and $\boldsymbol{\theta}^I$, $\boldsymbol{\theta}^J$ are the displacements of cells I and J respectively. The indices $\beta = N, M, L$ refer to a local system of reference defined by the local basis $\boldsymbol{n}_\beta^{IJ}$ where $\boldsymbol{n}_N^{IJ}$ is aligned with the line, of length l, connecting the centers of the two cells.

In the elastic regime, the facet stress traction components are proportional to the corresponding strains: $t_\beta = E_\beta \varepsilon_\beta$, where $E_N = E_0$ is the LDPM normal modulus and $E_M = E_L = \alpha E_0$ is the LDPM shear modulus. The LDPM elastic parameters, E_0 and α, are related to the macroscopic Young's modulus, E, and Poisson's ratio, ν

In the inelastic regime, LDPM simulates three different failure mechanisms: cohesive fracture, pore collapse, and frictional shearing.

For $\varepsilon_N > 0$, LDPM simulates cohesive fracture in tension and tension/shear. The effective stress $t = \sqrt[2]{t_N^2 + (t_M^2 + t_L^2)/\alpha}$. The effective stress-strain boundary $\sigma_{bt}(\varepsilon, \omega) = \sigma_0(\omega) \exp[-H_0(\omega), \frac{(\varepsilon_{max} - \varepsilon_0)}{\sigma_0(\omega)}]$, $\tan \omega = \frac{\varepsilon_N}{\sqrt{\alpha}\varepsilon_T}$, $\varepsilon_T = \sqrt{\varepsilon_M^2 + \varepsilon_L^2}$, $H_0(\omega) = 2E_N(\frac{2\omega}{\pi})^{n_t}/(\frac{l_t}{l} - 1)$ is the softening modulus governing the postpeak behavior, l_t is the tensile characteristic length, l is the tetrahedron edge length of the related facet, and n_t is the softening exponent. The effective strength σ, is defined as $\sigma_0(\omega) = \sigma_t r_{st}^2(-\sin(\omega) + \sqrt{\sin^2(\omega) + 4\alpha\cos^2(\omega)/r_{st}^2})/(2\alpha\cos^2(\omega))$, where $r_{st} = \sigma_s/\sigma_t$, σ_s is the shear strength, and σ_t is the tensile strength.

For $\varepsilon_N < 0$, LDPM simulates pore collapse and the subsequent material densification. The strain-hardening behavior in compression is simulated with the following strain-dependent boundary: $t_N = E_N \dot\varepsilon_N$, $-\sigma_{bc}(\varepsilon_D, \varepsilon_V) \leq t_N \leq 0$, $\sigma_{bc} = \sigma_{c0} + \langle -\varepsilon_V - \varepsilon_{c0}H_c(r_{DV})\rangle\varepsilon_{c0} = \sigma_{c0}/E_0$, $\varepsilon_V = (V - V_0)/V_0$ is the volumetric strain computed at the LDPM tetrahedral level, $\varepsilon_D = \varepsilon_N - \varepsilon_V H_c(r_{DV})$ is the initial hardening modulus.

In order to establish the connection between the solid and liquid phases, the principle of effective stress is used as follows.

$$t = t_0 - t_w \tag{2}$$

Where t is the total stress, t_0 is the facet stress vector applied to the solid phase, $t_0 = t_N n + t_M m + t_L l$, $t_w = \sigma_w \mathbf{n}$, and σ_w is the water pressure.

3.3 Constitutive equations for water fracture interaction

In the study, it is assumed that the specimen is completely saturated and the temperature is constant. With the two adjoining tetrahedrons being taken into account, the connection between two tet-points T_1 and T_2 is an FLE (Figure 3(c)). Its length is L. The FLE passes through a triangular face (Figure 3(f)) whose area is A. $T_i(i=1,2)$. And the triangular face makes up the control volume V_1 and V_2 respectively. L_1 and L_2 are the lengths of FLE associated with V_1 and V_2.

The mass conservation equation in the control volume V_i $(i = 1, 2)$ is:

$$\dot{M} = Q \tag{3}$$

Where, $\dot{M}$ is the derivative of water mass in V_i; and Q is the flow across the triangular face. $\dot{M}$ can be divided into two parts that in the uncracked domain $\dot{M}_u$ and that in cracked domain $\dot{M}_c$, i.e. $\dot{M} = \dot{M}_u + \dot{M}_c$. In the same way, $Q = Q_u + Q_c$. Q_u is the flow in the uncracked region, and Q_c is the flow in the cracked region.

For the uncracked region, the water mass $\dot{M}_u$ is as follows.

$$\dot{M}_u^i = \rho_{w0}(b\dot{\varepsilon}_i + \sigma_{wi}/M_b)V_i, (i = 1, 2) \tag{4}$$

Where, ρ_{w0} is water mass density, $\rho_{w0} = 1000 \text{ kg/m}^3$; ε_i is the volumetric strain; σ_{wi} is the water pressure in V_i; M_b is the Biot modulus.

For the uncracked region, according to Darcy's law, Q_u can be calculated as follows.

$$Q_u = \bar\rho_w \frac{\kappa_0}{\mu_w} A_n \frac{\sigma_{w1} - \sigma_{w2}}{L} \tag{5}$$

where κ_0 is the permeability coefficient of the material; μ_w is the water viscosity, $\mu_w = 8.9 \times 10^{-4}$ Pa; $\bar\rho_w$ is the average density of the fluid in V.

In the cracked region, $\dot{M}_c^i$ is as follows.

$$\dot{M}_c^i = \dot{\rho}_{wi} V_{ci} + \rho_{wi} \dot{V}_{ci} = \rho_{w0} \frac{V_{ci}}{K_w} \dot{\sigma}_{wi} + \rho_{wi} \dot{V}_{ci} \qquad (6)$$

Where, ρ_{wi} is the real water density, $\rho_{wi} = \rho_{w0}(1 + \frac{\sigma_{wi} - \sigma_{w0}}{K_w})$, and K_w is the water bulk modulus.

On the assumption that the water flow follows Poiseuille's formula in the crack, the crack width δ_β in each facet is $\delta_\beta = l\left(\varepsilon_\beta - t_\beta/E_\beta\right), (\beta = N, M, L)$ and Q_c can be calculated according to the following formula.

$$Q_c = \bar{\rho}_w \frac{\kappa_c}{\mu_w} A_n \frac{\sigma_{w1} - \sigma_{w2}}{L} \qquad (7)$$

Where, κ_c is the permeability of the cracked region, which is related to the crack length and width.

Finally, the discrete problem of flow is solved by the Crank-Nicolson method (Di Luzio & Cusatis 2009). The mechanical and water fracture modeling are coupled through a two-way coupling framework, i.e. the mechanical analytic process reckons the fluid pressure, whereas the crack opening and volumetric strain which are obtained from the mechanical calculation are used in the hydraulic fracture simulation.

3.4 Parameter identification

The model is first calibrated against a series of macroscopic experiments to identify the mesoscale LDPM parameters relevant to the mechanical behaviors of materials. The experiments include three-point bending, unconfined compression, and triaxial tests with different confining pressures. Parameters can be calibrated according to the following steps:

(1) The mix-design parameters, including cement content c, water to cement ratio w/c, aggregate to cement ratio a/c, and maximum aggregate size d_a, were set equal to the ones used in the experiments. The Fuller coefficient n_F was identified by fitting the fuller curve to the actual aggregate gradation. The minimum aggregate size was set to be one-quarter of d_a. So $c = 100$ kg/m^3, $w/c = 1.0$, $a/c = 21.3$, $d_a = 40$mm, $n_F = 0.65$, and $d_0 = 10$mm.
(2) The normal modulus E_0 and α were identified from the macroscopic Young's modulus, E, and Poisson's ratio, ν, with the formulas $E_0 = E/(1 - 2\nu)$, $\alpha = (1 - 4\nu)/(1 + \nu)$. That gives the value $E_0 = 1500$MPa and $\alpha = 0.168$.
(3) The tensile strength σ_t and the tensile characteristic length l_t were calibrated by simulating the Three-Point-Bending (TPB) tests. The tests were performed on 300 mm × 100 mm × 100 mm beams with a loading span of 240 mm. During the entire test run, the applied load and the load-line displacement were collected for comparison with the simulation data. The simulated load-displacement curves of best fit compared with the corresponding experimental data are presented in Figure 4(a). $\sigma_t = 0.70$MPa and $l_t = 90$mm can be calibrated.
(4) The shear strength ratio σ_s/σ_t is identified from the unconfined compression tests. The tests were performed using 150 mm × 150 mm × 150 mm cubes. The simulations mimicked exactly the experimental setup. Figure 4(b) reports the best fitting in terms of P/A vs. $\Delta L/L$ where the stress is calculated by the applied force P divided by the specimen cross-sectional area A and the strain is calculated by the length change of the specimen ΔL divided by the original length L. The value of $\sigma_s/\sigma_t = 2.7$ can be obtained.
(5) Finally, the compressive yielding strength σ_{c0} and the shear soft modulus ratio r_s were calibrated by fitting the triaxial tests with different confining pressures. The triaxial tests were performed on the cylinder with 300 mm diameter and 700 mm length. Four different confining pressure values, $\sigma_c = 300$ kPa, 600 kPa, 900 kPa, and 1200 kPa, were considered. The simulation results of best fit and the experimental data are compared in Figure 4(c). The numerical results are in excellent agreement with the experimental results. So $\sigma_{c0} = 20$MPa and $r_s = 0.01$.

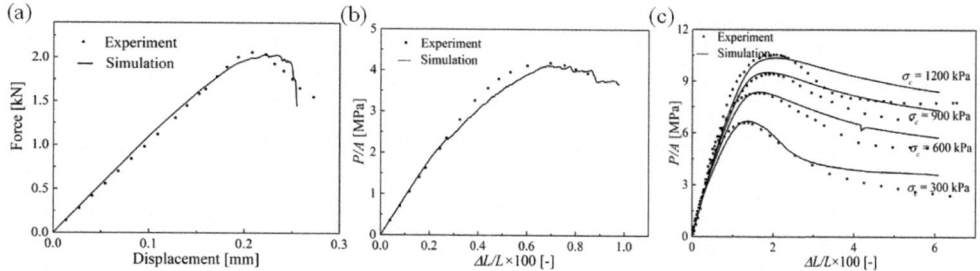

Figure 4. Results of experiments and simulations (a) Three-point bending tests; (b) Compression tests (c) Triaxial tests.

4 NUMERICAL MODELING

In order to analyze properly the results of the wedge splitting test, this section presents the numerical simulations of the tests by means of the so-called M-LDPM, which is briefly reviewed in the last section. Furthermore, specimens in different sizes are also simulated to study the size effect on water fracture properties of CSG material.

4.1 Experimental simulation

In simulations, the water pressure of all the boundaries except the notch is set to be zero. Permeability parameters are defined according to experiments, that is, $\kappa_0 = 7.2 \times 10^{-18}$ m^2 and Biot modulus $M_b = 5.0$ Gpa, providing that the maximum value of κ_c is 10^6 times that of κ_0.

Figure 5(a) and Figure 5(b) show the crack opening and pressure distribution for different CMOD ($\sigma_w = 0.02$ MPa). As the CMOD increases, crack width and length enlarge. Meanwhile, water was injected into the crack, which induces further crack propagation. The crack length under peak load for different water pressure is shown in Figure 5(c), which decreases as water pressure increases. It also indicates that FPZ becomes shorter for larger water pressure. Figure 5(d) shows the comparison of the peak load F_{max} between experimental and simulated results for different water pressures σ_w. As the results show, the M-LDPM simulations provide a very accurate prediction of the water fracture interaction behavior of the CSG material.

4.2 Size effect on water fracture properties

To study the size effect on water fracture properties of CSG material, specimens in different sizes are simulated. Their dimensions are 250 mm × 250 mm × 200 mm, 1000 mm × 1000 mm × 200 mm, and 2000 mm × 2000 mm × 200 mm. The heights (H) of the specimens are 250 mm, 1000 mm, and 2000 mm respectively. The initial notch is 0.4 of the height of different specimens.

Figure 6(a) shows the peak load for different size specimens. For a specific size, peak load decreases as water pressure increases. For a given value of water pressure, the larger the specimen's size, the bigger the peak load is. The fitting curve falls faster with the increase in specimen size. It indicates that a larger CSG structure is more likely to be affected by the change in water pressure.

The peak load is normalized with the method proposed by Bažant and Planas (1997). The nominal strength is defined as $\sigma_{Nu} = F_{max}/(DX)$, where, σ_{Nu} is the nominal strength in MPa, D is the thickness of the specimen in mm, and H is the height of the specimen in mm. According to Bažant's size effect law, $\sigma_{Nu} = \sigma_0/\sqrt{1 + H/H_0}$ where, σ_0 and H_0 are parameters depending on the size of the specimen, in MPa and mm respectively. Table 1 lists the value of σ_0 and H_0 for different water pressure.

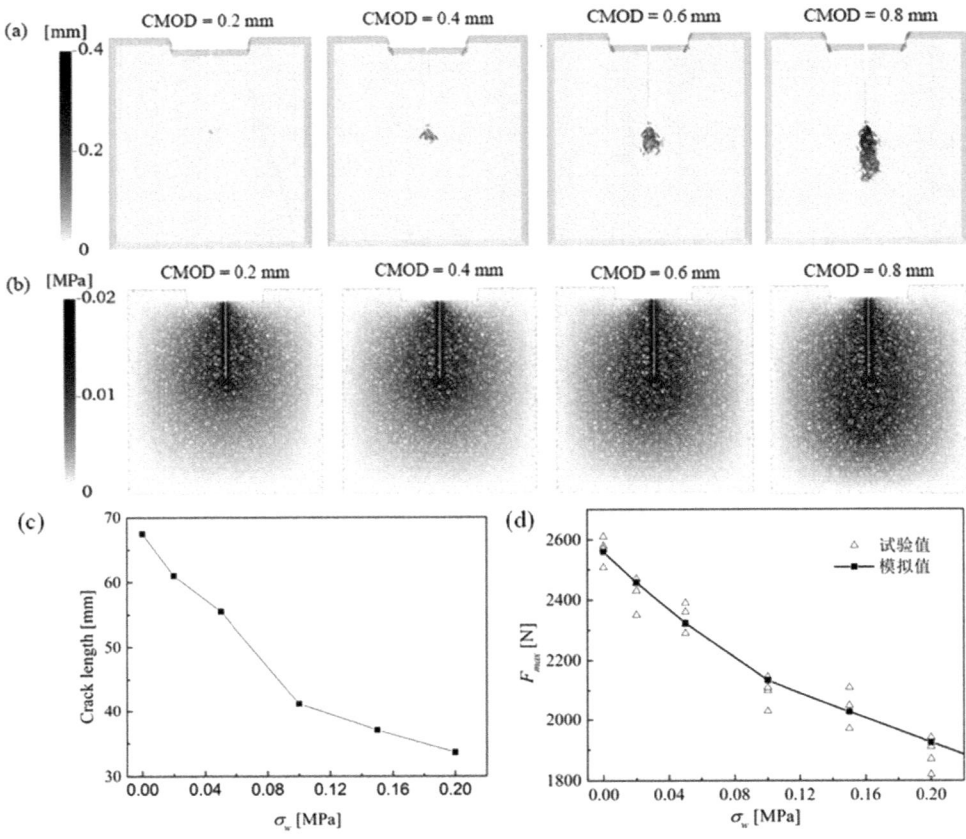

Figure 5. Simulation results (a) Crack opening ($\sigma_w = 0.02$ MPa); (b) Water pressure distribution ($\sigma_w = 0.02$ MPa); (c) Crack length under peak load; (d) Peak load of experimental and simulated results.

Table 1. The value of σ_0 and H_0 for different water pressure.

σ_w [MPa]	0	0.05	0.10	0.15	0.20
σ_0 [MPa]	0.0412	0.0478	0.0548	0.0557	0.0566
H_0 [mm]	291	152	89.6	74.2	66.0

For $\sigma_w = 0.1$ MPa, $\log(\sigma_{Nu})$ versus $\log(H)$ (Figure 6(b)) fit well with Bažant's size effect law. This means that the fracture behavior of CSG material is different from classical linear elastic fracture mechanics (LEFM), and it has the characteristics of quasi-brittle material. The same conclusion can be drawn for the analysis of different water pressure. According to Bažant's size effect law, $\sigma_0 = \left(E \cdot G_f / \left(c_f \cdot g_0' \right) \right)^{0.5}$, $H_0 = c_f g_0'/g_0$, E is the elastic modulus; G_f is the initial fracture energy in N/m; c_f is the effective fracture process zone length in mm; g_0 and g_0' are the initial dimensionless energy release rate and its derivative value. For wedging splitting test, $g_0 = (2 + \alpha_0)^2 \left(0.886 + 4.64\alpha_0 - 13.32\alpha_0^2 + 14.72\alpha_0^3 - 5.6\alpha_0^4 \right) / (1 - \alpha_0)^3$, α_0 is the length of the initial notch divide H. In this study, $\alpha_0 = 0.4$. The initial fracture energy G_f and the effective fracture process zone length c_f can be figured out. Figure 6(c) shows the value of G_f and c_f for different water pressure σ_w. It reveals that as water pressure increases, the initial fracture energy and effective fracture process zone length are all decreased.

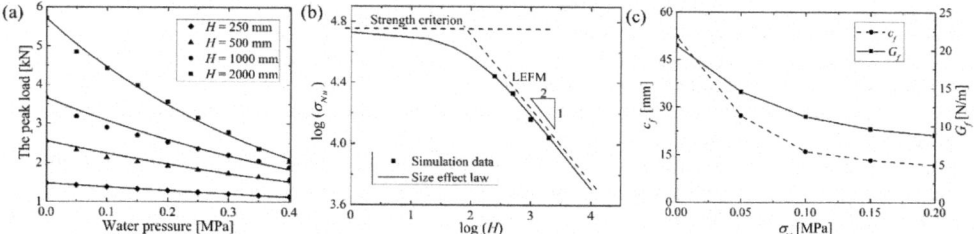

Figure 6. Size effect (a) Peak load vs water pressure curves for different size specimens; (b) comparison of simulation data with Bažant's size effect law; (c) The initial fracture energy and the effective fracture process zone length for different water pressure.

5 CONCLUSIONS

In this study, a wedge splitting test with water pressure and numerical simulations was carried out to investigate the water fracture interaction behavior of CSG material. The adopted numerical modeling approach is the Multiphysics Lattice Discrete Particle Model (M-LDPM) which can be effective in revealing the water pressure distribution and fracture propagation. Based on the obtained results, the following conclusions can be drawn.

(1) With the increase of water pressure, the initial fracture energy and the effective fracture process zone length are all decreased, and the failure mode tends to be more brittle. The initial fracture energy of CSG material is much smaller than that of normal concrete. If this material is to be used in the high dam, an impervious structure is needed.

(2) The comparison of the simulations with both laboratory tests shows that M-LDPM is an accurate computational tool for the simulation of water fracture interaction behavior of CSG materials. It provides a calculation model and reference for the failure behavior analysis of CSG dam and cofferdam under the hydromechanical coupling situation.

(3) CSG material has the characteristics of quasi-brittle material, and its water fracture behavior fits Bažant's size effect law. The size factor should also be taken into consideration if the results getting from experiments will be applied to practical engineering.

ACKNOWLEDGMENT

This work was supported by the Basic Science (Natural Science) Research Projects of Higher Education of Jiangsu Province (21KJB570002), the Research Fund of Nanjing Vocational Institute of Transport Technology (JZ2005), and the Innovation and Entrepreneurship Training Program for College students of Jiangsu Province (202112804010Y). This research was supported in part by the computational resources of the Quest high-performance computing facility at Northwestern University. Jiaojiao Chen expresses gratitude to Gianluca Cusatis for his timely help in the LDPM study during her visit to Northwestern University.

REFERENCES

Ashari, S.E., Buscarnera, G., Cusatis, G., 2017. A lattice discrete particle model for pressure-dependent inelasticity in granular rocks. *International Journal of Rock Mechanics and Mining Sciences* 91, 49–58.

Bažant, Z.P., Planas, J., 1997. *Fracture and size effect in concrete and other quasibrittle materials.* volume 16. CRC press.

Bruhwiler, E., Saouma, V.E., 1995a. Water fracture interaction in concrete–part I: Fracture properties. *ACI Materials Journal* 92, 296–303.

Bruhwiler, E., Saouma, V.E., 1995b. Water fracture interaction in concrete–part II: Hydrostatic pressure in cracks. *ACI Materials Journal* 92, 383–390.

Cusatis, G., Mencarelli, A., Pelessone, D., Baylot, J., 2011a. Lattice discrete particle model (LDPM) for failure behavior of concrete. II: Calibration and validation. *Cement and Concrete Composites* 33, 891–905.

Cusatis, G., Pelessone, D., Mencarelli, A., 2011b. Lattice discrete particle model (LDPM) for failure behavior of concrete. I: Theory. *Cement and Concrete Composites* 33, 881–890.

Cusatis, G., Rezakhani, R., Schauffert, E.A., 2017. Discontinuous cell method (DCM) for the simulation of cohesive fracture and fragmentation of continuous media. *Engineering Fracture Mechanics* 170, 1–22.

Di Luzio, G., Cusatis, G., 2009. Hygro-thermo-chemicalmodeling of high performance concrete. I: Theory. *Cement and Concrete Composites* 31, 301–308.

Jaworski, G.W., Duncan, J., Seed, H.B., 1981. Laboratory study of hydraulic fracturing. *J. Geotech. Eng. Div., Am. Soc. Civ. Eng.*;(United States) 107.

Lale, E., Zhou, X., Cusatis, G., 2017. Isogeometric implementation of high-order microplane model for the simulation of high-order elasticity, softening, and localization. *Journal of Applied Mechanics* 84, 011005.

Mao, R., Feng, Z., Liu, Z., Zhao, Y., 2017. Laboratory hydraulic fracturing test on large-scale pre-cracked granite specimens. *Journal of Natural Gas Science and Engineering* 44, 278–286.

Ohtomo, T., Ito, K., Hirakawa, K., Kusumi, M., 2005. Study on estimation for mixing action of CSG (Cemented Sand and Gravel) by gravity force and on numerical simulation method of that action. *Concrete Research and Technology* 16, 61–69.

Raphael, J.M., 1992. *The optimum gravity dam in Roller Compacted Concrete III*, ASCE. pp. 5–19.

Slowik, V., Saouma, V.E., 2000. Water pressure in propagating concrete cracks. *Journal of Structural Engineering* 126, 235–242.

Smith, J., Cusatis, G., 2017. Numerical analysis of projectile penetration and perforation of plain and fiber reinforced concrete slabs. *International Journal for Numerical and Analytical Methods in Geomechanics* 41, 315–337.

Yi, S.T., Hyun, T.Y., Kim, J.K., 2011. The effects of hydraulic pressure and crack width on water permeability of penetration crack-induced concrete. *Construction and Building Materials* 25, 2576–2583.

Frontiers in Civil and Hydraulic Engineering – Mohamed A. Ismail and
Hazem Samih Mohamed (Eds)
© 2023 The Authors, ISBN 978-1-032-38247-0

Effect of brick aggregate content on mechanical properties of recycled concrete

Ziyang Liao, Shenao Cui, Chenyu Du, Haipeng Zhang & Yufeng Xu
Department of Architectural Engineering, School of Civil and Architectural Engineering,
Shandong University of Technology, Zibo, Shangdong, China

Tian Su*
Department of Architectural Engineering, School of Civil and Architectural Engineering,
Shandong University of Technology, Zibo, Shangdong, China
Department of Architectural Engineering, School of Civil Engineering, Wuhan University,
Wuhan, Hubei, China
China Railway 11 Bureau Group Co., Ltd., Wuhan, Hube, China

ABSTRACT: The destruction mode, the 14-day compressive strength, and microstructure of recycled clay brick aggregate concrete are investigated in this paper. The results show that when the water-cement ratios are 0.58 and 0.68, the 14-day compressive strength shows a trend of increasing and then continues to increase with the rising replacement rate of brick aggregate. When the water-cement ratio is 0.48, the 14-day compressive strength shows a trend of increasing first but then decreases with the rising replacement rate of brick aggregate. The 14-day compressive strength of concrete decreases with the increasing water-cement ratio. And when the specimen is damaged, the natural coarse aggregate has no cracks while the recycled brick aggregate is cracked.

1   INTRODUCTION

With the acceleration of urbanization, a large amount of construction waste is generated in urban construction, increasing year by year (Su 2021). Construction waste accounts for 34% of the total urban waste, and waste concrete and waste clay bricks have become the primary pollution sources (Mohammed 2014). In addition, the development of the construction industry has led to an increasing demand for sand and gravel, which has seriously damaged the ecological environment (Su 2019). As an environmentally friendly building material, recycled concrete can reduce the mining of natural aggregates, protect the ecological environment, and solve the problem of environmental pollution (Su 2022).

Compared with natural aggregates, recycled brick aggregate have the characteristics of large porosity, low apparent density, high water absorption, and low strength, resulting in the mechanical properties of recycled aggregate concrete being lower than that of natural aggregate concrete (Khalaf 2006). Zheng (2018) concluded that the compressive strength of recycled brick aggregate concrete was lower than that of natural aggregate concrete, and the compressive strength decreased with the increased replacement of recycled brick aggregate. Yuan (2018) found that the brick aggregate replacement rate had a more significant effect on the compressive strength and splitting tensile strength of concrete. Uddin (2017) pointed out that the particle size of recycled brick aggregate affected its compressive strength. Chen (2017) found that when the content of waste brick aggregate was 20%-35%, the strength of recycled brick aggregate concrete was the largest. However, the effect

---

*Corresponding Author: sutian@sdut.edu.cn

DOI 10.1201/9781003344209-2

of different recycled brick aggregate replacement rates on the mechanical properties of concrete needs to be further investigated.

The purpose of this paper is: 1) to investigate the destruction mode and the compressive strength of recycled clay brick aggregate concrete with different water-cement ratios and brick aggregate replacement rates; and 2) to investigate the microstructure of recycled clay brick aggregate concrete. This research can provide a theoretical basis and technical support for the wide application of recycled aggregate concrete technology in engineering.

2 EXPERIMENT

2.1 *Raw materials*

The fine aggregate used in the experiment is natural river sand. And the fineness modulus of the fine aggregate is 2.64, which belongs to medium sand. The apparent density is 2514kg/m^3, and the mud content is 1.9%. Coarse aggregates are divided into natural coarse aggregates and recycled brick aggregates, and the basic physical properties are shown in Table 1.

This experiment uses Hailuo brand P.O 42.5 ordinary Portland cement, produced in Zibo Luzhong Cement Co., Ltd. The physical and mechanical properties of cement are shown in Table 2.

Table 1. Basic physical properties of coarse aggregates.

Type	Particle size (mm)	Apparent density (kg/m^{-3})	Bulk density/ (kg/m^{-3})	Crush indicator (%)	Water absorption (%)
NCA	5-20	2800	1580	6.0	2.0
RBA	5-20	2550	1120	23.7	10.7

Table 2. Physical and mechanical properties of cement.

Index	Ignition loss (%)	Specific surface area (m^2/kg)	Initial setting time (min)	Final setting time (min)	28-day strength (MPa) Flexural strength	28-day strength (MPa) Compressive strength
Measured value	2.8	345	211	321	8.5	52.1
Standard value	≤ 5.0	≥ 300	≥ 45	≤ 600	≥ 6.5	≥ 42.5

2.2 *Mix proportions design and specimen design*

The mixed proportions of concrete are shown in Table 3. Due to the high water absorption of recycled brick aggregate, additional water should be added during the mixing process. 150 mm × 150 mm × 150 mm cubic specimens were cast to investigate the compressive strength.

Table 3. Mix proportions (kg·m^{-3}).

	w/c	Water	cement	sand	natural coarse aggregate	recycled brick aggregate
NC-1	0.48	211	385	695	1200	0
RC-1-10	0.48	211	385	695	1080	105
RC-1-20	0.48	211	385	695	960	210
RC-1-30	0.48	211	385	695	840	315

(continued)

Table 3. Continued.

	w/c	Water	cement	sand	natural coarse aggregate	recycled brick aggregate
RC-1-40	0.48	211	385	695	720	420
RC-1-50	0.48	211	385	695	600	525
NC-2	0.58	185	319	720	1176	0
RC-2-30	0.58	185	319	720	823	308.7
RC-2-50	0.58	185	319	720	588	514.5
NC-3	0.68	185	272	738	1158	0
RC-3-30	0.68	185	272	738	811	303.98
RC-3-50	0.68	185	272	738	579	506.63

3 TESTS AND RESULTS

3.1 *Destruction characteristics*

The destruction mode of natural coarse aggregate concrete and recycled brick aggregate concrete is shown in Figure 1. When the natural coarse aggregate concrete was destroyed, the cracks around the specimen widened rapidly until the specimen was destroyed. The concrete around the specimen cracked and partially fell off, which is a typical brittle failure. The failure process of recycled brick aggregate concrete was the same as that of natural coarse aggregate concrete, showing apparent brittle failure. However, with the increase in recycled brick aggregate content, concrete damage was more obvious.

(a) Natural coarse aggregate concrete (b) Recycled brick aggregate concrete

Figure 1. Destruction mode.

3.2 *14-day compressive strength*

Figure 2 shows the 14-day compressive strength of concrete. It can be found that when the water-cement ratios were 0.58 and 0.68, the 14-day compressive strength showed a trend of increasing at first and then continuing to increase with the rising replacement rate of brick aggregate. When the water-cement ratio was 0.48, the 14-day compressive strength showed a trend of increasing at first but then decreased with the rising replacement rate of brick aggregate. When the brick aggregate replacement rate was 10%, the 14-day compressive strength was greater than that of natural coarse aggregate concrete; when the brick aggregate replacement rate was 20%, the 14-day compressive strength was minimum; when the brick aggregate replacement rate was larger than 20%, although the 14-day compressive strength increased, it was still lower than that of natural coarse aggregate concrete.

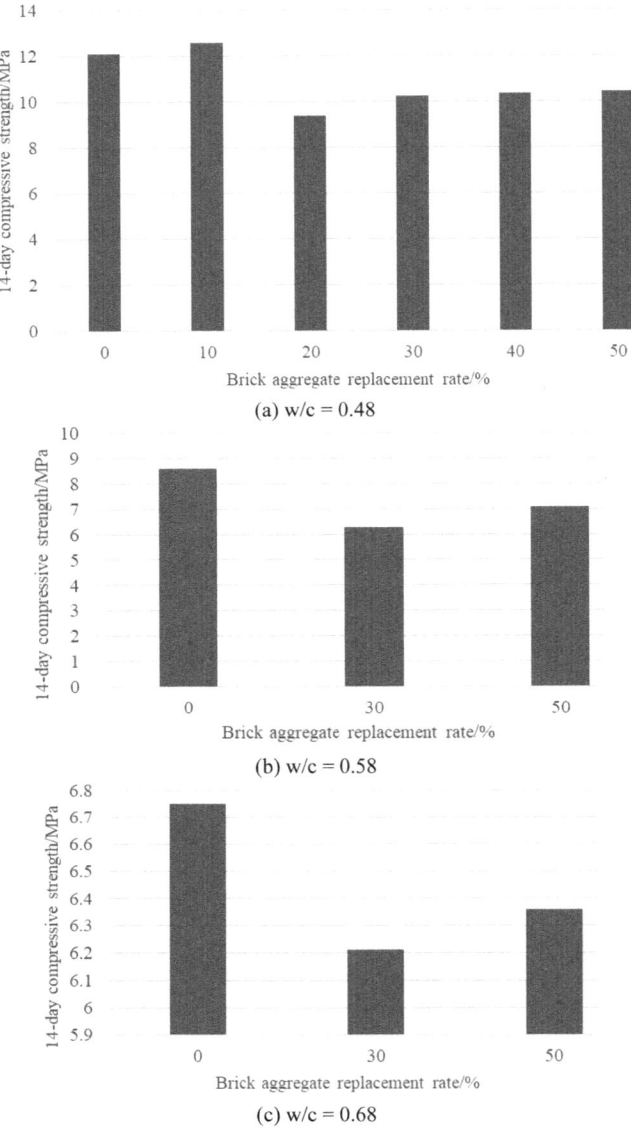

Figure 2.　14-day compressive strength of concrete.

The reasons for the decrease in the recycled brick aggregate concrete compressive strength are as follows. Firstly, the recycled brick aggregate strength is low, and the cement mortar is attached to the aggregate surface, resulting in weak adhesion between the brick aggregate and cement mortar. Secondly, the aggregate crushing process produces secondary damage to recycled brick aggregate, resulting in a large number of microcracks inside the brick aggregate. Thirdly, the high water absorption of brick aggregate leads to the deterioration of the workability of concrete during mixing, which reduces the cohesion between the brick aggregate and cement mortar.

The reasons for the rebound trend of recycled brick aggregate concrete compressive strength are as follows. On the one hand, the recycled brick aggregate has a large water absorption rate and a large water storage capacity when the concrete hydration is gradually released under the action of surface tension and the concrete hydration with improved strength. On the other, the surface porosity and roughness of recycled brick aggregate are relatively large, which makes the recycled aggregate and cement mortar better bond together, and has a particular effect on improving the concrete compressive strength.

In addition, it can be seen from Figure 2 that when the brick aggregate replacement rate was the same, the 14-day compressive strength of concrete decreased with the rising water-cement ratio. The reason is that with the increase of the water-cement ratio, the strength of the cement paste around the recycled aggregate is low, and the bonding force between the aggregate and the cement paste becomes weaker, which leads to a decrease in the concrete compressive strength (Zhang 2020).

3.3 *Microstructure*

The microstructure of natural coarse aggregate concrete and recycled brick aggregate concrete is shown in Figure 3 and Figure 4 respectively. It can be seen that there was no crack in the natural coarse aggregate, but there were cracks at the interface between the natural coarse aggregate and the mortar. There were cracks in the recycled brick aggregate, which was due to the stress of recycled brick aggregate concrete was first concentrated in the interface transition zone between aggregate and mortar, but the aggregate hindered the development of cracks, and the stress was transmitted to the surface of aggregate, resulting in the penetration of the brick aggregate or the failure of the aggregate surface (Liu 2020).

Figure 3. Microstructure of natural coarse aggregate concrete.

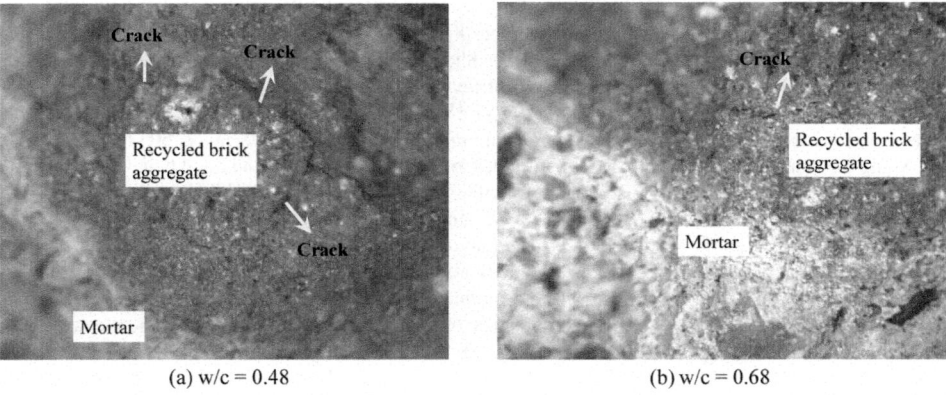

(a) w/c = 0.48   (b) w/c = 0.68

Figure 4. Microstructure of recycled brick aggregate concrete.

17

It can also be seen from Figure 4 that when the water-cement ratio was 0.68, the number and width of cracks on the aggregate surface were smaller than those on the aggregate surface when the water-cement ratio was 0.48. This is because the thickness and porosity of the interface transition zone increase with the rising water-cement ratio. When subjected to load, cracks are more likely to develop along the interface transition zone, thereby reducing the damage to the aggregate. Therefore, with the increase of the water-cement ratio, the failure of the transition interface became a key factor, and the influence of aggregate performance on the mechanical properties of concrete was gradually reduced.

4 CONCLUSIONS

This paper investigated the destruction mode, the compressive strength, and the microstructure of recycled clay brick aggregate concrete, which could provide theoretical basis and technical support for the wide application of recycled aggregate concrete technology in engineering. The main results of this paper are as follows.

(1) When the water-cement ratios were 0.58 and 0.68, the 14-day compressive strength showed a trend of increasing at first and then continued to increase with the rising replacement rate of brick aggregate;
(2) When the water-cement ratio was 0.48, the 14-day compressive strength showed a trend of increasing at first but then decreased with the rising replacement rate of brick aggregate;
(3) The 14-day compressive strength of concrete decreased with the rising water-cement ratio;
(4) When the specimen was damaged, the natural coarse aggregate had no cracks, while the recycled brick aggregate had cracks.

FUNDING INFORMATION

The study was carried out with the support of the National College Student Innovation and Entrepreneurship Training Program (202110433079) and the Doctoral Science and Technology Startup Foundation of Shandong University of Technology (420048).

REFERENCES

Khalaf F. (2006). Using crushed clay brick as coarse aggregate in concrete[J]. *Journal of Materials in Civil Engineering*, 18(4):518–526.
Liu Q., Xiao J., Poon C. & Li L (2020). Investigation of modeled recycled concrete prepared with recycled concrete aggregate and recycled brick aggregate[J]. *Journal of Building Structures*, 41(12): 133–140.
Mohammed T., Hasnat A., Awal M. & Bosunia S. (2014). Recycling of brick aggregate concrete as coarse aggregate[J]. *Journal of Materials in Civil Engineering*, 27 (7): B4014005.1-B4014005.9.
Su T. Wu J., Yang G., Jing X. & Mueller. A (2019). Shear behavior of recycled coarse aggregate concrete beams after freezing and thawing cycles[J]. *ACI Structure Journal*, 116(5): 67–76.
Su T., Wang C., Cao F., Zou Z., Wang C., Wang J. & Yi. H. (2021). An overview of bond behavior of recycled coarse aggregate concrete with steel bar[J]. *Reviews on Advanced Materials Science*, 60 (1):127–144.
Su T., Wang T., Wang C. & Yi H. (2022). The influence of salt-frost cycles on the bond behavior distribution between rebar and recycled coarse aggregate concretes[J]. *Journal of Building Engineering*, 45: 103568.
Uddin M., Mahmood A., Kamal M., Yashin S. & Zihan Z. (2017). Effects of the maximum size of brick aggregate on properties of concrete[J]. *Construction and Building Materials*, 134:713–726.
Yuan C., Li S., Zeng L. & Chen Z. (2018). Mechanical properties of brick and concrete mixed recycled coarse aggregate concrete[J]. *Bulletin of The Chinese Ceramic Society*, 37(2):398–402. (in Chinese)
Zhang L., Li H. & Ren J. (2020). Experimental study on compressive strength of the concrete with 100% recycled clay brick aggregates [J]. *Concrete*, (11): 79–88. (in Chinese)
Zheng C., Lou C., Du G., Li X., Liu Z. & Li L. (2018). Mechanical properties of recycled concrete with demolished waste concrete aggregate and clay brick aggregate[J]. *Results in Physics*, 9:1317–1322.

Frontiers in Civil and Hydraulic Engineering – Mohamed A. Ismail and Hazem Samih Mohamed (Eds)

Experimental study on the mechanical properties of steel fiber reinforced expansive concrete

Weixing Wang
Jiangsu Yanghu Construction Project Administration Co., Ltd., Changzhou, China

Sheng Ding & Guanyang Huang
Shanghai Mechanized Construction Group Co., Ltd, Shanghai, China

Longlong Niu
Nanjing Institute of Technology, Nanjing, China

Mengyu Li*
Chongqing Jiaotong University, Chongqing, China

Li Chen
Nanjing Institute of Technology, Nanjing, China

ABSTRACT: Steel fiber reinforced expansive concrete is a new building material, which adds the steel fibers into the expansion concrete to enhance the bond strength and improve the performance of the concrete. To study the influence of the steel fibers and the expansion agent on the mechanical properties of the concrete, steel fiber reinforced expansive concrete specimens were prepared, with the steel fiber volume contents being 0.0%, 0.5%, 1.0%, 1.5%, and 2.0% respectively, and the expansive agent mass contents being 0.0%, 0.8%, 1.0%, 1.5%, and 2.0% correspondingly. The mechanical properties of the specimens above were tested to reveal the influences of the contents of the steel fibers and the expansion agent on the compressive strength, the splitting tensile strength, the flexural strength, the impact resistance, and the ratio of flexural strength to compression strength of the concrete. It was shown by the results that all these properties can be improved by the steel fibers yet the expansion agent hardly has any effect on the performance of the concrete. When the steel fiber volume content is within 1%~1.5%, and the expansion agent mass content is within 0.8%~1%, the steel fiber reinforced expansive concrete has the best performance.

1 INTRODUCTION

Concrete is the most widely used building material in the world today. However, it is disadvantaged in many ways, such as low tensile strength, poor crack resistance, and poor impact resistance. Therefore, researches on improving the above mechanical properties of concrete are always a major concern. Steel fiber can hinder the formation and expansion of cracks in the concrete matrix and greatly improve the mechanical properties of concrete, so it is widely studied and applied (Abbass 2018; Amin 2018; Chanh 2017; Nataraja 1997). Expansive concrete has weak shrinkage and strong crack resistance (Hu 1999; Nguyen 2019; Suhara 2019; Wyrzykowski 2018). The steel fiber reinforced expansive concrete, which is a new building material with excellent performance,

---

*Corresponding Author: 1271911805@qq.com

DOI 10.1201/9781003344209-3

is developed by combining the steel fiber reinforced concrete with expansive concrete while using expansive concrete as the matrix and steel fiber as the dispersed constraint component (Cao 2017; Tian 2019; Wenling 2019; Yu 2018). It can make full use of the advantages of the two materials. The existing test results (Cui 2010; Li 2006) indicate that, when the content of steel fiber is fixed, the tensile strength of concrete increases at first and then decreases with increasing expansion agent; when the content of the expansion agent is fixed, the changing trend of the concrete tensile strength is consistent with the steel fiber content. In the meanwhile, the content of the steel fibers and that of the expansion agent should be well-proportioned. According to existing research (Huanan 2009; Zixiang 2002), the synergistic reinforcement effect of the steel fibers and the expansion agent is caused by the combined action of the expansion restriction and reinforcement effect of the steel fibers and the crack resistance and filling effect of the expansion agent.

It is of great significance to study the mechanical properties of steel fiber-reinforced expansive concrete. The influence of steel fiber and expansion agents on the mechanical properties of concrete was studied in this paper to promote the application of steel fiber-reinforced expansive concrete.

2 TEST MATERIALS AND METHODS

2.1 Raw materials

The adopted cement for the tests was the P·O 42.5-grade cement produced by Nanjing Zhonglian company. The fine aggregate was the river sand with the fineness modulus of 2.6. The coarse aggregate was the continuously graded gravel with a particle size of 5~15mm. The water-reducing agent was a polycarboxylate superplasticizer with a solid content of 18%. The water for mixing and curing was tap water. The steel fibers were 0.6mm in dimension and 33mm in length and were with an end hook. The expansion agent is UEA with good performance.

2.2 Concrete mix proportions

The mix proportion of the plain concrete was designed according to the design criterion (Li 2011). On the basis of the plain concrete, the steel fiber reinforced expansive concretes were prepared by replacing gravels with equal volume steel fibers. It should be noted that during the preparation, it was by way of adjusting the dosages of the water-reducing agent that the slump values were kept within 80 ~ 120mm. The following Table 1 shows the concrete mix proportions.

Table 1. Concrete mix proportions.

Specimen series identifier	Cement (kg/m^{-3})	Water (kg/m^{-3})	Sand (kg/m^{-3})	Gravel (kg/m^{-3})	Steel fibers (kg/m^{-3})	Expansion agent (kg/m^{-3})
PC	380	174	750	1100	0	0
SF0.5	380	174	750	1087	39	0
SF1.0	380	174	750	1073	78	0
SF1.5	380	174	750	1060	117	0
SF2.0	380	174	750	1046	156	0
CE0.8	380	174	750	1100	0	3.04
SF0.5E0.8	380	174	750	1087	39	3.04
SF1.0 E0.8	380	174	750	1073	78	3.04
SF1.5 E0.8	380	174	750	1060	117	3.04
SF2.0 E0.8	380	174	750	1046	156	3.04
CE1	380	174	750	1100	0	3.8
SF0.5E1	380	174	750	1087	39	3.8

(continued)

Table 1. Continued.

Specimen series identifier	Cement (kg/m^{-3})	Water (kg/m^{-3})	Sand (kg/m^{-3})	Gravel (kg/m^{-3})	Steel fibers (kg/m^{-3})	Expansion agent (kg/m^{-3})
SF1.0 E1	380	174	750	1073	78	3.8
SF1.5 E1	380	174	750	1060	117	3.8
SF2.0 E1	380	174	750	1046	156	3.8
CE1.5	380	174	750	1100	0	5.7
SF0.5E1.5	380	174	750	1087	39	5.7
SF1.0 E1.5	380	174	750	1073	78	5.7
SF1.5 E1.5	380	174	750	1060	117	5.7
SF2.0 E1.5	380	174	750	1046	156	5.7
CE2	380	174	750	1100	0	7.6
SF0.5E2	380	174	750	1087	39	7.6
SF1.0 E2	380	174	750	1073	78	7.6
SF1.5 E2	380	174	750	1060	117	7.6
SF2.0 E2	380	174	750	1046	156	7.6

PC: Plain concrete without steel fibers and expansive agents.
SF0.5: Only the steel fibers existed in the concrete, of which the volume content is 0.5%.
CE1: Only the expansive agent existed in the concrete, of which the mass content is 1.0%.
SF0.5E1: The steel fiber volume content and the expansive agent mass content in the concrete are 0.5% and 1.0% respectively.
The meaning of the specimen series identifiers can be analogized like above.

2.3 *Test methods*

The concrete specimens were cured in 20°C water for 28 days before the properties were tested.

The basic mechanical properties of the concretes were tested according to the test criterion (Leng 2019). There were three 100mm × 100mm × 100mm specimens in each specimen group for the compressive strength tests and the splitting tensile strength tests, and three 100mm × 100mm × 400mm specimens in each specimen group for the four-point bending strength tests.

The impact resistance properties of the concretes were tested according to the test criterion (Huang 2009). The height of the impact hammer was 1000mm, the weight of the impact hammer was 5.18kg, and the diameter of the impact ball was 63mm. For the impact tests, there were six 100mm × 100mm × 100mm specimens in each specimen group. When the first crack appeared on the surface of the tested specimen, the impact number was taken as the initial crack impact number N_1. Likewise, the impact number when the crack widens to the bottom of the specimen was taken as the penetrating crack impact number N_2.

3 RESULT ANALYSIS

3.1 *Compression stress*

Figure 1 shows the compression strengths of the different types of concrete specimens. Where E0, E0.8, E1.0, E1.5, and E2.0 represent the specimen series with 0, 0.8%, 1.0%, 1.5% and 2.0% expansive agent content respectively. It can be seen that, except for a few groups, with the increase of the steel fiber content or the expansion agent content, the compression strength of the concrete changes insignificantly, less than 10%. It should be noted that with the increase of steel fiber content, the steel fibers agglomerate gradually, and the compression strength begins to decrease when the steel fiber content is 2%. When the expansion agent is added in, the compression strength of the steel fiber reinforced concrete is not significantly improved, and the compression strength of the concrete with 2% steel fiber is still low.

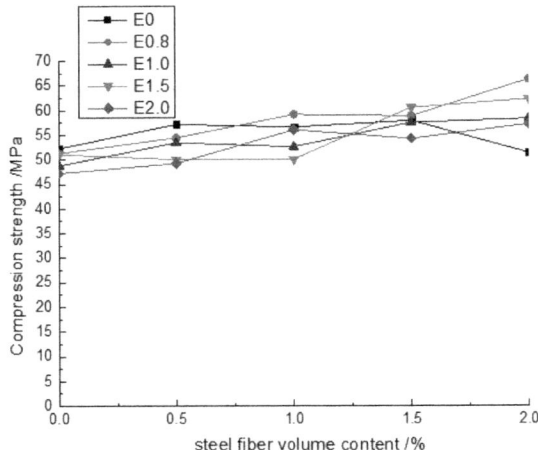

Figure 1.　Compression strengths of the different series of concrete specimens.

3.2　*Splitting tensile strength*

Figure 2 shows the splitting tensile strength of the different types of concrete specimens, which were E0, E0.8, E1.0, E1.5, and E2.0 respectively. It can be seen that with the increase of the steel fiber content, the splitting tensile strength of the concrete increases gradually by about 60%~90%. The content of steel fiber has a significant effect on the splitting tensile strength of the concrete. Yet with the increase of the expansive agent content, the splitting tensile strength of the concrete without steel fibers has no significant change. On the contrary, under the combined action of both the steel fibers and the expansive agent, the splitting tensile strength of the concrete increases speedily.

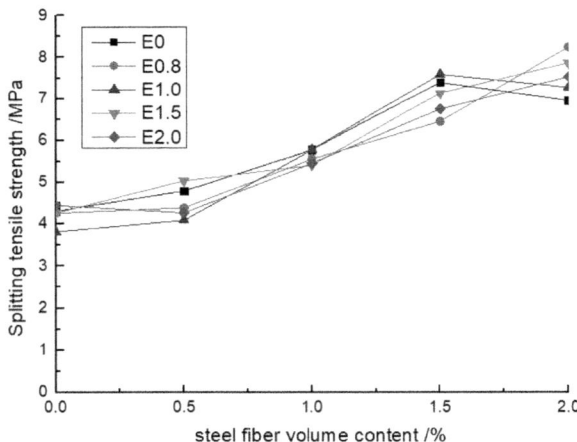

Figure 2.　Splitting tensile strengths of the different series of concrete specimens.

3.3　*Flexural strength*

Figure 3 is the flexural strength of the different types of concrete specimens, which were E0, E0.8, E1.0, E1.5, and E2.0 respectively. It can be seen that with the increase of the steel fiber content, the flexural strength of the concrete increases gradually by about 20%~50%. The content of steel fiber can significantly improve the flexural strength of the concrete. With the increase of the content of the expansive agent, the flexural strength of concrete has no significant change. However, under the combined action of both the steel fibers and the expansive agent, the flexural strength of the concrete increases speedily. When the content of the expansive agent is 0.8% or 1.0%, the flexural strength increases along with increasing steel fiber.

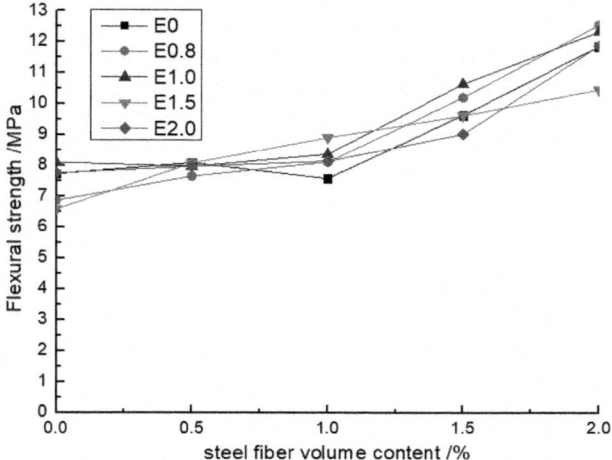

Figure 3. Flexural strengths of the different series of concrete specimens.

3.4 *Crack resistance*

The crack resistance of concrete is indicated by the ratio of flexural strength to compressive strength. Figure 4 shows the ratio of flexural strength to compression strength of the different types of concrete specimens, which were E0, E0.8, E1.0, E1.5, and E2.0 respectively. It can be seen that with the increase of the steel fiber content, the ratio of flexural strength to compression strength of the concrete increases gradually, and the steel fiber content has a significant effect on the ratio of flexural strength to compression strength of the concrete. However, under the combined action of both the steel fibers and the expansive agent, the ratio of flexural strength to compression strength of the concrete only increases a bit as the contents of the two increase. When the content of the expansive agent is 0.8% or 1.0%, the ratio of flexural strength to compression strength increases relatively rapidly with the increase of the steel fiber content.

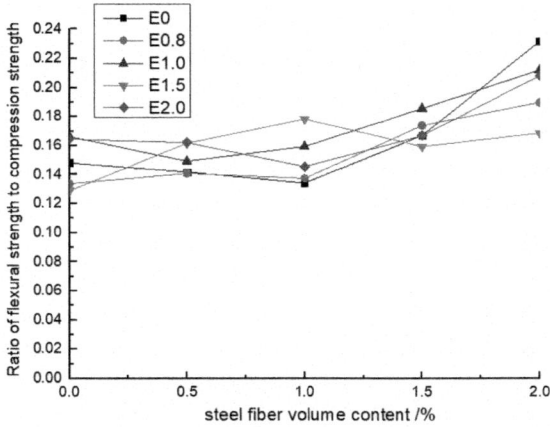

Figure 4. Ratios of flexural strength to compression strength of the different series of concrete specimens.

3.5 *Impact resistance*

Figure 5 displays the impact resistance of the different types of concrete specimens, which were E0, E0.8, E1.0, E1.5, and E2.0 respectively. It can be seen that with the increase of the steel fiber content, the impact resistance numbers of the concrete increase gradually, and the steel fiber content has a significant effect on the impact resistance of the concrete. The impact resistance of concrete is slightly improved by an increasing amount of expansion agents. Under the combined action of both the steel fibers and the expansive agent, the impact resistance of the concrete relates mainly to the content of the steel fiber, and the effect of the expansion agent is not obvious. When the steel fiber content is 1.5%~ 2%, the difference between the initial crack impact number and the penetrating crack impact number is relatively large. This indicates that after the concrete cracks, the penetrating crack impact number is mainly affected by the steel fibers and born by the steel fibers.

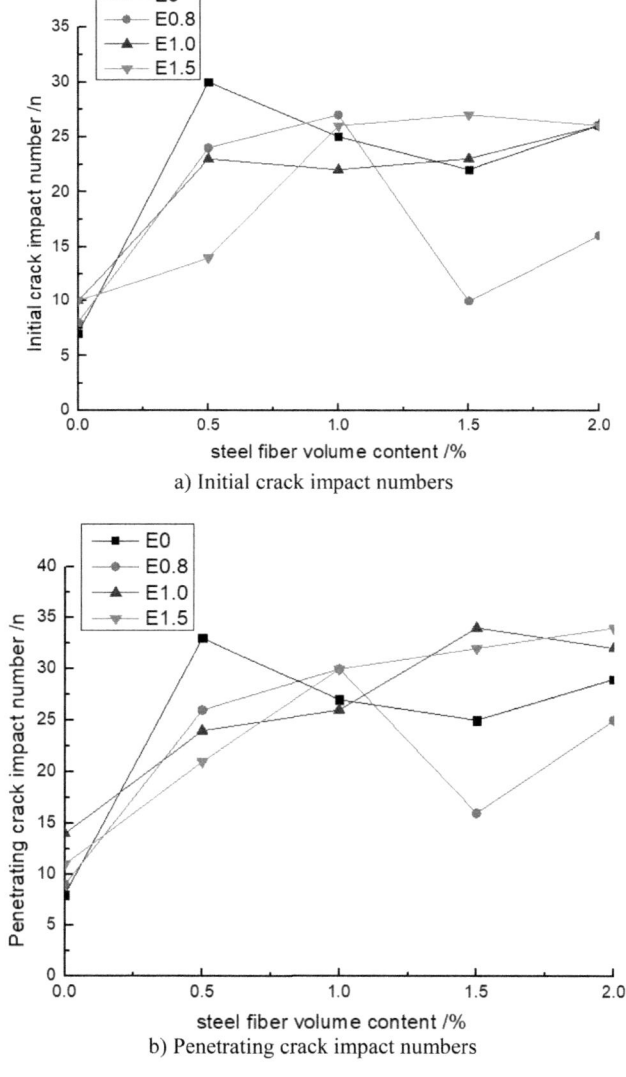

a) Initial crack impact numbers

b) Penetrating crack impact numbers

Figure 5.   Impact resistance of the different series of concrete specimens.

24

3.6 *Discussion*

Steel fiber netlike structures can be formed when the steel fibers are added to the concrete. With the increase of the load on the concrete, cracks appear in the concrete. The steel fibers hinder the expansion of cracks. With the expansion of the cracks in the concrete, especially after the matrix cracks, the steel fibers are pulled out from the matrix or broken. The steel fibers bear some stress so that the bearing capacity of the concrete increases. With the increase of the steel fiber content, the number of bridging fibers on the same length of the crack increases so as to better play its role in crack prevention and increase the fracture stress of the concrete. Therefore, the splitting tensile strength, flexural strength, and impact resistance of the concrete increase with the increase of the steel fiber content. On the other side, because the compression strength of concrete mainly depends on the compactness of the matrix and the roughness of the aggregate interface, the steel fibers have little effect on the compression strength of the concrete.

The changing trend in the improvement degree of the ratio of flexural strength to compression strength is consistent with that for the splitting tensile strength of the concrete. This manifests that the ratio of flexural strength to compression strength is mainly affected by the flexural strength of the concrete. With the increase of the steel fiber content, the flexural strength of the concrete increases gradually. Therefore, the ratio of flexural strength to compression strength of the concrete increases accordingly.

Adding expansive agents can also improve the density and the interface roughness of the concrete, and may cause certain prestress in the steel fiber reinforced concrete. But when the expansion agent and the steel fibers are both added to the concrete, there is hardly any synergy between them, which needs to be further researched.

4 CONCLUSIONS

(1) The variation for the compression strength of the steel fiber reinforced expansive concrete is less than 10% except for very few groups. The two materials added have no significant improvement in the compressive strength of concrete.

(2) With the increase of the steel fiber content, the splitting tensile strength, the flexural strength as well as the ratio of flexural strength to compression strength for the concrete increase gradually, and the initial crack impact number and the penetration crack impact number also increase significantly. But the expansion agent has no obvious effect on the concrete performance.

(3) The synthetical consideration of the performance of the steel fiber reinforced expansive concrete reveals that the ranges of the steel fiber content and the expansion agent are set as 1%~1.5% and 0.8%~1% respectively.

Based on the experimental research, the influence law of the content of steel fiber and expansive agent in steel fiber reinforced expansive concrete on multiple strength and toughness parameters of the material is obtained in this paper, and the suggestions on the mix proportion design of this material are put forward. Considering that the expansion agent does not play a good role in steel fiber reinforced expansive concrete, theoretical and experimental research should be further carried out on the mechanism of the expansion agent in steel fiber reinforced expansive concrete to better guide its application in steel fiber expansive reinforced concrete.

ACKNOWLEDGMENTS

The research is supported by the Construction System Science and Technology Guidance Project of Jiangsu (2017ZD131, 2017ZD132, 2018ZD182) which the sincere thanks go to.

REFERENCES

Abbass W., Khan M.I., Mourad S. Evaluation of mechanical properties of steel fiber reinforced concrete with different strengths of concrete [J]. *Construction & Building Materials*, 2018, 168(APR.20): 556–569.

Amin, Ali, Gilbert, et al. Instantaneous crack width calculation for steel fiber-reinforced concrete flexural members [J]. *Aci Structural Journal*, 2018.

Cao Q., Cheng Y., Cao M., et al. Workability, strength and shrinkage of fiber reinforced expansive self-consolidating concrete [J]. *Construction & Building Materials*, 2017, 131(JAN.30): 178–185.

Chanh N.V. *Steel Fiber Reinforced Concrete*[M]. Springer Singapore, 2017.

Guo-Xin L., Qin L., Jun-Fen Y., et al. Study on expansion admixture and steel fiber reinforcing high strength lightweight aggregate concrete[J]. *Concrete*, 2006(05): 40–42.

Hu S., Li Y. Research on the hydration, hardening mechanism, and microstructure of high-performance expansive concrete[J]. *Cement & Concrete Research*, 1999, 29(7): 1013–1017.

Huanan H., Jie Q., Chengkui H. Experimental research on long-term restrained deformation of steel fiber reinforced micro-concrete[J]. *China Concrete and Cement Products*, 2009(2): 41–44.

Huang C. CECS 13: 2009 *Standard test methods for fiber reinforced concrete*[S]. China Planning Press, 2009.

Leng F. GB/T 50081-2019 *Standard for test methods of concrete physical and mechanical properties*[S]. China Construction Industry Press, 2019.

Li Y. JGJ55-2011 *Specification for mix proportion design of ordinary concrete*[S]. Beijing: China Architecture& Building Press, 2011.

Nataraja M.C., Dhang N., Gupta A.P. Stress-strain curves for steel-fiber reinforced concrete under compression[J]. *Cement & Concrete Composites*, 1999, 21(5–6): 383–390.

Nguyen H.V., Nakarai K., Okazaki A., et al. Applicability of a simplified estimation method to steam-cured expansive concrete[J]. *Cement & Concrete Composites*, 2019, 95: 217–227.

Peng-Bo C., Qin-Yong M. Effects of expansive agents on mechanics properties of shrinkage-compensating steel fiber reinforced shotcrete[J]. *China Concrete and Cement Products*, 2010(5): 45–47.

Suhara K., Tsuji Y., Mashimo M., et al. Measureing methods and devices of length change for expansive concrete[J]. *Cement Science & Concrete Technology*, 2019, 72.

Tian W., Huang C., Zixiang L.I. Study on tension stress \| strain full curve for annular member of steel fiber reinforced expansive concrete[J]. *Journal of Dalian University of Technology*, 2000.

Wenling T., Shichun L., Chengkui H., et al. Experimental study on mechanical characteristics of steel fiber reinforced expansive concrete[J]. *Journal of Building Materials*, 2000, 3(3): 223–228.

Wyrzykowski M., Terrasi G., Lura P. Expansive high-performance concrete for chemical-prestress applications[J]. *Cement & Concrete Research*, 2018, 107: 275–283.

Yu H., Wu L., Liu W.V., et al. Effects of fibers on expansive shotcrete mixtures consisting of calcium sulfoaluminate cement, ordinary Portland cement, and calcium sulfate[J]. *Journal of Rock Mechanics and Geotechnical Engineering*, 2018, 10(2): 20–29.

Zixiang L., Wenling T. Experimental study on deformation performances of steel fiber reinforced expansive concrete[J]. *China Concrete and Cement Products*, 2002(6): 40–43.

Frontiers in Civil and Hydraulic Engineering – Mohamed A. Ismail and Hazem Samih Mohamed (Eds)
© 2023 The Authors, ISBN 978-1-032-38247-0

Study on the mechanical properties of nano-modified ceramsite concrete under dry-wet cycle and temperature

Peng Deng* & Yunkai Chen*
College of Civil Engineering and Architecture, Shandong University of Science and Technology, Qingdao, China

Sai Liu*
College of Civil Engineering and Architecture, Shandong University of Science and Technology, Qingdao, China
Taishan Fiberglass Inc., Tai'an, China

Yao Wang* & Yan Liu*
College of Civil Engineering and Architecture, Shandong University of Science and Technology, Qingdao, China

ABSTRACT: For the degradation law of mechanical properties of nano-modified ceramsite concrete under dry-wet cycle and temperature, nano-modified ceramsite concrete samples are tested at different temperatures (30°C, 40°C, and 50°C) and different dry-wet cycle times (15 times, 45 times, and 75 times). It shows that as the cycle times increase, the compressive strength changes with the typical three-stage law, and the splitting tensile strength shows the law of deterioration, strengthening, and secondary deterioration. With the temperature increasing, the compressive strength first increases and then decreases, and the splitting tensile strength decreases significantly. Under two factors, three failure modes of compression and splitting tensile are analyzed. Finally, based on the experimental data under the dry-wet cycle and temperature, the formula of mechanical property attenuation of nano-modified ceramsite concrete is obtained by fitting.

Chinese Library Classification: TU528.2   **Document Code**: A

1 INSTRUCTION

The deterioration and failure of marine concrete structures under the dry-wet cycle are serious, serving as one of the most adverse factors affecting its durability (Cody 2001) and hindering the development of marine engineering. As a lightweight artificial aggregate, ceramsite can improve the durability of concrete (He 2016), but there are many defects in the microstructure of concrete with ceramsite. Nanomaterials have a small size effect and surface effect (Li 2020), and their incorporation into concrete can improve the microstructure and durability. Nanomaterials are selected to be added to the ceramsite concrete further to improve the mechanical properties of the ceramsite concrete and better serve the complex environment.

Scholars have studied the durability of marine concrete. The experimental study of Ganjian E et al. (2015) shows that with the increase of the times of the dry-wet cycle, the durability of salt corroded concrete shows a deterioration law of first rising and then accelerating decline. Jiong Tan (2020) tested and studied that higher temperature produces tensile stress to crack the concrete surface under the dry-wet cycle, and finally, the measured compressive strength decreased by 10%

---

*Corresponding Authors: dengpeng1226@sdust.edu.cn, 1752030194@qq.com, liu-sai@sdust.edu.cn, 519379288@qq.com and ly966@sina.com

DOI 10.1201/9781003344209-4

Yuan Gao (2012) tested and studied that the free deformation of concrete with different water-cement ratios under the dry-wet cycle is different. The longer the dry-wet cycle is, the longer the free deformation will be. At the same time, the research on the modification of concrete with nanomaterials is relatively mature. Qing Ye et al. (2003) and Erhan Guneyisi et al. (2016) respectively discussed the influence of nano-SiO_2 on the durability of concrete, and the test showed that nano-SiO_2 promoted hydration, thereby improving its durability; Guhua Li et al. (2007) found that nano-SiO_2 greatly improved the performance of concrete later stage because it stimulated the activity of the materials. Most existing studies have discussed the durability of nano concrete under a single factor, but it is inaccurate to analyze it only from one side in actual marine engineering. Therefore, through comprehensively considering, the combination of dry-wet cycle and temperature is more in line with the actual working conditions, which is rarely reported.

Therefore, the research group performed experimental research on nano-SiO_2 modified ceramsite concrete under a dry-wet cycle (Wang 2021). By controlling different seawater temperatures (30°C, 40°C, and 50°C) and different dry-wet cycle times (15 times, 45 times, and 75 times) and keeping the seawater concentration unchanged, the research group explored the influence of temperature and dry-wet cycle times on the mechanical properties of nanomaterial modified concrete. Based on the parameters fitted of ceramsite concrete under dry-wet cycle times and temperature in a clear hygrothermal ocean environment, the attenuation formulas of concrete compressive strength and splitting tensile strength were obtained to provide a reference for their application in marine engineering.

2 TEST MATERIALS AND METHODS

2.1 Test materials and mix proportion

The cement is PO·42.5 composite Portland cement produced by a factory in Shandong. The fine aggregate is river sand provided by a company in Shandong, with fineness modulus greater than 2.5 and silt content less than 2%. The coarse aggregate is shale ceramsite with a nominal particle size of 5-20 mm, a density degree of 500 kg/m^3, and a bulk density of 470 kg/m^3. The mixing water is the standard water in Qingdao. Nano-SiO_2 is provided by a company in Jiangsu, with a particle size of 20 nm, a specific surface area of 200m^2/g, and a loose volume of 0.1g/cm^3. The artificial simulated seawater consists of water, sodium chloride, magnesium chloride, sodium sulfate, potassium chloride, and calcium chloride. The heating rod adopts stainless steel, explosion-proof, automatic constant temperature heating rod.

The mixing proportion of ceramsite concrete is designed according to JGJ/T12-2019 Technical Standard for Application of Lightweight Aggregate Concrete (Industrial Standard of the People's Republic of China 2021) and GB/T50082-2009 Standard for Test Methods of Long-term Performance and Durability of Ordinary Concrete (Industrial Standard of the People's Republic of China 2009). Ceramsite concrete with different strength grades (LC30 and LC35) is obtained by the mechanical property test of ceramsite concrete to optimize the best mix proportion of nano-SiO2 modified ceramsite concrete. Four groups of water-cement ratios (0.38, 0.35, 0.32, and 0.29) were designed to reduce the test error. 100 mm × 100 mm × 100 mm standard test block shall be used and subject to compressive strength test after 28d standard curing. Finally, the measured data selected the water-cement ratio of 0.35 and 0.29 as the test proportion. The mixing proportion is shown in Table 1.

Table 1. Mix proportion design of ceramsite concrete.

Strength	Water cement ratio	Cementitious materials	Cement	SiO$_2$ mixing amount	Water	Sand ratio	Ceramsite	Water reducer	Sand
LC30	0.35	343	336.14	6.86	120	40%	768	460	3.4
LC35	0.29	415	406.7	8.3	120	40%	768	460	4.2

2.2 *Test method*

First, the absorbed ceramsite, river sand, and Portland cement were successively put into the mixing tank. To prevent the aggregate from sinking, water was partially poured and mixed to bind the cement to the aggregate, and then the water and water reducers were added for wet mixing. In mixing, SiO_2 nanomaterials shall be evenly sprinkled on the mixture. Then, the mixed nano-SiO_2 ceramsite concrete was poured into the test block of 100 mm×100 mm×100 mm and 150 mm×150 mm×300 mm. After vibration compaction and flattening, the test block was put into a curing chamber with relatively stable temperature and humidity. The temperature was about 20°C, the relative humidity was more than 95%, and the curing age was 28 days.

According to the specifications, the dry-wet cycle tests were designed at different temperatures (30°C, 40°C, and 50°C) and different cycles (15 times, 45 times, and 75 times). The test piece was dried before the dry-wet cycle began. During the dry-wet cycle, the total time for the test piece to soak, put in the solution, and discharge the solution was 16 h. The drying was at about 80°C for 11 h, then cooling for one hour and wetting for one day. The total time of such a dry-wet cycle was 24 h. Such a dry-wet cycle system is more in line with the service environment of marine concrete, and the data obtained will be more representative. For the convenience of later experimental research, the naming rule is as follows. LC30-30-15 represents the strength of ceramsite concrete, the temperature of the soaking solution, and the times of the dry-wet cycle.

3   FAILURE MODE ANALYSIS UNDER THE COMBINATION OF TEMPERATURE AND DRY-WET CYCLE

Based on references (Cui et al. 2020; Jin et al. 2021; Liu et al. 2018; Pan et al. 2015; Sun et al. 2021; Wong et al. 2006), the failure mode of nano-modified ceramsite concrete is analyzed. During the compressive strength test of concrete test block, the bond failure between aggregate and cement, the splitting failure of aggregate, and the tensile failure of cement usually occur, as shown in Figure 1. The first failure is the vertical crack appearing. As the stress concentration, the crack extends and penetrates, and the failure will be more serious with the increase of the times of the dry-wet cycle. The second failure is that the concrete produces transverse and longitudinal cracks under pressure. The transverse cracks in the middle of the test piece reach the ultimate tensile strain of the concrete and extend through. The third failure is that there will be cracking when only the tensile stress of several parts is less than the bond strength in the initial stage. The crack penetration occurs in the weak parts and gradually expands to show brittle failure.

 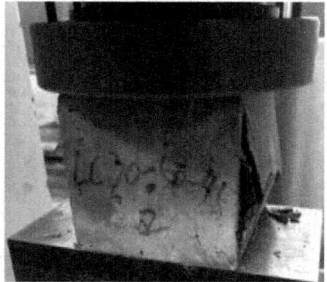

(a) Bond failure between aggregate and cement.   (b) Tensile failure of cement.

Figure 1.   The failure mode of the test piece.

The splitting failure mode of concrete generally includes central cracking failure, local crushing failure, and secondary crack failure. The failure mode of splitting tensile strength under the dry-wet cycle is a central cracking failure. In the initial stage, due to the fewer cycles, there is only peeling

29

on the corners and edges of concrete on the concrete surface. In the medium stage, the splitting tensile strength of concrete has a peak value, and the cracking surface is relatively complete, which confirms the increase of the splitting tensile strength macroscopically, as shown in Figure 2(a). After 75 cycles in the later stage, the splitting tensile strength of concrete shows a downtrend. The concrete surface is peeling, and the aggregate is exposed, with small cracks developing parallel to the edge, as shown in Figure 2(b).

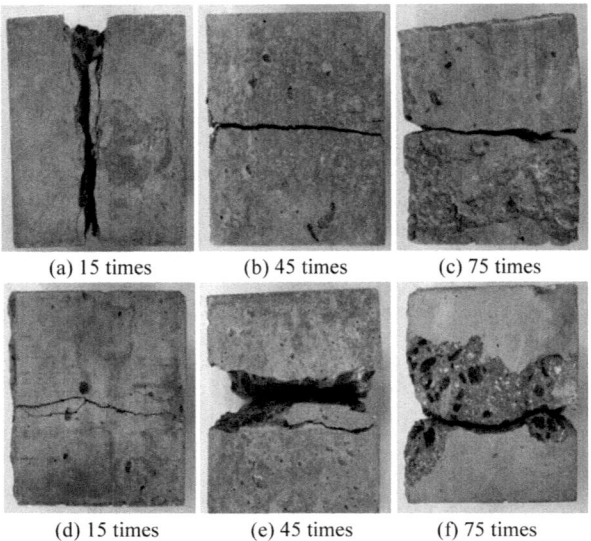

(a) 15 times	(b) 45 times	(c) 75 times
(d) 15 times	(e) 45 times	(f) 75 times

Figure 2. The failure mode of different dry-wet cycle times.

4 CONCRETE CUBE COMPRESSIVE STRENGTH ANALYSIS

4.1 *Concrete cube compressive strength test*

The 100 mm×100 mm×100 mm test blocks were soaked in artificial seawater at 30°C, 40°C, and 50°C, respectively, and the dry-wet cycles were 15 times, 45 times, and 75 times, respectively. The test blocks were not corroded. According to the requirements of the dry-wet cycle in Part 1, the test under planned working conditions was completed, and the dimension conversion coefficient was 0.95. The compressive strength of LC30 and LC35 obtained from the data processed are 31.9 MPa and 38.2 MPa, respectively. The strengths of other numbers are shown in Table 2:

Table 2. Compressive strength of nano-modified ceramsite concrete cube.

No. of the test piece	Sand ratio	Water-binder ratio	Cementitious materials	Strength (MPa)
LC30-30-15	40%	0.35	343	33.5
LC30-40-15	40%	0.35	343	37.4
LC30-50-15	40%	0.35	343	35.2
LC35-30-15	40%	0.29	415	39.3
LC35-40-15	40%	0.29	415	43.2
LC35-50-15	40%	0.29	415	41.6
LC30-30-45	40%	0.35	343	31.0

(*continued*)

Table 2. Continued.

No. of the test piece	Sand ratio	Water-binder ratio	Cementitious materials	Strength (MPa)
LC30-40-45	40%	0.35	343	36.0
LC30-50-45	40%	0.35	343	33.0
LC35-30-45	40%	0.29	415	37.1
LC35-40-45	40%	0.29	415	42.6
LC35-50-45	40%	0.29	415	39.5
LC30-30-75	40%	0.35	343	27.6
LC30-40-75	40%	0.35	343	31.5
LC30-50-75	40%	0.35	343	28.2
LC35-30-75	40%	0.29	415	31.9
LC35-40-75	40%	0.29	415	39.5
LC35-50-75	40%	0.29	415	34.4

4.2 *Effect of the combination of temperature and dry-wet cycle on the compressive strength of nanomaterial-modified ceramsite concrete*

The change law of compressive strength of nanomaterial-modified concrete under the combination of temperature and the dry-wet cycle is shown in Figure 3. At the same temperature, the compressive strength increases first and then accelerates deterioration as the dry-wet cycles increase. Taking the working condition of LC30-30 as an example, the compressive strength of LC30-30-15 reaches 33.5 MPa, which is 1.6 MPa (4%) higher than that of the test block without a dry-wet cycle. Under the working condition of LC30-30-45, the compressive strength decreases by 2.5 MPa, and the loss rate is 7% compared with 15 cycles.

With the increase in temperature, its intensity increases and then decreases under the same dry-wet cycles. Taking the working condition of LC35-30 as an example, the strength of LC35-30-15 is 1.1 MPa (2.8%) higher than that of the test block without a dry-wet cycle. The compressive strength of LC35-30-45 is 2.2 MPa (<1%) lower than that of 15 cycles, the compressive strength of LC35-30-75 is 5.2 MPa lower than that of 45 cycles, and the strength loss rate is 14%.

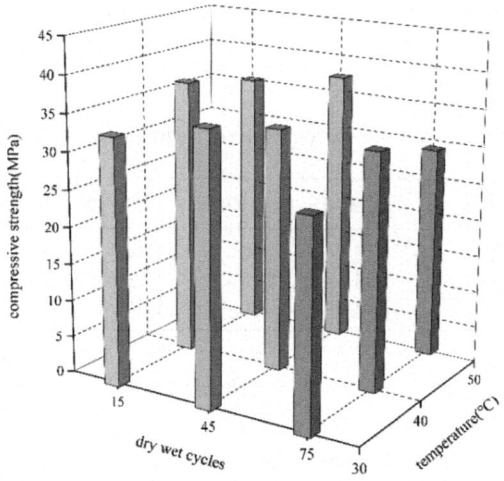

(a) The influence of dry-wet cycles on compressive strength

Figure 3. Influence of different factors on compressive strength.

31

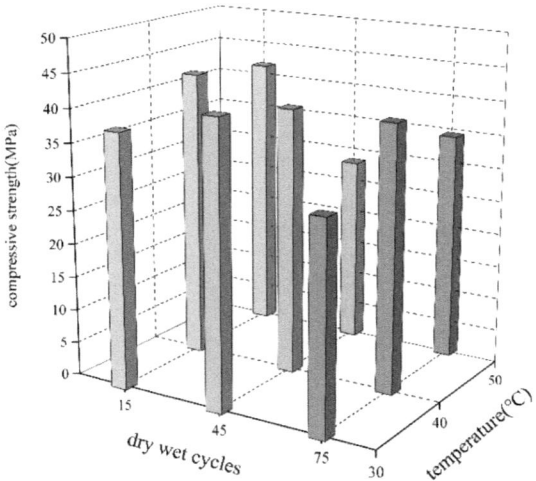

(b) The influence of temperature on compressive strength

Figure 3.   Continued.

Taking the compressive strength values of LC35-50-45 and lC30-30-75 as the benchmark values, the statistical summary strength ratio presents the influence of the times of the dry-wet cycle and temperature on the compressive strength, shown in Tables 3 and 4. The above figure shows that at the initial stage of the dry-wet cycle, the concrete with high strength grade has a large increase in strength, and the change in relative compressive strength is different due to different temperatures.

Table 3.   Normalized statistics of compressive strength under the influence of the times of the dry-wet cycle.

Times of dry-wet cycle	LC30-30	LC30-40	LC30-50	LC35-30	LC35-40	LC35-50
15	1.19	1.34	1.25	1.36	1.54	1.48
45	1.29	1.18	1.33	1.51	1.42	1.00
75	1.01	1.15	1.03	1.11	1.41	1.22

Table 4.   Normalized statistics of compressive strength under the influence of temperature.

Temperature	LC30-15	LC30-45	LC30-75	LC35-15	LC35-45	LC35-75
30	1.22	1.15	1.00	1.44	1.37	1.18
40	1.37	1.34	1.15	1.63	1.61	1.40
50	1.32	1.22	1.26	1.56	1.44	1.29

4.3   *Comparative analysis of compressive strength test data of nanomaterial modified ceramsite concrete*

For better analysis of the effect mechanism of the dry-wet cycle times, the test's compressive strength value is statistically compared with the relevant data (Chen et al. 2019; Huang et al. 2021; Li et al. 2009; Piao et al.2016; Shao et al. 2021; Wang et al. 2020; Zhang et al. 2019), as shown in Figure 4.

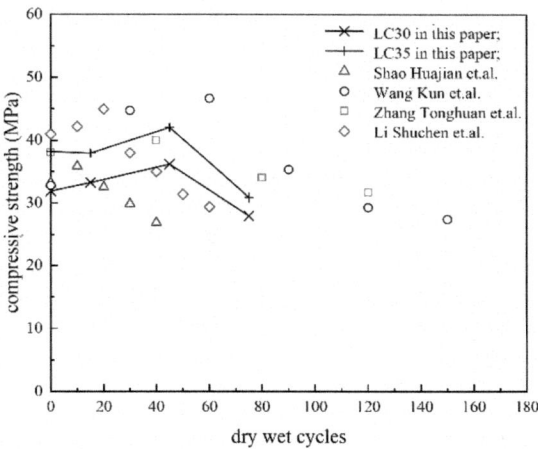

Figure 4.   Comparison of compressive strength and references.

Figure 4 shows that the change of concrete compressive strength shows a similar law, but the decline rate of the test in this paper is significantly faster than other test data because other tests only consider the influence of a single factor of the dry-wet cycle on the compressive strength of concrete, which ignores the possible role of other influencing factors, such as the coupling of temperature. On the one hand, the existing research has directly confirmed that temperature significantly influences the compressive strength of concrete, which is also reflected in the test data in this paper.

On the other hand, the decline rate of the compressive strength of the test data in this paper is even faster than that of other test data without nanomaterials, which directly proves that the dry-wet cycle and temperature will promote each other and accelerate the deterioration of concrete. Therefore, it is imperfect to consider single influencing factors.

4.4   *The attenuation law of compressive strength of nanomaterial modified ceramsite concrete under the combination of temperature and dry-wet cycle*

The research on marine concrete mostly stays under the influence of a single factor, but in the actual service process of the structure, it is often the combination of multiple factors. Therefore, to discuss the calculation method of the compressive strength of nanomaterial-modified ceramsite concrete, this paper uses the existing test and formula derivation and fitting method for reference.

In the existing studies, most scholars have confirmed that the dry-wet cycle is one of the important reasons affecting the durability of marine concrete. Shuchen Li et al. (2009) proposed a formula for compressive strength reduction of concrete with the times of the dry-wet cycle, as shown in Formula (1).

$$F = -0.0042N^2 - 0.1413N + 46.65 \tag{1}$$

where F is the compressive strength of concrete, and N is the times of the dry-wet cycle. This formula only considers the change under the dry-wet cycle at one intensity, so it is not universally applicable.

The above formula only considers the influence of a single factor on the compressive strength of concrete, which is often inaccurate for predicting strength attenuation in complex marine environments. Therefore, based on the test data and existing formulas, considering the influence of temperature, the times of dry-wet cycle, and the compressive strength of nano-modified ceramsite concrete without the action of temperature and dry-wet cycle, and based on the least square method and other analysis and fitting methods, this paper establishes the formula for the attenuation of the compressive strength of ceramsite concrete in the hygrothermal ocean environment:

$$f_{cu}^{T,C,\text{cu}} = (1.141T^{0.538} - 0.11T - 3.233) \times (1.397C^{0.318} - 0.063C + 6.771) \times (0.193f_{cu} + 0.09)$$

where $f_{cu}^{T,C,\text{cu}}$ are the compressive strength of ceramsite concrete, f_{cu} is the compressive strength of ceramsite concrete without the dry-wet cycle test, T is the temperature, and C is the times of the dry-wet cycle.

The fitting correlation coefficient R^2 of the fitting formula is 0.96. The comparison among the results obtained from the above fitting formula and the test data, as shown in Figure 5, shows that the error between the calculated value and the fitting value is within 5%, indicating that the fitting degree of the formula is high.

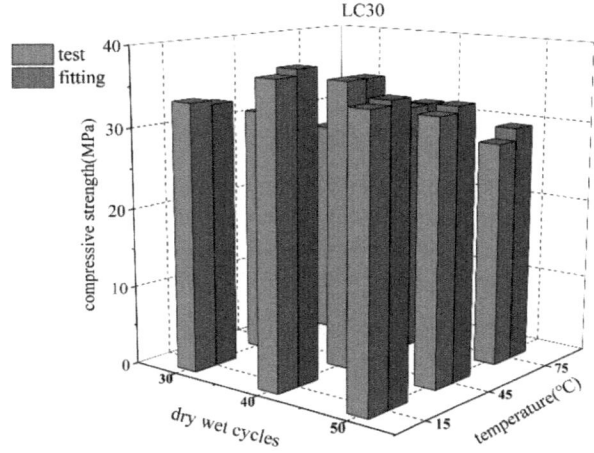

(a) Test and fitting comparison under LC30 strength

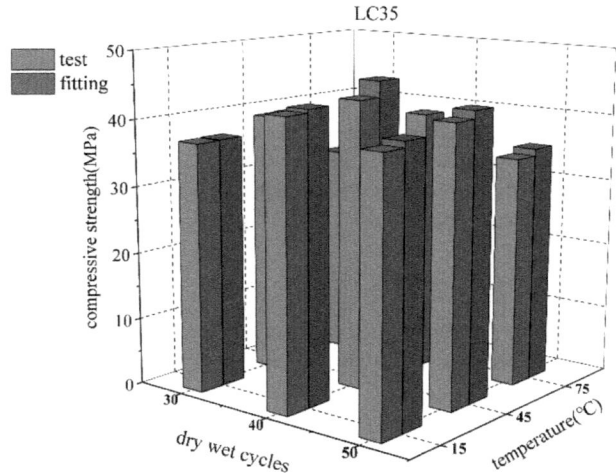

(b) Test and fitting comparison under LC35 strength

Figure 5. Comparison of fitting results.

5 CONCLUSIONS

1) Most of the failure of nano-modified ceramsite concrete is aggregate penetration failure, and a few test blocks have the phenomenon of section separation.
2) Under the combination of the dry-wet cycle and temperature, the increase of cycle times and temperature make the compressive strength of ceramsite concrete show a law of strengthening first and then deteriorating faster. The strength of ceramsite concrete with a large water-cement ratio increases greatly at the initial stage of the dry-wet cycle.
3) Under the combination of the dry-wet cycle and temperature, the increase of cycle times and temperature makes the splitting tensile strength of ceramsite concrete show a law of first strengthening and then deteriorating. The experimental data of compressive strength and splitting tensile strength of nano-modified ceramsite concrete under different cycles and temperatures is compared with the data in the references, which shows that the change law is consistent.
4) By the origin nonlinear fitting, the attenuation change law of the strength coupling of nano-modified ceramsite concrete with different dry-wet cycle times, different temperatures, and noncycling is obtained, of which the error of fitting result is small.

REFERENCES

Chen, J.W. (2019). *Mechanical properties of fiber nano-modified rubber concrete under high temperature.* Zhengzhou: Zhengzhou University. (in Chinese)

Cui, W., Liu, M.M. & Song, H.F. (2020). Influence of initial defects on deformation and failure of concrete under uniaxial compression. *J. Engineering Fracture Mechanics.* 234, 1–14.

Ganjian, E., Canpdat, F. & Claisse P. (2015). Special issue on sustainable construction materials. *J. Journal of Materials in Civil Engineering.* 23(7), B2015001.

Gao, Y., J. Zhang. & W. Sun. (2012). Concrete deformation and interior humidity during dry-wet cycles. *J. Journal of Tsinghua University* (Sciences and Technology). 52(2): 144–149. (in Chinese)

Guneyisi, E., Gesoglu, M., Azez, O.A. & Oz, H.O. (2016). Effect of nano-silica on the workability of self-compacting concrete having untreated and surface treated lightweight aggregates. *J. Construction and Building Materials.* (15), 371–380.

He, K.C., Guo, R.X. & Ma, Q.M. (2016). Experimental research on high-temperature resistance of modified lightweight concrete after exposure to elevated temperatures. *J. Advances in Materials Science and Engineering.* 206(16), 1–6.

Huang, Z.H. (2021). *Study on mechanical properties of modified rubber concrete subjected to the temperature.* Wuhan: Hubei University of Technology. (in Chinese)

Industrial Standard of the People's Republic of China. (2009). (Standard No. GB/T50082-2009). *The standard for Test Methods of Long-Term Performance and Durability of Ordinary Concrete.* Beijing: China Architecture and Building Press.

Industrial standard of the people's republic of China. (2021). (Standard No. JGJ51-2019). *Technical Standard for Application of Lightweight Aggregate Concrete.* Beijing: China Architecture and Building Press.

Jin, L., Yang, W.X. & Yu, W.X. (2021). Study on failure behavior and size effect of lightweight aggregate concrete based on meso-simulation. *J. Journal of Disaster Prevention and Mitigation Engineering.* 41(1), 91–99. (in Chinese)

Li, G.H. & B. Gao. (2007). Effect of nano-SiO_2 on the performance of concrete in dissolving/crystallizing cycles of salts. *J. Journal of Southwest Jiaotong University.* 42(1), 70–74. (in Chinese)

Li, S.C., Zhang, F. & Zhu, J.P. (2009). Test on the deterioration of mechanical properties of concrete under seawater erosion environment. *J. Journal of Highway and Transportation Research and Development.* 26(12), 35–38. (in Chinese)

Li, Z.D. & Meng, D. (2020). Analysis of macroscopic property and microscopic control mechanism of nano-sio_2 modified concrete. *J. Bulletin of the Chinese Ceramic Society.* 39(07), 2145–2153. (in Chinese)

Liu, T.J., Qin, S.S. & Zou, D.J. (2018). Mesoscopic modeling method of concrete based on a statistical analysis of CT images. *J. Construction Building Materials.* 192, 429–441.

Pan, Z.C., Chen, A.R. & Ruan, X. (2015). Spatial variability of chloride and its influence on the thickness of concrete cover: a two-dimensional mesoscopic numerical research. *J. Engineering Structures.* 95, 154–169.

Piao, Z.D. (2016). *Experimental study on mechanical properties of basalt fiber concrete under high temperature*. D. Zhengzhou: Zhengzhou University. (in Chinese)

Robert, D.C., Anita M.C., Paul, G.S. & Hyomin, L. (2001). *Reduction of concrete deterioration by ettringite using crystal growth inhibition techniques. Department of geological and atmospheric sciences.* Iowa State University. 111–116.

Shao, H.J. & Li, Z.L. & Xiao, S.P. et al. (2021). Mechanical properties and microstructure of concrete under drying-wetting cycles. *J. Bulletin of the Chinese Ceramic Society*. 40(09), 2949–2955. (in Chinese)

Sun, H. K., Gao, Y. & Zheng, X.Y. et al. (2021). The failure mechanism of concrete with precast defects based on image statistics. *J. Journal of Building Materials*. 24(6), 1154–1162. (in Chinese)

Tan, J. (2020). Study on the durability test of foam concrete subjected to drying-wetting cycles. *J. Journal of Transport Science and Engineering*. 36(03), 14–18. (in Chinese)

Wang, K. (2020). *Experimental study on the deterioration of mechanical properties of concrete sulfate erosion under drying-wetting and environments*. Zhengzhou: Zhengzhou University. (in Chinese)

Wang, Y. (2021). *Experimental study on mechanical properties and durability of modified ceramsite concrete in a hot and humid marine environment*. Qingdao: Shandong University of Science and Technology. (in Chinese)

Wong, T.F., Wong, R.H.C. & Chau, K.T. et al. (2006). Microcrack statistics, Weibull distribution and micromechanical modeling of compressive failure in rock. *J. Mechanics of Materials*. 38(7), 664–681.

Ye, Q., Zhang, Z.N. & Kong, D.Y. et al. (2003). Comparison of properties of high strength concrete with nano-SiO_2 and silica fume added. *J. Journal of Building Materials*. (04), 381–385. (in Chinese)

Zhang, T.H. (2019). *Study on the deterioration law of concrete performance under the coupling of sulfate attack and dry wet cycle*. Xi'an: Xijing University. (in Chinese)

Frontiers in Civil and Hydraulic Engineering – Mohamed A. Ismail and
Hazem Samih Mohamed (Eds)
© 2023 The Authors, ISBN 978-1-032-38247-0

Numerical research on water seepage in underground station gaps based on the CEL method

Tao Yang, Guodong Li & Hongna Yang
China Power Construction Municipal Construction Group Co., Ltd, Tianjin, China

Haotian Luo & Fengting Li
School of Civil Engineering, Shandong University, Jinan, China, China

ABSTRACT: Aiming at the problem of water seepage in underground stations, the explicit dynamic CEL method is used to analyze the development of groundwater seepage after groundwater construction is completed, and simulate the seepage process of liquid groundwater in the gaps of real stations. The research results show that the movement of groundwater in the hole is similar to a "fountain" after the construction is completed, and a thin layer of groundwater is distributed at the bottom of the bottom plate. In the process of groundwater infiltration, the seepage amount of the first floor and the second floor of the subway has a great difference. The seepage period is divided into the seepage spread period, the seepage pressure period, and the seepage stable period.

1 INTRODUCTION

During the construction of the underground station, a series of precipitation measures will be carried out to keep the surrounding soil above the water level during construction to ensure the quality and safety of the construction. When the construction is completed, the precipitation measures will be withdrawn and the groundwater level will be restored. Some underground stations will be surrounded by groundwater over time. Groundwater penetrates the interior through the concrete cracks of the subway station structure, and the phenomenon of water seepage inside the station occurs. This phenomenon is more obvious in cities with more rain in my country. Once a water leakage accident occurs in an underground station, the normal operation of the subway will be seriously affected. Residents' travel will also cause immeasurable economic and property losses. Therefore, it is of great significance for the urban development of our country to simulate the water seepage process of subway stations and explore its water seepage laws. Zhao Jiahui studied the overall influence of precipitation on the structure during the construction of the prefabricated subway station. It is found that in the process of backfilling, the pore pressure at the bottom of the ground connection wall has large water pressure, and the pore water pressure of the station floor was small. Most of the Groundwater flows to the interior of the station through the bottom of the ground connection wall (Zhao 2021). Park Dongsoo et al. tested the different properties of concrete structural cracks by changing the hydraulic pressure and observing the change in the leakage behavior of the specimens. Moreover, the waterproof performance of different waterproof materials was summarized (Park 2012). Based on the existing tunnel waterproofing situation, Chen Yunyao et al. observed the macroscopic seepage phenomenon of tunnel cracks, counted the seepage flow, analyzed the tunnel failure mode under unfavorable conditions through finite element analysis, and summarized the best waterproofing scheme (Chen 2019). Zhu Shufen studied the causes of cracks in the structural concrete of subway stations at different stages, made statistics on the occurrence probability of relevant cracks, and dealt with them with corresponding control measures to enhance the effect of concrete in the structure of subway stations (Zhu 2018). Yang

Tianhong studied the construction of subway stations when water seepage cannot be controlled, observed the main sources of seepage water, analyzed the changes in seepage paths and seepage amounts, and summarized a variety of deep water control methods at different construction stages, which aims to ensure the construction quality and waterproofing of concrete structures (Yang 2013).

The current research mainly simulates the process of groundwater level change through soil seepage, but some concrete cracks have macroscopic characteristics, and some cracks can reach 5~8mm in width. Conventional microscopic seepage research has certain limitations in this case, and it is difficult to simulate the Water seepage process in wider cracks. To this end, this study uses the CEL (Coupled Eulerian-Lagrangian analysis) method to simulate the real liquid groundwater seepage process. The Abaqus finite element software was used to simulate how the water seepage in the soil medium filled with pores, and to explore the influence on the groundwater seepage under different conditions.

1.1 *Numerical model building*

After the subway construction is completed, if the surrounding water level is higher than the subway floor, the pore water pressure on both sides of the subway retaining wall will be greater than that on the floor side, and seepage will occur. The traditional Lagrangian analysis method uses the finite element method to deform the element mesh to simulate the deformation of the material. If the deformation of the mesh is too large, the simulation will be distorted. The Euler analysis method effectively solves the extreme deformation and fluid flow problems that occur in the Lagrangian analysis method by fixing the mesh element nodes to make the material move between each element. However, in the fluid-solid coupling analysis such as the interaction between groundwater and subway stations in this study, although Eulerian analysis can effectively handle fluid flow analysis, it cannot analyze the interaction between groundwater and concrete structure cracks. At this time, the CEL method can be applied to solve it. The CEL method can effectively capture the interface between liquid and solid, simulate the interaction between groundwater and cracks in concrete structures through ABAQUS finite element general contact, and combine Lagrangian elements with Eulerian elements to simulate the crack seepage process.

In the numerical simulation model of groundwater seepage in the subway station, as shown in Figure 1, the depth of the retaining wall is 20m, the length of the subway floor is 20m, and the water level on both sides is 10m underground. The subway structure is divided into two layers, and each layer is 5m high. The crack width is 1 cm, and the cracks are respectively arranged at the bottom and the top of the first layer and the top of the second layer. The top of the subway is 5m from the ground. As shown in Figure 2, the Euler unit EC3D8R is used for the location where groundwater and groundwater may flow, and the number of units is 42.56 million. Because the groundwater flow has little influence on the force of the subway structure, the subway structure adopts the discrete rigid body element R3D4, which is ignored in the calculation. The deformation and internal force of the subway structure reduce the complexity of the model and increase the accuracy of the model.

Figure 1. Distribution of groundwater in subway stations.

Figure 2. Distribution of seepage grids in subway stations.

1.2 *Model mechanical parameters*

In simulating groundwater flow, water, as a fluid medium, is treated as an incompressible viscous laminar flow. The volume response of aqueous media is described by the Mie-Grüneisen equation of state in linear Hugoniot form (Wang 2018).

The usual form of the Mie-Grüneisen equation of state is:

$$p - p_H = \Gamma_\rho(E_m - E_H)$$

where: p_H is the Hugon-new pressure and E_H is the Hugon-new specific energy. Both are functions of density only; $\Gamma = \Gamma_\rho \rho_0 / \rho$, Γ_ρ is the material constant, and ρ_0 is the reference density.

Then the Mie-Grüneisen equation of state in linear $U_s - U_p$ Hugoniot form is obtained, which is expressed as:

$$p = \frac{\rho_0 c_0^2 \eta}{(1 - s\eta)^2}\left(1 - \frac{\Gamma_0 \eta}{2}\right) + \Gamma_0 \rho_0 E_m$$

In the formula, under the small nominal strain, $\rho_0 c_0^2$ is equivalent to the elastic bulk modulus $K = \rho_0 c_0^2$.

To sum up, based on the hydrodynamic state equation and the viscous shear model, the definition of the material properties of the water medium is completed. The basic parameters of the linear $U_s - U_p$ Mie-Grüneisen equation of state for real water media are as follows: density $\rho = 9.832 \times 10^2 \text{kg/m}^3$, $c_0 = 1.4506 \times 10^3 \text{m/s}$, $s = 0$, $\Gamma_0 = 0$, $\mu = 1 \times 10^{-3}$ Pa • s.

2 NUMERICAL RESULTS ANALYSIS

The Abaqus explicit dynamic method is used to simulate the groundwater variation on both sides of the subway station using the coupled Euler-Lagrangian analysis (CEL). As shown in Figure 3, by solving the fluid-solid coupling of water, after the construction of the subway station is completed, the pore water pressure on both sides of the side wall is higher, while the pore water pressure in the soil below the subway station floor is lower, and the seepage flow from the pore pressure is higher. At the same time, due to the influence of gravity, the water in the seepage first collects from both sides to the lower side of the center and then moves upward from the center of the bottom to the depression of the hole with the change of pore pressure. Then, it continues to move toward the

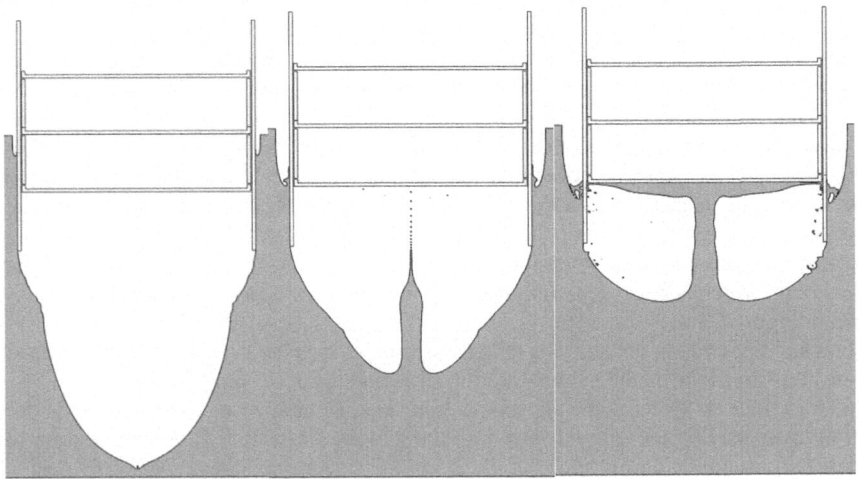

Figure 3. The performance of groundwater seepage on the floor of the subway station.

center of the bottom plate. The two sides of the subway move, and the groundwater movement is similar to a "fountain". A thin layer of groundwater is distributed at the bottom of the bottom plate, and the groundwater is less distributed when the bottom plate is buried deeper.

As shown in Figure 4, in the analysis of groundwater changes in the subway station, when the groundwater spreads to both sides of the subway station floor over time, the groundwater will flow up and down. Because the lower space is larger and the upper side wall gap is smaller, the groundwater will flow upwards and downwards. Under a certain water pressure, it moves vertically upward through the gap and continues to move upward under the push of groundwater at the bottom. During this period, there is no water seepage in the gap at the bottom of the first floor of the subway.

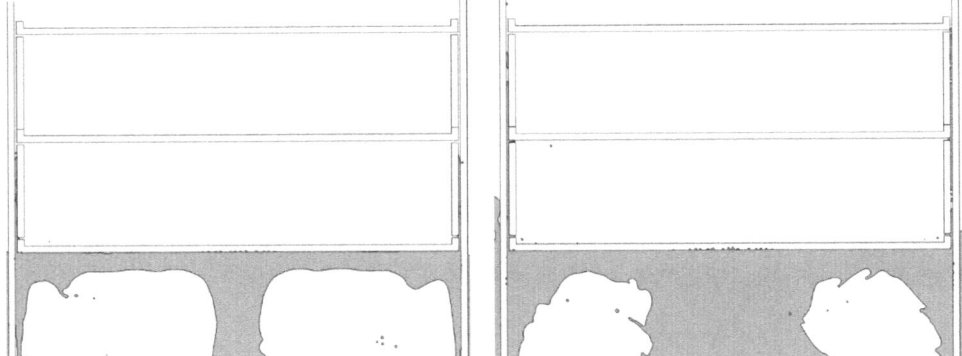

Figure 4. The performance of groundwater seepage on the side wall of the subway station.

Figure 5. The performance of groundwater seepage in the subway station.

As shown in Figure 5, in the analysis of groundwater changes in subway stations, the groundwater moves in the gaps on both sides of the side wall with the change of pore pressure and the inertia of groundwater movement. Even if groundwater has spread to the top of the second floor of the subway, groundwater, the main movement direction of the first layer is to spread upward, the bottom and top gaps of the first layer are horizontal gaps, and the groundwater flow direction is different. Therefore, the seepage of the first layer is not obvious, and the groundwater is not completely leaked, indicating that the direction of the gap affects the groundwater seepage. In the early stage, there was less water seepage, and it mainly spread and penetrated to distant places.

As shown in Figures 6 and 7, in the analysis of groundwater changes in subway stations, when groundwater has seeped to the upper part of the main subway structure, the pore pressure in the upper part is negative at this time. However, it continues to move to the upper part due to the action of the lower groundwater. At the same time, the groundwater located in the gap of the side wall of the first floor of the subway moves in all directions when the longitudinal pressure increases. It finally enters the subway station through the gap between the bottom and the top. The water seepage time is the same as that of the bottom in the first layer, and leakage will occur. When leakage occurs on the first floor of the subway station, there is no leakage on the second floor of the subway station, which is consistent with the phenomenon on the first floor of the subway station, as shown in Figure 5.

As shown in Figure 8, in the analysis of groundwater changes in subway stations, when seepage occurs on the first floor of the subway station, the amount of seepage in the side walls will be transferred to the first floor of the subway station. The groundwater in the side walls decreases discontinuously, leading the groundwater into the gap on the second floor. The station loses its upward thrust and no longer moves in all directions. The seepage phenomenon only occurs in the gap on the first floor of the subway, and the seepage phenomenon on the second floor is significantly weakened.

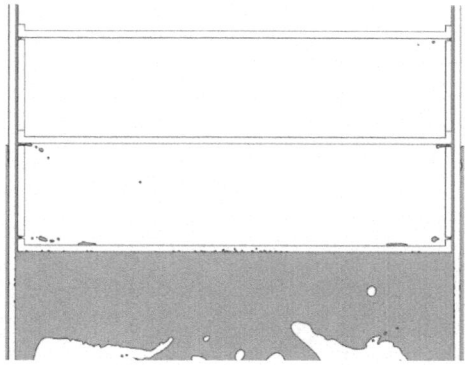

Figure 6. The performance of groundwater
seepage in the subway station.

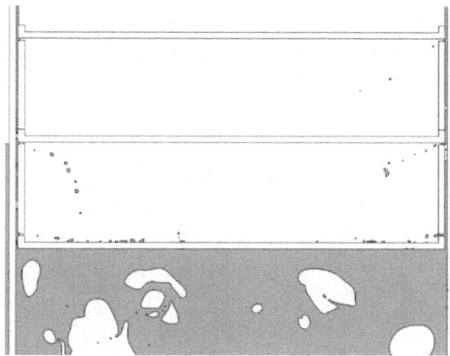

Figure 7. The performance of groundwater
seepage in the subway station.

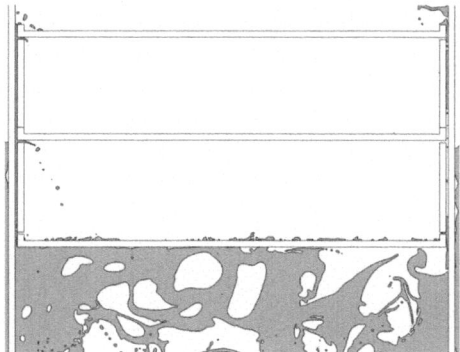

Figure 8. The performance of groundwater
seepage in the subway station.

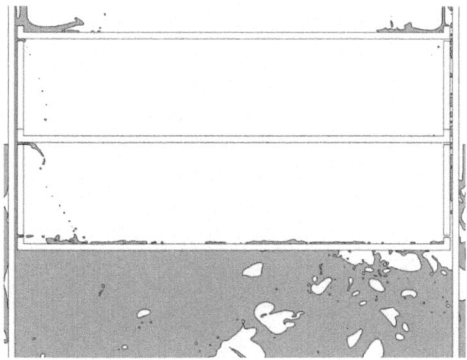

Figure 9. The performance of groundwater
seepage in the subway station.

As shown in Figure 9, in the analysis of the change of groundwater in the subway station, there are two possible situations. The groundwater on the left side wall of the subway station is continuous and has a certain pressure. Seepage occurs in each lateral gap. The water pressure decreases after the bottom seepage. The seepage flow of the gap on the second floor of the subway is significantly less. The groundwater on the right side wall of the subway station is discontinuous. The pressure among the groundwater is small, and it mainly moves in the vertical direction without passing through the horizontal gap. The groundwater does not occur seepage.

As shown in Figure 10, in the analysis of groundwater changes in the subway station, the groundwater on the right side of the subway station re-spreads to the side wall gap from bottom to top. The water seepage at the bottom of the first floor in the subway is less than that at the top of the first floor, indicating that the seepage first increases and then decreases when the pressure increases.

As shown in Figures 11 and 12, in the analysis of groundwater changes in subway stations, with the occurrence of seepage, the total seepage volume inside the subway station is divided into three stages. The first stage is the seepage spread period. It is proportional to the time, and the water seepage phenomenon on the second floor of the subway is not obvious at this time. The second stage is the seepage pressure period. When the groundwater pressure gradually increases, the seepage volume of the first floor of the subway increases parabolically, and the seepage volume of the second floor of the subway increases. It is proportional to the time. The third stage is the steady

41

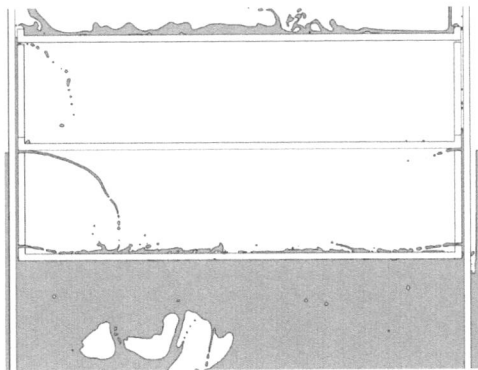

Figure 10. The performance of groundwater seepage in the subway station.

Figure 11. The performance of groundwater in the subway station.

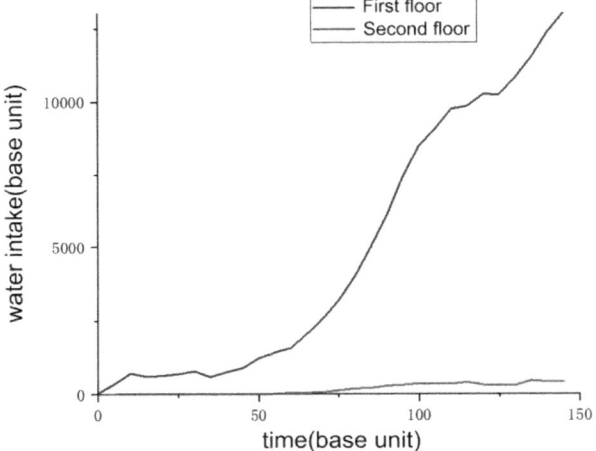

Figure 12.   The relationship between water inflow and the time of each layer in the subway.

water seepage period. After the groundwater pressure is constant, the seepage amount of the first floor of the subway is fixed, and the second floor of the subway does not have seepage.

As shown in Figure 12, there is a huge difference in the seepage amount between the first floor of the subway and the second floor of the subway, indicating that in the case of subway seepage, the seepage amount at different gaps is quite different. Where there is substantial seepage, precautions are designed.

3   CONCLUSION

Aiming at the problem of seepage in underground stations, the explicit dynamic CEL method of ABAQUS finite element software is used to analyze the development of groundwater seepage after groundwater construction is completed, and simulate the seepage process of liquid groundwater in the gaps of the real station. The main researches are as follows:

(1) After the construction is completed, the change of pore pressure moves up from the center of the bottom to the place where the pore is depressed, and then continues to move to both sides

42

of the subway after touching the middle of the bottom plate. At the same time, the bottom floor is buried deeper, but the groundwater distribution is less.

(2) Under a certain water pressure, the groundwater will move vertically upward through the gap and continue to move upwards under the push of the groundwater below. During this period, there is no water seepage in the gap at the bottom of the first floor of the subway.

(3) In the process of groundwater infiltration, the seepage amount of the first floor and the second floor of the subway has a great difference. The seepage period is divided into the seepage spread period, the seepage pressure period, and the seepage stable period. The seepage amount has different characteristics.

This paper mainly explores the law of groundwater seepage after subway station construction and summarizes the variation law of space and time of seepage water in underground stations through numerical simulation. The development of anti-leakage technology in stations is of great significance. Given the seepage law in the conclusion, it is suggested to further reduce the impact of water seepage on the underground station from the perspective of designing the waterproof structure and adopting new waterproof materials.

REFERENCES

Chen Yunyao, Zhang Junwei, Ma Shiwei, et al. Research on waterproof failure mode and improvement of shield tunnel segment gasket [J]. *Tunnel Construction* (Chinese and English), 2019, 39(6): 946–952. DOI:10.3973/j. issn.2096-4498.2019.06.007.

Park Dongsoo, Kwon Ki-joo. The experimental study on the development of durability performance test device for waterproofing material under hydrauiic pressure environment [J]. *Journal of the Architectural Institute of Korea Structure & Construction*, 2012, 28(06): 51–58.

Wang Xiaohui, Chu Xuesen, Feng Guang. Numerical study of sphere entry into water based on ABAQUS explicit CEL method [J]. *Ship Mechanics*, 2018, 22(7): 838–844.

Yang Tianhong. Water seepage control technology for waterproof construction of karst geological subway stations [J]. *Construction Technology*, 2013, 42(21): 81–84.

Zhao Jiahui. *Research on construction mechanical characteristics of prefabricated subway stations and performance of anti-seepage materials at joints* [D]. Shandong: Shandong University, 2021.

Zhu Shufen. analysis and control of water seepage cracks in subway station structural concrete [J]. *Smart City*, 2018, 4(17): 104–105.

Frontiers in Civil and Hydraulic Engineering – Mohamed A. Ismail and
Hazem Samih Mohamed (Eds)
© 2023 The Authors, ISBN 978-1-032-38247-0

Study of simple shear test between stale waste and pile surface with different roughness

Xinghao Hu*, Shuping Chen* & Xueqian Lou*
CCCC Fourth Harbor Engineering Institute Co., Ltd., Guangzhou, Guangdong, China

ABSTRACT: The rough and uneven surfaces of bored piles would affect the side friction as a foundation form, which is often used in landfill reused projects. In this paper, a regular saw-tooth surface was used to simulate the side surface of bored piles. At the same time, the influence of plastic fiber on stale waste was taken into consideration. A series of interface simple shear tests of stale waste were carried out with different roughnesses and different proportions. The test results show that the τ -ω curve of the interface occurs obvious strain hardening properties with the rough surface of the concrete. The shear strength of the stale waste interface is improved with the increasing of the roughness, and reduced with the existence of the plastic fiber. The deformation and failure modes are mainly determined by the relationship between interface strength and soil strength, which can be divided into the interface failure, the overall failure, and the shear zone failure in the test.

1 INTRODUCTION

As a common foundation form (Odud 2000; Yang 2005), the bored pile is often used in various engineering construction of abandoned landfill sites. Due to the small bearing capacity of waste soil foundation (Luo 2006), the bearing capacity of pile foundation in landfill is mainly provided by side resistance. The properties of side resistance are the key problem in the research of pile foundation bearing capacity in landfill. Meanwhile, engineering practice shows that the side surface of the bored pile is generally uneven and rough (Pando 2002; Pells 1980), which will inevitably affect the capacity of side resistance. Therefore, the influence of pile roughness should be considered in studying the side resistance of bored piles.

At present, the shear test of the soil-structure interface has become the main method to study the side resistance of pile foundations. The method of controlling the smoothness of the structure surface in the test has been widely used to study the influence of roughness on the properties of interfaces. For example, Chen et al. (Chen 2015) and other scholars have conducted experimental studies on the properties of the interface between different soil types and different roughnesses structural surfaces, including clay (Chen 2015), silty clay (Liu 2011), sand (Hu 2004), loess (Feng 2009), frozen soil (Zhao 2013), and macadam soil (Chen 2011), etc. The obtained results have improved the understanding of the pile-soil interface under different geological conditions. In addition, Gong et al. also studied the interaction among the stress history of clay (Gong 2011), the composition and thickness of coarse-grained soil crust (Peng 2011), different gradations of gravel (Zhou 2005), and the shear properties of interfaces with different roughnesses. The proposed stress-strain calculation method and conclusion further reveal the interface properties of soil structure.

---

*Corresponding Authors: 156441363@qq.com, 872510270@qq.com and 491259353@qq.com

DOI 10.1201/9781003344209-6

From the above, many results have been obtained in shear tests by controlling interface roughness, but different stress problems may be caused by the stale waste due to the special stress environment (Dan 2006). Therefore, it is necessary to study the properties of the interface between stale waste and concrete with different roughnesses. In addition, since the shear failure surface was not limited to the single shear test, it can truly reflect the deformation of soil during the shear process and engineering practice. Therefore, in this manuscript, the interface single shear test between stale waste and concrete was carried out. By adjusting the proportion of stale waste and concrete roughness, the mechanical and deformation properties and influencing factors of the interface between stale waste and concrete were comprehensively investigated. The research results can provide a reference for the research and design of pile foundations in landfill in the future.

2 TEST PREPARATION

2.1 *Test system*

A static interface single shear system was used in this test. The test device is shown in Figure 1. The sample with a diameter of 61.8mm and a height of 20.35mm was loaded into a stack ring. During the test, the cutting ring was placed on a concrete block with a size of 150mm×100mm×30mm. Displacement sensors are installed on the 1st, 3rd, 6th, and 10th layers of the bottom box, and the soil deformation was measured and recorded automatically during the shear process. The instrument can apply 50kPa-400kPa stable vertical stress, and its shear rate can be controlled between 0.01to 3mm/min.

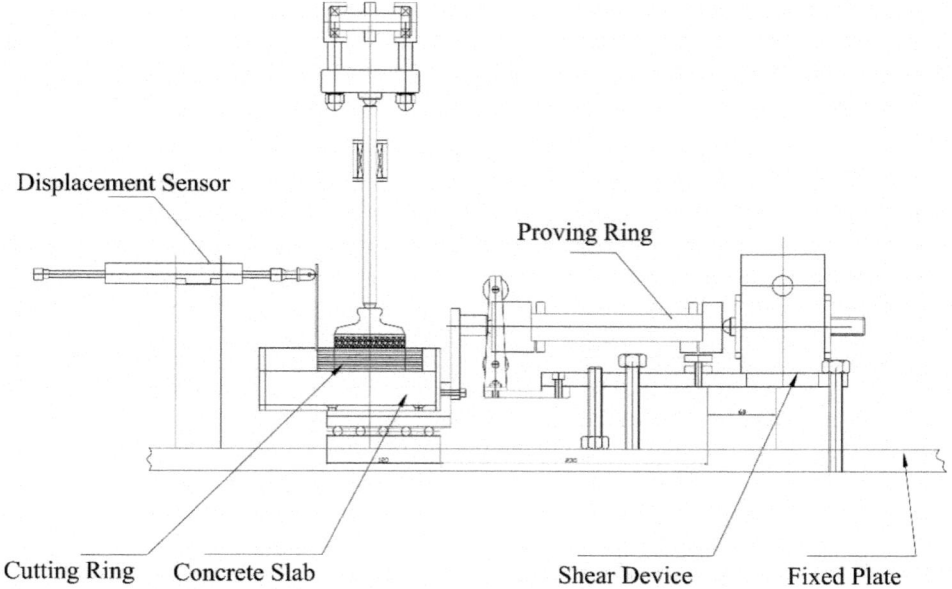

Figure 1.   Diagram of simple shear apparatus.

2.2 *Test materials and preparation*

2.2.1 *Preparation of stale waste materials*
The test parameters were obtained from the main components of the stale waste and the parameters of the foundation in a sealed landfill in Shenzhen (Gao 2011). The stale waste on site has been buried for more than 10 years and has been degraded stably. The degradation products are the fine

particles under the sieve of stale waste after removing plastic, broken glass, stones, animal bones, and other large particles. They are black, slightly smelly, and viscous soil-like substances with relatively uniform plasmids. Its composition is shown in Tables 1 and 2.

Table 1.   Composition of stale waste on site (%).

Composition	Plastic, cloth	Chip	Stone, rubber, metal, bone	Degradation products
Proportion /%	28.74	7.32	14.98	48.96
Specific materials	Floor leather, plastic bags	Chip	Gravel	The filter material of stale waste
Bulk density	1.51	1.45	2.62	2.36

Table 2.   Parameters of stale waste on site.

Moisture content	Void ratio	Dry density	Bulk density
ω/%	e	ρ_d/ g/cm^3	Gs
30	1.25	0.96	2.16

Before the test, the on-site proportion of stale waste and non-plastic fiber stale waste were prepared, and then comparative tests were conducted to investigate the influence of a large amount of plastic fiber (Dan 2006) in the waste on its interface shear properties. The on-site proportion of stale waste was prepared according to the material parameters in Tables 1 and 2, and the non-plastic fiber plastic stale waste was made by the on-site stale waste after removing plastic fibers. During sample preparation, the initial porosity proportion of the sample was kept unchanged at 1.25.

2.2.2   Preparation of concrete blocks with different roughnesses
According to the microscopic scanning results of pile body surface by Pando et al. (Odud 2002) (Figure 2) and the measured data of bored pile construction by Jiang et al. (Jiang 2006), it can be seen that the surface of bored pile is mainly a collection of sawtooth shapes of different sizes, and sawtooth shapes are the basic shapes that constitute the surface roughness of bored pile. Therefore, the roughness was set in the form of a regular sawtooth in the test. The concrete structural slabs were made by controlling the parameters (Joseph 2002) defined by ISO4287:1997 for sawtooth shape: sawtooth height Rt, sawtooth AngleΔ_a, and sawtooth spacing Sm. A total of 3 concrete slabs were made by using C35 concrete. The meaning of each parameter is shown in Figure 3, and the shape parameters of concrete blocks are shown in Table 3.

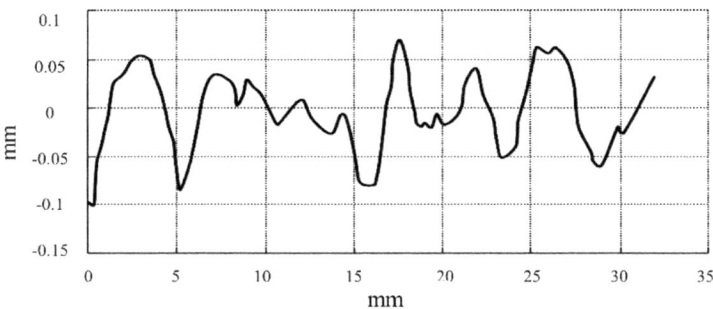

Figure 2.   Pile surface microscopic scan map.

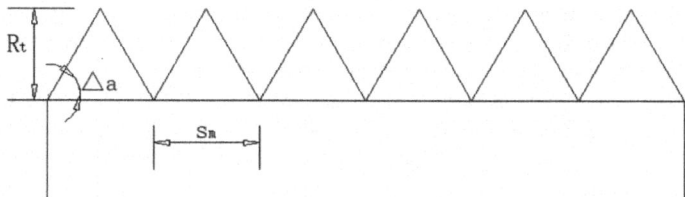

Figure 3.   Diagram for defining the parameters of the regular sawtooth.

Table 3.   The sawtooth parameter setting of each concrete block surface.

Block	Sawtooth spacing S_m (mm)	Sawtooth height Rt (mm)	Angle Δ_a (°C)
1	0	0	0
2	6	1	18.4
3	6	2	33.7

2.3   Test methods

The single shear test of the interface between non-plastic fiber stale waste and on-site proportion stale waste under different roughness conditions was carried out. In this test, the width of the concrete sawtooth was unchanged, the sawtooth height Rt and sawtooth Angle Δ_a were adjusted to control the roughness of the interface, and the influence of plastic fiber in stale waste was considered. The test scheme is shown in Table 4 below. Soil sample A was used to represent the on-site proportion of stale waste, and soil sample B was used to represent the non-plastic fiber stale waste. 24 groups of tests were carried out, in which the normal stress was set to 50kPa, 100kPa, 200kPa, and 400kPa four levels. The samples were pre-pressed before the test, and the shear began after the consolidation was stable. The shear rate was 0.8mm/min, and the test was stopped when the relative displacement reached 7mm.

Table 4.   Test plan.

Group	Sawtooth height R_t(mm)	Angle Δ_a(°)	Sample number	Plastic content
1	0	0	A	0
	0	0	B	28.74%
2	1	18.4	A	0
	1	18.4	B	28.74%
3	2	33.7	A	0
	2	33.7	B	28.74%

3   TEST RESULTS AND ANALYSIS

3.1   Shear strength analysis of interface

Figure 4 shows the τ-ω curve of the contact surface test between the two samples and concrete blocks with different roughnesses, from which it can be seen that:

(1) At the beginning stage, the slope of the τ-ω curve of each test is large, indicating that the shear stress of each interface increases rapidly with shear displacement at the initial stage of shear,

but when the displacement reaches a certain value, the slope of the curve gradually decreases. Subsequently, except for the τ-ω curve of the concrete with a flat surface (Rt=0), the strain hardening properties are shown in the other rough interfaces.

(2) The slope of the initial segment of the τ-ω curve decreased with the appearance of plastic fibers in the stale waste when the normal stress is low (50kPa, 100kPa). It indicates that the stale waste with plastic fibers requires a larger displacement to exert the same shear strength. However, when the normal stress increases, the initial segment of each τ-ω curve almost coincides, indicating that the larger the normal stress is, the smaller the influence of plastic fiber on the speed of shear strength development of contact surfaces with different roughnesses will be.

(3) Under each normal stress, the shear force of the interface in the two soil samples improves with the increase of the interface roughness. It shows that increasing pile side roughness will effectively increase pile foundation side resistance in the landfill. At the same time, under the same roughness condition, the interface shear force of soil sample A is smaller than that of soil sample B, indicating that plastic fiber will harm the shear strength of the stale waste interface. Therefore, the influence of plastic fiber should be paid more attention to in landfill reuse projects.

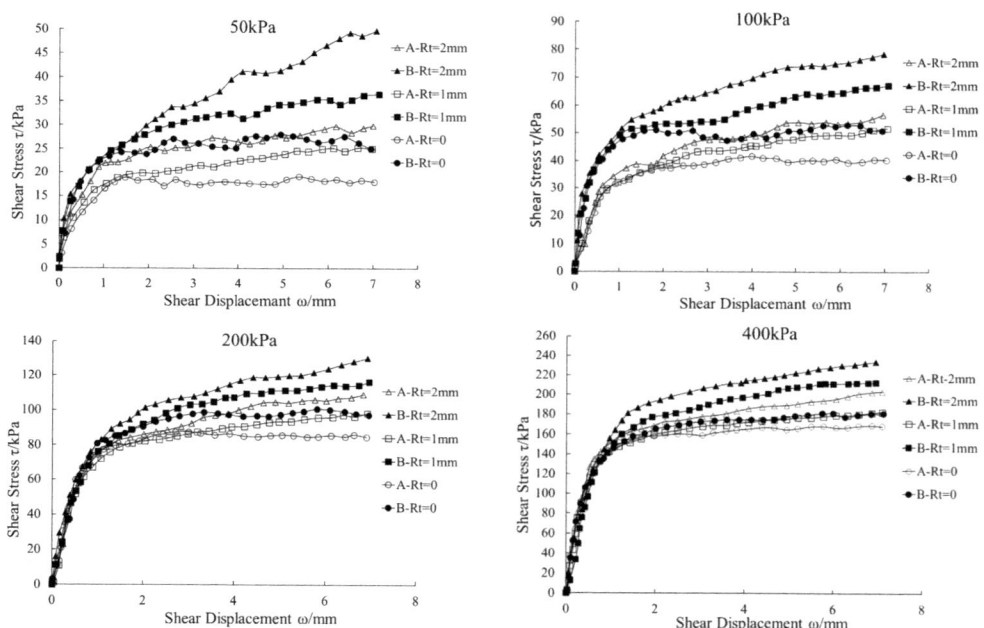

Figure 4. τ-ω curves of interface shear.

3.2   Analysis of deformation and failure of the interface

The deformation process of soil under shear and the soil contour after the test can be obtained by observing the displacement of each layer of soil during the single shear test on the interface. Through the statistical analysis of the deformation and failure modes of all the test soil samples, it can be concluded that the failure modes of the interface in this test can be divided into three types as shown in Figures 5.1–5.3 below interface failure (Table I), overall failure (Table II), and shear zone failure (Table III). The statistical results are shown in Table 5 and Table 6 below.

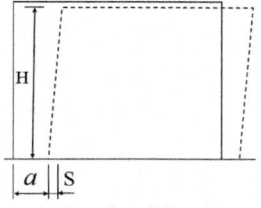
5.1 Interface failure (I)

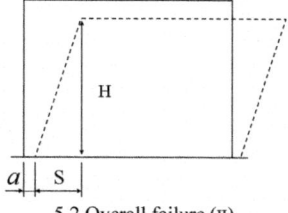
5.2 Overall failure (II)

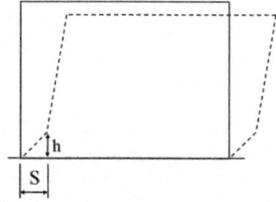
5.3 Shear zone failure (III)

Figure 5.　Three types of failure modes of the single shear interface.

Table 5.　Statistical table of failure form of soil A on the interface.

normal stress ＼ roughness	$R_t=0$	$R_t=1$mm	$R_t=2$mm
50kPa	I	II	II
100kPa	I	II	II
200kPa	I	II	II
400kPa	I	II	II

Table 6.　Statistical table of failure form of soil B on the interface.

normal stress ＼ roughness	$R_t=0$	$R_t=1$mm	$R_t=2$mm
50kPa	I	II	III
100kPa	I	II	III
200kPa	I	II	III
400kPa	I	II	III

3.2.1 *Interface failure*

Interface failure refers to the condition that interface dislocation displacement is ≥ 4mm and soil shear strain is $\leq 15\%$ in the test. Specifically, the maximum shear dislocation occurs among interfaces, while the deformation of soil is small, as shown in Figure 5.1. In this test, the interface failure occurred in the test of $Rt = 0$ concrete block with the lowest roughness and different soil samples. Figure 6 shows the soil displacement changes of each layer when $Rt = 0$ and normal stress $=100$kPa for the on-site proportion of stale waste.

In the figure, the soil in each layer moves with the concrete block at first, but the slope of the soil displacement curve changes abruptly after reaching a certain value. The soil displacement increment gradually decreases to 0, while the staggered deformation between the interfaces continues to increase, indicating that interfacial shear failure occurs. Figure 7 shows the comparison between the soil strength and the interface shear strength of the sample. It can be seen that the interface shear strength is significantly smaller than the soil strength when the interface failure occurs. In addition, the interface strength is controlled by the interface strength, and the interface failure occurs at the interface.

3.2.2 *Overall failure*

With the increase of the interface roughness, the shear strength of the interface increases, and the failure mode of the interface changes from interface failure to overall failure, as shown in Figure 5.2.

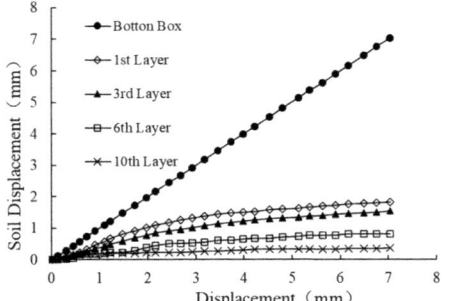
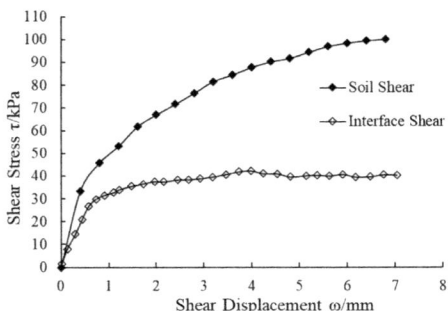

Figure 6. Displacement of each layer of soil under interface failure.

Figure 7. Comparison chart of interface strength and soil strength under interface failure.

In other words, when the interface displacement is ≤4mm, it occurs that the soil shear strain is >15%. Specifically, when the interface between soil and concrete staggers, there is also obvious dislocation between different heights of soil. The deformation of each layer of soil increases linearly from top to bottom without sudden change or inflection point. Figure 8 shows the soil displacement changes of each layer when Rt=2mm and normal stress =200kPa for the on-site proportion of stale waste. It can be seen that during the shear process, the displacement between the interface and the soil layer increases, and the slope of each curve does not appear sudden change. It shows that the interface dislocation and the deformation development of each layer of soil are stable during the shear process, which is the overall failure of the soil sample. The comparison between the strength of the soil sample and the interface shear strength is shown in Figure 8. It can be seen that although the interface strength has increased during the overall failure, it is still smaller than the strength of the soil.

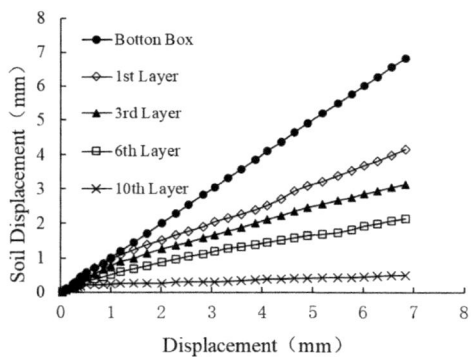
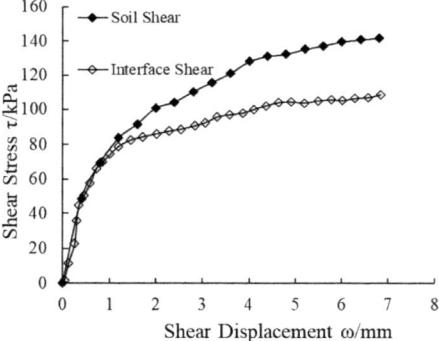

Figure 8. Displacement of each layer of soil under overall failure.

Figure 9. Comparison chart of interface strength and soil strength under overall failure.

3.2.3 Shear zone failure

Figure 5.3 shows that in the test of the interface between Rt = 2mm concrete slab and plastic-fiber stale waste, shear zone failure occurs due to the further increase of the strength of the interface. When the interface dislocation displacement is less than or equal to 4mm, a relatively concentrated shear zone appears near the interface. It is manifested as little interface dislocation in the shear process, uneven soil deformation, and inflection point on the soil contour. Figure 10 shows the soil displacement changes of each layer when Rt=2mm and normal force =100kPa for the non-plastic fiber stale waste. It can be seen that the soil displacement of the sample adjacent to the concrete block increases linearly with the movement of the lower shear box, indicating that the failure of the interface occurs in the soil.

According to the strength comparison diagram, the shear τ-ω curves of the interface and soil coincide, indicating that the strength of the interface is controlled by the soil strength. When the shear stress reaches or exceeds the soil shear strength, an obvious local strain is generated near the interface. The soil is a failure, forming a shear zone. The shear zone is generated between 1 and 3mm from the bottom box in this test.

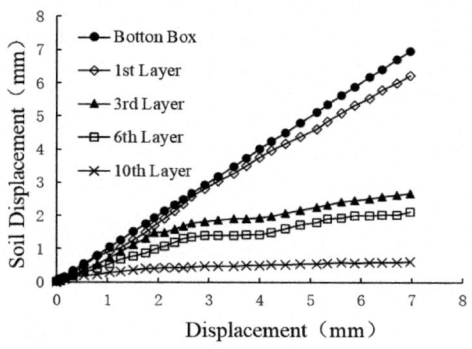

Figure 10. Displacement of each layer of soil under shear band failure.

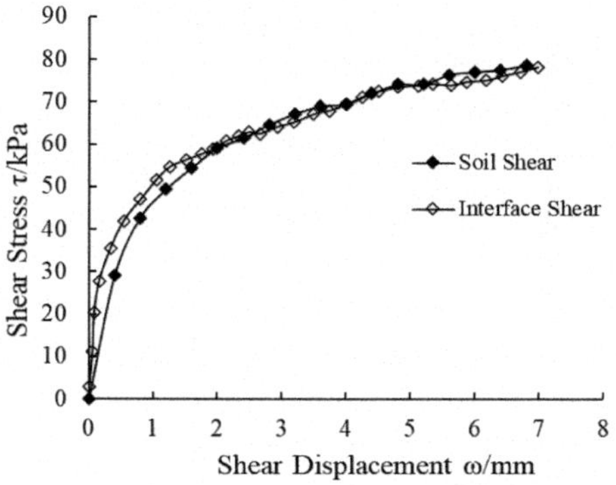

Figure 11. Comparison chart of interface strength and soil strength under shear band failure.

According to the statistics of the test results and the above analysis, with the increase of the shear strength of the contact surface, the failure mode of the interface will gradually change from interface failure to overall failure, and then to shear zone failure. It can be seen that the deformation and failure of soil continuously develop with the increase of interface strength. When the interface strength is small, the soil deformation will be small, and the interface failure occurs. When the interface strength increases, the upward shear force increases when the interface dislocation occurs, resulting in shear deformation of the soil. When the shear strain of soil reaches a certain value, the overall failure of the interface will occur. When the interface strength continues to increase greater than the soil strength, the soil will fail before the interface failure. Shear zone failure occurs, and the strength of the interface is controlled by the soil strength. It can be concluded that with the increase of pile side roughness, the interface strength is improved, and the deformation and failure form of the interface is changed as well. The relative dislocation failure among the pile-soil interface is transformed into the shear failure of the soil around the pile. It can be seen that the pile side friction resistance is not only controlled by the interface strength between pile and soil but also affected by the strength of soil around the pile.

4   CONCLUSION

In this manuscript, a regular saw-tooth surface was used to simulate the side surface of bored piles. Through the single shear test on the interface with considering the influence of plastic fiber in stale waste, the mechanics and deformation and failure properties of the interface between stale waste and concrete are studied. The main findings of this manuscript are summarized as follows:

(1) When the shear displacement is greater than a certain value, the interfaces of other rough concrete will show obvious strain hardening properties, except for the τ-ω curve of concrete with a flat surface, which will remain approximately horizontal.
(2) The side resistance of the pile foundation can be effectively improved by increasing pile side roughness since the shear strength of the interface of stale waste can be improved by increasing

the surface roughness of concrete. At the same time, plastic fiber will harm the shear strength of the stale waste interface. Therefore, the influence of plastic fiber should be paid more attention to in landfill reuse projects.

(3) According to the deformation and failure law of the stale waste in the single shear test, the deformation and failure of the interface can be divided into three types: interface failure, overall failure, and shear zone failure.

(4) The main factor of shear failure mode is determined by the relationship between stale waste strength and interface strength. With the increase in shear strength of the interface, the interface failure mode changes successively from interface failure to overall failure, as well as the shear zone failure. The increase of pile roughness not only affects the strength of the interface but also makes the deformation and failure mode of the interface change from the relative displacement failure between the pile and soil to the shear failure of the soil around the pile. The pile foundation side resistance is not only controlled by the interface strength of the pile and soil but also affected by the strength of the soil around the pile.

ACKNOWLEDGMENTS

Financial support from the Pearl River S&T Nova Program of Guangzhou (Project Nos. 201906010023).

REFERENCES

Chen, J., et al. (2011). Experimental study of mechanical properties of the interface between pile and macadam soil in an accumulative landslide in three gorges reservoir area. *Chinese Journal of Rock Mechanics and Engineering*, 30, 2888–2895.

Danzeng, Dan., et al. (2006). Strength characteristics of municipal solid waste. *Journal-Tsinghua University*, 46(9), 1538–1540.

Dove, Joseph. E. & Jarrett, J.B. (2002). The behavior of dilative sand interfaces in a geotribology framework. *Journal of Geotechnical and Geoenvironmental Engineering*, 128(1), 25–37.

Feng, J., et al. (2009). Study on characteristics of the interface between undisturbed loess and concrete by using large-sized single shear apparatus. *Journal of Water Resources and Architectural Engineering* 7(3), 120–122.

Gao, H., et al. (2011). Grouting improvement of the foundation of municipal solid waste. *Advances in Science and Technology of Water Resources*, 31(1), 58–61.

Gong, H., et al. (2011). Research on the effect of stress history on shear behavior of the interface between clay and concrete. *Chinese Journal of Rock Mechanics and Engineering*, 30(8), 1712–1919.

Hu, L. & Pu, J. (2004). Testing and modeling of the soil-structure interface. *Journal of Geotechnical and Geoenvironmental Engineering*, 130(8), 851–860.

Jiang, D.H. (2006). *Research on Quality of Drilled Holes for Cast-in-Place Piles in Shanghai District* (Doctoral dissertation, Tongji University).

Junhua, Chen., et al. (2015). Experimental research on mechanical characteristics of the cohesive soil-structure interface by considering its roughness. *Advanced Engineering Sciences*, 47(4), 22–30.

Kai, Peng., et al. (2011). Experiments on the influence of slurry kinds on the mechanical behavior of the interface between gravel and concrete. *Journal of Chongqing University* (Natural Science Edition), 34(1), 110–115.

Liu, F.C., et al. (2011). Study of shear properties of the silty clay-concrete interface by simple shear tests. *Chinese Journal of Rock Mechanics and Engineering*, 30(8), 1720–1728.

Ming-Liang., et al. (2005). Research on foundation and security of the edifice in Jinkou municipal refuse landfill. *Chinese Journal of Rock Mechanics and Engineering*, 24(4), 628–637.

Odud, C. (2000). *Current state of the practice of construction on closed landfill sites. In 16th International Conference on Solid waste technology and management*, PA, USA-December,12.

Pando, M.A., et al. (2002). Interface shear tests on FRP composite piles. In Deep Foundations 2002: *An International Perspective on Theory, Design, Construction, and Performance*, 1486–1500.

Pells, P.J.N., et al. (1980). *An experimental investigation into side shear for socketed piles in sandstone*. In Proceedings of the international conference on structural foundations on rock, Sydney, 1, 291–302.

Xing-Wen., et al. (2006). Experimental study on engineering mechanical properties of stale refuse [J]. *Chinese Journal of Geotechnical Engineering*, 28, 622–625.

Zhao L.Z., et al. (2013). Development and application of large-scale multi-functional frozen soil-structure interface cycle-shearing system. *Dam & Safety*, 707–713.

Zhou, X.W., et al. (2005). Large-scale simple shear test on mechanical properties of the interface between concrete face and gravel underlayer[J]. *Chinese Journal of Geotechnical Engineering*-Chinese edition-, 27(8), 876.

Frontiers in Civil and Hydraulic Engineering – Mohamed A. Ismail and Hazem Samih Mohamed (Eds)
© *2023 The Authors, ISBN 978-1-032-38247-0*

Direct shear test of root-containing loess under freeze-thaw conditions

Fanxing Meng*, Hui Li* & Ningshan Jiang
College of Civil Engineering, Qinghai University, Qinghai, China

Chengkui Liu
Qinghai Building and Materials Research Co., Ltd, Qinghai, China
The Key Lab of Plateau Building and Eco-community in Qinghai, China, Qinghai, China

Gencheng Liu
Zhongyu Hengxin Engineering Consulting Co., Ltd, Henan, China

ABSTRACT: To study the stability of root loess under freezing and thawing conditions, an indoor direct shear test was conducted at Qinghai University, Xining city, Qinghai Province. The roots of the three plants were Alkaline grass, North China Artemisia Sativa, and caragana. Through the conclusion of the normal temperature and freezing and thaw tests of the three samples under different root content, the relationship curve among shear strength, internal friction Angle, and cohesion is drawn, and the influence of different plant roots and freezing and thaw conditions on loess is analyzed, which strengthens the foundation of geological disaster prevention in the freezing and thaw period.

1   INTRODUCTION

Under the action of freezing and thawing, with the gradual rise of temperature, the melting of ice and frozen soil is easy to cause collapse and soil erosion. The main factor leading to soil instability is the lack of shear capacity. For such problems, it is best to use the management plan of planting vegetation, because the vegetation has the advantages of improving the ecological environment, reducing soil erosion, water conservation, fixing soil and slope, improving climate, and so on. Therefore, it is of great theoretical and practical significance to promote the mechanism of vegetation root soil fixation for the prevention of soil erosion and the construction of an ecological environment and to promote the development of basic research. Dong X.H. & A.J. Zhang. 2010. explored the changes in the shear strength of the reshaping loess in the Yangling area after the freezing and thaw cycle, and found that under the same moisture content, the cohesion of the loess would reduce to a minimum after 3 to 5 times, and gradually stabilized, but the internal friction angle did not change significantly. Mu, Y.H. 2011. analyzed the loess soil samples under different times of freezing and melting cycles as the research objects, and found that the porosity of the soil samples after freezing and melting cycles increased significantly, and the freezing and melting effect caused the weakening of the soil structure. Chen, H. 2019. studied the stability of vegetation roots and the loess effect of soil consolidation. The indoor direct shear test, model test and numerical simulation method of the vegetation root system can significantly improve the loess shear strength. With the increase of vegetation root number, the loess maximum shear strength and residual shear strength have increased significantly. Yang, Y.C. 1996. put forward the concept of "root soil complex", analyzed the mechanism of soil and water conservation efficiency of herbal plants, put forward the concept of "column soil grid" and "soil filter" at the connection of root

---

*Corresponding Authors: 1075377822@qq.com and 365329508@qq.com

                                                    DOI 10.1201/9781003344209-7

lotus of herbs, and studied the relationship between shear strength of soil and soil root content. Soil root content is positively correlated with value, not with the value itself. Yao, Y.L. 2015, Chen, Z.D. 2016, Mao, L.L. 2007. show that the more roots there are, the greater the cohesion of the soil body will be. The optimal root content makes the shear strength of the root and soil complex reach its peak. In the study of the influence of freezing temperature on the freezing and melting strength of soil, Wei, Y. 2018 found that the lower the freezing temperature is, the faster the unrestricted compressive strength of loess will decrease. With the decrease in freezing temperature, the cohesion force decreases, and the internal friction angle increases. Ziemer 1981 did large volume in-situ direct shear test of pine root soil. The test results showed that as 1kg of soil roots per unit volume increased, the shear strength of the root-soil complex increased by about 3.5kPa. Therefore, the relationship between the root content in soil and the increasing value of soil shear strength can be evaluated. Abernethy. 2011 used the remodeling root soil as the research object for the direct shear test, which showed that the roots existing in the soil can restrain the deformation of the soil body and improve the soil shear strength. Zhang, F. 2005 did soil and root-soil indoor straight shear test compared its shear strength, and found that the root density in a certain range of root-soil shear strength was enhanced with increasing root density. However, the root density in the soil has a peak, and more than the existence of the peak root system will aggravate soil homogeneity, with a negative effect on soil shear strength. Based on the indoor direct shear test, Hu, P. 2008. studied the anchorage effect of Manila grass root system, and established the formula of root anchorage effect by analyzing the test results.

This experiment analyzed the direct shear test of the root-containing loess under freezing and thawing to determine the influence of the root system and the freezing and thawing environment on the soil.

2 DIRECT SHEAR TEST OF ROOT-CONTAINING LOESS UNDER FREEZE-THAW CONDITIONS

2.1 *Physical property index of soil body*

The soil sample selected in this experiment is on the hillside outside the south gate of Qinghai University in Chengbei District, Xining City, Qinghai Province. The soil type is silt, and the soil property indexes are shown in Table 1.

Table 1. Soil property indicators.

Degree of depth /cm	Soil Density (g/cm³)	The average moisture content of (%)	Grading analysis (%)			
			0.25–0.075mm	0.075–0.05mm	0.05–0.005mm	<0.005mm
0–100	1.58±0.04	9.11	7.83	16.17	65.36	10.64

2.2 *Select Materials*

The selected slope element soil was transported back to the laboratory and dried at 106°C for 12h. It can be seen from the soil property index that the selected soil sample particles were mainly concentrated below 0.25mm, so the dried soil samples were prepared by the geotechnical sieve with an aperture of 0.25mm. The soil samples were selected through a geosieve with an aperture of 0.25mm, and the parameters can be used to establish the slope model with roots below (as shown in Figure 1).

The basic principles of screening the three kinds of plants are cold resistance. The root system is very developed, which can better absorb the deep water in the soil. According to this principle, three roots of plants, Alkaline grass, caragana, and North China Artemisia Sativa, were selected as test subjects (as shown in Figure 2).

The roots of the three plants purchased in the same cycle were cleaned and controlled at about 2cm when cutting, avoiding the final results caused by different lengths.

Figure 1.   Soil particle size with less than 0.25mm plain soil.

Figure 2.   Root systems of the three plants.

2.3   *Experimental methods*

Determination of the root content gradient: To study the shear strength of the root-soil complex among the three plants under different root content conditions, the soil with an average water content of 9.11% in the depth range of 0–100cm below was first selected to prepare the samples of plain soil and different vegetation root content. Three root content gradients with root content of 0.5%, 1%, and 1.5% were set, respectively. The difference value of adjacent root content was 0.5%.

Composite plate method of plain soil and vegetation:

Plain soil: The compaction cylinder with an internal diameter of 61.8mm and 125mm high is selected. A certain amount of root-free soil is taken, which is divided into four layers for compaction. The number of each layer is consistent. After finishing, the solid soil will be taken out. The ring knife blade with an inner diameter of 61.8mm and 20mm high is downward placed on the soil sample, which is pressed down vertically, making the soil out of the ring knife and leveling the ring knife on both sides. The ring wall residual soil particles are cleaned to keep the specimen intact, weighing the total mass of the ring knife and soil samples. Then, the above methods are repeated to prepare plain soil samples without a root system.

Root-soil composite system sample method: first, the plant, rice, and chicken root are out and rinsed, and then a certain amount of three plant roots are taken. 20mm root sections are evenly cut, and each root-soil complex sets root content (0.5%, 1%, 1.5%) among every 3 groups, which fully stirs with plain soil into the root-soil complex sample. The three test samples are shown in Figures 3–5.

The prepared plain soil and three root-soil complex samples with different root contents were sealed in plastic wrap together with the ring knife, which are put in a freeze-thaw circulation tank. It is found that the initial freeze-thaw cycle has the greatest impact on soil strength and internal structure. With the increase, the influence of the freeze-thaw cycle on soil generally becomes smaller and gradually stable after 3-5 times. Therefore, the number of freeze-thaw cycles was selected in this test, the temperature range was-15°C–15°C, and the time was 12h. Since the freezing and thawing effect can affect the shear strength of the root-soil complex, the plain soil and root-soil complex which do not participate in the freezing and thawing cycle, namely under normal-temperature conditions, are used as the reference group.

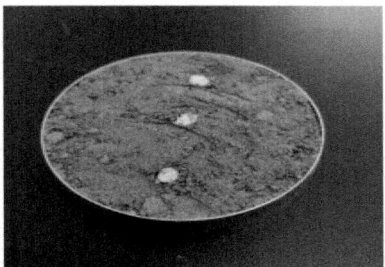

Figure 3.   Caragana sample.

Figure 4.   North China Artemisia Sativa sample.

Figure 5.   Alkaline grass sample.

2.4   *Direct shear test*

In this direct shear test, ZJ strain controlled direct shear instrument with a ring knife whose height is 20mm. The pressure was 50kPa, 100kPa, 150kPa, and 200kPa under four-grade vertical pressures. The shear strain rate was 2.4mm / min.

3 THE DATA OBTAINED FROM THE INDOOR STRAIGHT SHEAR TEST

3.1 *Relationship curves of different root content complexes under room temperature*

The blue line is the Alkaline grass sample, the yellow line is North China Artemisia Sativa sample, and the pink is the caragana sample (as shown in Figures 6–9).

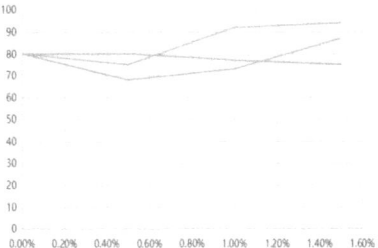

Figure 6. Shear strength at P = 100KPa under room temperature.

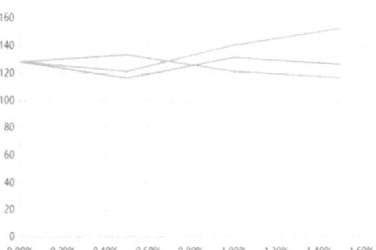

Figure 7. Shear strength at P = 200KPa under room temperature.

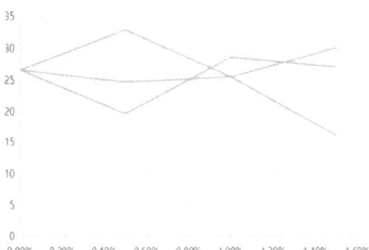

Figure 8. Internal friction Angle under room temperature.

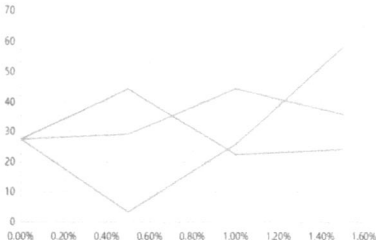

Figure 9. Cohesion under room temperature.

At room temperature, the shear strength and internal friction angle of the Alkaline grass sample increased with the increase of root content. However, the variation of cohesion increases first and then decreases.

At room temperature, the shear strength and cohesion of the North China Artemisia Sativa sample increased with the increase of root content at P = 100Kpa. At P = 200Kpa, the shear strength and internal friction angle decrease with the increase of root content.

At room temperature, the shear strength and cohesion of the caragana sample at P = 100Kpa decreased with the increase of root content. At P = 200Kpa, the shear strength and internal friction angle increase firstly and then decrease with the increase of root content.

3.2 *Relationship curves of different root content complexes under freeze-thaw*

The blue line is the Alkaline grass sample, the yellow line is North China Artemisia Sativa sample, and the pink is the caragana sample (as shown in Figures 10–13).

Under freezing and thawing conditions, the shear strength and cohesion of the Alkaline grass sample increased with the increase of root content. The internal friction angle decreases first and then rises.

Under freezing and thawing conditions, the shear strength and cohesion of the North China Artemisia Sativa sample increased with the increase of root content at P = 100Kpa. At P = 200Kpa, the shear strength and internal friction angle decrease with the increase of root content.

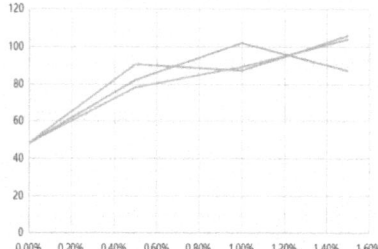

Figure 10. Shear strength
at P = 100KPa under freeze-thaw.

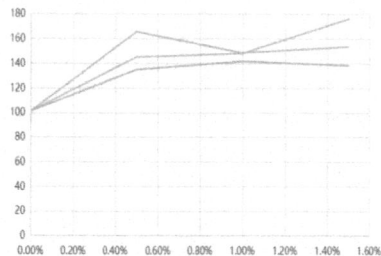

Figure 11. Shear strength
at P = 200KPa under freeze-thaw.

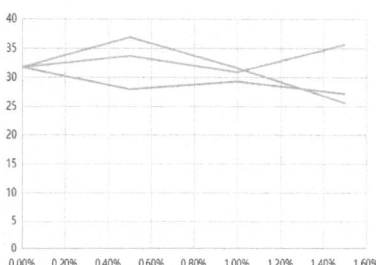

Figure 12. Internal friction Angle
under freeze-thaw.

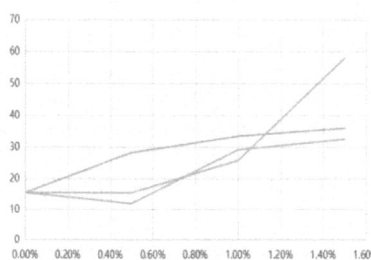

Figure 13. Cohesion under freeze-thaw

Under freezing and thawing conditions, the cohesion of caragana samples increased with the increase of root content. The shear strength and internal friction angle increase first and then decrease with the increase of root content.

4   CONCLUSION

(1) With the increase of vertical pressure, the shear strength of the three samples increases obviously.
(2) The root and soil complex of different root types had different mechanisms of strength change. With the increase in root content, the cohesion of shrub roots with larger diameters did not change significantly, and the internal friction angle tended to decrease. With the increase of root content, the cohesion of herb roots with smaller diameter increased obviously, and the internal friction angle also increased to a certain extent.
(3) Freezing and thawing can weaken the shear strength and cohesion of plain soil, and enhance the internal friction angle.
(4) The presence of roots reduces the weakening effect of freezing and thawing on the shear strength and cohesion of soil, and different roots have different effects on internal friction angles.
(5) Compared with the normal temperature situation, the internal friction angle and cohesion of the root-soil composite ratio are increased to a certain extent, and the enhancement effect of shear strength is more obvious.
(6) The values of internal friction angle and cohesion of the three samples with different root contents were brought into the slope numerical simulation established by ABAQUS. To study the influence of different roots on slope anti-sliding stability under hydraulic erosion, the stress-strain curve was drawn. The direct shear test under freezing and thawing in the laboratory and the numerical simulation of hydraulic erosion and scour slope were combined to explore the anti-sliding mechanism of loess slope under meltwater conditions.

REFERENCES

Abernethy, B. & I.D. Rutherford. The distribution and strength of riparian tree roots concerning river banker enforcement [J]. *Hydrol, Process*, 2001, 15: 63–79.

Chen, H. Influence of vegetation root system on loess slope stability and the effect of soil fixation [J]. *Hydropower and Energy Science*, 2019, 37 (10): 97–100.

Chen, Z.D. Study on strength properties of Fragrant Root Soil Complex [D]. *Central South University of Forestry and Technology*, 2016.

Dong, H.X. & A.J. Zhang (eds.). Experimental study on loess strength deterioration caused by long-term freeze-melting cycle [J]. *Journal of Engineering Geology*, 2010, 18(6): 887–893.

Hu, P. & X.G. Song (eds.). Experimental study on root-fixation effect of highway slope [J]. *Rock and soil mechanics*, 2008, (02): 442–444.

Mao, L.L. *Mechanical Analysis of Vegetation Root in Ecological Slope Protection* [D]. Wuhan University of Technology, 2007.

Mu, Y.H. & W. Ma (eds.). The microscopic quantitative study of the effects of freeze-thaw effects on the structure of compacted loess [J]. *Journal of Geotechnical Engineering*, 2011, 33(12): 1919–1925.

Wei, Y. & G.S. Yang (eds.). Study on the effect of frozen temperature on the mechanical characteristics of frozen-thaw loess [J]. *Journal of Yangtze River Academy of Sciences*, 2018, 35(8): 61–66.

Yang, Y.C. & Y.J. Mo(eds.). Experimental study on water erosion strength and shear strength of soil herbaceous vegetation root complex [J]. *Journal of China Agricultural University*, 1996, 02:31–38.

Yao, Y.L. *Study on Vegetation Protection Effect of Loess Cutting Slope in Shaanxi Highway* [D]. Chang'an University, 2015.

Zhang, F. & J.X. Chen (eds.). Study on the shear resistance of root-soil in slope ecological protection [J]. *Geotechnical Foundation*, 2005, (03): 25–27.

Ziemer, R.R. Roots and the stability of forested slopes[J]. *Inc association of hydrologic science*, 1981, 18(3): 343–362.

Frontiers in Civil and Hydraulic Engineering – Mohamed A. Ismail and Hazem Samih Mohamed (Eds)

Analysis of the influence of groundwater distribution characteristics on the impermeability of subway stations

Cao Wang* & Guodong Li
Power China Municipal Construction Group Co., Ltd, Tianjin, China

Wenbin Xiao* & Fengting Li*
School of Civil Engineering, Shandong University, Jinan, China

ABSTRACT: The urban groundwater environment has a great impact on the design and construction of subway stations, and it is also one of the important factors in the structural design of subway stations. To discuss the influence of groundwater distribution characteristics on the structure of subway stations, the numerical calculation and analysis method are used to study the distribution law of the seepage field of subway stations under different water pressure differences, which aims to elaborate on the influence effect of groundwater on the anti-seepage of subway station structure. The main conclusions are as follows. (1) As the groundwater level difference exceeds the station middle plate, and the water pressure is increasing with a first-order function, the influence of groundwater on the diaphragm wall is becoming more and more obvious. (2) When the pore pressure difference between the two sides of the station is large, the side with high groundwater level will be easier to bypass the bottom of the underground diaphragm wall and penetrate into the station floor. The more the pore pressure difference between the two sides of the station is, the more obvious this seepage phenomenon will be. (3) Although the underground water level monitoring has been included in the construction monitoring project of the metro station, the underground water level change caused by the artificial disturbance or extreme weather during the later operation period should also be paid attention to by the Metro builders.

1 INTRODUCTION

The urban subway project has become the way and method to solve the urban traffic congestion problem in large and medium-sized cities along the southeast coast of China. Meanwhile, the distribution of groundwater is high in these cities, and the subway station structure bears the dual role of soil pressure and groundwater pressure for a long time. Although the role of groundwater pressure is usually considered in the design process of subway stations, the renewal of the urban underground space and extreme weather bursts cause a large impact on local groundwater distribution, resulting in irregular changes in the groundwater distribution in local areas. Due to the change in groundwater distribution, it is easy to cause the subway station structure to face the reciprocal action of adding and unloading. The weak area reserved for the station structure may lead to its deformation and damage, causing groundwater infiltration. It affects the safety and normal operation of the subway station. Wang Xuwei (Wang 2021) took the subway station waterproofing project as an example to improve the quality management control system of material quality control, design quality control, construction quality control, and construction quality control and proposed specific measures and points for quality control of waterproofing projects in subway stations. Jia Zhuo (Jia 2021) designed a reasonable precipitation scheme based on hydrogeological pumping tests and used

---

*Corresponding Authors: 2587065175@qq.com, 1422539796@qq.com and 1422539796@qq.com

the groundwater analysis software Feflow to simulate the effect of single-well pumping tests. Precipitation wells were used to demonstrate the feasibility of sparse precipitation. Jiang Hao (Jiang 2022) developed a series of technical measures to apply the MJS method to stopping water at the joints of ground-linked walls of the enclosure structure, relying on the Jingyu Wudaogai subway station project of Harbin City Rail Transit Line 3. It solved the problem of precipitation and leakage of deep foundation pits in the construction of underground buildings. Xu Ruiyao (Xu 2020) conducted a quantitative analysis of the Chengdu Metro Line 8 Phase I project, and derived the congestion value of this project on urban groundwater, the ground settlement value caused by precipitation in the station pit, and the maximum surge volume of the station.

From the above analysis, it can be seen that the urban groundwater environment has a great influence on the design and construction of metro stations, which is also one of the important factors in the structural design of metro stations. To explore the influence of groundwater distribution characteristics on the structure of metro stations, numerical calculation and analysis methods are used to study the distribution law of the seepage field of metro stations under the effect of different water pressure differences, which aims to elaborate on the effect of groundwater on the structural impermeability of metro stations.

2 NUMERICAL CALCULATION MODEL AND ANALYSIS METHOD

2.1 *Numerical calculation model*

In this paper, it is assumed that the cross-sectional dimension of the subway station model is 20.8×13.5m, with 52m in length. The diameter of the station columns is 0.8m, the distance between columns is 8m, and there are 14 columns in total. The station is divided into two layers, namely the top slab, middle slab, bottom slab, and side walls. The thickness of the top and bottom slabs is 0.9m, the thickness of the middle slab is 0.4m, and the thickness of the side walls is 0.7m. The upper and lower sections of the columns are bound to the station respectively. This constraint can make the deformation between the two collaboratively, which is closer to the actual project. In ABAQUS, C3D8 cells are used to simulate each part of the station cells. The grid size of the station top, bottom, middle and side slabs is 1m, and the grid size of the station columns is 0.4m, with a total of 8560 cells.

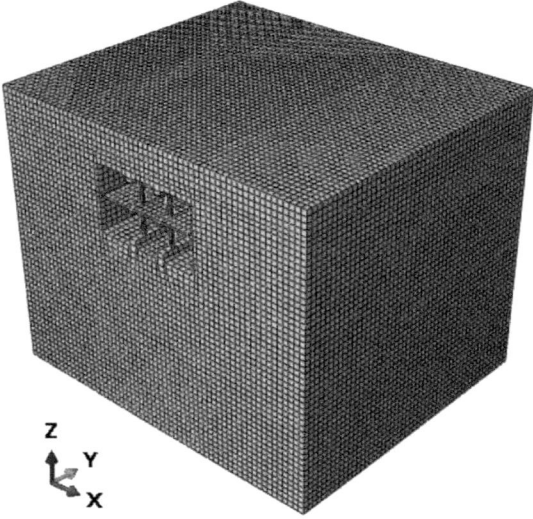

Figure 1. Numerical calculation model of the metro station.

2.2 *Numerical calculation and analysis method*

The station is modeled elastically. The following assumptions are used in the numerical simulation of the station model.

1. The uniform and isotropic materials are used for each part of the units in the model.
2. The station model is horizontal and represents an infinite distance.

The variability of groundwater is simulated by setting different groundwater level elevations around the subway station, and the station side walls and top slab are constrained to be displaced in the normal direction with water pressure differences of 0.00m, 4.00m, 5.62m, 11.97m, and 19.12m for a total of five conditions.

3   NUMERICAL CALCULATION RESULTS AND ANALYSIS

The loads on the ground connection wall of the station are mainly water pressure and earth pressure, followed by the self-weight of the surrounding buildings and other loads. Therefore, understanding the local groundwater level is a key issue. In the construction of open-cut stations, usually, the water head inside and outside the pit is different, and the groundwater seeps from the side with a high head to the side with a low head. The seepage of groundwater will have an impact on the lateral earth pressure of the underground diaphragm wall, which may lead to safety accidents such as sudden surges and sand flow in the station. Therefore, for the force influence of groundwater on the bottom slab of the station, based on numerical calculation analysis, the pore pressure distribution of the surrounding soil of the station under five kinds of water level conditions is discussed separately to reveal the influence law of groundwater on the station. As shown in Figure 2, the height difference

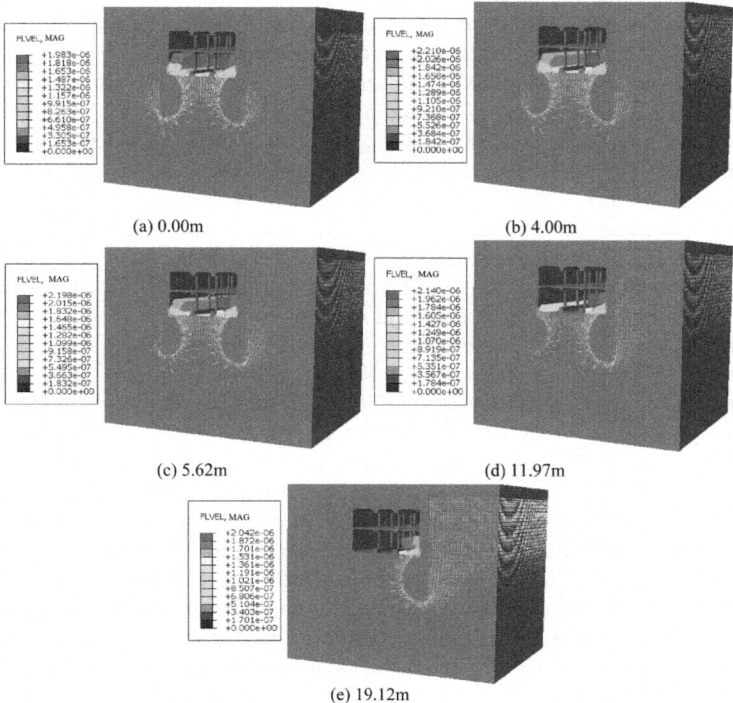

(a) 0.00m   (b) 4.00m

(c) 5.62m   (d) 11.97m

(e) 19.12m

Figure 2.   Station cross-sectional groundwater level vector map.

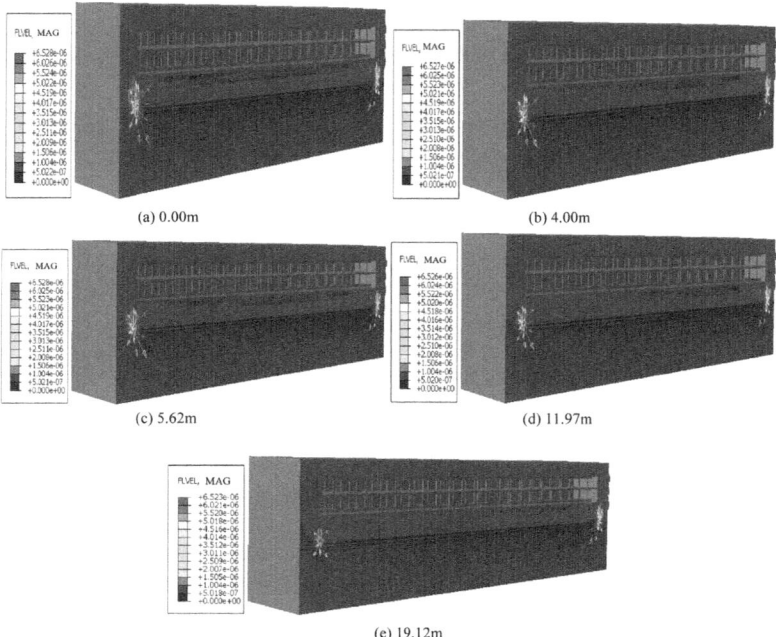

(a) 0.00m   (b) 4.00m

(c) 5.62m   (d) 11.97m

(e) 19.12m

Figure 3. Groundwater level flow rate variability in the longitudinal section of the station.

of the water pressure in the soil on both sides of the station cross-section is 0.00m, 4.00m, 5.62m, 11.97m, and 19.12m respectively, where 5.62m underground is the height of the station top slab position, 11.97m is the height of the station middle slab position, and 19.12m is the height of the station bottom slab position. As the groundwater level decreases, the pore pressure difference between the side walls at both ends of the station becomes more obvious. The pore pressure at the bottom slab location increases continuously, especially when the water pressure difference increases from 5.62m to 11.97m. By comparison, it can be seen that when the pore pressure on both sides of the station is at 4.00m and 5.62m, the influence of groundwater on the underground diaphragm wall will be small. In other words, the difference in water table height does not reach the top plate of the station. With the groundwater level difference exceeding the station mid-slab, the water pressure increases as a function of the law, and the influence of the groundwater on the underground diaphragm wall becomes more and more obvious.

As shown in Figure 2, the groundwater level pore pressure variability diagram of the longitudinal section in the station is shown. It is the same as the station cross-section. Based on the numerical calculation analysis, the stresses of the station floor under the five water level working conditions are discussed separately to reveal the law of the groundwater influence on the station floor. The height difference of the water pressure in the soil on both sides of the longitudinal section of the station is 0.00m, 4.00m, 5.62m, 11.97m, and 19.12m respectively, where 5.62m underground is the height of the top slab position of the station, 11.97m is the height of the middle slab position of the station, and 19.12m is the height of the bottom slab position of the station.

When the pore pressure difference between the two sides of the station is large, the water flow will be more likely to infiltrate into the station floor from the bottom of the diaphragm wall. The increase in the pore pressure difference will also lead to an increase in the lateral pressure of the diaphragm wall, causing the lateral deformation of the diaphragm wall. When the difference in pore pressure between the two sides of the station is large, the side with a high water table will be more likely to infiltrate into the station floor through the bottom of the diaphragm wall, and the more the difference in pore pressure between the two sides of the station is, the more significant this infiltration phenomenon will be. The more the difference in pore pressure between the two sides of the station is, the more significant this infiltration phenomenon will be. The station diaphragm wall is more susceptible to seepage at the bottom.

4 CONCLUSIONS AND RECOMMENDATIONS

Urban underground space renewal and extreme weather bursts cause a large impact on the local groundwater distribution, which are likely to cause the subway station structure to face plus or minus reciprocal action. The weak area reserved for the station structure may lead to its deformation and damage, causing groundwater infiltration and affecting the safety and normal operation of the subway station. To explore the influence law of the groundwater distribution characteristics on the subway station structure, numerical calculation and analysis methods are used to discuss the pore pressure distribution of the surrounding soil in the station under five water level working conditions, respectively, which aims to reveal the influence law of groundwater on the station. The main conclusions are as follows.

(1) With the groundwater level difference exceeding the station mid-slab, the water pressure increases in a primary function law, and the influence of the groundwater on the underground diaphragm wall becomes more and more obvious.
(2) When the difference in the pore pressure between the two sides of the station is large, it is easier for the side with a high groundwater level to infiltrate into the station floor through the bottom of the diaphragm wall. The more the difference in pore pressure between the two sides of the station is, the more significant this infiltration phenomenon will be.
(3) Although groundwater level monitoring has been included in the subway station construction monitoring project, groundwater level changes caused by artificial disturbance or extreme weather during later operation should also be paid attention to by the subway builders.

REFERENCES

Jia Zhuo. (2021) Research on groundwater control measures in the foundation pit of Qiligou subway station [J]. *Sichuan Cement*, 2021(02): 326–328.
Jiang Hao. (2022) Research on the design and construction technology of deep foundation pit precipitation for subway stations with a water-rich sand layer [J]. *Anhui Architecture*, 2022, 29(3): 40–42.
Ouyang Lina, Cao Guangyong, Zhang Feng, Wu Yaowu. (2021) Analysis of the effect of rainfall on the groundwater level in the stations of the southern extension of Metro 3 in Hefei [J]. *Journal of Jiamusi University* (Natural Science Edition), 2021, 39(4): 1–4.
Wang Xuwei. (2021) Research on the construction technology of waterproofing projects in metro stations [J]. *Construction Machinery and Maintenance*, 2021, (03): 238–240.
Xu Ruiyu. (2020) Groundwater impact analysis of Chengdu Metro Line 8 Phase I [J]. *Henan Science and Technology*, 2020, (7):122–125.

Frontiers in Civil and Hydraulic Engineering – Mohamed A. Ismail and
Hazem Samih Mohamed (Eds)
© 2023 The Authors, ISBN 978-1-032-38247-0

The influence of the different fiber content on the micro-structure characteristics of fiber concrete

Hailin Wu* & Jie Liu
College of Hydraulic & Environmental Engineering, China Three Gorges University, Yichang, Hubei, China

Ziqiang Pei
Wanjiazhai Town Power Station Administration District, Youguan County, Xinzhou City,
Shanxi Province, China

ABSTRACT: To study the micro-structural characteristics of fiber concrete and the mechanism of fiber concrete crack resistance, the micro-structural morphology of fiber concrete specimens was observed and analyzed in this paper by using the electron microscope scanning technique. Based on the research group's previous studies on axial tensile tests of steel fiber and nano-fiber concrete specimens with different fiber admixtures, samples were taken from the fiber concrete specimens with different fiber admixtures. This paper focuses on the effect of different steel fiber doping and nanofiber doping on the micro-structural characteristics of fiber concrete. The results show that the micro-structural compactness of the fiber concrete specimens with 1.5% steel fiber and 0.1% nanofiber doping is optimal, and the optimal doping rate is consistent with the doping rate with the optimal effect of crack width control in the fiber concrete axial tensile test.

1 INTRODUCTION

In the study of concrete, it is known that the microstructure of the concrete determines its macroscopic properties in a certain sense. Through the micro-structure of concrete, it is possible to understand the mechanism of its macroscopic properties and eventually improve the macroscopic properties by changing the micro-structure in an appropriate way. Generally speaking, the microscopic properties of concrete are compared to its macroscopic properties, and it is usually considered that the properties of the concrete within the invisible size of the naked eye (below 200 μm) can be considered the microscopic properties of concrete.

Studying the microscopic properties of the concrete, its microscopic properties consist of two main stages, namely the microscopic properties of the mix and the microscopic properties of the hardened concrete. They have completely different properties of the concrete structure and components. The microscopic properties of the mix are time-varying, and the micro-structure of the concrete mix is gradually changing as hydration proceeds. Although the microstructure of concrete also changes after hardening with environmental effects and other factors, the rate of change is slow compared with that of the mix. A large number of studies have shown that fibers will be uniformly distributed in the mortar matrix of concrete, forming an irregular but more stable mesh structure. It can effectively prevent shrinkage cracks arising from the internal setting and hardening of concrete. The uniformly distributed fibers can improve the stress distribution in the internal structure of concrete, improve the mechanical properties of the intersection between the aggregate and the mortar matrix, and improve the stability at the intersection.

At present, some scholars have studied the micro-structure of steel fiber concrete. For fiber concrete, the change of the micro-structure in the concrete by incorporating fibers of different

---

*Corresponding Author: 82667105@qq.com

DOI 10.1201/9781003344209-9

scales will have obvious differences, which will affect the cracking performance of the fiber concrete. Therefore, this paper carries out an axial tensile test of the fiber concrete. A combination of macroscopic tests and microscopic analyses is used to compare and analyze the microscopic cracking mechanism of the two different scales of fiber concrete.

2 DESIGN OF THE EXPERIMENTAL PROTOCOL

The tests were conducted with nanofibers and steel fibers as variables for fiber concrete axial tensile tests, and micro-structural observations were made after sampling the cracked fiber concrete specimens.

2.1 *Test material*

The raw materials used in the test include cement, gravel, river sand, nanofiber, steel fiber, high-efficiency water reducing agent, and water. The strength of the matrix concrete is C25, using Yichang Three Gorges P.O42.5 ordinary silicate cement with high-quality river sand. The gravel is with a particle size of 5mm~20mm. The water source is tap water, and the water-reducing agent is polycarboxylic acid mother liquor with a solid content of $40 \pm 1\%$. The raw material indexes meet the test requirements, and the fiber types mixed in are steel fiber and nanofiber. Nanofiber is cellulose with a filament diameter of less than 100nm, and its micro-structure is 30~40 cellulose molecules in a bundle stretching chain structure. The chemical composition is β-glycolic anhydride condensed glucose unit, and the macro state is a transparent gel with pseudoelasticity. The specific parameters and appearance of the steel fibers and nanofibers are as follows.

Table 1. Physical properties of steel fibers.

Fiber name	Aspect Ratio (L/D)	Type	Density (g/m^3)	Tensile strength (MPa)	Elongation (%)
Steel Fiber	80	Shear corrugated type	7.8	≥ 600	–

Figure 1. Steel fibers for testing.

Table 2. Physical properties of nanofibers.

Fiber name	Length (μm)	Width (nm)	morphological ratio	Appearance	Concentration	Tensile strength (MPa)
Nanofiber	>1	20–50	>30	Clear gel	4.85	222–233

Figure 2.   Experimental nanofibers.

2.2   *Mixing ratio design*

The tests were considered from two aspects of a single admixture of steel fibers and a single admixture of nanofibers, which aims to study the microstructure of fiber concrete. The volume rates of the steel fibers incorporated in this test were 0.5%, 1.0%, and 1.5%. The admixtures of nanofibers were 0.05%, 0.1%, 0.15% and 0.2% of the mass of cement. The design tests were a single admixture of steel fibers and a single admixture of nanofibers and plain concrete. They were in eight groups.

This test is based on the "Ordinary Concrete Proportioning Design Regulations: JGJ 55-2011" and "Steel Fiber Concrete: JG/T 472-2015. Due to the different amounts of the steel fiber admixture, the fit of each group of specimens was designed as shown in Table 3.

Table 3.   Test piece fitting ratio (unit: kg).

Cement	Gravel	Sand	Water	Water Reducer	Steel Fiber
321	1241	668	170	1.1	0
360	1122.6	748.4	180	1.1	39
376	1106.6	740.4	188	1.2	78
392	1090.6	732.4	196	1.2	117

2.3   *Sample preparation and maintenance*

100mm × 100mm × 700mm concrete specimens are made. After 36 hours of specimen casting, they are demoulded and numbered, which were then put into the standard curing room for maintenance. After the specimens are cured for 28 days, the axial tension specimens are then treated as follows.

1. Sand the surface of the test piece with sandpaper.
2. Use a two-inch small roller to brush the water-based paint evenly on the four sides of the axial pull test piece.

The test results show that when the single-doped steel fiber specimen steel fiber dose is 1.5%, the average crack width of the specimen is the smallest, and the crack limiting effect is more ideal. The steel fiber dose is 1.5% as the optimal dose. When the single-doped nanofiber specimen nanofiber dose is 0.1%, the average crack width of the specimen is the smallest, and the crack limiting effect is the best. The nanofiber dose is 0.1% as the optimal dose.

Figure 3.   Specimen surface treatment.

2.4   *Test methods and results*

The loading procedure was carried out by using a CMT5305 micro-controlled electronic universal testing machine to conduct an axial tensile test on steel-nanofiber concrete specimens. After the test, the specimen is removed, and a sample is taken at the cracked specimen.

Figure 4.   CMT5305 testing machine.

2.5   *Microstructure observation experiments*

The specimens were sampled and numbered, and micro-structural observations were made on the samples, which aim to observe the effects of the incorporation of steel fibers and nanofibers on the hydration products, microscopic crack structure, and crack generation in the concrete.

Samples were taken on the specimen block for the axial tensile test, and the sample size was controlled at about 10mm × 10mm × 10mm. After the sampling was completed, the taken specimens were immersed in anhydrous ethanol. The specimens were taken out after 48h, and the specimens were dried.

When doing the microscopic test, the prepared specimens were sprayed with gold, and then taken out and put into the scanning electron microscope to observe the microscopic morphology of each specimen respectively. A scanning electron microscope (SEM) of JSM-7500F was used to observe the microscopic morphology of the specimens.

Figure 5.   Gold spraying treatment.

3   MICROSTRUCTURE ANALYSIS OF FIBER CONCRETE

3.1   *Microstructure analysis of concrete with different steel fiber admixture*

The specimens of concrete corresponding to the nanofiber dose of 0 and the steel fiber volume dose of 0%, 0.5%, 1.0%, and 1.5% were taken for the microstructure analysis. It aims to study the role of the concrete structure corresponding to the micro-structure at different steel fiber doses. The microstructure of the concrete corresponding to them during the continuous change of the steel fiber admixture is shown in the following figures.

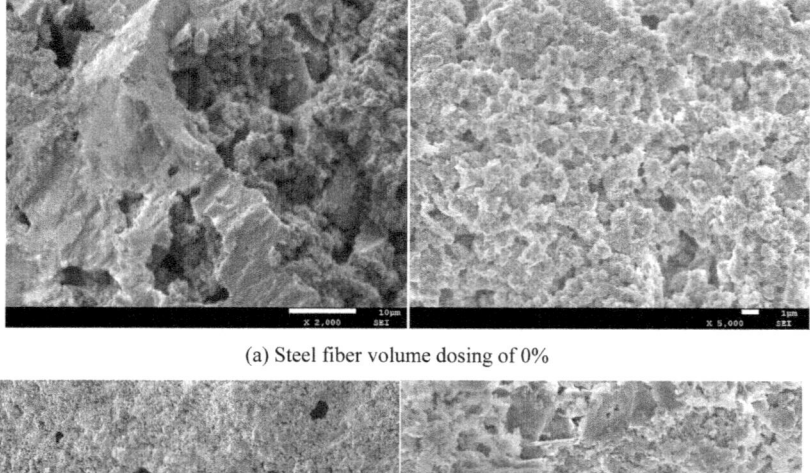

(a) Steel fiber volume dosing of 0%

(b) Steel fiber volume dosing of 0.5%

Figure 6.   Micro-structure morphology of the steel fiber concrete with different dosing levels.

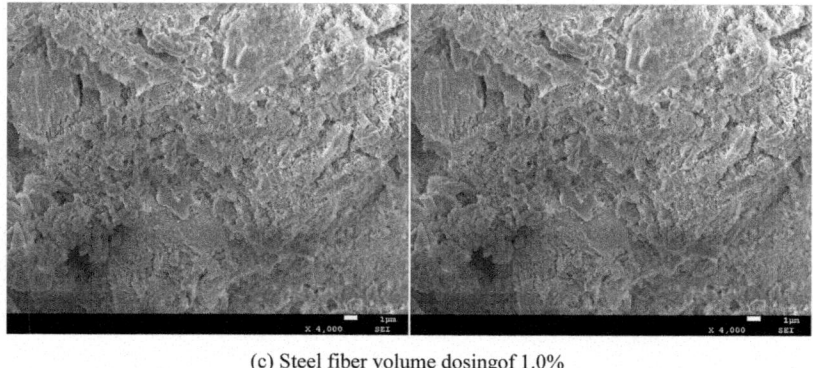

(c) Steel fiber volume dosingof 1.0%

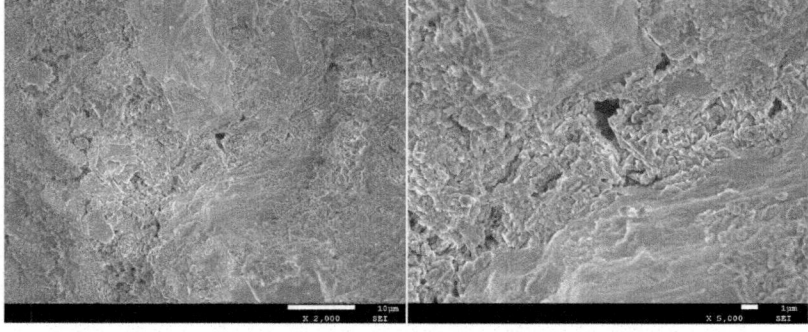

(d) Steel fiber volume dosing of 1.5%

Figure 6.   Continued.

Figure (a) is the microscopic morphology of the concrete without any fiber. It is obvious that there are many original pores inside, and the structure is also very loose. The hydration products are mainly irregularly layered $Ca(OH)_2$, and the internal structure is poorly compacted, which is reflected in the macroscopic mechanical properties as low strength. The effect of crack control is relatively unsatisfactory. From figure (b), it can be seen that for the steel fiber concrete, the number of holes in the matrix is high, which is due to the air mixed in the process of adding steel fibers to the concrete and the uneven mixing of steel fibers. However, it is obvious that the micro-structural denseness of the concrete has improved compared with the concrete without steel fibers, which is due to the disorganized distribution of steel fibers inside the concrete. To some extent, it prevents the generation and development of microcracks inside the concrete, making the internal structural denseness of the concrete matrix enhanced.

Figure (c) shows that the concrete matrix structure is relatively dense, and the hydration products are mainly calcium hydroxide and hydrated calcium silicate. Figure (d) shows the micro-structure morphology of the specimen with 1.5% steel fiber doping, which is the best hydration in comparison with the other doped steel fiber specimens mentioned above. The main hydration products are mainly calcium hydroxide and hydrated calcium silicate, and there is a large amount of mass hydration at the gap. Calcium silicate is filled with only a small amount of pores, which makes the internal structure of concrete better compactness, reflecting in the macro-mechanical properties. The mechanical properties of the concrete to improve the magnitude and the axial tensile specimens to limit the cracking effect are outstanding.

In summary, it can be seen that in the single-doped steel fiber concrete test, the steel fiber dosing at 1.5% is the optimal dosing, and the micro-structure of the concrete specimen is denser.

The mechanical properties of the concrete to improve the magnitude of the crack control effect on the axial tensile specimens is better.

3.2 *Microstructure analysis of the concrete under different nanofiber admixture*

The specimens of the concrete with 0 steel fiber admixture and 0.05%, 0.10%, 0.15%, and 0.20% of cement mass of nanofiber admixture were taken for micro-structure analysis to study their role in the micro-structure of concrete corresponding to different nanofiber admixture. The microstructure of their corresponding concrete micro-structures during the constant variation of nanofiber admixture is shown in the Figure.

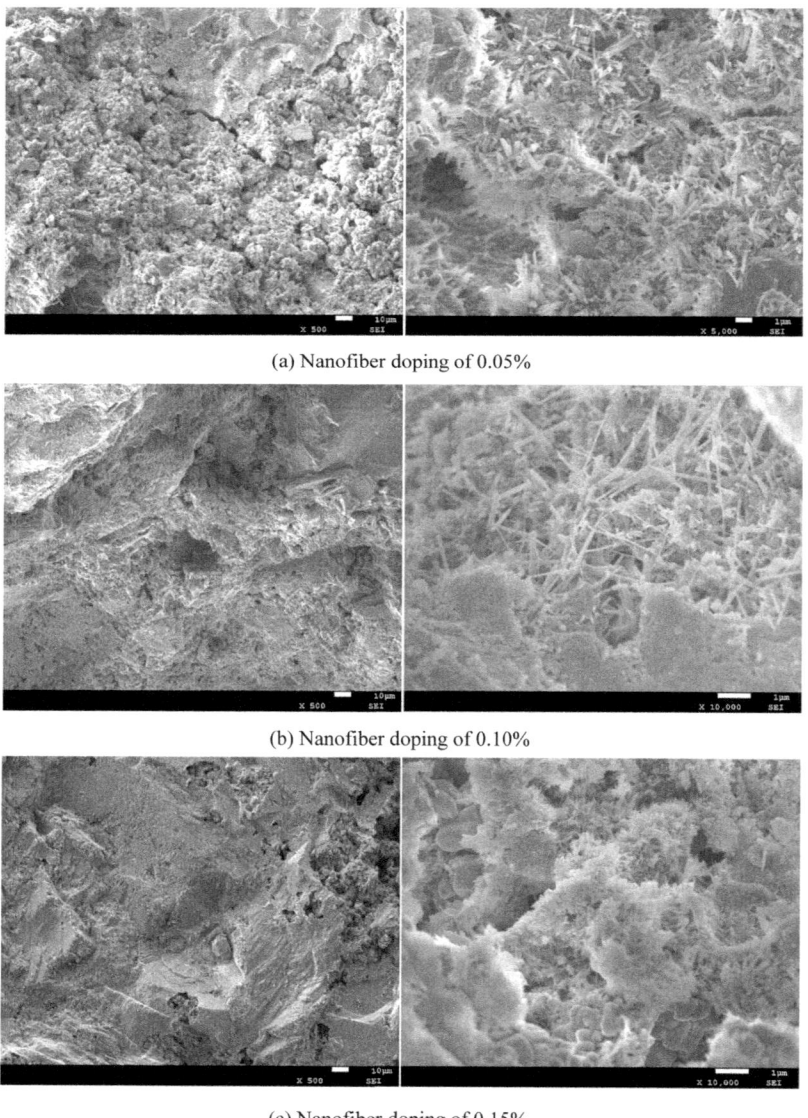

(a) Nanofiber doping of 0.05%

(b) Nanofiber doping of 0.10%

(c) Nanofiber doping of 0.15%

Figure 7.   Micro-structure morphology of the concrete with different dosing of nanofiber.

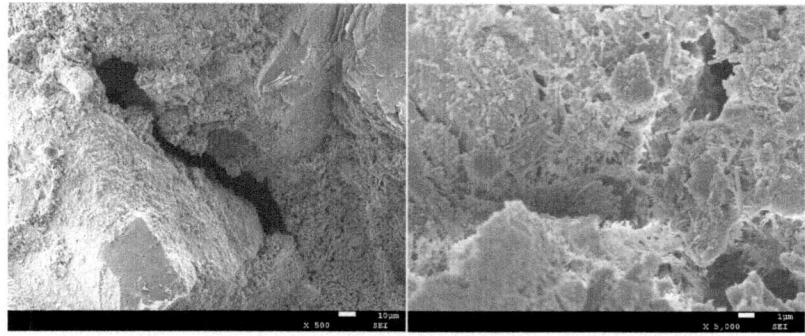

(d) Nanofiber doping of 0.20%

Figure 7.   Continued.

Figure (a) shows that the specimens mixed with 0.05% nanofiber are denser in micro-structure than the specimens without nanofiber, and the hydration products are mainly $C_3A.3CS.H_{32}$ crystals and hydrated calcium silicate. A large number of $C_3A.3CS.H_{32}$ crystals are filled in the gaps so that the gaps inside the concrete matrix can be greatly reduced. In Figure (b), when the nanofiber H_{32} crystals are inside the concrete, needle-like $C_3A.3CS.H_{32}$ crystals and agglomerated hydrated calcium silicate fill in the gaps, and no irregular $Ca(OH)_2$ crystals can be seen. The internal structure of the concrete is dense, and the macroscopic performance is reflected by an obvious increase in compressive strength, which has a better effect on the axial tensile specimens. H_{32} crystals and the pores inside the concrete cannot be filled well. The macroscopic mechanical properties are reflected by the decrease of compressive strength and the weakening of crack control.

From the above observation and analysis of the microscopic morphology of single-doped nanofiber concrete specimens, it is known that the incorporation of the nanofiber makes the number of the concrete hydration product $C3A.3CS.H_{32}$ crystals increase. The $C3A.3CS.H_{32}$ crystals and hydrated calcium silicate lap each other, and then the internal pores of the concrete are filled. This is due to the fact that nanocellulose is initially adsorbed on the surface of the cement particles and retained in the hydration product shell. Due to the hydrophilicity and wettability of nanocellulose, the pore solution water molecules slowly diffuse to the unhydrated cement nuclei through the nanocellulose channels, which promotes the hydration of the cement, thus generating more hydration products, namely the regular of $C3A.3CS.H_{32}$ crystals. In addition, since the size of the nanofibers is at the nano level, they can fill into the tiny pores inside the concrete and play a bridging role, which will make the concrete compactness further improve again.

Through the above analysis, it can be concluded that the nano-fiber doping amount of 0.10% is the optimal doping amount. When the microstructure of the concrete specimen is the densest, the crack control effect on the axial tensile specimen is also the best. When the nano-fiber doping amount continues to increase, there will be the phenomenon of fiber agglomeration, which will make the $C3A.3CS.H_{32}$ crystals less and less. Finally, the internal structure of concrete denseness is relatively reduced, resulting in a decrease in strength. This will lead to a decrease in the concrete strength.

4   CONCLUSION

In this paper, the axial tensile tests of plain concrete, steel fiber concrete, and nanofiber concrete are designed, and the effects of steel fiber admixture and nanofiber admixture on the micro-structure morphology of concrete are studied through the electron microscopy scanning (SEM) tests. The conclusions are as follows.

1) The steel fiber's admixture into concrete can make the internal structure of the concrete improve to a certain extent. The disordered distribution of the steel fiber inside the concrete can effectively prevent the generation and development of microcracks inside the concrete so that the internal structural denseness of the concrete matrix can be increased. The optimum admixture of steel fiber is 1.5% in this test.

2) When the nanofiber admixture is from 0 to 0.10%, the internal structural compactness of the concrete is higher. When the nanofiber admixture is from 0.1% to 0.2%, the internal structural compactness of the concrete decreases instead.

3) Nanofiber mixed into the concrete can promote the hydration reaction of cement so that its hydration products $C_3A.3CS.H_{32}$ crystals can increase in number. When a large number of $C_3A.3CS.H_{32}$ crystals and hydrated calcium silicate gel are filled in the gap, it will effectively improve the denseness of the internal structure of concrete. The test concluded that the optimal amount of nanofiber mixing is 0.1%.

4) The steel fiber dose is 1.5%, and the nanofiber dose is 0.1%. The best micro-structure of the specimen with higher denseness is reflected in the macroscopic performance of the specimen to limit the effect of good cracking. It is also consistent with the optimal fiber dose based on the relationship between the fiber dose and the axial tensile specimen crack width control study of the fiber concrete obtained by the subject group.

ACKNOWLEDGMENTS

The author received financial support from the National Natural Science Foundation of China (Grant No.51879146).

REFERENCES

Marsh, B.K. *Relationship between engineering properties and micro-structural properties of hardened cement paste containing PFA as a partial cement replacement.* Ph. C thesis, Hatfield Polytechnic, 1984.

Bai Min, Niu Ditao, Jiang Lei, Miao Yuanyao. Study on the improvement of mechanical properties and microstructure of concrete by steel fibers [J]. *Silicate Bulletin*, 2013, 32(10): 2084–2089.

Cao Y.Z., Tian N.N., Bahr D., et al. The influence of nanocrystals on the micro-structure of cement paste [J]. *Cement & Concrete Composites*, 2016, 74:164.

Cao Y.Z., Zavattieri P., Youngblood J., et al. The relationship between cellulose nanocrystal dispersion and strength[J]. *Construction and Building Materials*, 2016, 119:71.

Dong Xiang, Shen Zheng. Effect of fiber species and admixture on the frost resistance and micro-structure of concrete[J]. *Journal of Nanjing Forestry University* (Natural Science Edition), 2010, 34(05): 91–95.

Flores J., Kamali M., Ghahremaninezhad A. An investigation into the properties and microstructure of cement mixtures modified with cellulose nanocrystal [J]. *Materials*, 2017, 10(5): 2.

Jiao Huazhe, Han Zhenyu, Chen Xinming, Liu Zilu, Chen Fengbin, PETER Hughes. Mechanisms of basalt fibers on the mechanical properties and microstructure of shotcrete[J]. *Journal of Composites*, 2019, 36(08): 1926–1934.

Jin Shanshan. Study on the relationship between microstructural characteristics and macroscopic properties of concrete [D]. Beijing Institute of Technology, 2010.

Klemm D., Kramer F., Moritz S., et al. Nanocelluloses: A new family of nature-based materials [J]. *Angewandte Chemie-International Edition*, 2011, 50(24): 5438.

Lin Jiafu. Study on mechanical properties and micro-structure of basalt fiber concrete based on SEM [J]. *Construction Technology*, 2018(18): 97–101.

Lin Lian, Li Yuhai, Deng Yonggang Research on the microstructure of high-performance fiber concrete and its compressive strength [J]. *Journal of Shenyang University of Technology*, 2021, 40(01): 45–49+59.

Lv Zhiheng, Cheng Ming, Jiang Xisheng, Zhou Yihui, Jia Yanmin. Study on the improvement of concrete micro-structure by glass fiber and polypropylene fiber [J]. *Sino-Foreign Highway*, 2020, 40(06): 267–270.

Petersson L, Kvien I, Oksman K. Structure and thermal properties of poly(lacticacid)/cellulose whiskers nanocomposite materials [J]. *Composites Science and Technology*, 2007, 67: 2535–2544.

Frontiers in Civil and Hydraulic Engineering – Mohamed A. Ismail and
Hazem Samih Mohamed (Eds)
© 2023 The Authors, ISBN 978-1-032-38247-0

Experimental study on the interfacial bonding behavior between geopolymer concrete and old concrete in corrosive environment

Peng Deng

Shandong Provincial Key Laboratory of Civil Engineering Disaster Prevention and Mitigation, Shandong University of Science and Technology, Qingdao Shandong, China
College of Civil Engineering and Architecture, Shandong University of Science and Technology, Qingdao Shandong, China

Qi Zheng & Meng Cao

College of Civil Engineering and Architecture, Shandong University of Science and Technology, Qingdao Shandong, China

Yan Liu*

Shandong Provincial Key Laboratory of Civil Engineering Disaster Prevention and Mitigation, Shandong University of Science and Technology, Qingdao Shandong, China
College of Civil Engineering and Architecture, Shandong University of Science and Technology, Qingdao Shandong, China

ABSTRACT: In this paper, the interfacial bonding behavior between Slag-Fly ash Geopolymer Concrete (SFGPC) and old concrete in a corrosive environment was investigated by splitting tensile test. The effects of SFGPC slag-ash ratio (0.25, 0.43, 0.67), old concrete strength (C25, C35, C45), and corrosive environment (acid, saline) on the interfacial bonding behavior of the new and old concrete were analyzed. The test results show that the interfacial bond strength of the new and old concrete increases with the increase of the SFGPC slag-ash ratio. The interfacial bond strength of the new and old concrete increases with the increase of the strength of old concrete. The bond strength of the specimen with the same SFGPC slag-ash ratio and old concrete strength is reduced by about 0.26 MPa in the acid environment compared with that in the saline environment.

1 INSTRUCTION

The concrete structures located in acid rain and saline environments are prone to damage. To prolong the life of the building and ensure the safety of the structure, it is necessary to repair the damaged parts. The junction of the old and new concrete is the weakest part of the component, and it is easy to cause bond failure under the combined action of load and corrosion. Therefore, it is necessary to study the bonding behavior of the interface between new and old concrete.

In recent years, the academic community has made a systematic analysis of the factors affecting the bonding behavior of the interface between new and old concrete (Baloch 2021; Fathy 2022; Zhang 2017). First, the strength of new and old concrete is an important factor affecting the interfacial bonding behavior. Wang (2011) found that the bonding tensile strength of the interface increased by about 0.17 MPa for every 10 MPa increase in the compressive strength of the old concrete through the pull-out test. Gao (2019) found through the cylinder splitting tensile test, when the average compressive strength of new concrete increases from 39.9 MPa to 44.9 MPa, the average splitting tensile strength of the interface increases by 0.09 MPa. Secondly, Pejman (2017)

---

*Corresponding Author: ly966@sina.com

DOI 10.1201/9781003344209-10

pointed out that the corrosive medium produced chemical degradation at the interface of the new and old concrete, resulting in tensile stress at the interface, and weakening its bond strength. Gao (2019) corroded two different types of cement-based concrete for 120 d respectively. Through the cylinder splitting test, it was found that the average splitting tensile strength of the two decreased by 0.54 MPa and 0.69 MPa, respectively. Compared with cement-based materials, SFGPC has better corrosion resistance, which is more suitable as a material for repairing concrete structures in corrosive environments. The current research on concrete repair is mainly for traditional Portland cement and other composite materials. Therefore, it is necessary to study the bonding properties of the interface repaired by SFGPC in corrosive environments.

In this paper, the effects of SFGPC slag-ash ratio (0.25, 0.43, 0.67), old concrete strength (C25, C35, C45), and corrosive environment (acid, saline) on the interfacial bonding behavior between new and old concrete were studied.

2 SUMMARY OF THE EXPERIMENT

2.1 *Raw materials*

The raw materials for preparing Slag-Fly ash Geopolymer Concrete (SFGPC) include the slag of S95, the fly ash of Class I, the crushed stone with a particle size of 10–20 mm, and Qingdao local medium river sand and alkaline activator with a modulus of 2.80. S95 slag was produced by Henan Hengnuo Company, with a density of 2.84 g/cm^3 and a specific surface area of 472 cm^3/kg. Class I fly ash was produced by Hebei Yousheng Company with a density of 2.1 g/cm^3. Alkaline activators are solid sodium hydroxide and liquid sodium silicate produced by Henan Hengnuo Company. In addition to the above-mentioned crushed stone and river sand, the raw materials for preparing old concrete also include P-O 42.5 grade cement with an apparent density of 3.05 g/cm^3.

2.2 *Design of mix ratio*

In this experiment, SFGPC with slag-ash ratios of 0.25, 0.45, and 0.67 and Portland cement concrete with strengths of C25, C35, and C45 were designed respectively. The compressive strength of the concrete was measured by the specimens of 100 mm × 100 mm × 100 mm. The experimental process was based on the *Standard for test method of concrete physical and mechanical properties (GB/T 50081-2019)* (Ministry of Housing and Urban-Rural Development of the People's Republic of China 2019). Among them, the design mix ratios of old and new concrete are shown in Tables 1 and 2.

Table 1. Design mix ratio of old concrete.

Strength grade of concrete	Water-cement ratio	Cement/ $kg \cdot m^{-3}$	Sand/ $kg \cdot m^{-3}$	Stone/ $kg \cdot m^{-3}$	Water/ $kg \cdot m^{-3}$	Water reducer/ $kg \cdot m^{-3}$	Compressive strength/ MPa
C25	0.76	250	885	1125	190	0	28.5
C35	0.58	325	815	1120	190	0	39.5
C45	0.48	395	745	1120	190	1.19	47.5

2.3 *Design of specimens*

In this test, the bonding behavior of the interface between new and old concrete was studied by splitting tensile strength test. A total of 18 groups of specimens were designed. To ensure the reliability of the test, each group was designed with 3 specimens. The experimental design is shown in Table 3. Among them, *L* represents the splitting tensile specimen of the new and old concrete.

Table 2.   Design mix ratio of SFGPC.

Water cement ratio	Slag/ kg·m^{-3}	Fly ash/ kg·m^{-3}	Sodium silicate/ kg·m^{-3}	NaOH/ kg·m^{-3}	Water/ kg·m^{-3}	Sand/ kg·m^{-3}	Stone/ kg·m^{-3}	Compressive strength/ MPa
0.33	100	400	164.80	16.50	76.93	454.55	1061.62	29.6
0.33	150	350	210.25	21.03	52.62	454.55	1061.62	38.5
0.33	200	300	225.95	22.60	44.23	454.55	1061.62	48.1

0.25, 0.45, and 0.67 represent the slag-ash ratio of SFGPC, respectively. 2 and 7 represent the pH value of the corrosive medium, respectively. C25, C35, and C45 represent the strength grade of the old concrete, respectively.

Table 3.   Design of specimens.

Group number of specimens	Corrosive medium	SFGPC slag-ash ratio	Strength grade of old concrete
L0.25-2-25	Acid (pH=2)		C25
L0.25-7-25	Saline (pH=7)		
L0.25-2-35	Acid (pH=2)		C35
L0.25-7-35	Saline (pH=7)	0.25	
L0.25-2-45	Acid (pH=2)		C45
L0.25-7-45	Saline (pH=7)		
L0.43-2-25	Acid (pH=2)		C25
L0.43-7-25	Saline (pH=7)		
L0.43-2-35	Acid (pH=2)		C35
L0.43-7-35	Saline (pH=7)	0.43	
L0.43-2-45	Acid (pH=2)		C45
L0.43-7-45	Saline (pH=7)		
L0.67-2-25	Acid (pH=2)		C25
L0.67-7-25	Saline (pH=7)		
L0.67-2-35	Acid (pH=2)		C35
L0.67-7-35	Saline (pH=7)	0.67	
L0.67-2-45	Acid (pH=2)		C45
L0.67-7-45	Saline (pH=7)		

2.4   Preparation and corrosion of specimens

The thin plywood was placed vertically in the middle of each 100 mm × 100 mm × 100 mm mold. The freshly mixed concrete was poured. After 2 days of pouring, the molds were removed, and two old concrete specimens can be obtained from each mold. Immediately after removing the molds, the specimens were placed in a curing box at (20±2)°C and cured for 28 days.

An old concrete specimen with an artificially chiseled surface was placed on the side of the mold. The freshly stirred SFGPC are poured into the reserved mold space. After 2 days of pouring, the molds were removed, and then the specimens were placed in a curing box at (20±2) °C and cured for 28 days.

The immersion method was used for the corrosion treatment of the specimens. The acid medium was an oxalic acid solution with pH=2, and the saline medium was a NaHCO$_3$ solution with pH=7. When soaking, the liquid level was 2 mm higher than the specimens, and the distance between the specimens was not less than 2 mm. The upper surface of the plastic water tank was covered with a plastic film to avoid the volatilization of water during the test. During the soaking period, the pH values of the two corrosive environments were monitored in real-time by a pH meter, and the corresponding solutes were added in time. After soaking for 120 d, the specimens were taken out and dried.

2.5 Test method

The splitting tensile strength test was based on the *Standard for test method of concrete physical and mechanical properties (GB/T 50081-2019)* (Ministry of Housing and Urban-Rural Development of the People's Republic of China 2019). During the test, the load was continuously and uniformly applied. Specimens with different strengths had different loading speeds. The loading speed of C25 was 0.05 MPa/s, the loading speed of C35 was 0.06 MPa/s, and the loading speed of C45 was 0.07 MPa/s.

3 RESULT

3.1 Appearance of the specimens in a corrosive environment

The specimens were placed in two corrosive environments, and the apparent changes in the specimens were observed during the period. In the acid environment, all the old concrete parts of the specimens did not change significantly before 60 d. At 60 d, the edge skins began to fall off, but the surfaces were still intact. At 120 d, the edges and corners were corroded and fell off obviously, and rough surfaces began to appear. All the SFGPC parts appeared with a large number of bubbles when they were just immersed in the corrosive medium, and then the bubbles gradually decreased. At 60 d, a large number of bubble pits appeared on the surfaces. At 120 d, a small number of edges and corners began to fall off. At this time, the solution changed from colorless to dark brown.

In the saline environment, all the old concrete parts had a small amount of corrosion pit on the surfaces before 100 d, and the edges and corners began to fall off at 120 d. All SFGPC parts had no obvious change before 100 d, and all surfaces had a small amount of corrosion pit at 120 d. A layer of white corrosion appeared on the entire surface of the specimens, and the color of the solution did not change significantly.

3.2 Results of test

Through the splitting tensile strength test, it was found that the failure position of all the specimens was at the bond between the new and old concrete, namely the interface between the old concrete and SFGPC. After a fracture, concave points appeared on the old concrete, and convex points appeared on the corresponding part of SFGPC. The reason for this phenomenon is as follows. From a macroscopic point of view, the interface between the old and new concrete is the weakest position of the entire specimen. From a microscopic point of view, the water absorption of the old concrete makes a water film to be formed on the interface, resulting in the appearance of more ettringite. Calcium hydroxide crystals make the aggregate and calcium silicate hydrate (Calcium Silicate Hydrate, C-S-H) gel not fully contacted, resulting in reduced interface bond strength (Cui 2018).

4 ANALYSIS OF THE RESULTS

4.1 Effect of SFGPC slag-ash ratio on interfacial bonding behavior

It can be seen from Figure 1 that with the increase of the slag-ash ratio of SFGPC, the splitting tensile strength of the specimen increases, and the increased range of the splitting tensile strength also increases gradually. Compared with specimens L0.25-2-25, L0.25-2-35, L0.25-2-45, L0.25-7-25, L0.25-7-35 and L0.25-7-45, the splitting tensile strengths of specimens L0.43-2-25, L0.43-2-35,

L0.43-2-45, L0.43-7-25, L0.43-7-35 and L0.43-7-45 increased by 13.7%, 18.8%, 25.6%, 22.5%, 16.3% and 12.7% respectively. Compared with specimens L0.43-2-25, L0.43-2-35, L0.43-2-45, L0.43-7-25, L0.43-7-35 and L0.43-7-45, the splitting tensile strengths of specimens L0.67-2-25, L0.67-2-35, L0.67-2-45, L0.67-7-25, L0.67-7-35 and L0.67-7-45 increased by 22.9%, 31.6%, 36.1%, 31.2%, 27.3% and 19.5% respectively.

The reason why the splitting tensile strength increases with the increase of the slag-ash ratio of SFGPC may be that the bleeding water of slag-based geopolymer concrete is stimulated by alkali decreases, which makes the relative water-cement ratio at the interface of new and old concrete lower, thus enhancing the interfacial bond strength between the new and old concrete. The C-S-H gel generated in the geopolymer reaction continuously enters the capillary pores of the old concrete, making the material structure more compact (Fan 2021). Therefore, more C-S-H gels enter the capillary pores of the old concrete, and the increase of the splitting tensile strength will be greater.

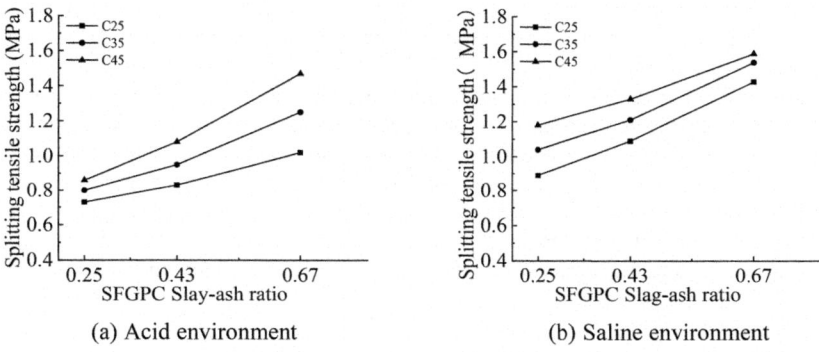

(a) Acid environment   (b) Saline environment

Figure 1.   Effect of SFGPC slag-ash ratio on the splitting tensile strength.

4.2   *Effect of the old concrete strength on interfacial bonding behavior*

It can be seen from Figure 2 that with the strength of the old concrete increases, the splitting tensile strength of the specimen increases. Compared with specimens L0.25-2-35, L0.43-2-35, L0.67-2-35, L0.25-7-35, L0.43-7-35 and L0.67-7-35, the splitting tensile strengths of specimens L0.25-2-45, L0.43-2-45, L0.67-2-45, L0.25-7-45, L0.43-7-45 and L0.67-7-45 increased by 9.6%, 14.5%, 22.5%, 16.9%, 11.0% and 7.7% respectively. Compared with specimens L0.25-2-25, L0.43-2-25,

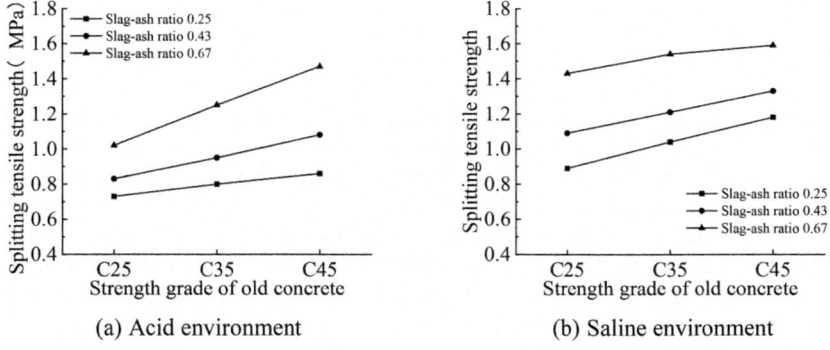

(a) Acid environment   (b) Saline environment

Figure 2.   Effect of the old concrete strength on splitting tensile strength.

L0.67-2-25, L0.25-7-25, L0.43-7-25 and L0.67-7-25, the splitting tensile strengths of specimens L0.25-2-35, L0.43-2-35, L0.67-2-35, L0.25-7-35, L0.43-7-35 and L0.67-7-35 increased by 7.5%, 13.7%, 17.6%, 13.5%, 10.0% and 3.2% respectively. The reason for this phenomenon may be as follows. With the decrease in the water-cement ratio of the old concrete, the strength of the old concrete increases, the thickness of the water film generated at the interface decreases, and the ettringite and calcium hydroxide crystals generated in the water film also decrease, which makes the aggregate fully contact with the C-S-H gel (Cui 2018), finally resulting in the increase of the splitting tensile strength of the interface between the new and old concrete.

4.3 *Effect of corrosive environment on interfacial bonding behavior*

It can be seen from Figure 3 that the splitting tensile strengths of specimens L0.25-2-35, L0.43-2-35, and L0.67-2-35 under an acid environment are reduced by 0.24 MPa, 0.26 MPa, and 0.29 MPa, respectively compared with specimens L0.25-7-35, L0.43-7-35 and L0.67-7-35 under saline environment. Similarly, the splitting tensile strengths of specimens L0.43-2-25, L0.43-2-35, and L0.43-2-45 under an acid environment are reduced by 0.27 MPa, 0.26 MPa, and 0.25 MPa, respectively compared with specimens L0.43-7-25, L0.43-7-35 and L0.43-7-45 under saline environment. In other words, under the same SFGPC slag-ash ratio and strength of the old concrete, the weakening degree of the splitting tensile strength of the new and old concrete interface in the acid environment is greater than that in the saline environment. This is mainly due to the reaction of Ca(OH)$_2$ in the outermost layer of the specimen with an acid solution, which leads to the continuous formation of calcium oxalate in the specimen. The increase in the volume of the product will produce expansion stress. When the expansion stress exceeds a certain value, expansion cracks will occur, resulting in an decrease in the bonding strength between the new and old concrete interface (Ma 2017).

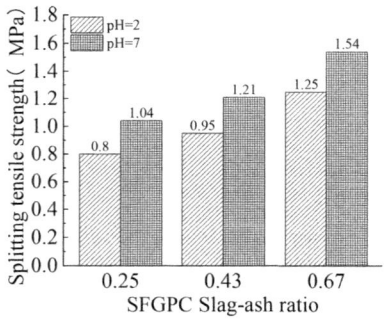
(a) The strength grade of old concrete is C35

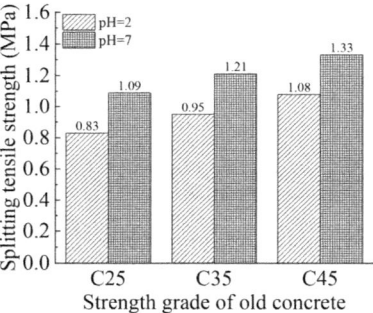
(b) The slag-ash ratio of SFGPC is 0.43

Figure 3.   Effect of the corrosive environment on splitting tensile strength.

5   CONCLUSION

1) With the increase of the SFGPC slag-ash ratio, the splitting tensile strength of the new and old concrete increases, and the increase of splitting tensile strength also increases. The splitting tensile strengths of the specimens with the SFGPC slag-ash ratio of 0.43 increased by 12%–26% compared with the specimens of 0.25. The splitting tensile strengths of the specimens with the SFGPC slag-ash ratio of 0.67 increased by 19%–37% compared with the specimens of 0.43.
2) As the compressive strength of the old concrete increases, the splitting tensile strength of the new and old concrete increases. The splitting tensile strengths of the specimens with the old concrete strength grade of C35 increased by 9%–23% compared with the specimens of C25. The splitting tensile strengths of the specimens with the old concrete strength grade of C45 increased by 3%–18% compared with the specimens of C35.

3) The splitting tensile strength of the specimens with the same SFGPC slag-ash ratio and old concrete strength is weakened by about 0.26MPa in the acid environment compared with the saline environment.

REFERENCES

Baloch, W.L., Siad, H., Lachemi, M. & Sahmaran, M. (2021). A review on the durability of concrete-to-concrete bond in recent rehabilitated structures. *J. Journal of Building Engineering*. 44, 103315.

Cui, H. (2018). *Influencing factors and mechanism of bonding between new and old concrete under the action of new interfacial agents*. D. Zhengzhou: North China University of Water Resources and Electric Power.

Fan, J. & Zhang, B. (2021). Repair of ordinary Portland cement concrete using alkali-activated slag/fly ash, Freeze-thaw resistance and pore size evolution of adhesive interface *J. Construction and Building Materials*. 300, 124334.

Fathy, A., Zhu, H. & Kohail, M. (2022). Factors affecting the fresh-to-hardened concrete repair system. *J. Construction and Building Materials*. 320, 126279.

Gao, S., Jin, J., Hu, G. & Qi, L. (2019). Experimental investigation of the interface bond properties between SHCC and concrete under sulfate attack. *J. Construction and Building Materials*. 217, 651–663.

Gao, S., Zhao, X., Qiao, J., Guo, Y. & Hu, G. (2019). Study on the bonding properties of Engineered Cementitious Composites (ECC) and existing concrete exposed to high temperature. *J. Construction and Building Materials*. 196, 330–344.

Ma, B. (2017). *Experimental study on mechanical properties of concrete in acid rain environment D*. Xian: Chang'an University.

Ministry of housing and Urban-Rural development of the people's republic of China (2019). *The standard for test method of concrete physical and mechanical properties*: GB/T 50081-2019 S. Beijing: China Architecture and Building Press.

Pejman, A. & Rishi, G. (2017). Electrical resistivity of concrete for durability evaluation: a review *J. Advances in Materials Science and Engineering*. 2017, 1–30.

Wang, B. (2011). *Ultrahigh toughness cementitious composite: Its bond behavior with the concrete and structural application for flexural strengthening of RC beams D*. Dalian: Dalian University of Technology.

Zhang J. & Li Y. (2017). Research summary on factors influencing the strength of the interface between new and old concrete *J. Concrete*. (10), 156–159.

*Frontiers in Civil and Hydraulic Engineering – Mohamed A. Ismail and
Hazem Samih Mohamed (Eds)
© 2023 The Authors, ISBN 978-1-032-38247-0*

An experimental study on the optimal proportion of foam concrete with flow

Dandan Shi*

School of Civil Engineering, Xi'an Traffic Engineering Institute, Xi'an, Shaanxi, China

ABSTRACT: Flow concrete is widely used in construction engineering, especially in self-compacting concrete. In this paper, the mixing ratio of the flow state concrete is designed. According to the performance test results of the concrete mixture, the concrete ratio is optimized and the optimal ratio is finally given. The design scheme meets the requirements for applicability, durability, and economy.

1 INTRODUCTION

In the construction project, many special-shaped working surfaces are involved in the concrete construction, which puts forward higher requirements for the vibration and compactness of concrete, and the use of flow concrete can effectively solve similar problems. To ensure that the flow state of concrete construction meets the standard of construction quality, the durability of concrete is considered. Sand and high-performance water-reducing agents are mixed with fly ash and mineral powder minerals in concrete in, for example, the configuration C40 flow state. Through the design of the flow state of the concrete mixture ratio, the proportions of raw materials and mixture schemes of concrete are given. Finally, the experimental results are analyzed to optimize the concrete mixture ratio and identify the best mixture ratio.

2 SELECTION OF RAW MATERIALS

(1) Cement: P • O ordinary Portland cement, strength grade 42.5, density 3100 kg/m^3;
(2) Fine aggregate: 1# machine-made sand with stone powder content of about 5%, medium sand, density 2600 kg/m^3;
(3) Coarse aggregate: Two kinds of single gravels with a size of 5 mm to 10 mm, 10 mm–20 mm, and a density of 2650 kg/m^3. The gravel with a particle size of 5 mm–10 mm is 30% of the total gravel, and the gravel with a particle size of 10 mm–20 mm is 70% of the total gravel, which meets the grading requirements.
(4) Admixture: Fly ash, with a density of 2200 kg/m^3; slag powder, with a density of 2950 kg/m^3; High-efficiency water reducing agent, polycarboxylic acid high-performance water reducing agent.

---

*Corresponding Author: xjy@xjy.edu.cn

DOI 10.1201/9781003344209-11

3 MIX RATIO DESIGN METHOD

3.1 *Determine the strength of concrete configuration*

When the design strength grade is less than C60, the configuration strength (Ke, 2012) is calculated by following the formula:

$$f_{cu,o} \geq f_{cu,k} + 1.645\sigma \qquad (1)$$

Where $f_{cu,o}$ is the configured strength of concrete (MPa); $f_{cu,k}$ is the standard value of compressive strength of concrete cube (MPa); σ is the standard deviation of concrete strength (MPa). The standard deviation of concrete strength can be determined according to Table 1.

Table 1. Standard deviation.

Standard deviation of concrete strength	$\leq$ C20	C25 $\sim$ C45	C50 $\sim$ C55
σ	4.0	5.0	6.0

3.2 *Determine the water-binder ratio*

When the strength grade of concrete is not greater than C60, the water-binder ratio of concrete (Qian 2008) is calculated as follows:

$$\frac{W}{B} = \frac{\alpha_a f_b}{f_{cu,0} + \alpha_a \alpha_b f_b} \qquad (2)$$

where α_a and α_b are regression coefficients and are valued according to Table 2. f_b is the strength of cementing material (MPa) for 28d. When the mineral admixture is fly ash and slag powder, f_b can be calculated according to the formula below:

$$f_b = 1.1\gamma_f \gamma_s f_{ce} \qquad (3)$$

where γ_f, γ_s is the influence coefficient of fly ash and slag powder, ranging from 0.85 to 1, and the value varies according to the dosage; f_{ce} is the 28d compressive strength of cement mortar (MPa), which is determined according to the following formula:

$$f_{ce} = \gamma_e f_{ce,g} \qquad (4)$$

where γ_e is the affluence coefficient of cement strength grade value, which is valued according to Table 3; $f_{ce,g}$ is the cement strength grade value (MPa).

Table 2. Regression coefficient selection table.

Coefficient/coarse aggregate variety	gravel	pebble
α_a	0.53	0.49
α_b	0.20	0.13

Table 3. Affluence coefficient of cement strength grade values.

Cement strength grade value/MPa	32.5	42.5	52.5
affluence coefficient	1.12	1.16	1.10

3.3 *Determine the amount of water per unit*

Unit water consumption combined with construction conditions, according to the requirements of concrete slump value and the selection of aggregate type, maximum particle size to select and determine. In this study, coarse aggregate is gravel with a maximum particle size of 20 mm, and the standard water consumption is 215 kg. For the water consumption per unit of concrete with the flow state, based on the water consumption per unit of concrete with a standard slump of 90 mm, the water consumption per unit of water consumption should be increased by 5 kg for each increase in the slump of 20 mm. When the slump increases to more than 180 mm, water consumption increases with the decreased slump (Qiao 2017). In addition, the unit water consumption of concrete with admixture is calculated as follows:

$$m_{wo} = m'_{wo}(1 - \beta) \tag{5}$$

where m'_{wo} is the water consumption (kg) of concrete per cubic meter that meets the slump requirement presumed without admixture, and β is the water reduction rate (%) of admixture.

3.4 *Determine the amount of cementing material, mineral admixture, and cement*

For cementitious material per cubic meter of concrete, the dosage of mineral admixture and cement dosage can be calculated according to the following formulae (Jiang 2018):

$$m_{b0} = \frac{m_{wo}}{W/B} \tag{6}$$

$$m_{f0} = m_{b0}\beta_f \tag{7}$$

$$m_{k0} = m_{b0}\beta_k \tag{8}$$

$$m_{c0} = m_{b0} - m_{f0} - m_{k0} \tag{9}$$

where m_{b0} is the amount of cementitious material (kg) in concrete per cubic meter of mixture ratio, m_{f0} is the amount of fly ash (kg) in each cubic meter of concrete, m_{k0} is the amount of slag powder per cubic meter of concrete calculated by mixing ratio (kg), and m_{c0} is the amount of cement (kg) per cubic meter of concrete in the calculation of mixture ratio.

3.5 *Determine the amount of admixture*

The amount of admixture per cubic meter of concrete is calculated as follows:

$$m_{a0} = m_{b0}\beta_a \tag{10}$$

where m_{a0} is the amount of admixture (kg) in concrete per cubic meter of mixture ratio.

3.6 *Determine the rate of sand*

According to the water-binder ratio, aggregate technical index, concrete mixture performance, and construction requirements, the sand ratio in this study is 38%~42% (Shi 2020).

3.7 *Determine the amount of coarse and fine aggregate*

In this study, the coarse and fine aggregate amounts of concrete are calculated by the volume method.

$$\frac{m_{c0}}{\rho_c} + \frac{m_{f0}}{\rho_f} + \frac{m_{k0}}{\rho_k} + \frac{m_{g0}}{\rho_g} + \frac{m_{s0}}{\rho_s} + 0.01\alpha = 1 \tag{11}$$

$$\beta_s = \frac{m_{s0}}{m_{s0} + m_{g0}} \times 100\% \tag{12}$$

where ρ_c is the density of cement, ρ_f is the density of fly ash, ρ_k is the density of slag powder, ρ_g is the density of gravel, ρ_s is the density of sand, and α is the percentage of gas content of concrete. When no air-entraining admixture is used, α can be 1.

4 MIX RATIO OPTIMIZATION DESIGN

Combined with the relevant indicators of raw materials and according to the mix ratio design method, the mix ratio of four groups of concrete with flow state is given as follows.

4.1 *First group*

The concrete slump is required to be at a size of 180–220 mm, the mineral admixtures are blended with 20% fly ash, and the water-reducing agent water reduction rate is 20%. According to the relevant technical indicators of raw materials, the concrete mix ratio is designed, and the raw material consumption per cubic meter of concrete is given as follows: Cement 342.05 kg, gravel 1068.02 kg (gravel with a particle size of 5–10 mm is 30% of the total gravel, gravel with a particle size of 10–20 mm is 70% of the total gravel), sand 712.02 kg, water 192.4 kg, fly ash 85.51 kg, and admixture 8.55 kg. The concrete mixing process is shown in Figure 1. The test results are as follows:

(1) State: There are bubbles, sticky plate phenomenon, and slight bleeding phenomenon.
(2) Detection results: The lump is at a size of 206 mm; the expansibility ranges from 650 mm to 620 mm, with an average of 635 mm; the stack height is 80 mm, and the apparent density is 2380 kg/m^3, with an error rate of 0.8%.
(3) Adjustment method: The sand rate increases from 40 to 42%.
(4) After adjustment, the detection result: the size of the slump is 207 mm; the extensibility ranges from 640 mm to 615 mm, with an average of 628 mm; the stack height is 80 mm; the apparent density is 2378 kg/m^3, with an error rate of 0.9%.

Figure 1. Concrete mixing with flow state.

4.2 Second group

The slump of concrete is required to be 180–220 mm, and the mineral admixture is fly ash and slag powder with the dosage of 20% and 30%, respectively. Ordinary water-reducing agent is used, the water-reducing rate of water-reducing agent is 20%, and the dosage is 1.5%. According to the relevant technical indicators of raw materials, concrete mix ratio design is carried out, and the raw material consumption per cubic meter of concrete is given as follows: 240.05 kg of cement, 995 kg of gravel, 663 kg of sand, 192.4 kg of water, 96.2 kg of fly ash, 144.3 kg of slag powder, and 6.71 kg of admixture. The test results are as follows:

(1) Status: Sticky plate phenomenon (admixture quantity is too much).
(2) Detection results: The slump is at a size of 230 mm; the expansibility ranges from 700 mm to 650 mm, with an average of 675 mm; the stack height is 60mm, and the apparent density is 2390 kg/m^3, with an error rate of 0.4%.
(3) Adjustment method: Increase sand and stone by 5%.
(4) After adjustment, the detection result: slump 215 mm; The expansibility ranges from 525 mm to 500 mm, with an average of 513 mm; the stack height is 50 mm, and the apparent density is 2386 kg/m^3, with an error rate of 0.5%.

4.3 Third group

The slump of concrete is 180–220 mm, and the mineral admixture is fly ash and slag powder, which are 20% and 30%, respectively. A high-performance water-reducing agent with a water reduction rate of 25% is mixed by a mixture ratio of 1.3%. According to the relevant technical indicators of raw materials, concrete mix ratio design is carried out, and the raw material consumption per cubic meter of concrete is given as follows: cement 225.9 kg, gravel 1010.53 kg, sand 673.7 kg, water 180.75 kg, fly ash 90.38 kg, slag powder 135.56 kg, and admixture 5.87 kg. The test results are as follows:

(1) State: The admixture is too saturated, the flow rate is too slow, the material is too sticky and the bubble is too much, the admixture quantity is too large.
(2) Detection results: Slump 206 mm; the extension is 650mm, 620mm, and the average is 635mm; Stack height 80mm, apparent density 2380 kg/m^3, error 0.8%.
(3) Adjustment method: Increase sand rate from 40% to 42%.
(4) After adjustment, the detection result is: the slump is at a size of 207 mm; the expansibility ranges from 640 mm to 615 mm, with an average of 628 mm; the stack height is 80 mm, and the apparent density is 2378 kg/m^3, with an error rate of 0.9%.

4.4 Fourth group

The concrete slump is at a size of 200–240 mm, mineral admixture with fly ash and slag powder at a proportion of 25% and 15%, respectively. Using high-performance water-reducing agent, water reduction rate of 25% (consider 27%), and mixing amount of 1.1%. According to the relevant technical indicators of raw materials, concrete mix ratio design is carried out, and the raw material consumption per cubic meter of concrete is given as follows: cement 247.11 kg, gravel 1006.41 kg, sand 729.28 kg, water 177.1 kg, fly ash 102.97 kg, slag powder 61.78 kg, and admixture 4.53 kg. The test results are as follows:

(1) State: Good.
(2) Detection results: The slump is at a size of 200 mm; the expansibility ranges from 550 mm to 535 mm, with an average of 540 mm; the stack height is 110 mm; and the apparent density is 2356 kg/m^3, with an error rate of 1.8%.

5 CONCLUSIONS

In this paper, based on the study of the mixture design method of flow mode foam concrete, the raw material dosage is given, and the concrete mixing is carried out. Finally, combined with the workability determination method, the performance of five groups of concrete mixing is tested. The main test indexes are slump, expansion, apparent density, and pile height. According to the performance test results of the concrete mixture, the concrete ratio is optimized and the optimal ratio is finally given, which is of great significance to the engineering application of flow-foamed concrete.

ACKNOWLEDGMENTS

This paper is supported by the Xi'an Traffic Engineering Institute 2021 Young and Middle-aged Fund Project "the foam concrete mechanics performance influence factors analysis" (project number: 21KY-24).

REFERENCES

Jiang, Y. (2018) *Principle of Engineering Structure Design*. Nanjing Southeast University Press, Nanjing.
Ke, G. (2012) *Civil Engineering Materials*. Peking University Press, Beijing.
Qian, X. (2008) *Building materials*. China Architecture and Building Press, Beijing.
Qiao, W. (2017) Effect of vibration mixing on the performance of high strength flow concrete. *Concrete World*, 12: 90–95.
Shi, D. (2020) Characteristics and application status of foamed concrete. *Ju She*, 35: 25–26.

Frontiers in Civil and Hydraulic Engineering – Mohamed A. Ismail and Hazem Samih Mohamed (Eds)

Research on ecological embankment structure of navigation junction

Chunxin Zhong*

College of Harbour, Coastal and Offshore Engineering, Hohai University, Nanjing, China

ABSTRACT: Hydrodynamic characteristics on river banks vary between different water levels, which is important for revetment structure. In this paper, according to the characteristics of different water levels, the bank revetment is divided into 4 zones: bottom-protection zone, key-protection zone, hydrophilic zone, and landscape zone. The results showed that the bottom-protection zone is perennially submerged and vulnerable to water erosion, so construction should be chosen to protect the toe of the slope; the key-protection zone is enduring inundation and erosion, therefore permeable structure with strong scouring resistance and good durability is optimal; hydrophilic zone, of which the hydrodynamic characteristic is short-term submerged, plus slope runoffs, is suitable for both safe and ecological revetment forms; landscape zone is not flooded, but scoured by runoffs, so it is preferable to select appropriate plant types and distributions. Finally, the revetment project of the Naji navigation junction is taken as an example to apply the method proposed in the paper.

1   INTRODUCTION

Revetment work is the most popular and effective method to protect bank stability and control river regimes. In the 1990s, while enjoying the huge economic and social benefits brought by the large-scale development and utilization of rivers, western countries also felt the damage to the river ecosystem. People are gradually aware that rigid concrete revetment has played a certain role in bank maintenance, soil erosion prevention, flood control, and drainage; however, it produced negative impacts on landscape, environment, and ecology to a variable extent, against the sustained and healthy development of river ecosystems. To meet the needs of protecting the environment and closing to nature, ecological rebuilding of existing bank slopes or new construction of revetment has become a trend (Bai 2020).

Researchers have done lots of work on ecological revetment techniques and constructions, mainly employing vegetation to increase the ecology of the embankment (Jiang 2022; Zhao 2017). In Germany, gravel paving, dry placer techniques plus some plants have been adopted in the regulation of Rhine; Vetiver grass was used by Vietnam, and coconut fiber volume was used by England to mitigate the impact of ship waves. Besides, ecology concepts have been introduced in the waterway project of Grand Canal, China, with the use of the already lush green reeds and other green plants as the main channel slope protection materials. In addition, Japan has put forward the "hydrophile" concept, and structures like wood piles and wooden frames with stones have been taken to stabilize the riverbed and improve the ecological environment. Precast concrete interlocking blocks pavement has been adopted in Yangzhou urban sections of Grand Canal, China, with structures mutually engaged and cavity grass within the block, to ensure water exchange. Although the new materials and techniques might suggest lessons for increasing the ecology of slopes, previous ecological revetment paid little attention to the hydrodynamic characteristics of bank slopes; furthermore, the zoning and selection of structures are without basis for hydrodynamic characteristics.

In this paper, based on the hydrodynamic characteristics of a shipping hub, the bank slope is zoned according to different feature water levels. Finally, suitable revetment constructions are recommended.

---

*Corresponding Author: cxzhong@hhu.edu.cn

   DOI 10.1201/9781003344209-12

2 REVETMENT ZONING OF A SHIPPING HUB

A shipping hub is a navigation project to improve natural rivers. The characteristic water levels of which can be briefly summarized as protective high water level, normal pool level, and dead water level. The dead water level is the minimum level of the reservoir under normal operation. The normal pool level is the reservoir's storage level that meets water conversation demand in normal conditions. A protective high water level is the highest normal operating elevation of the reservoir or flowage in the case of downstream protective objects. Considering different water levels, hydrodynamic characteristics, and landscape features, the embankment can be divided into a bottom-protection zone, a key-protection zone, a hydrophilic zone, and a landscape zone (Zhang 2007), as shown in Figure 1.

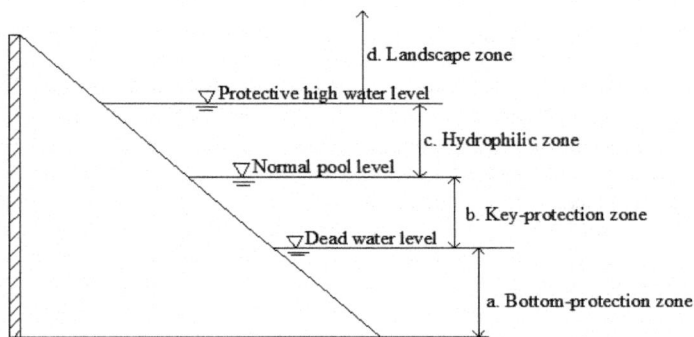

Figure 1.   Revetment zoning in the protective section of reservoirs.

2.1   Bottom-protection zone

Bottom-protection zone is from the dead water level of the embankment down to the riverbed, of which the hydrodynamic characteristic is perennial submerged. The intensity of bank soil decreases due to prolonged immersion. During flood period, the navigation junction's gate is wide open to discharge flood, then the flow rate is comparatively large, and it may crash the toe of slope and threaten the stability of the bank when mainstream close to the slope. Therefore, foot care must be done in the first place. Bottom-protection zone's primary function is protecting the toe of the slope from scouring and safeguarding slope stability.

2.2   Key-protection zone

Key-protection zone is from normal pool level to dead water level. Water level fluctuations caused by ship waves or storms affect the upper and lower boundary. Key-protection zone is the focused protective area in the ecological revetment works. There are two major patterns of bank deformation failure: One is the revetment structure damage with loss of soil, which is caused by ship or wind waves, and long-term scouring mainstream of river; The other is bank collapse, since lateral water pressure of riverbank cannot be released, owing to the lag on water levels drop of soil in the dry season. Key-protective zones' protective structures should give priority to the security and stability of the bank, and take into account ecological functions.

2.3   Hydrophilic zone

The hydrophilic zone is above normal pool level to a protective high water level, of which the hydrodynamic characteristic is short-term submerged, and only in flood season, bank slope will suffer flood soaking and water erosion plus slope runoffs. The bank is not affected by river force, but slope wash caused by rainfall runoff should be considered under normal pool level or in the dry season. Primarily, the hydrophilic zone is an important transition section of safety and ecology, aiming at creating sound ecological hydrophilicity.

2.4 *Landscape zone*

The landscape zone is above a high water level, of which the hydrodynamic characteristic is affected only by runoff erosions, instead of flows. Ecological revetment must be fitted into the culture and environment in the surroundings, with consideration to the antierosion properties of the slope. Based on the topography of the district, it can be divided into straight slopes and a ramp. The straight slope is more stable, but in the ramp, slope wash caused by rainfall runoff should be considered.

3 STRUCTURE FORMS OF REVETMENT ZONE

3.1 *Revetment structures of bottom-protection zone*

Protection works of the zone play a vital role in the entire revetment project. There are three revetment forms: one is reinforcement measures only focused on the lower part of the slope, one is foot reinforcement combined with bottom protection, and the other is the direct use of bottom protection. Foot reinforcements support the bank, prevent collapse, and serve as a smooth transition between bottom protection and embankment. Meanwhile, bottom protections stabilize the riverbed and prevent soil erosion (Guan 2014). The two are independent of each other but make up an indivisible whole. Because the bottom-protection zone suffers perennial erosion and immersion, foot reinforcement materials should have the advantages of resisting wave scouring and long-term soaking, good integrity, and the ability to adapt to the riverbed deformation, to avoid the loss of soil at the foot of the slope and ensure safety. In the case of poor ground conditions or steep slopes, slope failure is the major riparian damage, and foot reinforcement is recommended. In the corners of the concave bank and near thalweg shore waters, underwater topography is complex, and bottom protection or combined with foot reinforcement is utilized. When the slope gradient is small and foundation conditions are stable, erosion is the main bank damage, only bottom protection is adopted to guarantee riverbeds from soil erosion.

Currently, foot reinforcements consist of concrete, dry rubble foundation trench, wire gabion foundation trench, and other forms. Traditional concrete foot reinforcement demands much higher ground conditions, which means that the foundation trench is excavated to a solid base when the water level in dry weather is comparatively low. By contrast, the latter two can be adapted to inhomogeneous deformation of riverbed bottoms for ages, which has a broader scope. Therefore, geological conditions, construction conditions, and economic rationality of the reservoir should be taken into account in selecting suitable foot reinforcement. Meanwhile, some materials that can resist long-term water immersion and adapt to the inhomogeneous deformation of the riverbed are used in the bottom protection works, such as scattered-cast stone, combed fascine raft, concrete-caking soft mattresses, and geotextile soft mattresses. The layout in bottom-protection zone is shown in Figure 2.

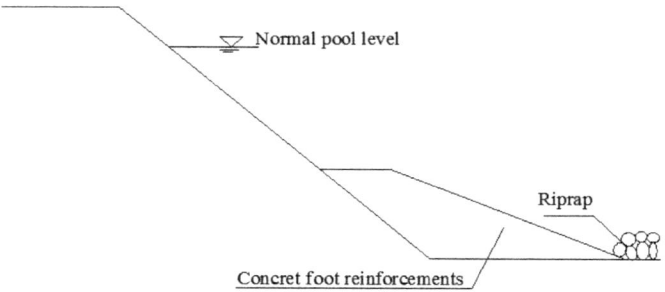

Figure 2. Layout in bottom-protection zone.

3.2 Revetment structures of key-protection zone

The key-protection zone is the most vital to ensure slope safety and stability. Rigid material like stone masonry is typically adopted but shortcomings of such revetment structures become increasingly obvious over time. First, the resource consumption of quality stone is large. Second, grouting surface protection blocks water exchange between rivers and banks and does great damage to the ecological environment as well. Third, stone masonry revetment often appears to soil erosion at the back of the slope, owing to porous grout, and small and thin face stone, which requires long-term maintenance in the dry season. Therefore, the selection of revetment structures of the key-protection zone should achieve a balance between slope revetment and ecological protection, so that they can: meet requirements of both structural strength and stability; enhance the antierosion ability of surface cover; overcome the disadvantages of conventional revetment's "crude" and vegetation revetment's narrow applicable range at the same time. Ecological revetment structures can be summarized as natural ecoengineering, semi-natural ecoengineering, and artificial ecoengineering, as discussed in detail below.

3.2.1 Examples
Natural ecoengineering adopts natural materials, including plants, dry stone, and timber lattice.

3.2.2 Semi-natural ecoengineering
Semi-natural ecoengineering adopts part of artificial stuff combined with natural materials, like wire gabion revetment and half-masonry revetment.

3.2.3 Artificial ecoengineering
Artificial ecoengineering is a new type of revetment, also an ecological transformation of the original artificial materials or constructions, mainly including permeable retaining wall revetment (for instance, permeable steel sheet pile retaining wall, permeable prefabricated concrete caissons, and permeable concrete scission block), and eco-organic materials revetment (for instance, eco-concrete and high-performance soil curing agent).

Each ecological revetment structure has its own advantages and application scope. Natural ecoengineering can be adopted when the slope is gentle and flow velocity is small so that local materials can be used and the environment is aesthetically pleasing. Semi-natural ecoengineering has good flexibility to adapt to ground deformation, and can timely discharge pore water from the soil. It can reduce water pressure behind the wall, accelerate the natural consolidation of the backfill, be conducive to the security and stability of the structure, and can also dissipate impact forces of ship waves, which has a wide range of applications. Artificial ecoengineering can withstand rapid flows of the curve and has no negative impact on the environment or landscape below the normal pool level, but it demands much higher ground and construction conditions and costs more than the previous two, which is often applied to the major hydroelectric project with the large flow velocity.

3.3 Revetment structures of hydrophilic zone

The hydrophilic zone is inundated for a short time during the flood season. Although affected by wind and waves, the hydrodynamic of the area is much weaker than the key-protection zone. Revetment structures of plants, geotextile mats, and reinforced eco-bags are mainly adopted.

Natural ecoengineering is more suitable, especially for vegetation measures, since the flexible leaves of the plants scattered in the water can absorb and dissipate water energy, and reduce the impact of waves on the dike. Another important role the vegetation acting as is reinforcement, for the spread of turf roots and nodule branches fixing soil between the roots. But vegetation revetment, eco-bags, or other natural ecoengineering should be engineered to increase their integrity and impact resistance of slope runoffs or water erosion. Local mesophytes with a higher survival rate, lower project cost and easily forming of a forest landscape are recommended; in addition, landscaping mesophytes are planted on the top, which can enhance ecological and landscape effects under the normal pool level, and act as a warning sign to the ship closing to the shore.

3.4 *Revetment structures of landscape zone*

The landscape zone is not flooded but scoured by runoffs. The case of flat slope or ramp is separately analyzed.

Typically, pure vegetation technology is used on the flat slope because there are fewer restrictions on plant planting. Submergence-resistance is not required in the zone, so herbs, shrubs, and trees that are strongly-rooted and inter-crossed, which have good mechanical properties, large coverage, and some ornamental value, are employed in the zone.

On the contrary, a combination of vegetation and reinforcement technology is optimal because the ramp suffers from runoff erosion. Living-brunch-layer-planted technology, or ecoengineering composite technology, such as three-dimension vegetative net, geotextile, alloyed steel net-pad, and other flexible reinforced sod revetment, is available, which has the advantages of low cost, convenient operation, and desirable maintenance results.

4 NAJI RESERVOIR ECOENGINEERING WORKS

4.1 *Revetment zoning of naji*

Naji navigation junction is located in River Youjiang, Baise, Guangxi Province, which is the anti-regulating reservoir of the Baise hydro-junction. Naji is a shipping hub that is also equipped with power generation capabilities. This is another example of effective comprehensive utilization engineering, where the normal water level is 115.0 m, the dead water level is 114.4 m, the designed flood standard is defined as occurring once every 500 years, and the checked flood level is 118.53 m. Besides, when the flow exceeds 2670 m^3/s, the gate is wide open to discharge flood following the operating rules of a reservoir.

According to the water level data of the Naji project: the normal water level is 115.0 m, the dead water level is 114.4 m, the designed flood level is 109.87 m, and the checked flood level is 118.53m. Although the checked flood level is relatively high, its occurrence is very rare, and the designed water level is under the normal water level, even the designed low water level. Therefore, the hydrophilic zone and the landscape zone are merged into a hydrophilic-landscape zone in the Naji project. Besides, in view of fluctuations caused by ship or wind waves, the water level of the key-protection zone ranges from 114.2 m to 115.5 m (Figure 3).

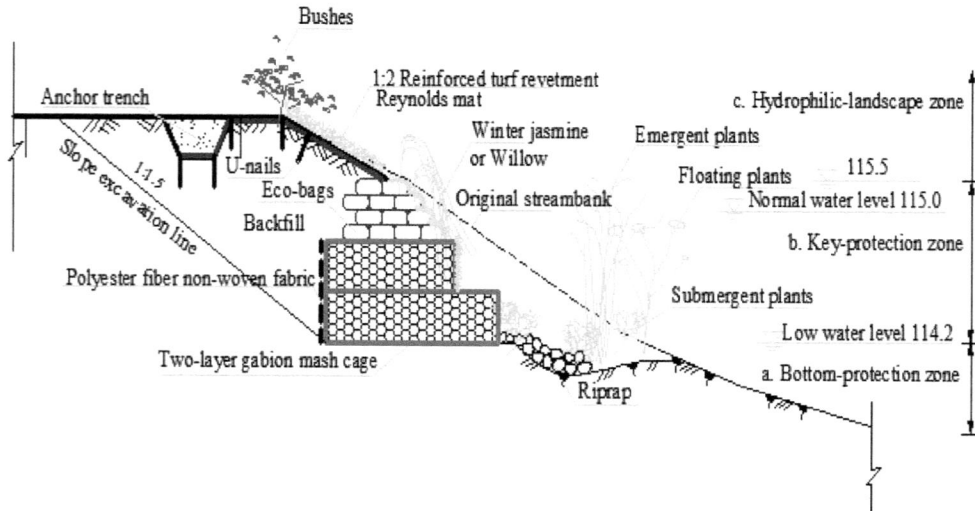

Figure 3.   Schematic diagram of ecological revetment in Naji navigation junction.

4.2 Revetment structures of naji

The original protected range of revetment structures of Naji is small because no bottom protection is designed in front water except stakes. Streambank is easy to collapse and retreat, and cultivated emergent macrophytes in the littoral zone failed to survive too, with actions of scouring, and abrasion to rear solids, produced by the rise and fall of water levels, ship, or wind waves. Also during maintenance, selected shrubs in the ramp above the normal water level are insufficient to stabilize and protect slopes, resulting in bare surface soil erosion by rainfall runoff. Therefore, the treatment of the hydrophilic-landscape zone of Wantang is necessary.

4.2.1 Bottom-protection zone

Bottom-protection zone is from the dead water level down to the designed flood level, while the designed flood level down to the riverbed also needs bottom protection when mainstream scouring the bank. So riprap is commonly installed, which is a simple construction and cheaper cost.

4.2.2 Key-protection zone

Key-protection zone ranges from 114.2m to 115.5m. Below the normal pool level, a two-layer gabion mash cage filling stone is established to withstand water waves, and polyester fiber non-woven fabric is also equipped behind the retaining wall as a reversed filter. Near the normal pool level, an eco-bags retaining wall (consisting of shrubs like winter jasmine and weeping willow) is constructed, to buffer water dissipation, guarantee ecology and provide visual aesthetics.

4.2.3 Hydrophilic-landscape zone

In the hydrophilic-landscape zone, ramps are herbal-based reinforcement, specifically, the arrangement of a three-dimensional vegetation net plus reinforced turf revetment, and the sowing of a mix of Bermuda grass (occupying 75%) and Tall Fescue (occupying 25%), which is easy to tolerate flooding and reproduce. While flat slopes are vegetation restoration due to fewer restrictions, Vitex negundo or Lantana camara is planted on the shoulder.

5 CONCLUSIONS

(i) Hydrodynamic characteristics are varying between different water levels on the river bank slope, which revetment structures should be based on, to maintain safety, ecology, and sight of the embankment.

(ii) According to the characteristic water level of the navigation channel, the embankment will be divided into a bottom-protection zone, a key-protection zone, a hydrophilic zone, and a landscape zone. The bottom-protection zone is perennially submerged, which is susceptible to erosion when mainstream close to the shore. The key-protection zone suffers enduring inundation, the main concern of which is that it is vulnerable to water erosion and affected by ship or wind waves. The hydrophilic zone suffers short-term submerging, plus slope runoffs. The landscape zone is affected only by runoff erosions, instead of flows.

(iii) Suitable revetment structures are selected according to the hydrodynamic characteristics, geological conditions, construction technique, and project efficiency. Structures of the bottom-protection zone should be adapted to bed deformation, and prevent toe scouring. Structures of the key-protection zone should be permeable, and also have abilities of antierosion and protecting slope stability. Structures of the hydrophilic zone should be both safe and ecological.

(iv) For the revetment project of Naji, Guangxi, China, revetment zoning is based on normal pool levels and low water levels, and revetment structures are chosen according to hydrodynamic conditions. Bottom-protection zone takes measures of riprap protection, the key-protection zone uses a gabion retaining wall structure to ensure safety while focusing on water exchange, hydrophilic-landscape zone introduces eco-bags and reinforced turf revetment.

REFERENCES

Bo Bai & Shuguang Xin (2020). Study on ecological revetment engineering technology of river. *J. Inner Mongolia Water Resources*, 47–49

Chenkun Jiang & Lushan Li (2022). Application of ecological revetment technology in river embankment prevention and treatment engineering. *J. Technology and Economy of Changjiang* 6(1), 36–44

Chunman Guan & Guirong Zhang (2014). Development trend and hot issues of ecological bank protection technology for small and medium-sized rivers. *J. Hydro-Science and Engineering* 4, 75–81

Peijian Zhao & Hongchun Ma (2017). River ecological revetment technology and design. *J. Building Technology Development* 44(4), 102–103

*Frontiers in Civil and Hydraulic Engineering – Mohamed A. Ismail and
Hazem Samih Mohamed (Eds)*
© 2023 The Authors, ISBN 978-1-032-38247-0

Response of microalgae cells to bacterial signal molecules during water treatment

Yiling Fang, Kang Xie*, Xianggan DuanMu & Zhiyong Zhou
School of Civil Engineering and Architecture, University of Jinan, Jinan, China

ABSTRACT: Bacteria algae symbiotic membrane bioreactor (BAS-MBR) has good application potential in sewage treatment and resource recovery. Microalgae widely exist in various water bodies, and can efficiently assimilate and utilize nitrogen and phosphorus nutrient elements in sewage. Microalgae are introduced into MBR to form bacteria algae symbiotic MBR (BAS-MBR), which effectively combines the efficient uptake of nitrogen and phosphorus by microalgae with the degradation ability of bacteria to organic matters, and at the same time overcomes the difficulty of difficult recovery and utilization of microalgae, so it has great application potential. But membrane pollution is still an important factor restricting its development. In recent years, the influence of the group sensing effect between bacteria and algae species on the expression of extracellular polymers and the copolymerization behavior of bacteria and algae has attracted wide attention, which provides a new idea for BAS-MBR research. In this study, under the action of bacterial signal molecules C6-HSL, C8-HSL, C10-HSL, C12-HSL, 3-oxo-C8-HSL, and 3-oxo-C12-HSL with a concentration of 50ng/L, the aggregation state, extracellular polymer content, nitrogen, and phosphorus removal effect and zeta potential change of Chlorella liquid were explored to clarify the response mechanism of microalgae to bacterial quorum sensing effect and provide theoretical and technical reference for stable operation of BAS-MBR and efficient energy conservation.

1 INTRODUCTION

With the continuous growth of China's population, rapid economic development, and accelerating urbanization, the water consumption of various industries in the society is rising, the sewage discharge is also increasing, and the eutrophication of water body is becoming more and more serious, which has become one of the core problems of surface water pollution prevention and control (Van Loosdrecht & Brdjanovic 2014). The bacteria algae symbiosis system (ABS) has been widely concerned by researchers at home and abroad because of its high energy recovery potential, low aeration energy consumption, and high-efficiency nutrient removal effect compared with traditional nitrogen and phosphorus removal methods. The application of ABS Technology in sewage treatment has been verified in practice, and it is considered as one of the most potential means to eradicate the eutrophication of water (Craggs et al. 2012). Algae were light-energy autotrophs that can airborne carbon dioxide, releasing oxygen to increase dissolved oxygen in water bodies. And its biomass is synthesized by absorbing nutrients such as nitrogen and phosphorus in the surrounding environment through assimilation for proliferation, to reach the role of purifying water bodies (Craggs et al. 2012; Wang et al. 2010). In addition, algal biomass and its metabolites were an important bioresource that can serve as raw materials for many renewable products (Brennan & Owende 2010; Chen et al. 2009; Milledge 2011). These biofuels are regarded as the most promising alternatives to fossil fuels in the future (Rawat et al. 2011).

---

*Corresponding Author: cea_xiek@ujn.edu.cn

At present, bacteria alga symbiotic MBRs have been widely studied in the field of wastewater treatment, but membrane fouling is a limiting condition for MBRs to be widely used (Liu et al. 2021). Studies have shown that membrane fouling and quorum sensing phenomenon were necessarily linked, and the quorum sensing phenomenon between bacteria has been explored by many scholars, while whether bacterial signaling molecules can have an effect on the cellular behavior of microalgae was not very clear (Byers et al. 2002; Rudner et al. 1991; Thomson et al. 2000). Therefore, it is of great practical importance to explore the effect of bacterial signaling molecules on the extracellular polymers as well as aggregation behavior of microalgae.

2 MATERIALS AND METHODS

2.1 Selection of algal species

The experimental algal species used were cropped into FACHB-10 *(proteinaceous nuclear)* *Chlorella*, a unicellular organism belonging to the *phylum chlorophytes, chroococcales*, and *Chlorella* at the Freshwater Algal Stock Bank of Chinese Academy of Sciences (Wuhan, China). The algal cells were often oval with 1 prominent proteinaceous nucleus, a thin cell wall, and a cell diameter of 3–5 μm. Proteinaceous Chlorella is a freshwater alga that was first shown more than 2 billion years ago by related studies to belong to the earliest group of life on earth. Synechocystis proteolytica has high photosynthetic efficiency, environmental tolerance, rapid growth and reproduction, and extremely wide distribution.

2.2 Formulation of growth solutions

In order to avoid the influence of bacterial pollution, the laboratory uses the BG11 medium as the simulated wastewater of microalgae. The composition of the culture medium is shown in Table 1.

Table 1. BG11 Media Formulation.

Sequence	Components	Consumption	Mother liquor concentrations
1	NaNO$_3$	10ml/L	15g/100ml dH$_2$O
2	K$_2$HPO$_4$	10ml/L	2g/500ml dH$_2$O
3	MgSO$_4$·7H$_2$O	10ml/L	3.75g/500ml dH$_2$O
4	CaCl$_2$·7H$_2$O	10ml/L	1.8g/500ml dH$_2$O
5	Citric Acid	10ml/L	0.3g/500ml dH$_2$O
6	Ferric Ammonium Citrate	10ml/L	0.3g/500ml dH$_2$O
7	EDTANa$_2$	10ml/L	0.05g/500ml dH$_2$O
8	Na$_2$CO$_3$	10ml/L	1g/500ml dH$_2$O
9	A5(Trace Mental Solution) *	1ml/L	

	Components	Concentrations	Components	Concentrations
A5	H$_3$BO$_3$	2.86g/L dH$_2$O	ZnSO$_4$·7H$_2$O	0.22g/L dH$_2$O
	MnCl$_2$·4H$_2$O	1.86g/L dH$_2$O	Na$_2$MoO$_4$·2H$_2$O	0.39g/L dH$_2$O
	CuSO$_4$·5H$_2$O	0.08g/L dH$_2$O	Co (NO$_3$)$_2$·6H$_2$O	0.05g/L dH$_2$O

*A5 was prepared by a mixture of trace elements, as detailed in this Table.

2.3 Experimental procedures

Seven 250 ml Erlenmeyer flasks were used as reactors, in which 200 ml of algal fluid with a concentration of 300 ± 30 mg/L was added to each reactor, the blank control was pure algal fluid without signal molecules, and the other six groups were spiked with 50 ng/L of C6-HSL, C8- HSL, C10- HSL, C12- HSL, 3-oxo-C8- HSL, 3-oxo-C12- HSL, which were used to study the effects of different kinds of signaling molecules on microalgae. The seven groups of reactors were placed in a constant temperature and light incubator with 12 h light and 12 h dark at 25°C, and the pH of the algal liquid in the reactor was 7.0–7.2, and the laboratory used incandescent lamps and light illumination was 4 KLX.

2.4 *Analytical methods*

The contents of TN and TP in the water quality index were determined spectrophotometrically. Microalgal aggregation was assessed using light microscopy, employing high-speed centrifugation to extract extracellular polymers. Detection of proteins and polysaccharides in the extracellular polymers was performed by the BCA protein method (Frølund et al. 1996) and anthrone-sulfuric acid spectrophotometer (Li & Yang 2007), respectively. Zeta potential values were monitored using a zeta potentiometer.

3 RESULTS AND DISCUSSION

3.1 *Effects of bacterial signaling molecules on microalgal aggregation*

We could see the change in aggregation morphology of microalgae after adding signal molecules for five days from Figure 1. We could see that microalgae have an obvious aggregation state from the first day to the fifth day of culture. On the first day, microalgae cells were in a normal dispersion state, while on the fifth day, microalgae cells were in an obvious aggregation state, and the aggregation effect of adding different signal molecule groups was slightly different. This showed that bacterial signal molecules can affect the colony behavior of microalgae cells to a certain extent, and microalgae cells can accept bacterial signal molecules and respond to them. Zhou et al. also reached a similar conclusion (Zhou et al. 2017).

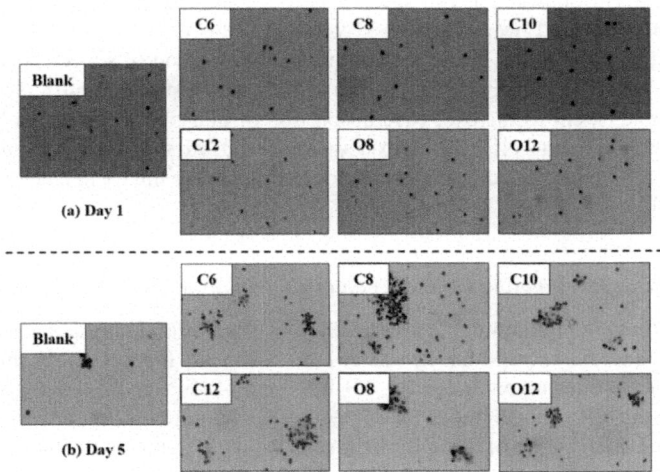

Figure 1. Microalgal aggregation on Day 1 (a) and Day 5 (b) after the addition of different signal molecules under the microscope.

3.2 *Effect of bacterial signal molecules on the removal of nitrogen and phosphorus from microalgae*

Figure 2a showed the removal effect of Chlorella on TN within 5 days of a mini reactor cycle, and it can be seen that the TN removal effect within the 7 groups of Erlenmeyer flasks exhibited a similar change trend throughout the process. The removal effect of Chlorella solution added with signal

97

molecule group on TN was better than the blank control group. It can be specifically seen from Figure 1a that the TN removal rates of the blank control group and the group with added signal molecules for 5 days were 2.48%, 10.07%, 9.43%, 12.26%, 15.62%, 17.42%, 15.70%, respectively. Studies have shown that Chlorella can take up and assimilate nitrogen elements into its own cell body (Uggetti et al. 2014), so the blank control group also had a certain removal effect on TN. Control experiments illustrated the addition of signaling molecules, which may have facilitated the uptake and utilization of nitrogen elements by Chlorella cells.

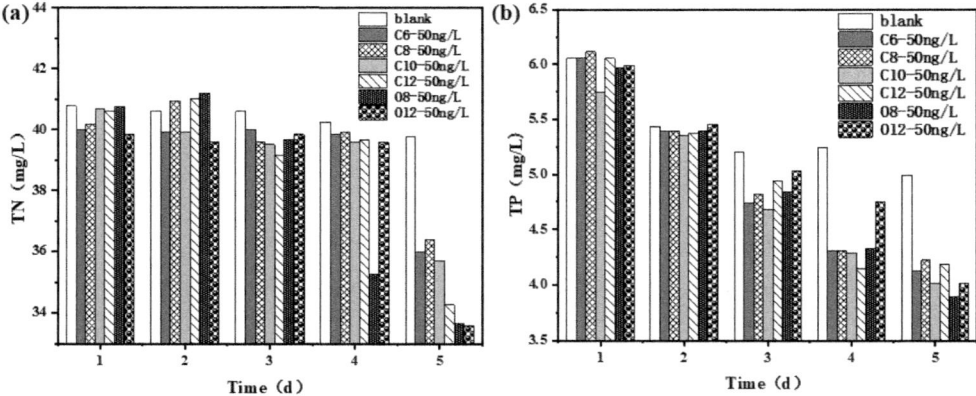

Figure 2. The variation law of nitrogen and phosphorus contents in the reactor.

The change of TP in the simulated sewage was shown in Figure 2b, and it was known from the figure that the TP removal rates in seven sets of Erlenmeyer flasks at influent phosphorus concentrations of 5.5–6 mg/L were 17.49%, 31.94%, 31.00%, 30.06%, 30.90%, 34.77%, 32.94%, respectively. Compared with the blank control group, the addition of signal molecules improved the removal efficiency of total phosphorus. This illustrated that signaling molecules may have contributed to some extent to the excessive absorption of phosphorus elements by Chlorella sp. Algae can take up phosphorus elements in excess and can absorb them from the environment to be stored inside the cell in the form of polyphosphates as used by the microalgal cells for their growth (Boelee et al. 2014).

3.3 *Effects of bacterial signaling molecules on extracellular polymers in microalgae*

It is well known that the production of extracellular polymers is an important cause of membrane fouling in membrane bioreactors during wastewater treatment. Figures 3a, b, and c showed the changes in polysaccharide content in S-EPS, LB-EPS, and TB-EPS in the seven groups of reaction systems respectively. EPS has an important impact on the hydrophobicity, adhesion, flocculation, sedimentation, and dehydration of flocs. It can be seen from the figure that with the extension of the reaction time, the polysaccharide concentration in EPS of the reactor broth of each group increased to a certain extent. Among them, the polysaccharide content in SMP was the highest, from 22–25 mg/gMLSS on the first day to 25–26 mg/gMLSS on the fourth day. The polysaccharide content in LB-EPS increased from 3 ± 0.5 mg/gMLSS to 3.5 ± 0.5 mg/gMLSS. The polysaccharide content in TB-EPS increased from 1 ± 0.5 mg/gMLSS to 3.5 ± 0.5 mg/gMLSS. On the whole, the content of TB-EPS increased the most. The analysis shows that the addition of AHLs signal molecules may affect the intercellular information exchange of microalgae, affect the biological metabolism of Chlorella, and then lead to the aggregation of Chlorella. Coupled with the increased secretion of extracellular polysaccharides, polysaccharides can be used as an adhesive between Chlorella cells (Parise et al. 2007), which will cause aggregation between cells when Chlorella cells move and collide.

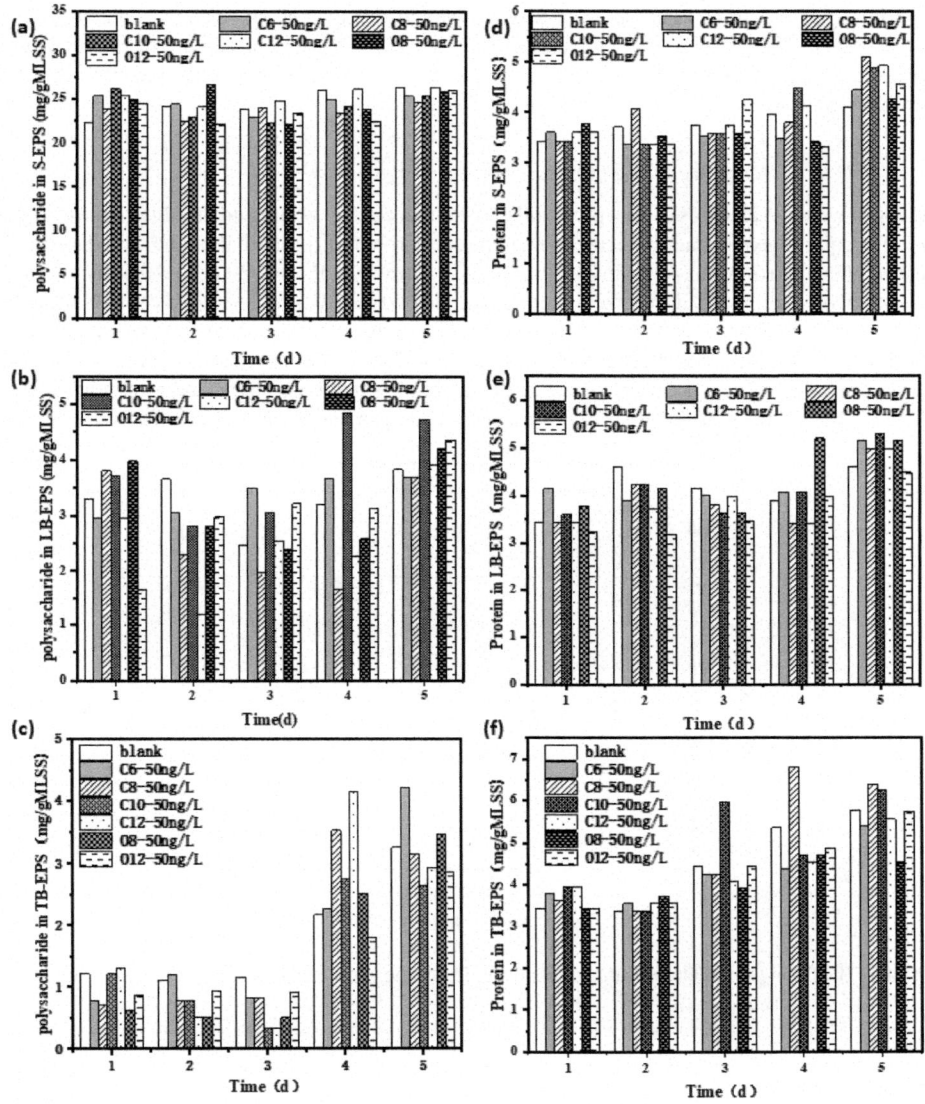

Figure 3. The variations of polysaccharide content (a, b, c) and protein content (d, e, f) in algal liquor.

Figures 3d, e, and f respectively showed the changes in protein content in S-EPS, LB-EPS, and TB-EPS in the seven groups of reaction systems. From Figure 3, it can be seen that the protein changes in the blank control group within 5 days are relatively small compared with the addition of signal molecular groups, fluctuating between 3.5–5.5mg/gMLSS. The protein in SMP, LB-EPS, and TB-EPS increased from the initial 3.5 ± 0.5 mg/gMLSS to 4.5 ± 0.5 mg/gMLSS, 4.5 ± 0.5 mg/gMLSS, and 5.5 ± 0.5 mg/gMLSS after five days, respectively from the perspective of protein and polysaccharide, both of which showed the fastest growth rate in TB-EPS. Studies have shown that cells widely use protein outer fibers to establish a fulcrum of surface contact, so as to promote the formation of biofilm (Anon 1974). At the same time, the increase in protein content also improved the surface hydrophobicity and ensures the stability of aggregates (Wang et al. 2014).

The change of extracellular polymer also explained the change of nitrogen and phosphorus removal effect. The increase of extracellular polymer leads to the corresponding increase of its adsorption on nitrogen and phosphorus, and the removal effect of nitrogen and phosphorus in wastewater was enhanced.

3.4 *Effects of bacterial signaling molecules on microalgal zeta*

Figure 4 showed the changes of zeta potential in the blank control group and the algae solution added with different kinds of signal molecules. It can be seen from the figure that on the first day, the zeta potential of each group was-20.8mv,-20.5 mv,-21.7 mv,-21.9 mv,-20 mv,-20.5 mv and-20.7 mv, respectively. On the fifth day, zeta potential values were-20.5 mv,-19.4 mv,-19.8 mv,-18.1 mv,-18.3 mv,-17.9 mv and-19.1mv, respectively. The zeta potential value of each group increased to a certain extent. Zeta potential value represents the stability of colloidal dispersion. The larger the absolute value of zeta potential is, the more stable the system is. Therefore, after adding signal molecules, the stability of the algal liquid decreases and gradually shows aggregation.

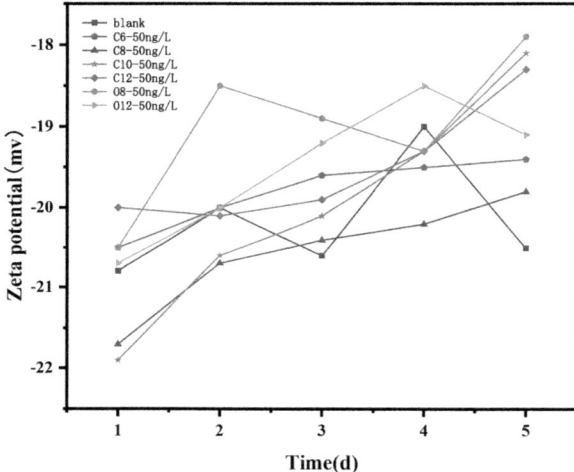

Figure 4.   The Changes in Zeta potential of algal fluid after the addition of different signal molecules.

4   CONCLUSION

In this paper, *Chlorella pyrenoidosa* was selected as the research object, and a 250ml Erlenmeyer flask was used as the reactor to explore the effects of bacterial signal molecules on extracellular polymers and the aggregation of *Chlorella pyrenoidosa*. It can be seen from the experimental results that bacterial signal molecules would contribute to the aggregation of microalgae and promote the secretion of extracellular polymers of microalgae to a certain extent. The addition of signal molecules would reduce the absolute value of the zeta potential of algal fluid, reduce the stability of algal fluid, aggravate the collision of algal fluid, and accelerate the aggregation of microalgae cells under the action of extracellular polymers. This conclusion can be used as a reference to slow down the membrane pollution in the subsequent operation of bacteria algae MBR for wastewater treatment. However, the influence of different signal molecules on the algal liquid was not obvious, which may be due to the insufficient concentration of signal molecules or the limited influence of individual signal molecules on the algal liquid. Further research is needed. This study further promotes the construction of an efficient and stable operation technology system of bacteria algae symbiosis MBR and has important theoretical value.

ACKNOWLEDGEMENTS

I would like to express my sincere thanks to my teachers, classmates, friends, and relatives who have given me guidance and help in the research process of this paper. In addition, I would like to thank project ZR2020ME227 supported by Shandong Provincial Natural Science Foundation, and project No. 2021SFGC0204 supported by the Shandong Provincial Key Research and Development Program of China support for this paper.

REFERENCES

Anon. 1974. "Book Reviews." *Journal of Child Psychology and Psychiatry* 15(4):335–40. doi: 10.1111/j.1469-7610.1974.tb01258.x.

Boelee, N.C., H. Temmink, M. Janssen, C.J.N. Buisman, and R.H. Wijffels. 2014. "Balancing the organic load and light supply in symbiotic microalgal-bacterial biofilm reactors treating synthetic municipal wastewater." *Ecological Engineering* 64:213–21. doi: 10.1016/j.ecoleng.2013.12.035.

Brennan, Liam, and Philip Owende. 2010. "Biofuels from microalgae-A review of technologies for production, processing, and extractions of biofuels and co-products." *Renewable and Sustainable Energy Reviews* 14(2):557–77. doi: 10.1016/j.rser.2009.10.009.

Byers, Joseph T., Claire Lucas, George P.C. Salmond, and Martin Welch. 2002. "Nonenzymatic turnover of an erwinia carotovora quorum-sensing signaling molecule." *Journal of Bacteriology* 184(4):1163–71. doi: 10.1128/jb.184.4.1163-1171.2002.

Chen, Paul, Min Min, Yifeng Chen, Liang Wang, Yecong Li, Qin Chen, Chenguang Wang, Yiqin Wan, Xiaoquan Wang, Yanling Cheng, Shaobo Deng, Kevin Hennessy, Xiangyang Lin, Yuhuan Liu, Yingkuan Wang, Blanca Martinez, and Roger Ruan. 2009. "Review of the biological and engineering aspects of algae to fuels approach." *International Journal of Agricultural and Biological Engineering* 2(4):1–30. doi: 10.3965/j.issn.1934-6344.2009.04.001-030.

Craggs, Rupert, Donna Sutherland, and Helena Campbell. 2012. "Hectare-scale demonstration of high rate algal ponds for enhanced wastewater treatment and biofuel production." *Journal of Applied Phycology* 24(3):329–37. doi: 10.1007/s10811-012-9810-8.

Frølund, Bo, Rikke Palmgren, Kristian Keiding, and Per Halkjær Nielsen. 1996. "Extraction of extracellular polymers from activated sludge using a cation exchange resin." *Water Research* 30(8):1749–58. doi: 10.1016/0043-1354(95)00323-1.

Li, X.Y., and S.F. Yang. 2007. "Influence of loosely bound extracellular polymeric substances (EPS) on the flocculation, sedimentation and dewaterability of activated sludge." *Water Research* 41(5):1022–30. doi: 10.1016/j.watres.2006.06.037.

Liu, Qiang, Jiayao Ren, Yongsheng Lu, Xiaolei Zhang, Felicity A. Roddick, Linhua Fan, Yufei Wang, Huarong Yu, and Ping Yao. 2021. "A review of the current in-situ fouling control strategies in MBR: biological versus physicochemical." *Journal of Industrial and Engineering Chemistry* 98:42–59. doi: 10.1016/j.jiec.2021.03.042.

Van Loosdrecht, Mark C.M., and Damir Brdjanovic. 2014. "Anticipating the next century of wastewater treatment." *Science* 344(6191):1452–53. doi: 10.1126/science.1255183.

Milledge, John J. 2011. "Commercial application of microalgae other than as biofuels: A brief review." *Reviews in Environmental Science and Biotechnology* 10(1):31–41. doi: 10.1007/s11157-010-9214-7.

Parise, Gina, Meenu Mishra, Yoshikane Itoh, Tony Romeo, and Rajendar Deora. 2007. "Role of a putative polysaccharide locus in bordetella biofilm development." 189(3):750–60. doi: 10.1128/JB.00953-06.

Rawat, I., R. Ranjith Kumar, T. Mutanda, and F. Bux. 2011. "Dual role of microalgae: phycoremediation of domestic wastewater and biomass production for sustainable biofuels production." *Applied Energy* 88(10):3411–24. doi: 10.1016/j.apenergy.2010.11.025.

Rudner, D.Z., J.R. LeDeaux, K. Ireton, and A.D. Grossman. 1991. "The SpoOK locus of bacillus subtilis is homologous to the oligopeptide permease locus and is required for sporulation and competence." *Journal of Bacteriology* 173(4):1388–98. doi: 10.1128/jb.173.4.1388-1398.1991.

Thomson, N.R., M.A. Crow, S.J. Mcgowan, A. Cox, and G.P.C. Salmond. 2000. "Pak-Iran to sign FTA after resumption of banking channels." *Molecular Biology* 36(3):539–56.

Uggetti, Enrica, Bruno Sialve, Eric Latrille, and Jean Philippe Steyer. 2014. "Anaerobic digestate as substrate for microalgae culture: The role of ammonium concentration on the microalgae productivity." *Bioresource Technology* 152:437–43. doi: 10.1016/j.biortech.2013.11.036.

Wang, Bin Bin, Qing Chang, Dang Cong Peng, Yin Ping Hou, Hui Juan Li, and Li Ying Pei. 2014. "A new classification paradigm of extracellular polymeric substances (EPS) in activated sludge: Separation and characterization of exopolymers between floc level and microcolony level." *Water Research* 64:53–60. doi: 10.1016/j.watres.2014.07.003.

Wang, Liang, Yingkuan Wang, Paul Chen, and Roger Ruan. 2010. "Semi-continuous cultivation of chlorella vulgaris for treating undigested and digested dairy manures." *Applied Biochemistry and Biotechnology* 162(8):2324–32. doi: 10.1007/s12010-010-9005-1.

Zhou, Dandan, Chaofan Zhang, Liang Fu, Liang Xu, Xiaochun Cui, Qingcheng Li, and John C. Crittenden. 2017. "Responses of the microalga chlorophyta sp. to bacterial quorum sensing molecules (N-Acylhomoserine Lactones): Aromatic protein-induced self-aggregation." *Environmental Science and Technology* 51(6):3490–98. doi: 10.1021/acs.est.7b00355.

Frontiers in Civil and Hydraulic Engineering – Mohamed A. Ismail and
Hazem Samih Mohamed (Eds)
© 2023 The Authors, ISBN 978-1-032-38247-0

Particle crushing mode during shearing of calcareous sand

Xu Shi
School of Earth Sciences and Engineering, Sun Yat-sen University, Guangdong, Zhuhai, China
Southern Marine Science and Engineering Guangdong Laboratory (Zhuhai), Guangdong, Zhuhai, China

Wenlong Li*
Guangzhou Urban Planning & Design Survey Research Institute, Guangzhou, China

Yan Gao
School of Earth Sciences and Engineering, Sun Yat-sen University, Guangdong, Zhuhai, China, China
Southern Marine Science and Engineering Guangdong Laboratory (Zhuhai), Guangdong, Zhuhai, China

ABSTRACT: As the main foundation material for filling islands and reefs, calcareous sand is characterized by the extremely irregular particle shape, abundant internal pore, being easy to be broken, etc. The particulate crushing mode of calcareous sand during the shearing has an important influence on the strength and deformation of the foundation. Based on the tailor-made transparent model box and the MTS testing machine, the particulate crushing mode of calcareous sand during the shearing was explored. Results have shown that during shearing, the particles have obvious directional arrangement; the crushing modes of particles with different shapes are quite different, in which the branched particles are mainly fractured, the bulk particles are mainly disintegrated, and the shell-shaped particles are generally not broken.

1 INTRODUCTION

With the advancement of China's national maritime strategies, construction of various large-scale island reef projects followed, native islands and reefs can no longer meet the development needs, and the reclamations of islands and reefs have become more and more important for future large-scale construction projects. As an island reef filling material, Calcareous sand has different engineering and mechanical properties from terrigenous siliceous sand. The slippage, fragmentation and rearrangement of soil particles are the main ways of compressive deformation of calcareous sand foundations, and the elastic deformation of the soil particles can be ignored. Calcareous sand is characterized by its irregular shape, high porosity and easily broken. The particle crushing is one of the important factors which causes its deformation (Datta et al. 1979; Liu et al. 1998).

The crushing mode of calcareous sand particles can be divided into three types: cracking, crushing and grinding (Guyon et al. 1994). The strength of silica sand particles is high, and the experimental results of its particle crushing show that its crushing form is mainly grinding of the corners of particles (Bastidas et al. 2016; Nakata et al. 2001; Vesic et al. 1968). The particle strength of calcareous sand is much lower than that of siliceous sand. Due to the irregular shape of particles and strong bite force between particles, the test results still show that under relatively low pressure, the crushing form of calcareous sand is mainly the grinding of sharp corners of particles (Coop et al. 2004; Peng et al. 2019; Wu et al. 1997; Zhang et al. 2008). Wang et al. (2018) conducted a large-scale triaxial test on calcareous sand and found that the number of the large-sized particles decreased obviously, that is, the phenomenon of the disintegration of large-sized particles and the wear of sharp corners of particles were obvious during the particle crushing process of coral reef gravel, indicating that the particle size has a prominent influence on the particle crushing form.

---

*Corresponding Author: liwl26@mail2.sysu.edu.cn

In order to explore the particulate crushing mode of calcareous sand during the shearing process, the biaxial shear tests on calcareous sand and siliceous sand were carried out based on the tailor-made transparent model box in this paper. Combined with image post-processing technology and particle image velocimetry, the effects of particle shape, internal porosity, particle size and other factors on particle crushing ability and mode in the shearing process were revealed.

2 TEST METHODS

2.1 Test equipment

In order to facilitate the application of image post-processing analysis technology and particle image velocimetry technology after the test, this paper used the tailor-made unconfined transparent pressurized model box. The model box is made of acrylic plates bonded together. The thickness of the acrylic plate of the model box is 25 mm, the upper part is a piston-type loading plate, and the two sides are rubber films with a thickness of 1mm. The sample size is 100×100×200 mm. Axial loading was performed with an MTS testing machine, and the axial loading rate was 4 mm/min, that was, the axial strain was 2%/min until the axial strain reached 20%. During the test, the sampling frequency of the high-speed camera was 1 Hz. The volumetric strain ε_v of the sample in the test was measured by the drainage method, that was, the entire shear box was placed in a drainage box with a scale line, and the liquid level rise scale can indicate the volume change of the sample (Figure 1).

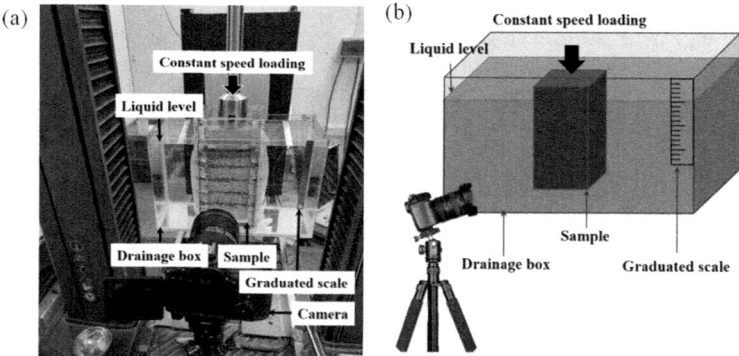

Figure 1.   Experimental device: (a) model diagram; (b) physical diagram.

2.2 Test arrangement

To avoid the influence of particle surface salinity, the calcareous sand sample was filtered and washed before the test. When the mass ratio of water to sand was 1:1 and the particle salinity content was less than 0.1%, it was regarded as filtered and washed, and the test was carried out after drying. In this paper, three groups of unconfined shear tests (Table 1) were carried out

Table 1.   Experimental groups.

Group	Category	Particle size	Relative density
A	Calcareous sand	2~5mm	70%
B	Siliceous sand	2~5mm	
C	Calcareous sand	>5mm	

according to the type and particle size of sand. The initial relative compactness of the samples was 70%, which was in dense sand condition.

3 STRESS AND DEFORMATION CHARACTERISTICS

Figure 2 shows the stress-strain and volume-strain curves of three groups of samples with different particle sizes. As can be seen from the figure, all three groups of samples show the behavior of "dense sand" subjected to shearing. The stress drop of the sample in group C (>5 mm calcareous sand) was not obvious. The reason is that the particle crushing of group A and B samples could be neglected, and their particle crushing potential Br values were 0.047 and 0.002, respectively, while that of group C samples was 0.229. The obvious particle crushing in group C causes its stress drop to be not obvious, and its dilatancy decreases for which the volume strain was obviously lower than that of the samples in groups A and B. Compared with group B, the particle shape of group A was very irregular and its internal friction angle was larger, so its peak strength was obviously higher than that of siliceous sand. In addition, the irregularity of particle shape also led to the increase of voids between particles during particle movement, and the final volume strain was larger than that of the siliceous sand sample.

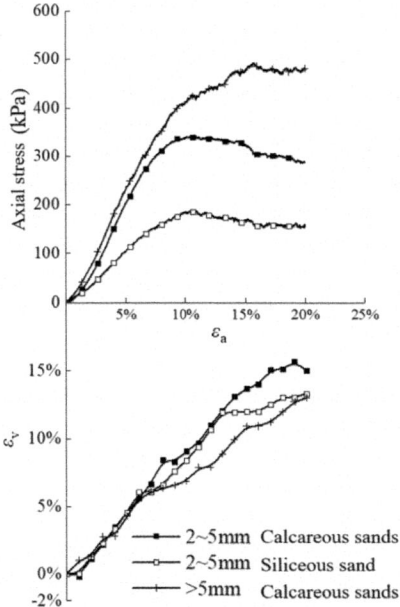

Figure 2.   Stress-strain curve and volumetric strain curve.

4 MICROSCOPIC MOTION CHARACTERISTICS

Three axial strain characteristic points (before peak stress, at peak stress and after peak stress) of each group of samples were selected and analyzed by image post-processing and particle image velocimetry (PIV) technology to explore the particle movement in the samples. Figures 3–5 are the particle motion contours based on PIV,which shows that before the peak stress, the sample was mainly compressed as a whole, and the deformation of the sample was mainly due to the compaction of pores between particles. After the peak stress, it can be seen from the contour

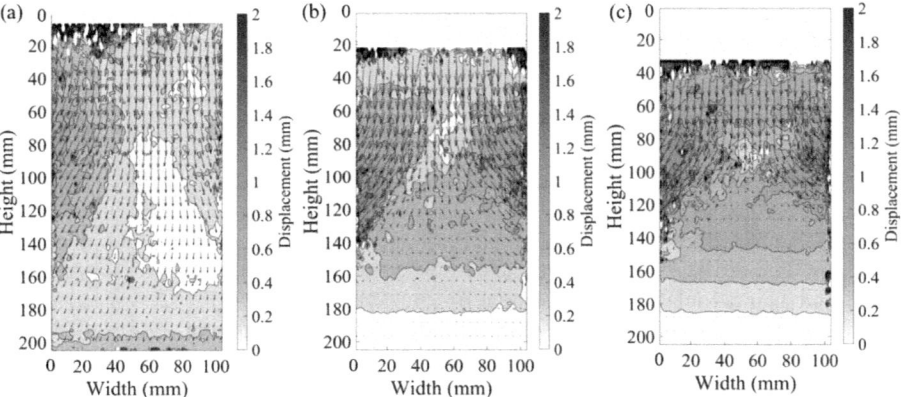

Figure 3.　Contour map of particle motion in group A (2~5 mm calcareous sand): (a) Before the peak intensity; (b) Peak intensity point; (c) after the peak intensity.

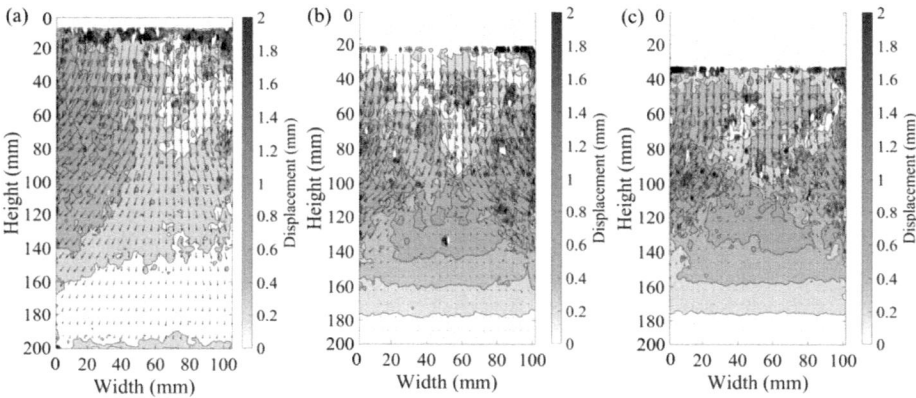

Figure 4.　Contour map of particle motion in group B (2~5 mm Siliceous sand): (a) Before the peak intensity; (b) Peak intensity point; (c) after the peak intensity.

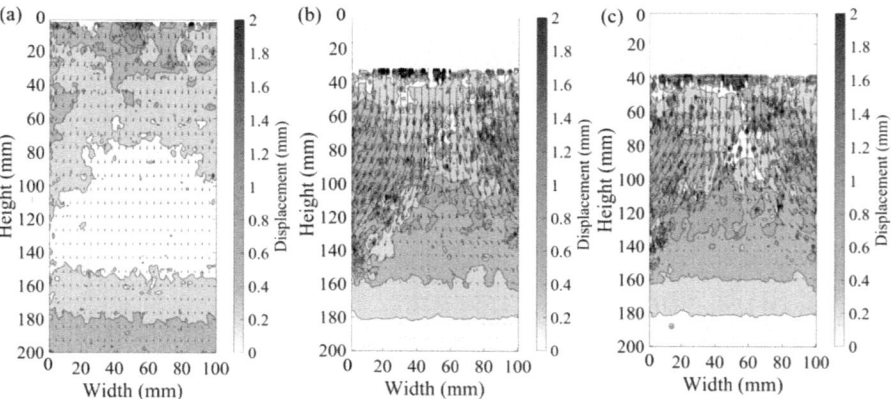

Figure 5.　Contour map of particle motion in group C (>5 mm Siliceous sand): (a) Before the peak intensity; (b) Peak intensity point; (c) after the peak intensity.

distribution of particle motion and the direction vector of particle motion in the figure that the relative motion between particles mainly occurs in the "X"-shaped area where the height of the sample is 40~140 mm. The "X"-shaped shear zone is gradually generated and developed in the sample. The deformation of the sample mainly comes from the relative motion between particles, which leads to the enlargement of pores in the shear zone and the sample exhibited a dilatancy trend. During shearing, the movements of particles in calcareous and siliceous sand samples are similar, and the relative movement between particles mainly occurs in the shear zone and is mainly distributed at the rubber membranes on both sides, and the particles at the bottom of the sample hardly move. In addition, the calcareous sand samples with a particle size larger than 5 mm have obvious particle crushing during shearing, also mainly concentrated in the shear zone.

5 PARTICULATE CRUSHING MODE

Understanding the particle crushing mode of calcareous sand is of great significance to improve the strength and deformation characteristics of calcareous sand. Therefore, in this paper, the initial state and three typical axial strain points (before peak stress, at peak stress) are studied. And after peak stress) of the calcareous sand sample with particle size larger than 5 mm with obvious particle crushing were selected for particle morphology extraction and analysis as shown in Figure 6 and Table 2. The results illustrate that the particle crushing mainly occurs after the peak stress and is concentrated in the shear zone. The particle crushing mode has a good correlation with particle shape. Calcareous sand particles can be divided into massive particles, branched particles and bioclastic particles according to their morphology (Table 2). The shape of the massive particles is relatively regular, and the force is relatively uniform and stable, but due to the relatively developed internal pores, when the force of the particles is greater than their crushing strength, the particles will be broken into multiple calcareous sand fragments (for example, particle 1 in Figure 6) or broken into 2–3 debris along the channels developed along the pores (for example, particle 2 in Figure 6). The dendritic particles have the most irregular particle shape, and the positions of the acting points of the particles are quite different, which leads to the fracture of the particles (for example, particles 3 and 4 in Figure 6). Bioclastic particles are mainly formed after shells and conch

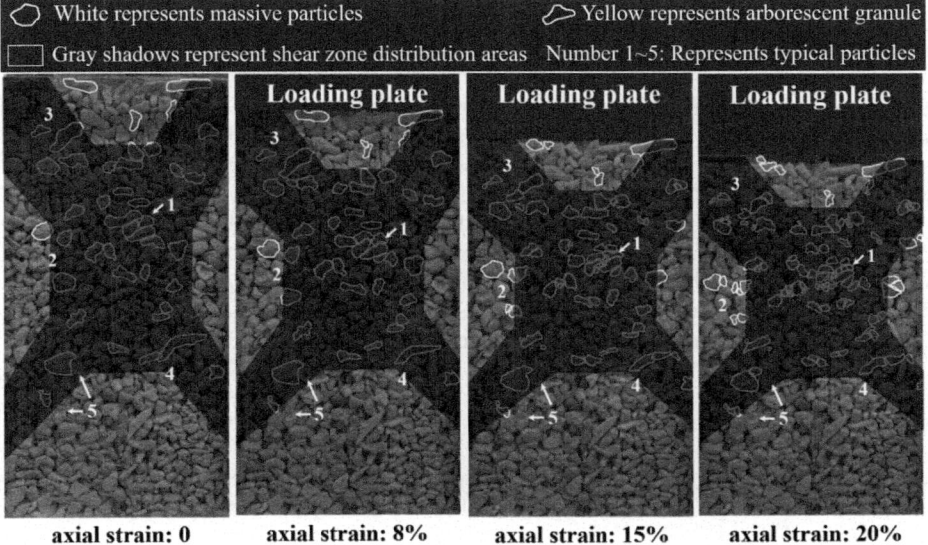

Figure 6. Variation of particle breakage with axial strain.

Table 2. Particle breakage mode.

| Particle shape | | Particle breakage mode | Breakage diagram | Observations | | The particle in Figure 6 |
				Before breakage	After breakage	
Massive particles	Internal pore development	Monolithic breakage or breaks along pore development channels				1, 2
Arborescent particles	Typical arborescent shape	Breaks along the branches and leaves				3
	Long strip shape	Fracture				4
Bioclastic particles		Barely break	–	–	–	5

shells were broken through edges and corners, and their shape is approximately disc or conch shape, with good overall strength and generally does not break (for example, particle 5 in Figure 6). Hence, one of the effective methods to improve the strength and deformation characteristics of calcareous sand foundation is to control the degree of particle crushing. That is, the calcareous sand material can be pre-pressed to reduce particle breakage during the use of the project so as to improve the long-term stability of the calcareous sand foundation.

6 CONCLUSION

Based on the biaxial shear test and image post-processing technology, compared with siliceous sand, it is found that the particle crushing of calcareous sand is much more obvious, and the particle movement mainly occurs in the shear zone at the upper part of the sample. In the process of shearing, particles with different shapes have different particle crushing modes. Branched particles are mainly broken, and shell particles are generally not broken, while blocky particles with well-developed internal pores are mostly broken as a whole or along the pore development channel. This is important for understanding the roles of particle shape and size of calcareous sand on particle crushing characteristics and also provides a certain reference value for the explanation of the strength and deformation resistance of calcareous sand from microscopic aspects.

REFERENCES

Bastidas P.A.M. *Ottawa F-65 sand characterization*[D]. California: University of California at Davis, 2016.
Coop M.R., Sorensen K.K., Bodas T., et al. Particle breakage during shearing of a carbonate sand[J]. *Géotechnique*, 2004, 54 (3):157–163.
Datta M., Gglhati S.K., Rao G.V. *Crushing of calcareous sands during shear*[C].//Offshore Technology Conference, 1979, OTC3525:1459–1467.

Guyon É., Troadec J.P. Du sac de billes au tas de sable[M]. *Odile Jacob Sciences*, France, 1994. (in France)

Liu Chong-qua, Wang Ren. Preliminary research on physical and mechanical properties of calcareous sand [J]. *Rock and Soil Mechanics*,1998(1):32–37+44. (in Chinese)

Nakata Y., Hyodo M., Hyde A.F.L., et al. Microscopic particle crushing of sand subjected to high pressure one-dimensional compression[J]. *Soils and Foundations*, 2001, 41(1):69-82.

Peng Yu, Ding Xuan-ming, Xiao Yang, Chu Jian, Deng Wei-ting, Study of particle breakage behaviour of calcareous sand by dyeing tracking and particle image segmentation method[J]. *Rock and Soil Mechanics*, 2019, 40(7): 2663–2672. (in Chinese)

Vesic A.S., Clough G.W. Behavior of granular materials under high stresses[J]. *Journal of Soil Mechanics and Foundation Division*, ASCE, 1968, 94(SM3):661–688.

Wang Gang, Ye Qin-guo, Zha Jing-jing. Experimental study on mechanical behavior and particle crushing of coral sand-gravel fill. *Chinese J. Geot. Eng.*, 2018, 40(5): 802–810. (in Chinese)

Wu Jingping, Chu Yao, Lou Zhigang. Influence of particle breakage on deformation and strength properties of calcareous sands. *Chinese J. Geot. Eng.*, 1997, 19(5): 51–57. (in Chinese)

Zhang Jia-ming, Zhang Lin, Jiang Guo-sheng, Wang Ren. Research on particle crushing of calcareous sands under triaxial shear[J]. *Rock and Soil Mechanics*, 2008, 29(10): 2789–2793. (in Chinese)

Frontiers in Civil and Hydraulic Engineering – Mohamed A. Ismail and
Hazem Samih Mohamed (Eds)
© 2023 The Authors, ISBN 978-1-032-38247-0

Research on key construction parameters of double-side heading method for newly-added four-lane tunnel with super large section

Jiangling He & Min Liu
Guangdong Expressway Co., Ltd., China

Xinghong Jiang
China Merchants Chongqing Traffic Research and Design Institute Co. LTD, China

Langping Xiong*
School of Civil Engineering, Chongqing Jiaotong University, China

ABSTRACT: Super-large cross-section four-lane tunnels often have the characteristics of large cross-section span, deep and shallow burial, and poor stability of surrounding rock. Taking the example of the Shenshan West Tunnel as the engineering background, this paper simulates and analyzes the key parameters of the construction of the double-side heading method in the excavation of a super large section tunnel, and studies the effects of different process parameter schemes on the deformation and stability of the tunnel surrounding rock during the tunnel excavation process. The results of the study show that the double-side heading method is used to construct a four-lane tunnel with a super large cross-section of V-level surrounding rock. The deformation and mechanical characteristics of the surrounding rock of the tunnel are comprehensively analyzed. The radius of curvature is 45° and the height of the steps on the pilot tunnel is 1/2 of the tunnel height.

1 INTRODUCTION

In recent years, with the continuous improvement of people's living standards, more and more road tunnels have caused traffic jams and other problems due to limited lanes and heavy traffic. In order to meet the increasingly saturated demand of traffic volume, and to overcome the influence of topography and geological conditions, the reconstruction and expansion of expressways will become the trend of tunnel construction in the future (Yang 2010).

The reconstruction and expansion of the tunnel have the characteristics of large cross-section span, flat shape and complex engineering geological conditions (Zhang 2014). The integrity of the surrounding rock at the entrance and exit of the tunnel is mostly low, and the surrounding rock is mostly low. The tunnel excavation process is often accompanied by stress concentration. It causes large deformation of the surrounding rock (Li 2011), which makes the surrounding rock unstable. In order to reduce the large deformation of surrounding rock caused by tunnel excavation, based on the conditions of grade V surrounding rock and the characteristics of large section tunnels, tunnel excavation is usually carried out by the double-side heading method (Li 2011). Li Hao and others carried out numerical simulations for the construction of two-lane tunnels with small cross-sections and analyzed that a reasonable shape of the pilot pit and the location of the pilot pit can reduce the displacement of the vault (Zhang 2009). Fang Qian (Fang 2010) used numerical calculations to

---

*Corresponding Author: 1757147657@qq.com

DOI 10.1201/9781003344209-15

compare and study the construction process of the double-side-wall pilot method, and proposed an improved double-wall pilot-pit method. However, in the study of the numerical simulation scheme of the tunnel, most of the analysis focuses on the influence of different construction methods on the tunnel construction; and the study of the comprehensive consideration of the vertical and horizontal parameters for the double-side heading method is rare.

Based on the engineering geological conditions of the Shenshan West Tunnel, this paper compares the deformation and displacement of the tunnel under different schemes by numerically calculating the position and shape of the next wall in the double-side piloting method of the large-section flat tunnel and the height of the middle pilot tunnel. The research results provide a basis for the subsequent design and optimization of construction parameters of large-section tunnels and also provide references for the design and construction of similar projects in the future.

2 PROJECT OVERVIEW

The Shenshan West Tunnel is a two-hole and eight-lane highway tunnel. The overall form is a hidden mountain tunnel + an open-cut sunken tunnel. It is located at mileage YK56+334~YK58+137, with a total length of 1803 m, and a designed tunnel of 120 km/h. The clear width of the single-hole four-lane tunnel is 19.25 m, and the clear height of the tunnel is 5.0 m. The overall single-tunnel excavation span is 21.66 m and the height is 13.6 m. The outline of the tunnel is shown in Figure 1.

The tunnel passes through the hilly mountainous area and the plain landform +- area. The topography undulates. The tunnel body has multiple sets of joints, fault gouge and a large number of fragmented quartz veins. The trend is basically distributed along the valley, landslides and collapse are developed, and the strata are fragmented. The surrounding rocks on both sides of the fault are the second subgroup of the Lower Jurassic, argillaceous siltstone, sandstone and shale. Affected by this, the surrounding rock joints and fissures in the vicinity of the fracture are extremely developed, the rock mass is broken, and the surrounding rock integrity is poor. The maximum buried depth is 69 m. The tunnel body is dominated by Grade V surrounding rock, and the double-side pilot excavation method is adopted.

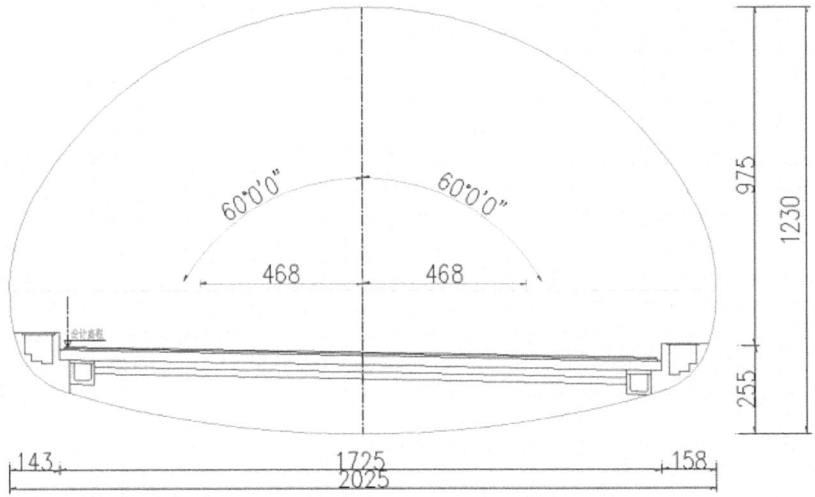

Figure 1. The outline of the four-lane tunnel.

3 DYNAMIC NUMERICAL SIMULATION OF LARGE CROSS SECTION TUNNEL CONSTRUCTION

3.1 Calculation model and parameters

3.1.1 Calculation model

Based on the initial survey report of the Shenshan West Tunnel, considering that the tunnel has the characteristics of large cross-section span, excavation seepage, deep and shallow buried stratum, poor surrounding rock stability, and more landslides at the entrance of the tunnel, this article has conducted a large number of literature investigations. And referring to various industry standards (94–94 Technical code for building pile foundation 1995), it can be seen that the double-side wall construction method is applicable to various types of shallow buried/biased crushed rock masses, soft soil surrounding rocks, water-rich, karst, and natural formations. The calculation section is selected to be buried 60 m deep (level V surrounding rock). According to the influence range of tunnel excavation, [] the three-dimensional calculation model is the upper boundary to the ground surface, the lower boundary should not be less than 5 times the height of the tunnel, and the left and right sides are taken respectively 50 min order to reduce the boundary effect. In the numerical analysis, a length of 90 m is taken along the excavation direction, and the FLAC3D numerical software is used to establish the double-side-wall heading method model. The calculation model is shown in Figure 2 below.

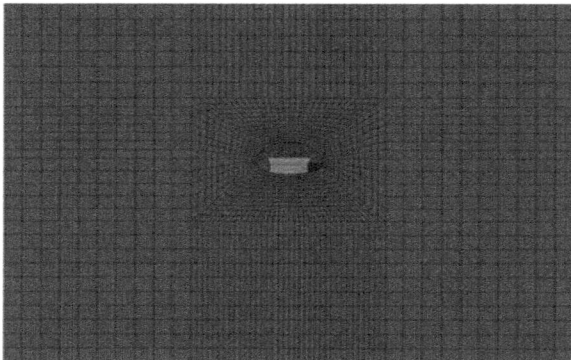

Figure 2.   Three-dimensional numerical calculation model.

3.1.2 Parameter selection

The vertical displacement constraint is applied to the bottom surface of the model, and the lateral displacement constraint is applied to the left and right sides. According to the plane strain model, the calculation and analysis of the partial excavation construction process are carried out. The rock mass around the tunnel body is finitely dispersed with hexahedral elements. The M-C theoretical constitutive model is used for calculation. Only the self-weight stress is considered, and the influence of groundwater on the tunnel excavation is not considered. In order to consider the supporting effect of the construction method itself, the effect of the system bolt and the grouting of the leading small pipe are not considered. The initial support is simulated by solid elements and obeys the linear elastic stress-strain relationship. The physical and mechanical parameters are shown in the table below.

Table 1.   Parameter table of computational physical mechanics.

level	Severe $\gamma(kN/m^3)$	Deformation modulus E (GPa)	Poissn's ratio μ	Internal friction angle $\psi(°)$	Cohesion c(MPa)
V	19.75	1.3	0.35	23.5	0.1250

The inner wall of the tunnel adopts a section steel frame as temporary support, and its function adopts the equivalent stiffness method to convert it into a concrete area for conversion. The equivalent method of section steel frame can be expressed as:

$$E_1A_1 + E_2A_2 = E_1A_3$$

Where: E1 is the modulus of elasticity of concrete; E2 is the modulus of elasticity of steel bar or I-steel; A1 is the cross-sectional area of the sprayed layer of the original design initial support or temporary support, and A2 is the original design initial support or temporary support,the cross-sectional area of the inner steel bar or I-beam, A3 is the cross-sectional area of the initial support or temporary support spray layer after equivalent.

3.2 *Simulate the excavation process*

According to the engineering geology of the tunnel body, the double-side-wall pilot method is adopted for excavation, and the finite element software is used to simulate the tunnel excavation steps. The excavation process is shown in Figure 4 below. Aiming at the excavation characteristics of the double-side pilot tunnel method, the deformation mechanical state of the surrounding rock and lining structure during the excavation process can be reflected in a timely and effective manner by arranging monitoring points during the numerical analysis process(94–94 Technical code for building pile foundation 199). Seven displacement monitoring points are set on the side walls of the upper and lower steps of the vault and side wall pilot pits, including 3 measurement points for the settlement of the left, middle and right pilot pits, and the upper and lower side walls of the side wall pilot pits, four horizontal displacement measuring points. The layout of monitoring points is shown in Figure 3.

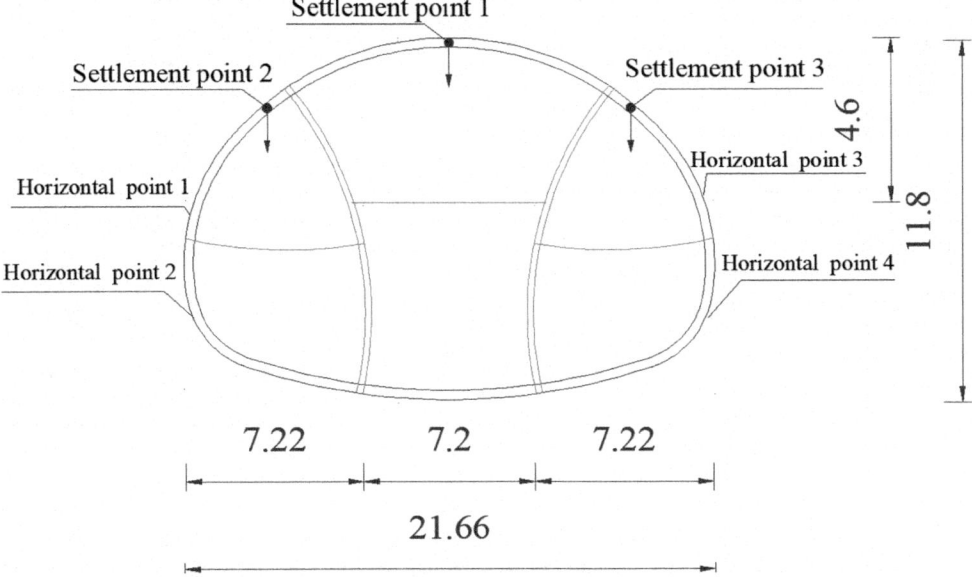

Figure 3.   Layout of measuring points.

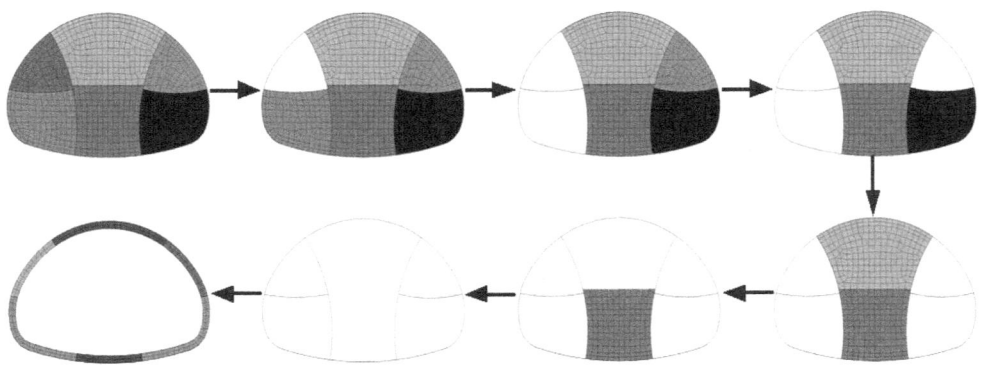

Figure 4.   Schematic diagram of excavation.

3.3   *Key construction parameter calculation scheme*

According to relevant industry technical specifications [], the shape of the sidewall pilot pit should be close to an elliptical section, and the section size of the pilot pit should not exceed 1/3 of the maximum span of the section. Relying on the originally designed excavation plan for the hidden excavation section of the Shenshan West Tunnel, the process parameters in this paper are mainly based on the number of excavation sections, the size of the pilot pit, the curvature and position of the vertical temporary support, etc., to study the influence of different calculation schemes on the tunnel excavation. The calculation scheme of each process parameter is shown in the table below.

Table 2.   The calculation scheme of the process parameters of the double-side-wall pilot pit method.

Process parameters	Calculation conditions	Calculation model
The middle wall position L	① L=1/3 B ② L=1/4 B ③ L=1/5 B	B is the maximum excavation span of the tunnel L is the position from the side wall to the middle wall
The radius of curvature of the partition wall R	①R=15.7m∠45° ②R=10.35m∠60° ③R=7.37m∠75°	The central angle corresponding to the arc length of the middle partition
The height of the upper steps of the pilot hole h1	①H=1/2H (two steps) ②h_1=1/3H (two steps) ③h_1=3/5H (two steps)	H is the height at the central axis of the tunnel h_1 is the height of the steps on the pilot hole h_2 is the height of the steps on the pilot hole

The control variable method [] is used to study the influence of three process parameters on the construction of large-section tunnels, and the conclusions obtained are verified by the mechanical parameters of surrounding rock. Therefore, it is necessary to analyze separately on the basis of fixing the irrelevant quantities of L, R and h1. The influence of L, R and h1 on the tunnel. The optimization flowchart is shown in Figure 5.

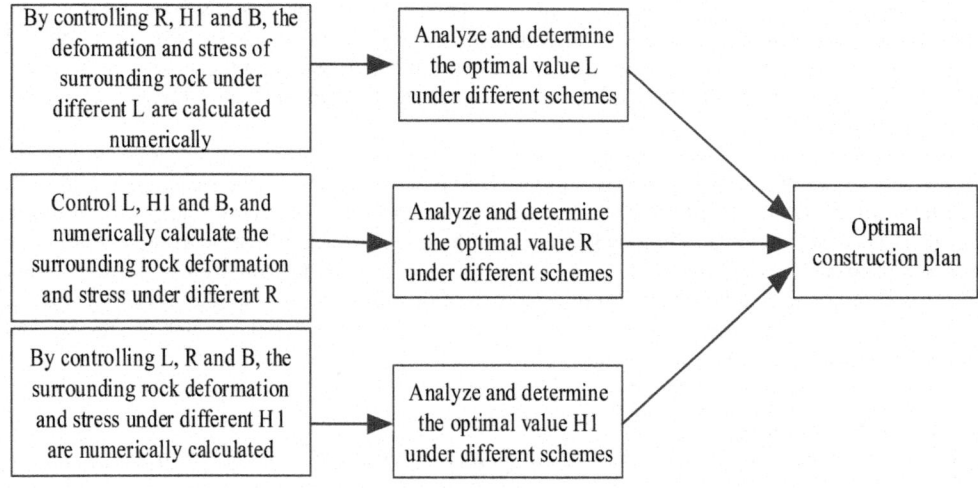

Figure 5. Optimization flow chart.

4 ANALYSIS OF CALCULATION RESULTS

Tunnels with super-large cross-sections are often located at the entrance section and have the characteristics of poor geological conditions, many procedures, and strong interference []. In order to ensure the stability of the surrounding rock under construction, this paper combines the above-mentioned key construction parameter calculation schemes in the numerical calculation process. The plan's vault settlement, horizontal convergence and internal force of the supporting structure are compared and analyzed.

4.1 *Next door position*

4.1.1 *Surrounding rock deformation*
Figure 6 shows the change curve of settlement and horizontal displacement with the position of the middle wall under the condition of V-grade surrounding rock. It can be seen from the figure that the dome settlement of the left, middle, and right pilot pits changes with the position of the

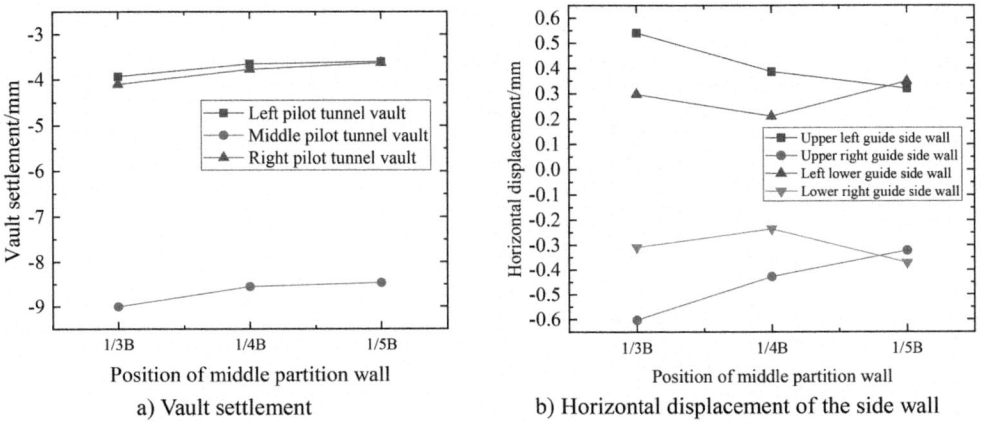

| a) Vault settlement | b) Horizontal displacement of the side wall |

Figure 6. Displacement curve of grade V surrounding rock versus position of the middle wall.

middle wall, and the changing trend is not obvious. The maximum settlement value is 9mm for the dome of the middle pilot tunnel; at the same time, the excavation process starts with the left tunnel excavation. After the excavation of the rear right tunnel, it can be seen that the sinking trend of the left and right pilot tunnels is roughly the same, and the left pilot tunnel excavation has less impact on the right pilot tunnel.

From the perspective of surrounding rock deformation, as the width L of the sidewall pilot pit decreases, the settlement of the vault and the horizontal displacement of the side wall are down to a certain extent. At the same time, the upper pilot pit excavation contributes a large proportion to the deformation of the surrounding rock. As the width of the sidewall pilot pit becomes smaller, the radius of curvature has less influence on the deformation of the surrounding rock of the tunnel, and the deformation magnitude of the surrounding rock of the tunnel is very small, indicating that the three kinds of sidewall pilot pit spans can meet the requirements of the deep-shanxi large-section tunnel Construction safety requirements.

4.1.2 *Surrounding rock stress*

Table 3 shows the initial support stress cloud diagram under L=1/3B working conditions of grade V surrounding rock. The table also shows that the maximum compressive stress value under L=1/3B working condition is 11.36 MPa, and the maximum tensile stress value is 0.044 MPa; the maximum

Table 3. Cloud diagram of initial support stress of grade V surrounding rock.

Working condition	Minimum principal stress distribution cloud chart
L=1/3 B	

Working condition	Maximum principal stress distribution cloud map
L=1/3 B	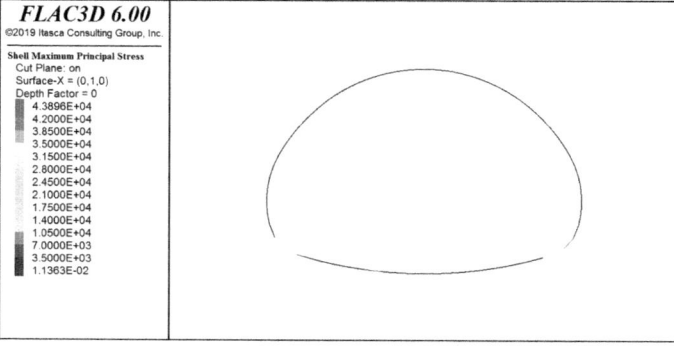

compressive stress value under L=1/4B working condition is 11.36 MPa, and the maximum tensile stress is 11.36 MPa. The stress value is 0.043 MPa; the maximum compressive stress value under L=1/5B working condition is 12.13 MPa, and the maximum tensile stress value is 0.041 MPa.

Therefore, the maximum compressive stress under Grade V surrounding rock is the largest at the side wall, and the maximum stress can reach 12.51 MPa. There is a maximum tensile stress at the temporary supports at the bottom of the lower steps of the left and right guide pits and the arch toe, and the maximum stress can reach 0.09 MPa. With the decrease of the position of the middle partition, the maximum compressive stress under the three working conditions increases and the maximum tensile stress decreases correspondingly but the magnitude is small, indicating that the spans of the three sidewall pilot pits can meet the requirements of the large section of the western deep section. Tunnel construction can meet safety requirements, but considering the requirements of construction safety, economic cost, construction period and other aspects, the L=1/3B working condition is more reasonable in the tunnel excavation process.

4.2 Radius of curvature

4.2.1 Surrounding rock deformation
In order to find the optimal radius of curvature under the three schemes, Figure 7 shows the change curve of the settlement and horizontal displacement with the radius of curvature under V-level surrounding rock conditions.It also shows that the settlement of the dome changes with the radius of curvature, and the maximum settlement value is 9 mm for the dome of the middle pilot tunnel; at the same time, the excavation process first excavates the left tunnel and then the right tunnel. The picture shows the left and right pilot tunnels under the vault The sinking trend remains roughly the same, and the excavation of the left pilot tunnel has less impact on the right pilot tunnel.

From the perspective of surrounding rock deformation, with the increase of the radius of curvature, the settlement of the vault and the horizontal displacement of the side wall are reduced to a certain extent, and the excavation of the upper pilot pit contributes a large proportion to the deformation of the surrounding rock. When R=10.35 m∠600, the calculation condition is optimal, that is, when R=10.35 m∠600, the restraint deformation effect of support is relatively better. The maximum horizontal displacement value is 0.45 mm, but the magnitude is small and the difference is not obvious. The three kinds of sidewall heading spans can meet the construction safety requirements of the deep-shanxi large-section tunnel.

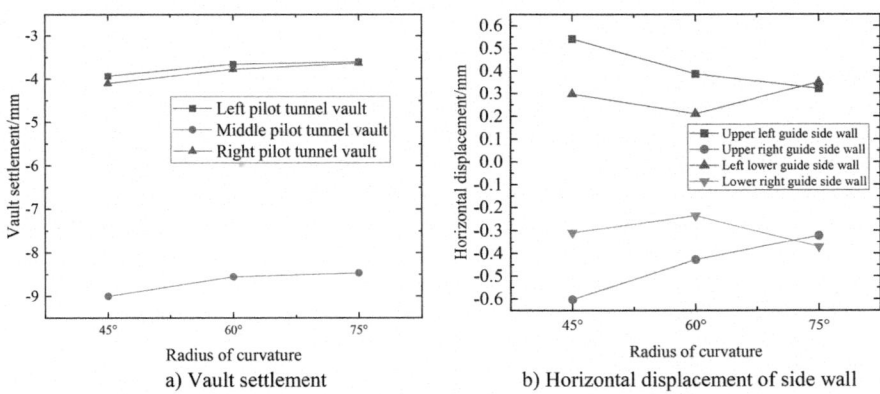

a) Vault settlement   b) Horizontal displacement of side wall

Figure 7.   Curve diagram of displacement of grade V surrounding rock and position of diaphragm wall.

4.2.2 Surrounding rock stress
Table 4 shows the initial support stress cloud diagram under the R=15.7 m ∠45° working conditions of grade V surrounding rock. It also shows that the maximum compressive stress under Grade V surrounding rock is the largest at the side wall and the arch waist, and the maximum stress can reach

Table 4. Cloud diagram of initial support stress of grade V surrounding rock.

Working condition	Minimum principal stress distribution cloud chart
R=15.7m∠45°	

Working condition	Maximum principal stress distribution cloud map
R=15.7m∠45°	

12.13 MPa. The maximum tensile stress exists at the temporary support at the bottom of the lower step of the left and right guide pits and the arch foot, and the maximum stress can reach 0.042 MPa. With the decrease of the position of the middle partition, the maximum compressive stress under the three working conditions increases and the maximum tensile stress decreases correspondingly but the magnitude is small, indicating that the spans of the three sidewall pilot pits can meet the requirements of the large cross-section of Shenshan West Tunnel construction safety requirements, but considering the requirements of construction safety, economic cost, construction period and other aspects, the R=15.7 m ∠45° working condition is more reasonable in the tunnel excavation process.

4.3 Block size of the pilot hole

4.3.1 Surrounding rock deformation

In order to find the block size of the middle pilot tunnel, the final calculation results of the three calculation conditions are compared below. Figure 8 shows the deformation trend of the surrounding rock in the left. It can been seen that middle and right pilot tunnels remains the same, and the impact of the left pilot tunnel excavation on the right pilot tunnel is smaller, the excavation of the upper pilot pit is a key construction step, which contributes a large proportion to the deformation of the surrounding rock. The three methods of dividing the middle pilot tunnel have little effect on the overall mechanical properties of the tunnel, that is, the middle pilot tunnel is excavated in two blocks, and the block position is placed at 1/2 height.

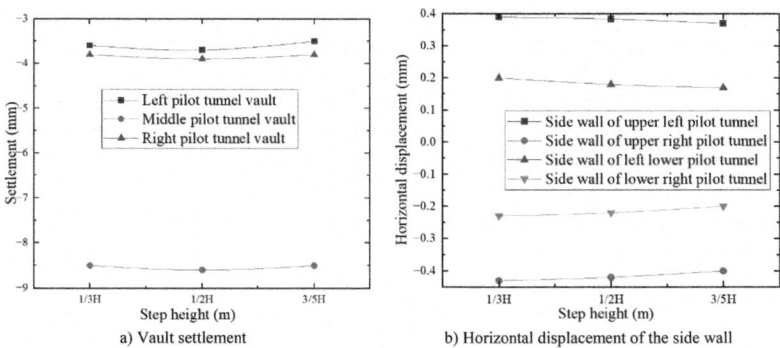

| a) Vault settlement | b) Horizontal displacement of the side wall |

Figure 8. Curve diagram of displacement of grade V surrounding rock and position of diaphragm wall.

4.3.2 *Surrounding rock stress*

Table 5 shows the initial support stress cloud diagram under three conditions of grade V surrounding rock. It can be seen from the table that the maximum compressive stress under Grade V surrounding

Table 5. Cloud diagram of initial support stress of grade V surrounding rock.

Working condition	Minimum principal stress distribution cloud chart
$h_1 = 1/2\ H$	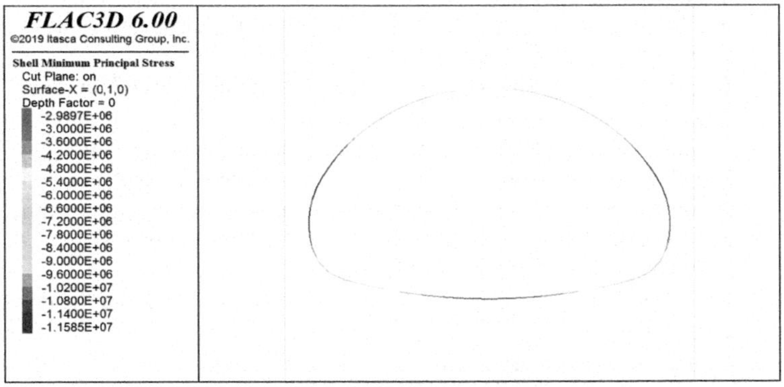

FLAC3D 6.00
©2019 Itasca Consulting Group, Inc.

Shell Minimum Principal Stress
Cut Plane: on
Surface-X = (0,1,0)
Depth Factor = 0

-2.9897E+06
-3.0000E+06
-3.6000E+06
-4.2000E+06
-4.8000E+06
-5.4000E+06
-6.0000E+06
-6.6000E+06
-7.2000E+06
-7.8000E+06
-8.4000E+06
-9.0000E+06
-9.6000E+06
-1.0200E+07
-1.0800E+07
-1.1400E+07
-1.1585E+07

Working condition	Maximum principal stress distribution cloud map
$h_1 = 1/2\ H$	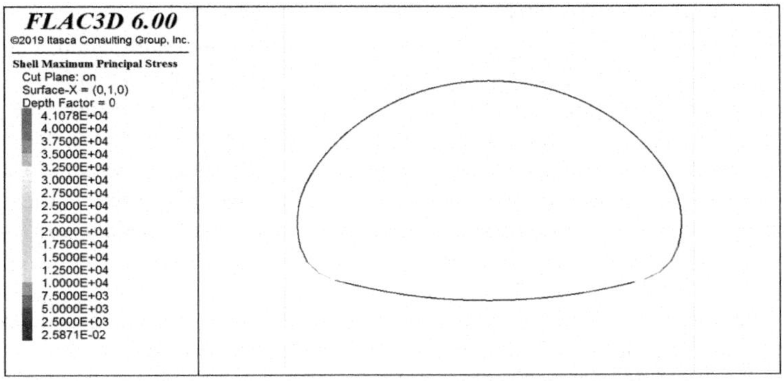

FLAC3D 6.00
©2019 Itasca Consulting Group, Inc.

Shell Maximum Principal Stress
Cut Plane: on
Surface-X = (0,1,0)
Depth Factor = 0

4.1078E+04
4.0000E+04
3.7500E+04
3.5000E+04
3.2500E+04
3.0000E+04
2.7500E+04
2.5000E+04
2.2500E+04
2.0000E+04
1.7500E+04
1.5000E+04
1.2500E+04
1.0000E+04
7.5000E+03
5.0000E+03
2.5000E+03
2.5871E-02

rock is the largest at the side wall and the arch waist, and the maximum stress can reach 11.6 MPa. There is maximum tensile stress at the temporary supports at the bottom of the lower steps of the left and right guide pits and the arch toe, and the maximum stress can reach 0.04 MPa. With the decrease of the pilot tunnel, the maximum compressive stress under the three working conditions increases and the maximum tensile stress decreases correspondingly, but the magnitude is small, indicating that the spans of the three sidewall pilot pits can meet the requirements of the deep-shanxi large-section tunnel Construction safety requirements, but considering the randomness of the actual project selected, the working condition of h1=1/2H is more reasonable in the tunnel excavation process.

5 IN CONCLUSION

By comparing the different schemes under the double-side pilot pit method, the following conclusions can be drawn:

(1) The position of the middle wall has a greater influence on the deformation and mechanical state of the surrounding rock of the tunnel. Under the condition of grade V surrounding rock, during the construction of the large-section tunnel using the double-side heading method, when the width of the side heading pit increases, that is, when the cross-sectional area of the heading pit increases, the surrounding rock deforms during the tunnel excavation. With the decrease of the position of the middle partition, the maximum compressive stress under the three working conditions increases and the maximum tensile stress decreases correspondingly but the magnitude is small. In order to reduce the large deformation of the surrounding rock, the width of the side pilot pit can be L=1/3B.

(2) The shape of the pilot pit can be determined by the radius of curvature, which also has a greater impact on the deformation and mechanical state of the surrounding rock of the tunnel. Under the condition of grade V surrounding rock, in the construction process of the large-section tunnel using the double-side heading method, the larger the radius of curvature, the smaller the influence on the deformation of the surrounding rock of the tunnel, but the stress changing trend is not obvious. In order to reduce the large deformation of the surrounding rock, the radius of curvature can be R=15.7 m∠450.

(3) The height of the steps of the pilot tunnel has little effect on the deformation and mechanical state of the surrounding rock of the tunnel. Under the condition of grade V surrounding rock, during the construction of the large-section tunnel using the double-side pilot method, with the increase of the upper step of the pilot tunnel, the deformation and mechanical state of the surrounding rock of the tunnel have the same changing trend, which has a greater impact on the tunnel. Taking into account the requirements of construction safety, economic cost, construction period and other aspects in the process of tunnel excavation, the height of the stairs can be h1=1/2H.

(4) During the excavation of large-section tunnels, the stress field and displacement field will change with different working procedures. The sinking trend of the left and right pilot tunnel arches is roughly the same, and the left pilot tunnel excavation has less impact on the right pilot tunnel. The excavation of the upper pilot tunnel on the left and right sides makes a great contribution to the horizontal displacement of the tunnel and has a decisive influence on the deformation of the surrounding rock of the tunnel. Therefore, in the process of excavating the pilot pit, the parameter scheme should be selected reasonably. During the excavation process of the upstairs, the initial support should be done in time.

Taking into account the requirements of construction safety, economic cost, construction period, etc., the location of the middle wall is L=1/3B, the radius of curvature is 450, and the upper step block size of the pilot tunnel h1=1/2H is more important in the tunnel excavation process.

REFERENCES

94–94 *Technical code for building pile foundation*[S]. Beijing: China Architecture and Building Press, 1995. (in Chinese).

Fang Qian. *Study on the relationship between high-speed railway tunnel support and surrounding rock* [D] Beijing: Beijing Jiaotong University, 2010.

Li Hao, Han Lijun, Meng Qingbin, et al. Optimization analysis of double-side heading method for V-level two-lane surrounding rock tunnel [J]. *Modern Tunnelling Technology*, 2015, v.52; No.365(06):118–125.

Li Tingchun. Construction control technology of Maoyushan tunnel with high ground stress and large deformation in soft rock[J]. *Modern Tunnelling Technology*, 2011, 48(2): 59–67.

Yang Yongbo, Liu Minggui, Zhang Guohua, et al. Optimization analysis of construction parameters of a new large-section tunnel adjacent to an existing tunnel[J]. *Rock and Soil Mechanics*, 2010, 31(4): 1217–1226.

Zhang Guisheng, Feng Wenwen, Liu Xinrong, et al. Discussion on the reasonable clear distance of tunnels with small clear distance under V-level surrounding rock[J]. *Chinese Journal of Underground Space and Engineering*, 2009, 5(3): 582–586.

Zhang Yalong, Wang Gaoyan, Li Zhiqing, et al. Analysis of parameters influencing the construction of deep-buried large-section tunnels[J]. *Construction Technology*, 2014(S2): 116–119.

Frontiers in Civil and Hydraulic Engineering – Mohamed A. Ismail and
Hazem Samih Mohamed (Eds)
© 2023 The Authors, ISBN 978-1-032-38247-0

Application of magnetotactic bacteria to treat heavy metal wastewater

Yingjie Dou, Kang Xie*, Zhiyong Zhou & Xianggan Duanmu
School of Civil Engineering and Architecture, University of Jinan, China

ABSTRACT: Magnetotactic bacteria are anaerobic or facultative anaerobic bacteria with unique magnetic field motion characteristics and biological characteristics, which can synthesize magnetic nanoparticles with large particle sizes and crystal shapes, namely magnetosomes. In vivo magnetotactic bacteria of magnetometers can be used as magnet-fixed carriers and magnetic memory materials for bioactive substances. At the same time, magnetosomes can specifically adsorb paramagnetic metal ions to achieve their accumulation outside cells or cell masses, and under the action of an external constant magnetic field, magnetotactic bacteria carrying metal ions migrate to the fixed surface to aggregate and realize the removal of heavy metals. Therefore, Magnetotactic bacteria have great potential in the treatment of heavy metal wastewater. In this paper, the discovery, types, and characteristics of magnetotactic bacteria and their treatment of heavy metal wastewater containing iron, manganese, zinc, nickel, chromium, and other heavy metals are elaborated and their future prospects are expected.

1 INTRODUCTION

Nearly with the deepening of the "made in China 2025" strategy, China's industrial production technology is constantly updated, satisfying people's material and cultural needs in the industrial products category. With huge production, the resulting industrial wastes present new characteristics: Water pollutants increase, the processing difficulty is aggravated, and the damage after pollutants are discharged into the environment is intensified (Zhang 2020). Industrial wastewater is characterized by complex types, difficult treatment, and great harm, and wastewater treatment faces many challenges (Liu 2020).

As the main pollutant in industrial wastewater, heavy metals generally refer to metals with atomic weights between 21 and 83. Heavy metal wastewater is the wastewater containing heavy metals discharged from mining and metallurgy, machinery manufacturing, chemical industry, electronics, instrumentation, and other industrial production processes, including cadmium, nickel, mercury, zinc, and other elements. Some metal elements are essential trace elements for the human body, but excessive intake of trace elements will also cause damage to human health. For example, excessive Zn will reduce human immunity, excessive Fe will damage the liver, and excessive Cu will cause metabolic disorders (Deng 2021). The discharge of heavy metal wastewater into the ecological environment not only affects our intake of trace elements and causes damage to our health, but also causes ecological damage and economic loss. Scientific and reasonable removal of heavy metal ions from wastewater or conversion to low toxic products is an urgent problem to be solved in environmental protection.

Generally, heavy metals in wastewater cannot be decomposed and destroyed, but can only be transferred to different locations and physical and chemical forms. At present, there are three main methods for treating wastewater containing heavy metal ions (Tao 2022): (1) The dissolved heavy metals are transformed into insoluble or insoluble metal compounds, to remove them from water;

---

*Corresponding Author: cea_xiek@ujn.edu.cn

DOI 10.1201/9781003344209-16

(2) adsorption, concentration, and separation without changing the chemical form of heavy metals; (3) Methods to remove heavy metals from wastewater by means of microbial or plant flocculation, absorption, accumulation, and enrichment.

With the progress and development of society, people's requirements for water quality are gradually increasing, and constantly exploring how to reduce the cost of treatment, minimize secondary pollution, and realize the biological treatment of heavy metal contaminated wastewater has become a research hotspot.

2 DISCOVERY AND STUDY OF MAGNETOTACTIC BACTERIA

Magnetotactic bacterium is a general category of microorganisms that can move along the direction of a magnetic field. Under the action of an external magnetic field, it can do directional movement and form nano-magnetic particles in the body. Magnetosomes are mainly distributed in soil, lakes, oceans, and other underwater sludges (Brenner 2013; Richard 2009). In 1963, Salvatore Bellin, an Italian scholar, discovered magnetotactic bacteria in a freshwater lake near Pavia and observed a large number of consistent and single bacteria swimming toward the n-pole of the magnetic field. It was speculated that the magnetic behavior of cells was due to the internal "magnetic compass" (Salvatore 2009; Salvatore 2009). In 1974, Richard P. Blakemore observed by electron microscope and found that there were chains of Fe_3O_4 nanoparticles in bacterial cells. This chain of magnetic bodies acts as a compass and a guide, enabling the bacteria to carry out a magnetotactic movement, which first proved Salvatore Bellin's guess: "There are magnetic magnetosomes in magnetotactic bacterial cells" (Blakemore 1975).

The morphology of magnetotactic bacteria is diverse. At present, the single cell morphology of magnetotactic bacteria has been detected as spherical, rod-shaped, arc-shaped, helical, and multicellular aggregates (Figure 1) (Dziuba 2016). The synthesized magnetosomes are usually hexagonal prismatic, cubic octahedral, bullet-shaped, and irregular. The arrangement modes of magnetic bodies in cells are also different, mainly including single-strand, double-strand, multi-strand, beam arrangement, and dispersed distribution (Figure 2) (Fuduche 2015; Fang 2019). In the same magnetotactic bacteria, the synthesis of magnetosomes is regulated by genes, and the particle size, composition, and shape of magnetotactic bacteria are relatively uniform. However, a few magnetotactic bacteria can simultaneously synthesize Fe_3O_4 and Fe_3S_4 magnetosomes in cells. Magnetotactic bacteria are mostly micro-aerobic organisms. Guided by intracellular magnetosomes and driven by their own flagella, magnetotactic bacteria carry out a magnetotactic and oxygen-tactic movement to reach the suitable habitat.

At present, the known magnetotactic bacteria belonging to Pro-Teobacteria include alpha-Proteobacteria, β-proteobacteria, γ-(gamma-proteobacteria-RIA), Alphaproteobacteria, Beta proteobacteria, and Gamma proteobacteria. δ (Delta-Proteobacteria, Eta-Proteobacteria, nitrosourea, Planctomycetes, Om-Nitrophica phylum, and suspected Latescibacteria phylum (Xiao 2019). Most of the common magnetotactic bacteria belong to the genera Magnetocous, Magnetospira, and Magnetovibrio.

3 APPLICATION OF MAGNETOTACTIC BACTERIA IN THE TREATMENT OF HEAVY METAL POLLUTED WASTEWATER

Magnetotactic bacteria contain magnetosomes, which are easily removed in magnetic fields, and are highly adsorbent to heavy metals. Some heavy metals, such as Fe, Ni, Mg, and Cu, can also be absorbed by magnetotactic bacteria and used for their own reproduction and metabolism. Magnetotactic bacteria and heavy metal ions have a mutual attraction. If the metal ions in sewage are paramagnetic, they will be continuously attracted by magnetotactic microbial cells or cell clusters, and realize their accumulation outside the cells or cell clusters. Subsequently, under the action of a constant external magnetic field, magnetotactic bacteria will swim along a certain magnetic

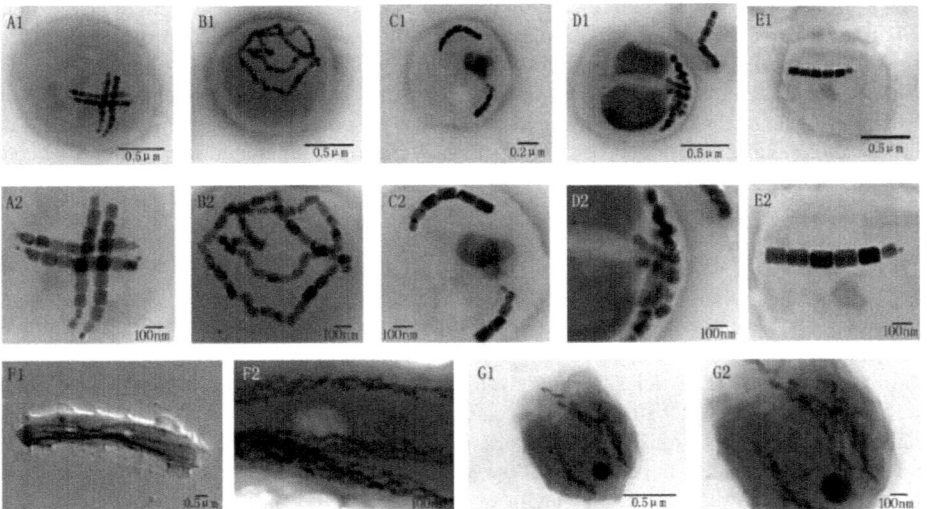

Figure 1.   Transmission Electron Micrographs of Magnetotactic Bacteria and Magnetosomes (Wang 2020).

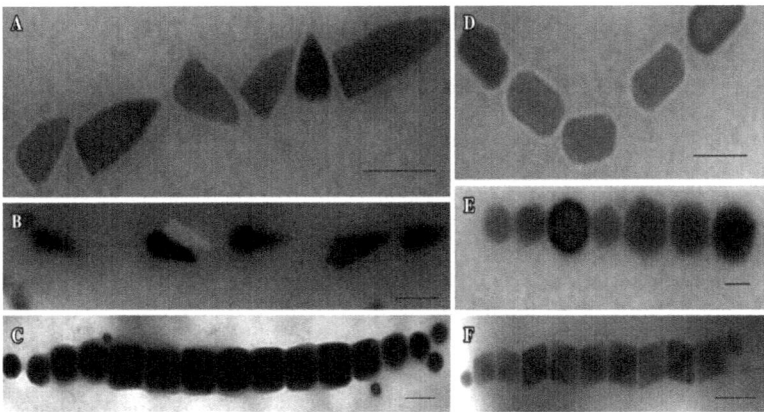

Figure 2.   Electron Micrograph Determination of Different Crystal Forms of Magnetosomes in Different Magnetotactic Bacteria (Fuduche 2015). A, B Bullet or Tooth, C Prism, D Rectangle, E Cube Octahedron, F Octahedron.

field direction, and the magnetotactic bacteria carrying metal ions migrate to the fixed surface to accumulate, thus completing the removal of heavy metal ions in sewage solution (Lin 2006).

The ion adsorption of heavy metals is divided into 2 stages, the first stage is the passive adsorption, including extracellular accumulation, precipitation, and cell surface adsorption and complexation, mainly in the cell surface area of extracellular polysaccharide (EPS), the cell walls of phosphoric acid roots, carboxyl, thiol, and amino groups and intracellular chemical groups as well as combination between the metals. It is fast and reversible and does not depend on energy metabolism. The second stage is active absorption, that is, the metal ions adsorbed on the cell surface area in combination with some enzymes on the cell surface and transferred to the cell. It includes transport and deposition with slow and irreversible speed, and it is related to cell metabolism.

Since the discovery, studies on the use of magnetotactic bacteria to treat heavy metal wastewater have begun abroad. For example, Professor Bahaj (Bahaj 1998) from the University of Southampton studied the influence of different heavy metal ions on the activity of magnetotactic and found that

Cu, Co, Cd, and Zn plasmas can inhibit the activity of magnetotactic bacteria. However, Fe, Mg, Mn, Al, and Cr can be effectively adsorbed and removed, so they can be used to treat specific heavy metal wastewater. It is found that the removal efficiency of the above ions by magnetotaxis bacteria is very high, and the concentration can be reduced from $(10–100) \times 10–6$ mg/L to $(10–100) \times 10–9$ mg/L in general, which is far below the international emission standards. Using magnetotactic bacteria to treat wastewater containing heavy metals, the wastewater can be moved through a fixed magnetic field to remove heavy metal ions without the need of additional power or the addition of other drugs. Therefore, it can greatly reduce the cost of wastewater treatment and has an exclusive advantage in economic feasibility.

Qu et al., from Tianjin University (Qu 2014) used a new culture method to study the effects of temperature, contact time, pH value, microbial concentration, initial concentration of Cr^{6+} and coexisting ions (Co^{2+} and Cu^{2+}) on the removal rate of Cr^{6+}. Under the pH value of 6.00 and temperature of 29°C, the removal efficiency of Cr^{6+} by living cells reached 77% within 10 min, indicating that Co^{2+}, Cu^{2+} and applied electric field promoted the removal of Cr^{6+}, while the applied magnetic field hindered the removal. Ren M M et al. (Ren 2004) studied the effects of factors such as pH value, temperature, time, and microbial biomass on the adsorption of Cr^{3+} by magnetotactic bacteria, and selected the corresponding adsorption model. The removal rate of Cr^{3+} by magnetotactic bacteria can reach more than 80%, and the magnetotactic bacteria can successfully remove Cr^{6+}, which is of great value to improve the process of heavy metal removal by magnetotactic bacteria. Wu Z H et al. (Wu 2007) studied the optimal adsorption conditions of four precious metal ions Pd^{2+}, Au^{3+}, Pt^{4+} and Ag^+ in precious metal waste liquid by using magnetotactic MTB as the adsorbent. The results showed that the adsorption capacity of magnetotactic bacteria on precious metal ions was much higher than that of other common metal ions, indicating that in the multiple systems, magnetotactic bacteria have good selective absorption of precious metal ions. Liu Jun (Liu 2013) studied the removal effect of Cu^{2+} and Zn^{2+} in sewage by the combined technology of magnetic field and magnet-seeking bacteria. The results showed that under the environmental temperature of 25°C, adsorption duration of 1 h, and magnetic field strength of 4000 Gs, compared with magnetotactic bacteria alone, the removal rate of Cu^{2+} and Zn^{2+} increased by 44.7% and 43.5%, respectively, and the high-efficiency separation and recycling of heavy metal ions by magnetotactic bacteria could be realized. In addition, Shu H H et al. (Shu 2005) pointed out in the study of combining a magnetic field with magnetotactic bacteria in the treatment of Ni^{2+} wastewater that the creation of an external magnetic field can improve the adsorption performance of magnetotactic bacteria on Ni^{2+}. When the magnetic field intensity is 100 Gs and the metal wire frame in the magnetic separator is perpendicular to the direction of a magnetic field, the separation efficiency of magnetotactic bacteria is up to 98% under the room temperature 20°C, pH value of 5.0 and the number of magnetotactic bacteria of 80 g /L for 1 h, and the synchronous separation of heavy metal ions in sewage is also achieved.

Based on the biomineralization mechanism of magnetotactic bacteria (MTB), Diaz-Alarcon, J. A. et al. (Diaz-Alarcón 2019) proposed an alternative biological method to remove soluble elements Fe^{2+} and Mn^{2+} in groundwater. Figure 3A shows the removal of iron and manganese in 10 mL of groundwater samples in the UPTC well. The average removal efficiency of Fe^{2+} and Mn2+ is 56.97% and 23.44%, respectively. With 20 mL of groundwater samples (Figure 3B), the average removal efficiency of Fe^{2+} and Mn^{2+} is 38.75% and 7.08%, respectively. In addition, zinc ion concentrations at two volumes were measured to achieve the removal rate of 75% in 10 mL of groundwater samples; It reached 18.56% in 20 mL of groundwater samples. The average removal efficiency of Fe^{2+}, Mn^{2+}, and Zn^{2+} was 47.86%, 15.26%, and 46.78%, respectively. In addition, Sumana Sannigrahi et al. (Sannigrahi 2019) studied the adsorption and recovery of heavy metals such as cadmium, copper, nickel, lead, and zinc in the diode and resistor of the printed circuit board by five magnetostrophic bacterial strains (RJS2, RJS5, RJS6, RJS7, and Mar-1) and the consortium MASG (RJS2, RJS5, and MSR-1). The results showed that RJS2, MSR-1, and RJS6 had the highest recoveries of cadmium, lead, and nickel, which were 97%, 100%, and 99%, respectively. RJS2 (copper -89%) and RJS6 (zinc -88%) reached the highest recovery in terms of resistance. The

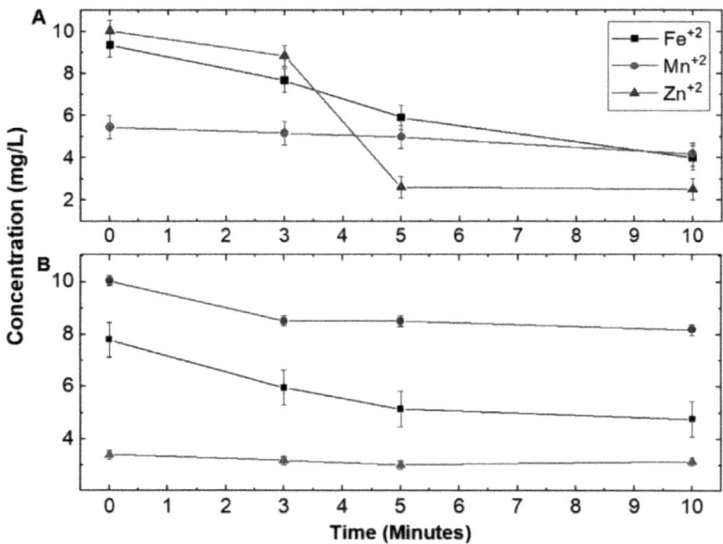

Figure 3.   Removal Rates of Fe^{2+}, Mn^{2+}, and Zn^{2+} by (A)10 mL of Groundwater Sample and (B) 20 mL of Groundwater Sample (Diaz-Alarcón 2019).

overall average recovery of cadmium (80%) and lead (66%) was higher than that from the treated diodes. Similarly, copper (45%) and lead (40%) are recovered from resistors. MAG1 recovers nickel (100%) and zinc (75%) from the diode, as shown at the left side in Figure 4. Cadmium (90%), nickel (22%), and zinc (47%) are recovered from the resistor, as shown at the right side in Figure 4. Compared with a single strain, MAG1 achieves a better recovery rate.

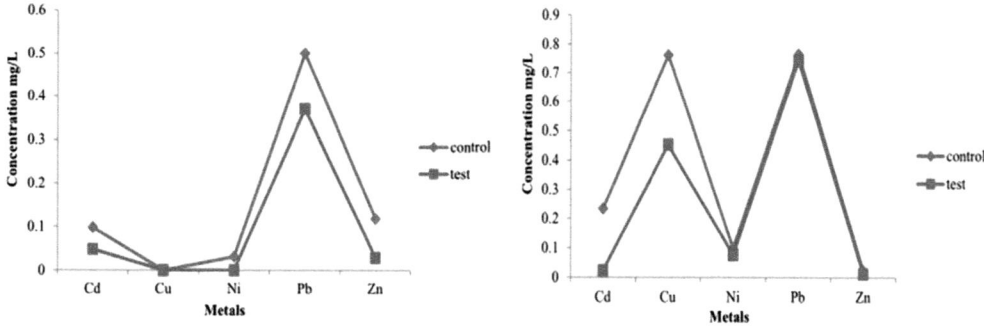

Figure 4.   Atomic Absorption Spectrum Analysis of Diodes in Waste PRINTED Boards Treated and Resistance in Waste Circuit Board Processed with MAG1 (Sannigrahi 2019).

4   CONCLUSIONS

Magnetotactic bacteria are diverse and rich in resources in nature. Their own magnetotactic ability to adsorb metal ions provides a new, effective, and low-cost method to improve the effect of wastewater treatment, and also provides a new solution to membrane pollution in water treatment. Their internal magnetosomes are also valuable in magnetic separation, immobilized enzyme, food detection, environmental monitoring, medical diagnosis, magnetic resonance imaging, magnetic hyperthermia, and targeted therapy.

5 APPLICATION PROSPECTS OF MAGNETOTACTIC BACTERIA

At present, the main bottleneck restricting the application of magnetotactic bacteria and magne-tosomes is the separation and purification of magnetotactic bacteria, large-scale pure culture, and magnetosome synthesis mechanism. The study of magnetotactic bacteria diversity and biominer-alization is crucial to understanding the evolution of magnetic induction in higher organisms based on free ions, and considering the complex composition of magnetotactic bacteria in nature or labo-ratory microspheres. Next, it is necessary to link targeted 16S RNA genes to specific magnetotactic bacterial morphologic types at the single-cell level (Li 2020; Liu 2021). With the deepening of research work, the application of magnetotactic bacteria and magnetosomes will show a better prospect, and it is believed that they will be better used in people's lives in the future.

ACKNOWLEDGEMENTS

This work was supported by the School of Civil Engineering and Architecture, University of Jinan. This research was supported by the Shandong Provincial Key Research and Development Program of China (No. 2021SFGC0204) and Shandong Provincial Natural Science Foundation project ZR2020ME227.

REFERENCES

Bahaj, A.S. & I.W. Croudace, et al. (1998). *Continuous radionuclide recovery from wastewater using magnetotactic bacterial Presented in part at ICM'97 in Cairns*, Australia.1.

Blakemore, R. (1975). Magnetotactic bacteria. *Science* **190** (4212): 377–9.

Brenner, D.J. & Garrity, G.M. et al. (2013). Bergey's manual of systematic bacteriology: Volume Two: The proteobacteria, Part A introductory essays. *Asian Australasian Journal of Animal Sciences* **26** (9): 1237–46.

Deng, S. (2021). Research on the present situation and solution of industrial wastewater treatment. *Resource Conservation and Environmental Protection* (02): 75–76.

Diaz-Alarcón, J.A. & Alfonso-Pérez, M.P. et al. (2019). Removal of iron and manganese in groundwater through magnetotactic bacteria. *Journal of Environmental Management* **249**: 109381.

Dziuba, M. & Koziaeva, V. et al. (2016). Magnetospirillum caucaseum sp. nov., Magnetospirillum marisnigri sp. nov. and Magnetospirillum moscoviense sp. nov., freshwater magnetotactic bacteria isolated from three distinct geographical locations in European Russia. *International Journal of Systematic & Evolutionary Microbiology* **66** (5): 2069.

Fang, Y. & Zhang, T. et al. (2019). Research progress on diversity and application of magnetotactic bacteria. *Chinese Journal of Biological Engineering* **39** (12): 73–82.

Fuduche & Maxime, et al. (2015). Diversity of Magnetotactic Bacteria from a French Pristine Mediterranean Area. *Current Microbiology: An International Journal* **70** (4): 499–505.

Li, J. & Liu, P. et al. Magnetotaxis as an adaptation to enable bacterial shuttling of microbial sulfur and sulfur cycling across aquatic oxic-anoxic interfaces. *Journal of Geophysical Research: Biogeosciences*.

Li, W. & Ti, X. et al. (2006). Advances in magnetotactic bacterial magnetosomes. *Bulletin of Microbiology* (03): 133–137.

Liu, J. & Zhou, P. et al. (2013). Study on the application of magnetotactic bacteria in the treatment of wastewater containing Cu^{2+} and Zn^{2+}. *Environmental Science and Technology* **26** (04): 20–24.

Liu, Y. (2020). Research on current status and development trend of heavy metal wastewater treatment technology. *Leather Manufacturing and Environmental Technology* **1** (16): 66–71.

Qu, Y. & Zhang, X. et al. (2014). Removal of hexavalent chromium from wastewater using magnetotactic bacteria. *Separation and Purification Technology* **136**: 10–17.

Ren, M. & Wang, Y. et al. (2004). Adsorption of Cr(3+)-containing wastewater by magnetotactic bacteria. *Journal of Kunming University of Science and Technology* (Science and Technology) (01): 97–99+107.

Richard, B. et al. (2009). The discovery of magnetotactic/magnetosensitive bacteria. *Chinese Journal of Oceanology & Limnology*.

Salvatore & Bellini, et al. (2009). Further studies on "magnetosensitive bacteria". *Chinese Journal of Oceanology and Limnology* **27** (1): 7.

Sannigrahi, S. and Suthindhiran, K. (2019). Metal recovery from printed circuit boards by magnetotactic bacteria. *Hydrometallurgy* **187**.

Shu, H., Wang, Y. & Sun, J. et al. (2005). *Study on the treatment of nickel-containing wastewater by magnetotactic bacteria and magnetic field*. Ion Exchange and Adsorption (04): 365–369.

Tao, W. (2022). Discussion on heavy metal wastewater treatment technology and recycling. *Resource Conservation and Environmental Protection* (02): 98–101.

Wang, H. & Zhang, L. et al. (2020). Diversity of magnetotactic bacteria in spring water system of Jinan in winter. *Shandong Agricultural Sciences* **52** (04): 79–85.

Wu, Z. (2007). *Study on treatment of waste liquid containing noble metal ions by magnetic field – magnetotactic bacteria*, Tianjin University: 86.

Xiao, T. & Pan, H. et al. (2019). Studies on magnetotactic bacteria in the deep sea. *Journal of Microbiology* **39** (02): 1–10.

Zhang, T. & Li, Z. et al. (2020). The present situation of industrial wastewater treatment and the countermeasures of pollution prevention and control. *Water Supply and Drainage* **56** (10): 1–3+18.

*Frontiers in Civil and Hydraulic Engineering – Mohamed A. Ismail and
Hazem Samih Mohamed (Eds)*

Study on seepage deformation characteristics of soil-structure interface under confining pressure

Hualin Sun & Lin Meng
*Haihe River, Huaihe River, and Xiaoqinghe River Basin Water Conservancy Management and
Service Center of Shandong Province, Jinan, China*

Xuesen Zhang & Quanyi Xie*
School of Qilu Transportation, Shandong University, Jinan, China

ABSTRACT: The interface between soil and structure widely exists in all kinds of engineering
structures. It is a key part of engineering structures. Once the seepage failure condition is reached,
it will lead to disaster accidents such as structural instability. According to the seepage deformation characteristics of the soil-structure interface under confining pressure, a fluid-solid coupling
numerical calculation model is constructed, and the deformation laws of the interface under the confining pressure are analyzed in this paper. The main conclusions are as follows: With the increase
in the cohesion, interface stiffness, and confining pressure, the deformation of the interface shows
a nonlinear growth trend, and the critical hydraulic gradient shows a linear growth trend. With the
increase of the confining pressure, the influences of soil sample cohesion and interface stiffness
decrease gradually.

1   INTRODUCTION

The interface between soil and structure is very common in practical engineering, such as the
interface between the face slab and cushion of the face slab dam, the interface between concrete
foundation and foundation, and the interface between the pile and surrounding soil. The soil-
structure interface is the key and weak part of the project. Once the critical hydraulic gradient
is reached, it will have a significant impact on the stability of the structure. Many infrastructure
buildings in the world have been subjected to major accidents due to seepage damage at the soil-
structure interface. In 1963, the Baldwin Hills dam in the United States crashed due to seepage
damage at the interface between blanket and foundation (Scott 1987, Leps 1987). In 1976, the Teton
dam in Idaho State of the United States broke due to the interfacial erosion between the impervious
wall and the bedrock (Chen 2006; Penmana 1977; Seed 1987; SEED 1981; Sherard 1987). In
1990, due to the interfacial erosion between the core wall of the main dam and the bedrock layer
of Songshan reservoir in China, the downstream slope of the earth dam experienced leakage (Zhan
2009). In 1993, the Gouhou reservoir in China suffered from reservoir leakage due to interfacial
erosion between the bottom plate of the retaining wall of the earth dam and the foundation (Chang
2012; China Water & Power Press 2006).

---

*Corresponding Author: xiequanyi@sdu.edu.cn

In the construction of water conservancy projects, research on the soil and structure interface has always been an important topic. A large number of interface shear equipment have been developed to study the shear failure behavior of interfaces, such as direct shear equipment (Zhang 2005), single shear equipment (Desai 2005), and torsional shear equipment (Yoshimi 1981). At the same time, in terms of interfacial erosion, seepage devices such as horizontal permeameter (Beguin 2013), vertical permeameter (Zhu 2016), cylindrical permeameter (Kim 2018), and semi-cylindrical permeameter (Xie 2019) have also been developed for exploring interfacial erosion.

However, there are relatively few studies on the erosion characteristics of the soil-structure interface under normal pressure. Therefore, the FLAC numerical simulation software is used to build the fluid-solid coupling numerical model of the soil-structure interfacial erosion, and the interfacial deformation characteristics under the action of the normal pressure are studied. The research results of this paper are of great significance to the construction of water conservancy projects and the development of the water conservancy industry.

2 NUMERICAL ANALYSIS

2.1 Numerical calculation model

In this paper, FLAC 3D is used to establish a three-dimensional numerical calculation model of the soil structure interface, as shown in Figure 1. The size of the model sample is 150 mm in diameter and 170 mm in height. The model adopts a three-dimensional solid element structure, and the mesh element is an octagonal hexahedron element. The whole model has 14158 calculation elements and 16859 nodes. The cylindrical boundary of the model is set as the impervious boundary, and the bottom boundary is the permeable boundary. The vertical displacement of the bottom boundary of the model is fixed, and the water pressure is applied. The confining pressure is applied to the cylinder boundary of the model. In this paper, the M-C model is used in the numerical calculation model of soil. The M-C shear model is adopted for the soil structure interface, which can reflect the slip, separation, or closure of the interface under the stress.

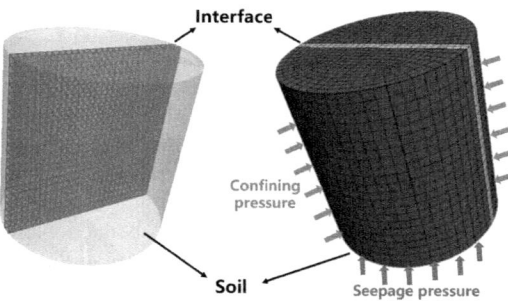

Figure 1.   Numerical calculation model.

2.2 Numerical calculation conditions

In this paper, the numerical calculation conditions are designed according to the cohesion, normal pressure, and roughness of the interface, which affect the erosion characteristics of the soil-structure interface. Among them, the cohesion and normal pressure of soil samples are set at three levels respectively. The friction angle, elastic modulus, and Poisson's ratio of soil samples are 30°, 3.2 e^7Pa, and 0.34, respectively. By changing the normal stiffness and tangential stiffness of the interface, this paper simulates the interface with different roughness values and sets up three kinds of interfaces. A total of 15 calculation conditions are set in this paper, as shown in Table 1.

Table 1. Numerical calculation conditions.

	Parameters of soil			Parameters of interface				
Number	Density (kg/m^3)	Cohesion (Pa)	Permeability coefficient (cm/s)	Normal stiffness (N/m)	Tangential stiffness (N/m)	Cohesion (Pa)	Friction angle (°)	Confining pressure (kPa)
N-1								50
N-2	1650	$3.1e^4$	$7.7e^{-4}$	$8e^8$	$8e^8$	$2.5e^4$	20	100
N-3								150
N-4								50
N-5	1635	$4.2e^4$	$6.9e^{-4}$	$8e^8$	$8e^8$	$2.5e^4$	20	100
N-6								150
N-7								50
N-8	1611	$5.5e^4$	$6.5e^{-4}$	$8e^8$	$8e^8$	$2.5e^4$	20	100
N-9								150
N-10								50
N-11	1611	$5.5e^4$	$6.5e^{-4}$	$8.6e^8$	$8.6e^8$	$2.5e^4$	26	100
N-12								150
N-13								50
N-14	1611	$5.5e^4$	$6.5e^{-4}$	$9e^8$	$9e^8$	$2.5e^4$	30	100
N-15								150

3 EFFECT OF MATERIAL COHESION

Figures 2–4 show the interfacial deformations under different soil cohesion conditions. It can be seen from the figures that with the increase of the water head at the bottom of the soil sample, the deformation of the interface presents a nonlinear increasing trend under the same confining pressure and interface roughness. The higher the cohesion of the soil sample, the smaller the deformation of the interface between soil and structure. The hydraulic slope with abrupt deformation of the interface is regarded as the destructive hydraulic slope, as shown in Figure 5. Under confining pressures at all levels, the hydraulic slope of contact seepage failure shows a linear growth trend with the increase in soil cohesion. With the increase of confining pressure, the slope of the cohesion-hydraulic slope curve shows a decreasing trend, indicating that with the increase of the confining pressure, the influence of soil cohesion on the contact seepage failure shows a decreasing trend.

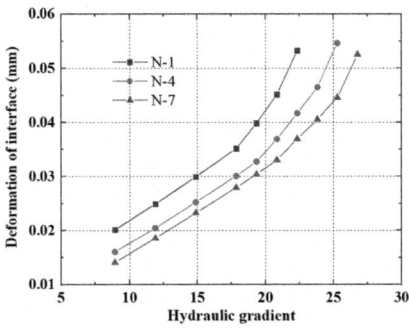

Figure 2. Interfacial deformation for different soil samples under confining pressure of 50 kPa.

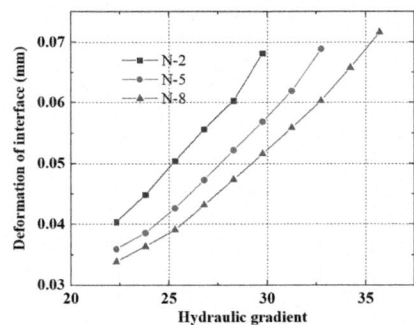

Figure 3. Interfacial deformation for different soil samples under confining pressure of 100 kPa.

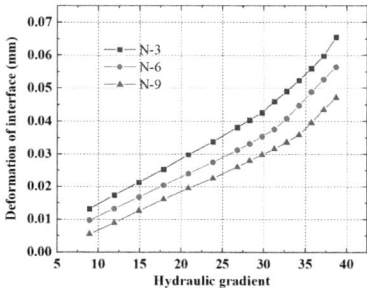

Figure 4.   Interfacial deformation for different soil samples under confining pressure of 150 kPa.

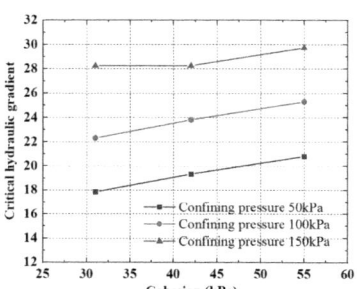

Figure 5.   Critical hydraulic gradient for different soil samples.

4   INFLUENCE OF INTERFACE ROUGHNESS

Figures 6–8 show the interfacial deformation laws under different interface conditions. It can be seen that the stiffness of the interface has a very significant influence on the deformation of the interface between soil and structure under different confining pressures. The larger the stiffness of the interface, the smaller the deformation of the interface between soil and structure. It indicates that with the increase of the stiffness of the interface, the roughness of the interface increases, so does the resistance against the soil deformation at the interface under the action of seepage. The hydraulic failure gradient under different interfaces is shown in Figure 9. As seen from the figure, with the increase in the stiffness of the interface, the hydraulic failure slope presents an increasing linear trend. It indicates that the erosion process of soil particles on the interface can be delayed by increasing the roughness of the interface, such as gouging and spraying.

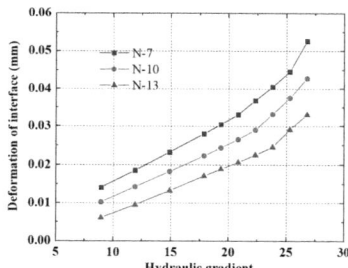

Figure 6.   Interfacial deformation for different interfaces under confining pressure of 50 kPa.

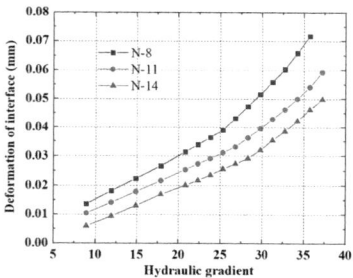

Figure 7.   Interfacial deformation for different interfaces under confining pressure of 100 kPa.

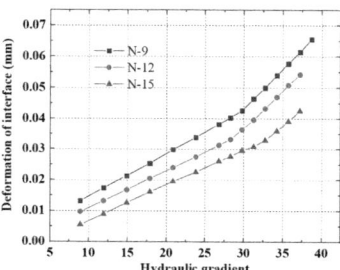

Figure 8.   Interfacial deformation for different interfaces under confining pressure of 150 kPa.

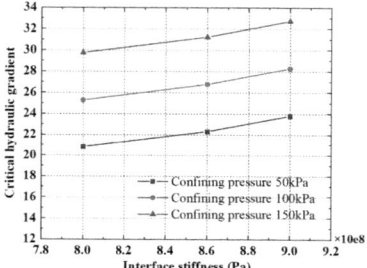

Figure 9.   Critical hydraulic gradient for different interfaces.

5   ANALYSIS OF INFLUENCE OF CONFINING PRESSURE

Figure 10 and Figure 11 show the interfacial deformation and hydraulic gradient under different confining pressures. As can be seen from the figures, the larger the confining pressure, the smaller the deformation of the interface between soil and structure. When the hydraulic gradient is slight, the deformation gap at the soil-structure interface under different confining pressures is tiny. With the gradual increase of hydraulic slope, the deformation gap of the soil-structure interface under other confining pressures increases gradually. At the same time, with the confining pressure increasing, the hydraulic failure slope shows an increasing linear trend. This is mainly because with the increase of confining pressure, the interface between the soil and the structure is more closely bonded, and the ability of the interface to resist deformation is more vital.

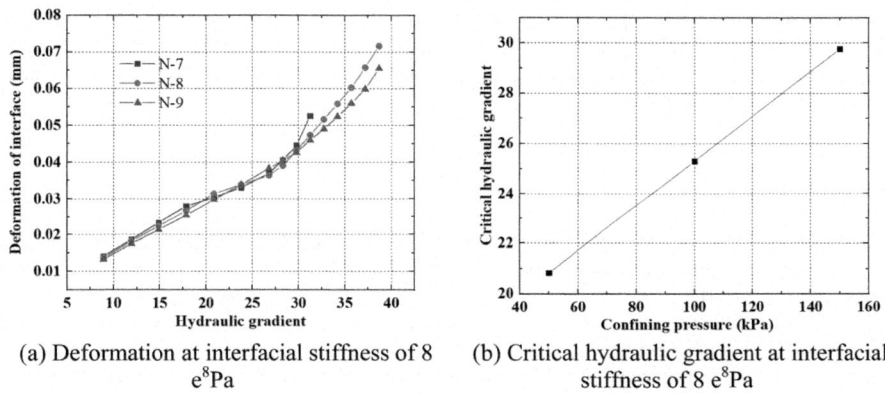

(a) Deformation at interfacial stiffness of 8 e^8Pa

(b) Critical hydraulic gradient at interfacial stiffness of 8 e^8Pa

Figure 10.   Influence of confining pressure under interface stiffness of 8 e^8Pa.

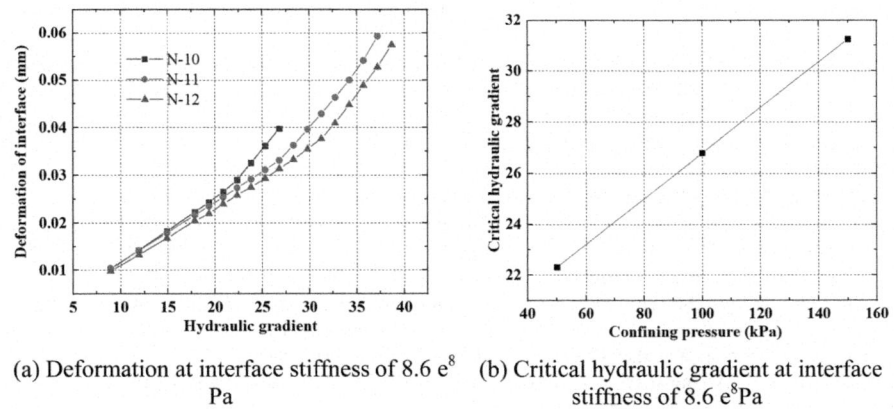

(a) Deformation at interface stiffness of 8.6 e^8 Pa

(b) Critical hydraulic gradient at interface stiffness of 8.6 e^8Pa

Figure 11.   Influence of confining pressure under interface stiffness of 8.6 e^8Pa.

6   CONCLUSION

In this paper, according to the seepage deformation characteristics of the soil-structure interface under the confining pressure, a fluid-solid coupling numerical calculation model is constructed, and the deformation law and its influencing mechanism of the interface under the confining pressure are analyzed. The main conclusions are as follows:

(1) The deformation of the interface shows a nonlinear growth trend with the increase in soil cohesion, interface stiffness, and confining pressure.
(2) The hydraulic slope of interface failure shows a linear growth trend with an increase in soil cohesion, interface stiffness, and confining pressure. The influences of soil cohesion and interface stiffness gradually decrease with the confining pressure.

Data Availability Statement: The data used to support the findings of this study are included within the article. Any additional data related to the paper may be requested from the corresponding author.

Conflicts of Interest: The authors declare that there are no conflicts of interest regarding the publication of this paper.

Funding: This work was supported by the Regional Project of the National Natural Science Foundation of China (No. 42162022), the Science and Technology Plan of Shandong Transportation Department (No. 2021B52) and the Youth Foundation of Shandong Natural Science Foundation of China (No. ZR2021QE279).

REFERENCES

Beguin R., Philippe P., Faure Y.H. Pore-scale flow measurements at the interface between A sandy layer and A model porous medium: Application to statistical modeling of contact erosion [J]. *Journal of Hydraulic Engineering ASCE*, 2013, 139(1): 1–11

Chang L.Y., Chen Q. Progress in contact scouring research [J]. *Advances in Science and Technology of Water Resources*, 2012, 32(02): 79–82.

Chen L.Y. *Analysis of Embankment Dam Seepage Problem and Prototype Observation* [D]. Zhengzhou University, 2006.

Desai S., Pradhan K., Cohen D. Cyclic testing and constitutive modeling of satu-rated sand-concrete interfaces using the disturbed state concept [J]. *International Journal of Geomechanics*, 2005, 5(4): 286–294.

Jansen R.B. A review of the baldwin hills reservoir failure [J]. *Engineering Geology*, 1987, 24 (1-4): 7–81.

Kim H., Park J., Shin J. Flow behavior and piping potential at the soil–structure interface [J]. *Geotechnique*, 2018, 69 (1): 1–6.

Leps T.M. Failure of Baldwin Hills reservoir, 1963—Interpretation of step-by-step failure sequence [J]. *Engineering Geology*, 1987, 24 (1–4): 83–88.

National flood control and Drought Relief Headquarters, Science and Technology Department of the Ministry of water resources. *Gouhou reservoir gravel faced dam design, construction, operation, and accident* [M]. China Water & Power Press, Beijing, 2006.

Penmana D.M. The failure of Teton Dam [J]. *Ground Engineering*, 1977, 10(6): 18–27.

Scott R.F. The Baldwin Hills reservoir failure [J]. *Engineering Geology*, 1987, 24 (s 1–4): 155–159.

Seed H.B., Duncan J.M. The failure of Teton Dam [J]. *Engineering Geology*, 1987, 24(1–4): 173–205.

SEED, B. The Teton dam failure a retrospective review [J]. *Proc. of Inter. conf. smfe*, 1981, 3.

Sherard J.L. Lessons from the Teton Dam failure [J]. *Engineering Geology*, 1987, 24 (1–4): 239–256.

Xie Q.Y., Liu J., Han B., Li H.T., Li Y.Y., Jiang Z.Q. Experimental investigation of interfacial erosion on the culvert-soil interface in earth dams [J]. *Soils and Foundations*, 2019, 59 (3).

Yoshimi Y., Kishida T. A ring torsion apparatus for evaluating friction between soil and metal surfaces[J]. *Geotechnical Testing Journal*, 1981, 4(4): 145–452.

Zhan M.L., Gao F., He S.Y., Sheng J.C., Xu X.C. Experimental study on contact scouring of clay under normal stress [J]. *Journal of Liaoning Technical University* (Natural Science), 2009, 28(S1): 206–208.

Zhang G., Zhang J.M. Reversible and irreversible dilatancy of the soil-structure interface [J]. *Rock and Soil Mechanics*, 2005, (05): 699–704.

Zhu Y.J., Peng J., Chen Q. Contact scouring tests on sandy gravel and cohesive soil [J]. *Chinese Journal of Geotechnical Engineering*, 2016, 38(S2): 92–97.

Frontiers in Civil and Hydraulic Engineering – Mohamed A. Ismail and
Hazem Samih Mohamed (Eds)
© 2023 The Authors, ISBN 978-1-032-38247-0

Research on ground settlement control based on slurry indicators during shield tunneling

Xiangchuan Yao* & Wen Liu*
CCCC Second Harbor Engineering Co., Ltd., Wuhan, Hubei, China

Yang Chen*
School of Civil and Hydraulic Engineering, Huazhong University of Science and Technology, Wuhan, Hubei, China

ABSTRACT: Good slurry performance is a requisite for ground settlement control during shield tunneling. When the ground settlement is under strict control, the slurry used for tunneling generally has a large pressure, high gravity, and high viscosity (thick). The increased slurry pressure is beneficial to control the settlement of the soil mass in front of the cutter head. Thick slurry helps increase the shear strength of surrounding soil mass and reduce the consolidation settlement caused by stratum disturbance. However, it needs to be noticed that although the thick slurry at the shallow soil layer is good for avoiding mud spillovers, the excessive slurry pressure is still likely to cause mud spillovers. Therefore, how to control the ground settlement by adjusting the slurry indicators is one of the challenges in research in the engineering field. In this paper, we took the Nanjing Weisan Road River-Crossing Project as an example to study the impact of slurry indicators on ground settlement in the process of slurry balance shield tunneling. We analyzed the relationship between the excavation face stability and slurry indicators during shield tunneling and looked into the impacts of slurry permeability on the increase in the formation strength and the impacts of thick slurry on splitting pressures. Through related tests, we obtained the quantitative measurements, providing reference and experience for subsequent similar engineering projects and being of certain guiding significance.

1 INTRODUCTION

In recent years, China has surpassed the U.S in underground development in cities, thus being one of the important players in tunneling works over the world. The shield method, as the main approach for underground development and construction, is widely used in tunnel projects. As urban underground tunnel projects have complex construction conditions and large risks, in the process of shield tunneling, ground settlement control is essential for measuring construction safety and quality. The root causes for ground settlement are the stratum loss caused during shield tunneling and the reconsolidation after the surrounding strata of the tunnel are subject to disturbance and shear failure. When the ground settlement is under strict control, the slurry used for tunneling generally features large pressure, high gravity, and high viscosity (thick). The increased slurry pressure is beneficial to control the settlement of the soil mass in front of the cutter head. Thick slurry helps increase the shear strength of surrounding soil mass and reduce the consolidation settlement caused by stratum disturbance. However, it needs to be noticed that although the thick slurry at the shallow soil layer is good for avoiding mud spillovers, the excessive slurry pressure is still likely to cause

---

*Corresponding Authors: 605866412@qq.com, 379047775@qq.com and 824890254@qq.com

DOI 10.1201/9781003344209-18

mud spillovers. Therefore, how to control the ground settlement by adjusting the slurry indicators is one of the challenges in research in the engineering field.

Currently, some theories and numerical methods for analyzing the stability of slurry-shied heading faces have been established over the world. For the estimation of the stability, no matter the upper and lower solutions, limit equilibrium method, or finite element analysis, they all employ "thin film models", that is, slurry pressure only imposes on the heading face in the form of surface force without considering the impact of slurry permeability on the stability of the heading face (Gao 2010; Liu 2019; Wu 2021). In practice, however, under complex geographic conditions, in the case of special geographic features, such as a local sand layer, high-quality films may not be formed in the process of shied tunneling because of failure to adjust the slurry parameters timely, thus the slurry permeability cannot be avoided. As the slurry penetration into the soil mass in front of the tunnel heading face leads to excessive pore water pressure, such pressure increase may result in the drop of the effective slurry pressure and thus affect the stability of the heading face (Lu 2020). In this case, using this "film model" is irrational; on the contrary, the wedge model considering the slurry permeability should be used to analyze the stability of the heading face (Zhou 2021). In China, the research on the related fields is in the bud and mainly relies on engineering experience. The understanding of the stability mechanism of the slurry shield heading face is quite limited (Lv 2019; Zhu 2015).

In this paper, we took the Nanjing Weisan Road River-Crossing Project as an example to study the impact of slurry indicators on ground settlement in the process of slurry balance shield tunneling. We analyzed the relationship between the excavation face stability and slurry indicators during shield tunneling, looked into the impacts of slurry permeability on stratum strength increase and the impacts of thick slurry on splitting pressures, summarized some empirical equations, and obtained measured data from related tests, which can be used as guidance for other slurry shied projects to some degree.

2 STABILITY MECHANISM OF EXCAVATION FACE OF SLURRY BALANCE SHIELD

In terms of water pressure, the slurry pressure imposed on any point on the excavation face is always greater than the groundwater pressure, so a hydraulic gradient spreading toward the periphery of the excavation face is formed in the soil mass near the excavation face. This is a basic condition for keeping the excavation face stable in the process of slurry balance shield tunneling (Editorial Department of China Journal of Highway and Transport 2022).

2.1 Basic requirements for slurry membrane formation

From the slurry balance shield theory, when the slurry shield advances, it is vital to form a non-permeable slurry membrane on the excavation face. A slurry membrane is formed only when the following basic requirements are met:

(1) Maximum particle size of slurry

   A slurry membrane is formed based on clogging and bridging effects, so the maximum particle size of the slurry has a great impact on the formation of a slurry membrane. In the soil layers with different permeability coefficients K, the slurry may have different maximum particle sizes. These two elements must correspond to each other.

(2) Particle gradation

   Particle gradation also affects the slurry membrane formation a lot. The optimal particle size distribution of slurry must be determined through numerous tests. Some research proposed that high-quality slurry contains the following: gel: 42–88%; silty soil: 7–46; sand: 5–13%. After the particle components in the soil mass peeled off by the cutter head are fully used to mix with new slurry in construction, the resulting suspended slurry will form a slurry membrane with a certain tension, to balance the water-earth pressure in front of the excavation face.

(3) Slurry density

The slurry density is closely related to the slurry viscosity, bleeding degree, and slurry ratio. Higher slurry density means higher yield value and better stability of slurry membranes. Generally, the slurry density is controlled within 1.15–1.30 g/cm3, being slightly lower in cohesive soil layer or fine sand, and higher in coarse sandy soil layer or shallow overburden where the shield passes through, or the places where the ground buildings are protected, generally reaching 1.3g/cm3.

(4) Slurry pressure

Although the permeation volume increases as the slurry pressure rises, such an increase is far less than the pressure rise, and the slurry pressure rise will affect the effective supporting pressure of the excavation face. Therefore, under high-quality slurry conditions, slurry pressure rise will enhance the stability of the excavation face.

2.2 *Relationship between the slurry membrane and tunneling speed*

When the slurry pressurized shield is advancing normally, its cutter head does not peel off the soil mass directly but cuts the formed slurry membrane opposite to the cutter head. A new slurry membrane will be formed immediately after cutting. The cutter head runs at a certain speed, and the maximum capacity of the shield at the tunneling speed is limited to some degree, therefore, the tunneling speed relates to the depth of the cut soil mass instead of slurry membranes. On the contrary, when the slurry pressurized shield is not advancing normally, especially when the slurry quality and slurry pressure are not up to the design requirements, the slurry membrane may be formed until a long period, which restrains the tunneling speed. Generally, with high-quality slurry, it may take 1–2 s to form a slurry membrane.

3 LIMIT ANALYSIS OF HEADING FACE CONSIDERING SLURRY PERMEABILITY

In practice, the excessive supporting pressure caused by slurry permeability might be four times the supporting pressure without considering slurry permeability. Therefore, it is essential to consider the slurry permeability in the limit analysis of the heading face of a tunnel.

3.1 *Slurry permeability*

When the soil mass features coarse particles and large porosity, high-quality slurry membranes cannot be formed, and the slurry will penetrate the soil mass to a certain depth. The typical slurry permeability is shown in the following figures. From Figure (a), the formed slurry membrane is of high quality, the slurry does not penetrate the soil mass in front of the heading face, and the slurry membrane transfers the supporting pressure in the slurry chamber completely to the soil skeletons in front of the heading face. Figure (b) shows that when the slurry quality is poor, no high-quality slurry membrane is formed in front of the heading face, and the slurry penetrates a large distance in front of the heading face. This is the most adverse case for slurry shield tunneling. Figure (c) shows the practical projects, especially in the highly water-bearing sand and gravel strata. With poor quality, the slurry penetrates into the soil mass in front of the heading face of the tunnel, causing excessive pore water pressure.

The groundwater analysis is simplified as appropriate for limit equilibrium analysis of the heading face stability. Firstly, the groundwater analysis is broken down into a single-degree-of-freedom (SDOF) problem, and each layer of soil mass is described as a semi-confined aquifer. Secondly, for each isotropic or anisotropic soil mass, the SDOF analysis needs to be modified as needed to adapt to the flow rate calculation of groundwater in front of the heading face. Relevant methods need to be employed to calculate the excess pore water pressure on the basis of the simplified groundwater analysis.

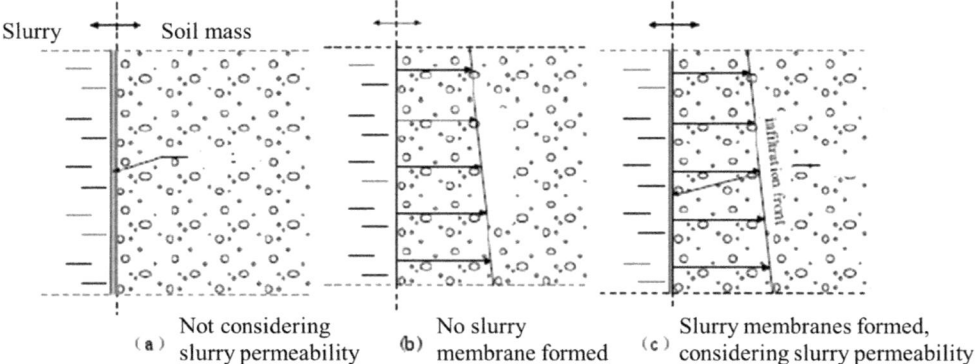

Figure 1. Three cases of slurry permeability.

The following figure illustrates the semi-confined aquifer at a given SDOF. This aquifer is defined as a permeable geological unit that can transmit a large amount of water under a normal hydraulic gradient. The dash area in the figure indicates the overlay weakly permeable soil layer. The excessive pore water pressures in the semi-confined aquifer are distributed as shown in Equation 1.

$$\Delta p(x,z,t) = p(x,z,t) - p_0(z) = \Delta p_s \frac{t}{a+t} \exp\left(-(x - e(t))/\lambda\right) \tag{1}$$

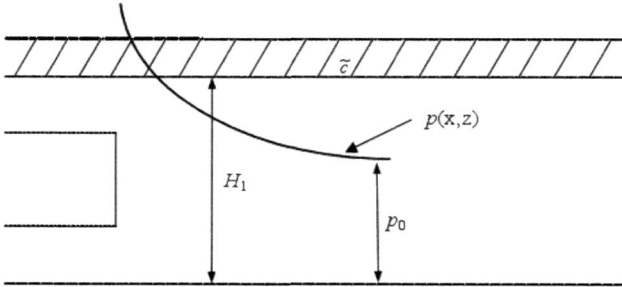

Figure 2. Semi-confined aquifer.

3.2 Layered wedge considering slurry permeability

In this paper, the wedge mass and the overburden on the wedge mass in front of the heading face are divided into several layers, for calculation in two parts. Besides, a wedge model for stabilizing the heading face of the slurry balance shield tunnel is established in consideration of the impacts of slurry permeability and groundwater seepage. The modified model is shown in the following figure.

To establish a wedge model for stabilizing the heading face under complex geological conditions, the unstable wedge is divided into several layers with different soil mass parameters and thicknesses. The soil mass parameters and conditions vary at different layers, including the included angle $\theta(i)$ between the unstable plane and horizontal plane at each layer of the soil mass. The layered wedge model is shown in the following figure.

Considering that the slurry permeability may cause excessive pore water pressure, the analytic expression of the supporting pressure for stabilizing the heading face of the slurry shield tunnel is as below:

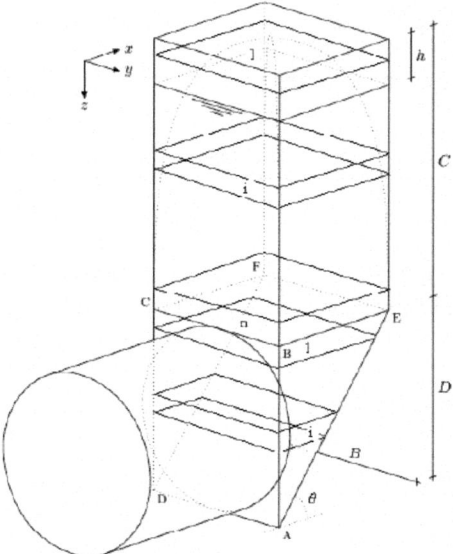

Figure 3. Modified 3D wedge model.

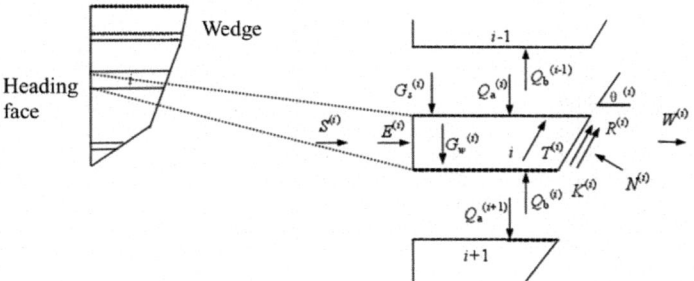

Figure 4. Layered wedge model.

$$S = -\frac{(G_s + G_w)\,\xi_- + (2T + K)}{\xi_+} + D \int_{z\text{bot}}^{Z\text{top}} \left(p_0(z) + \Delta p_s \frac{t}{a+t} \exp\left(-(x - e(t))/\lambda \right) \right) dz \quad (2)$$

4   RESEARCH ON THE IMPACTS OF THICK SLURRY ON SPLITTING PRESSURE

The splitting failure caused by the slurry shield tunneling is a circumstance where the slurry breaks through the overburden and merges into river water when the slurry pressure is excessively large to severely impair the soil mass strength in the process of slurry shield tunneling. The slurry splitting closely relates to the soil mass strength, hydrostatic pressure, overburden thickness, and slurry performance, among others.

Considering the operability of the site and the similarity to the soil layers the tunnel passes through, the test site is about 100 m in section K1+348–K1+449 of the connecting road. To measure the in-situ static earth pressure at the splitting test holes, the in-situ flat spade test is carried out directly at the original drills. The flat spade test results are shown in Table 1.

Table 1. Summary of flat spade test results.

Test hole No.	A1	A2	A3	B1	B2	B3
Test hole depth/m	15	10	5	5	10	15
Lateral pressure coefficient	0.74	0.61	0.71	0.70	0.59	0.73
Horizontal lateral pressure/kPa	199.8	109.8	63.9	63.0	106.2	197.1

To explore the impacts of different slurry properties on splitting performance, two slurry samples are proposed in the test for comparison. For those two slurry samples, the preparation results are shown in Table 2.

Table 2. Sample preparation.

Sample group No.		Component				Slurry quality	Property	
		Clay	Bentonite	CMC additive	Water		Viscosity	Gravity
Sample 1	1	–	140 g	–	930 g	1070 g (1000 mL)	18 s	1.07
	2	–	140 g	–	930 g	1050 g (1000 mL)	17 s	1.05
	3	–	140 g	–	930 g	1060 g (1000 mL)	18 s	1.06
	4	–	80 g	0.75 g	919.25 g	1030 g (1000 mL)	20 s	1.03
	5	–	80 g	0.75 g	919.25 g	1020 g (1000 mL)	19.5 s	1.02
	6	–	80 g	0.75 g	919.25 g	1025 g (1000 mL)	20 s	1.02
Sample 2	7	200 g	80 g	0.75 g	719.25 g	1150 g (1000 mL)	37 s	1.15
	8	200 g	80 g	0.75 g	719.25 g	1137 g (1000 mL)	35 s	1.14
	9	200 g	80 g	0.75 g	719.25 g	1143 g (1000 mL)	34.5 s	1.14
	10	220 g	80 g	1.00 g	699 g	1141 g (1000 mL)	40 s	1.14
	11	220 g	80 g	1.00 g	699 g	1160 g (1000 mL)	42 s	1.16
	12	220 g	80 g	1.00 g	699 g	1153 g (1000 mL)	42 s	1.15

To perform the stand test of the prepared samples, the mixed-ready sample is poured into a 500 mL graduated cylinder and let aside, and its precipitation effect is observed. Among these samples, the standing effects of Samples 1–5 are illustrated in Figure 5. It can be observed that, except for a little precipitation from the beginning, there is no sign of precipitation increase within 24 hours.

0 hour   1 hour   24 hours

Figure 5. Slurry stand test.

Upon comparison, the gravity and viscosity of Sample 1 are within the empirical value scope, and the slurry in the mix ratio of bentonite: CMC: water=8:0.075:91.925 (Group A) is more cost-effective, so this type of slurry is used for splitting test, with a gravity of 1.02 and a viscosity of 20 s. The slurry in the mix ratio of clay: bentonite: CMC: water=20:8:0.075:71.925 (Group B) is used for the contract test, with a gravity of 1.14 and a viscosity of 35 s.

For a more accurate measurement of the anti-splitting pressure of the mucky clay, the upper 1 m thick fill was excavated. The mucky clay below the water table was reserved, with an about 5 cm thick overlying aquifer, easy for observing mud spillovers during splitting. The splitting results at the A2 hole are shown in the following figure.

Figure 6.   Mud spillover during splitting test.

The 5 m deep B1 hole is excavated to expose the splitting path of the slurry splitting strata. Manual excavation is employed for protecting the split face from damages. The excavation starts 2 m from the split face, and each excavation shall be not deeper than 2 cm. The excavated fissure surface is shown in the following figure. The 3D split face is shown in Figure 7.

(a) At the 0.3 m depth of Hole B1 (b) At the 0.6 m depth of Hole B1 (c) At the 1.2 m depth of Hole B1

Figure 7.   Split fissure surface.

The splitting test results are shown in the table. The result analysis shows that in Group A with the mix ratio of bentonite: CMC: water=8:0.075:91.925, the splitting values of the 15 m Hole A1, 10 m Hole A2, and 5 m Hole A3 are 250 kPa, 155 kPa, and 80 kPa, respectively. In Group B with the mix ratio of clay: bentonite: CMC: water=20:8:0.075:71.925, the splitting values of the 5 m Hole B1 and 5 m Hole B3 are 82 kPa and 255 kPa, respectively. Mori Akira and Masaya Tamura conducted a laboratory test over the splitting during the shield tail grouting, and obtained the relationship between the splitting pressure with the slurry viscosity and soil strength (Mori 1990), i.e.:

$$P_f = \sigma_3 + \alpha q_u \tag{3}$$

The α values of different measuring holes can be calculated according to the above equation. As the actual overburden thickness is more than 10 m in calculations, the 10 m and 15 m holes are selected to obtain the α values. For safety, $\alpha=1.8$ was taken as the basis for theoretical analysis of slurry splitting at clayey soil layers.

141

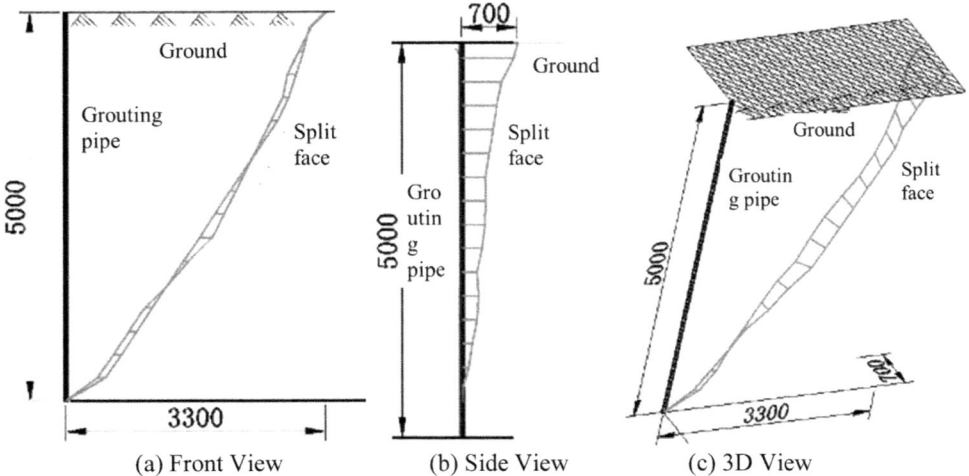

Figure 8.   Front view, side view, and 3D view of fissure surface of hole B1 (Unit: mm).

Table 3.   Summary of splitting test results.

	Group A			Group B	
Test hole No.	A1	A2	A3	B1	B3
Overburden thickness/m	15	10	5	5	15
Overlaying hydraulic pressure/kPa	270	180	90	90	270
Horizontal lateral pressure/kPa	199.8	109.8	63.9	63.0	197.1
Splitting pressure/kPa	250	155	80	82	255
Unconfined compressive strength of soil mass/kPa	27.8	27.8	27.8	27.8	27.8
α value	1.80	1.62	0.58	0.68	2.08
Type of slurry used	Bentonite: CMC: water = 8:0.075:91.925			Bentonite: CMC: water = 20:8:0.075:71.925	

5   CONCLUSION

In this paper, we took the Nanjing Weisan Road River-Crossing Project as an example to study the influences of slurry indicators on ground settlement in the process of slurry balance shield tunneling. We designed a proven and effective test upon the detailed survey on the stability of the excavation face of the slurry balance shield, in combination with the actual engineering geological conditions on the site, and conducted tests over multiple groups on the site to generate adequate sample data. Upon comparative analysis of the test results, we made the following conclusions:

(1) The slurry in the slurry chamber of the shield permeates into the soil mass in front of the heading face, resulting in excessive pore water pressure, so the calculation of slurry pressure not considering the slurry permeability is unsafe in practice if the slurry permeates seriously.
(2) In shallow soil layers, the slurry splitting path runs sidewise and upwards. In the section in the river, the splitting, if any, will go upwards vertically.
(3) In Group A with the mix of bentonite: CMC: water, the splitting values of the 15 m Hole A1, 10 m Hole A2, and 5 m Hole A3 are 250 kPa, 155 kPa, and 80 kPa, respectively. In Group B, the splitting values of the 5 m Hole B1 and 5 m Hole B3 are 82 kPa and 255 kPa, respectively.

For safety, $\alpha=1.8$ was taken as the basis for theoretical analysis of slurry splitting at clayey soil layers.

REFERENCES

Editorial Department of China Journal of Highway and Transport. Review on China's Traffic Tunnel Engineering Research: 2022 [J]. *China Journal of Highway and Transport*, 2022, 35(04): 1–40.

Gao J., Zhang Y.T. Face stability analysis of tunnel considering shield advance velocity [J]. *Rock and Soil Mechanics*, 2010, 31(07): 2232–2237+2240.

Liu X.Y., Wang F.M., Yuan D.J., Fang H.Y., Zhang S.L. Range of support pressures for slurry shield and analysis of its influence factors [J]. *Chinese Journal of Geotechnical Engineering*, 2019, 41(05): 908–917.

Lu K.D. Study on distribution characteristics of pore water pressure in the cross-river of shield tunnel [J]. *Building Structure*, 2020, 50(S2): 759–763.

Lv X.L., Zeng S., Wang Y.P., Ma S.K., Huang M.S. Physical model tests on stability of shield tunnel face in saturated gravel stratum [J]. *Chinese Journal of Geotechnical Engineering*, 2019, 41(S2): 129–132.

Mori A, Tamura M. *Hydraulic fractures phenomenon of shield construction in cohesive soil grounds (No.2) — hydraulic fractures phenomenon with back filling*[C]// Proceedings of the 24th Symposium on Geotechnical Engineering. Tokyo: Japan Society of Geotechnical Engineering, 1990: 1773–1774.

Wu B., Liu W., Shi B.X., Fu C.Q. Two-dimensional spiral failure model of the heading face of shield tunneling [J]. *Rock and Soil Mechanics*, 2021, 42(03): 767–774.

Zhou L.J., Zhang M.X., Wang W., Jia W.R., Zhang X.Q. Stability mechanism of the excavation face for shield tunneling in soft and hard composite ground [J]. *Journal of Shanghai University* (Natural Science Edition), 2021, 27(06): 1094–1105.

Zhu J.M., Lin Q.T., Kang Y. Research on dip angle of sliding surface of wedge model in shield tunneling [J]. *Rock and Soil Mechanics*, 2015, 36(S2): 327–332.

*Frontiers in Civil and Hydraulic Engineering – Mohamed A. Ismail and
Hazem Samih Mohamed (Eds)*
© 2023 The Authors, ISBN 978-1-032-38247-0

Experimental study on oil spill adsorption on sandy boundary of river

Xiang Wang & Sichen Tong*
School of River and Ocean Engineering, Chongqing Jiaotong University, Chongqing, China

Changyao Li
CCCC Second Harbor Consultants Co., Ltd., Wuhan, China

Guoxian Huang
Chinese Research Academy of Environmental Sciences, Beijing, China

Pingfeng Jang
School of River and Ocean Engineering, Chongqing Jiaotong University, Chongqing, China

ABSTRACT: Oil spill accidents from ships in inland rivers often occur. In order to understand the adsorption characteristics of oil spills on the sandy beach boundaries of inland rivers, a generalized adsorption test was carried out in the flume. This paper firstly establishes the thickness-grayscale relation of oil spilled oil film based on the hydrostatic calibration test. By carrying out a generalized experiment on sandy river bank oil spill adsorption, the quantitative relationships among the oil spill adsorption amount, adsorption rate, and time per unit length of the sandy boundary were established. The results show that with the oil spill expansion and adsorption, the amount of oil spilled on the boundary of unit length has an exponential relationship with time, and the adsorption rate has a logarithmic relationship with time. The parameters in the fitting formula are analyzed. The research can provide a certain basis for oil spill expansion adsorption research and accident handling in inland waterways.

1   INTRODUCTION

In recent years, water resources and inland shipping have been rationally developed and utilized (You et al 2014). With the increase in the number of ships, the risk of oil spill accidents has become very prominent. Previous studies have mainly focused on marine oil spill accidents, but the research on inland waterways is relatively lacking. The adsorption module in the existing oil spill model is not perfect. In an experimental study, for instance, Zhao et al (2012) simply divided the oil spill adsorption into complete absorption and complete reflection when dealing with the adsorption of oil spills on the boundary; when establishing an oil spill model of a complex river network, Wang Peng et al (2021) considered the adsorption of the river bank by specifying the adsorption probability coefficient of oil particles. These processing methods only roughly analyze the adsorption of oil by the boundary and do not consider the maximum value of the medium on the bank. Factors such as adsorption amount and adsorption rate are quite different from the actual oil spill adsorption, so the research on oil spill adsorption on the river bank boundary needs to be carried out urgently. Based on the hydrostatic calibration experiment and the generalized experiment on sandy bank boundary oil spill adsorption. And the adsorption capacity formula and the adsorption rate decay formula of sandy bank boundary were obtained. This study provides some reference and theoretical references for exploring the oil spill adsorption mechanism and accident handling.

---

*Corresponding Author: 81441849@qq.com

DOI 10.1201/9781003344209-19

2 TEST LAYOUT AND SCHEME

2.1 *Test arrangement*

This experiment was carried out in the National Inland Waterway Improvement Engineering Technology Research Center of Chongqing Jiaotong University. A thickened beef tendon plastic basin with a size of 800⊃37003280 mm was selected as the test container and placed in a horizontal tank and the generalized sand boundary adsorption test layout is shown in Figure 1. A shadowless light strip is arranged above the sink to provide a stable light source. The whole process of the test was recorded by a high-definition camera, and the relevant physical quantities of oil spill adsorption were obtained through image processing and data analysis in the later stage (Xiao et al 2020).

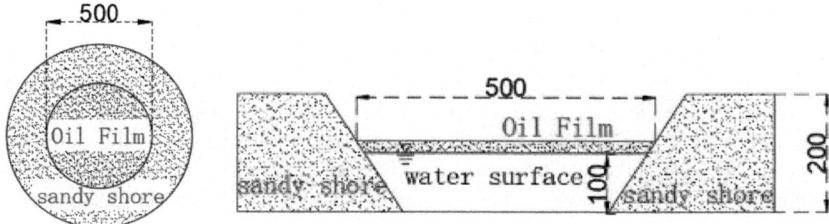

Figure 1.   Generalized experiment on sand boundary oil spill adsorption.

2.2 *Oil selection*

After the oil spill occurs on the water surface, it will spread around under the combined action of surface tension, inertial force, gravity, and viscous force. The viscosity of the oil spill has a significant impact on the expansion whether it is the initial stage of gravity expansion or the subsequent shear expansion stage (Liu et al 2006). Lubricating oil is a common oil in oil spill accidents. In this experiment, a large number of pre-experiments were carried out when selecting oil products, and finally, No. 100 hydraulic oil was selected as the test oil. The oil parameters are shown in Table 1.

Table 1.   Oil parameters for the generalized experiment on oil spill adsorption.

Oil and properties	Density (g/cm^3)	Kinematic viscosity (mm^2/s)
No. 100 hydraulic oil	0.82	99.40

In order to accurately observe the grayscale change of the oil film, oil-soluble nigrosine is used to colorize the lubricating oil. In this test, the configuration of the test oil is carried out according to the ratio of 200:1. After measuring the configured lubricating oil, the measured kinematic viscosity is 99.91 mm^2/s and the density is 0.81 g/cm^3, indicating that the coloring has little effect on the properties of the oil.

2.3 *Sandy boundary selection*

In this experiment, natural river sand with a median particle size of 0.66 mm was used to construct the river bank boundary for the oil spill adsorption experiment. The sediment has a large specific surface area. The gradation composition of sand particles is shown in Figure 2.

2.4 Experimental content

2.4.1 Calibration experiment of oil film grayscale-thickness relationship

In this experiment, a fixed-point one-time oil spill is carried out in a beaker. During the expansion process of the oil film on the free water surface, the expansion speed is constantly attenuated (Bymiltonvandyke E et al. 1970), its area and gray level are also constantly changing, and the thickness of the oil film at the end of the expansion is about 0.01 cm (Lehr W et al 2022). The amount of oil dripping is determined by weighing the difference in the quality of the beaker before and after oil dripping, and the oil density ρ is known. At this time, the instantaneous thickness of the oil film can be determined by Formula (1).

$$h_t = \frac{m}{\rho \times A_t} \tag{1}$$

where h_t = instantaneous oil film thickness (cm), m = oil drop determined by mass difference method (g), ρ= oil density (g/cm^3), A_t = instantaneous area of oil film at a certain time (cm^2).

In order to determine the area of the oil film, ImageJ software is used to convert the collected image into a grayscale image. Luo Hao et al (2021) used ImageJ software to analyze the microstructure of loess. By measuring the oil film area under different gray levels, the average thickness of the oil film is obtained, and the relationship between the oil film gray level and thickness is established.

2.4.2 Oil spill adsorption test on sandy river bank boundary

In order to study the adsorption characteristics of oil spills on sandy river bank boundaries, this experiment intends to carry out adsorption experiments on four different oil spill volumes under the annular sandy boundary with a thickness of 10 cm. The diffusion and adsorption process of the oil film on the static water sandy boundary under different oil spill volumes is collected by the camera, and the area and the gray level of the oil film are calculated by ImageJ software, and then the corresponding oil film is calculated by the oil film gray level-thickness formula obtained from the calibration experiment. Subsequently, the adsorption amount and adsorption rate of the boundary are determined, and their relationships with time are analyzed.

3 TEST RESULTS

3.1 Calibration test of oil film grayscale-thickness relationship

The grayscale-thickness relationship of the oil film was calibrated according to the variation range of oil film thickness between 0.07-0.18 cm, and the relationship formula was fitted to obtain the empirical Formula (2).

$$h = -0.015x + 2.3 \tag{2}$$

where x = oil film grayscale, h = oil film thickness (cm).

It has been verified that the deviation between the experimental value and the calculated value is within 5%, and the oil film thickness and grayscale have a good linear relationship with a correlation coefficient of 0.9986 (Figure 3).

3.2 Oil spill adsorption test on sandy river bank boundary

In this experiment, the oil spill amount was used as the variable. In order to study the oil spill adsorption on the sandy boundary under different oil spill amounts, four different oil spill amounts were drawn up, namely 70 g, 90 g, 110 g, and 120 g respectively.

The camera recorded the diffusion and adsorption process of oil spilled on the sandy boundary at different times, as shown in Figure 4.

The image collected at a certain time interval was used to calculate the oil film area and gray level by ImageJ software, and then the average thickness of the oil film in each period was calculated by

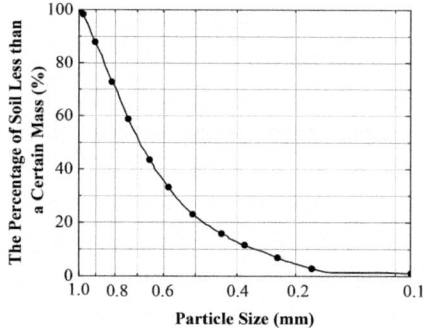

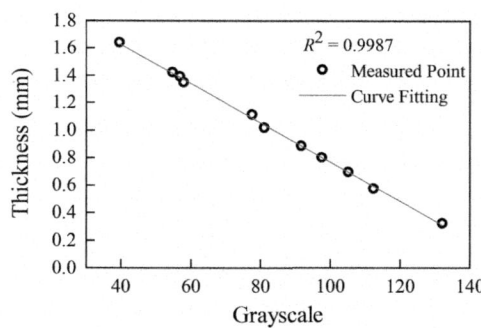

Figure 2.   Gradation curve of natural river sand.

Figure 3.   Calibration of gray-thickness relationship.

using Equation 2.1. From $m_a = m - Ah\rho g$, the oil spill volume m, oil film area A, average thickness h, and density ρ are known, and the oil spill adsorption amount m_a of the annular boundary at each time period can be obtained. Then, the variation relationship of the adsorption amount with time is obtained, and the variation relationship of the oil spill adsorption amount with time under each working condition is shown in Figure 5. The variation trend of the adsorption capacity of the sandy boundary under different oil spillages is consistent. With the advancement of the adsorption process, the adsorption capacity of the boundary presents an increasing trend.

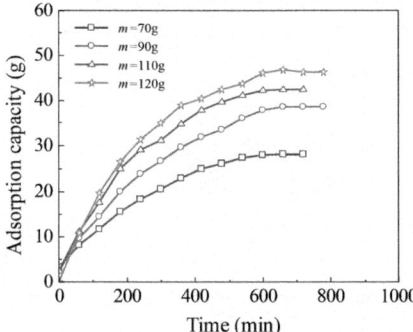

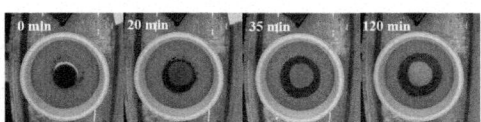

Figure 4.   Sand boundary oil film expansion and adsorption process.

Figure 5.   Variation of adsorption capacity with time under various working conditions on sandy boundary.

4   ANALYSIS OF TEST RESULTS

4.1   *Analysis of oil spill adsorption*

Since the thickness of the oil film in contact with the boundary is different under different oil spill masses, in order to consider the influence of the oil film thickness change on the adsorption capacity, the concept of generalized unit oil spill body is introduced (Figure 6). The mass M_0 of the generalized unit oil spill body is calculated by Formula (3).

$$M_0 = V_0\rho \tag{3}$$

$$V_0 = abd_t \tag{4}$$

147

where V_0 = generalized unit volume of the oil spill (m^3), d_t = oil film thickness (m), a = length of contact between unit oil body and river bank (1m), b = length of the unit oil spill body in the direction perpendicular to the river bank (1m).

After introducing the generalized unit oil spill body, the adsorption amount m_a of the annular river bank in the experiment is extended to the adsorption amount M_A of the unit river bank length, as shown in Equation (5).

$$M_A = \frac{m_a}{l_0} \times l_1 \tag{5}$$

where M_A = adsorption capacity per unit length of the river bank (kg), m_a = measured adsorption capacity on circular banks (kg), l_0 = perimeter of the circular bank in the experiment (m), l_1 = length of oil spill adsorption per unit (1m).

It can be seen from Figure 7 that with the progress of oil film diffusion and adsorption, the adsorption amount per unit length of boundary presents an increasing trend with time. According to the measured results, the relationship between the adsorption capacity per unit boundary length (M_A/M_0) of the sandy river bank and the time (t/t_0) can be fitted to obtain the Equation (6). The adsorption capacity per unit boundary length of sandy banks (M_A/M_0) has a negative exponential relationship with time (t/t_0), and the correlation coefficient is 0.9723, which is a good agreement. The comparison results are shown in Figure 7.

$$\frac{M_A}{M_0} = 0.107 \times \left(1 - e^{-\frac{1.37t}{t_0}}\right) \tag{6}$$

where M_A = adsorption capacity of 1 m long river bank (kg), M_0 = the mass of the generalized unit oil spill body at the corresponding time (kg), t = adsorption time (s), t_0 = time for adsorption to be stable being 600 min for 10 cm thick sandy boundary (s).

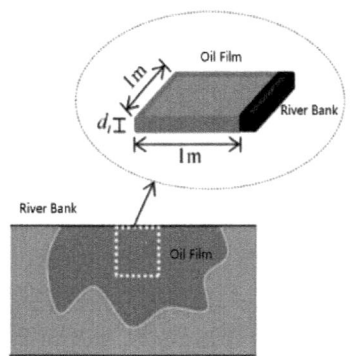

Figure 6. Schematic diagram of a generalized unit oil spill body.

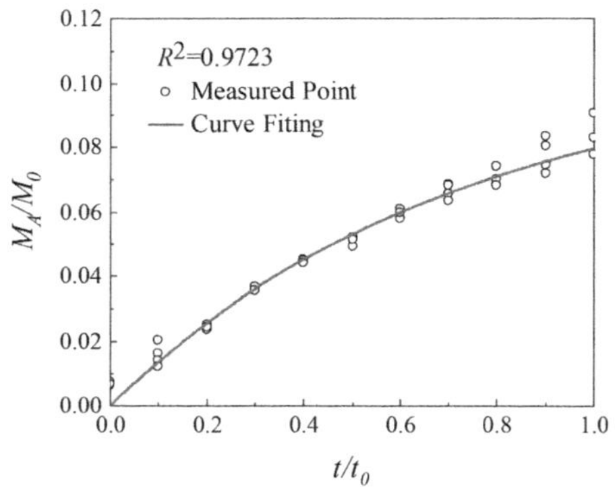

Figure 7. Comparison results of adsorption capacity formulas per unit length of sandy bank boundary.

4.2 Analysis of the change process of oil spill adsorption rate

The variation process of the adsorption rate with time is shown in Figure 8. By fitting the relationship between the oil spill adsorption rate and time under different working conditions, Formula (7) can be obtained, and it is found that the oil spill adsorption rate has a natural logarithmic relationship with time. The correlation coefficient R^2 is shown in Table 7, and R^2 is greater than 0.90, indicating that the measured value is in good agreement with the calculated value of the calibration formula.

$$y = aln(t) + b \tag{7}$$

where y = adsorption rate of the oil spill by a sandy boundary (g/min), t = adsorption time (min).

148

Since the oil spill adsorption rate is affected by the oil spill quality and boundary types, a and b are parameters related to the oil spill quality and boundary types, respectively. The values of a and b under different working conditions are shown in Table 2. From the parameter values in the table, it can be seen that the oil spill adsorption rate y has a natural logarithmic relationship with the time t, indicating that the initial adsorption rate is larger.

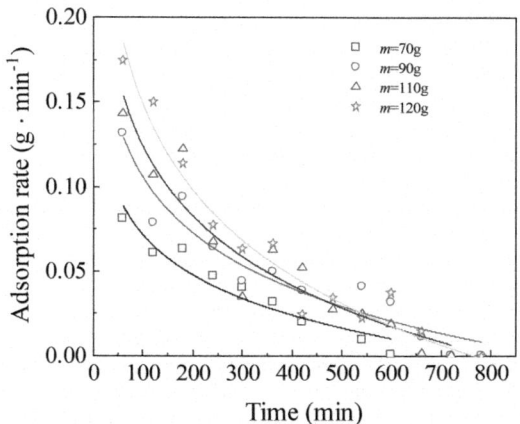

Figure 8. Variation of adsorption rate with time under various working conditions on sandy boundary.

Table 2. Relevant parameter values in the fitting formula of oil spill adsorption rate and time.

Serial Number	Working Condition	a	b	R^2
1	$m = 70$ g	−0.034	0.2280	0.92
2	$m = 90$ g	−0.047	0.3224	0.92
3	$m = 110$ g	−0.059	0.3957	0.90
4	$m = 120$ g	−0.073	0.4834	0.96

5 CONCLUSIONS

The sandy boundary is an important part of the inland river boundary. The generalized oil spill adsorption experiment was carried out with different oil spill amounts in the sandy boundary, and the rationality of the experiment was verified. The results of this test preliminarily reveal the characteristics and laws of the diffusion and adsorption of oil spills under the sandy boundary, which can provide a basis for the calibration and verification of oil spill adsorption coefficients in oil spill numerical simulations. Through the generalized sandy boundary oil spill adsorption experiment, the relationship between the adsorption amount and the oil spill quality and the change process of the adsorption rate with time are studied, and the following conclusions are obtained:

(1) According to the oil spill still water calibration test, the relationship between oil film grayscale and thickness can be established. It is found that the oil film grayscale and thickness have a linear relationship and a good correlation.

(2) According to the experimental results of the oil spill adsorption model on the generalized sandy boundary, the quantitative relationship between the oil spill adsorption amount per unit boundary length (M_A/M_0) and the time (t/t_0) is preliminarily established, that is, the adsorption

process expands with the oil spill. With the development of the oil spill, the adsorption per unit length of sandy boundary (M_A/M_0) shows an increasing trend, and has a negative exponential relationship with time (t/t_0).

(3) According to the experimental results of the oil spill adsorption model on the generalized sandy boundary, the quantitative relationship between the adsorption rate and time is established, that is, the larger the oil spill volume, the larger the initial adsorption rate and the slower the adsorption rate decay. The oil spill adsorption rate shows a decreasing trend and has a natural logarithmic relationship with time.

REFERENCES

Bymiltonvandyke E., et al. (1970). *Annual review of fluid mechanics: v.2. M. Annual review of fluid mechanics: v.2.*

Lehr W., Jones R., Evans M., et al. (2022). Revisions of the ADIOS oil spill model. *J. Environmental Modelling & Software*, 17(2), 189–197.

Liu, D., Lin, W.Q., Zhong, B.C., et al. (2006). Characteristic experiment of spread and transport of oil spill in tidal river. *J. Chinese Journal and Hydrodynamics*, (06), 744–751.

Luo, H., Wu, F.Q., Chang, J.Y., et al. (2021). Pore characteristics of Malan loess—A case study at Zhaojia'an landslide. *J. Journal of Engineering Geology*, 29(05), 1366–1372.

Wang, P., Shen, X., Wang, C.H., et al. (2021). Development of the oil particle model for complex river network and numerical simulation of oil spill pollution. *J. Journal of Hydraulic Engineering*, 13(12), 1–8.

Xiao, H., Yang, Z.W. (2020). 2D flume experimental study on the movement characteristics of oil spill from pipeline under water. *J. Journal of Waterway and Harbor*, 41(5), 531–534.

You, S.B., Ren, J.R., Peng, D.F. (2014). Research on the Interaction Impact of Yangtze River Shipping and Basin Economy. *J. Regional Economic Review*, 04, 64–70.

Zhao, Y.X., Wang, Y.G., Zhang, W.S., et al. (2012). Two-dimensional numerical simulation model for oil spill in river. *J. Yangtze River*, 43(15), 81–84.

*Frontiers in Civil and Hydraulic Engineering – Mohamed A. Ismail and
Hazem Samih Mohamed (Eds)
© 2023 The Authors, ISBN 978-1-032-38247-0*

Stability analysis of dam foundation of extra-large hydropower station under complicated geological conditions—Taking Xiangjiaba hydropower station as an example

Bin Zhang*

Changjiang Spatial Information Technology Engineering Co., Ltd, Hubei Water Conservancy Information Perception and Big Data Engineering Technology Research Center, China

Shuangping Li*

Changjiang Spatial Information Technology Engineering Co., Ltd, Hubei Water Conservancy Information Perception and Big Data Engineering Technology Research Center, School of Geodesy and Geomatics Wuhan University, China

Yonghua Li*

Changjiang Spatial Information Technology Engineering Co., Ltd, Hubei Water Conservancy Information Perception and Big Data Engineering Technology Research Center, China

ABSTRACT: The Xiangjiaba Hydropower Station dam site is located in the NEE direction of the eastern part of the shortest anticlinal part of Tang Fang Wan. There are NW-shaped vertical coal knee flexures, more weak intercalations, compression zones, small faults, and joint fracture structures, the geological structure at the dam site is very strange and complex, with more unfavorable anti-sliding stability and seepage stability of the dam foundation. Based on these unfavorable factors, a series of engineering measures have been adopted to ensure the stability of the dam foundation. In this paper, based on the deformation and seepage monitoring data of Xiangjiaba Hydropower Station, the influence of the water level rise on the dam's foundation in terms of the settlement, seepage, and slide-resistant stability after all previous impoundments is analyzed. The results show that the deformation and seepage of the dam's foundation are within normal limits, and the uplift pressure generally meets the requirements of the code. The practice has proved that under complex geological conditions, effective engineering measures can ensure the stable operation of the dam foundation.

1 INTRODUCTION

The Xiangjiaba Hydropower Station is the last stage planned for the lower reaches of the Jinsha River. It is located at the junction between Yibin City in Sichuan province and Shuifu City in Yunnan Province, which is 33 km away from Yibin City and 1.5 km away from Shuifu City. The development task of the project is mainly to generate electricity and improve shipping conditions at the same time considering flood control, irrigation, sands blocking, and reverse regulation of Xiluodu Hydropower station. The project is a first-class large-scale project, consisting of a dam, plant, ship lift, and other buildings, and the dam is a concrete gravity dam. From the left to the right, the dam retaining structure is composed of the left bank non-overflow dam section, sand flushing hole dam section, ship lift dam section, left bank powerhouse dam section, discharge dam section, and right bank non-overflow dam section. The elevation at the top of the dam is 384.00 m,

*Corresponding Authors: cjwjcgs_zb@126.com; lishuangping@cjwsjy.com.cn and 1051158892@qq.com

the maximum dam height is 162 m, and the length at the top of the dam is 892.26 m. The power plant is located underground on the right bank and behind the left bank dam, and each is equipped with 4 machines. The capacity of each machine is 800 MW, and the total capacity of the machine is 6400 MW.

Currently, the conventional storage water level of Xiangjiaba Hydropower station is 380 m, which has gone through three stages to reach the conventional storage water level. The first water storage period was from October 10, 2012, to October 16, 2012, with the water level rising from 280 m to 354 m. The second water storage started on June 26, 2013, and ended on July 5, 2013, with the water level increasing from 354 m to 370 m. The third water storage period was from September 7, 2013, to September 12, 2013, and the water level reached 370 m to 380 m.

2 INTRODUCTION TO ENGINEERING GEOLOGY

The overall shape of the bed bedrock at the dam site of Xiangjiaba Hydropower station is high in the lower reaches and low in the upper reaches, and high in the middle and low reaches on both sides. That is, the bed bedrock faces a slight dip upstream, and there are coherent grooves on both sides. The bedrock of the dam site is mainly sandstone and mudstone interbedded with coal lines, which are deposited in the river, lake, and swamp facies of the upper Triassic Xujiahe Formation, and the lithology and lithofacies change greatly (Peng et al. 2012). And the main structural planes include flexural core fracture zone, compression zone, soft interlayer, small fault, and joint fissure. The site conditions of the dam foundation are complicated, and there are major technical problems such as deep anti-sliding, dam deformation, and foundation leakage (He et al. 2015).

According to the analysis of specialized exploration drilling holes, the fractured rock mass at the flexural core can form a fractured rock belt in space from 0–65 above the dam to 0+132 below the dam. The core samples of drilling holes are mostly fragmentary, with a few short columns, generally with more than one section of clastic sandstone and mudstone. The top and bottom interfaces of the broken rock belt are irregular and undulating so the thickness of the broken rock belt varies greatly. The vertical thickness mostly ranges from 10 m to 602 m, and the distribution range and thickness are large. The quality of the rock mass is poor, and even the distribution of multi-layer sandstone and mudstone with clastic structures has a negative impact on the stability of the dam foundation and the stability of the slope excavated by the tooth groove (Zeng 2011).

3 INTRODUCTION OF GEOLOGICAL DEFECT TREATMENT AND SEEPAGE CONTROL CONSTRUCTION OF DAM FOUNDATION

3.1 *Geological defect treatment of dam foundation*

Bad geological rock masses such as extrusion zone, soft rock interlayer, broken zone, weathered interlayer, and cystic inclusion may produce uneven deformation and may cause stress concentration. In order to improve the integrity, bearing capacity, and surface impermeability of rock masses, it is necessary to excavate and replace such geological defects. The section type and depth of excavation and displacement treatment are determined according to the location of unfavorable geological bodies and foundation planes.

The dam foundation of the discharge dam section and the workshop dam section is excavated around the above geological defects. Finally, grooves are set in the shipping dam section, the workshop dam section, and the Spillway 1–3 dam section. And the elevation of the bottom of the tooth groove is from 203.00 m to 215.00 m, which gradually climbs to the left bank. The Spillway 4–7 dam section is provided with diagonal troughs across the dam foundation, and the elevation at the trough bottom is 203.00 m. The geological condition of the foundation of the Spillway 8–13 dam section is satisfactory, and the excavation and displacement are mainly carried out in the dampening pool, and the elevation at the displacement concrete bottom is 210.00 m. In the process of excavation, the downstream area of the dam section from Spillway 9 to Spillway 11 has been further excavated to an elevation of 203.00 m according to the revealed geological conditions (Du 2007; Pan et al. 2017; Yu et al. 2010; Zhang et al. 2015).

3.2 *Foundation seepage control project*

The seepage control scheme of Xiangjiaba Hydropower Station dam foundation is implemented by combining the grouting curtain with the drainage curtain. Specifically, the seepage control design mainly considers three requirements: Firstly, it is necessary to reduce the leakage around the dam foundation and prevent it from adversely affecting the stability of the dam foundation and the slope on both sides. Secondly, it is necessary to prevent weak interlayer and structural fracture from producing seepage damages under the seepage action. Finally, the seepage pressure of the dam foundation needs to be controlled within an allowable range (Fan et al. 2015; Feng et al. 2017).

For the first phase of the project, the crushed zone of the anti-seepage line of the foundation of the riverbed section has been dug out through open digging, and the crushed zone of the foundation of some sections has been replaced by concrete through the excavation of the cave. In the second phase of the project, the extrusion and flexural fracture zones of the impermeable line upstream of the dam foundation have been dug out by opening digging. The impervious wall is adopted as the main basic scheme in the center of the riverbed and in other parts of the dam foundation such as the compressional fracture zone and flexural fracture zone, which are partially overlying and shallowly buried and treated by combining the conventional cement grouting with chemical grouting.

4 DAM FOUNDATION DEFORMATION AND SEEPAGE MONITORING ARRANGEMENT

4.1 *Deformation monitoring of dam foundation*

The dam foundation deformation monitoring mainly includes horizontal displacement monitoring and vertical displacement monitoring. In addition, the inclination monitoring of the dam foundation can also be carried out by combining vertical displacement measuring points in upstream and downstream directions.

Horizontal displacement monitoring of dam foundation is mainly done by combining inverted vertical with wire alignment. A total of 25 inverted vertical lines are arranged in the dam retaining line from the left bank to the right bank, respectively in the left bank 14, left bank 7, left bank 1, hang 1 left, hang 1 right, Powerhouse 8, Powerhouse 6, Powerhouse 4, Spillway 1, Spillway 4, Spillway 6, Spillway 10, and Spillway 13 dam sections. In order to monitor the anti-sliding stability of foundations at different depths, Powerhouse 8, Powerhouse 6, Powerhouse 4, Spillway 4, Spillway 6, and Spillway 10 are all inverted groups. The wire alignment of the foundation part is arranged in Powerhouse 8 to Spillway 1 dam section of the corridor factory at elevation 243, and a wire alignment measuring point is arranged in each dam section.

The vertical displacement of the dam foundation is mainly monitored by precision leveling. Forty leveling points are established at the 210 m elevation of the deep groove in the foundation corridor, which are arranged in the front groove of the 8# dam section of the left plant to the rear groove of the Spillway 10 dam station, and seventy-nine leveling points are arranged in the foundation corridor with an elevation of 243m, which are arranged in each dam section of each row of the longitudinal corridor from the 8# dam section of the left plant to the 2# dam section of the right plant. Considering that the horizontal observation results of the foundation corridor at an elevation of 243 m are relatively representative, the vertical displacement analysis in this paper mainly focuses on the observation results at the horizontal points of this part.

4.2 *Seepage and seepage pressure monitoring of dam foundation*

The number of piezometric tubes installed in the dam foundation corridor is 219. A set of piezometric tubes are respectively arranged in each dam section of the upstream curtain grouting corridor, each dam section of the downstream curtain grouting corridor, each dam section at the 210 m elevation of the main drainage corridor, and each dam section in the first and second rows of the auxiliary drainage corridor.

Xiangjiaba drainage system is composed of two drainage systems: The drainage system of the dam foundation and the drainage system of stilling pool. The dam foundation drainage system includes: right non-dam section drainage corridor, 210 m drainage corridor and 238 m–245m drainage corridor in the discharge dam section, 210 m drainage corridor and 243 m–245 m drainage corridor in the workshop dam section, 226 m drainage corridor in the workshop behind the dam, and left non-dam section drainage corridor in the first stage of the left bank. The drainage system of the stilling basin consists of the drainage gallery of the left basin, the drainage gallery of the right basin, the drainage gallery of the middle guide wall, and the drainage gallery of the tail dam.

5 STABILITY ANALYSIS AND EVALUATION OF DAM FOUNDATION

5.1 *Deformation stability analysis of dam foundation*

5.1.1 *Horizontal displacement of dam foundation*
The horizontal displacement of the dam foundation is mainly based on the observation results of the inverted vertical of the foundation, and the measured results show that the downstream deformation of the spillway dam section is large after the water storage, which is ascribed to the knee-shaped flexural core of Limewan that is inclined to leak through the dam section, and the main geological structures include flexural core fracture zone, left bank compression zone, 11 weak interlayers, and 1 small fault. The flexural core fracture zone flows through the dam section diagonally and extends to the dampening pool.

From the monitoring results, the deformation of the Spillway 4, Spillway 6, and Spillway 10 dam sections located in the riverbed is relatively large, all exceeding 7 mm. In particular, the displacement of the measuring point on the inverted vertical line downstream of the Spillway 6 dam section is the largest. The position is arranged with two inverted verticals to form the inverted group, with a deep hole of 90 m and a shallow hole of 30 m, wherein, the maximum displacement of the deep hole to the downstream is 9.87 mm while the maximum displacement of the shallow hole to the downstream is 0.6 mm, indicating that the deformation mainly occurs in the deep flexural core fracture zone with poor geological conditions, and the displacement of the rest of the dam is less than 7 mm downstream.

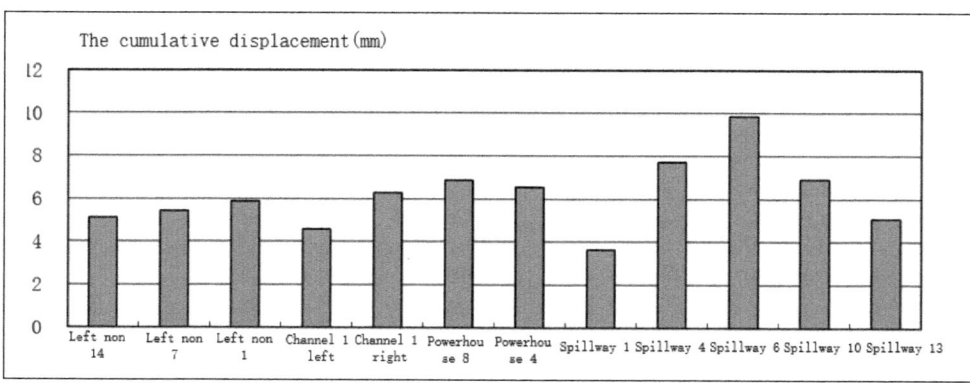

Figure 1. Cumulative displacement distribution diagram of foundation inverted vertical in upstream and downstream directions.

From the analysis of the displacement change process, the displacement of the Spillway 6 dam section mainly occurs during the period of 354 m water storage, and the displacement of the measuring point downstream is 6.63 mm during the upstream reservoir water level changes from 280 m to 354 m, and before 354 m to 370 m (from October 10, 2012, to June 20, 2013), and the downstream displacements of 370.0 m and 380.0 m are 1.07 mm and 1.54 mm, respectively. When the upstream reservoir water level reaches 380.0 m, the deformation of the dam section tends to be stable, and the process line of the measuring point is shown in Figure 2. In other dam sections, the displacement of the foundation in upstream and downstream directions is similar, and the displacement of the foundation mainly occurs during the first impinging period and tends to slow down during the impinging period of 370 m and 380 m.

It can be seen that channeling and replacing the damaged rock mass of the dam foundation dam site and the downstream resistance body with concrete can not only improve the comprehensive strength of the sliding shear surface under the double-sliding-surface mode but also greatly improve the strength and stiffness of the resistance body under the single-sliding-surface mode.

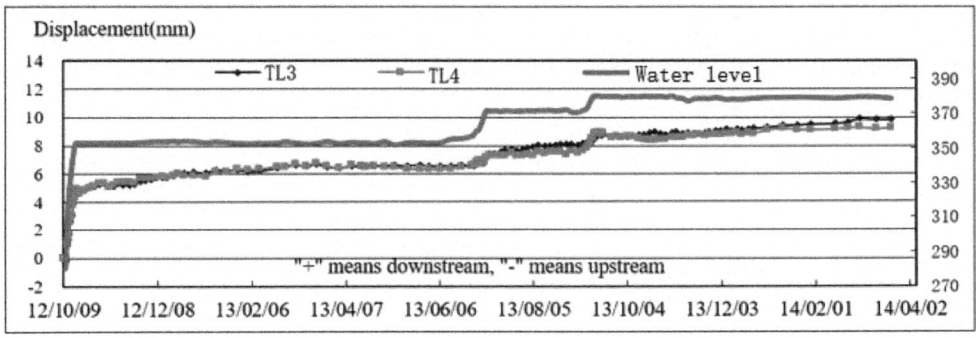

Figure 2. Displacement process diagram of upstream and downstream direction of measuring point on inverted vertical of spillway 6 dam section.

5.1.2 *Vertical displacement of dam foundation*
Considering that the 210 m corridor is only equipped with leveling points at the tooth slot, and the 243 m foundation corridor is equipped with leveling points at all the longitudinal corridors, this paper mainly analyzes the observation results of the vertical displacement of the 243 m foundation corridor.

The measured cumulative vertical displacement of measuring points in each dam section is between 10.81 mm and 36.91 mm, and all dam sections show a trend of settlement deformation. The maximum settlement displacement occurs in the Powerhouse 2 dam section, the vertical displacement increases gradually from the bank slope to the river bed, the uneven settlement of adjacent dam sections is small, and the settlement is mainly affected by the increase in the dam pouring load. The dam section of Powerhouse 2 is located in the middle of the riverbed. The main structural planes distributed in the dam section of the whole plant are the left bank compression zone, 10 weak interlayers, and 2 small faults, and the dam base strata are $T3^{2-4} - T3^{2-6-2}$, mainly being sandstone, which accounts for more than 90%.

After water impinging at an elevation of 354 m (September 2012–June 2013), the vertical displacement of the dam heel varies from 0.21 mm to 2.01 mm, and the vertical displacement of the dam toe varies from -1.90 mm to -0.07 mm, and the dam heel is slightly subsided and the dam toe is slightly raised, both of which have small magnitudes. After impinging at 370 m (June 2013–March 2014), the vertical displacement of the dam heel varies from 2.27 mm to 5.65 mm, and the vertical displacement of the dam toe varies from 5.82 mm to 9.54 mm, both of which show a trend of settlement deformation. Due to the thrust of the reservoir pressure on the upstream surface of the dam, the dam foundation has a slight tendency of inclining to the downstream, that is, the settlement at the toe of the dam is greater than that at the heel of the dam, which is consistent with the physical deformation characteristics of the concrete dam. In general, in each impoundment, the deformation regularity of the dam foundation is good. The first impoundment is mainly manifested as the initial increase of load, and the dead weight of water exerts pressure on the reservoir area, leading to the settlement and slight uplift of the dam toe. After impoundment of 370 and 380 m, due to the high water level, the dead weight of water not only exerts pressure on the reservoir area, but also produces a thrust on the upper part of the dam body, resulting in an increase in the compressive stress of the dam toe, and its settlement is slightly greater than that at the dam toe. After impoundments, the uneven settlement between the adjacent dam sections is less than 1 mm, and there is almost no uneven settlement. The displacement distribution of the dam heel and toe after each impoundment is shown in Figures 3 and 4.

According to the above data analysis, the vertical displacement of the dam foundation mainly occurs before the first impoundment, which is ascribed to the deformation caused by the dead weight of the concrete dam, and there is no abnormal deformation of the dam foundation after impoundments. It is beneficial to the settlement stability of the dam foundation by excavating large grooves at the dam heel and truncating the left bank extrusion zone and the weak interlayer with a shallow burial depth, which strengthens the integrity and consistency of the foundation settlement deformation.

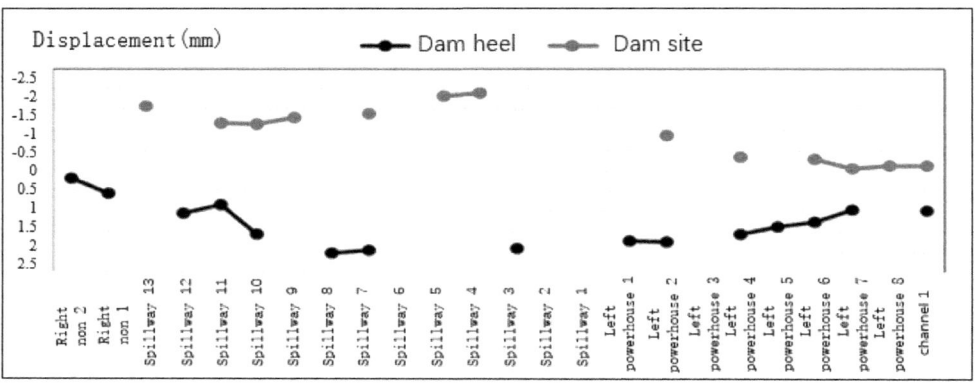

Figure 3.   Distribution of vertical displacement of dam heel and dam site after 354 m impoundment.

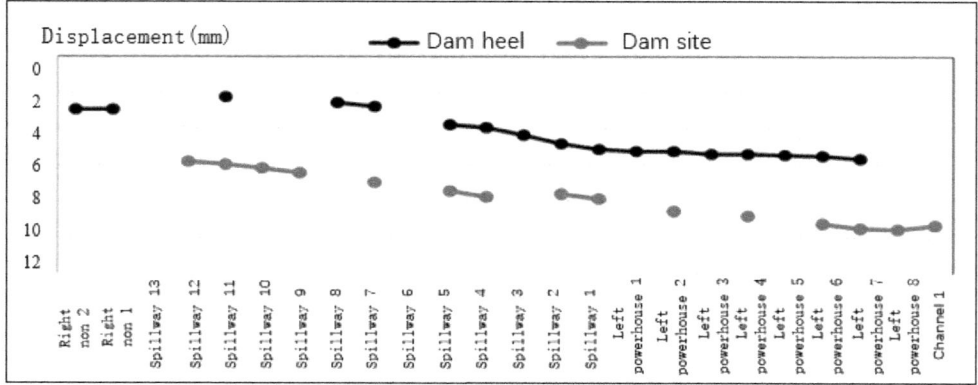

Figure 4.   Distribution of vertical displacement of dam heel and dam site after 370 m impoundment.

5.2 Seepage stability analysis of dam foundation

5.2.1 Leakage monitoring of dam foundation

In the process of 354 m water storage, the total number of water outlets and drainage holes change from 742 to 861 with an increment of 119. The flood spillway section and the first phase of the left bank increase more greatly by 107 and 32, respectively. The total water output from the drainage hole increases from 13.8 m^3/min to 17.66 m^3/min, with an increment of 3.86 m^3/min and an increased rate of about 27.9%. In the first phase of the left bank, the total water output increases from 2.64 m^3/min to 5.31 m^3/min, with an increment of 2.67 m^3/min and an increased rate of about 101.2%. For the dam section of the workshop, it increases from 0.91 m^3/min to 1.45 m^3/min, with an increment of 0.54 m^3/min, and an increased rate of about 58.8%. After water storage, the number and displacement of drainage holes are still increased. With an elevation of 243 m as the base plane, considering that the measured lift pressure reduction coefficient is relatively rich, the water output of the drainage hole has been adjusted three times. After the adjustment, the outlet location and the water yield from a single hole tend to be dispersed and balanced, and the uplift pressure of the dam foundation is within the allowable range of the code.

Before the impoundment to 370 m, some drainage holes with muddy water are sealed. After the water level is raised, the total number of drainage holes in the process of water storage changes from 642 to 704, with an increment of 62, and the water output of drainage holes increases from 6.85 m^3/min to 7.43 m^3/min, with an increment of 587.3 L/min and an increased rate of about 8.58%. Among them, the increment in the sand flushing 1 dam section to the left non-overflow 5 dam section is the largest, being 298.7 L/min, which accounts for 50.8% of the increase in the displacement. After the impoundment, the drainage holes of the dam foundation are adjusted for the fourth time, and all the drainage holes with a single hole displacement greater than 40 L/min are adjusted to 40 L/min. In this adjustment, a total of 20 drainage holes are adjusted. After the adjustment, the number of water outlet holes increases from 704 to 722, and the water output from water outlet holes increases from 7.43 m^3/min to 7.7 m^3/min. The water outlet location and single holes tend to be dispersed and balanced, and slightly increased.

In the process of 380 m water storage, after the water level is raised, the total number of outlet and drainage holes changes from 679 to 686, with an increment of 7 drainage holes. The water output from the drainage hole increases from 7.22 m^3/min to 8.00 m^3/min, with an increment of 780.3 L/min and an increased rate of about 10.81%. Among them, the increase of 423.3 L/min from the sand flushing 1 dam section to the left non-overflow 5 dam section is the largest, accounting for 54.2% of the displacement increment.

According to the above data analysis, the increment of the drainage hole in the dam foundation during the first impoundment period is large, and the water output of the drainage hole in the subsequent impoundment dam foundation is decentralized, balanced and gradually decreased until it is stable, which is mainly attributed to two aspects: First, the sediment deposition of the reservoir after impoundments exceeds the cover weight of the dam site, which compresses or even blocks the seepage channel. Second, the controllable regulation of the displacement can effectively prevent water from gushing and carrying sand, so that the seepage channel gradually decreases with the silt deposition.

5.2.2 Pressure monitoring of dam foundation

(1) Uplift pressure of upstream curtain grouting corridor: At present, a total of 18 piezometric tubes (UP1-21–UP1-36) are arranged in the main curtain grouting corridor, and the actual measured converted water level ranges from 209.471 m to 275.8 m. A total of 8 piezometric tubes (UP1-37–UP1-44) are arranged in the 245# elevation curtain grouting corridor, and the actual measured converted water level is 242.751 m–259.894m, and the current converted water level of the curtain grouting corridor (bank slope dam section) of the main body on the left bank is 224.79 m–367.06 m. Figure 5 shows the distribution of osmotic pressure and converted water level of the pressure gauge in each dam section of the upstream curtain grouting corridor. Generally speaking, the seepage pressure in the right dam 0+049.00 (Spillway 3 dam section) to the right dam 0+125.00

Table 1. Statistical table of displacement changes of water storage dam foundation and drainage hole of stilling pool.

Part	354 m water storage		370 m water storage		380 m water storage	
	Water output increment (L/min)	Increase amplitude (%)	Water output increment (L/min)	Increase amplitude (%)	Water output increment (L/min)	Increase amplitude (%)
Phase I dam on the right bank	2586.5	117.29	548.5	16.68	390.60	11.34
Phase II dam on the right bank	1028.04	22.93	158.1	6.14	58.6	1.97
Dam foundation subtotal	3614.54	54.04	706.6	12.05	449.2	7.00
Stilling pool	218.3	3.27	377.5	35.99	50	4.51
Summation	3832.84	28.67	1084.1	15.69	499.2	6.63

(Spillway 7 dam section) is larger, the seepage pressure in other dam sections is smaller, and the converted water level of the piezometric tube is basically located in the corridor floor elevation. Among them, the measured converted water level of Powerhouse 4 dam section is relatively high before the storage of 380 m, and the actual water level is 300.4 m, indicating that there may be a micro-seepage channel. In order to verify the actual seepage situation of this part, two pressure measuring pipes UP1-26-1 and UP1-26-2 are drilled in the range of about 1.5 m. After drilling, the converted water level of UP1-26 pressure measuring pipe drops to 261.512 m, and the current converted water level of the newly installed pressure measuring pipe is 257.039 m and 223.971 m, respectively, with a large difference in the converted water level. The high converted water level of UP1-26 can only reflect the local situation. With the blockage of the micro-seepage channel, the overall uplift pressure in this part decreases significantly and tends to be stable.

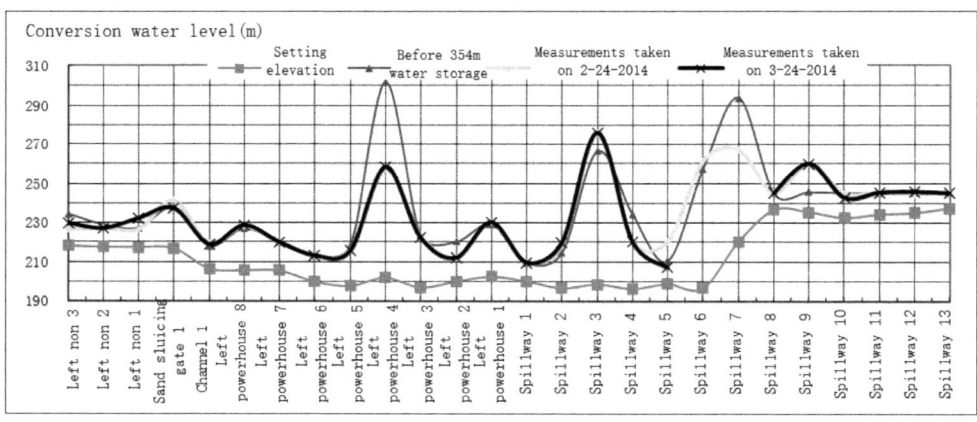

Figure 5. Distribution diagram of seepage pressure and reduced water level of the pressure gauge in each dam section of the upstream curtain grouting corridor.

Table 2 shows the observed seepage pressure value and uplift pressure reduction coefficient of the dam foundation corridor in late March, 2014. It can be seen from the table that the reduction coefficient of uplift pressure at the dam foundation surface (calculated based on 240 m elevation) is between 0.000 and 0.259. Among them, only the reduction coefficient of the Spillway 3 dam section exceeds the allowable value of 0.2, and the specific reduction coefficient of uplift pressure is 0.259. By observing the water yield of the piezometric tubes, the water yield is very small, being about 2 drops/second. The calculation results of the whole uplift pressure in the dam section show that the large uplift pressure reduction coefficient is only a local phenomenon, and the rest of the uplift pressure reduction coefficient is within the allowable range of the code. The measured uplift pressure on the dam foundation and the deeper part is less than that calculated according to the code, thus meeting the requirements of the code.

Table 2. Calculation table of observed seepage pressure value and uplift pressure reduction coefficient of dam foundation corridor.

Dam section	Instrument number	Water level in front of dam/m	Elevation of foundation surface/m	Converted water level/m	Measured lift pressure reduction coefficient ($\alpha 1$)
Sand sluicing gate 1	up1-18	377.9	222	237.3	0.098
Left non 3	up1-15	377.9	222	229.7	0.049
Left powerhouse 8	up1-22	377.9	215	228.7	0.084
Left powerhouse 6	up1-24	377.9	240	213.1	0.000
Left powerhouse 4	up1-26	377.9	240	258.2	0.132
Left powerhouse 4	up1-26-1	377.9	240	256.4	0.118
Left powerhouse 4	up1-26-2	377.9	240	227.8	0.000
Left powerhouse 2	up1-28	377.9	240	212.0	0.000
Spillway 1	up1-30	377.9	240	209.7	0.000
Spillway 2	up1-31	377.9	240	219.8	0.000
Spillway 3	up1-33 new	377.9	240	275.8	0.259
Spillway 4	up1-34	377.9	240	220.1	0.000
Spillway 5	up1-35 new	377.9	240	220.2	0.000
Spillway 6	up1-36	377.9	240	261.4	0.155
Spillway 7	up1-37 repair	377.9	240	267.1	0.196
Spillway 8	up1-38	377.9	240	245.0	0.036
Spillway 10	up1-40	377.9	240	245.2	0.037
Spillway 12	up1-43	377.9	240	245.7	0.041
Right non 3	up1-46	377.9	240	245.1	0.037

The above data analysis shows that although there are many geological defects in the dam foundation and abundant underground water, concrete replacement is carried out after the excavation of the tooth groove at the dam heel, and the basic scheme of the impervious wall is adopted. For the compressional fracture zone and the flexural fracture zone covered and buried shallowly beneath the dam foundation, the combination of conventional cement grouting (such as wet grinding cement grouting) and chemical grouting can control the uplift pressure reduction coefficient of the dam foundation within the allowable range of the code, and the overall uplift pressure in the deep part of each dam section is much greater than the calculated value.

6 CONCLUSION

There are many risks in the construction of super-large hydropower stations under complicated geological conditions, and thus the anti-slip and seepage stability of the dam foundation are the most important. In this paper, according to the complex geological conditions of Xiangjiaba Hydropower Station, a series of engineering technical measures and means are designed, including concrete replacement of the dam foundation, curtain grouting in various ways, reasonable design of drainage curtain, and controllable dam foundation drainage system. The monitoring results show that the horizontal and vertical displacements of the dam foundation are slightly larger than that of other similar projects in the initial stage of impounding [the horizontal displacement of the dam foundation on the right bank of TGP is less than 3 mm and the vertical displacement is less than 15 mm (Wu et al. 2010)], but after the impoundment to the normal water level, the horizontal displacement and vertical displacement of the dam foundation are stable and have no obvious increasing trend.

The seepage pressure of the dam foundation is within the allowable range of the code, the displacement of the dam foundation changes steadily, and the working behavior of the dam is normal. Through the analysis of monitoring results, the stability of Xiangjiaba Hydropower Station is good after adopting these measures, such as cleaning measures along the bottom of the extrusion zone, concrete replacement measures of the dam heel tooth groove, dam foundation and damping pool, adjustment of the part of the curtain as the seepage wall, chemical grouting of the bad geological body deep in the dam foundation and the strengthening of other seepage control measures. There is no adverse sliding and abnormal seepage, and the dam and its foundation are safe and controllable.

REFERENCES

Du Y.H. *Determination of sliding stability boundary and parameter evaluation of complex rock mass foundation for high concrete gravity Dam—A case study of Xiangjiaba Hydropower Station* [D]. Chengdu University of Technology, 2007.

Fan K, Zou Y.S, Wang L.T. Design and dynamic optimization of seepage control system for Dam foundation of Xiangjiaba Hydropower Station [J]. *Yangtze River*, 2015(2):89–93.

Feng S.R, Jiang Z.M, Zhong H.Y, et al. *Journal of Hydraulic Engineering*, 2017, 48(1):21–30.

He L.H, Zhang Y.T, Zhou H.B, et al. Study on seismic safety of Xiangjiaba Hydropower Station dam [J]. *Yangtze River*, 2015(2):94–97.

Pan J.Y, Feng S.R, Zhang Y.T, et al. Deformation control and anti-seepage and anti-slip treatment of Xiangjiaba Hydropower Station dam foundation. *Hydroelectric Power*, 2017, 43(2):60–66.

Peng G., Zhang H.Q. Main technical characteristics, difficulties and countermeasures of Xiangjiaba Project [J]. *China Three Gorges*, 2012(11):50–56.

Wu X, Zhang W.S. Analysis on the deformation law of the dam on the right bank of the Three Gorges Project [J]. *Yangtze River*, 2010, 41(020):19–22.

Yu S, Chen Z.Y., Jia Z.X., et al. *Journal of China Institute of Water Resources and Hydropower Research*, 2010, 8(1):11–17.

Zeng X.X. *Study on mechanical Properties and Parameter Values of Dam Foundation rock of Xiangjiaba Hydropower Station* [D]. Chengdu University of Technology, 2011.

Zhang Y.T., Zeng X.X., Shi Y. Design of dam foundation treatment for Xiangjiaba Hydropower Station [J]. *Yangtze River*, 2015(2):76–80.

*Frontiers in Civil and Hydraulic Engineering – Mohamed A. Ismail and
Hazem Samih Mohamed (Eds)*
© 2023 The Authors, ISBN 978-1-032-38247-0

The influence of the size of the first step on the hydraulic characteristics of the combined energy dissipator

Jianqing Tang*
Yunnan College of Business Management, China

Jurui Yang*
Kunming University of Science and Technology, China

ABSTRACT: In high-head hydraulic structures, the energy dissipator combined with a wide tail pier, stepped chute, and stilling basin have good hydraulic characteristics. However, it still has serious problems such as cavitation and cavitation erosion. The size of the first step has an obvious influence on the hydraulic characteristics of the combined energy dissipator. Based on the Ahai Hydropower Station, six groups of first transition steps of different sizes are tested by the hydraulic model test. By comparing these hydraulic characteristics, it is found that the first transition step has a positive effect on the hydraulic characteristics of the dissipator, which is mainly reflected in the aerated area and the height-width ratio. The cavitation erosion on the step surface of the energy dissipator will be reduced by appropriately increasing the aerated area and the aspect ratio. The maximum time average pressure at different positions of the energy dissipater will also be appropriately reduced by appropriately increasing the aeration area and reducing the height-width ratio.

1 INTRODUCTION

In recent years, with the development of hydraulic structures such as wide tail piers and stepped overflow weirs, the joint energy dissipator combining the wide tail pier, stepped overflow dam and stilling basin has made remarkable progress. It has been widely used in discharge structures with high head and large unit width discharge (Hu 2006; Tian 2009; Wei 2012). Combined energy dissipation is widely used in the following hydropower stations, such as Guangxi Baise hydropower station, Yunnan Dachaoshan Hydropower Station, Fujian Shuidong hydropower station, and Guizhou Suofengying Hydropower Station. The energy dissipation mechanism, hydraulic characteristics, aeration characteristics, and the influencing factors of the aeration cavity of the combined energy dissipator have been tested and numerically simulated, and have been widely used in the actual engineering of water release structures in China (Liang 2009, Yin 2007). For Suofengying Hydropower Station, the negative pressure at the junction of the vertical surface of the first step and the overflow dam surface is determined by numerical simulation and physical tests (Zhang 2007). The numerical simulation method is used to analyze Ahai Hydropower Station, and the conclusion that the first step has the greatest influence on the energy dissipation effect of the stepped overflow dam is obtained (Wang 2016). Through the analysis of Ahai Hydropower Station through physical tests, it is found that under transition steps of three steps, the transition steps of different sizes have their own advantages in hydraulic characteristics (Dong 2018). However, the surface of the spillway of the actual hydropower station still has serious cavitation damage. For example, less than half a year after the operation of Ahai Hydropower Station, cavitation damage

*Corresponding Authors: 894360316@qq.com and yangjurui@kmust.edu.cn

occurred on the stepped pressure surface. When the unit width discharge of Shuidong Hydropower Station is 90 m²/s, the spillway also has slight damage. The main reason is that under large unit width discharge, the water depth on the stepped dam surface is too large, and the bottom lacks sufficient aeration conditions, resulting in excessive negative pressure on the stepped dam surface (Hu 2009; Tian 2012). Therefore, by reducing or eliminating the negative pressure on the stepped dam surface, aeration conditions are created for the bottom of the water tongue. By increasing the aeration cavity and aeration concentration, the cavitation degree of the overflow dam surface can be reduced. These are of great significance for joint energy dissipation to promote the discharge structure with high head and large unit width discharge.

In this paper, relying on the Ahai Hydropower Station and using the method of physical model test, the influences of the WES Weir surface curve section and the first-stage transition steps with different sizes on the negative pressure of the first stage step surface, the flow pattern in the stilling basin, the time average pressure of the stilling basin bottom plate and the energy dissipation rate are studied. The research results are of great significance to the development of combined energy dissipators.

2 TEST MODEL AND SCHEME

2.1 Model design

The maximum dam height of Ahai Hydropower Station is 138 m. At all levels of flood flow, the water level difference between the upstream and downstream of the dam is 62–110 m. Its maximum flood discharge power is 884.2 MW, which is a typical problem of flood discharge and energy dissipation with high head and large flow. Ahai Power Station is a five-hole outlet with a symmetrical structure. Due to the limitations of the experimental site, the overall model cannot be made. Therefore, the same length scale is used in three directions of space, and the model is completely geometrically similar to the prototype. The model adopts a normal semi-integral model with two holes discharging and a symmetrical structure. It is designed according to the gravity similarity criterion. According to the requirements of similarity criteria, combined with the site and water supply capacity, the scale of the model is taken as 1:60. The similar conditions shown in Table 1 are selected according to the gravity similarity criteria, and the continuous similar conditions of flow.

Table 1. The model of main scale relationship.

Type	Relationship	Model	Prototype
Geometric scale/ λ_L	λ_L	1	60.000
Flow scale / λ_Q	$\lambda_L^{2.5}$	1	27885.480
Velocity scale/ λ_V	$\lambda_L^{0.5}$	1	7.746
Pressure scale / λ_P	λ_L	1	60.000
Roughness scale /λ_n	$\lambda_L^{1/6}$	1	1.979
Time scale /λ_t	$\lambda_L^{0.5}$	1	7.746

2.2 Test method

19 measuring sections are set at the center line of the 1# surface hole of the model and the dam axis respectively. Here, four measuring points are set at 0+043.812 m–0+066.042 m, three measuring points are set at 0+069.924 m–0+097.800 m and 12 measuring points are set at 0+105.760 m–0+256.970 m. The negative pressure of the first step surface is measured by the CY200 digital pressure sensor. The measuring range of the instrument is 0–50 kPa, and the comprehensive accuracy is 1‰. The measuring points are located at No. 1, 2, and 3 measuring points on the facade of the first transition step (0+043.812) and No. 4, 5, and 6 measuring points on the plane, totaling 6 measuring points (see Figure 1). The flow pattern in the stilling basin is mainly represented

by hydraulic elements such as water surface profile and flow velocity. The water surface line is measured with a measuring needle, and the accuracy of the instrument is 0.1 mm. The location of the measuring points is 12 cross-sections with a chainage of 0+105.760 m–0+256.970 m. The flow velocity is mainly measured near the bottom. The pitot tube is used to measure the flow velocity near the bottom. The location of the measuring points is 12 cross-sections with a chainage of 0+105.760 m–0+256.970 m. The position of each measurement section of the bottom flow rate is about 12 cm away from the stilling basin bottom plate (the model is 0.2 cm). The time average pressure is mainly measured by the glass tube pressure measuring row with an inner diameter of 12 mm, and its accuracy is 1 mm. The location of the measuring points is 18 cross-sections with a chainage of 0+050.682 m–0+256.970 m. The layout of each measuring point is shown in Figure 1.

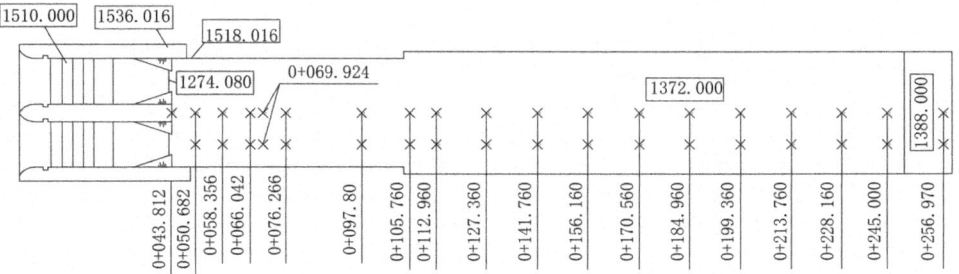

NOTE: THE ELEVATIONS IN THE DRAWING ARE ALL IN M; × INDICATES PRESSURE MEASURING POINT; STAKE NUMBER UNIT:M.

Figure 1. The plan of model measuring points.

2.3 *Test scheme*

In order to study the hydraulic characteristics of the joint energy dissipator with the transition step size of the first step, the test scheme is determined according to the original design scheme of Ahai Hydropower Station. Scheme 1 is to set the first two uniform steps of 1 m×0.75 m as the first transition step, which is the original working condition. Scheme 2 is to change two uniform steps of 1 m×0.75 m into one large uniform step of 2 m×1.5 m. Scheme 3 is to add a step of 1.5 m×0.375 m at the first stage of the transition step on the basis of Scheme 2 to form the first stage of the transition step of 0.5 m×0.375 m. In schemes 1, 2, and 3, the height-width ratio of the wide tail pier size does not change, but the aeration area of the three schemes is different. In scheme 4, a 1 m×1.125 m first-stage transition step is added on the basis of Scheme 2 to form a 1 m×1.125 m first-stage transition step. In scheme 5, a first stage transition step of 0.5 m×0.75 m is added on the basis of Scheme 2 to form a first stage transition step of 0.5 m×0.75 m. The first step of scheme 6 is the first step of 0.5 m×1.125 m by adding a 1.5 m×1.125 m step on the basis of Scheme 2. The six different sizes of the first transition step scheme are shown in Figure 2. The check flood level with a return period of five thousand years is taken for the test, the water levels of the upstream and downstream reservoirs are taken as 1507.23 m and 1445.23 m, respectively, and the unit width flow is taken as 269.47 m^2/s.

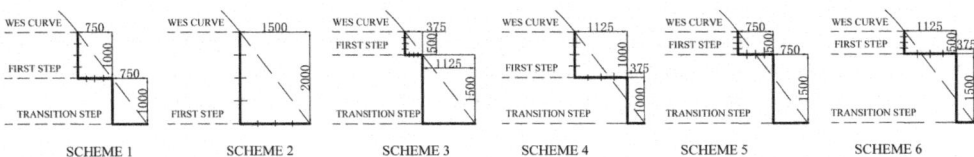

Note: all dimensions in the drawing are prototype dimensions, in mm, Bold lines indicate ladders

Figure 2. The type of first-stage steps.

3 TEST RESULTS AND ANALYSIS

3.1 *Negative pressure distribution*

According to Table 2, the measuring points of negative pressure on the first step surface of the six schemes are different. When the height-width ratio is 0.75, the negative pressure appears at measuring points 1, and 5 respectively, and the larger the aeration area is, the greater the negative pressure value is. Scheme 2 has the largest aeration area and the largest negative pressure. The negative pressure values of No. 1 and No. 5 measuring points are -6.30 kPa and -14.27 kPa, respectively. In Scheme 2, there is a large amount of gas to be supplemented, and the ability of the letdown flow to entrain gas is limited. Because the rate of external gas supplement to the first step is basically the same, the maximum negative pressure on the first step surface appears in scheme 2. In Schemes 4, 5, and 6, negative pressure is measured at the No. 4 measuring point on the step surface in Schemes 4 and 5. Among them, negative pressure in Scheme 4 is -5.22 kPa, and that in Scheme 5 is -2.36 kPa. The aeration area of Schemes 4 and 5 is 1.5 m², and the difference between them is the aspect ratio. Since the height-width ratio of Scheme 4 is 0.9, while that of Scheme 5 is 1.5, Scheme 4 may have a small barrier effect on the drainage tongue, making the water flow on the step plane, thus affecting the negative pressure value of the step surface. The aeration area of Scheme 6 is 1.125 m², and the height-width ratio is 0.4. The stepped surface is flooded with water, and there is no negative pressure. The results show that, on the one hand, the size of the first transition step can be appropriately changed to ensure sufficient aeration area so that there is negative pressure on the surface of the first step to facilitate the entrainment of the gas. On the other hand, the height-width ratio of the first transition step can be changed, and the height-width ratio can be appropriately increased, which is conducive to the aeration of the dam surface and is of great significance to the erosion reduction of the ladder surface.

Table 2. The first-step negative pressure value of each scheme.

Scheme	Scheme 1	Scheme 2	Scheme 3	Scheme 4	Scheme 6
Vertical measuring point 1	-3.00	-6.30	2.34	6.72	8.16
Vertical measuring point 2	4.80	14.31	4.13	12.74	1.65
Vertical measuring point 3	10.20	29.15	7.14	-1.44	11.06
Plane measuring point 4	7.80	-9.14	5.23	-5.22	16.84
Plane measuring point 5	-7.80	-14.27	-3.66	1.62	7.62
Plane measuring point 6	15.00	2.63	1.12	3.28	8.81

3.2 *Bottom flow velocity*

Among the six schemes, the maximum value of the temporary bottom flow velocity in Scheme 4 is located at stake 0+112.960 m, and the minimum value is located at stake 0+228.160 m, which is 2.17 m/s. According to the analysis, the difference in the height-width ratio between various cases has a restraining effect on the water strand emitted from the wide tail pier and then affects the discharge flow. Therefore, in Scheme 4, in the WES reverse arc section and the stilling basin, the flow pattern of the water body is the most chaotic, and the efficiency of velocity attenuation near the bottom is the highest.

Comparing Figures 3 and 4, there is a partial difference in the bottom flow velocity between the center line of 1# surface hole and the dam axis, which is mainly manifested in the low bottom flow velocity at 0+069.924m of the dam axis, and in the reverse arc section, the rising trend of the bottom flow velocity is more obvious than that at the center line of 1# surface hole.

3.3 *Time average pressure*

Comparing Figures 5 and 6, it can be seen that the time average pressure of Schemes 1, 2, and 3 is the maximum at the stake No. 0+097.800 m. The time average pressure of Scheme 2 is the

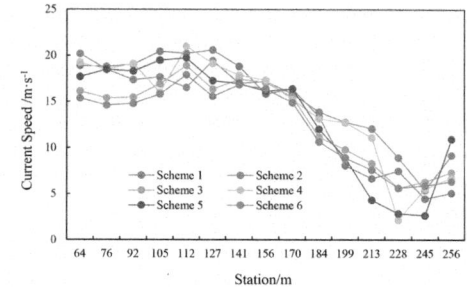

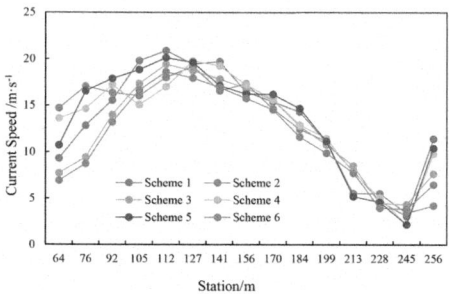

Figure 3. The velocity distribution of the
1# surface outlet.

Figure 4. The velocity distribution of the
dam axis.

maximum, and its value is 445.72 kPa, and that of Scheme 1 is the smallest, being 431.61 kPa. It shows that the different first transition steps have positive significance for the reduction of time average pressure. Compared with Scheme 2, the maximum hourly average pressure of Scheme 1 is reduced by 3.2%. The larger the aeration area is, the more obvious the reduction of hourly average pressure is. In Schemes 4, 5, and 6, similarly, the time average pressure reaches the maximum at stake 0+097.800 m. The hourly average pressure of Scheme 6 is the largest, with a value of 435.73 kPa, and that of Scheme 4 is the smallest, being 425.73 kPa. Compared with Scheme 6, the maximum hourly average pressure of Scheme 4 decreases by 2.3%, and the smaller the aspect ratio, the higher the reduction level of hourly average pressure.

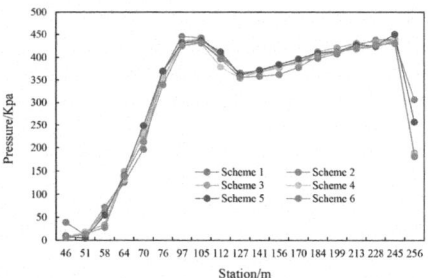

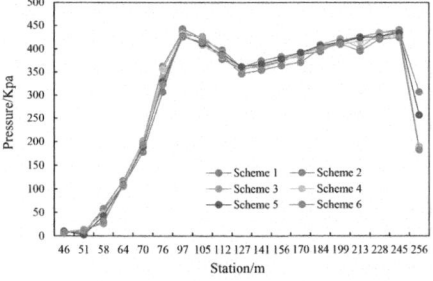

Figure 5. The time-average pressure of the
1# surface outlet.

Figure 6. The time-average pressure of the
dam axis.

3.4 *Energy dissipation rate*

Energy dissipation rate is a key index to evaluate the overall performance of water release structures, so the calculation of energy dissipation rate is very important. In order to study the energy dissipation rate of the combined energy dissipation of the first transition step with different sizes, the energy equations are established for the upstream transition section and the downstream transition section of the stepped overflow dam. Here, the ratio of the energy difference between the upstream and downstream overflow sections of the stepped overflow dam and the energy of the upstream overflow section of the stepped overflow dam is used as the energy dissipation rate:

$$\eta = \frac{\Delta E}{E_1} \times 100\% = \frac{E_1 - E_2}{E_1} \times 100\% \tag{1}$$

$$E_1 = Z_1 + H_1 + \frac{\alpha_1 v_1^2}{2g} \tag{2}$$

$$E_2 = Z_2 + H_2 + \frac{\alpha_2 v_2^2}{2g} \tag{3}$$

165

where Z_1 and Z_2 are the elevations of the upstream and downstream overflow sections relative to the stilling basin bottom plate, respectively. H_1 and H_2 are the water depths of upstream and downstream overflow sections, respectively. v_1 and v_2 are the average flow velocities of the upstream and downstream cross-sections, respectively. α_1 and α_2 are the kinetic energy correction coefficients of the upstream and downstream cross-sections, respectively, which depend on the uneven distribution of the velocity on the upstream and downstream cross-sections. α_1 and α_2 are approximately 1.

Table 3. Comparison of energy dissipation rate of each scheme.

Scheme	E_1	E_2	ΔE	η
Scheme 1	104.15	46.12	58.03	55.72%
Scheme 2	104.15	45.73	58.42	56.09%
Scheme 3	104.15	46.36	57.79	55.49%
Scheme 4	104.15	43.89	60.26	57.86%
Scheme 5	104.15	48.25	55.90	53.67%
Scheme 6	104.15	45.31	58.87	56.50%

From Table 3, the energy dissipation rate of Scheme 4 is the highest, with a value of 57.86%. The energy dissipation rate of Scheme 5 is the lowest, with a value of 53.67%. The difference in the energy dissipation rate between them is 4.19%. Scheme 4 has a larger height-width ratio and a larger aeration area than the first transition step of Scheme 5. This is conducive to changing the flow pattern of the discharge water and increasing the aeration concentration of the discharge water, which is of great significance to improving the energy dissipation rate. The results of six test schemes show that the aerated area and height-width ratio of the step are the main factors affecting the energy dissipation rate of the first transition step with different sizes. The smaller the aspect ratio is, the larger the aeration area is, which is beneficial to the energy dissipation rate of combined energy dissipation.

4 CONCLUSIONS

In this paper, six different sizes of first transition steps are selected, and the joint energy dissipation test of wide tail pier + stepped overflow dam + stilling basin is carried out. The hydraulic characteristics of the negative pressure on the first step, the flow pattern in the stilling basin, and the time average pressure and energy dissipation rate of the stilling basin floor are studied. The main conclusion is that the aeration area and the height-width ratio of the step are the two main factors affecting the negative pressure distribution on the step surface. Negative pressure is mainly distributed at the lower edge of the first step facade and the front end of the first step plane. The increase in the aeration area and height-width ratio can make the first step surface have negative pressure, which is convenient for gas entraining and is of great significance to the corrosion reduction of the step surface. However, excessive negative pressure is detrimental to the safety of the first step and is prone to cavitation damage. Therefore, it is necessary to control the size of the first step to achieve a relatively better aeration effect. The change in the size of the first step will lead to a difference in the aeration area and height-width ratio. If the aeration area is appropriately increased and the height-width ratio is appropriately reduced, the maximum time-averaged pressure at different pile numbers of the stilling basin will also be appropriately reduced.

REFERENCES

Dong L.Y., Yang J.R., Li S.Z. (2018). Influence of transitional ladder on the hydraulic characteristics of integrated energy dissipator. J. *Journal of Hydroelectric Engineering*. 37(3): 50–58.

Hu Y.H., Wu C., Lu H. et al. (2006). Study on the hydraulic structure of flaring gate piers locating at upstream of the stepped spillway. *J. Journal of Hydroelectric Engineering*. 25(5): 37–41.

Hu Y.H., Wu C., Zhang T., et al. (2009). Aeration of skimming flow on stepped spillway combined with flaring gate piers. *J. Journal of Hydraulic Engineering*. 40(5): 564–568.

Liang Z.X., Yin J.B., Zheng Z., et al. (2009). Aeration of skimming flow on stepped spillway combined with flaring gate piers. *J. Journal of Hydraulic Engineering*. 40(5): 564–568.

Tian J.N., Yasuda Youichi, Li J.Z. (2009). Energy dissipation on stepped sluicing structure. *J. Journal of Hydroelectric Engineering*. 28(2): 96–100.

Tian J.N., Zhao Q., Fan L.M. (2012). Pressure characteristic in stilling basin of stepped spillways. *J. Journal of Hydroelectric Engineering*. 31(4): 113–118.

Wang Q., Yang J.R., Wu Z.Z., et al. (2016). Impact of transition step with different number of steps on dam surface pressure and energy dissipation of stepped spillway. *J. Journal of Hydroelectric Engineering*. 35(5): 84–93.

Wei W.L., Lv B., Liu Y.L. (2012). Numerical simulation of flow on stepped spillway combined with flaring gate piers and stilling basin. *J. Journal of Hydrodynamics*. 27(4):442–448.

Yin J.B., Liang Z.X., Gong H.L. (2007). Experimental study on application and development of X type flaring gate piers. *J. Journal of Hydroelectric Engineering*. 26(4):36–39, 35.

Zhang T., Zhou Q., Wu C., et al. (2007). Numerical simulation of energy dissipator with new style flaring gate pier and stepped spillway *J. Journal of Fuzhou University* (Natural Science Edition), 35(1): 111–115.

Frontiers in Civil and Hydraulic Engineering – Mohamed A. Ismail and Hazem Samih Mohamed (Eds)
© 2023 The Authors, ISBN 978-1-032-38247-0

Analysis of the influence of different excavation sequences of foundation pit above existing tunnels

Ting Bao*
Publisher, Chinese, Zhejiang Mingsui Technology Co., Ltd, Zhejiang, China

Xiaobo Sun*
Chinese, China Railway 14th Bureau Group Mege Shield Construction Engineering Co., Ltd, China

Jin Pang*, Lingchao Shou* & Lifeng Wang*
Chinese, Zhejiang Mingsui Technology Co., Ltd, Zhejiang, China

ABSTRACT: Based on a foundation pit project in Hangzhou, two different excavation sequences of the foundation pit are simulated and analyzed by Plaxis 2D. The results show that the influence of different excavation sequences of the foundation pit on the subterranean tunnel is quite different. Compared with the excavation method of excavating the inner soil first, excavating the outer soil first can reduce the impact of foundation pit excavation on the subterranean tunnel, but the deformation of the foundation pit itself is relatively large. In the project, the appropriate excavation method should be selected according to the characteristics of the project.

1   INTRODUCTION

With the acceleration of urbanization, the excavation of foundation pits over existing subway tunnels is becoming more and more common, which will inevitably affect the safety of subway operations (Liu 2021; Zheng 2016). In order to ensure the smooth excavation of foundation pits and the safe operation of subways, a reasonable excavation sequence is very important.

At present, many experts and scholars have carried out certain research on the influence of the excavation sequence of foundation pits. Bi S Q et al. (Bi 2022) analyzed the influence of foundation pit excavation construction on the subterranean tunnel in combination with the actual project. The results show that the initial stage of the tunnel's upward deformation is greatly affected by the excavation sequence. Ye J F et al. (Ye 2017) used PLAXIS to establish a two-dimensional model of adjacent foundation pit excavation and analyzed that the adjacent foundation pit spacing, excavation sequence, and support method are important influencing factors of foundation pit deformation. Yan C (2019) studied the adjacent foundation pits and found that the lateral displacement of the foundation pit enclosure structure during the later excavation construction will shift towards the direction of the first excavation foundation pit, and considered that the simultaneous excavation construction of adjacent foundation pits is preferred. Wu S Y et al. (2012) used Midas/GTS to study the influences of the internal force and displacement deformation of the tunnel structure when the deep foundation pits on both sides of the existing tunnel were constructed in different excavation sequences.

The excavation sequence of the foundation pit soil directly affects the stress release of the soil, which has a great impact on the envelope structure and the surrounding environment (Zhou 2021). Based on the research background of an upper-span operation subway foundation pit project in

---

*Corresponding Authors: 1542528435@qq.com; 114938824@qq.com; pangj00@163.com; slczust@163.com and wanglfzust@163.com

DOI 10.1201/9781003344209-22

Hangzhou, this paper used the method of numerical simulation to analyze the surface subsidence and uplift caused by different excavation sequences, the deformation of the enclosure structure, and the deformation characteristics of the subterranean tunnel, so as to provide a reference for similar projects.

2 ENGINEERING BACKGROUND

2.1 Project overview

The Genshan East Road crossing the river tunnel project is designed as an urban expressway, including the river crossing tunnel, comprehensive pipe gallery, ground connecting line project, and supporting ancillary projects. The total length of the project is 4612.26 m (including the ground connecting line road), and the total length of the tunnel is 4462.26 m (including the shield section of about 3210 m). Among them, the open-cut section near YK0+700 on the Xiasha side is orthogonal to the existing tunnel, and the foundation pit of the open-cut section is shown in Figure 1. It is located about 6 m below the proposed tunnel project.

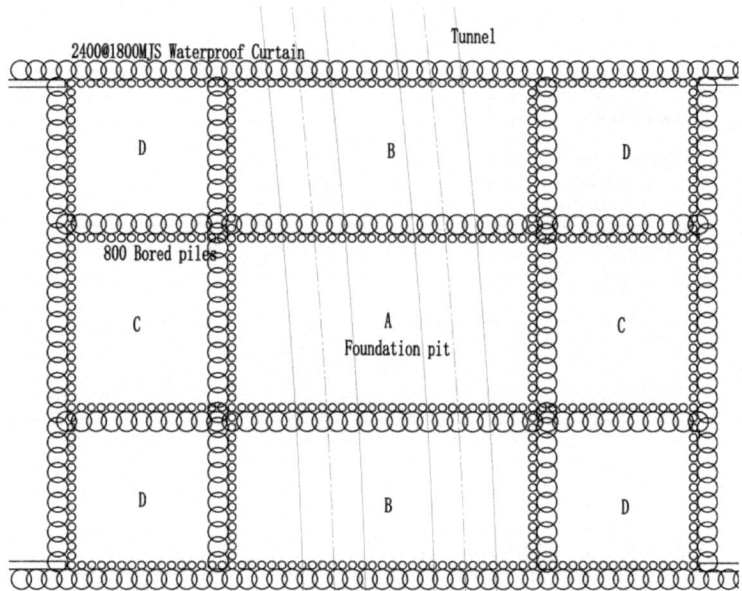

Figure 1.   Plane position relationship between the foundation pit and the subway tunnel.

2.2 Foundation pit support structure

For the foundation pit enclosure, bored cast-in-place piles are adopted with a diameter of 800 mm and a spacing of 1000 mm, of which double-row piles are used directly above the subway, and the waterproof curtain consists of 2400@1800MJS rotary jet piles. A bored cast-in-place pile with a diameter of 1200 mm is installed in the foundation pit as an anti-uplift pile. The vertical distance between the bottom of the enclosure pile directly above the shield tunnel and the top of the tunnel is greater than 2.0 m, and the elevation of the bottom of the remaining enclosure piles is –21.2 m. The two sides of the shield tunnel of Metro Line 1 are reinforced with portal bodies. MJS rotary jet piles are used for the sake of reinforcement. The depth of the foundation pit is reinforced to –21.2 m. The sub-pits are separated by bored piles, and the overall construction sequence is to excavate areas A, B, C, and D of the foundation pit above in sequence.

3 CALCULATION RESULETS AND ANALYSIS

3.1 *Establishment of finite element model*

According to the engineering geological characteristics of the Genshan East Road crossing the river tunnel and combining the foundation pit design with the construction plan, this paper selects the excavation sections of areas A and C in Figure 1, and uses PLAXIS 2D finite element software to establish a numerical model, and analyzes the impact of different excavation sequences in area A and area C on the surrounding environment and the underground operation of the subway. The small strain soil hardening constitutive (HSS) structure is used for the soil, and the soil layer model is established according to the geological survey report.

The mesh of the calculation model is shown in Figure 2. The excavation depth of the foundation pit in the model is 9.8 m. Since the foundation pit is geometrically symmetrical along the width direction, a 1/2 foundation pit width model is established. According to the principle of Saint-Venant, the impact depth of foundation pit excavation is generally 3-5 times that of the foundation pit excavation depth, so the soil layer boundary is 100 m wide and 50 m deep. Fully fixed constraints are applied to the bottom of the geometric model, and horizontal constraints are applied to the left and right edges. In addition, 20 kPa of the load is applied near the excavation surface of the foundation pit to simulate the construction load. The calculation parameters of the soil layer are shown in Table 1.

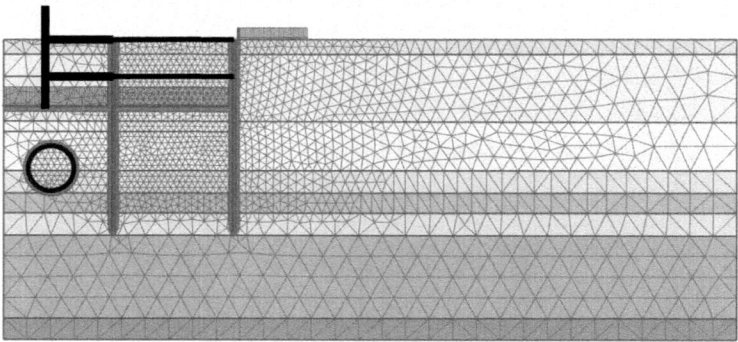

Figure 2.   Model meshing diagram.

Table 1.   Calculation parameters of the soil layer.

Name of soil layers	γ/(kNm^{-3})	$E_{50}{}^{ref}$/MPa	$E_{oed}{}^{ref}$/MPa	$E_{ur}{}^{ref}$/MPa	c/kPa	ϕ /(°)	G^{ref}/MPa	$\gamma_{0.7}$/10^{-4}
Miscellaneous fill	17.71	3	3	9	10	12	27	2
Sandy silt	19.5	19	19	57	6	25	171	2
Silt	19.7	20.31	20.31	60.93	4	32	180	1
Sandy silt	19.4	13.39	13.39	40	7	23	120	2
Silty clay	18.7	4.9	4.9	22	14	21	110	2
Silty clay	18.3	3.9	3.9	20	13	11	100	2
Silty clay	18.62	5.1	5.1	15.3	22	15	45.9	2

3.2 *Analysis of calculation results*

The two different excavation sequences are: (1) excavating the middle first and then excavating the outer side (Scheme 1); (2) excavating the outer side first and then excavating the middle (Scheme 2). The calculation steps of the two schemes of the model are shown in Table 2.

Table 2. Construction steps of foundation pit excavation.

Step	Working condition description of Scheme 1	Working condition description of Scheme 2
Phase 0	Equilibrium stress field	Equilibrium stress field
Phase 1	Tunnel construction	Tunnel construction
Phase 2	Ground wall construction	Ground wall construction
Phase 3	Construction of the first concrete support in the middle	Construction of the first concrete support on the outside
Phase 4	Middle excavation to the second support bottom	Outside excavation to the second support bottom
Phase 5	Construction of the middle second steel support	Construction of the second steel support on the outside
Phase 6	Middle excavation	Excavation to the bottom
Phase 7	Intermediate floor casting	Outer base plate pouring
Phase 8	Construction of the first concrete support on the outside	The first concrete support construction in the middle
Phase 9	Outside excavation to the second support bottom	Middle excavation to the second support bottom
Phase 10	Construction of the second steel support on the outside	Construction of the middle second steel support
Phase 11	Excavation to the bottom	Middle excavation
Phase 12	Outer base plate pouring	Intermediate floor casting

3.2.1 *Deformation analysis of the enclosure structure*

After the excavation of the foundation pit is completed, the deformation curves of the horizontal displacement of the connecting wall under the two excavation sequences are extracted, as shown in Figure 4.

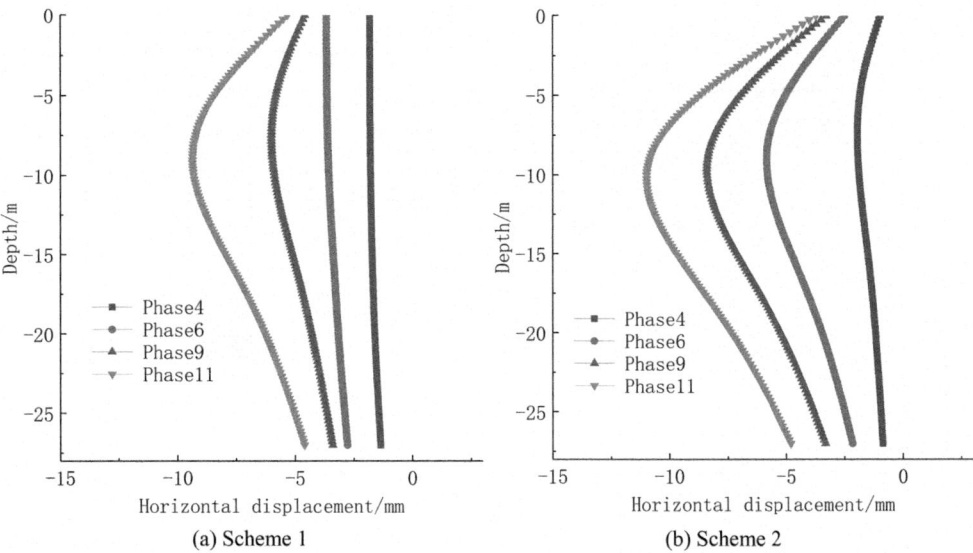

(a) Scheme 1   (b) Scheme 2

Figure 3. Horizontal displacement curve of the ground connecting wall.

It can be seen from Figure 3 that different excavation sequences have a greater impact on the deformation of the ground connecting wall, and a reasonable excavation sequence can effectively reduce the horizontal displacement of the ground connecting wall. The excavation of the middle part of the soil in Scheme 1 has little effect on the horizontal displacement of the ground connecting wall. The maximum displacement occurs at a distance of 9.21 m from the top of the pile, and the maximum displacement value is 9.37 mm; middle soil excavation continues to deform the ground connecting wall. The maximum displacement occurs at 10.01 m from the top of the pile, and the maximum displacement is 10.98 mm, which is 1.61 mm (17.18%) larger than that of Scheme 1.

3.2.2 *Analysis of surface subsidence and pit bottom uplift*

After the excavation of the foundation pit is completed, the final vertical displacement cloud map of the ground surface under two different excavation sequences is shown in Figure 4.

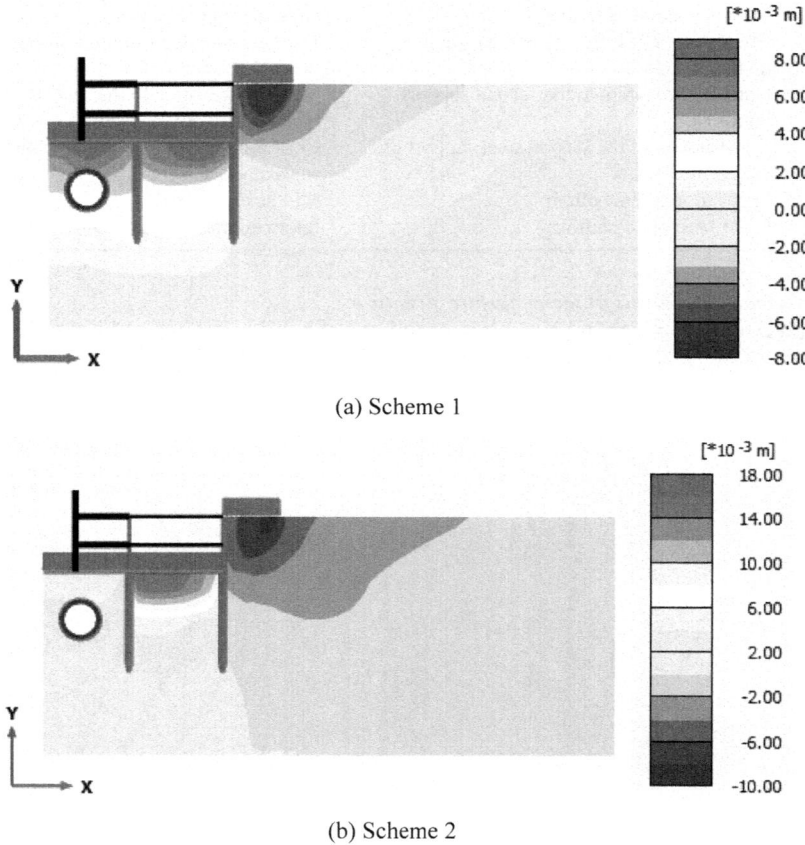

(a) Scheme 1

(b) Scheme 2

Figure 4.   Vertical displacement.

The maximum value of the surface settlement behind the wall in Scheme 1 is about −7.04 mm, and the maximum uplift of the inner and outer bases is about 8.09 mm and 7.91 mm, respectively; the maximum value of the surface settlement behind the wall in Scheme 2 is about −8.30 mm, and the maximum uplift of the inner and outer bases is about −8.30 mm. The values are about 5.56 mm and 16.09 mm. Comparing the two excavation sequences, it can be seen that the excavation of the outer soil body first reduces the uplift of the inner base but greatly increases the uplift of the outer base, and the surface settlement behind the wall also increases accordingly.

3.2.3 *Deformation analysis of subway tunnel*

The subterranean tunnel is located under the soil inside the foundation pit, so the excavation of the foundation pit will cause the uplift and deformation of the tunnel. After the excavation of the foundation pit is completed, the deformation of the tunnel under two different excavation sequences is shown in Figure 5.

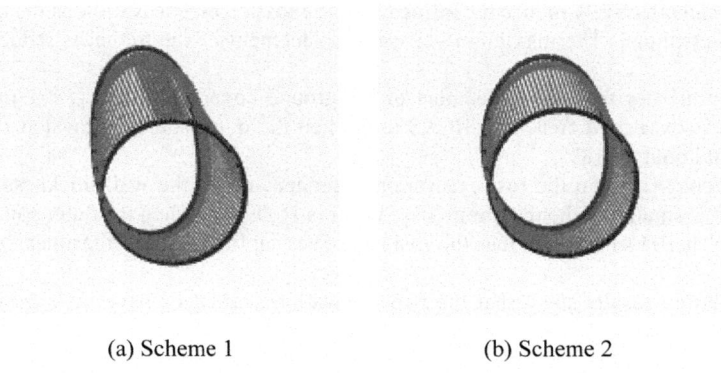

(a) Scheme 1   (b) Scheme 2

Figure 5.   Displacement map of subway tunnel.

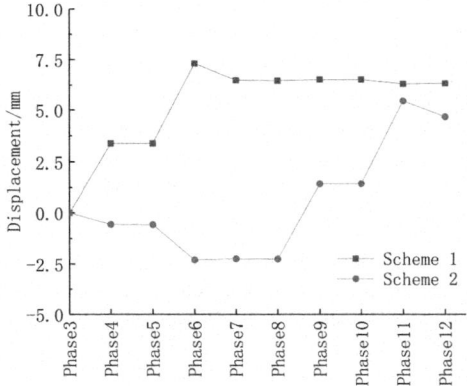

Figure 6.   Variation of the maximum vertical displacement of the tunnel under various working conditions.

Figure 6 reflects the change in the vertical displacement of the tunnel caused by each excavation step. It can be seen that because the tunnel is under the internal soil, the vertical displacement of the tunnel is mainly affected by the internal soil excavation and is less affected by the external soil excavation. The timely pouring of the bottom plate can reduce tunnel deformation. The maximum vertical displacement of the tunnel in Scheme 1 occurs when the internal soil is excavated to the bottom, and the maximum value is about 7.31 mm. After all the excavation is completed, the vertical displacement of the tunnel is 6.34 mm; when the external soil is excavated first in Scheme 2, the external base is uplifted greatly, and the internal soil also rushes to the outside, so the overall tunnel is slightly displaced downward, and the maximum vertical displacement of the tunnel also occurs when the internal soil is excavated to the bottom. The maximum value is about 5.46 mm. After all excavation works are completed, the vertical displacement of the tunnel is 4.68 mm, which is 26.18% lower than that of Scheme 1.

4 CONCLUSION

Through the numerical simulation analysis of different soil excavation sequences in the foundation pit of Genshan East Road crossing the river tunnel in Hangzhou, the following conclusions can be drawn:

(1) The outer soil is excavated first, followed by the inner soil, that is, the far-to-near excavation sequence can effectively reduce the influence of foundation pit excavation on the displacement of the lower tunnel. The maximum vertical displacement of the tunnel is reduced by about 26.18%.
(2) The maximum horizontal displacement of the ground connecting wall is 9.37mm when the inner soil is excavated first, and 10 .98 mm when the outer soil is excavated first, with an increase of about 17.18%.
(3) The difference between the two excavation sequences under the wall thickness and surface settlement is small, but the maximum base uplift is 16.09 mm when the outer soil is excavated first, which is 103.41% higher than the 7.91 mm of base uplift caused by the inner soil excavation first.
(4) The simulation results show that the two excavation sequences have their own advantages and disadvantages. The first excavation of the outer soil can better protect the safety of the subway operating below, but the foundation pit itself has a large deformation. A reasonable excavation plan is an important guarantee of engineering safety. In the actual construction process, a reasonable excavation sequence should be selected according to the characteristics of the project to reduce the impacts on the foundation pit and surrounding environment.

REFERENCES

Bi, S.Q., Gan, B.L. & Liang, Y.H. et al. (2022). Measured analysis of the influence of foundation pit excavation on existing short-range subterranean tunnels. *J. Science Technology and Engineering*. 22(3), 1198–1204.
Liu, Y. & Zhang, L.H. (2021). Analysis of the influence of deep foundation pit excavation on adjacent subway safety based on Midas/GTS. *J. Journal of North China Institute of Science and Technology*. 18(04), 74–79.
Wu, S.Y., Yang, X.P. & Liu, T J. (2012). Analysis of the influence of double-sided deep foundation pit construction on the deformation of adjacent subway tunnels. *J. Chinese Journal of Rock Mechanics and Engineering*. 31(S1), 3452–3458.
Yan, C. (2019). Analysis of the influence of foundation pit excavation on the surrounding environment and adjacent foundation pits under different excavation conditions based on PLAXIS-2D. C. Proceedings of the Exchange Conference on New Materials, *New Technologies and Engineering Applications in Civil Engineering* Volume. 970–974.
Ye, J.F., Lin, H. & Yan, G.Y. (2017). Two-dimensional character analysis of mutual influence of adjacent double foundation pit excavation. *J. Journal of Fuzhou University* (Natural Science Edition). 45(02), 190–198.
Zheng, G., Du, Y.M. & Diao, Y. et al. (2016). Research on the influence area of adjacent existing tunnels caused by excavation of foundation pits. *J. Chinese Journal of Geotechnical Engineering*. 38(04), 599–612.
Zhou, W. (2021). Analysis of excavation sequence of single-layer soil blocks adjacent to double foundation pits. *J. Railway Construction Technology*. (09), 152–157+176.

*Frontiers in Civil and Hydraulic Engineering – Mohamed A. Ismail and
Hazem Samih Mohamed (Eds)*
© 2023 The Authors, ISBN 978-1-032-38247-0

Analysis of characteristics and control measures of water leakage in Shanxi Mountain tunnel lining

Dasheng Liu*
Chongqing Quantong Engineering Construction Management Co., Ltd., Chongqing, China

Lihua Sun* & Jiong Zhu*
China Merchants Chongqing Communications Technology Research and Design Institute Co., Ltd., Chongqing, China

ABSTRACT: In recent years, a survey has found that most highway tunnels in China operate not for a long time. Due to complex geological conditions and construction technology for the cross section of road tunnels, the design and calculation theory and method are not yet mature, operation management and maintenance technology is relatively backward, and the disease of highway tunnels in China is also more serious, especially the seepage. In this paper, based on the investigation of water leakage in the lining of the mountain tunnel in Shanxi, the surface characteristics are analyzed and the data of various water leakage phenomena are statistically analyzed. The distribution law of water leakage and the remediation of water leakage are obtained by combining the phenomena. The research results can provide a reference for the pipe maintenance of highway tunnels in China, and then gradually form the standardized technology of water leakage remediation of tunnels and effectively improve the level of tunnel maintenance in our country.

1 INTRODUCTION

With the strategies of "new urbanization" and "building national strength in transportation" progressively advanced, development opportunities are being created increasingly for tunnels and underground works, which are the main component of infrastructure construction. A large number of tunnel projects have been completed and put into operation. However, the flourishing tunnel construction has also created some major challenges. Tunnels are built in different periods and are subject to different design standards and construction processes. In addition, the geological conditions in tunnel-passing areas differ greatly. Most tunnels are found with leakage, cracked lining, hollowing, and other problems to different degrees. Leakage of tunnel lining is one of the most common structural hazards in tunnel engineering, which can lead to short-circuited power systems and wet and slippery roads. Leakage, if not treated in time, will cause concrete erosion, lining cracking, frost heave, and other hazards. The significant reduction in structural safety will lay huge risks for tunnel operation.

In 1990, Japan Road Association organized a safety evaluation for 4,307 highway tunnels in operation. The investigation showed that as many as 84% of the tunnels were hazardous, including 60% found with leakage, which was generally related to lining cracking, and 24% with other hazards. In the United States, Germany, France, and the U.K., tunnel hazards have occurred to different degrees due to various reasons. Back at the end of the last century, some foreign experts and scholars had highly valued the repair and maintenance of highway tunnels and had developed some new methods and technologies to move up highway repair and maintenance management to a new

---

*Corresponding Authors: 27076661@qq.com; Slh18223661823@163.com and zhujiong@cmhk.com

DOI 10.1201/9781003344209-23

level. For example, they discussed the application of NDT to concealed structures— tunnels—and adopted TST, multispectral analysis, and other methods to quickly carry out preliminary investigations on tunnels. In 1986, Japan conducted an investigation of over 100 railway tunnels and 10 highway tunnels at home, aiming to systematically study the probability and causes of tunnel hazards. Lining leakage has attracted the attention of experts, who have carried out a series of studies. Xue (2016) discussed the mechanism of and countermeasures for lining leakage based on statistics on leakage of highway tunnels in Shanxi Province. Ying (2015) carried out a statistical analysis of tunnel lining leakage and threw out suggestions on controlling leakage hazards in different forms. Zhou (2013), based on detection data obtained from tunnel sites, summarized the distribution characteristics and types of cracks and leakage hazards and comprehensively analyzed the causes of lining leakage from several aspects.

As a result, this paper aims to specially study the leakage hazards occurring in mountain tunnels in Shanxi Province, find out the quantity, surface characteristics, and distribution pattern of tunnel leakage based on leakage phenomena, and draw conclusions on measures against leakage occurring to mountain tunnels in Shanxi Province.

2 TUNNEL OVERVIEW

At present, Shanxi Province has 602 expressway tunnels in operation, with a total length of 535 km; among these tunnels, there are 50 extra-long tunnels and 84 long tunnels. There is a large number of highway tunnels that are found with severe leakage hazards. For mountain tunnels in Shanxi Province, the groundwater leakage and dripping in summer when rainfall is heavy and icing in winter, especially in the northern Shanxi region where winter frost heave, spring thawing, leakage, and other hazards can easily occur to mountain tunnels. These phenomena will create great potential hazards for tunnel structure and traffic safety.

3 ANALYSIS OF CHARACTERISTICS OF LEAKAGE HAZARDS

3.1 Leakage site

To clearly present the leakage phenomena of mountain tunnels in Shanxi Province, it is necessary to describe the parts of leakage, which are divided by tunnel structure, tunnel construction & operation environment, and control method as shown in Figure 1.

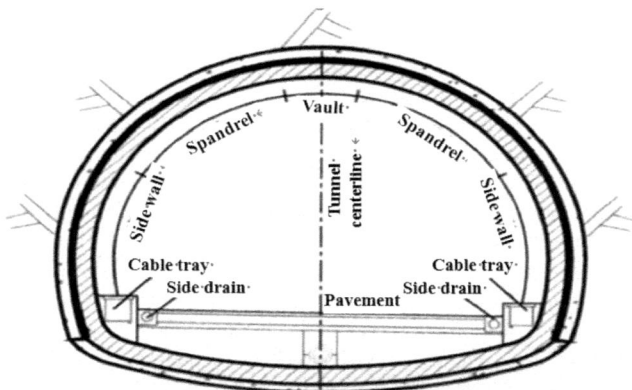

Figure 1. Division of water leakage parts in tunnels.

Parts of the tunnel found with leakage include the construction joint vault, spandrel, side wall, arch springing, construction joint, and equipment box. The research-based statistics on all parts with leakage of mountain tunnels in Shanxi Province are shown in Figure 2. According to Figure 2, most parts of the tunnel lining with a leakage are at the side wall, followed by the spandrel and arch springing.

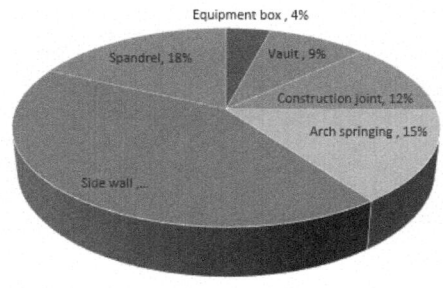

Figure 2.   Statistics of various parts of mountain tunnel leakage in shanxi province.

3.2   Type of leakage

The leakage phenomena of tunnels are divided into the following 5 categories according to acceptance criteria and practices in tunnel engineering: wet stains, leakage, water droplets, drips, and line leaks.

The author summarized and analyzed the results of extensive research and studies on tunnels with leakage hazards that are currently in operation in Shanxi Province. The results show that 4 types of leakage phenomena exist: wet stains, leakage, water droplets, and drips, which are shown in Figures 3 and 4.

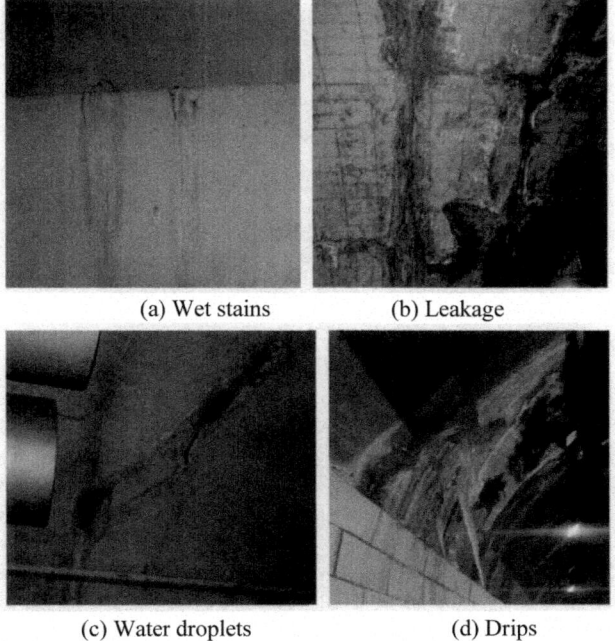

(a) Wet stains   (b) Leakage

(c) Water droplets   (d) Drips

Figure 3.   Water leakage type of mountain tunnel in shanxi province.

177

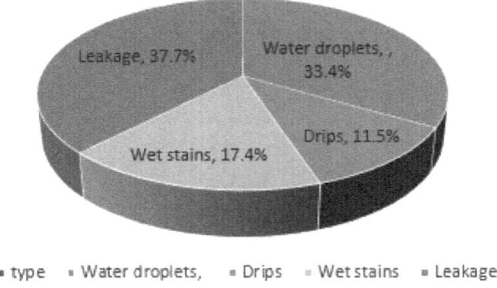

Figure 4. Statistical chart of water leakage types of mountain tunnel in shanxi province.

These figures show that the leakage phenomena found in tunnels in operation in Shanxi Province are mostly leakage and water droplets. Only 11.5% of the leakage phenomena are drips. The reason is that after a period of operation, the cracked lining and the aged water stop made of waterproofing materials provide certain conditions for tunnel leakage.

3.3 Analysis of characteristics of leakage at cracks

3.3.1 Impact of crack width on leakage

During the use of concrete members, only 20% of the cracks are formed under the action of loading, including the joint action of deformation and loading. More than 80% of the cracks in concrete structures are related to non-load deformation. Therefore, non-load factors have a great impact on crack formation. Cracks in concrete structures caused by non-load factors are divided into three types: 0.2–1 mm, 1–3 mm, and >3 mm, which are respectively named microcracks, small cracks, medium cracks, and large cracks.

Lining cracks of mountain tunnels in Shanxi Province are counted by width, as shown in Table 1.

Table 1. Statistical table of crack width of mountain tunnels in Shanxi.

Width (mm)	Percentage (%)
<1	80.74
1–3	17.64
>3	1.63

Statistical results show that the lining cracks detected from mountain tunnels in Shanxi Province are generally less than 1 mm in width. The number of such cracks is 2,512, accounting for 80.74% of the total number of cracks. This is followed by medium cracks, which account for 17.64% of the total number of cracks. For cracks with a width of < 1.0 mm, the impact on the structural safety of tunnels is minor because they affect the bearing capacity less, which may not be a key concern in the study. It can be seen that for cracks with a width > 1.0 mm, the greater the width is, the higher the probability for leakage in the crack will be, and the greater the load-bearing hazard of the tunnel lining structure will be. Figure 5 shows the leakage rate for most cracks. In microcracks, the leakage rate is as low as 0.8%, which is usually caused by drying shrinkage of concrete and will not greatly affect the safety of the lining structure. In large cracks, however, the leakage rate is as high as 23%, which is usually caused by structural loading; some are penetrating cracks featuring large-area leakage. As a result, special attention must be paid to tunnel cracks.

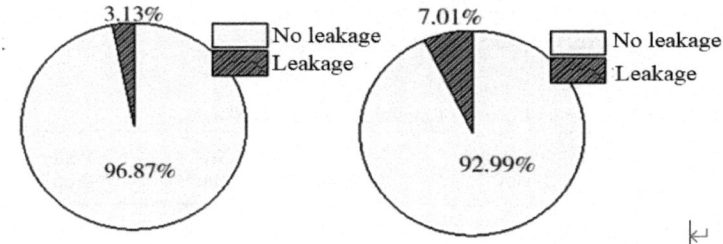

Figure 5.   Crack leakage ratio statistics.

3.3.2 *Impact of crack orientation on leakage*

In tunnels, there are longitudinal, circumferential, and diagonal cracks. Circumferential cracks have a minor impact on the lining structure under normal loading; the longitudinal and diagonal cracks in the vault and side wall damage structural integrity and can cause severe hazards.

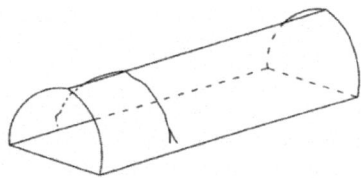

Figure 6.   Schematic diagram of circumferential cracks in lining.

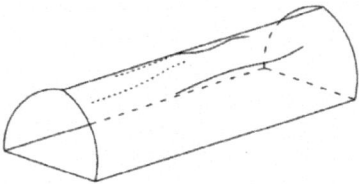

Figure 7.   Schematic diagram of longitudinal cracks in lining.

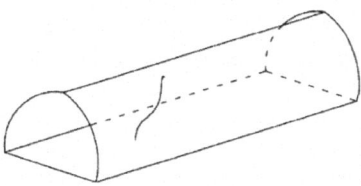

Figure 8.   Schematic diagram of oblique cracks in lining.

The statistical cracks and leakage of the lining are shown in Table 2. It can be seen that most lining cracks of mountain tunnels in Shanxi Province are longitudinal and circumferential; diagonal cracks are relatively uncommon. In general, tunnel cracks show a pattern of circumferential cracks > longitudinal cracks > diagonal cracks. Circumferential cracks exist in quantity and can cause major hazards, which must be paid much attention to. Longitudinal cracks account for 1/3 of the total number of cracks, which are less hazardous. From the perspective of operation management and control, the priority for crack prevention and control is longitudinal cracks > diagonal cracks > circumferential cracks. Regarding crack orientation, the main leakage phenomena are leakage and wet stains, which usually do not affect the safety of tunnel operation. The uncommon phenomena of water droplets and drips indicate that, although the water protection and drainage systems of the tunnels are locally defective, they are not completely failed.

Table 2. Statistical table of crack direction of mountain tunnels in Shanxi.

Tunnel type	Percentage (100%)	Leakage	Wet stains	Water droplets	Drips
Circumferential cracks	43.86	48.89	23.55	15.36	12.5
Longitudinal cracks	38.08	21.37	46.52	23.56	8.55
Diagonal cracks	18.06	58.23	30.01	5.88	5.88

4 ANALYSIS OF HAZARD CAUSES AND COUNTERMEASURES

4.1 *Causes of leakage*

Leakage is often related to lining cracking. Among the numerous causes of leakage occurring to tunnel linings, there are generally three causes: natural conditions, construction conditions, and the waterproofing material used in tunnel engineering.

Natural conditions: The greater the drop in ambient temperature in a tunnel is, the greater the shrinkage deformation due to temperature changes will be, and the more likely cracks will appear; the impact on the tunnel lining of the entrance section is the greatest, which is most susceptible to cracking; in winter, cold shrinkage of concrete reaches the maximum. With other unfavorable factors, cracks may occur in the entrance linings of short and medium tunnels. The lower the ambient humidity is, the greater the shrinkage deformation of tunnel lining concrete will be, and the more likely cracks will appear in the lining. Leakage is very likely to occur once cracks appear.

Construction conditions: After a mountain tunnel is excavated, a surrounding rock loosened zone is formed within a certain area of the tunnel. In this area, the surrounding rock is deformed due to changes in the original ground stress of the formation. Cracks open up, allowing groundwater to flow into the tunnel along the open cracks. In areas farther from the tunnel excavation line, the formation remains undisturbed, and the groundwater still runs off following the pattern.

Deficiencies in tunnel water control and drainage designs: For a long time, the lack of awareness of the leakage hazard in tunnels has led to a tendency to emphasize the structure, and light waterproofing in tunnel design. Designers calculate and test the strength, stiffness, and stability of the structure repeatedly, but in addition to the gutter section size, few of them calculate the progress of water control and drainage. Even when calculating water pressure in sections with high water pressure, they only think about the requirements for structural design.

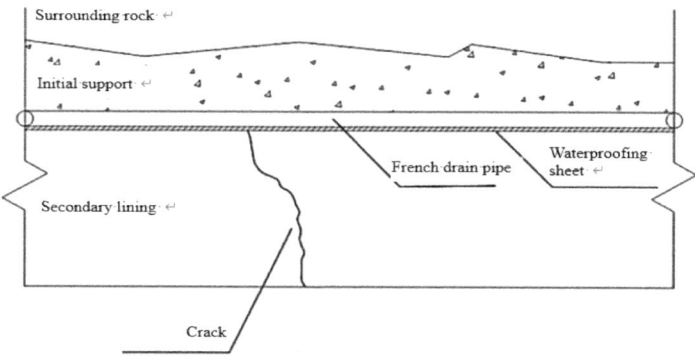

Figure 9. Drainage system plan.

4.2 Leakage countermeasures

4.2.1 Countermeasures for cracks

For mountain tunnels in Shanxi Province, cracks with a width of < 0.2mm are treated by surface sealing; cracks with a width ≥ 0.2 mm are treated by pressure grouting. In case of small tunnel clearance, cracks are usually reinforced using steel strips. In case of cracks in thin and weak linings or linings with local damages, shotcrete reinforcement is an effective way.

4.2.2 Countermeasures for leakage

To treat leakage, it is necessary to carefully analyze the causes of leakage, take pertinent measures and follow the principles of "combination of drainage and blocking, rigid and flexible measures, consideration of local conditions, and comprehensive management". For tunnels in operation in Shanxi Province, leakage and water droplets are the main leakage phenomena. Considering the small leakage amount, it is recommended to apply SWF concrete sealant to the lined surface. This material contains a special compound water-based solution with strong seepage force and induces an alkaline reaction with the concrete. The gel it generates will fill the capillaries and block the seeping channels. The percentage of drips in total leakage phenomena is just 11.5%. However, a large number of drips can easily lead to pavement ponding and icing in winter, which will cause great threats to traffic safety. Therefore, drips must be timely treated by slurry plugging, buried pipe drainage, etc.

5 CONCLUSIONS

(1) For mountain tunnels in Shanxi Province, most parts of tunnel lining with a leakage are at the side wall, accounting for 42.1% of the total leakage, followed by the spandrel and arch springing; the leakage phenomena are mostly leakage and water droplets, respectively accounting for 37.7% and 33.4%. Only 11.5% of the leakage phenomena are drips. As to the width of cracks at lining leakage, most cracks are small; most lining cracks of mountain tunnels are longitudinal and circumferential; diagonal cracks are relatively uncommon. Tunnel cracks show a pattern of circumferential cracks > longitudinal cracks > diagonal cracks. When building a leakage hazard evaluation and handling system, different types of leakage must be considered comprehensively instead of being graded just by percentage or severity.

(2) General research and investigation were carried out on the causes and phenomena of lining leakage hazards of mountain tunnels in Shanxi Province. The distribution pattern of different types of leakage in terms of parts, scales, and forms was summarized. Measures were proposed pertinently to treat crack leakage.

REFERENCES

Cui J.T. Discussion on treatment measures of water seepage in tunnel [J]. *Building Safety*, 2020, 35(04):38–40.
Ding W.Q. Analysis of water prevention and drainage system of a large span tunnel in Shenzhen [J]. *Highway Traffic Technology*, 2018, 34 (S1) :76+82.
Lin C.J. *Studies on Lining Water Leaking Mechanism and Grouting Governing in Operational Tunnels* [D]. Shangdong University, 2017.
Lei D.J. Causes analysis and remediation measures of water leakage in a tunnel lining during operation period [J]. *Fujian Construction Science and Technology*, 2022(01):93–95.
Li S.T., Feng X.B. Causes of water leakage in tunnel lining and remediation measures [J]. *Guangdong Building Materials*, 2021, 37(06):73–75.
Liang F. *Study on the Influence of groundwater on mountain tunnel Construction and Prevention Measures* [D]. Chongqing University, 2007.
Li X.Q., Zhang S.L., Bao T., et al. Analysis on characteristics, causes and control measures of water leakage disease of highway tunnel in operation [J]. *Shandong Communications Science and Technology*, 2020(05):56–59+76.

Ma P.D. Research on water seepage control measures of tunnel lining [J]. *Enterprise Science and Development*, 2022(02):93–95.

Tian J.J. Cause analysis and solution of water leakage in a highway tunnel [J]. *Traffic World*, 2020(Z2):140–141+187.

Xue X.H., *Study on water Seepage Prevention and control technology of Operating tunnel in Shanxi Province*. Shanxi Transportation Research Institute, Shanxi Province, February 26, 2016.

Ying W., Case study on causes and treatment measures for water leakages of highway tunnels in Taizhou, Zhejiang Province[J]. *Tunnel Construction*, 2015(S2):220–223.

Zou Y.L., He C., Zhou Y., et al. Statistics and cause analysis of leakage diseases in operating expressway tunnels in Chongqing[J]. *Journal of Highway and Transportation Research and Development*, 2013, 30(01):86–93+101.

Zhang C. Study on comprehensive prevention technology of water Seepage in tunnel[J]. *Shandong Transportation Science and Technology*, 2018(05):70–71+78

Zhang S.L., Feng Q.Z., Ying G.G., et al. Cause of water seepage in highway tunnel and field test of new drainage material [J]. *Highway Traffic Science and Technology*, 2013, 30 (10): 86–91.

*Frontiers in Civil and Hydraulic Engineering – Mohamed A. Ismail and
Hazem Samih Mohamed (Eds)*
© *2023 The Authors, ISBN 978-1-032-38247-0*

Research on the mechanism and influencing factors of geysers in deep tunnel drainage systems

Bin Wang*, Wenbin Lv, Rurong Chen & Jinming Zhang
Pearil River Water Resources Research Institute, Guangdong, China

Chun Li
Graduate School of Hohai University, Nanjing, China

ABSTRACT: The geysering phenomenon is a serious accident that can occur in deep tunnel drainage systems. Many geyser incidents have occurred in the actual operation of deep tunnel drainage tunnels. In this paper, a physical model is used to imitate the occurrence process of the geyser phenomenon. The mechanism of geysers is studied, and the key factors and influencing patterns of geysers are analyzed. The conclusion provides a reference for the safe operation of the deep tunnel drainage system.

1 GENERAL INSTRUCTION

Global climate changes have caused extreme weather frequently in recent years, The contradiction between the demand for urban drainage and sewage and the insufficient flow capacity of the existing drainage network is constantly intensifying, accompanied by prominent sewage overflow and waterlogging problems. In order to settle serious urban overflow pollution and rainstorm waterlogging problems, many cities (Hong Kong, Tokyo, Japan, Edogawa, Singapore, Chicago, Mexico, Paris, and London) have developed deep tunnel drainage systems as an important part of the urban drainage system. However, many geysering accidents have occurred during the operation of the deep tunnel system, which can be described as gas-entrained water flow blowing out of the shaft like geysering. The gas-water mixture can reach up to tens of meters, and cause serious damage to the manhole cover. With the construction and operation of deep tunnels in major cities in the world, the geysering problem will become the major problem affecting the safe operation of the tunnel.

Figure 1.   Cities with geysering-damaged structures.
(a) St. Paul, Minnesota, United States 1982 (b) Minneapolis, Minnesota, U.S. 1997 (c) Chicago, Illinois 2010.6.23 (d) Quebec City, Montreal 2011.7.18 (e) Novosibirsk, Russia (f) Structure damaged by geysering.

*Corresponding Author: 278273213@qq.com

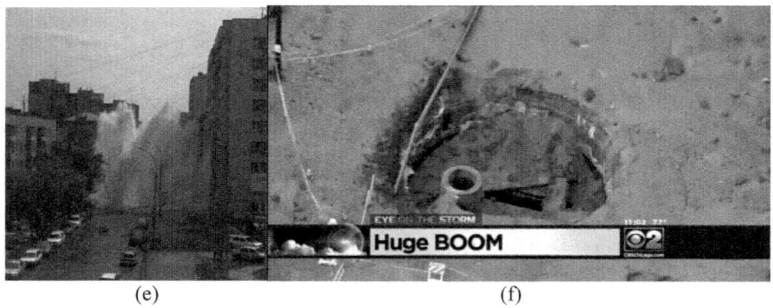

(e)   (f)

Figure 1. Continued.

The geysering phenomenon is defined as the violent and intermittent gushing phenomenon of the water-gas mixture from the shaft (e.g., Leon 2006; Leon et al. 2006). The gas-water mixture can reach up to tens of meters. Wright, Lewis, and Vasconcelos (2011) studied the geysering phenomenon in Minneapolis drainage tunnels, recorded 9 geysering processes, each of which consisted of several eruptions, and each eruption lasted 10–25 seconds, with an interval of about 75–90 s. Many researchers have studied the hydrodynamic characteristics of the geysering phenomenon since the 1980s (Cardle et al. 1989; Hamam & McCorquodale 1982; Valentin 1981; Zech et al. 1984; and recently Aimable & Zech 2003; Zhou et al. 2002, 2004; and Vasconcelos and Wright 2003, 2005). Due to the complexity of the problem, most research starts with physical modeling studies. Guo and Song (1991) discussed the hydrodynamic characteristics of the small-scale shaft model, and they assumed that the inflow rate of the shaft increases rapidly and the water flow in the shaft is uniformly mixed, indicating that a strong surge in the main tunnel is one of the sources of the geysering phenomenon. Vasconcelos and Wright (2008, 2009), Gui Z N et al. (2006), and Vasconcelos and Leite (2012) found that the free-surface-pressurized flow could cause irregular changes in water depth, pressure, flow rate, and air cavity in the tunnel, which will lead to structural damage and overflow in severe cases. Song et al. (1983), and Guo and Song (1990) believed that the rapid fluctuating inflow into the tunnel will cause the free-surface-pressurized flow, which can create an air cavity in the tunnel, and make the flow regime in the tunnel extremely complex, and the air cavity migration and pressure fluctuations will lead to the geysering phenomenon. Steven J. Wright (2016) indicated that negative pressure may exist in tideway deep tunnels, which can cause instability of the tunnel structure, and lead to the air cavity in the tunnel. Li and McCorquodale (1999) concluded that the geysering phenomenon will endanger the operation of the deep tunnel drainage system.

Abad et al. (2009) pointed out that drop shafts are important facilities that lead inflow from the ground into the tunnel, exhausting air while consuming inflow energy. The ideal operating state of the drop shaft should ensure the minimization of the intake air volume, while the inflow rate reaches the maximum. In order to make the geyser reach a height of tens of meters (Wright et al. (2011a, b), the flow velocity in the shaft should also be large enough. Vasconcelos (2005, 2017) and Wright et al. (2008) studied the transient flow in the tunnel and proposed that the floating air cavity in the shaft will lead to geysering as well. Zhou et al. (2002) found that the poor ventilation of the shaft and pressure increase in the tunnel are also direct reasons for the geysering phenomenon. Vasconcelos and Wright (2003) believed that the air pressure is not the only cause of manhole cover blowout and structural damage, and the water volume in the shaft is also a necessary factor. Wang J (2018) recorded the fluctuating pressure process in drop shafts during geysering.

Previous research results show that two cases could cause the geysering phenomenon: The first is the air cavity floating from the ventilation shaft or the drop shaft (Vasconcelos, J.G., and Wright, S.J. 2011). The air cavity migrates at the crown with the pressurized flow in the tunnel. When the air cavity reaches the ventilation shaft, it drives the water into the surface, which causes the geysering phenomenon. The water-air mixture released from the ventilation shaft is the second possibility. A surge will be generated in the interface between the open channel flow and the pressurized flow in the fast-filling tunnel. When the open channel flow reaches the ventilation shaft, the pressurized water-air mixture creates the geysering phenomenon in the shaft.

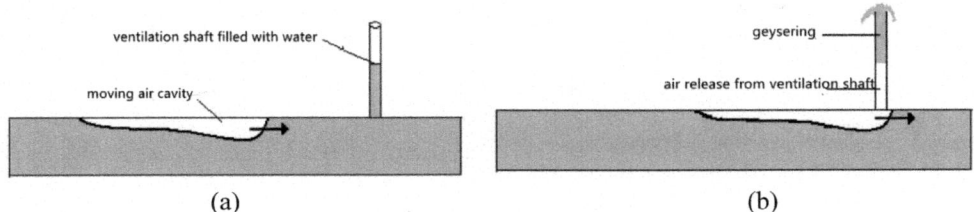

Figure 2. Air cavity overflow from the ventilation shaft.
(a) Air Cavity Transport Along the Tunnel (b) Air Cavity Overflow from the Ventilation Shaft.

Researchers have concluded the preliminary understanding of the geysering mechanism: With hydraulic surge, buoyancy, and water-air interaction in the tunnel system, the air cavity is transported from the tunnel to the ventilation shaft and released, which causes the geysering phenomenon. When multiple air cavities are released intermittently, continuous geysering may be caused. Although it is known that the shaft diameter, shaft water level, air cavity volume, and pressure will affect the occurrence of geysering, and its classification and occurrence mechanisms, and various influencing factors remain unknown.

This paper focuses on the hydraulic characteristic and process during geysering, categorizes geysering types, analyzes the geysering mechanism, and discusses the influences of various factors on geysering by changing the shaft diameter, shaft water level, and air cavity volume.

2   MODEL DESIGN

The model setup in this paper includes upstream pool, downstream pool, pipe, ventilation shaft, and backwater facility, among which the size of the downstream pool is 2.5 m×2.5 m×4.0 m, and that of the upstream pool is 2.0 m×1.0 m×2.5 m. A 10 m long plexiglass pipe is erected between the two pools, with a slope of 0.1% and the inner diameter of 0.27 m; the upstream and downstream valves are located at both ends of the pipe, the ventilation shaft is located in the middle of the pipeline to simulate geysering, and 4 pressure transmitters are set to record the instantaneous pressure change during geysering; in order to change the volume of the air cavity and switch the experimental operation mode, 4 air valves are arranged in the pipe, with an interval of 0.8 m. 2 industrial cameras are used to record the geysering process.

In order to explore the geysering phenomenon in both the closed conduit pipe and open flow pipe, 2 models are set in this experiment. In the experiment model A, namely the closed conduit pipe, the downstream valve is always closed, $H_{shaft}=H_{up}$, by adjusting air valves 1#-4#, and different air cavity volumes can be combined. In the experiment, the air valve 1# will be opened quickly, and the release process of the air cavity can be recorded.

In the experiment model B, namely the connected pipe, the downstream valve is always opened by adjusting air valves 1#-4#, and different air cavity volumes can be combined. In the experiment, the air valve 1# will be opened quickly, and the release process of the air cavity can be recorded.

The fluctuating pressure is measured with a ϕ10 mm miniature pressure sensor at the signal sampling frequency of 500 Hz. The gas geysering process is recorded by an industrial camera (pixel: 2560*1024).

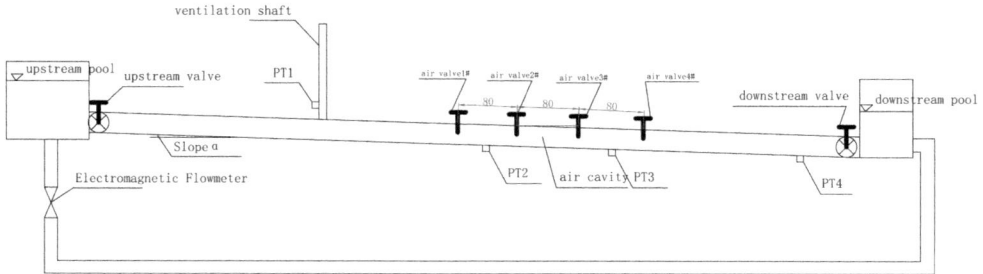

Figure 3.　Layout of experiment devices.

Figure 4.　Model photo.

Figure 5.　Ventilation shaft photo.

Figure 6.　Fluctuating pressure sensor and acquisition equipment.

3　TEST CONDITIONS

Multiple experimental configurations are created by changing three parameters. shaft diameter (0.075, 0.1, and 0.15 m), air cavity volume (0.046, 0.092, and 0.137 m^3), and shaft water level, i.e., initial upstream water level, as seen in Tables 1 and 2. Each experimental run is repeated 15 times.

Table 1. Test conditions of experiment model A (Closed Conduit Model).

Shaft diameter D (m)	Air cavity volume (m³)	Upstream water level H_0 (m)	Shaft water level H_{shaft} (m)
0.075/0.1/0.15	0.046/0.092/0.137	0.31/0.45/0.61/0.75	Same as the upstream water level

Table 2. Test conditions of experiment model B (Closed Conduit Model).

Shaft diameter D (m)	Air cavity volume (m³)	Upstream water level H_0 (m)	Downstream water level H_{down} (m)	Shaft water level H_{shaft} (m)
0.075/0.1/0.15	0.046/0.092/0.137	0.31/0.45/0.61/0.75	0.45/0.61/0.75/0.9	Same as the upstream water level

4 ANALYSIS OF EXPERIMENT RESULTS

The geysering intensity can be reflected by the water crushed level (H_{cr}) and velocity (V_{cr}). The former is defined as the level at the moment when the air column breaks, and the latter is defined as the maximum flow rate of geysering. If the air column breaks in the shaft, H_{cr} and V_{cr} can be measured. If the air column breaks outside the shaft, H_{cr} and V_{cr} are obtained through industrial camera photos.

4.1 Geysering mechanism

Two types of geysering are found in the experiment: gas-flow geysers and surge-type geysers, and the mechanism and performance characteristics of the two types are different.

1. Mechanism of gas-flow geysers
   When the air cavity in the tunnel is transported to the shaft, it is driven by buoyancy and external pressure. When the trapped air cavity in the tunnel is transported to the shaft position, the air cavity enters the shaft to form an air column under the action of the air cavity buoyancy. The gas

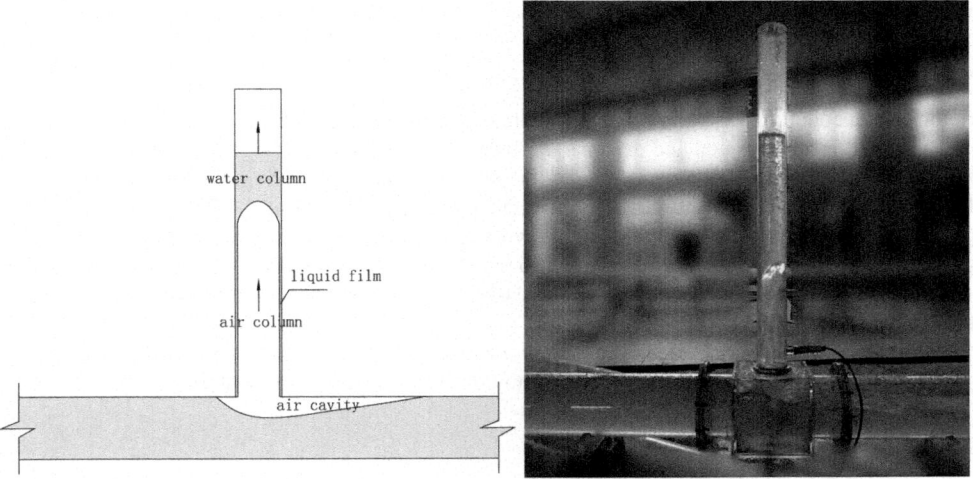

Figure 7. Schematic diagram of gas-flow geysers.

column rises rapidly under the action of external pressure, and the water-air mixture spews out to form the gas-flow geysers. After the continuous release of water-air mixture in the shaft, the pressure in the tunnel system is gradually reduced. The two necessary conditions for gas-flow geysers are: the large volume air cavity and external pressure of the tunnel system.

2. Mechanism of surge-type geysers

Surge-type geysers generally follow gas-flow geysers. After the trapped air cavity in the tunnel is released, a low-pressure zone is formed in the shaft under the high pressure from upstream and downstream of the shaft. The water flow fills the low-pressure zone instantly. The surge effect causes instantaneous high pressure in the tunnel and squeezes the air near the shaft into a high-pressure air cavity. When the water-air mixture is expelled from the shaft, the surge-type geysers are formed. The surge-type geysers are the result of the water inertial oscillation and the compression effect of the air cavity in the tunnel. The higher the external pressure and local pressure at the shaft contributes to the high intensity of surge-type geysers.

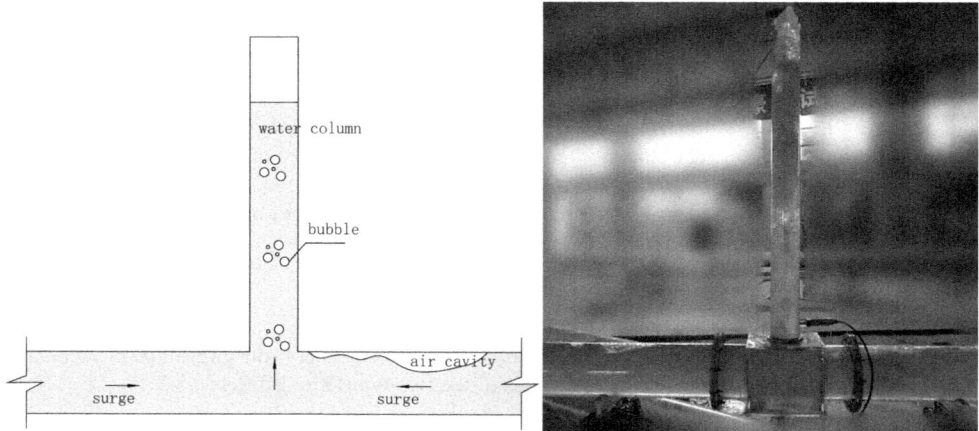

Figure 8.　Schematic diagram of surge-type geysers.

Comparing the two types of geysering, the intensity of gas-flow geysers is relatively small, while that of surge-type geysers is much larger, but surge-type geysers generally follow gas-flow geysers, which is the same as the phenomenon recorded by Wright et al. (2011).

4.2　*Pressure characteristics of geysering*

In the closed conduit pipe experiment A (Figure 9), with the effect of the upstream pressure head of the system, the compression and expansion effects of the air cavity are significantly enhanced. The system pressure oscillation is accelerated with the occurrence of geysering. The pressure patterns of PT1 and PT2 are basically the same. In the connected pipe experiment B (Figure 10), the external pressure head exists both upstream and downstream of the shaft, and pressure oscillations are more complex. Due to the difference between the upstream and downstream water levels, the flow velocity of the water flow in the tunnel increases, and the compression and expansion effects of the air cavity are weakened.

Under the pressure surge effect, the system pressure rises sharply. Due to the instantaneous high pressure, the water-gas mixture is ejected from the shaft and forms surge-type geysers.

4.3　*Influence of water column length*

Figures 11 and 12 display the relationship between the water level in the shaft, and the water crushed level (H_{cr}) and velocity (V_{cr}). Experiments show that both the water crushed level (H_{cr})

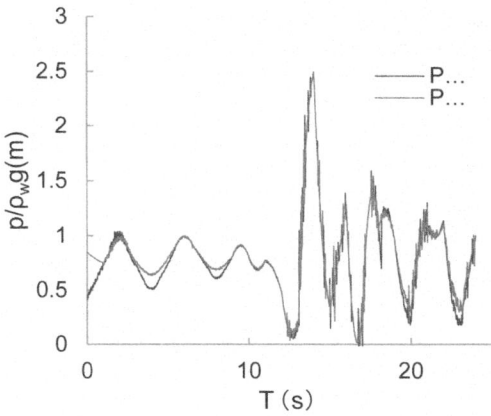

Figure 9.   Transient pressure process diagram for expA.

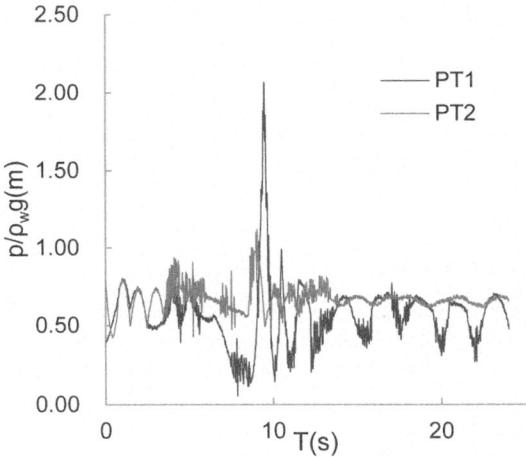

Figure 10.   Transient pressure process diagram for ExpB.

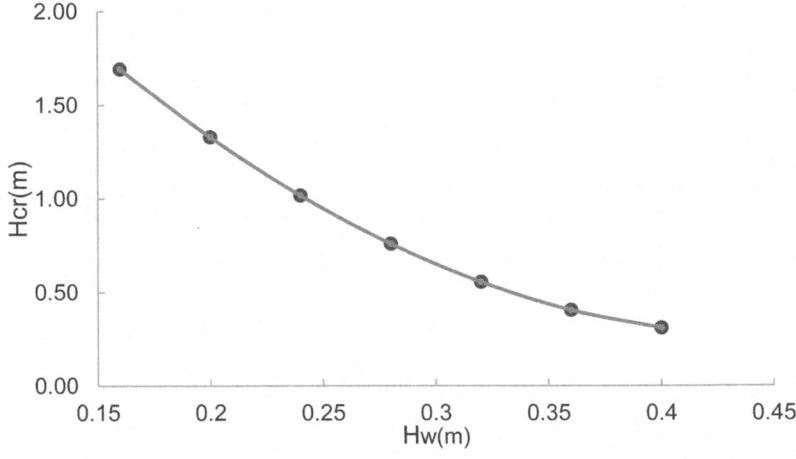

Figure 11.   The relationship between water level in the shaft and the water crushed level.

and velocity (V_{cr}) decrease with the increase in the water level in the shaft. The variation rule of the water crushed level (H_{cr}) and velocity (V_{cr}) under different shaft diameters are basically the same.

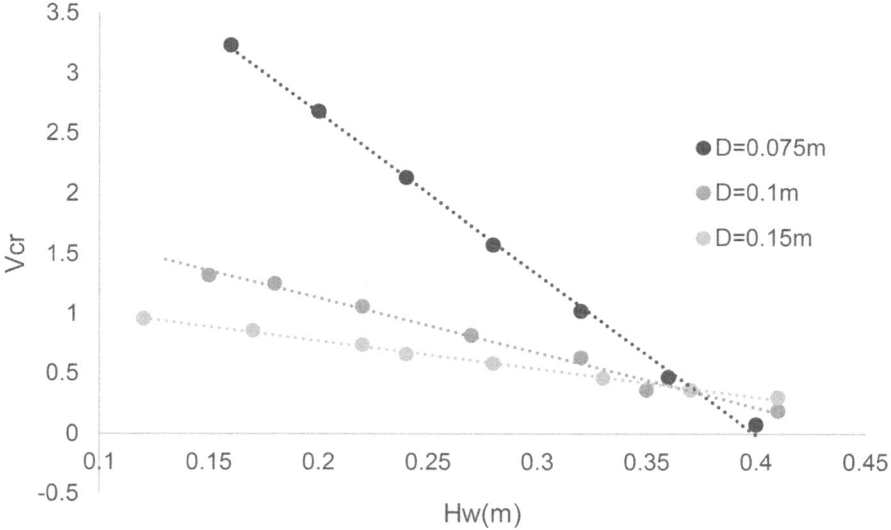

Figure 12. Relationship between the water level in the shaft and the water crushed velocity.

4.4 Influence of upstream water level

Figures 13 and 14 present the relationship between the upstream water level, and the water crushed level (H_{cr}) and velocity (V_{cr}). When the shaft diameter is 0.075 m in Experiment A with different air cavities, the variation rules of the water crushed level (H_{cr}) and velocity (V_{cr}) under different

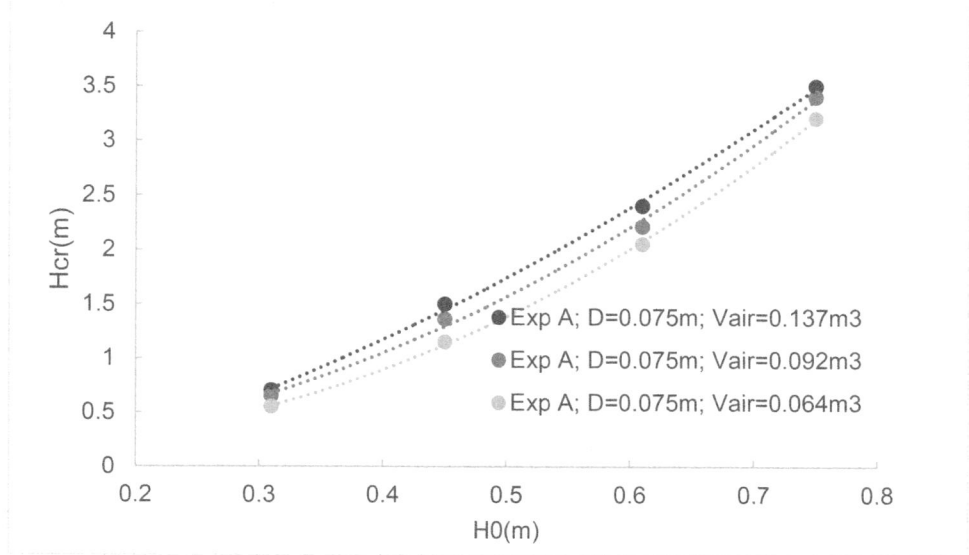

Figure 13. Relationship between the upstream water level and the water crushed level.

shaft diameters are basically the same. Under small air cavities, when the shaft air column and the tunnel air cavity are not connected, the water crushed level decreases significantly; for large air cavities, the relationship between the water crushed level and the external pressure is more regular, and the water crushed level (H_{cr}) is proportional to the external pressure.

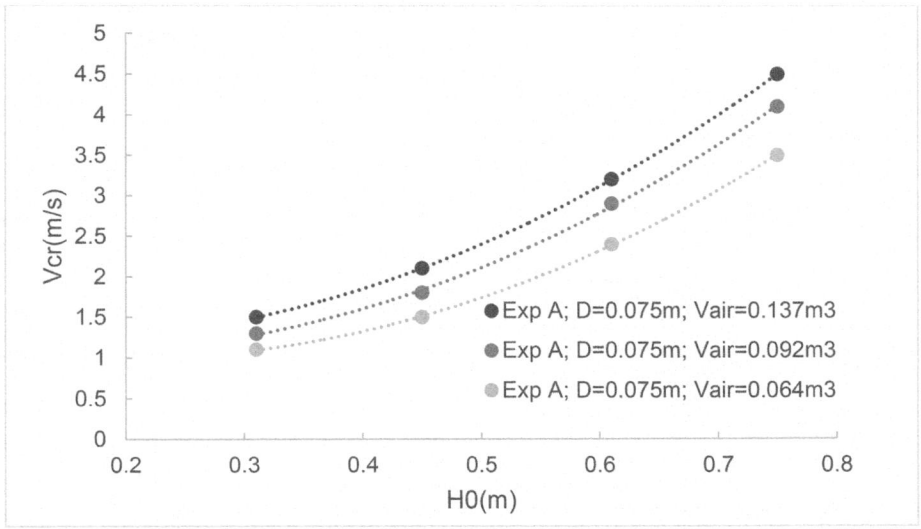

Figure 14.   Relationship between the upstream water level and the water crushed velocity.

4.5   *Influence of air cavity volume*

Figures 15 and 16 show the relationship between the air cavity volume, and the water crushed level (H_{cr}) and velocity (V_{cr}). When the shaft diameter is 0.075 m in Experiment A with a long enough

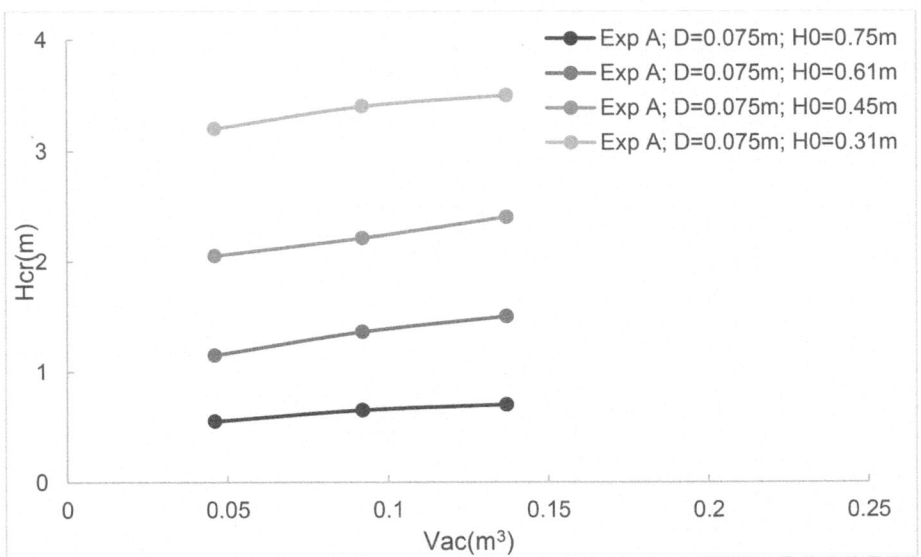

Figure 15.   Relationship between the air cavity volume and the water crushed level.

air cavity (i.e., the shaft air column is connected to the tunnel air cavity), the water crushed level (H_{cr}) is stable, and the H_{cr} changes little with the increase in the air cavity volume.

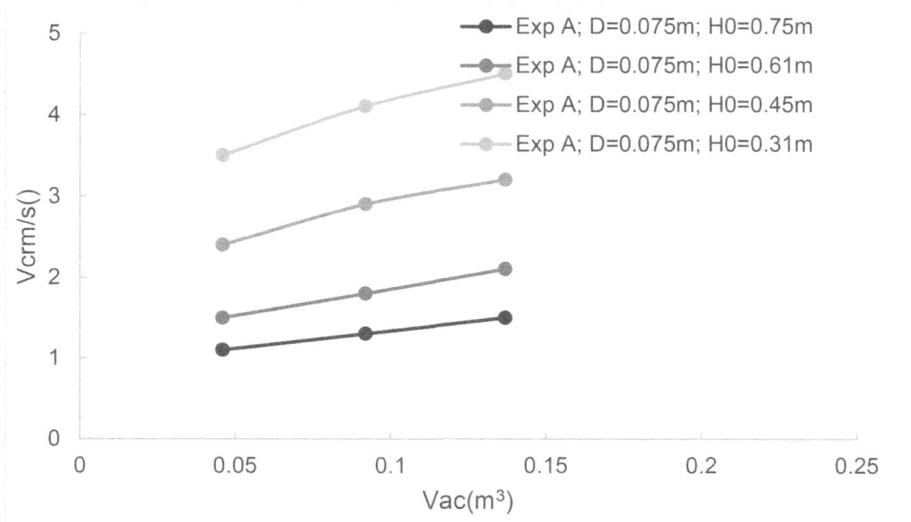

Figure 16.   Relationship between the air cavity volume and the water crushed velocity.

4.6   *Influence of the shaft diameter*

Figures 17 and 18 exhibit the relationship between the drop shaft diameter, and the water crushed level (H_{cr}) and velocity (V_{cr}). When the air cavity volume is 0.137 m^3 in Experiment A, both the water crushed level (H_{cr}) and velocity (V_{cr}) decrease significantly with the increase in the shaft

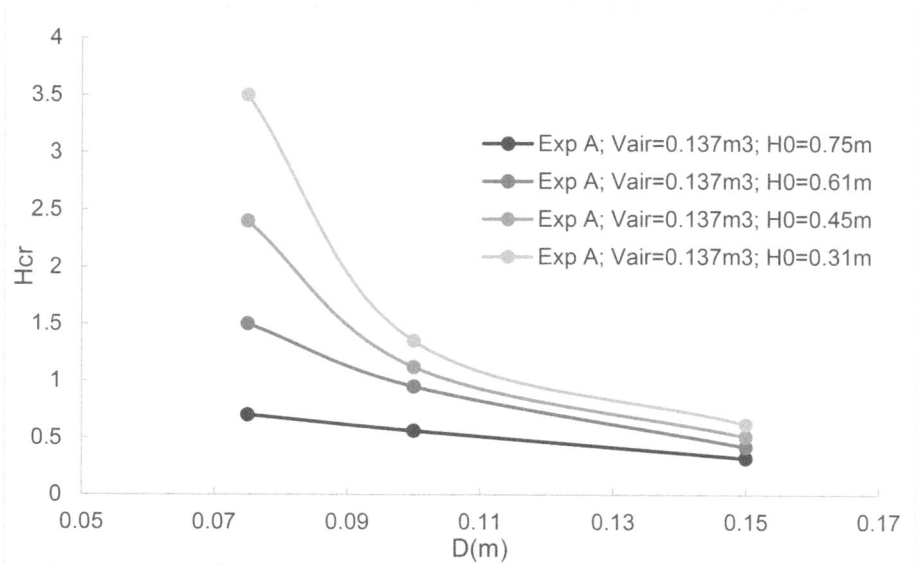

Figure 17.   Relationship between the drop shaft diameter and the water crushed level.

diameter. When the shaft diameter is larger than 0.1 m, the geysering intensity gradually remains the same. When the shaft diameter reaches 0.15, the geysering phenomenon cannot be observed in the experiment.

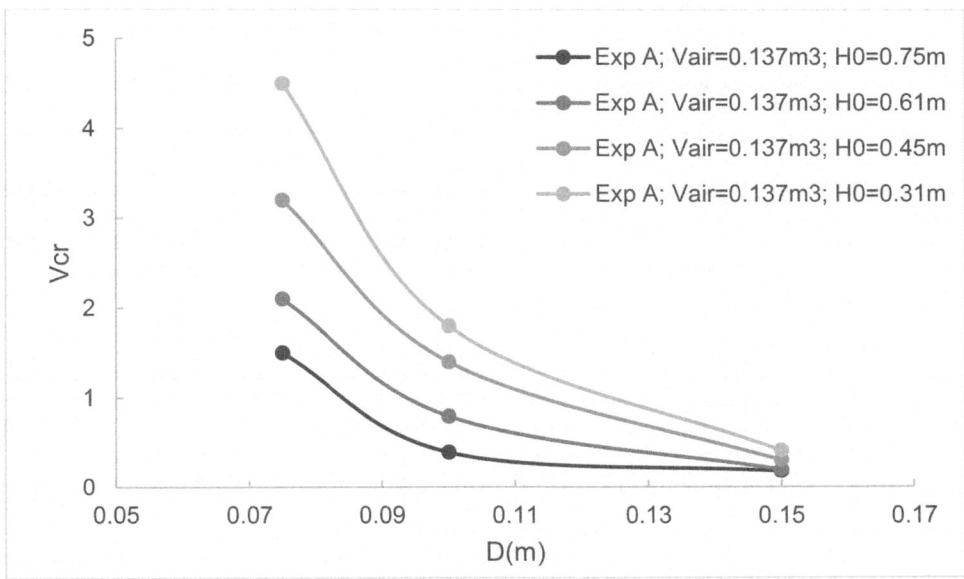

Figure 18. Relationship between the drop shaft diameter and the water crushed velocity.

5 CONCLUSION

A physical model is used to imitate the occurrence process of the geyser phenomenon. In this paper, the geysering mechanism is studied, the key influencing factors and influencing patterns of gaysers are analyzed. Conclusions are as follows:

1. Two types of geysering are found in the experiment: gas-flow geysers and surge-type geysers. Gas-flow geysers are driven by the buoyancy and external pressure. The two necessary conditions for gas-flow geysers are: the large volume air cavity and external pressure of the tunnel system.
2. The intensity of gas-flow geysers is relatively small, while that of surge-type geysers is much larger, but surge-type geysers generally follow gas-flow geysers.

FUNDING

This study was financially supported by the Science and Technology Project of Guangzhou (Research on the mechanism and influencing factors of geysers in the deep tunnel drainage system with the grant number of 201904010335).

REFERENCES

Arturo S. Leon. (2019) Mechanisms that lead to violent geysers in vertical shafts. *Journal of Hydraulic Research* 57:3, 295–306.
Besharat M., Tarinejad R., Ramos H.M. The effect of water hammer on a confined air pocket towards flow energy storage system [J]. *Aqua,* 2016, 65(2).

Cataño-Lopera Y.A., Tokyay T.E., Martin J.E., et al. Modeling of a transient event in the tunnel and reservoir plan system in Chicago, Illinois [J]. *Journal of Hydraulic Engineering*, 2015, 140 (9): 05014005.

Cong J., Chan S.N., Lee J.H.W. Geyser formation by release of entrapped air from horizontal pipe into vertical shaft [J]. *Journal of Hydraulic Engineering*, 2017, 143(9).

Eng. Appl. *Computation Fluid Mech.*, 5(1), 127–140.

Hatcher T.M., Vasconcelos J.G. Peak pressure surges and pressure damping following sudden air pocket compression [J]. *Journal of Hydraulic Engineering*, 2016.

Huber, W.C., and Dickinson, R. (1992). *Storm water management model, version 4: User's manual*, 2nd Ed., U.S. Environmental Protection Agency, Athens, GA.

Izquierdo, J., Fuertes, V.S., Cabrera, E., Iglesias, P.L., and García-Serra, J. (1999). Pipeline start-up with en-trapped air. *J. Hydraul. Res.*, 37(5), 579–590.

Jiachun Liu, Shuangqing Zhang, Biao Huang, David Z. Zhu. (2021) Numerical study on effects of chamber design and multi-inlet on storm geyser. *Water Science and Technology* 142.

Jiachun Liu, Yu Qian, David Z. Zhu, Jian Zhang, Stephen Edwini-Bonsu, Fayi Zhou. (2022) Numerical study on the mechanisms of storm geysers in a vertical riser-chamber system. *Journal of Hydraulic Research* 60:2, pages 341–356.

Jorge Molina, Pablo Ortiz. (2022) Propagation of large air pockets in ducts. Analytical and numerical approaches. *Applied Mathematical Modelling* 110, pages 633–662.

Kerger, F., Archambeau, P., Erpicum, S., Dewals, B.J., and Pirotton, M. (2011). An exact Riemann solver and a Godunov scheme for simulating highly transient mixed flows. *J. Comput. Appl. Math.*, 235(8), 2030–2040.

Klaver P., Collins D., Robinson K., et al. Modeling of Transient Pneumatic Events in a Combined Sewer Overflow Storage Tunnel System[J]. *Journal of Water Management Modeling*, 2016.

León, A.S., Ghidaoui, M.S., Schmidt, A.R., and García, M.H. (2010). A robust two-equation model for transient-mixed flows. *J. Hydraul. Res.*, 48(1), 44–56.

León, A.S., Oberg, N., Schmidt, A.R., and García, M.H. (2011). *The Illinois transient model: A state-of-the-art model for simulating the flow dynamics in combined storm-sewer systems*. Cognitive Modeling Urban Water Systems, monograph 19, W. James, ed., CHI, Guelph, ON.

Li, J., and McCorquodale, A. (1999). *Modelling mixed flow in storm sewers*. J. Hydraul. Eng., 10.1061/(ASCE)0733-9429(1999)125: 11(1170), 1170–1180.

Pasadena, CA. Nielsen, K. D., and Davis, A. L. (2009). *Air migration analysis of the Terror Lake tunnel. Proc., of the congress Int. Association for Hydraulic Research (IAHR)*, Water Engineering for a Sustainable Environment, Madrid, Spain, 262–268.

Taha, T., and Cui, Z.F. (2006). CFD modelling of slug flow in vertical tubes. *Chem. Eng. Sci.*, 61(2), 676–687.

Taher Chegini, Arturo S. Leon. (2020) Numerical investigation of field-scale geysers in a vertical shaft. *Journal of Hydraulic Research* 58:3, pages 503–515.

Trindade, B.C., and Vasconcelos, J.G. (2013). Modeling of water pipeline filling events accounting for air phase interactions. *J. Hydraul. Eng.*, 10.1061/(ASCE)HY.1943-7900.0000757, 921–934.

Ubbink, O. (1997). *Numerical prediction of two fluid systems with sharp interfaces. Ph.D. thesis*, Imperial College of Science, Technology and Medicine, London.

Vasconcelos J.G., Wright S.J. Anticipating transient problems during the rapid filling of deep stormwater storage tunnel systems [J]. *Journal of Hydraulic Engineering*, 2017, 143(3).

Vasconcelos, J.G., and Wright, S.J. (2005). Experimental investigation of surges in a stormwater storage tunnel. *J. Hydraul. Eng.*, 10.1061/ (ASCE)0733-9429(2005)131:10(853), 853–861.

Vasconcelos, J.G., and Wright, S.J. (2006). Mechanisms for air pocket entrapment in stormwater storage tunnels. Proc., *World Environment and land Water Resources Congress*, ASCE, Reston, VA.

Vasconcelos, J.G., and Wright, S.J. (2011). Geysering generated by large air pockets released through water-filled ventilation shafts. *J. Hydraul. Eng.*, 10.1061/(ASCE)HY.1943-7900.0000332, 543–555.

Wang J., Vasconcelos J.G. Manhole cover displacement caused by the release of entrapped air pockets [J]. *Journal of Water Management Modeling*, 2018, 26.

William L. Peirson, John H. Harris, Richard T. Kingsford, Xi Mao, Stefan Felder. (2021) Piping fish over dams. *Journal of Hydro-environment Research* 39, pages 71–80.

Wright S.J., Klaver P. Assessment of pressure transients due to trapped air compression in rapidly filling combined sewer overflow tunnels [J]. *Journal of Water Management Modeling*, 2016.

Wright S.J., Vasconcelos J.G., Lewis J.W. Air–water interactions in urban drainage systems[J]. *Engineering & Computational Mechanics*, 2017, 170(3).

Wright, S.J., Lewis, J.W., and Vasconcelos, J.G. (2011a). Geysering in rapidly filling storm-water tunnels. *J. Hydraul. Eng.*, 10.1061/(ASCE) HY.1943-7900.0000245, 112–115.

Wright, S.J., Lewis, J.W., and Vasconcelos, J.G. (2011b). Physical processes resulting in geysers in rapidly filling storm-water tunnels. *J. Irrig. Drain. Eng.*, 10.1061/(ASCE)IR.1943-4774.0000176, 199–202.

Wright, S.J., Vasconcelos, J.G., Creech, C.T., and Lewis, J.W. (2008). Flow regime transition mechanisms in rapidly filling stormwaters storage tunnels. *Environ. Fluid Mech.*, 8 (5–6), 605–616.

Yan Zhang, Yaohui Chen, Shangtuo Qian, Hui Xu, Jiangang Feng, Xiaosheng Wang. (2022) Experimental study on geysers in covered manholes during release of air pockets in stormwater systems. *Journal of Hydraulic Engineering* 148:5.

Yu Qian, David Z. Zhu, Bert van Duin. (2022) Design considerations for high-speed flow in sewer systems. *Journal of Hydraulic Engineering* 148: 9.

Zhou, F., Hicks, F.E., and Steffler, P.M. (2002). Transient flow in a rapidly filling horizontal pipe containing trapped air. *J. Hydraul. Eng.*, 10.1061/(ASCE)0733-9429(2002)128:6(625), 625–634.

Zhou, L., Liu, D., and Karney, B. (2013). Investigation of hydraulic transients of two entrapped air pockets in a water pipeline. *J. Hydraul. Eng.*, 10.1061/(ASCE)HY.1943-7900.0000750, 949–959.

Zhou, L., Liu, D., and Ou, C. (2011). *Simulation of flow transients in a water filling pipe containing entrapped air pocket with VOF model.*

Frontiers in Civil and Hydraulic Engineering – Mohamed A. Ismail and
Hazem Samih Mohamed (Eds)
© 2023 The Authors, ISBN 978-1-032-38247-0

Creep analysis of rocky slope

Qingyun-Hou*

China Investment Consulting Co., Ltd. Beijing, China

ABSTRACT: Rock creep is one of the important causes of slope instability. For projects with a long service life, the long-term stability analysis is required in addition to the stability during construction. Based on the foundation pit of a Yangtze River Bridge in Chongqing, the numerical simulation of the foundation pit slope is carried out using the finite difference program FLAC with the Cvsic model to analyze the deformation creep characteristics. First, the stress and strain at the foot of the slope after slope excavation is calculated, the creep analysis of the horizontal and vertical displacements at different positions on the slope within 10 years is performed, and then the horizontal and vertical displacements of the slope after 10-year use of anchor supports are calculated, with the creep characteristics of the axial force of the anchor also considered. The results show that the horizontal stress at the foot of the slope tends to decrease after 10 years. The shear stress and vertical stress are gradually stabilized. The horizontal displacement near the foot of the slope increases slowly after 2 years. The horizontal displacement at other locations basically tends to be stable after 2 years, and the horizontal displacement mainly occurs in soft rocks. The vertical displacements of each point on the slope surface are small and tend to be stable in the second year. After the anchor support is added, the horizontal displacement of the slope is significantly reduced. The anchor rod plays a larger role near the slope foot.

1 INTRODUCTION

Rock rheology is one of the important reasons for the large deformation and even instability of rock underground works, building foundations, slopes, and landslides. For some projects with a long service life, it is necessary to consider the creep characteristics of the rock and conduct the long-term stability analysis in addition to the stability during construction (Cai 2002). In recent years, some progress has been made in the study of rock creep characteristics. In terms of theory, a unified rheological model is proposed by Xia C C et al. (Xia 2008), along with the method of identifying the rheological model corresponding to the rheological state of the rock from the loading and unloading creep test results under different stress levels. Qi (2008) put forward a strain-triggered nonlinear sticky pot in series on the Nishihara model, and proposed an improved Nishihara model. Xu (2014) proposed a fracture plastic element to describe the change law of the instantaneous plastic strain in rock rheological tests, and combined this element with the traditional Burgers model to form an improved Burgers model. Zhao (2016) combined the Burgers model and the nonlinear M-C plastic element in series to establish a new BNMC creep damage model. In terms of experiments, Ding (2005) studied the creep failure characteristics of different types of soft and hard rocks by conducting indoor rock compression creep tests, and established a rheological constitutive model of the rock. Cheng (2009) concluded that the long-term strength of red-bed soft rocks should be about 75% of the short-term strength through indoor shear creep tests of weak interlayer samples at two typical red-bed soft rock slope sites. Jiang (2015) used the rock automatic servo triaxial rheological test equipment to carry out the creep and elastic after-effect tests of the typical sandstone samples in the Three Gorges Reservoir area by graded loading and unloading, obtained the aging deformation of the rock samples under different stress levels, and

---

*Corresponding Author: 18800126955@163.com

DOI 10.1201/9781003344209-25

deduced the creep and elastic after-effect constitutive equations of the Burgers model under a three-dimensional stress state. Because the creep test is time-consuming and expensive, and it is difficult to operate in practice, many scholars have begun to carry out numerical simulations to study the creep characteristics of rocks. Wang (2009) used the geotechnical analysis software FLAC3D to establish a test model and conduct simulation research on indoor creep tests. Xu (2012) conducted a numerical analysis of typical mud and sandstone interbed slopes in Chongqing using the finite element program. Gao (2015) established a siltstone uniaxial compression creep numerical test model based on FLAC3D, called a custom constitutive model to conduct uniaxial creep numerical simulation tests, and discussed the creep characteristics of siltstones with different stress levels under the uniaxial compression. The above studies, in forms of theoretical studies, experiments, or numerical simulations, all show that the creep characteristics of rocks cannot be ignored in rock underground works, structure foundations, slopes, landslides, etc. Therefore, in the slope stability analysis, it is necessary to consider the creep properties of rock slopes. Based on the anchorage foundation pit project of Xintian Yangtze River Bridge, the finite difference program FLAC is used to analyze the creep characteristics of the foundation pit slope, and the influence of the support structure on the slope creep characteristics after the foundation pit excavation is discussed.

2 PROJECT OVERVIEW

There are two levels of sandstone cliffs developed on the peak bank of a Yangtze River Bridge in Chongqing. The height of the next level of sandstone cliffs is relatively low, and no dangerous rocks are found. The dangerous rock belt is mainly distributed in the K13+000–K13+500 section. In the upper cliff area, the overall cliff section is distributed in an east-west direction with a belt-like distribution and a total length of about 520 m, and the overall trend of the cliff is about 96°. The elevation of the top of the cliff is 254–328 m, the elevation of the bottom of the cliff is about 240–302 m, and the relative height difference is about 14–26 m. The overall terrain is a combination of gentle slope platform-cliff-slope. The lithology is mudstone. Because the upper part is a huge sandstone mass, under the long-term self-weight of the upper sandstone, the lower weak mudstone will undergo compression deformation, resulting in the vertical displacement of the upper brittle sandstone mass, accompanied by fissures in the direction of the cliff. The rock formation in the dangerous rock area is gentle. Under the cutting of weathering fissures and structural fissures, the single-layer rock mass of the sandstone layer falls down and the rock cavity expands. The continuous development of this process forms the roof-type probe rock mass in the middle and upper parts of the cliff.

Figure 1. Bridge renderings and dangerous rocks on site.

The allowable bearing capacity of the anchorage foundation base is not less than 1.35 MPa, the minimum slope of the foundation pit slope is 1:0.3, and the maximum slope is 1:1. The design elevation of the bottom of the foundation pit is 198 m (the bottom of the anchor block). The bottom area of the anchorage foundation pit is 1026 m². The area of the side slope is 5633 m², and the excavation area of the foundation pit is 5958 m². The whole process of the excavation of the foundation pit shall meet the requirements of dry operation. The on-site bridge effect drawing and dangerous rocks are shown in Figure 1.

3 NUMERICAL MODEL

3.1 Modeling

The shape of the foundation pit studied in this paper is irregular, and the section adjacent to the dangerous rock is selected for the numerical simulation using the finite difference program FLAC, and the creep characteristics and long-term stability of the slope after the excavation of the foundation pit are analyzed. The plane strain model is established according to the geological profile model, and the geological profile model and numerical calculation model are shown in Figures 2 and 3. Among them, the length of the bottom of the slope is 200 m, the total height is 156 m, the boundary adopts a fixed displacement boundary, the left and right sides are fixed horizontal displacement, and the bottom is the fixed horizontal and vertical displacement.

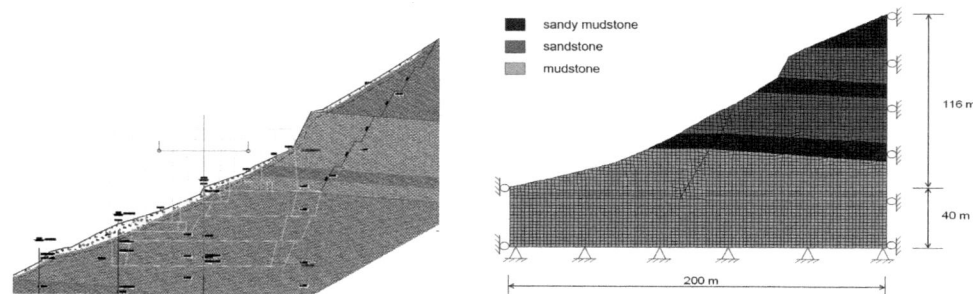

Figure 2. Geological Profile of the Foundation Pit.   Figure 3. Numerical Model Grid diagram.

3.2 Model and parameter selection

According to different simulation processes, different models are selected for the simulation. After the model is established, the initial stress balance is carried out first. Among them, the sandstone, sandy mudstone and mudstone adopt the Mohr-Coulomb model. The specific parameters are shown in Table 1. After the excavation is completed, the creep calculation of the slope is carried out. The

Table 1. Mohr-coulomb model parameters.

Medium	Elastic modulus E/GPa	Poisson's ratio μ	Bulk modulus K/GPa	Shear modulus G/GPa	Cohesion c/MPa	Internal friction angle ϕ/°	Weight γ/ kN/m³	Tensile strength σ^t/MPa
Sandstone	1.754	0.2	0.974	0.731	1.01	33.15	25.6	0.7
Sandy mudstone	8.3	0.24	5.321	3.347	2.17	37.23	25.1	0.5
Mudstone	1	0.34	1.042	0.373	0.43	30.23	25.8	0.4

creep model adopts the Cvisc model, including Maxwell elements, Kelvin elements, and Mohr-Coulomb plastic elements, as shown in Figure 4, in which the Maxwell elements and Kelvin elements are used to simulate the viscosity, and the Mohr-Coulomb plastic elements are used to simulate plasticity. The creep calculation parameters are shown in Table 2.

Table 2. Calculated creep parameters.

Medium	G^K/GPa	η^K/(GPa·d)	G^M/GPa	η^M/(GPa·d)
Sandstone	26.84	8.84×10^3	2.1	6.93×10^6
Sandy mudstone	18.87	5.78×10^3	1.19	5.96×10^6
Mudstone	2.47	8.47×10^2	0.5	3.40×10^4

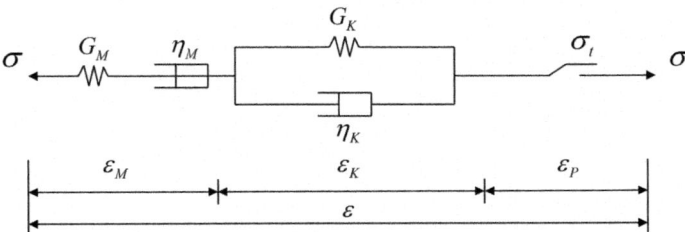

Figure 4. Creep Components of the Cvsic Model.

4 RESULT ANALYSIS

4.1 *Unsupported structure*

After the initial stress of the foundation pit slope is balanced, the front part of the slope is excavated from a slope height of 96 m. The slope model after the excavation is completed is shown in Figure 5. After the excavation is completed, the creep analysis of the slope is carried out to monitor the horizontal stress, shear stress, and vertical stress at the toe of the slope (p1), as well as the horizontal and vertical displacements at different heights of the slope. Figures 6–8 show the change in the stress at the toe of the slope in 10 years. It can be seen that the horizontal stress shows a downward trend as a whole, and the shear stress and vertical stress do not change much. Figure 9 shows the change in the horizontal displacement at different positions on the slope with time. After 2 years, the

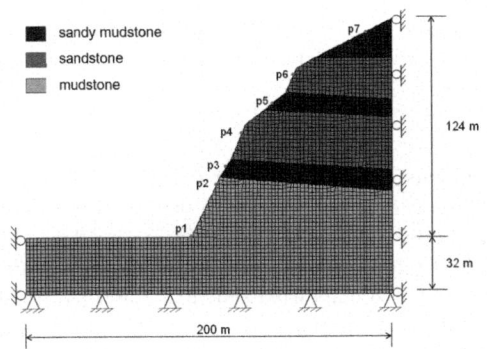

Figure 5. Numerical model grid diagram (after Excavation).

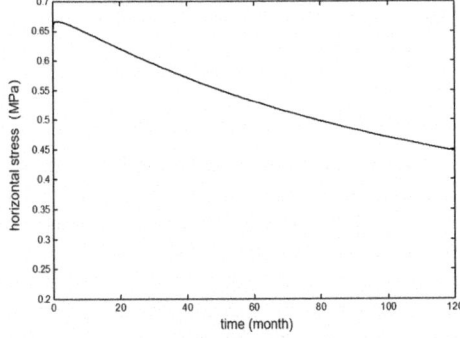

Figure 6. Variation of horizontal stress with time at P1.

horizontal displacement at points p4–p7 tends to be stable. Although the horizontal displacement at points p1–p3 is still increasing, the increase rate decreases significantly. After 10 years, the maximum displacement of the rear slope is 8.3 mm, which occurs in the mudstone in the middle and lower part of the slope. Figure 10 shows the vertical displacement of different positions on the slope with time. The vertical displacement at points p1–p4 does not change after 2 years, and the vertical displacement at points p5–p7 still increases after 2 years, but little change is shown overall. It can be found that the upward vertical displacement occurs at the points p1–p3. This is because after the excavation of the foundation pit is completed, the points p1–p3 become a free surface, and the slope is inclined to the shear displacement, which leads to the vertical displacement of the slope. After 10 years, the maximum vertical displacement of the downward slope is 6.6 mm, and the maximum vertical displacement of the upward slope is 6.3 mm.

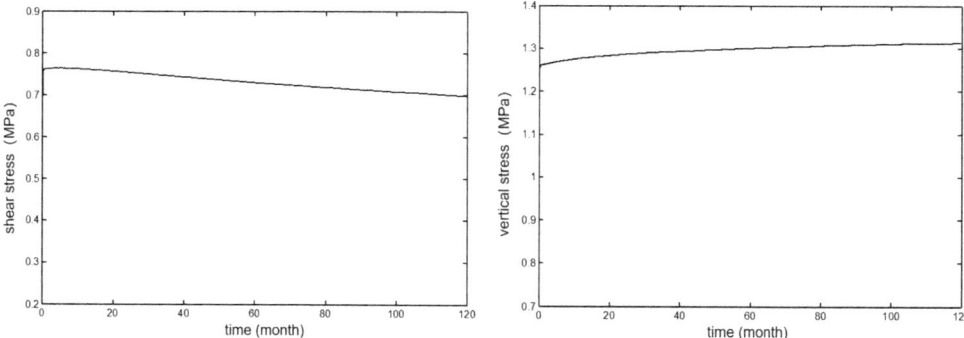

Figure 7. Variation of shear stress with time at P1.   Figure 8. Variation of vertical stress with time at P1.

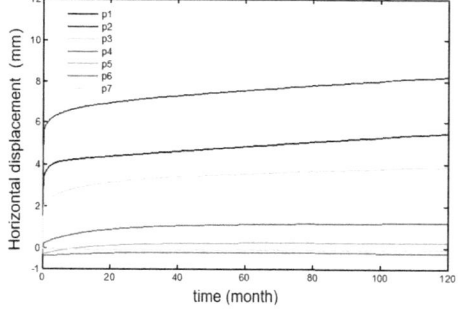
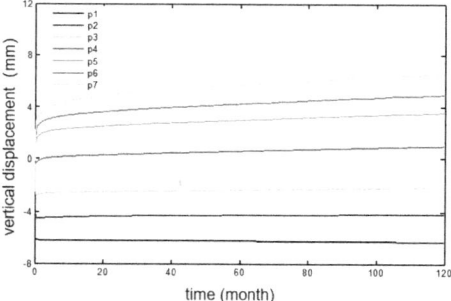

Figure 9. Variation of horizontal displacement of the slope with time.   Figure 10. Variation of vertical displacement of the slope with time.

4.2 *Grouted rock bolts timbering*

After the excavation of the foundation pit is completed, the bolt support is carried out at the excavation site. The arrangement of the bolt from the excavation position to the toe of the slope is shown in Figure 11. The inclination angle of the bolt is set to 30°, and the length is 40 m. The anchor rod is simulated by the cable element in the finite difference program FLAC, where the prestress is 560 kN. Figure 12 shows the change in the horizontal displacement at points p1–p3. It can be seen that the bolt support can effectively reduce the horizontal displacement of the slope. Figure 13 shows the change in the axial force of the anchor rod with time. The first layer is the anchor rod at the toe of the slope, and the second and third layers are the two layers of anchor rods from the toe of the slope upward in turn along the slope height. It can be seen that the bolt tension remains basically unchanged after 2 years. Figure 14 shows that the bolt tension near the toe of the slope is larger, which is attributed to the large deformation of the slope near the toe of the slope.

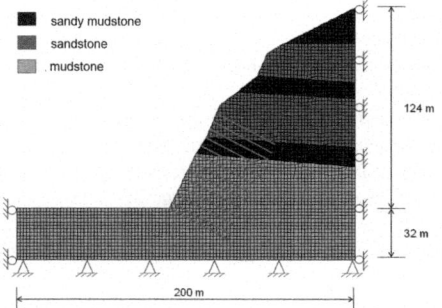

Figure 11. Numerical model grid diagram (Grouted Rock Bolts Timbering).

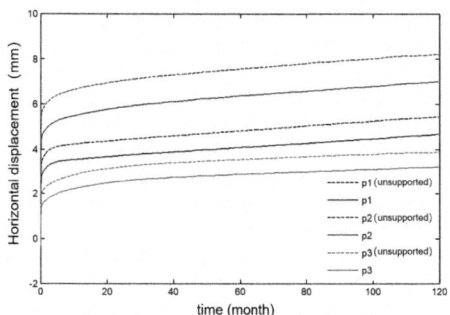

Figure 12. Variation of horizontal displacement of the slope with time at p1–p3 (Grouted Rock Bolts Timbering).

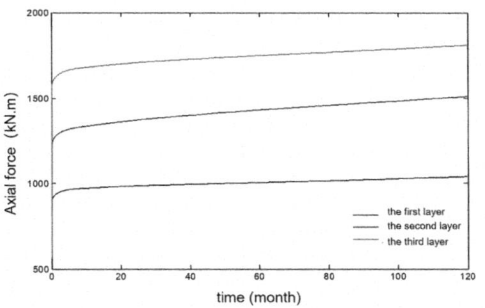

Figure 13. Variation of axial force with time.

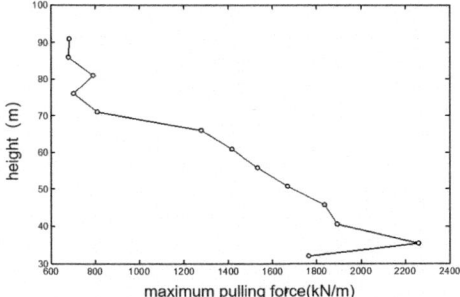

Figure 14. Distribution of maximum axial force along the slope height.

5 CONCLUSION

Based on the anchorage pit of a Yangtze River Bridge in Chongqing, this paper uses the Cvisc creep model of the finite difference program FLAC to study the creep characteristics of the rock slope after the excavation of the foundation pit, and analyzes the effect of bolt support on the creep of the slope. The research results show that the excavation of the foundation pit has a great influence on the horizontal stress, and the horizontal stress shows a downward trend as a whole in 10 years, while the vertical stress and shear stress change little. The horizontal displacement near the foot of the slope increases slowly after 2 years, the horizontal displacement at other positions is basically stabilized after 2 years, and the horizontal displacement mainly occurs in soft rocks. The vertical displacement basically tends to be stable after 2 years. Under the influence of the shear deformation, there is an upward vertical displacement in the middle and lower part of the slope. The displacement caused by the creep cannot be ignored in the study of long-term stability of the slope. The bolt support can effectively reduce the horizontal deformation of the slope and reduce the creep of the slope. The anchor rod plays a greater role near the toe of the slope.

REFERENCES

Cai M.F., He M.C., Liu D.Y. *Rock Mechanics and Engineering* [M]. Beijing: Science Press, 2002.

Cheng Q., Zhou D.P., Feng Z.J. Study on shear creep properties of soft interlayers in typical red-bed soft rocks [J]. *Chinese Journal of Rock Mechanics and Engineering*, 2009, 28(S1):3176–3180.

Ding X.L., Fu J., Liu J., et al. Research on creep characteristics and stability analysis of soft-hard interbedded slope rock mass[J]. *Chinese Journal of Rock Mechanics and Engineering*, 2005, 24(19):3410–3410.

Gao W.H., Liu Z., Zhang Z.M. Numerical simulation of siltstone compression creep test based on FLAC3D [J]. *Chinese Journal of Civil Engineering*, 2015(3):96–102.

Jiang Y.Z., Wang R.H., Zhu J.B., et al. Experimental study on creep and elastic after-effect characteristics of sandstone[J]. *Chinese Journal of Rock Mechanics and Engineering*, 2015, 034(010):2010–2017.

Qi Y.J., Jiang Q.H., Wang Z.J., et al. Three-dimensional creep constitutive equation and parameter identification of improved Nishihara model[J]. *Chinese Journal of Rock Mechanics and Engineering*, 2012, 31(2):347–355.

Wang X.D, Xu X.M. Numerical simulation study of compression creep mechanical test[J]. *Chinese Journal of Engineering Geology*, 2009(04):103–107.

Xia C.C., Wang X.D., Xu C.B., et al. Method and example of identifying rheological model with unified rheological mechanics model theory[J]. *Chinese Journal of Rock Mechanics and Engineering*, 2008(08): 86–92.

Xu P., Yang S.Q., Chen G.F. Improved Burgers rheological model of rock and its experimental verification [J]. *Chinese Journal of Coal*, 2014, 39(10):1993–2000.

Zhao Y.L., Tang J.Z., Fu C.C., et al. Rheological test and creep damage model of viscoelastic-plastic strain separation in rock[J]. *Chinese Journal of Rock Mechanics and Engineering*, 2016, 035(007):1297–1308.

*Frontiers in Civil and Hydraulic Engineering – Mohamed A. Ismail and
Hazem Samih Mohamed (Eds)*
© 2023 The Authors, ISBN 978-1-032-38247-0

Study on dam seepage characteristics of Changtan reservoir

Jincheng Du*

Changtan Reservoir Affairs Center, Taizhou City, Zhejiang Province, China

ABSTRACT: Seepage is the most common and harmful typical disease in dam engineering. The seepage channel is located inside the dam, so it is difficult to intuitively find the location and time of seepage. Taking Changtan Reservoir as an example, this paper analyzes the correlation between the reservoir water level and the seepage pressure water level from January 1, 2015, to December 30, 2021, to calculate the seepage discharge per unit width of the dam. The single-width seepage discharge of the dam is calculated. At the same time, two-dimensional finite element modeling is used to calculate the seepage characteristics of the normal water level, design water level, and check water level. The results show that the correlation is not obvious between the seepage water level and the reservoir water level, and the reservoir water loss caused by the seepage is very small, thus proving the rationality of anti-seepage measures and providing the experience for similar projects.

1 INTRODUCTION

As important water conservancy projects, dams can be divided into concrete dams and earth rock dams (Adamo et al. 2022; Omofunmi et al. 2017). Of all kinds of dams in China, earth rock dams account for more than 92% (Wang et al 2018). Seepage is the most common and harmful typical disease in earth dam projects (Turkmen et al. 2003). When the seepage force reaches a certain degree, it may lead to major seepage accidents such as dam slope sliding, seepage control breakdown, pipe inrush in the dam foundation, and flow soil, which directly affect the safety of the dam (Lottermoser et al 2005; Rice et al. 2010). Therefore, the dam seepage analysis and the safety evaluation are of great significance to ensuring the safe operation of dams.

This paper takes the Changtan reservoir as the research object and analyzes the correlation between the reservoir water level and the osmotic pressure water level over the years. The single-width seepage discharge of the dam foundation over the years is calculated. At the same time, the two-dimensional finite element analysis is used to calculate the theoretical seepage state at three different water levels of the typical section of the dam. The seepage of the dam is systematically analyzed, and the seepage safety of the dam is evaluated effectively.

2 OVERVIEW OF THE STUDY AREA

2.1 *Geography of the study area*

Changtan Reservoir is located in the west of Taizhou City, Zhejiang Province, China. It is a large-scale water conservancy project with comprehensive benefits of flood control, irrigation, and water supply. The normal water level of the reservoir is 36.00 m, and the corresponding storage capacity is 4.57×108 m^3. The designed flood level is 39.39 m, the checked flood level is 43.01 m, and the total storage capacity is 7.32×108 m^3. The dam is a typical earth-rock dam with a crest length of 480 m, a width of 7.0 m, and a crest elevation of 44.00 m. The spillway is located on the north side of Fuhu Mountain at 400 m from the left end of the dam. The spillway tunnel passes through

---

*Corresponding Author: 187369867@qq.com

DOI 10.1201/9781003344209-26

Fuhu Mountain on the left bank of the dam and is located on the left side of the power generation irrigation tunnel (Figure 1).

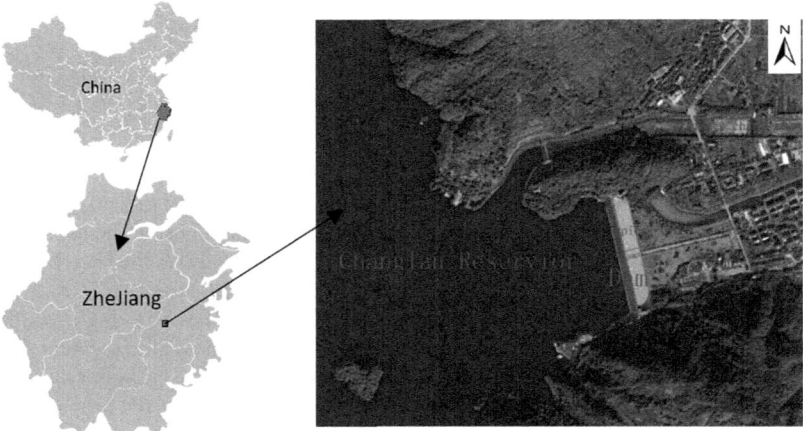

Figure 1.   Geographical location of the study area.

2.2   *Engineering overview and testing layout*

The dam face is protected by precast concrete blocks, and the back dam face is protected by granite coarse stones. The dam foundation is the deep quaternary alluvial sand and gravel layer, the shallow layer is the extremely permeable layer, and the permeability coefficient decreases gradually with the increase in depth. The seepage control system of the dam foundation includes a clay intercepting flume, clay blanket, clay grouting curtain, and plastic concrete seepage control wall. A monitoring section is arranged for the seepage flow of the dam body, and three seepage gauges are buried. Two pressure measuring pipes are set in the riverbed beneath the dam.

3   RESERVOIR WATER LEVEL AND SEEPAGE PRESSURE WATER LEVEL

The reservoir water level is the main factor affecting the change in the seepage pressure of the dam. The osmotic pressure decreases with the decrease in the reservoir water level. The statistical characteristic values of the seepage pressure and reservoir water level at the measured points of the dam are shown in Figure 2. The characteristic value of the reservoir water level is 28.37–38.17,

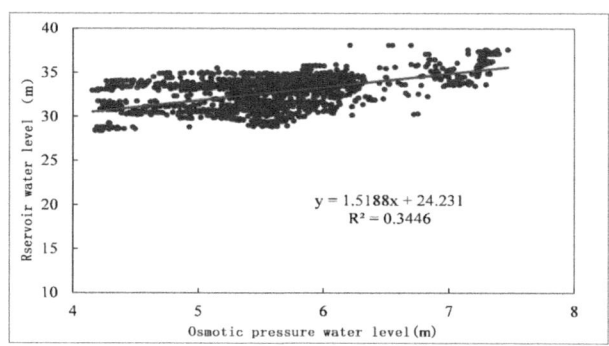

Figure 2.   Correlation between the seepage water level and reservoir water level.

while the seepage pressure water level is 4.18–7.47 m. The osmotic pressure water level is obviously lower than the reservoir water level in the same period, and the change is stable over the years. The correlation between the seepage level and the reservoir level is not obvious ($R^2 = 0.3446$), indicating that the seepage of the dam foundation is stable.

4 SINGLE-WIDTH PERMEABILITY FLOW ANALYSIS

When there is no water downstream of the reservoir and the seepage flow observation facilities are not installed, the seepage flow cannot be directly observed. In general, the dam foundation is equivalent to the finite uniform foundation, and the seepage flow of dam foundation is converted by the water level of the pressure measuring pipe downstream of the dam.

Darcy's law, which is based on the motion of pressure seepage, can be widely applied to study the motion of free surfaces without any pressure seepage. Its expression is as follows:

$$v = -k\frac{dy}{dx} \tag{1}$$

where: k is the permeability coefficient of medium; $\frac{dy}{dx}$ is the osmotic slope; v is the osmotic flow rate.

The flow per unit width is:

$$q = vy = -ky\frac{dy}{dx} \tag{2}$$

Separating variable integrals can be obtained:

$$qx = -\frac{1}{2}y^2 + C \tag{3}$$

So, we plug in the boundary conditions, namely $y = h_1$ when $x = 0$; when $x = 1$ and $y = h_2$, the single-width flow can be obtained as follows:

$$q = \frac{k}{2l}\left(h_1^2 - h_2^2\right) \tag{4}$$

The permeability coefficient of the dam foundation is $k = 8.83 10^{-4}$ m/s, and the calculated depth is -10 m. Figure 3 shows the relationship between the seepage discharge per unit width, and the reservoir water level and time. It can be seen from the Figure 3 that seepage flow is affected by the reservoir water level. In 2021, the seepage volume is between 0.28 L/s and 1.18 L/s, and the variation of the seepage volume is small. The seepage flow is generally stable, and the anti-seepage effect of the dam is good.

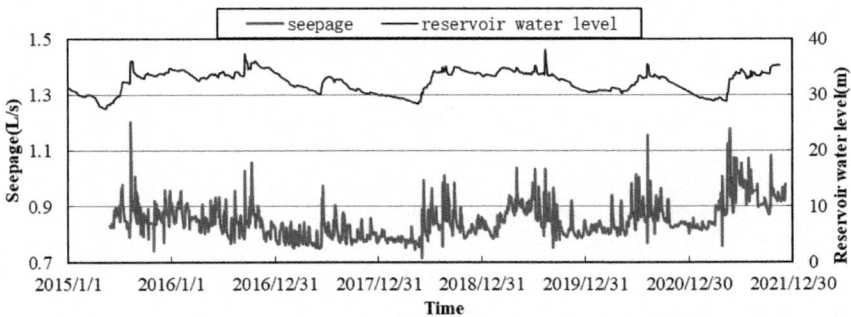

Figure 3.   Changes in single-width seepage discharge and reservoir water level with time.

5 ANALYSIS OF TWO-DIMENSIONAL FINITE ELEMENT MODEL FOR DAM SEEPAGE

The finite element analysis of the dam seepage for the typical section is mainly based on the distribution of engineering geological exploration strata, the longitudinal section of the dam, and the transverse section of the typical section. The initial value of permeability coefficient for each section is mainly selected according to geological prospecting data, test data, and general engineering experience. The specific parameters are shown in Table 1.

Table 1. Inversion Permeability Coefficients of Materials in Each Typical Section.

Material number	Division of permeability coefficient		Permeability coefficient (cm/s)	
			k_x	k_y
1	Geomembrane		1.46×10^{-9}	1.46×10^{-9}
2	Concrete diaphragm wall		7.70×10^{-8}	7.70×10^{-8}
3	Original clay inclined wall		5.23×10^{-5}	5.29×10^{-5}
4	New clay inclined wall		3.90×10^{-7}	3.90×10^{-7}
5	Filter transition zone		1.00×10^{-2}	1.00×10^{-2}
6	Original Cement Clay Curtain		2.31×10^{-3}	2.31×10^{-3}
7	Downstream drainage		1.00×10^{-1}	1.00×10^{-1}
8	Sand pebble of dam shell		9.26×10^{-1}	9.26×10^{-1}
9	Foundation sandy cobble layer	-10.00 m-7.50 m	8.83×10^{-2}	8.83×10^{-2}
		-22.00 m-10.00 m	5.79×10^{-4}	5.79×10^{-4}
		Below about 22.00 m	1.16×10^{-5}	1.16×10^{-5}

According to the actual situation of the Changtan reservoir and the structural stability analysis, it can be seen that the anti-seepage system composed of clay inclined wall and geomembrane also

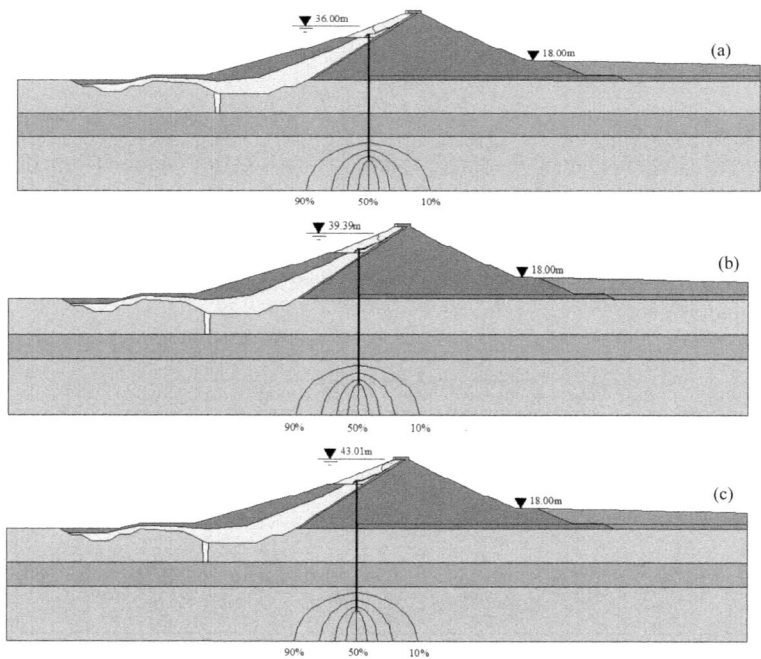

Figure 4. Theoretically calculated stable seepage fields at three different water levels (a indicates the normal storage water level, b indicates the designed flood water level, and c indicates the checked flood water level).

plays a good anti-seepage effect. The water level has been substantially reduced in front of the concrete cutoff wall (Figure 4). The concrete cutoff wall goes deep into the bedrock, forming a closed seepage control system, and the seepage slope of each part of the upstream and downstream dam is basically 0.

The dam foundation is a sandy pebble layer. And the center of the dam is a clay core wall. In the dam with a small permeability coefficient, there is obvious hysteresis between the water level of the seepage and the water level of the reservoir. The water level of the osmotic pressure gauge is the instantaneous state result of the unsteady seepage under varying reservoir water levels. Therefore, according to the measured infiltration line of the section, the measured seepage field map is obtained by analyzing and calculating the measured seepage field of the three water levels (Figure 5).

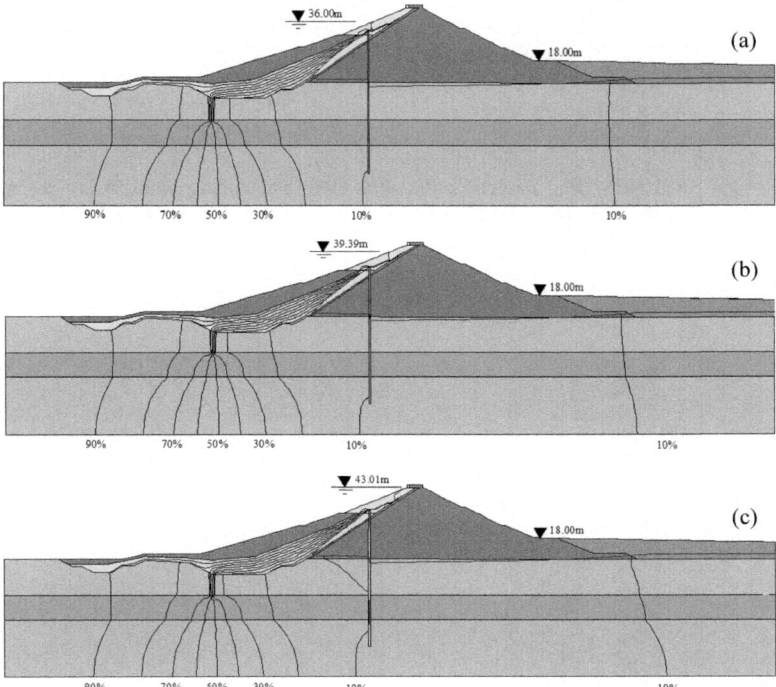

Figure 5. Measured stable seepage fields at three different flood water levels (a indicates the normal storage water level, b indicates the designed flood water level, and c indicates the checked flood water level).

When the flood level is 43.01 m, the theoretically calculated value of the single-width seepage is 0.38 m³/m × d, and the actual calculation value of the single-width seepage is 0.042 m³/m × d. The length is calculated as 470 m. The annual seepage discharge of the dam at the checked flood level is about 6.52×104 m³ (theoretically calculated value) and 7.21×103 m³ (measured value), accounting for about 0.09‰ and 0.01‰ of the total storage capacity, respectively. It can be seen that the reservoir water loss caused by the seepage is very small.

6 CONCLUSION

The seepage level of the Changtan Reservoir changes stably over the years, and the correlation between the reservoir water level and the seepage level is not obvious. According to the Darcy's law, the seepage variation over the years is small. The seepage is generally stable, and the anti-seepage effect of the dam is good. A two-dimensional finite element model is used to calculate the theoretical and measured annual seepage discharge of the Changtan reservoir under the checked flood level. The seepage rate accounts for 0.09‰ and 0.01‰ of the total storage capacity, respectively. It is proved that the anti-seepage system of the dam can meet the seepage requirements of the dam, which can provide a good reference for the anti-seepage of similar projects.

REFERENCES

Adamo N., Al-Ansari N., Sissakian V., et al. Dam safety problems related to seepage[J]. *Journal of Earth Sciences and Geotechnical Engineering*, 2020, 10(6): 191–239.

Lottermoser B.G., Ashley P.M. Tailings dam seepage at the rehabilitated Mary Kathleen uranium mine, Australia[J]. *Journal of Geochemical Exploration*, 2005, 85(3): 119–137.

Omofunmi O.E., Kolo J.G., Oladipo A.S., et al. A review on effects and control of seepage through earth-fill dam[J]. *Current Journal of Applied Science and Technology*, 2017, 22(5): 1–11.

Rice J.D., Duncan J.M. Findings of case histories on the long-term performance of seepage barriers in dams[J]. *Journal of Geotechnical and Geoenvironmental Engineering*, 2010, 136(1): 2–15.

Turkmen S. Treatment of the seepage problems at the Kalecik Dam (Turkey)[J]. *Engineering Geology*, 2003, 68(3–4): 159–169.

Wang S, Xu Y, Gu C, et al. Monitoring models for base flow effect and daily variation of dam seepage elements considering time lag effect[J]. *Water Science and Engineering*, 2018, 11(4): 344–354.

*Frontiers in Civil and Hydraulic Engineering – Mohamed A. Ismail and
Hazem Samih Mohamed (Eds)*

Variations and characteristics of deposition on the tidal landforms: The role of vegetation

Lin Chong*, Yunze Shen* & Zhilin Sun*
College of Architecture and Civil Engineering, Zhejiang University, Hangzhou, China

ABSTRACT: Vegetation is a geophysical driver influencing estuarine tidal landforms. To investigate the deposition characteristics of tidal landforms under the influence of vegetation, a field survey has been conducted in the Yangtze Estuary. The distribution characteristics of sediment samples collected from *bare flat along vegetated regions (BFV)* and *bare flat in non-vegetated regions (BFNV)*, and marshes with *Scirpus mariqueter (SM)*, *Spartina alterniflora (SA)*, and *Reed (RE)* were studied, the results of which were compared to that before the Three Gorges Project and a slightly coarse trend of deposition was found. The deposition in the *BFV* was similar to that in the vegetated marshes which presented better sorting performance and was finer than that in the *BFNV*. The roles of *SM*, *SA*, and *RE* were similar in the retention of fine sediments. The sorting performance and the symmetry of frequency distribution curves of sediment samples were ranked as *SM>SA>RE*.

1 INTRODUCTION

Scirpus mariqueter (SM), Spartina alterniflora (SA) and *Reed (RE)* are common plant species in the Yangtze Estuary. These species have been widely used in ecological restoration to stabilize the sand, protect banks against erosion and slumping, improve water quality, and create habitats for organisms (Camporeale et al. 2013; Carpenter & Lodge 1986; Demars et al. 2012; Schulz et al. 2003). *SA* has been introduced to the newly-formed wetlands on the Jiangsu coast, China, and serves as an ecological engineer to reduce near-bed shear stress, trap fine-grained sediments and protect the coast from wave-induced erosion (Wang et al. 2021). The complex morphology and flexibility of natural vegetation physically change flow conditions and deposition patterns, generating a variety of river landforms (Meng et al. 2021). The tidal landforms including marshes, tidal flats, and subtidal areas have been formed in the estuary with the interaction of vegetation and sediment transport (Da Lio et al. 2013). The effect of natural emergent vegetation on deposition in tidal landforms is significant but rarely studied by the field survey.

The ecological restoration engineering of tidal landforms is the focus of both ecologists and hydrologists. The construction of the Three Gorges Reservoir impacts downstream sediment transport and so influences the estuarine tidal flats. The drop of sediment load at Datong (seaward most station) was 128.37×10^6 t/ year due to the impoundment of Three Gorges Dam in 2003–2012 and 27.28×10^6 t/ year due to new cascade dam impoundment in 2013–2017 (Tian et al. 2021). As a result, the tidal environment was changed. However, there has been a lack of follow-up reports about the impact of the ecological restoration project on deposition in the estuarine tidal flats recently.

In this study, surficial sediment samples collected from different areas and times were compared to investigate the impact of vegetation on the spatial distribution of deposition and to study the temporal changes of deposition in the tidal landform, Yangtze Estuary, after the Three Gorges Project. The general grain-size characteristics of deposition were compared as follows: bare flats VS. vegetated marshes, *bare flats in non-vegetated regions (BFNV)* VS. *bare flats along vegetated*

---

*Corresponding Authors: 11812073@zju.edu.cn, 12012014@zju.edu.cn and oceansun@zju.edu.cn

DOI 10.1201/9781003344209-27

regions (BFV), comparison in *SM, SA,* and *RE* marshes, as well as 1990s VS. 2019. The results provide helpful references for estuarine ecological restoration engineering.

2   MATERIALS AND METHODS

2.1   *Sampling and experimental analysis*

The field sampling was conducted by using the line transect method in the tidal flat, south of Fushan tidal channel, Yangtze Estuary, in October 2019 [Figures 1(a) and (b)]. There is an irregular semidiurnal tide in the Fushan tidal channel, which is mainly affected by tidal forces. The current velocity and sediment concentration are asymmetric during flood and ebb tide. The average velocity in the north of Fushan tidal channel was greater than that in the south. The flood and ebb velocities were 0.183 m/s and 0.075 m/s respectively at spring tide while those were 0.096 m/s and 0.058 m/s respectively at neap tide in 2016. The sediment concentration in the channel at flood tide is 0.08–0.10 kg/m^3, which is greater than that (0.04–0.05 kg/m^3) at ebbing tide (Zhang et al. 2011). The suspended sediment concentration around Jiuduansha shoal was measured to be 0.1–0.7 kg/m^3 in the surface water (Yang 1999).

There are three species of plants, *Scirpus mariqueter (SM), Spartina alterniflora (SA),* and *Reed (RE),* growing on the tidal marsh [Figure 1(d)]. Five survey lines *A* to *E* were set to take sediment samples and the distribution of plants was recorded [Figure 1(d)]. The typical environment where sediment samples were taken was classified into five categories, i.e., marsh such as *Scirpus mariqueter (SM), Spartina alterniflora (SA), Reed (RE), bare flat along vegetated regions (BFV),* and *bare flat in non-vegetated regions (BFNV),* respectively. The grain size and its probability density distribution curve of 50 samples were measured for further study by using a laser diffraction grain

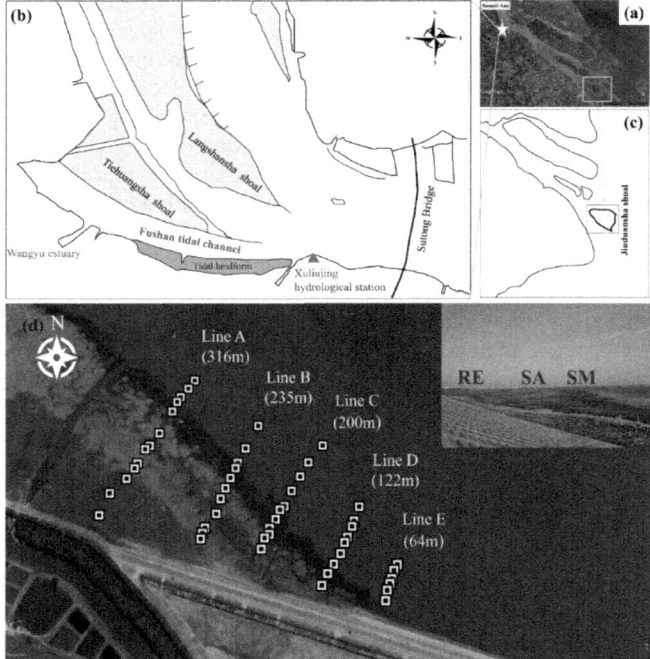

Figure 1.   Geographical location of the Study Site: (a) the study area in the Yangtze Estuary. (b) the Fushan tidal channel. (c) the location of Jiuduansha shoal (Yang 1998). (d) the sampling tidal flat and vegetation distribution. *SM, SA* and *RE* represent *Scirpus mariqueter, Spartina alterniflora,* and *Reed,* respectively.

size analyzer (BT-9300Z). Moreover, by comparing the deposition in the south bank of the Fushan tidal channel in 2019 with that in the Jiuduansha shoal, the variation of deposition in the tidal environment under the influence of the Three Gorges Project, Yangtze Estuary, was studied [Figure 1(c)].

2.2 *Size parameters of sediment samples*

The parameters used to describe particle size distribution are divided into four main groups (Blott & Pye 2001): (a) the average or median size, (b) the spread of the sizes around the average (sorting coefficient or standard deviation), (c) the symmetry or preferential spread to one side of the average (skewness), and (d) the degree of concentration of the grains relative to the average (kurtosis). $\varphi_{50} = -log_2 D_{50}$, where D is grain size in mm, being used as a measure of the central tendency in this study. The logarithmic graphical measures were proposed (Folk & Ward 1957), and the sorting coefficient (standard deviation) σ is expressed as,

$$\sigma = \frac{\varphi_{84} - \varphi_{16}}{4} + \frac{\varphi_{95} - \varphi_5}{6.6} \tag{1}$$

where the subscript, i.e., 84, indicates the φ-size at which 84% by weight is smaller.

The skewness S of the collected samples is estimated as,

$$S = \frac{\varphi_{16} + \varphi_{84} - 2\varphi_{50}}{2(\varphi_{84} - \varphi_{16})} + \frac{\varphi_5 + \varphi_{95} - 2\varphi_{50}}{2(\varphi_{95} - \varphi_5)} \tag{2}$$

The kurtosis K of the collected samples is estimated as,

$$K = \frac{\varphi_{95} - \varphi_5}{2.44(\varphi_{75} - \varphi_{25})} \tag{3}$$

3 RESULTS AND DISCUSSION

3.1 *Multivariate characteristics of samples*

As for the sorting coefficient σ, it can be seen from Equation (1) that small sorting coefficients indicate well sorted (more uniform-sized) sediment samples whereas poorly sorted sediment samples show large sorting coefficients. The sorting coefficient σ varies with φ_{50} [Figure 2(a)] Compared to the samples from *BFV*, the particle size of *BFNV* samples was finer with the φ_{50} larger than 4 ($D_{50} < 63\mu m$), below which are silt and clay. It is found that the sorting coefficient σ of *BFNV* was larger than that of *BFV*, ranging from 1.75 to 1.9. For the sediment samples from vegetated marshes, the range of the sorting coefficient was from 1.5 to 1.9, the variation trend of which was consistent with φ_{50}. As the value of φ_{50} increased (D_{50} decreases), the sorting performance became better. The sorting coefficient σ of sediment samples from *SM* was larger than that of *SA* and *RE*, indicating a poorer sorting performance. Comparing samples from different vegetated marshes to bare tidal flats *BFNV* and *BFV*, the sorting performance became poorer with increasing content of coarse particles in the samples.

Regarding the skewness S of sediment samples, the positive skewness corresponds to the coarser sediment component in the sample, while the finer component in the collected sample leads to the negative skewness. The zero skewness of the collected sample indicates that the grain size distribution curve is completely symmetrical on average. The variation trend of the skewness S with φ_{50} is shown in Figure 2 (b). With the increase of φ_{50} (the mean particle size decreases), the skewness tends to decrease until it was close to 0. That is, the frequency distribution curve became more symmetrical from bare tidal flat *BFNV* and *BFV* to different vegetated marshes with the median particle size being smaller.

As for kurtosis K, a smaller or larger value corresponds to the platykurtic or leptokurtic size distribution, respectively. The larger the coarse particle composition is, the more obvious the tail of the frequency distribution curve is, and the greater the kurtosis value is. The variation trend of the kurtosis K with φ_{50} is shown in Figure 2 (c). With the change in the median particle size φ_{50}, the kurtosis K of *BFNV* and *BFV* samples has a slight variation ranging from 0.90 to 1.25. The kurtosis K of most vegetated marsh samples was different from that of bare tidal flats *BFNV* and *BFV* samples with values increasing to nearby 1 as φ_{50} increased, indicating that the distribution curves became more concentrated with a less significant tail.

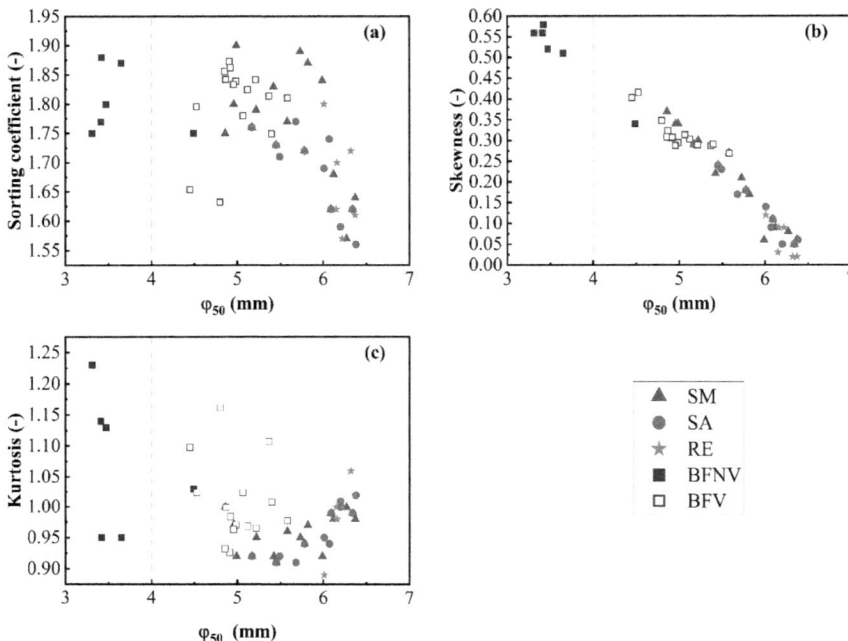

Figure 2.   Grain Size Parameters of Samples, including (a) sorting coefficient (b) skewness and (c) kurtosis.

3.2   *Comparison of deposition in the 1990s to 2019*

In the 1990s, the influence of marshes on sedimentation was examined in Jiuduansha Shoal, Yangtze Estuary (Yang 1998). Comparative pairs of measurements were conducted respectively in marsh environments and their adjacent bare flats for data on the grain size of deposits. The trapping effect of vegetation on water flow was found in the marshes where the ratios of mean current velocity to its adjacent bare flat were 0.18–0.84. The average ratio of suspended sediment concentration over the vegetated marsh to the bare flat was 0.71. The characteristics of deposition in the tidal flats in the 1990s and 2019 were compared (Figure 3), manifesting that the influence of marshes on hydrodynamics and deposition was related to plants and the location of the measurement.

Deposition became finer from the tidal flat to *SM* through *SA* marshes in the 1990s, which was similar to that in 2019. Also, marsh sediments were much finer than those on the adjacent flat with φ_{50} of 8.27 in *SA* marshes, 7.45 in *SM* marshes, and 5.83 on the bare flat in the 1990s, compared with those (6.12, 5.83, and 4.99, respectively) in 2019 [Figure3(a)]. The observation supported that sediment sorting is functionally related to grain size. The positive correlation between mean grain size and sorting coefficient σ was also found in the 1990s when the best sorting occurred at a median grain size of 2.5–3 compared to that (5.5–6.5) in 2019. There was also a variation in deposition in 2019 compared to that in the 1990s. The sediment deposited in the marsh showed a coarsening trend over the past 20 years with the silt and clay content decreasing from 99.92% to 88.53% as the decreased sediment concentration [Figure 3(b)].

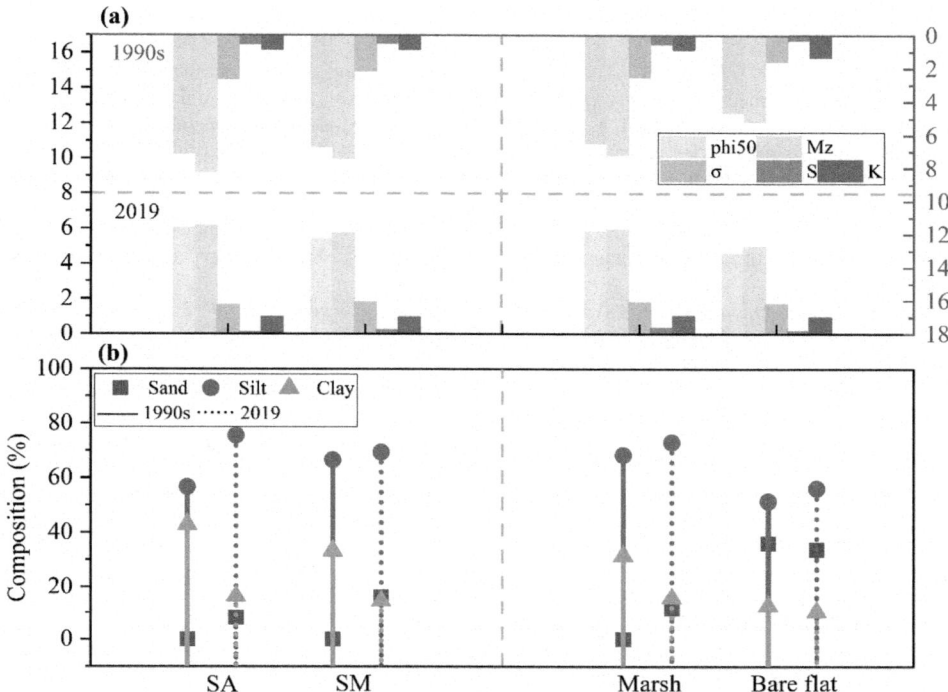

Figure 3. Comparison of Deposition in the Tidal Flat, Yangtze Estuary, in the 1990s and 2019: (a) size parameters and (b) composition of deposition. *Mz* presents the mean particle size.

4 CONCLUSION

The role of emergent vegetation on deposition in the tidal landforms, Yangtze Estuary, was studied by the field survey. The distribution curves of deposition in the vegetated marshes with less significant tails were more concentrated nearby silt composition, compared to that in the bare tidal flat. Also, the distribution of deposition was different between *BFNV* and *BFV*, and between *SM, SA,* and *RE* marshes. The deposition of *BFV* with φ_{50} larger than 4 was finer and the sorting performance was poorer than that of *BFNV.* This is because the fine sediment retained in the vegetated marsh was deposited on the nearby tidal flat under the influence of flooding and ebbing tide. The sorting performance of sediment samples from different vegetated marshes was ranked as *SM> SA> RE*, which was the same as the symmetry of frequency distribution curves. Comparing the deposition in the 1990s to that in 2019, a slightly coarsening trend was found, although the roles of marshes with *SM, SA,* and *RE* were similar in the retention of fine sediments. The bio-deposition by vegetation under different hydrodynamic conditions can be studied by experimental analysis in the future. The grain size of deposition in the marshes and tidal flats is also affected by other factors including current velocity, wave height, and suspended sediment concentration, which should be considered in further study.

ACKNOWLEDGMENTS

This study was financially supported by the key project of Three Gorges (12610100000018J129-06) and the Natural Science Foundation of China (91647209).

REFERENCES

Blott, S.J., Pye, K., 2001. GRADISTAT: A grain size distribution and statistics package for the analysis of unconsolidated sediments. *Earth Surf. Process. Landforms* 26, 1237–1248. https://doi.org/10.1002/esp.261

Camporeale, C., Perucca, E., Ridolfi, L., Gurnell, A.M., 2013. Modeling the interactions between river morphodynamics and riparian vegetation. *Rev. Geophys.* 51, 379–414. https://doi.org/10.1002/rog.20014

Carpenter, S., Lodge, D., 1986. Effects of submersed macrophytes on ecosystem processes. *Aquat. Bot.* 26, 341–370. https://doi.org/10.1016/0304-3770(86)90031-8

Da Lio, C., D'Alpaos, A., Marani, M., 2013. The secret gardener: Vegetation and the emergence of biogeomorphic patterns in tidal environments. *Phil. Trans. R. Soc. A.* 371, 20120367. https://doi.org/10.1098/rsta.2012.0367

Demars, B.O.L., Kemp, J.L., *Friberg, N., Usseglio-Polatera, P., Harper, D.M., 2012. Linking biotopes to invertebrates in rivers: Biological traits, taxonomic composition, and diversity. Ecol. Indic.* 23, 301–311. https://doi.org/10.1016/j.ecolind.2012.04.011

Folk, R.L., Ward, W.C., 1957. Brazos River bar: A study in the significance of grain size parameters. *J. Sediment. Petrol.* 27, 3–26.

Meng, X., Zhou, Y., Sun, Z., Ding, K., Chong, L., 2021. Hydraulic characteristics of emerged rigid and submerged flexible vegetations in the riparian zone. *Water* 13, 1057. https://doi.org/10.3390/w13081057

Schulz, M., Kozerski, H.-P., Pluntke, T., Rinke, K., 2003. The influence of macrophytes on sedimentation and nutrient retention in the lower river spree (Germany). *Water Research* 37, 569–578. https://doi.org/10.1016/S0043-1354(02)00276-2

Tian, Q., Xu, K.H., Dong, C.M., Yang, S.L., He, Y.J., Shi, B.W., 2021. Declining sediment discharge in the Yangtze River from 1956 to 2017: Spatial and Temporal Changes and Their Causes. *Water Resour. Res.* 57, e2020WR028645. https://doi.org/10.1029/2020WR028645

Wang, D., Gao, S., Zhao, Yangyang, Chatzipavlis, A., Chen, Y., Gao, J., Zhao, Yongqiang, 2021. An eco-parametric method to derive sedimentation rates for coastal saltmarshes. *Sci. Total Environ.* 770, 144756. https://doi.org/10.1016/j.scitotenv.2020.144756

Yang, S.L., 1998. The role of scirpus marsh in attenuation of hydrodynamics and retention of fine sediment in the Yangtze Estuary. *Estuarine, Coastal and Shelf Science* 47, 227–233. https://doi.org/10.1006/ecss.1998.0348

Yang, S.L., 1999. *Sedimentation on a Growing Intertidal Island in the Yangtze River Mouth. Estuarine, Coastal and Shelf Science* 49, 401–410. https://doi.org/10.1006/ecss.1999.0501

Zhang, S.Z., Xia, Y.F., Wu, D.W., 2011. *Preliminary study on the regulation and utilization of Fushan waterway in Chengtong reach of the Yangtze River.* Proceedings of the 15th China Ocean (Shore) Engineering Symposium (middle), 2011:551–554 (in Chinese).

*Frontiers in Civil and Hydraulic Engineering – Mohamed A. Ismail and
Hazem Samih Mohamed (Eds)
© 2023 The Authors, ISBN 978-1-032-38247-0*

A study on surface settlement of diversion tunnel in the sandy loam composite stratum constructed by shield tunneling

Zichang Ma*

College of Water Conservancy and Hydropower Engineering, Hohai University, Nanjing, China

ABSTRACT: Shield tunneling is used more frequently in water diversion projects. The prediction and control of surface settlement are particularly important when shield tunneling through the composite stratum. The numerical simulation method can be used to effectively predict the surface settlement caused by shield construction. A two-dimensional finite element model is established based on a shield construction diversion tunnel project passing through a sandy loam composite stratum. The orthogonal test with the surface settlement as the index is designed. The sensitivity analysis of internal friction angle, elastic modulus, density, and elastic modulus of the lining of the sandy soil layer is carried out. The results show that the surface settlement is the most sensitive to the change in the friction angle in the sand layer. Then, taking the monitored value of surface settlement as the target value, the improved genetic algorithm is used to inverse the elastic modulus of the sandy soil layer, internal friction angle, and the elastic modulus of the lining. The inversion results are substituted into the finite element model. The results show that the calculated results are in good agreement with the measured values. On this basis, the surface settlement of shield construction under different burial depths and different loam thicknesses is calculated. The results show that when the shield depth increases, the surface settlement will decrease, and the width of the "settlement tank" will increase. When the thickness of the loam layer increases, the surface settlement will increase, and the width of the "settlement tank" will decrease.

1 INTRODUCTION

Shield construction has the characteristics of a small impact on the surrounding environment, a high degree of automation and easy management. It plays an important role in water diversion projects. In the sandy loam composite stratum, shield construction also has many risks, such as excessive surface deformation, so the prediction and control of surface settlement are particularly important.

Numerical simulation is an important method for studying the surface settlement caused by shield construction. In the aspect of numerical simulation, a three-dimensional fluid-solid coupling finite element calculation model is established (Luo 2020). The changes in surface settlement and pore water pressure caused by the shield construction of Chengdu Metro Line 4 are calculated. The conclusion is that the changing trend of soil pore water pressure is similar to that of the surface settlement, and the pore water pressure increases and tends to be stable after the shield passes through. A finite difference model of double tunnel excavation is established (Do 2014). The influence of double tunnel excavation on surface settlement and lining structure is calculated. It is concluded that the construction sequence of the two tunnels has a great impact on the surface settlement. The one-dimensional leakage element is embedded into the three-dimensional finite element model to simulate the leakage at the joint of the shield tunnel (Wu 2020). The results show that the leakage at the vault has little impact on the tunnel settlement and has a greater impact on the surface settlement. The surface settlement caused by shield construction is simulated (Bao 2020). The feasibility and accuracy of using a two-dimensional numerical model to calculate the surface

*Corresponding Author: 1121269271@qq.com

DOI 10.1201/9781003344209-28

settlement caused by shield construction are verified. The influence of the seepage erosion of the vault on the surface settlement is studied (Ye 2018). The results show that the increase in erosion range will cause an increase in surface settlement. The influence of shield tunnel spacing and depth on surface settlement and distribution is studied, and the law that the settlement decreases with the increase in the buried depth and spacing is obtained (Cheng 2021). The influence of shield tunneling on the existing tunnel and ground deformation is analyzed, and the conclusion that grouting in the shield can effectively control the settlement caused by excavation is drawn (Yin 2018).

The above research rarely involves the situation where the shield is buried up to 30 m. The following aspects are less considered: (1) During field sampling of the sandy soil layer, due to sampling disturbance and other reasons, the actual mechanical parameters such as elastic modulus, internal friction angle, and density of sandy soil deviate from the test results, so it is difficult to accurately determine the field parameters of sandy soil layer from the test results. (2) Under the pushing and extrusion of the shield machine, the joint gap between the lining sheets is continuously compacted, resulting in a wide range of changes in the overall elastic modulus of the lining material, which is difficult to determine its size. Therefore, it is necessary to determine the size of its parameters through inversion. (3) When shield tunneling through sandy loam composite stratum, the influence of tunnel buried depth and loam layer thickness on surface settlement.

In this paper, a two-dimensional finite element model is established based on a shield construction diversion tunnel project passing through the sandy loam composite stratum. The sensitivity analysis of sand and lining parameters is carried out. The improved genetic algorithm is used to inverse these parameters. The surface settlement of shield construction under different buried depths and different loam thicknesses is calculated. The research results with engineering practical value are obtained.

2 CALCULATION MODEL AND METHOD

2.1 Parameter sensitivity

To explore the parameters that have a great impact on surface settlement, the orthogonal test is designed. The surface settlement under different parameter combinations is calculated. Range analysis is used to analyze the sensitivity of different parameters.

2.2 The range method

Let $A, B...$ represent different factors respectively, and t is the number of factor levels, A_i is the value of factor A at level i ($i = 1, 2, \ldots, t$), and Y_k is the result of different groups of experiments ($k = 1, 2 \ldots n$). The calculated statistics are as follows:

$$K_{ij} = \frac{1}{n}\sum_{k=1}^{n} Y_k - \overline{Y} \tag{1}$$

where K_{ij} is the average value of each test result of factor j at level i, n is the number of tests of factor j at level i, Y_k is the k_{th} test index value, is the average of all test results.

The evaluation index of range analysis is the range value R_j.

$$R_j = \max\{K_{1j}, K_{2j}, ...\} - \min\{K_{1j}, K_{2j}, ...\} \tag{2}$$

The greater the range value R_j of the factor, the greater the sensitivity of the factor.

2.3 Genetic algorithm model for parameter inversion

A genetic algorithm is an optimization algorithm that simulates natural evolution proposed by Holland. The mathematical model of parameter inversion can be described as searching for appropriate values in the parameter interval to minimize the square sum of errors between the calculated value and the measured value:

$$\min \sum_{i}^{n} (S_i - S_i')^2 \qquad (3)$$

where S_i represents the measured value of measuring point i, and S_i' represents the calculated value of measuring point i. There are n surface settlement monitoring points, and the measured value at monitoring point i is $S_i (i = 1, 2 \ldots n)$, and the calculated value at monitoring point i is S_i'.

$$S_i' = g_i[\gamma_{sand}, E_{sand}, \phi_{sand}, E_{lining}] \qquad (4)$$

Where g_i refers to the settlement value of the i_{th} monitoring point calculated by the finite element model.

Individual fitness is used to describe the probability of individual genes being retained. The following expression is selected as the fitness function:

$$\begin{aligned} f(k) &= temp(k) - \min (temp(k)) + c \\ temp(k) &= 1/\sum_{i}^{n} (S_i - S_{ik})^2 \end{aligned} \qquad (5)$$

Where $f(k)$ refers to individual fitness, k refers to the individual number ($k = 1, 2 \ldots l$), $temp(k)$ refers to the reciprocal of the sum of squares of errors, and i refers to the point i in n monitoring points, $S_i(t)$ represents the measured value of measuring point i, and S_{ik}' represents the calculated value of individual k at measuring point i.

2.4 Improved genetic algorithm

The crossover probability and mutation probability of each generation of genetic algorithm are fixed, resulting in poor local search ability of the algorithm. When the objective function is multimodal, it may fall into local optimization. It lacks intergenerational competition, resulting in the "degradation" phenomenon that the offspring are inferior to the parents.

To improve the convergence speed and avoid the result falling into local optimization, the following improvement measures are implemented:

a. The optimal retention strategy is adopted. The optimal individual of the parent is allowed to enter the offspring directly without crossing and mutation.
b. Variable crossover probability and mutation probability are adopted. In the process of crossover and mutation, individuals are sorted according to their fitness. Two individuals with similar fitness values were combined. If the average fitness of this combination is larger in modern times, it will be given a smaller crossover probability and a larger mutation probability.

The improved genetic algorithm and the interface between the algorithm and ABAQUS are written. MATLAB is used for programming. The calculation process of the algorithm is as follows:

Step 1: According to the range and accuracy of parameters to be retrieved, the length of the binary code and the number of the populations are set. The initial population is generated.
Step 2: The parameters to be inverted are decoded. The binary code of each individual is converted into decimal inversion parameters. The decoded parameters are written into the INP file of the corresponding individual.
Step 3: The INP file is submitted to ABAQUS for calculation. The displacement values of each monitoring point are extracted. They are used to calculate individual fitness. According to the fitness value, individuals are sorted. The optimal individual is extracted. It does not participate in natural selection. The rest of the individuals are naturally selected according to their fitness values.

Step 4: According to the average fitness of two individuals, they are given crossover probability and mutation probability. According to the probability of crossover and mutation, the coding of individuals is changed.

Step 5: If the evolutionary algebra reaches the maximum algebra, the algorithm process is terminated. If the evolution does not reach the maximum algebra, the optimal individual is added to the individual after crossover and mutation. The initial population of the next generation is formed. The process is transferred to Step 2.

3 ENGINEERING CASE

3.1 Project overview

A diversion tunnel constructed by shield tunneling located in Henan Province is buried 15~30 m deep. The thickness of the loam soil layer is 5~20 m, and the thickness of the sandy soil layer exceeds 60 m. The engineering geology is shown in Figure 1. The section of the shield with a buried depth of 30 m and a loam layer thickness of 11 m is modeled. The finite element calculation model is shown in Figure 2. The height of the model is 60 m. The three-level wavelet decomposition is used to deal with the surface monitoring point data of the section. The variation of surface settlement with time at 38 days after the shield passing is shown in Figure 3. After the shield machine passes through for 30 days, the surface settlement tends to be stable. The maximum surface settlement is 4.8 mm.

The model calculation parameters are shown in Table 1. The Mohr-Coulomb constitutive model is used for the soil. The elastic constitutive model is used for lining. The stress relief factor is taken as 0.1. The normal displacement and tangential displacement on the bottom of the model are constrained. The normal displacement on both the left and right sides of the model is constrained.

The finite element calculation process is as follows:

Step 1: Set the initial conditions and balance the stress.

Step 2: Extract the nodal force of the excavation unit on the soil element at the excavation boundary. Remove the excavation elements. The extracted node force is applied to the node of the excavation boundary in the form of concentrated force.

Step 3: Release the concentrated force on the excavation boundary by 0.1.

Step 4: Activate the lining unit and completely release the concentrated force on the excavation boundary.

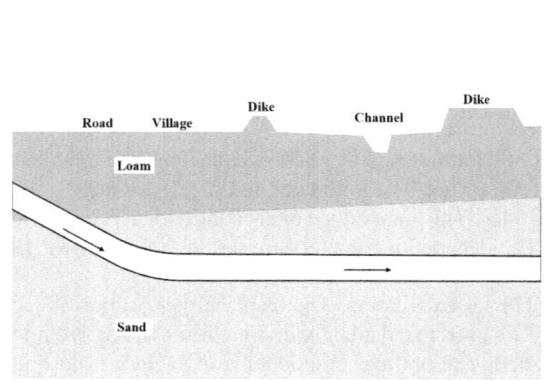

Figure 1. Overview of engineering geology.

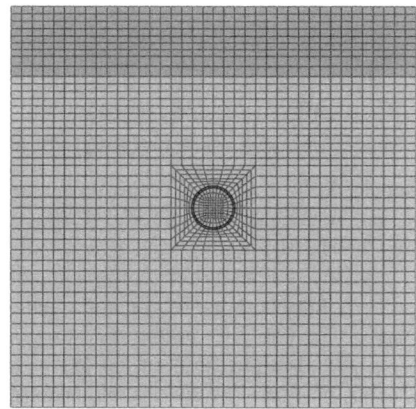

Figure 2. Finite element calculation model.

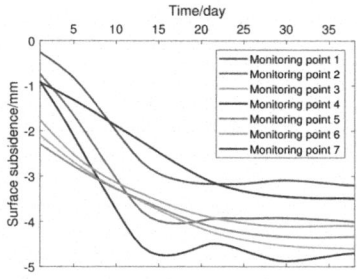

Figure 3. Monitoring point settlement changes over time.

Table 1. Calculation parameters of the finite element model.

Material	Thickness	ρ	E	μ	ϕ	c
Loam	11 m	1.9 t/m³	4.9 MPa	0.35	16°	25 kPa
Sand	49 m	*	*	0.3	*	0
Lining	0.35 m	2.5 t/m³	*	0.2	–	–

*represents the parameters to be inverted.

3.2 Parameter sensitivity analysis

3.2.1 Test factors and factor levels

In the process of shield construction, the elastic-plastic state of the soil layer and the elastic modulus of the lining material have an impact on the surface settlement.

In this paper, the cohesion of the sandy soil layer is considered as 0. The internal friction angle of the sand layer ϕ_{sand}, the elastic modulus of the sand layer E_{sand}, the density of the sand layer ρ_{sand}, and the elastic modulus of the lining E_{lining} in the model are selected as sensitivity analysis factors. The parameters of a sand layer are based on the results of laboratory tests. The elastic modulus of the lining is set as the reference level according to relevant engineering experience. In the sensitivity analysis, the 10% increase and decrease of the parameter benchmark level are taken as the test level. The sensitivity analysis factors and the level of each factor are shown in Table 2.

3.2.2 Orthogonal test results

According to the number and level of experimental factors, the L_9 (3^4) orthogonal table is selected. The elements in the orthogonal table are replaced with corresponding factors and levels. The maximum settlement S_{max} of surface monitoring points in each test scheme is calculated and filled in the last column of Table 3. The range analysis method is used to analyze the sensitivity of various parameters to surface settlement. The average value K and sensitivity R of different levels of each factor are shown in Table 4. The sensitivity analysis results show that the elastic modulus of sand has the greatest influence on the surface settlement, followed by the elastic modulus of the lining. And the internal friction angle and density of the sand layer influence the surface settlement.

Table 2. Factor levels for sensitivity analysis.

Level of factor	$\rho_{sand}/$ t/m³	$E_{sand}/$ MPa	ϕ_{sand}	$E_{lining}/$ GPa
1	1.782	13.5	25.2	13.5
2	1.98	15	28	15
3	2.187	16.5	30.8	16.5

Table 3. Orthogonal test scheme.

Test plan	ρ_{sand}	E_{sand}	ϕ_{sand}	E_{lining}	$S_{max}/$ mm
1	1.782	13.5	25.2	13.5	5.63
2	1.782	15	28	15	5.08
3	1.782	16.5	30.8	16.5	4.63
4	1.98	13.5	28	16.5	5.62
5	1.98	15	30.8	13.5	5.08
6	1.98	16.5	25.2	15	4.63
7	2.187	13.5	30.8	15	5.62
8	2.187	15	25.2	16.5	5.08
9	2.187	16.5	28	13.5	4.63

Table 4. Results of range analysis.

Range analysis	ρ_{sand}	E_{sand}	ϕ_{sand}	E_{lining}
K_1	7.97×10^{-5}	0.514	3.97×10^{-5}	2.50×10^{-3}
K_2	-2.33×10^{-5}	0.032	-2.55×10^{-4}	-5.07×10^{-5}
K_3	-7.73×10^{-5}	-0.481	-2.15×10^{-4}	2.50×10^{-3}
R_j	1.57×10^{-4}	0.995	4.71×10^{-4}	5.00×10^{-3}
Sensibility	$E_{sand}>E_{lining}>\phi_{sand}>\rho_{sand}$			

Table 5. Parameter value range and parameter inversion results.

Algorithm parameters	E_{sand}	ϕ_{sand}	E_{lining}
Upper limit	16.5	30.8	16.5
Lower limit	13.5	25.2	13.5
Accuracy	0.001	0.01	0.001
Code length	13	10	12
Inversion results	15.627	30.01	14.610

3.3 Inversion calculation

The population number of the improved genetic algorithm is 40, and the maximum algebra is 100 generations. As the sand layer density is the least sensitive to surface settlement, it is taken as 1.98 t/m³ from its inner test results. Highly sensitive parameters are endowed with high calculation accuracy. The value range, calculation accuracy, coding length of the parameters to be inverted, and the inversion results of each parameter are listed in the last row of Table 5. The parameter inversion results are used in the forward analysis of the finite element model, and the settlement calculation values of each point are shown in Figure 4. The difference between the finite element calculation value and the measured value is small, and the changing trend is relatively consistent.

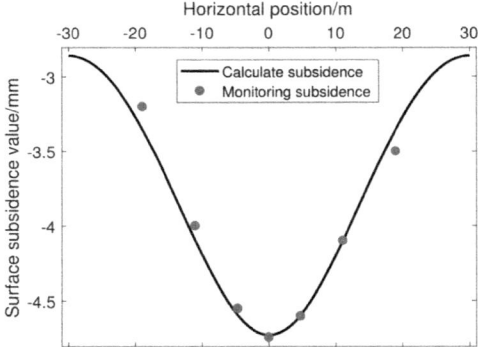

Figure 4. The calculated value of the settlement.

3.4 Surface settlement analysis

The surface settlement curve caused by shield excavation is generally called "settlement trough". The two important indicators of the "settlement trough" are the maximum value of the surface settlement and the "width of the settlement trough". The maximum value of the land settlement is located at the midpoint of the settlement curve. "Settlement trough width" is the distance from the center point of the settlement curve to the inflection point. In the modified Peck formula (Rankin 1988):

$$S = S_{\max} \exp\left[\frac{-y^2}{2i^2}\right] \qquad (6)$$

where y refers to the horizontal position, the settlement value of any point on the ground can be calculated according to the maximum value of surface settlement S_{max} and "settlement tank width" i.

Based on the parameters obtained from the inversion, this paper studies the influence of the shield depth and the thickness of the upper loam layer on the surface settlement curve of shield construction. According to the finite element model of shield excavation established in Section 3.1, the maximum surface settlement (Figure 5) and the width of settlement trough (Figure 6) of the shield under the burial depth of 15 m, 20 m, 25 m and 30 m and different soil layer thicknesses are calculated.

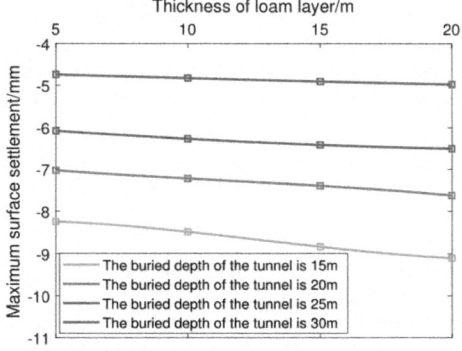

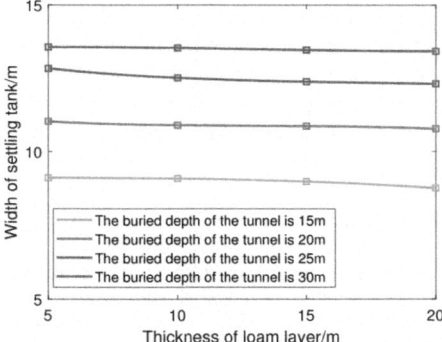

Figure 5. Maximum surface settlement under different burial depths of the tunnel.

Figure 6. Width of the sinking tank with different burial depths of the tunnel.

The results show that the maximum surface settlement will decrease with the increase of shield depth. Under the same tunnel depth, the thicker the loam layer is, the greater the maximum surface settlement is. Compared with the change in soil layer thickness, the change in shield depth causes the change in the maximum surface settlement more obvious. The smaller the shield depth is, the more significant the change in the maximum surface settlement caused by the increase in the thickness of the loam layer is. With the increase of shield depth, the width of the surface settlement trough will increase. Under the same shield depth, the width of the settlement trough will increase with the increase of soil layer thickness. However, the change is within the range of 0.5 m, and the change is not significant. Changing the depth of the shield causes the width of the settlement trough to change more significantly.

4 CONCLUSIONS

A two-dimensional finite element model is established according to a shield tunneling project through a sandy soil layer. An orthogonal test with surface settlement as the index is designed to analyze the sensitivity of sand and lining parameters. Based on the surface monitoring data, the improved genetic algorithm is used to inverse the material parameters. Based on this, the influence of different shield depths and different soil layer thicknesses on surface settlement is calculated. The main conclusions are as follows:

(1) The result of the orthogonal test is analyzed by range analysis. The results show that the change of the elastic modulus of the sand layer where the shield is located has the greatest impact on the surface settlement, followed by the elastic modulus of the lining and the friction angle in the sand layer, and the change of the density of the sand layer has little impact on the surface settlement. The results show that the change of elastic modulus of the sand layer has the greatest influence on the surface settlement. The changes in the elastic modulus of the lining and the friction angle in the sand layer have a secondary effect on the surface settlement. The change of sand layer density has the least effect on surface settlement. In the numerical calculation of shield tunneling in the sandy loam stratum, the elastic modulus of the sandy soil layer and the elastic modulus of the lining should be reasonably selected.

(2) Based on the measured settlement values of the surface monitoring points, the elastic modulus of the sand layer, the internal friction angle of the sand layer and the elastic modulus of the

lining are inverted by using the improved genetic algorithm. The inversion parameters are substituted into the finite element model, and the calculated results are in good agreement with the measured values.

(3) The maximum value of the surface settlement and the width of the settlement trough of the shield under different buried depths and different thicknesses of the soil layer are calculated. The results show that the shield depth has the greatest influence on the surface settlement curve. When the buried depth of the shield increases, the maximum settlement of the surface settlement curve will decrease, and the width of the "settlement trough" will increase. When the thickness of the soil layer increases, the maximum settlement of the surface settlement curve will increase, and the width of the "settlement trough" will decrease.

REFERENCES

Bao, H. & Y. Xue (2020). A study on surface deformation induced by shield tunneling with multi-layered soil [J]. *Chinese Journal of Underground Space and Engineering*, 16 (S1): 431–436.

Cheng, J. & Y. Jiang (2021). A study on influence characteristics of shield tunnel depth and spacing on settlement in clay strata [J]. IOP conference series. *Earth and Environmental Science*, 651 (3): 32041.

Do, N. & D. Dias (2014). Three-dimensional numerical simulation of mechanized twin tunnels in soft ground [J]. *Tunnelling and Underground Space Technology*, 42: 40–51.

Luo, Z. & Z. Li (2020). Three-dimensional fluid-soil full coupling numerical simulation of ground settlement caused by shield tunnelling [J]. *European Journal of Environmental and Civil Engineering*, 24 (8): 1261–1275.

Rankin, W. (1988). Ground movements resulting from urban tunnelling: Predictions and effects.

Wu, H. & S. Shen (2020). Three-dimensional numerical modelling on localised leakage in segmental lining of shield tunnels [J]. *Computers and Geotechnics*, 122: 103549.

Ye, Z. & H. Liu (2018). Mechanism and countermeasure of segmental lining damage induced by large water inflow from excavation face in shield tunneling [J]. *American Society of Civil Engineers*.

Yin, M. & H. Jiang (2018). Effect of the excavation clearance of an under-crossing shield tunnel on existing shield tunnels [J]. *Tunnelling and Underground Space Technology*, 78: 245–258.

Zhao, Y. & J. Zhao (2020). Study on surface settlement caused by shield tunneling in water-rich fine sand stratum [J]. *Chinese Journal of Underground Space and Engineering*, 16 (S2): 918–924.

*Frontiers in Civil and Hydraulic Engineering – Mohamed A. Ismail and
Hazem Samih Mohamed (Eds)*
© 2023 The Authors, ISBN 978-1-032-38247-0

Experiment on the external reinforcement of prestressed concrete cylinder pipes

Lijun Zhao*

*Earthquake Engineering Research Center, China Institute of Water Resources and Hydropower Research
(IWHR), Beijing, China*

Tiesheng Dou

Division of Materials, China Institute of Water Resources and Hydropower Research (IWHR), Beijing, China

Chunlei Li

*Earthquake Engineering Research Center, China Institute of Water Resources and Hydropower Research
(IWHR), Beijing, China*

ABSTRACT: The bearing capacity of the prestressed concrete cylinder pipe (short for PCCP)
is very important for pipeline safety. External reinforcement with strands is an efficient way to
restore the pipe to the original carrying capacity. The advantage of this reinforcement is actively
compensating for the prestress loss in an economic way and there is no need to empty the pipes.
To evaluate the external reinforcement effect, a full-scale test of a PCCP was conducted. The test
apparatus was mainly constituted by two PCCPs with an external diameter of 2330 mm. The pipe
after strengthening can sustain the internal hydraulic pressure of design and does not leak. The
external reinforcement of the PCCP with strands is capable to meet the strengthening requirement.

1   INTRODUCTION

A prestressed concrete cylinder pipe (PCCP) has been adopted for more than 70 years [Roller, J.J
2013] since it was invented in 1942. PCCP failures (such as cracks and fractures of prestressed
wires) may lead to a catastrophic loss without warning due to the pipes' dimensions and high
internal hydraulic pressure [S.Q. Ge 2015]. To restore the deteriorating pipe's ability to undertake
the internal hydraulic pressure, a lot of studies were conducted by S. Rahman [S. Rahman 2012],
Michael K. Kenny [Michael K. Kenny 2014], Michael Ambroziak [Michael Ambroziak 2010],
Mehdi S. Zarghamee [2013], and Dou Tiesheng [Dou Tiesheng 2017]. There is no need to take the
pipes out of service and the reinforcement can compensate for the prestress loss in an economic
way. This reinforcement is an economic way to ensure long-term service of the pipes with wires-
breakage in a good state. However, few researches have been carried out on the strengthening
effect of the external reinforcement of PCCP. A full-scale test under changing internal hydraulic
pressure was conducted for a full-scale PCCP. The structural behaviors of the pipe and the external
strengthening effect are studied in this paper.

2   EXPERIMENT PLANNING

2.1   *Experiment materials*

Table 1 gives several critical parameters of the full-scale PCCP. The working pressure of the
water inside the pipe is 0.6 MPa, and the design hydraulic pressure is 0.9 MPa according to
the specifications.

---

*Corresponding Author: zhaolj@iwhr.com

DOI 10.1201/9781003344209-29

Table 1. Critical variables of the specimen.

Critical Variables		Critical Variables	
Inner diameter of concrete/mm	2000	Standard compressive strength of concrete /(N/mm^2)	55
Thickness of core concrete/mm	140	Modulus of concrete/(N/mm^2)	2.786×10^4
Outer diameter of cylinder/mm	2103	Compressive strength of mortar /(N/mm^2)	45
Thickness of cylinder/mm	1.5	Modulus of mortar/(N/mm^2)	2.535×10^4
Diameter of wires/mm	6	Modulus of cylinder/(N/mm^2)	2.069×10^5
Spacing between each wire/mm	22.1	Modulus of wire/(N/mm^2)	1.93×10^5

2.2 Experiment apparatus

A full-scale test was carried out in an assembled device (Figure 1) and the device consisted of two pipes and a steel fitting. The sealing at the bell and spigot was specially designed for this full-scale test of external reinforcement. The inspection of the entire device is significant. It is essential to ensure that the assembled device do not leak after the installation.

Figure 1. Design sketch in 3D of the test apparatus.

2.3 Procedures of test

The procedures of the test mainly consisted of 5 phases.

Phase 1: increase the internal hydraulic pressure from 0 to the working pressure 0.6 MPa.

Phase 2: cut the prestressed wires manually under the working pressure of 0.6 MPa, and stop until the cracks propagate in the concrete core.

Phase 3: decrease the internal hydraulic pressure from the working pressure of 0.6 MPa to the artesian pressure of 0.2 MPa.

Phase 4: wrap the strands externally around the pipe under the artesian pressure of 0.2 MPa, and then perform the tensioning process. The target tensile strength of strands is 1172 MPa (Ma, Z. 2017; Ye, Z. 2012).

Phase 5: increase the internal hydraulic pressure from 0.2 MPa to the designed value of 0.9 MPa.

To ensure a steady increase in internal hydraulic pressure, the test increased the internal hydraulic pressure step by step. Macro-cracks in the pipe were observed and particular information about the macro-cracks (such as the position, the width, etc.) was recorded during the whole process.

3 TEST RESULTS

If the strain of core concrete was up to $1.5\varepsilon_t' = 208\mu\varepsilon$, where ε_t' represented the elastic strain when the stress of the core concrete reached the designed tensile strength, the micro-cracking in the concrete core began to propagate. When the strain reached $11\varepsilon'_t = 1523\mu\varepsilon$, macro-cracks were about to appear.

Macro-cracks were first discovered when the percentage of broken wires was up to 20.18% through field inspection. The location of the macro-cracks were on the surface of the mortar coating at the orientation of 3 o'clock or 9 o'clock. The cracks continued to propagate while the operation of breaking wires lasted. Inter-facial separation between the outer concrete core and the mortar coating occurred at the orientation of 3 o'clock or 9 o'clock. The widest cracks occurred on the outer surface of the concrete core. The maximum width of these cracks was 2.2 mm at the orientation of 3 o'clock.

When the process of wire breakage was finished, in the subsequent load phase, the internal hydraulic pressure decreased from the working value of 0.6 MPa to the artesian pressure of 0.2 MPa. The macro-cracks on the surface of the pipe showed a slight closure during the decreasing process of the internal hydraulic pressure. The maximum width of cracks in the outer concrete core at the orientation of 3 o'clock decreased from 2.2mm to 1.2 mm.

The fourth phase was the tensioning process of the external strands, which aimed to actively compensate the prestress loss of the pipe. The strains of the core concrete showed an obvious decrease during this phase. The maximum width of cracks in the outer concrete core at the orientation of 3 o'clock decreased from the former 1.2 mm to 0.1 mm (Figure 2). Most macro-cracks tended to close and were eventually difficult to be observed with the naked eyes.

The final phase was to increase the internal hydraulic pressure from 0.2 MPa to the design value of 0.9 MPa, aiming to testify the effect of the external strengthening. The maximum width of cracks in the outer concrete core at the orientation of 3 o'clock almost remained unchanged at about 0.1 mm during this phase, which means that the prestressed strands have the ability to constrain the propagation of the cracks in the pipes. What's more, the concrete core returned to the compressive situation. It was found that the tested pipe did not leak during the field inspection.

The strains of the inner core concrete and the outer concrete core were all less than $207\mu\varepsilon$. This means the areas where the strain gauges were attached did not coincide with the area where the cracks continued to propagate.

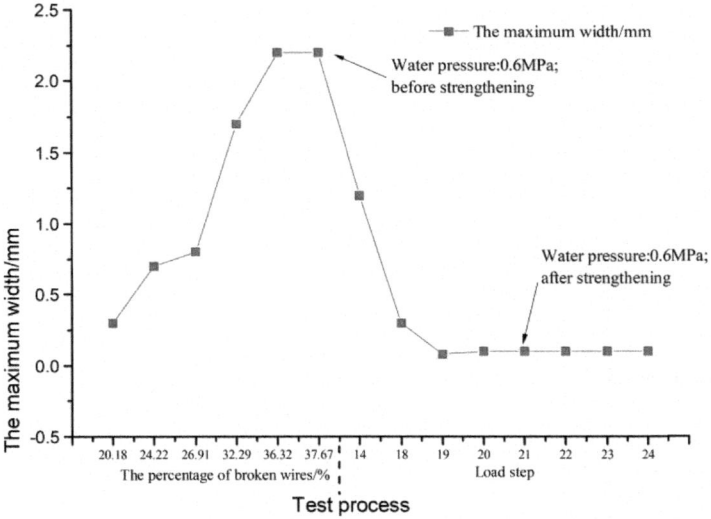

Figure 2.   The width of macro-cracks observed.

The actual situation of pre-stressed strands after the tensioning process is as shown in Figure 3. The strains of each prestressed strands were monitored during the whole process (Figure 4). The nominal tensile strain was $\varepsilon_S = \frac{\sigma_s}{E_{st}} = \frac{1860MPa}{1.95\times10^5 N/m^2} = 9538.46 \times 10^{-6}$.

Figure 3.   An actual situation of the strengthened pipe.

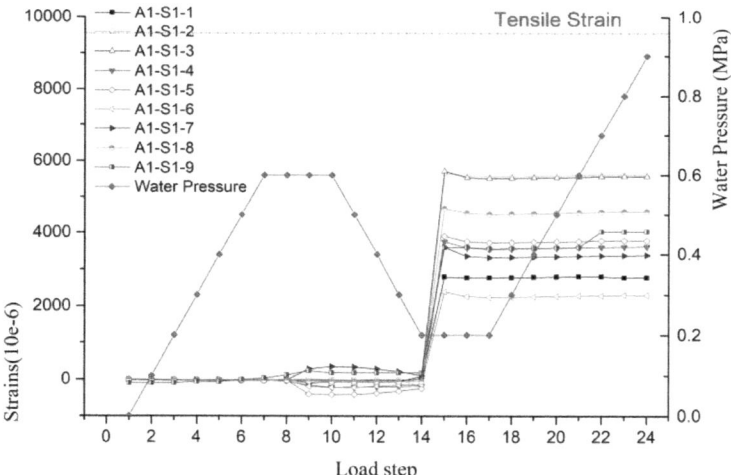

Figure 4.   Monitored strains of external pre-stressed strands.

4   CONCLUSIONS

A full-scale test was conducted to investigate the effect of the external strengthening with pre-stressed steel strands on a PCCPE. According to the test results, the following conclusions can be drawn:

(1) Wires breaking led to significant prestress loss. And the prestress resumed partially due to the bond quality of the mortar coating.

(2) When the tensioning process of the pre-stressed strands finished, the strands were able to compensate for the prestress loss. The crack propagation in the core concrete can be constrained by the prestressed strands. The strengthened pipe with external strands was able to bear the designed internal hydraulic pressure and did not leak according to the field observation. The maximum width of the macro-cracks in the outer concrete core at the orientation of 3 o'clock was reduced from the 1.2 mm to 0.1 mm. The effect of external strengthening with external prestressed strands was evident.

(3) The strains of each steel strand were all below the value of the tensile strain. The external strengthened way of PCCPs with steel strands is able to reach the strengthening demand of the strengthening and is an effective method for strengthening PCCP. The study results provide support for future application into the external reinforcement of PCCP with prestressed strands.

REFERENCES

Ambroziak, M.& Kelso, P.E.B. (**2010**) Development and Construction of the Nations Largest Water Main Rehab Project. In Proceedings of the Pipelines 2010: Climbing New Peaks to Infrastructure Reliability: Renew, Rehab, and Reinvest, August 28-September 1. 51–60.

AWWA C304 Standard for Design of Prestressed Concrete Cylinder Pipe; American Water Works Association, 2015.

Dou, T.S. & Cheng, B.Q. (**2017**). The experimental study on CFRP renewal of PCCP with broken wires. *China Concr. Cem. Prod.* 12, 35–40.

Elnakhat, H. & Raymond (**2006**) R. Repair of PCCP by post tensioning; pipelines 2006, held in Chicago, Illinois, July 30-August 2; 1–5.

Feng, X.Z (**2008**). Comparison of ASTM concrete compressive strength and China's concrete strength grade. *Northwest Hydropower*, 3, 65–67.

Ge, S.Q. & Sunil, S. (**2015**) Effect of mortar coating's bond quality on the structural integrity of prestressed concrete cylinder pipe with broken wires. *J. Mater. Sci. Res.* **2015**, 4, 59–75.

Kenny, M.K. & Rahman, S. (**2014**). *San Diego County Water Authority Aqueduct Protection Program since 1992: Evolution in Design and Construction of Steel Cylinder Relining of PCCP*. In Proceedings of the Pipelines 2014: From Underground to the Forefront of Innovation and Sustainability, Portland, OR, USA, 3–6 August.

L-Hacha, R.E. & Elbadry, M. (**2006**) *Strengthening Concrete Beams with Externally Prestressed Carbon Fiber Composite Cables: Experimental Investigation.*

Ma, Z. (**2017**) Study on the application of CFRP prestressed tendons in precast segmental bridge. *J. Transp. Sci. Eng.* 33, 50–55.

Ojdrovic, R. (**2008**) Inspection, failure risk analysis, and repair of cooling-water lines in one outage. In Proceedings of the Pipelines 2008: Pipeline Asset Management: Maximizing Performance of Our Pipeline Infrastructure, 22–27 July; 1–10.

Roller, J.J. (**2013**) *PCCP Risk Management-State of the Art and Strategy*; American Water Works Association.

Rahman, S. (**2012**). *Rehabilitation of Large Diameter PCCP: Relining and Sliplining with Steel Pipe. In Proceedings of the Pipelines 2012: Innovations in Design, Construction, Operations, and Maintenance-Doing More with Less*, August 19-22, 494–504.

Ye, Z. (**2017**) The external prestressing and carbon fiber reinforced joint concrete continuous beam research; *Jilin Architecture and Civil Engineering Institute*, 2012.

Zarghamee, M.S. & McReynolds, M. (**2013**). *Retrofit of CFRP Installation to Meet Current Design Standard*. In Proceedings of the Pipelines 2013: Pipelines and Trenchless Construction and Renewals, 23–26 June; 1258–1267.

Frontiers in Civil and Hydraulic Engineering – Mohamed A. Ismail and
Hazem Samih Mohamed (Eds)
© 2023 The Authors, ISBN 978-1-032-38247-0

Study on the evolution law of water–sediment relationship of the yellow river in Gansu section, China

Rui Zhang* & Xiaoxia Lu*
Gansu Agricultural University School of Water Conservancy and Hydropower Engineering, China

Zhe Cao*
Soil and Water Conservation Center of Gansu Provincial Department of Water Resources, China

Heping Shu*
Gansu Agricultural University School of Water Conservancy and Hydropower Engineering, China

Stan*
Soil and Water Conservation Center of Gansu Provincial Department of Water Resources, China

Xiaoyan Zhang* & Yanting Gao*
Gansu Agricultural University School of Water Conservancy and Hydropower Engineering, China

ABSTRACT: The influence of human activities on water and sediment has been increasing on the Yellow River in recent years, and their relationship is constantly changing. Therefore, the Gansu section of Yellow River was selected as the object. The data of precipitation, runoff, and sediment transport at Maqu station, Lanzhou station, and Anningdu station have been collected from 1956 to 2020. Then regression analysis and double accumulation curve and Kendall rank correlation test were used to obtain the evolution law of water-sediment and the impact of human activities on water-sediment relationship. The results showed that the runoff and sediment transport in the Gansu section of the mainstream of the Yellow River showed a decreasing trend in terms of grade changes. The runoff of each station was most affected by human activities, and its contribution rate was 52.16% (Maqu station in 1990s), 88.99% (Lanzhou station in 1980s) and 95.50% (Anningdu station in 1990s), respectively. Meanwhile, there were differences in the maximum period and the contribution rate of sediment transport at each station affected by human activities. They were 80.05% (2000s), 98.73% (2010s) and Anningdu 99.10% (2010s), respectively. These results indicated that human activities were the main factor in the reduction of water-sediment in the studied area. This study can provide a scientific basis for ecological protection and high-quality development in the Yellow River.

1 INTRODUCTION

The watershed is a relatively independent area and an important place for human activities (Downs et al. 1991). The core issue affecting the ecological protection and high-quality development of the watershed is the relationship between water and sediment, which is affected by a series of climate and human activities (Syvitski & Kettner 2011; Wohl 2006). The changes of water and sediment and its influencing factors have received extensive attention and have become one of the hotspots in hydrological scientific research (Quan et al. 2004; Wang et al. 2014). Built dams, built embankments and returned farmland to forests have caused great changes in the water and sediment

---

*Corresponding Authors: 873850358@qq.com, 487564607@qq.com, 34535193@qq.com, shuhp@lzu.edu.cn, wq-c3h3@163.com, gyt830717@gsau.edu.cn and 304084908@qq.com

DOI 10.1201/9781003344209-30

situation in the Yellow River Basin since 1960 (Zheng et al. 2021). Meanwhile, these have greatly changed the complex sediment transport relationship in the basin (Zhang et al. 2017), so that the relationship of water and sediment were significantly altered (Shen et al. 2020). Therefore, studying the relationship between water and sediment in the Yellow River Basin is the basis for comprehensive management of the Yellow River Basin (Fan et al. 2022), which can provide a basis for ecological protection and high-quality development of the basin. Scholars studied the relationship between water and sediment in the Yangtze River Basin, and found that different human activities have different effects on runoff and sediment. The sediment retention in the reservoir is the main factor for the change of sediment transport in the Basin, and the water conservation measures are an important factor. The contribution rate of the reservoir and soil-water conservation measures were 64 % and 36%, respectively (Wang et al. 2020). Engineering and vegetation measures have caused surface changes based on the analysis of the driving forces of water and sediment in the Yellow River, resulting in a 76% water reduction (Wang et al. 2017). Liu et al. studied the relationship between water and sediment in the Huangfuchuan watershed, and found that the annual increase in human activities in the watershed is the main factor that drives the reduction of runoff and sediment transport (Liu et al. 2021). Xie et al. used the linear regression cumulative distance equality method to analyze the changes of water and sediment in the interval between Tangnaihai and Gaocun and the main stream of the Yellow River, (Xie et al. 2021). The above studies have expounded on the main driving factors and their contribution rates of the changes in the Yellow River's water and sediment, but they are mainly concentrated on the scale of a large watershed or a small watershed.

In view of this, this paper uses the monitoring data of precipitation, runoff, and sediment transport in the Gansu section of the Yellow River from 1956 to 2020, and uses regression analysis, double cumulative curve and Kendall rank correlation method to analyze the mainstream of the Yellow River. The evolution law of water and sediment in the Gansu section of the main stream of the Yellow River was analyzed, hoping to provide a scientific basis for the restoration of the ecological environment and high-quality development of the entire Yellow River Basin.

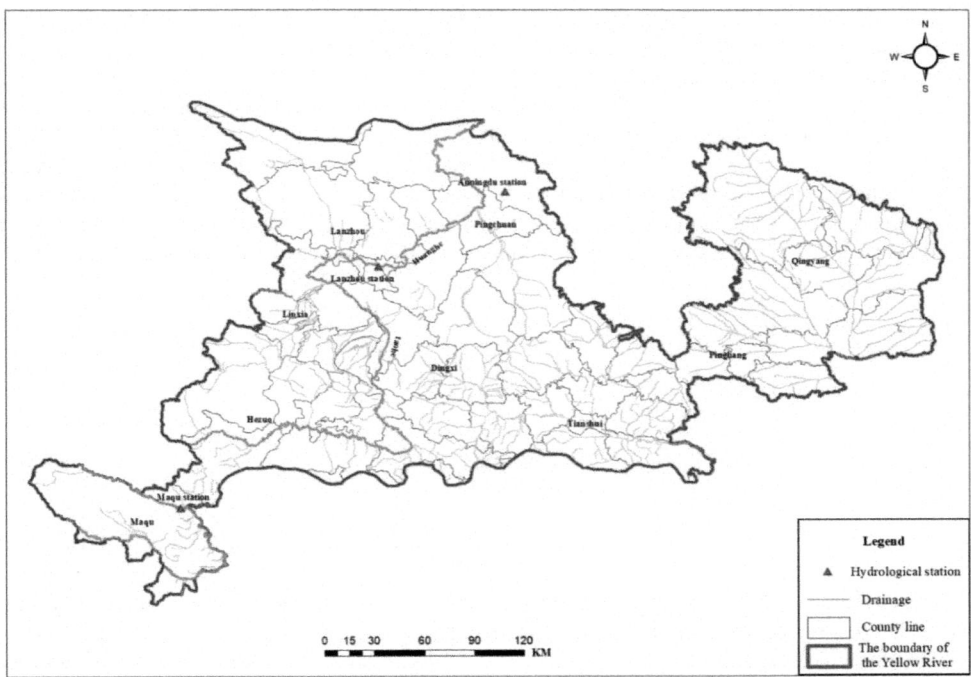

Figure 1.   Water system of the Gansu section of the mainstream of the Yellow River.

2 OVERVIEW OF THE STUDY AREA

The Yellow River is the second largest Yellow River in China. It originates from the Yogu zonglie Basin and flows through 9 provinces (including the autonomous regions) (Shandong, Henan, Shanxi, Shaanxi, Inner Mongolia, Ningxia, Gansu, Sichuan and Qinghai) with a total length of about 5464 kilometers. The Gansu section of the upper and middle reaches of the Yellow River was selected as the main research object, and its geographical location was between $32°11' \sim 42°57'$ north latitude and $92°13' \sim 106°46'$ east longitude. To study the changes of water and sediment in the upper and middle reaches of the Yellow River in recent years, the main control stations in the upper and middle reaches of Lanzhou station, Maqu station and Anningdu station were selected to analyze the evolution of water and sediment (Figure 1).

3 DATA AND METHODS

3.1 *Data*

The data series of rainfall, runoff, and sediment transport in Maqu station, Lanzhou station and Anningdu station from 1956 to 2020 were used.

3.2 *Research methods*

(1) Regression analysis method is to establish the regression relationship function expression between dependent variables and independent variables (Zhang et al. 1986). Mathematical statistical methods based on a large number of observation data are used. Regression analysis can be used to judge the fitting degree between two variables.

(2) The double cumulative curve method is to test the consistency and change of the relationship of two parameters (Easterling et al. 2000). It can describe the change relationship between the cumulative annual runoff and cumulative annual sediment discharge, so that the various characteristics of water and sediment was reflected in the Basin.

(3) Kendall rank correlation test

For the sequence, $X_1, X_2, \cdots, X_n$, the number of occurrences in all dual values was determined $(X_i, X_j, j > i)X_i < X_j$ (set to P). The (i,j) subset of the order is: $(i = 1, j = 2, 3, 4, \cdots, n)$, $(i = 2, j = 3, 4, 5, \cdots, n)$, $\cdots$, $(i = n - 1, j = n)$. If the values of advancing in order are all greater than the previous value, this is an uptrend, $p = (n + 1) + (n + 2) + \cdots + 1$, is an arithmetic progression, and the sum is $\frac{1}{2}(n - 1)n$. If the series are all reversed, then $p = 0$, and it is a downtrend. Therefore, this is a series without a trend, P mathematical expectations is $E(P) = \frac{1}{4}n(n - 1)$.

Statistics for this test:

$$U = \frac{\tau}{[V_{ar}(\tau)]^{\frac{1}{2}}} \tag{1}$$

Where:

$$\tau = \frac{4p}{n(n - 1)} - 1 \tag{2}$$

$$V_{ar}(\tau) = \frac{2(2n + 5)}{9n(n - 1)} \tag{3}$$

When n increases, U is quickly converged to a standardized normal distribution. The null hypothesis is that there is no trend when a significant level is given α, and the critical value in the normal distribution table is found $U_{\alpha/2}$. At that time, $|U| < U_{\alpha/2}$, the null hypothesis is accepted, and the trend is not significant. At that time, $|U| > U_{\alpha/2}$ the null hypothesis is rejected, and the trend is significant (Fu et al. 1992).

4 RESULTS AND ANALYSIS

4.1 *Basic characteristics of water-sediment changes*

The runoff changes and trend analysis of each hydrological station were obtained in the Gansu section of the mainstream of the Yellow River from 1956 to 2020 (Figure 2). And we found that the annual runoff of Maqu, Lanzhou, and Anningdu stations are 14.44 billion m^3, 31.03 billion m^3

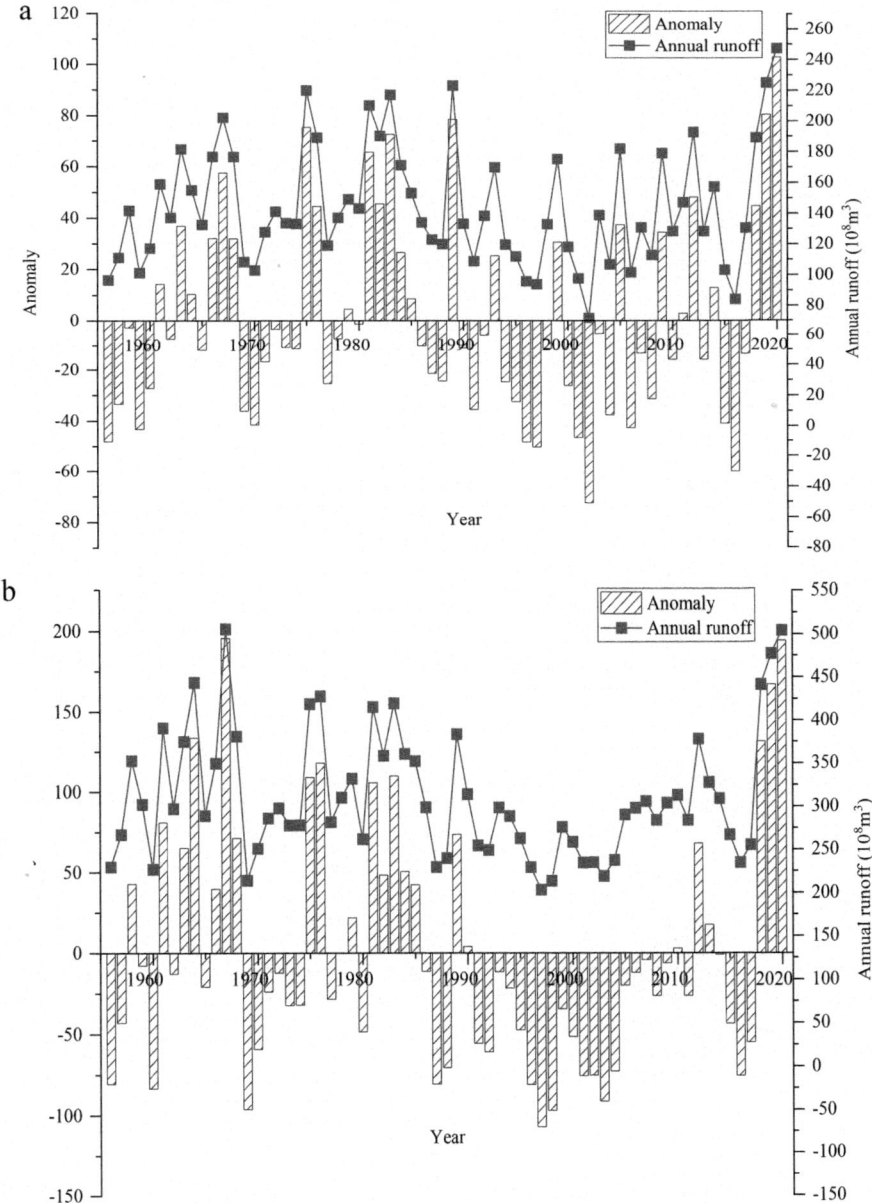

Figure 2.   Runoff anomaly and change curve, Maqu (a) station, Lanzhou (b) station, Anningdu (c) station.

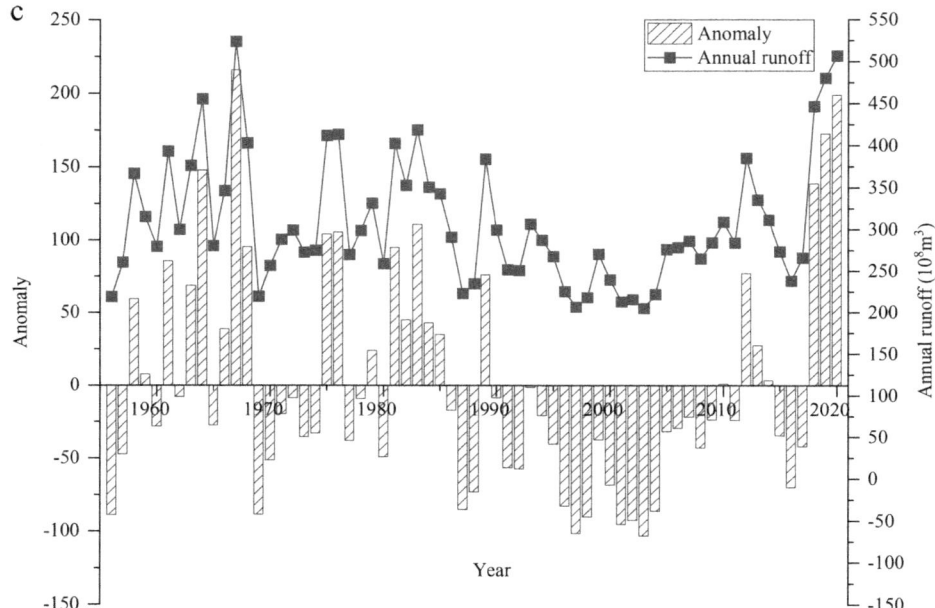

Figure 2. Continued.

and 30.82 billion m³, respectively. A maximum runoff and a minimum runoff were 24.74 billion m³ (time was 2020) and 7.19 billion m³ (2021) in Maqu station, respectively. With respect to Lanzhou station, they were 50.60 billion m³ (2020) and 21.42 billion m³ (1969). As for Anningdu stations, the maximum and minimum values were 52.41 billion m³ (1967) and 21.93 billion m³ (1956). On the other hand, there was a slightly increasing trend on the annual runoff of Maqu station (Figure 2), but it was not significant. As for Lanzhou station and Anningdu station, they were significantly gradually decreasing. These results indicated that the runoff of Yellow River was obviously decreasing in the Gansu section.

Table 1. Trend test results of runoff and sediment transport at each station.

| Name | Trend equation | Site name | Kendall rank correlation test | | |
			|U|	$U_{\alpha/2}$	Trend significance
Runoff (10^8 m³)	$y = 0.0063x + 144.95$	Maqu	0.0392	1.96	Non-significant
	$y = -0.2368x + 318.12$	Lanzhou	0.5257	1.96	Non-significant
	$y = -0.4102x + 321.74$	Anningdu	1.5547	1.96	Non-significant
Sediment load (10^4 t)	$y = 0.7893x + 446.99$	Maqu	0.2944	1.96	Non-significant
	$y = -149.51x + 1063$	Lanzhou	5.2089	1.96	Significant
	$y = -252.91x + 19630$	Anningdu	5.3114	1.96	Significant

The change and trend of sediment transport were analyzed (Figure 3, Table 1), and results presented that the annual sediment transport of Maqu, Lanzhou, and Anningdu stations were 4.73 million tons, 56.26 million tons, and 111.57 million tons, respectively. The maximum sediment discharge of Maqu station was 37.20 million tons in 1992, and its minimum value was 5.60 million tons in 2002. With respect to Lanzhou station, the maximum and minimum values were 266.10 million tons (1967) and 8.86 million tons (2017). As for Anningdu station, the maximum value was 644.60 million tons (1997) and the minimum was 12.11 million tons (1976). There was no significantly increase at Maqu station. Nevertheless, the value significantly decreasing at Lanzhou station and Anningdu station. Therefore, a significantly decreasing trend was presented in the studied area about the sediment transport.

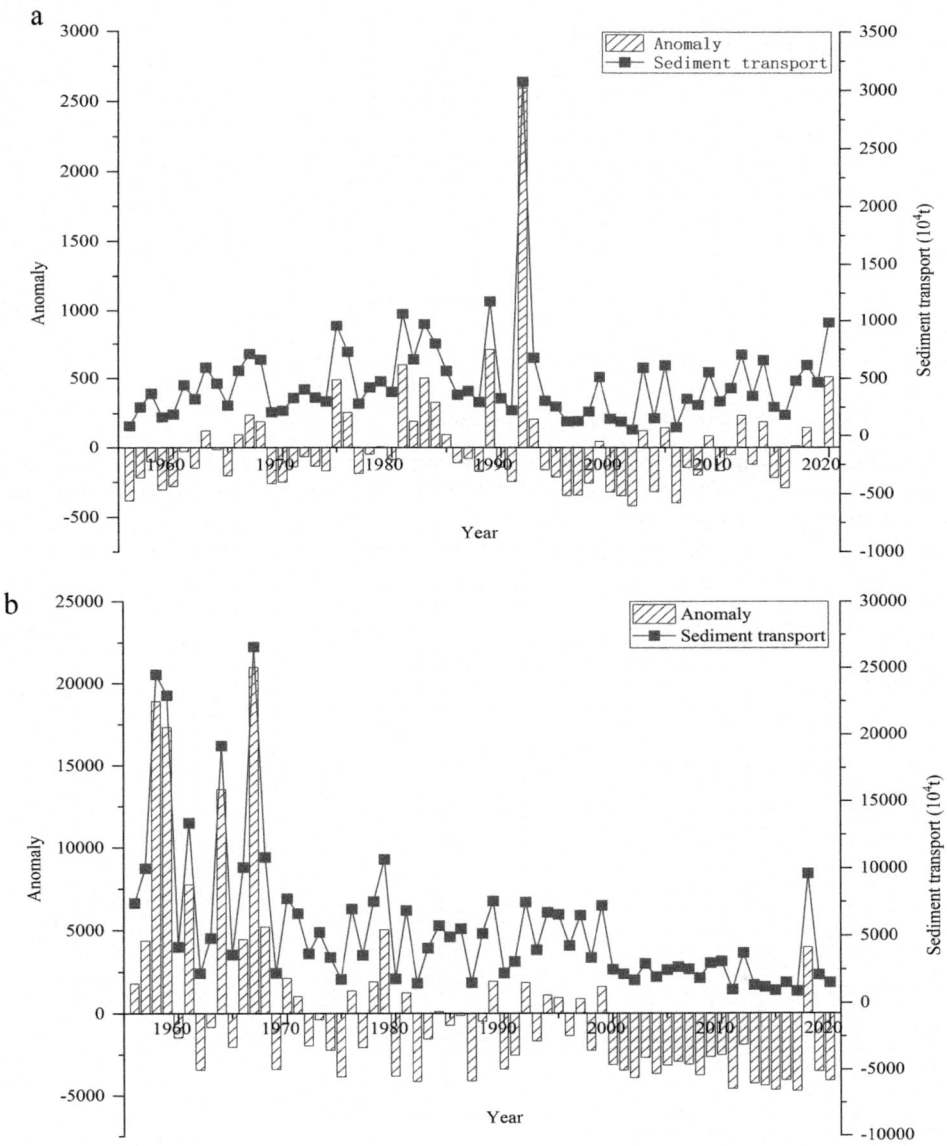

Figure 3. Sediment transport anomaly and change curve of Maqu (a), Lanzhou (b) and Anningdu (c) stations.

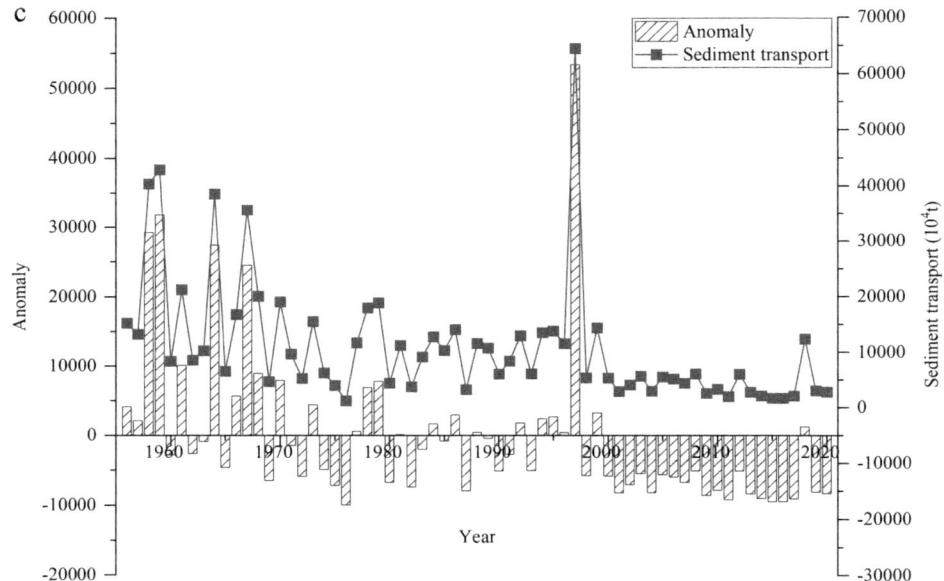

Figure 3. Continued.

4.2 *The relationship of accumulated water-sediment*

The double accumulation curve of annual runoff and sediment transport at each station was obtained in the Gansu section of Yellow River (Figure 4). We found that the slope of the double accumulation curve was significantly decreased in 1992 at Maqu station. This phenomenon illustrated the sediment transport was obviously decreased around 1992, and this was related to the implementation of

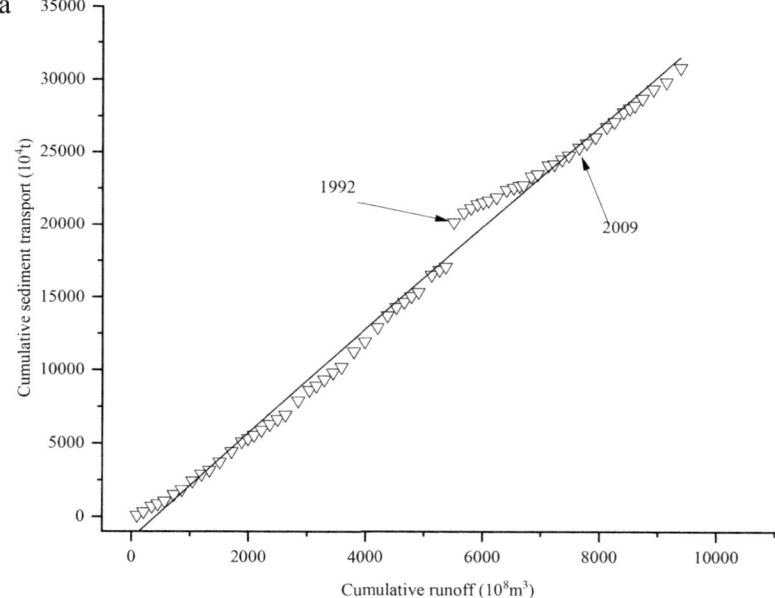

Figure 4. Double accumulation curves of annual runoff and annual sediment transport in Maqu (a), Lanzhou (b) and Anningdu (c) stations.

234

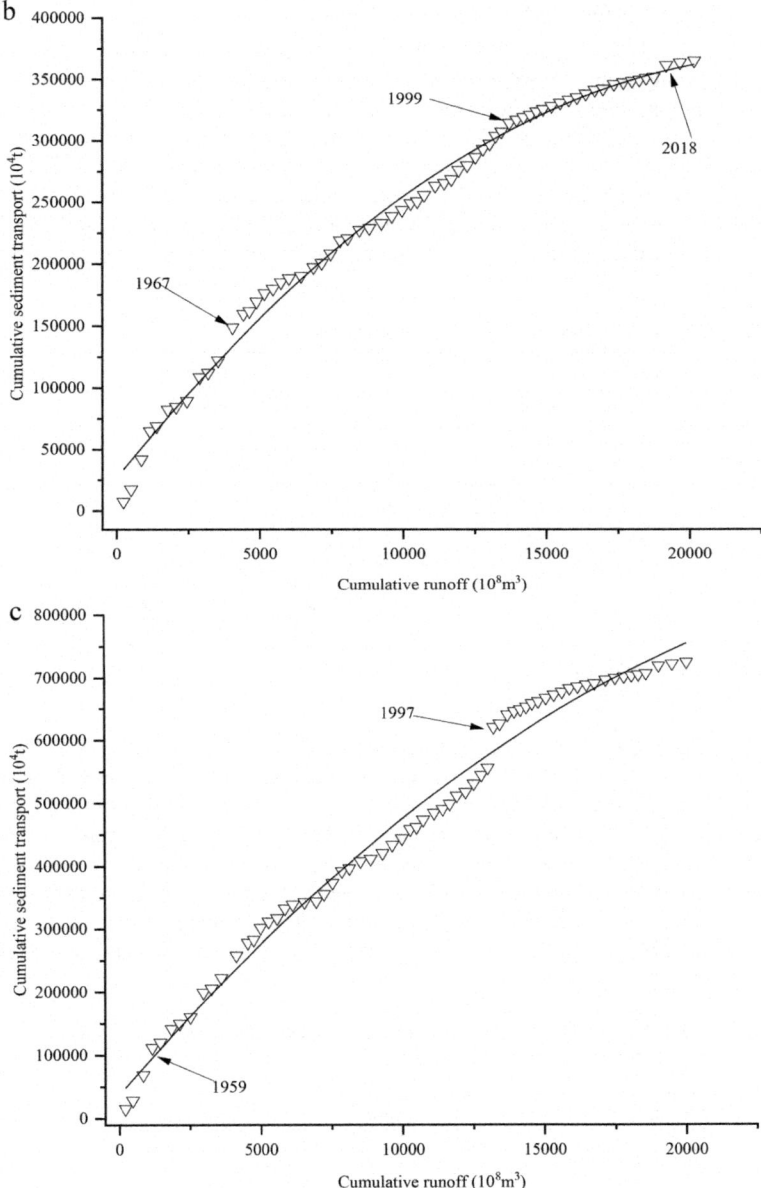

Figure 4. Continued.

soil and water conservation measures. The slope was rapidly increased after 2009, and it indicated that the sediment transport was significantly increased. This may be relate to the torrential rain that occurred in Maqu County on 21st July, 2009. The slope was decreased in 1967 at Lanzhou Station, and it was represented by a trend of decrease on sediment transport, and a sharp decrease was presented in sediment transport in 1999. Thereafter, the slope was quickly increased in 2018 due to a flood occurred on the 20th July 2018 in Lanzhou. The sediment transport of Anningdu station was decreased in 1959, and then there was an obvious decrease in sediment transport in 1997. This was related to the implementation of soil and water conservation measures.

4.3 *Analysis of contribution rates in different years*

Human activities have had a great impact on the relationship of water-sediment in the Basin (Wang et al. 2020). The study found that the runoff of Maqu, Lanzhou, and Anningdu stations were all affected by human activities, and the degree of impact was different. The impact rate of human activities was between 3.6% and 52.2% at Maqu station. As for Lanzhou station, the value was ranged from 34.5% to 89.0%. Regarding Anningdu station, it was between 34.3% and 95.5% (Table 2). These results indicated that the runoff of Maqu station is less affected by human activities. In addition, Maqu station was greatly affected by human activities in the 1990, and Lanzhou station was slightly affected in the 1970 and 1980, Anningdu station had the greatest influence between1970 and 1990.

Table 2. Effects of human activities on runoff in different years.

	Maqu				Lanzhou				Anningdu			
Time	Measured runoff ($10^8 m^3$)	Calculate runoff ($10^8 m^3$)	human activity impact ($10^8 m^3$)	Human activity impact rate (%)	Measured runoff ($10^8 m^3$)	Calculate runoff ($10^8 m^3$)	human activity impact ($10^8 m^3$)	Human activity impact rate (%)	Measured runoff ($10^8 m^3$)	Calculate runoff ($10^8 m^3$)	human activity impact ($10^8 m^3$)	Human Human impact rate (%)
1956–1960	112.6	112.6			288.2	288.2			290.9	290.9		
1970	154.5	140.2	14.3	34.1	347.6	270.7	77.0	81.5	358.1	277.8	80.3	86.0
1980	145.2	137.5	7.70	23.6	316.4	284.2	32.2	89.0	331.9	271.5	60.4	75.7
1990	168.4	139.4	29.0	52.2	332.2	238.0	94.2	65.2	326.1	289.2	36.9	95.5
2000	128.0	1275	0.6	3.6	259.1	264.6	5.5	34.5	258.5	423.6	165.1	55.5
2010	124.1	133.9	9.9	31.7	266.9	251.0	16.0	69.1	248.6	263.1	14.5	34.3
2020	157.7	145.7	12.0	26.7	345.0	245.7	99.2	70.0	348.8	327.4	21.4	37.0

Note: The base period is 1956-1960s. Same below.

Moreover, sediment transport has been significantly decreased due to influence of human activities between 1956 and 2020 (Table 3). The reduction rate of sediment transport was between 14.1% and 80.1% at Maqu station under the influence of human activities. This indicated that it was greatly affected. However, the sediment transport was hardly affected by human activities in 2010. Human activities have always greatly influenced on the Lanzhou station except for 1970 and 1980, and the value exceeded 80.0% between 1970 and 1980. The rest values were over 90.0%. As for Anningdu station, the value was 63.5% in 2000, and in the other periods the value was more than 80.0%. These results illustrated that the sediment transport of Maqu station was the least affected by human activities.

Table 3. Effects of human activities on sediment transport in different years.

	Maqu				Lanzhou				Anningdu			
Time	Measured sediment transport ($10^4 t$)	Calculate sediment load ($10^4 t$)	human activity impact ($10^4 t$)	Human activity impact rate (%)	Measured sediment transport ($10^4 t$)	Calculate sediment load ($10^4 t$)	human activity impact ($10^4 t$)	Human activity impact rate (%)	Measured sediment transport ($10^4 t$)	Calculate sediment load ($10^4 t$)	human activity impact ($10^4 t$)	Human activity impact rate (%)
1956–1960	221.8	221.8			16222.5	16222.5			27975.0	27975.0		
1970	444.3	321.7	118.9	54.3	9701.0	17306.9	7605.9	87.5	17091.0	29314.6	12223.6	90.1
1980	449.2	315.6	133.6	58.8	5717.0	18173.5	12456.5	86.5	10965.0	28654.3	17689.3	96.3
1990	670.7	320.0	350.8	73.1	4455.0	15221.5	10766.5	91.5	9140.0	30514.5	21374.5	89.4
2000	587.9	292.5	295.4	80.1	5138.0	16919.7	11781.7	94.4	15665.0	44699.2	29034.2	63.5
2010	293.4	307.4	14.0	14.1	2374.0	16046.8	13672.8	98.7	4476.0	27764.3	23288.3	99.1
2020	493.7	334.4	159.4	58.6	2451.6	15713.3	13261.66	96.3	3633.6	34550.5	30916.9	82.5

To sum up, the impact degree of human activities on the runoff and sediment transport at three stations was ranked below: Lanzhou station, Anningdu station, and Maqu station. The specific performance is as follows: the runoff of Maqu station was most affected by human activities in the 1990s, its sediment transport was most affected by human activities in the 1990s and 2000s, and the impacts on runoff and sediment transport by human activities were basically continuous and consistent. This shows that the comprehensive start of the project of returning farmland to forest have changed the relationship between water and sediment at Maqu station. In the 1980s, the runoff of Lanzhou Station was most affected by human activities. The amount of sediment transport was greatly affected by human activities after the 21st century. Especially a large number of water conservancy projects were constructed such as Liujiaxia Reservoir and Yanguoxia Reservoir. They played a role in water storage and diversion and adjusting the relationship between water and sediment in the Basin. The runoff of Anningdu station was most affected by human beings in the 1990s. The implementation of small water conservancy projects such as check dams and terraces has changed the relationship between water and sediment. The sediment transport of Anningdu station was most affected by human activities in the 2010s. Anningdu station is mainly a water area, and the amount of sand coming from Anningdu is less. Therefore, the amount of sand deposited along the route reduces the amount of sand along the route.

5 CONCLUSION

The runoff and sediment transport were gradually decreasing between 1956 and 2020. The sediment transport of Maqu station was decreased in the 1990, and it was increased in the 2000. This phenomenon was related to the occurrence of severe rainstorms. The sediment transport of Lanzhou station was decreased in the 1960 and 1990, and it was increased in 2018. This was caused by floods. The sediment transport of Anningdu station was decreased in the 1950 and 1990, and it was related to the implementation of soil and water conservation measures.

The impact degree of human activities on the runoff and sediment transport was ranked from Lanzhou station, Anningdu station, to Maqu station. With respect to runoff, Maqu station and Anningdu station were affected in the 1990, and Lanzhou station was in the 1980. As for sediment transport, Maqu station was affected in the 2000, Lanzhou station and Anningdu station were in the 2010. In addition, water conservancy projects and soil-water conservation measures have obvious impacts on water-sediment reduction, and human activities were the main reason for the changes in water-sediment in the studied area.

ACKNOWLEDGEMENTS

This work was funded by Gansu Province Water Conservancy Scientific Experiment Research and Technology Promotion Project (No. 22GSLK047), Water Resources Fee Project of Gansu Provincial Department of Water Resources (Gansu water resources [2021] No. 105), and Water Resources Fee Project of Gansu Provincial Department of Water Resources (Gansu water resources [2000] No. 94).

REFERENCES

Downs P.W., Gregory K.J., Brookes A. How integrated is river basin management? [J]. *Environmental Management*, 1991, 15(3): 299–309.

Easterling D.R, Meehl G.A, Parmesan C., et al. Climate extremes: observations, modeling, and impacts [J]. *Science*, 2000, 289(5487): 2068–2074.

Fan Junjian, Zhao Guangju, Mu Xingmin, et al. Variation of water and sediment in the upper Yellow River from 1956 to 2017 and its driving factors[J]. *China Soil and Water Conservation Science* (Chinese and English), 2022, 20(03): 1–9.

Fu Congbin, Wang Qiang. Definition and detection method of climate change[J]. *Atmospheric Science*, 1992, 16(4): 482–493.

Liu Qiang, Yu Feihong, Chang Kangfei, et al. Variation characteristics of water and sediment in the Huangfuchuan Basin and its influencing factors [J]. *Arid Region Research*, 2021, 38(06): 1506–1513.

Quan-xi X.U., Guo-yu S.H.I., Ze-fang C. Analysis of recent changing characteristics and tendency runoff and sediment transport in the upper reach of Yangtze River[J]. *Advances in Water Science*, 2004, 15(4): 420–426.

Shen Lianmian, Jiang Xiaohui, Lei Yuxin. Variation of river base flow and its driving factors in the Kuye River Basin[J]. *Journal of Water Resources Research*, 2020, 9: 373.

Syvitski J.P.M., Kettner A. Sediment flux and the Anthropocene[J]. *Philosophical Transactions of the Royal Society A: Mathematical, Physical and Engineering Sciences*, 2011, 369(1938): 957–975.

Wang S., Fu B., Liang W., et al. Driving forces of changes in the water and sediment relationship in the Yellow River[J]. *Science of the Total Environment*, 2017, 576: 453–461.

Wang Y.G., Chen Y. The influence of human activity on variations in basin erosion and runoff-sediment relationship of the Yangtze River[J]. *ISH Journal of Hydraulic Engineering*, 2020, 26(1): 68–77.

Wang Y.G., Liu X., Shi H.L. Variation and influence factors of runoff and sediment in the lower and Middle Yangtze River[J]. *Journal of Sediment Research*, 2014, 5(2020): 38–47.

Wohl E. Human impacts to mountain streams[J]. *Geomorphology*, 2006, 79(3–4): 217–248.

Xie Fabing, Zhao Guangju, Mu Xingmin, et al. Changes in the relationship between water and sediment in the mainstream of the Yellow River in the past 70 years [J]. *China Soil and Water Conservation Science (Chinese and English)*, 2021, 19(05): 1–9.

Zhang J., Zhang X., Li R., et al. Did streamflow or suspended sediment concentration changes reduce sediment load in the middle reaches of the Yellow River? [J]. *Journal of Hydrology*, 2017, 546: 357–369.

Zhang Mingnian. Regression analysis and experimental design[J]. *Xinjiang Agricultural Science*, 1986, 3: 33–36.

Zheng H., Miao C., Jiao J., et al. Complex relationships between water discharge and sediment concentration across the Loess Plateau, China[J]. *Journal of Hydrology*, 2021, 596: 126078.

Frontiers in Civil and Hydraulic Engineering – Mohamed A. Ismail and
Hazem Samih Mohamed (Eds)
© 2023 The Authors, ISBN 978-1-032-38247-0

Numerical simulation and mechanism analysis of shield cutting steel bar

Xupeng Liu
China Railway Tunnel Group Road & Bridge Engineering Co. Ltd, Tianjin, China

Mengxuan Kan*
Center for Urban Construction and Underground Space Engineering, Anhui Jianzhu University,
Anhui Hefei, China

Zhitao Liu & Quanquan Sun
China Railway Tunnel Group Road & Bridge Engineering Co. Ltd, Tianjin, China

Yang Liu
Center for Urban Construction and Underground Space Engineering, Anhui Jianzhu University,
Anhui Hefei, China

ABSTRACT: In order to better understand the interaction mechanism between steel and cutting tools in shield cutting reinforced concrete piles, taking the shield cutting of two bored piles in the Yaohai Park Station-Hefei Station section of Hefei Rail Transit Line 1 as the engineering background, the three-dimensional simulation model of cutting steel bar is established by finite element software, and the effects of rake angle, tool width and cutting depth on cutting effect and cutting force of steel bar are studied. The results show that the cutting process can be divided into four stages, including initial extrusion, uplift, shear fracture, and chip formation. The tangential force of the cutting steel bar is much larger than the penetration force, and both of them show a trend of "increasing first and then decreasing" in general while the lateral force is almost zero. Penetration force decreases with the increase of the cutting front angle, and the tangential force is positively correlated with the cutting front angle, cutting depth, and cutter width. However, when the parameters are increased to a certain value, the cutting force will change abruptly. Therefore, the blade with a negative rake angle should be selected as far as possible and the tool parameters should be strictly controlled to prevent the large wear of the tool due to excessive force and affect the construction efficiency in the construction.

1 INTRODUCTION

In recent years, with the rapid development of China's social economy and the continuous improvement of urban infrastructure, the construction of the subway has become an important measure to ease the urban traffic pressure (Ye & Feng 2021). The subway lines are mostly located below the urban roads. When the subway tunnel is constructed in downtown areas, the shield machine passes through the reinforced concrete pile foundation from time to time (Ge 2016; Zhang 2017; Zhao & Liu 2011). Generally, the ground is used to directly remove the pile foundation, and the method of the direct cutting pile is rarely used. Although the method of direct cutting pile foundation by the shield machine has the advantages of a short construction period, small environmental impact on the surrounding environment and high economic benefits, the cutter head of the traditional shield machine cannot directly cut the pile foundation, especially when cutting the steel bar, it is easy to cause the cutter head to be hooped by the steel bar, which affects the working performance of the

*Corresponding Author: kanmengxuan@stu.ahjzu.edu.cn

shield machine. Therefore, the study on the mechanism of shield cutter cutting reinforced concrete pile foundation has profound guiding significance for improving cutter configuration.

Scholars have also conducted a lot of research on the cutting mechanism of shield cutters. The early cutting mechanism of hob is to use the calculation model of coal rock planning and the cutting force formula of linear cutting coal rock to derive the cutting force (Evans 1965), and then the CSM model provides a basis for the calculation of the force of hob (Rostami 1993). The most representative scraper cutting models are the flow model (Fuzhaojiro 2001; Zhinoshiro 1980) and the Mckyes-Ali model (Ibarra et al. 2005). With the development of computer simulation technology, the numerical simulation method has been widely used in the study of cutting pile foundations by shield cutterhead due to its convenience and economy. Research shows that shield cutting large diameter reinforced concrete pile foundation is feasible (Li 2020; Wang 2013, 2014), and with the increase of cutting speed and tool spacing (Guo 2021; Jiang 2021), the maximum value of cutting force fluctuation range increases, and increasing the cutting depth can further improve the efficiency of shield tunneling.

The existing literature mainly focuses on the overall influence of shield cutterhead cutting pile foundation on stratum and reinforced concrete pile foundation, and rarely explores its cutting mechanism according to the micro-model of cutter cutting steel. Therefore, this paper takes the shield cutting of two bored piles in the Yaohai Park Station – Hefei Station section of Hefei Rail Transit Line 1 as the engineering background, and uses ABAQUS finite element software to establish the three-dimensional simulation model of cutter cutting reinforcement, so as to explore the influence of blade rake angle, blade width and cutting depth on the cutting effect and cutting force of reinforcement.

2 ESTABLISHMENT OF SIMULATION MODEL

In order to explore the mechanism of cutter cutting steel bars, ABAQUS software was used to simulate the process of shield cutter cutting piles. In Hefei rail transit line 1 Yaohai park station-Hefei station section under the Hefei station abandoned shed bored pile foundation engineering, pile diameter 500 mm, pile length 30 m, pile foundation steel diameter 22 mm. The cutter size is modeled according to the actual size of the shield cutter in the actual project. The cutter height is 110 mm and processed according to the rigid body. The specific parameters are shown in table 1. The cutting reinforcement simulation models of cutters with different rake angles ($-45°$, $-25°$, $25°$, $45°$), cutter widths (50, 60, 70, 80 mm), and cutting depths (2, 3, 4, 5 mm) were established.

Table 1. Physical parameters of cutter and steel bar.

Material	Internal friction angle	Poisson's ratio	Elastic modulus/MPa
Steel bar	0.3	7.85	210000
Cutter	0.22	14.5	640000

2.1 *Constitutive model*

The Johnson-Cook model is a common constitutive model of metal materials, which can reflect the characteristics of metal changes in the cutting process. Therefore, the Johnson-Cook model is used in this paper. Since the material parameters of steel bars are similar to those of 45 steel (Chen 2005), the material parameters of reference are shown in Tables 1 and 2.

Table 2. Johnson-Cook constitutive model parameters and failure parameters of steel bars.

Constitutive model parameters					Failure parameters				
A/MPa	B/MPa	n	C	m	d_1	d_2	d_3	d_4	d_5
506	320	0.28	0.064	1.06	0.1	0.76	1.57	0.005	-0.84

2.2 *Analysis steps, boundary conditions, and load setting*

The Explicit display algorithm of ABAQUS is used to realize the simulation process of cutting steel bars, and the calculation results of 0.10 s are output in the post-processing. The cutting of steel bars by cutter is a penetration problem. The interaction between the cutter and steel bar is simulated by the contact algorithm of the erosion surface with a symmetrical penalty function, and the friction coefficient is 0.15. The shear failure criterion of the J-C model is used as the cutting separation criterion. In the boundary conditions, the degrees of freedom of all nodes in the lower half and side of the reinforcement are constrained; the radius of the subway shield tunnel is 3m, and the cutter speed is 0.5–1.0r /min, in order to facilitate modeling, the cutting speed is set to 200mm/s.

2.3 *Establishment of cutting steel bar model*

In this project, the diameter of the main reinforcement of the pile foundation is 22mm, so the diameter of the reinforcement in the simulation is set to 22mm, and the edge length of the reinforcement

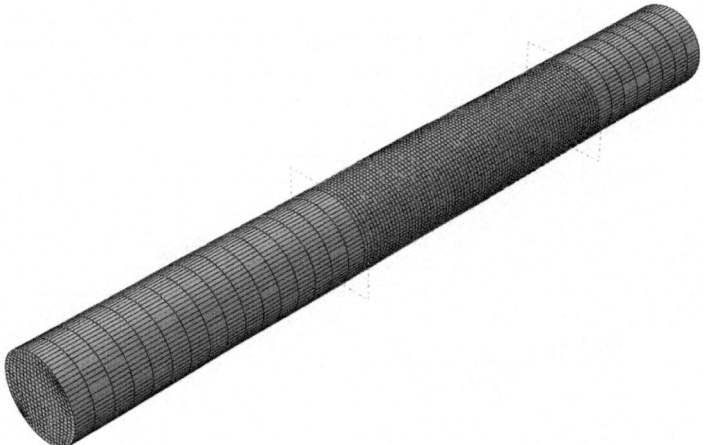

Figure 1.   Reinforcement.

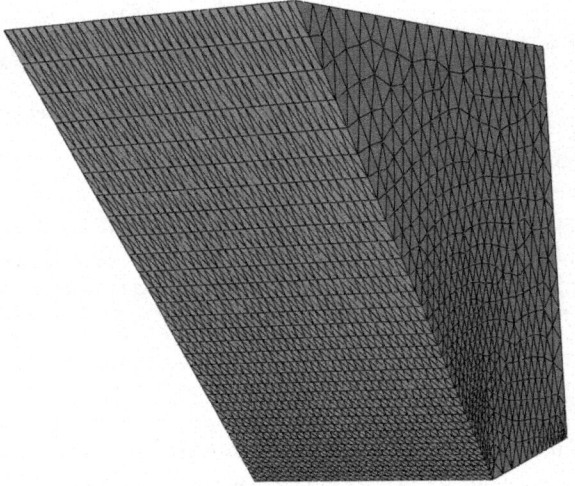

Figure 2.   Cutter with a negative rake angle of −45°.

241

unit in the cutting part is set to 1mm, and the mesh division of the reinforcement and the negative rake angle-45° is shown in Figures 1 and 2.

3 PROCESS AND CHARACTERISTICS OF STEEL BAR CUTTING BY CUTTER

3.1 *Cutting effect analysis of cutter*

Taking the negative rake angle −45° blade and the cutting depth of 3 mm as an example, the cutting process of the three-dimensional blade is analyzed, as shown in Figure 3. The cutting process of steel bars by the blade can be basically summarized into four stages. The first stage is the initial extrusion stage, and a small amount of compression occurs in the interaction area between the steel bar and blade. In the second stage, the extruded part of the steel bar has a certain degree of uplift; the third stage is the shear fracture stage, which is manifested by the tearing of the area where the steel bar and the left and right sides of the cutter contact each other, and the fracture surface appears. With the advance of the cutting edge and the further extrusion of the front steel bar, the cutting force exceeds the shear strength of the steel bar. The steel bar section below the tooltip contact surface is shear cracked and peeled off and finally enters into the fourth stage of chip formation to complete the whole cutting process. During the whole cutting process, the maximum Mises stress reaches 1.15 GPa, and the value becomes smaller and smaller with the shear fracture of steel bars. The stress is mainly distributed in the contact part between the steel bar and the cutter and the area that is about to be cut ahead.

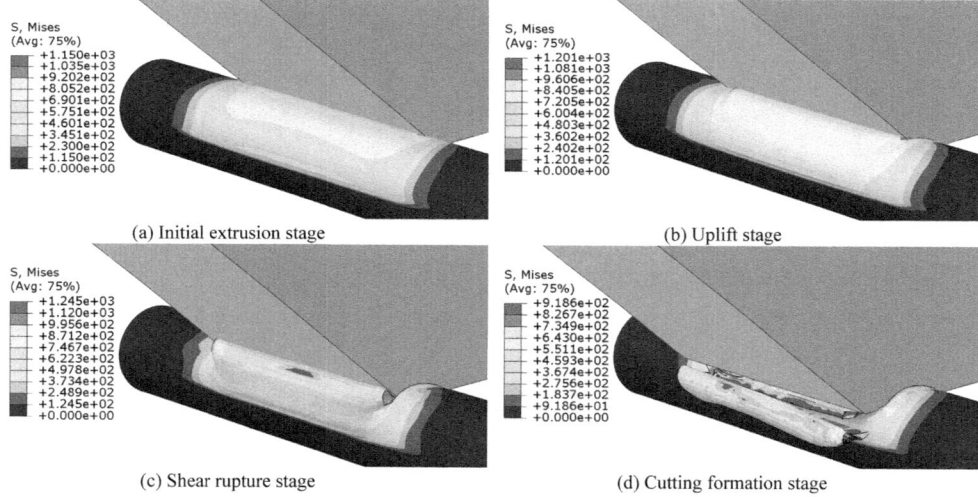

Figure 3.　Effect diagram of cutting steel bar with negative rake angle −45°.

Among them, the chip volume formed by cutting the steel bar by the blade is significantly smaller than that of the steel bar removed by the blade notch. This is because the unit automatic deletion function of the shear failure criterion will delete the unit that has reached the failure strain value in the component, and the strong extrusion effect of the blade will lead to the reduction of the steel bar volume.

3.2 *Cutting force analysis of cutter*

The cutting force of the cutter on the steel bar in the tunneling process is mainly manifested as the support reaction force along the cutting direction and the penetration force along the tunneling direction. The influencing factors mainly include the depth of cut, the rake angle of the cutter, and the width of the cutter. Under the condition of certain steel bars and cutters, the influence of the rake angle of the cutter is studied and discussed. The cutting tools with a cutting depth of 3 mm and rake angles of −45°, −25°, 25°, and 45° are simulated. The variation curve of cutting force is shown in Figure 4 (negative cutting along the X axis). It can be seen from the figure that when the rake angle of the blade is negative, the cutting force in the X direction mainly fluctuates between 400 kN and 500 kN, the penetration force in Y direction mainly fluctuates around 200 kN, and the direction is downward. When the rake angle of the blade is positive, the cutting force in the X direction is small, mainly fluctuating between 200 and 300 kN, and the Y direction is upward, fluctuating near 100 kN.

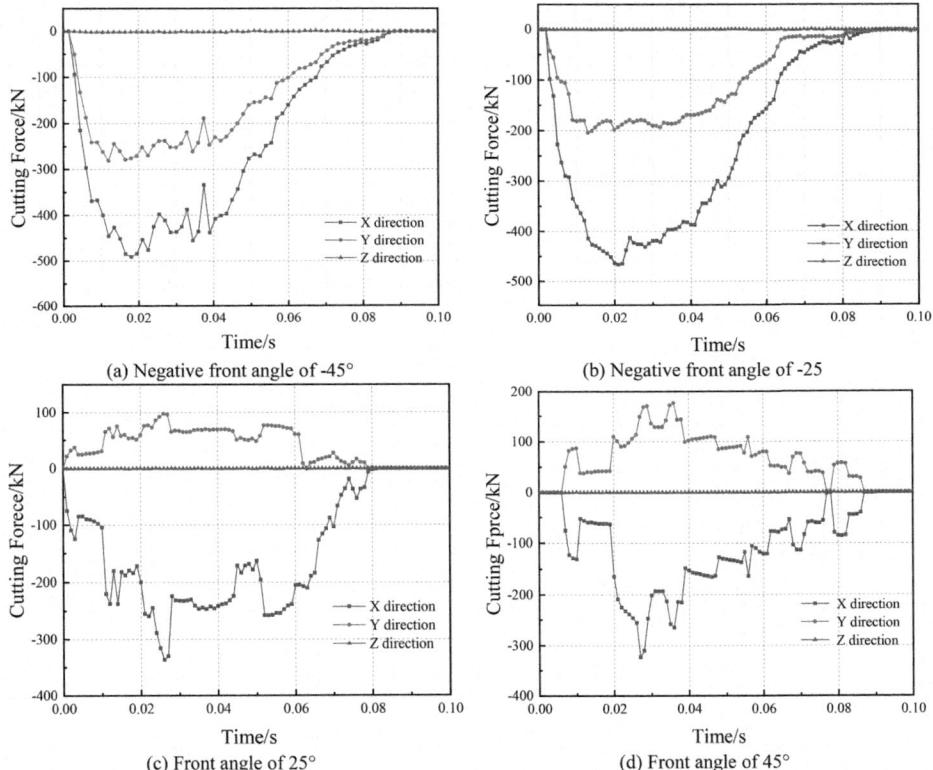

Figure 4.   Cutting force curve of cutting steel bar with the blade.

During the whole cutting process, the tangential force in the X direction is always greater than that in the Y direction, and the lateral pressure in the Z direction is almost zero. It can be seen from the curve in the figure that the initial stage curve of the rigid contact between the tool and the steel bar will change suddenly, which is because the tool has a certain loading speed, and when it begins to contact and cut the steel bar, it will form a large impact, resulting in a load mutation (Cui 2012). With the continuous advancement of the cutterhead, the cutting value gradually fluctuates within a certain range, and the cutting force falls back to 0 after the steel bar is gradually peeled off.

4   ANALYSIS OF CUTTING INFLUENCE PARAMETERS

After comprehensively considering the cutter parameters in the actual dynamic cutting process, this paper uses finite element software to study and analyze the influence of blade rake angle, cutter width, and cutting depth on the cutting force and cutting effect by using the single variable method.

4.1 Analysis of the influence of blade rake angle on cutting reinforcement

The blade angle is an important parameter to evaluate the cutting performance of the tool. The width of the cutter used in the construction site is 70 mm. Therefore, the width of the cutter is assumed to be 70 mm in the simulation model, and the influence of four different blade rake angles ($-45°$, $-25°$, $25°$, $45°$) on the cutting of analyzed.

As shown in Figure 5, the penetration force is negatively correlated with the rake angle of the blade, that is, the larger the rake angle is, the smaller the penetration force is (according to the positive and negative values). For the positive rake angle tool, since the chip is located in front of the blade, the vertical force on the blade is downward (that is, along the positive direction of the Y axis in the simulation model); for the blade with a negative rake angle, the cutting force generated by cutting steel bars increases with the increase of rake angle. However, when the negative rake angle of the blade is greater than $-25°$, the cutting force decreases with the increase of the rake angle. The maximum cutting force is 371 kN, and the direction is opposite to the cutting direction, that is, along the negative direction of the X axis. This is because the larger the negative rake angle of the cutter is, the blunter the blade is, and its penetration effect on the cutting of steel bars is more obvious than that of forwarding shear.

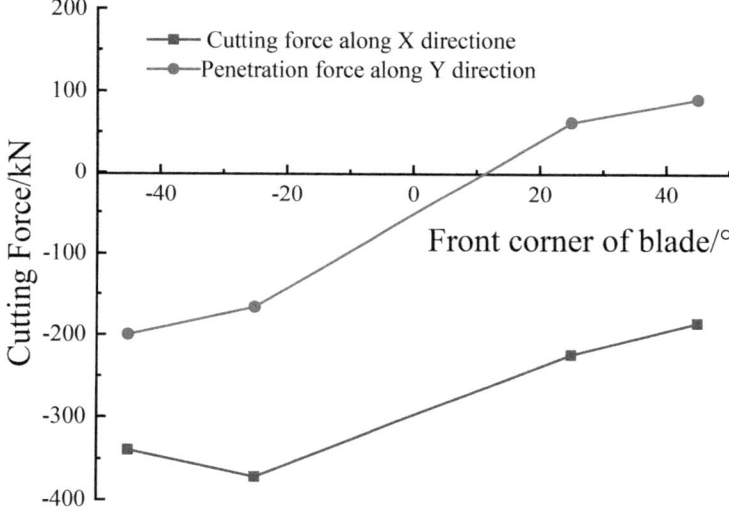

Figure 5.   Relationship between cutting force and rake angle.

4.2 Analysis of the influence of blade width on cutting reinforcement

According to the actual engineering parameters, the negative rake angle of the blade is determined to be $-45°$, the cutting depth is 3 mm, the cutting speed is 200 mm/s, and the time is 0.10 s. The influence of different blade widths (50, 60, 70, and 80 mm) on the cutting effect and cutting force is simulated. Figure 6 and Figure 7 are the cutting effect diagrams of different blade widths on the reinforcement and the corresponding cutting force diagrams.

According to the effect diagram of cutting steel bars in Figure 6, it is found that with the increase of tool width, the forward shear and downward penetration of the cutter on the steel bar are more obvious, and the formed chips are more compressed, so the volume is smaller. At the same time, the wider the cutter is, the larger the contact area between the cutter and the steel bar is, and the larger the influence area of the interaction is.

The relationship between the cutting force and the blade width in Figure 7 shows that the cutting force is positively correlated with the blade width, and the larger the blade width, the greater the

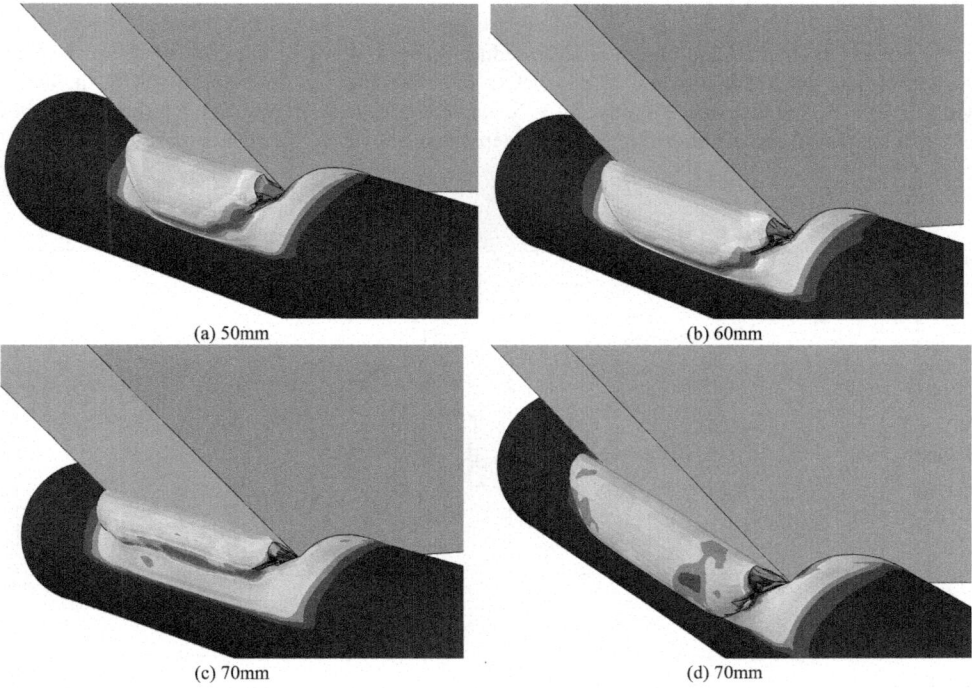

(a) 50mm   (b) 60mm

(c) 70mm   (d) 70mm

Figure 6.   Effect of cutting steel bar with different width.

maximum average cutting force growth, which can also be confirmed by the cutting effect diagram. However, when the cutter width increases to a certain value, the slope of the curve changes, and the influence on the cutting force is further deepened. Therefore, in the actual cutting engineering, it is necessary to strictly control the blade width to prevent the tool from large wear due to excessive force, which affects the construction efficiency.

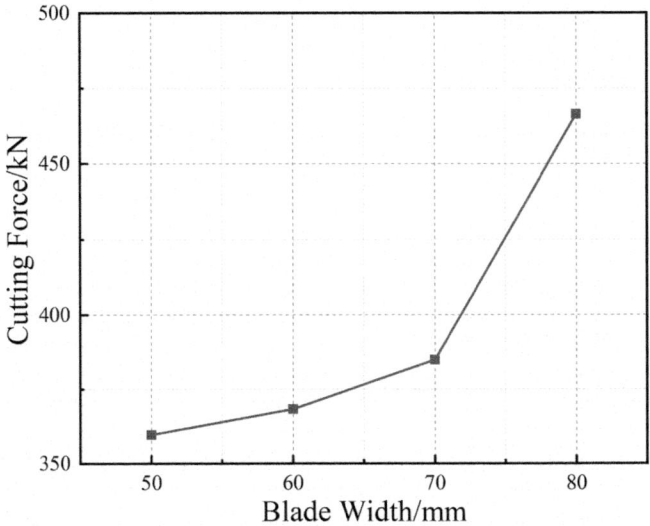

Figure 7.   Relationship between cutting force and blade width.

4.3 *Effect of cutting depth on cutting steel bar*

In the process of shield tunneling, in order to reduce the tool wear rate and prolong its service life, the appropriate cutting depth should be selected. The rake angle of the cutter is −45°, the cutter width is 70 mm, and the cutting depths are 2, 3, 4, and 5 mm, respectively. The relationship curves between tangential force and different cutting depths are shown in Figure 8 and Figure 9.

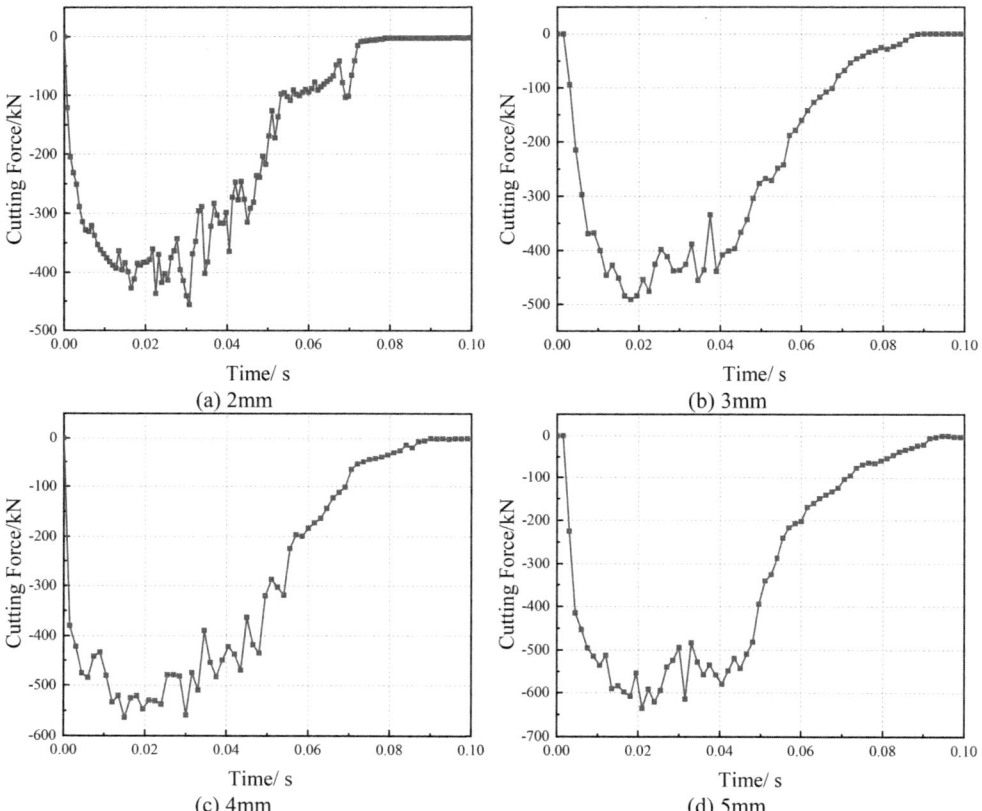

(a) 2mm (b) 3mm

(c) 4mm (d) 5mm

Figure 8. The variation of cutting force along X direction with time under different cutting depths.

It can be seen from Figure 8 that when the cutting depth is 2 mm, the cutting force in the X direction fluctuates between 350 kN and 450 kN. When the cutting depth is3mm, the cutting force fluctuates between 400–500 kN. When the cutting depth is 4mm, the cutting force fluctuates around 550 kN. When the cutting depth is 5 mm, the cutting force fluctuates between 550 and 650 kN, so the X-direction cutting force increases with the increase of cutting depth. When the shear failure occurs in the cutting part of the steel bar, the X-direction cutting force gradually decreases and tends to zero.

Cutting steel bars by cutter is a dynamic cutting process. Therefore, the average cutting force of the stable cutting section is taken as the tangential force corresponding to the cutting depth. It can be seen from Figure 9 that the tangential force is linearly related to the cutting depth, that is, the greater the cutting depth is, the greater the tangential force is.

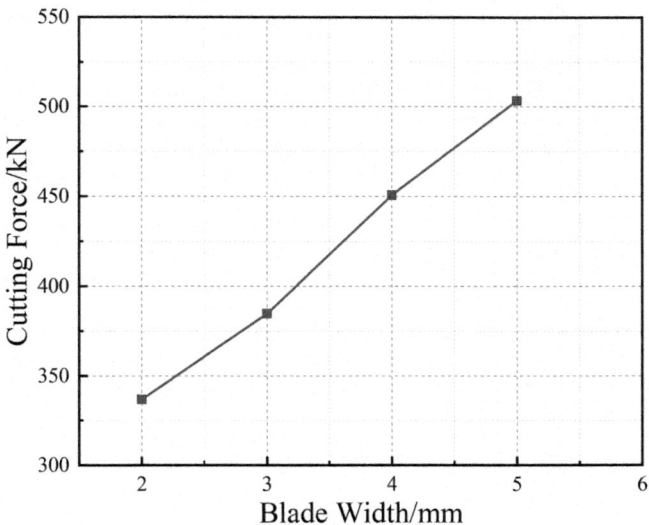

Figure 9.   Relationship between cutting force and cutting depth.

5   CONCLUSIONS

In this paper, ABAQUS finite element software is used to simulate the dynamic process of steel bar cutting by a shield cutter. The main conclusions are as follows:

(1) The cutting process of steel bar by the blade can be divided into four stages: initial extrusion, uplift, shear fracture, and chip formation;
(2) In the cutting force curve of a cutting steel bar, the tangential force in the X direction is always greater than the penetration force in the Y direction, but both generally show a trend of "increasing first and then decreasing", and the lateral force is almost zero;
(3) The penetration force decreases with the increase of the blade rake angle; the tangential force is basically positively correlated with the rake angle, cutting depth, and blade width. However, when the parameters increase to a certain value, the cutting force will change abruptly.

Therefore, the blade with a negative rake angle should be selected and other parameters of the cutter should be strictly controlled when the cutter is cutting steel bars to prevent large wear of the cutter due to excessive force and affect the construction efficiency.

REFERENCES

Chen, G., Chen, Z.F., Tao, J.L. (2005). Investigation and validation of plastic constitutive parameters of 45 steel. *Explosion and Shock waves* 25(5), 6.
Cui, J. (2012). *Numerical simulation for soil cutting process and calculation method of cutter head torque of shield machine.* Dalian University of Technology.
Fuzhaojiro. (2001). Construction machinery. Tokyo: Kajima Institute Press Conference.
Ge, J., Xu, Y. & Cheng, W. (2016). Key construction technology of shield tunneling crossing underneath a railway. *Infrastructures* 1(1), 4.
Guo Z.G., Wang J., Lv S., et al. (2021). *Numerical simulation analysis of coal rock crushed by disc cutter based on ABAQUS*[C]//IOP Conference Series: Materials Science and Engineering. IOP Publishing 1043(4), 042010.
Ibarra, S. McKyes, E. & Broughton, R. (2005). A model of stress distribution and cracking in cohesive soils produced by simple tillage implements. *Journal of Terramechanics* 42(2), 115–139.

Ivor. & Evans. (1984). A theory of the cutting force for point-attack picks. *Geotechnical & Geological Engineering* 2(1), 63–71.

Jiang, H., Zhang, J.X., Tang, F.P., et al. (2021). Numerical analysis of the cutting process of shield tunneling in Beijing sandy pebble formation. *Journal of China Coal Society* 46(S1), 539–548.

Li, H.B. (2020). Feasibility study on the direct cutting of reinforced concrete pile foundation with ϕ25 mm main reinforced bar by shield. *Tunnel Construction* 40(12), 1808–1816.

Wang, F. (2014). *Study on shield cutting large-diameter reinforced concrete piles directly*. Beijing Jiaotong University.

Wang, F., Yuan, D.J., Dong, C.W., Han, B., Nan, H.R. (2013). Test study of shield cutting large-diameter reinforced concrete piles directly. *Chinese Journal of Rock Mechanics and Engineering* 32(12), 2566–2574.

Ye, X. & Feng, A.J. (2021). Data statistics and development analysis of urban rail transit in China in 2020. *Tunnel Construction* 41(05), 871–876.

Yoshino S. (1980). *Shield construction method*. Tokyo: Kajima Institute Press Conference.

Zhang, L. & Liu, B.L. (2017). Analysis of the impact of construction control for subway shield tunneling through the traveling track groups in the beijing-guangzhou line. *Journal of Information Technology in Civil Engineering and Architecture* 9(01), 106–112.

Zhao, Q.W. (2011). *Study on Ground Settlement Prediction and Control Technology of Shield Undercrossing Zhengzhou Station of Beijing-Guangzhou Railway*. Central South University.

Numerical simulation and technical optimization of construction engineering

*Frontiers in Civil and Hydraulic Engineering – Mohamed A. Ismail and
Hazem Samih Mohamed (Eds)*
© 2023 The Authors, ISBN 978-1-032-38247-0

Parametric analysis of blind bolted end-plate connection joints under the condition of middle column-loss

Shaopeng Lei*, Yongshuai Li & Zike Jiao
Xi'an University of Architecture and Technology, Xi'an, China

ABSTRACT: Based on the experimental research on the CFST column-steel beam blind bolted end-plate connection joint, this paper uses ABAQUS to establish the condition of the CFST column-steel beam blind bolted end-plate connection joint under the failure mode of the middle column. The effects of the thickness of the steel tube wall and end plate, anchoring method, steel beam section, and stiffener on its mechanical properties were also studied respectively.

1 INTRODUCTION

The blind bolts can be installed and tightened from the unilateral side of the steel tube section with no need to open installation holes and on-site welding, which greatly improves the efficiency and speed of construction and saves labor and construction. Based on the existing blind bolts, many scholars have made improvements to them. To improve the anchoring effect of blind bolts in concrete, British scholars proposed EHB (Tizani 2003), extended the Hollo-Bolt's shank, and installed the anchor nut at its tail.

After the plane crash of the World Trade Center building in New York in 2001, the progressive collapse resistance of the structure has attracted extensive attention from academic and engineering circles. Joints are the critical locations of structural stress, and beam-column joints play a crucial role in load redistribution after structural column failure in designing progressive collapse resistance (Khandelwal 2007).

The square CFST column-composite beam frame gives full play to the advantages of the CFST column's good flexibility and strong lateral movement resistance (Han 2014). The composite beam also bears a great many advantages, such as small section height, light weight, high rigidity, good flexibility, short construction period, and easy installation. Therefore, the CFST-composite beam frame structure system has been widely used in multi-story and high-rise buildings.

2 FINITE ELEMENT MODEL

2.1 Material properties

2.1.1 Steel constitutive model

When the trilinear constitutive model (Bahaari 2000) is used, its mathematical expression is as follows. Under the multiaxial stress state, the Von Mises yield criterion is used to judge whether the steel has reached a yield or not.

$$\sigma = \begin{cases} E_s \varepsilon & (\varepsilon \leq \varepsilon_y) \\ f_y + 0.01E_s(\varepsilon - \varepsilon_y) & (\varepsilon_y < \varepsilon \leq \varepsilon_u) \\ 0 & (\varepsilon > \varepsilon_u) \end{cases} \tag{1}$$

---

*Corresponding Author: 596480835@qq.com

2.1.2 Concrete constitutive model

The plastic damage model (CDP) is adopted for the concrete constitutive in previous research (Kmiecik 2011). Due to the confinement effect of steel tube columns on the core concrete, the mechanical properties of the core concrete change. This paper adopts the uniaxial stress-strain relationship model of the core concrete (Han 2005), which introduces the constraint coefficient ξ to reflect the interaction between steel pipe and concrete, and the calculation of ξ is as follows.

$$\xi = \frac{A_s f_y}{A_c f_{ck}} = \alpha \frac{f_y}{f_{ck}} \tag{2}$$

2.2 Contact and restraint

The normal behavior of surface-to-surface contact adopts the "hard contact" method, which does not limit the contact pressure between the contact surfaces. The tangential behavior adopts the Coulomb friction model and takes the effect of interfacial bonding into consideration.

The end plate and the beam end; the stiffener, the beam end, and the end plate adopt Tie to bind the contact surfaces together. Coupling is applied to the ends of the CFST column and beam to keep the point deformations consistent on each coupling section.

2.3 Element selection and meshing

All parts of the finite element model adopt the first-order reduced-integration element (C3D8R). For the critical components with complex stress and many contact relationships, a denser mesh is used, and other areas of beams and columns far from the core area of the node are divided into a sparser mesh.

2.4 Loads and boundary conditions

The boundary conditions at both ends of the beam are hinged, the translation and rotation of the column are constrained at the same time, and only the vertical displacement is released. The Bolt Load in ABAQUS is used to apply the pre-tension to the blind bolt, and then the vertical displacement is applied to the column end.

2.5 Finite element model verification

Based on the comparison of the moment-rotation curve of the CFST column-steel beam blind bolted connection joints in the literature (Wang 2020), the necessary verification of the finite element model is carried out. The comparison between the finite element calculation results and the experimental results is shown in the following Figure 1. The curves obtained by simulation and experiment are in good agreement, which verifies the rationality and calculation accuracy of

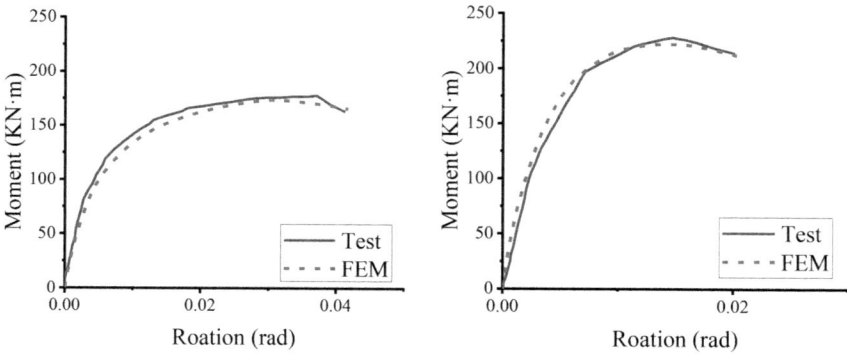

Figure 1.   Comparison of joined Moment-Rotation curves.

the finite element model and provides a basis for further using the model to study the working mechanism and parametric analysis of joints.

3 PARAMETER ANALYSIS

With the relevant specifications and the range of common engineering parameters serving as the basis, 8 CFST columns and beams extended hollo bolt extended end plate connection joints are designed. The test specimen is numbered SJ-1~SJ-8, and the main parameters of the test specimen are shown in the following Table 1. Models SJ-1~SJ-5 use extended end-plates with a cross-sectional size of 570×200mm, as shown in Figure 2. And specimens SJ-6~SJ-8 use extended end-plates with a cross-sectional dimension of 520×200mm, as shown in Figure 3. Models SJ-1~SJ-2 use square steel columns with a cross-sectional size of 200×200×5mm, and specimens SJ-3~SJ-8 use square steel columns with a cross-sectional size of 200×200×13mm. The blind bolts of the specimens were made of M16 Hollo-Bolt high-strength bolts with a specification of 8.8. Except for specimen SJ-5, the anchoring effect of the bolts was enhanced by extending the bolt rod and installing an 8.8-grade nut at the tail. Specimen SJ-5 only extends the threaded rod without installing a nut at the tail. The CFST columns are made of self-compacting concrete with a strength grade of C40. Except for the steel beams in the specimens SJ-6~SJ-8, which are made of Q235 steel, the rest of the steel beams are made of Q345 steel.

Table 1. Details of the specimens.

Specimen Number	Column Section (mm)	Column Length (mm)	Beam Section (mm)	Beam Length (mm)	End-Plate Thickness (mm)	Anchorage Method	Stiffener
SJ-1	250×5	1736	HN350×175×7×11	1400	12	Nut	None
SJ-2	250×5	1736	HN350×175×7×11	1400	24	Nut	None
SJ-3	250×13	1736	HN350×175×7×11	1400	12	Nut	None
SJ-4	250×13	1736	HN350×175×7×11	1400	24	Nut	None
SJ-5	250×13	1736	HN350×175×7×11	1400	12	None	None
SJ-6	250×13	1736	HN300×150×6×9	1400	12	Nut	None
SJ-7	250×13	1736	HN300×150×6×9	1400	24	Nut	None
SJ-8	250×13	1736	HN300×150×6×9	1400	12	Nut	With

The design of the test piece is summarized as follows. Specimen SJ-1 and SJ-3, specimen SJ-2 and SJ-4 serve to study the effect of steel tube wall thickness; specimen SJ-1 and SJ-2, and specimen SJ-3 and SJ-4 for the impact of end plate thickness; specimen SJ-3 and SJ-5 for the effect of anchoring methods; specimens SJ-3 and SJ-6, specimens SJ-4 and SJ-7 for the influence of the section of the steel beam; specimens SJ-6 and SJ-8 for the effect of stiffeners.

To improve the operation speed and convergence of the model, this paper uses the simplified model of the Hollo-Bolt bolt (Ataei 2013), as shown in Figure 2. The schematic diagram of the finite element model of the overall node is shown in Figure 3 below. The beams are hinged at both ends, and the columns only move vertically. At first, the blind bolts are pre-tension, and then the axial pressure is applied to the end of the CFST column. The influence of each parameter on the Moment-Rotation curve of the joint is shown in Figure 4 below.

As shown in Figure 4, the initial stiffnesses of models SJ-3 and SJ-1, and models SJ-4 and SJ-2 are almost the same, indicating that the steel tube wall thickness has little effect on the improvement of the initial stiffness. Compared with SJ-1, the compressive bearing capacity of model SJ-3 is improved, and the compressive bearing capacity of SJ-4 is improved compared with SJ-2, indicating that increasing the thickness of the steel tube wall can improve the compressive bearing capacity. The initial stiffnesses of models SJ-2 and SJ-1 and the initial stiffnesses of models SJ-4 and SJ-3

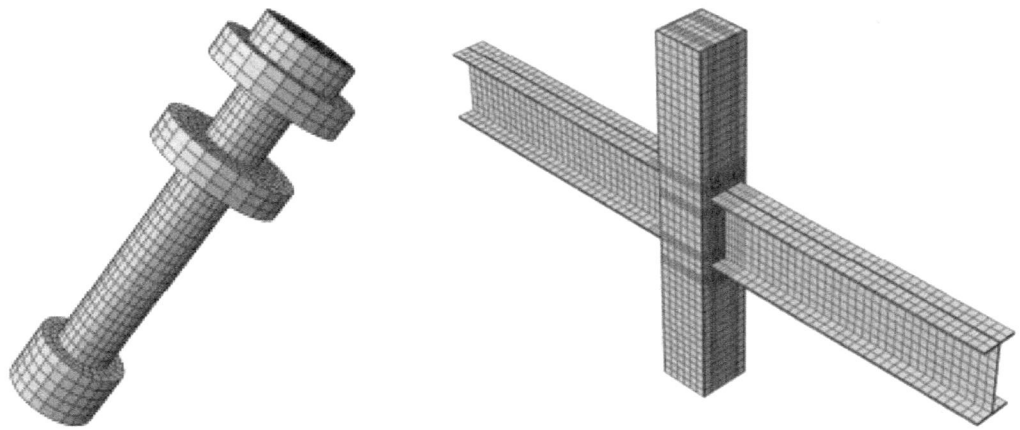

Figure 2.   Extended hollo-bolt.   Figure 3.   Model.

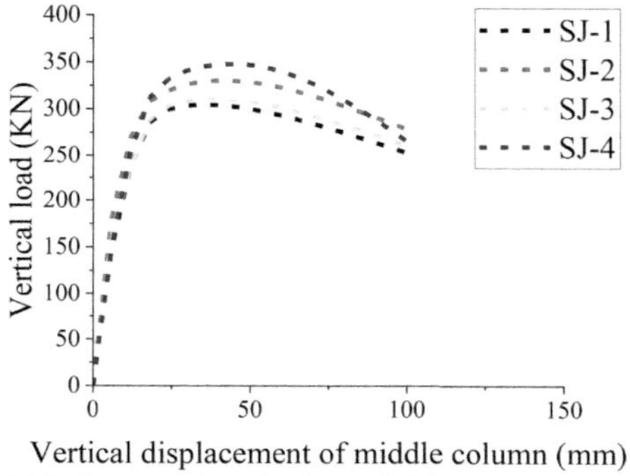

(a) Influence of steel tube wall thickness and end plate thickness

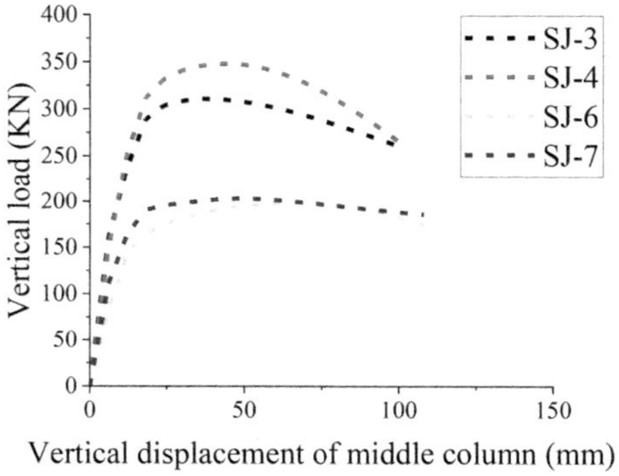

(b) The effect of the anchorage method

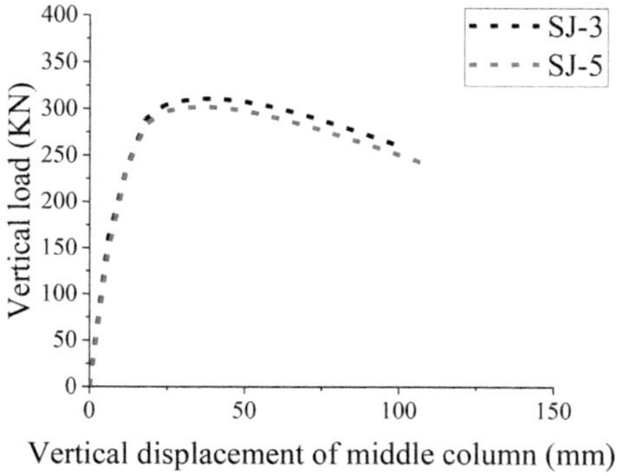

(c) Effect of the steel beam section on joint performance

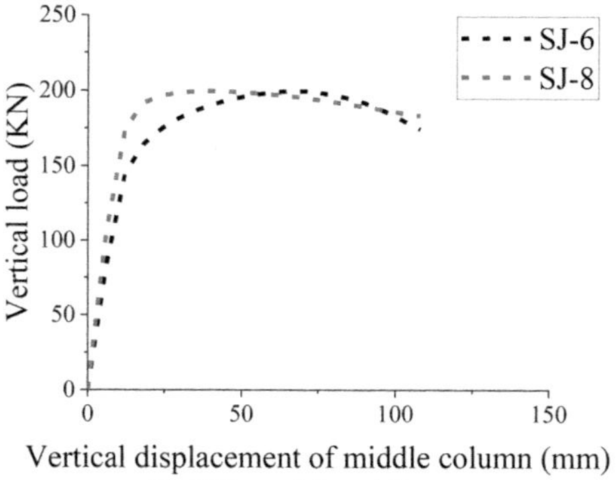

(d) Effect of the stiffeners

Figure 4.   Vertical load-middle column displacement curve comparison.

are almost the same, manifesting that the end-plate thickness has little effect on the initial stiffness. Compared with SJ-1, the compressive bearing capacity of model SJ-2 is significantly improved, and the compressive bearing capacity of SJ-4 is also improved considerably compared with that of SJ-3, revealing that the increase of end plate thickness can improve the compressive bearing capacity. Model SJ-3 has almost the same initial stiffness and compressive capacity compared to SJ-5, indicating that changes in the anchoring method (with or without anchor nuts) have little effect on the initial stiffness and compressive capacity. Compared with SJ-6, the initial stiffness and compressive bearing capacity of the model SJ-3 are significantly improved, showing that increasing the steel beam's section size can improve the joint's initial stiffness and compressive bearing capacity. Compared with the model SJ-6, the initial stiffness of the model SJ-8 is improved, and the compressive bearing capacity has no change, proving that the stiffener can improve the initial stiffness and has no effect on the compressive bearing capacity.

4 CONCLUSION

Through modeling and analysis of eight blind bolted end-plate connection joints with different parameters after the failure of the central column, the influence of each parameter on the mechanical properties of the joint is studied, and the conclusions are as follows.

(1) Increasing the steel tube's wall thickness can increase the joint's compressive capacity but has little effect on the initial stiffness.
(2) Increasing the end plate's thickness can improve the joint's compressive bearing capacity and has little effect on the initial stiffness.
(3) Changes in the anchoring method (with or without anchor nuts) have little effect on the initial stiffness and compressive capacity.
(4) Increasing the section size of the steel beam can significantly improve the initial stiffness and compressive bearing capacity.
(5) Stiffeners can improve the initial stiffness but have little effect on the compressive bearing capacity.

REFERENCES

Ataei A. & Bradford M.A. (2013). FE modeling of semi-rigid flush end plate joints with concrete-filled steel tubular columns[J]. *Research and Applications in Structural Engineering, Mechanics and Computation,* 62(4): 181.
Bahaari M.-R. & Sherbourne A.-N. (2000). Behavior of eight-bolt large capacity endplate connections[J]. *COMPUTERS & STRUCTURES,* 77(3): 315–325.
Han L.H., Yao G.H. & Zhao X.L. (2005). Tests and calculations for hollow structural steel (HSS) stub columns filled with self-consolidating concrete (SCC)[J]. *Journal of Constructional Steel Research,* 61(9): 1241–1269.
Han, L., Li, W. & Bjorhovde, R. (2014). Developments and advanced applications of concrete-filled steel tubular (CFST)structures: Members. *J. Constr. Steel Res.* 100, 211–228.
Khandelwal, K. & Ei-Tawil, S. (2007). Collapse behavior of steel special moment resisting frame connections, *J. Struct. Eng.* 133(5): 646–655.
Kmiecik P. & Kamiński M. (2011). Modeling of reinforced concrete structures and composite structures with concrete strength degradation taken into consideration[J]. *Archives of civil and mechanical engineering,* 11(3): 623–636.
Tizani, W. & Ridley, E. (2003). *The Performance of a New Blind-Bolt for Moment-Resisting Connections;* Taylor & Francis: Abingdon, UK.
Wang Y., Wang Z. & Pan J. (2020). Seismic behavior of a novel blind bolted flush end-plate connection to strengthened concrete-filled steel tube columns[J]. *Applied Sciences,* 10(7): 2517.

Frontiers in Civil and Hydraulic Engineering – Mohamed A. Ismail and
Hazem Samih Mohamed (Eds)
© 2023 The Authors, ISBN 978-1-032-38247-0

Knowledge map analysis of typhoon wave numerical simulation

Zhiyuan Li
Zhejiang Ocean University, Zhoushan, China
Zhejiang University of Water Resources and Electric Power, Hangzhou, China

Dongfeng Li*
Zhejiang University of Water Resources and Electric Power, Hangzhou, China

ABSTRACT: The 431 articles in the Web of Science core library were visually analyzed by Cite Space. This paper draws the knowledge structure map of core authors, research institutions, and research hot spots. The results show that the numerical simulation of typhoon waves has experienced three slow-development stages, steady development, and rapid development; the main research institution is the Chinese Academy of Science (CAOS); the future development trend of wind and wave numerical simulation will change from a single deterministic simulation to a probabilistic simulation, and adapt to more complex wind and wave changes.

1 INTRODUCTION

When a typhoon is generated by tropical cyclones at sea, typhoon waves are often formed while transporting water vapor and heat. Typhoon waves will pose a serious threat to coastal engineering and construction. Therefore, a large number of scholars around the world have carried out numerical simulation research on typhoon waves. Due to the irregularity of the wind above sea level, the shape of wind waves is also irregular, with steep wave surfaces, short wavelengths, and small periods (Du 2020). At the same time, the increase of water in the offshore area caused by typhoon waves will also bring property hazards to people living along the coast (Zhu & Xiong 2020). To reduce the losses caused by typhoon waves, it is of great significance to carry out high-precision and large-scale numerical simulations of typhoon waves (Hu 2021).

This paper uses CiteSpace6.1. R2 knowledge map software and bibliometric analysis to visually analyze the literature related to typhoon wave numerical simulation through Web Of Science core database retrieval, depicting the number of published papers, authors, publishing institutions, and literature co-citations as well as keyword analysis and other knowledge maps. The purpose is to objectively reveal the dynamics, development process, and evolution trend of typhoon wave numerical simulation research and explore the theoretical and practical research of typhoon wave numerical simulation. These fronts and hot spots are expected to provide scientific references for the numerical simulation of typhoon waves.

2 RESULTS AND ANALYSIS

2.1 *Time analysis of published volume*

Statistics of published articles from 1992 to 2021 are shown in Figure 1. It is found that the distribution of published records of numerical simulation-related research on typhoon waves generally

---

*Corresponding Author: lidf@zjweu.edu.cn

DOI 10.1201/9781003344209-33

shows an increasing trend. Generally speaking, it can be roughly divided into three stages: (1) 1992-2004 was the initial stage, and there was little research on typhoon wave numerical simulation. The number of published papers in this period is relatively small, with an average of only 1 paper per year, and the research field has certain limitations. The theoretical development and practical research are insufficient mainly due to the limitation of computer performance. (2) In the development period from 2005 to 2016, the number of documents has increased year by year, and countries have paid more attention to the prevention of typhoon wave disasters. Ya, T et al. (Yamashita, Kyeongok, Nishiguchi, et al. 2005) used the simple typhoon model of the gradient wind field established by the Schloemer atmospheric pressure distribution model to simulate the wind field and then used the mesoscale typhoon model MM5 to simulate the surface wind field and rainfall field caused by the typhoon. Since then, the research on numerical simulation of typhoon waves has begun to receive widespread attention, and relevant results have continued to emerge. (3) The period from 2016 to 2021 shows rapid growth, and the growth rate of literature is doubled compared with the previous period. Hu et al. (Hu et al. 2020) extracted actual data from the climate forecast system (CFSV2) and ERA5 data sets in 2020 and used two data sets for each. These data were combined with the parametric typhoon model as the meteorological conditions driving the coupled-wave circulation model. By simulating at different spatial and temporal resolutions, the accuracy of the typhoon model under the data coupling of different systems was studied.

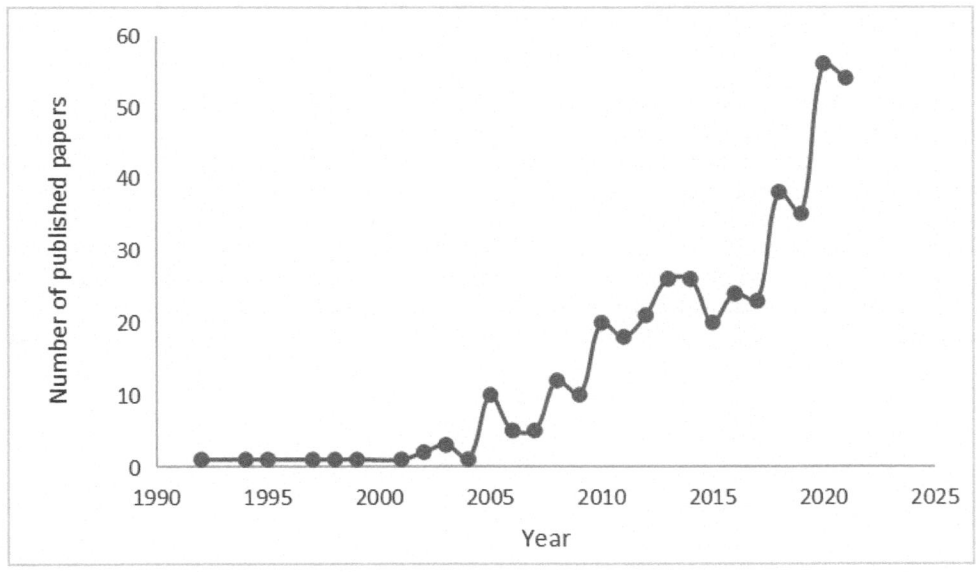

Figure 1.　Document issuing record of wind wave numerical simulation research.

2.2 *Analysis of the research team*

Cite Space is used to analyze the authors of the paper on the numerical simulation of typhoon waves, as shown in Figure 2. The author cooperation graph is obtained, which has 417 nodes, 551 links, and a network density of 0.0054. The color of the node circle layer indicates the change in the author's posting time. The brighter the color is, the closer the author will post, and the larger the node will be, indicating that the author has published more articles. The darker the color of the connecting line is, the closer the cooperation between the authors will be. As can be seen from Figure 2, the scholars with the largest number of papers are CHEN J, with a frequency of 12; LI S, with a frequency of 10; WANG Y, with a frequency of 9. Other high-profile authors include LI J, HOU Y, CHEN W, WADA A, etc.

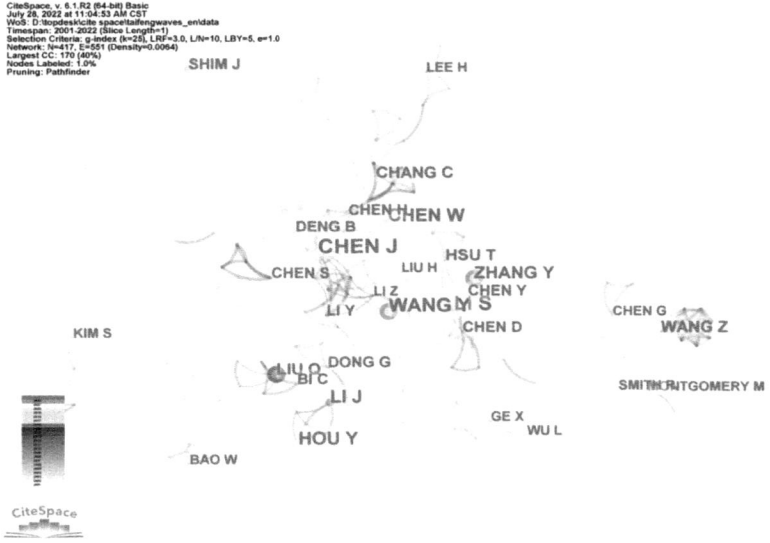

Figure 2. Map of authors of typhoon wave numerical simulation literature.

To explore the main research institutions of typhoon wave numerical simulation, Cite Space is used to visually analyze the document issuing institutions, as shown in Figure 3. Each node in the visual graph represents an institution, and the size of the node represents the volume of its publications. The issuing organization map has 291 nodes and 489 connections, and the network density is 0.0116. Chinese Academy of Science (CAOS), Nanjing University of Information Science Technology, Meteorological Research Institute Japan, and other institutions have the largest number of published papers, with 50, 24, and 22 papers respectively. As can be seen from the figure, CAOS is very closely connected with other institutions. It not only publishes the most papers but also cooperates with institutions around the world. In 2009, Yin, BS (Progress in Natural Science

Figure 3. Chart of document issuing institutions for typhoon wave numerical simulation.

2009) of CAOS published a wind field model for the coupling of three main driving forces: wave radiation stress, combined wave-current bottom shear stress, and surface wind stress related to the wave state. Uncoupled models can better simulate typhoons. The main publishing agencies are distributed globally, and most of them are located in Asia. Among them, China accounts for the majority of documents, and Florida and Hawaii in the United States publish more documents.

2.3 *Analysis of hot topics in typhoon wave numerical simulation research*

Keyword clustering can deepen the co-occurrence relationship of keywords and calculate a group of keywords with a closer relationship, thereby forming topic clusters containing multiple groups of words. To further investigate the knowledge structure of numerical simulation of typhoon waves and explore its research hotspots, this paper conducts cluster analysis of keywords, runs the software, and obtains the relevant clustering map (Figure 4). It can be observed from the cluster map that Modularity (Q value) is 0.555>0.3, which means the clustering result is significant; the mean silhouette (S value) is 0.8226>0.7, which means that the clustering reliability is high.

Scholars have researched the numerical simulation of typhoon waves, and a total of 14 research hot spots on keyword clustering have been formed. The clusters were sorted and analyzed, and two research hot topics were formed: (1) Sediment movement under the action of typhoon waves, including clusters #0 sediment transport, #1 flow, #2 surge, #9 tube-sinker, and #10 gravity wave. Scholars have found that although the original gentle slope equation (BERKOFFJCW, 1972/RADDERAC, 1979) can simulate wave refraction, diffraction, reflection, bottom friction, and breakage, its numerical method of refraction-diffraction is based on linear theory, and it describes a single frequency. With wave propagation motion, the Boussinesq equation (Nowguo 1993) is a plane two-dimensional nonlinear shallow water wave equation, including shallow water deformation, refraction, diffraction, reflection, and other mechanisms of waves, and is only suitable for a small area, dynamic spectrum balance equation(RISRC et al. 1994). An apparent wave is a random wave, which comprehensively and reasonably considers wave shallowing, refraction, bottom friction, breaking, white waves, wind energy input, and nonlinear effects of waves. It is suitable for large, medium, and small-scale waters. (2) Coupled models are used to simulate and verify typhoons, including clusters #6 atmosphere-wave-ocean coupled model, #4 storm surge, #7 model, and #11 typhoon size. At present, the wind field model usually used in the research on typhoon waves is the meteorological numerical prediction model (Li et al. 2016), such as MM5, WRF; the wave model

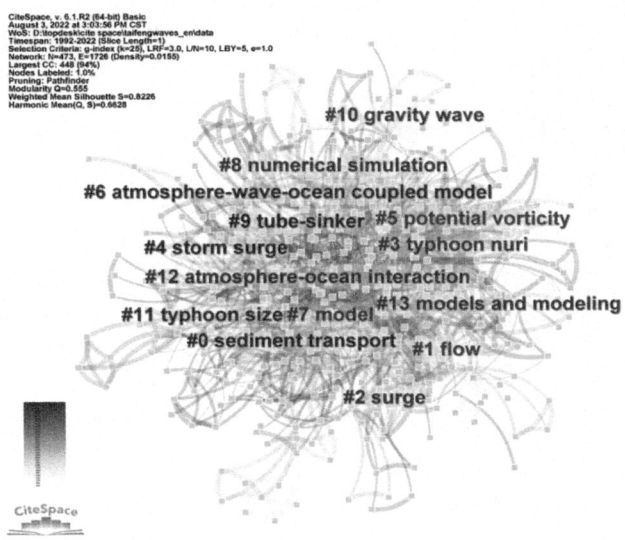

Figure 4.   Keyword clustering map of typhoon numerical simulation.

usually adopts the SWAN model of the third generation wave model. The typhoon wave model obtained by coupling is closer to the actual value.

3 RESEARCH TRENDS AND PROSPECTS

Emergent keywords refer to keywords with a sudden increase in frequency during a certain period, which are used to study the dynamic development and potential problems of a certain field. They are suitable for testing new trends and sudden changes in the development of disciplines (Chen 2009). It can be seen from Figure 3 that the research on typhoon waves has gradually evolved from analyzing the causes of typhoon waves to showing how to accurately predict the intensity of typhoon waves. The research and outlook statements are as follows:

1. Typhoon research is divided into two categories: forecast and hindsight. The forecast is to simulate and estimate the upcoming typhoon; hindsight is to study and analyze the past typhoon. The actual observational data of marine meteorology is very scarce, and the only on-site observational data is not coherent in time and space, so it is necessary to supplement the marine hydrological data with numerical post-reporting.
2. With the help of the SWAN model which considers the physical process comprehensively, the numerical simulation study is carried out on the spatial and temporal distribution characteristics of typhoon waves during the typhoon process. This has important significance both in theory and reality and provides a necessary basis for future engineering design and construction.
3. With the in-depth study of typhoon waves, several research directions for the evolution and distribution characteristics of typhoons have been gradually formed, and certain research results have been achieved. However, due to the uncertainty of typhoons and the complex and changeable process of theoretical analysis and derivation, even if the propagation of typhoon waves is replayed and predicted with the help of computers, the early warning and forecasting of typhoon waves still need further research and exploration.
4. With the emergence of ensemble wave forecasting, this forecasting method turns the single deterministic forecast in the traditional sense into an uncertain forecast. Fully considering the uncertainty of the physical process in the numerical model of ocean waves, the traditional deterministic forecast is transformed into an uncertain forecast. The forecast is extended to the field of probabilistic forecasting, thereby providing additional information that cannot be provided by pure deterministic forecasting, making the resulting wave forecasts include all possible states of future wave changes as much as possible, and achieving the purpose of improving wave forecasting skills.

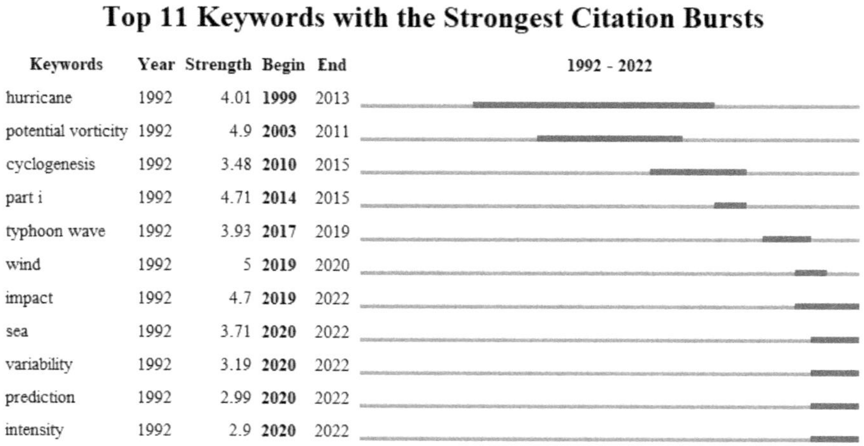

Top 11 Keywords with the Strongest Citation Bursts

Keywords	Year	Strength	Begin	End	1992 - 2022
hurricane	1992	4.01	1999	2013	
potential vorticity	1992	4.9	2003	2011	
cyclogenesis	1992	3.48	2010	2015	
part i	1992	4.71	2014	2015	
typhoon wave	1992	3.93	2017	2019	
wind	1992	5	2019	2020	
impact	1992	4.7	2019	2022	
sea	1992	3.71	2020	2022	
variability	1992	3.19	2020	2022	
prediction	1992	2.99	2020	2022	
intensity	1992	2.9	2020	2022	

Figure 5.   Keywords emergent in numerical simulation of typhoon waves.

4 CONCLUSION

Taking the literature on numerical simulation of typhoon waves in the core database of Web of Science from 1992 to 2021 as sample data, Cite space software was used to generate a map of the number of published papers, author groups, research institutions, hot keywords, etc. The analysis led to the following conclusions. First, the numerical simulation of typhoon waves has experienced a stage from slow to rapid development, and numerical simulation has entered a rapid development stage since 2016; second, the main research institutions are still CAOS, and CAOS is closely linked with other institutions; third, at present, the main research mode of typhoon wave numerical simulation is to make the working conditions under the coupling of different wind fields and wave models close to the actual data; fourth, the future development trend of typhoon wave numerical simulation will change from a single deterministic simulation to a probabilistic simulation, which can adapt to more complex wind and wave changes.

ACKNOWLEDGMENTS

This research was supported by the Funds Key Laboratory for Technology in Rural Water Management of Zhejiang Province (ZJWEU-RWM-202101), the Joint Funds of the Zhejiang Provincial Natural Science Foundation of China (No. LZJWZ22C030001, No. LZJWZ22E090004), the Funds of Water Resources of Science and Technology of Zhejiang Provincial Water Resources Department, China (No.RB2115, No.RC2040), the National Key Research and Development Program of China(No.2016YFC0402502), and the National Natural Science Foundation of China (51979249).

REFERENCES

BERKOFFJCW. *Computation of combined refraction diffraction* [A]. Proc. Of the 13th Conference Coastal Eng [C]., 1972.471490.

Chen Chaomei. Cite Space II: Identification and visualization of new trends and dynamics in the scientific literature [J]. *Chinese Journal of Information*, 2009, 8(3): 401–421.

Du Yan. *Simulation of wave distribution characteristics in the Yellow and East China Sea under the forcing of typhoon Canhong* [D]. Nanjing University of Information Science and Technology, 2020.

Hu Y., Shao W., Wei Y., et al. Analysis of typhoon-induced waves along typhoon tracks in the Western North Pacific Ocean, 1998–2017 [J]. *Journal of Marine Science and Engineering*, 2020, 8(7): 521.

Hu Yuyi. *Research on typhoon waves based on synthetic aperture radar and numerical simulation* [D]. Zhejiang Ocean University, 2021. DOI: 10.27747/d.cnki.gzjhy.2021.000076.

Li Xue, Wang Zhifeng, Wu Shuangquan, Dong Sheng, Jia Jing, Zhang Xiaoshuang, Wu Hao. Application of coastal wave-storm surge coupling model in Tianjin coast [J]. *Ocean Bulletin*, 2016, 35(06): 657–665.

NOWGUO. An alternative form of Boussinesq equations for nearshore wave propagation [J]. J. of Waterway, Port, *Coastal and Ocean Engineering*, 1993, 119(6): 618–638.

RADDERAC. On the parabolic equation method for water wave propagation [J]. *J.fluid Mech.*, 1979(1): 159–176.

RISRC, HOLTHUIJSENLH, BOOIJN. Aspectral model for waves in the nearshore zone [J]. *Coastal Engineering*, 1994, 1:6878.

Simulating a typhoon storm surge in the East Sea of China using a coupled model [J]. *Progress in Natural Science*, 2009, 19(01): 65–71.

Yamashita T., Kyeongok K., Nishiguchi H., et al. Numerical simulation of surface wind and rainfall fields caused by a typhoon [J]. *WIT Transactions on The Built Environment*, 2005, 78.

Zhu Zhixia, Xiong Wei. Three-dimensional sediment numerical simulation under the action of typhoon waves and storm surges [J]. *Journal of Harbin Engineering University*, 2020, 41(03): 332–339.

Frontiers in Civil and Hydraulic Engineering – Mohamed A. Ismail and
Hazem Samih Mohamed (Eds)
© 2023 The Authors, ISBN 978-1-032-38247-0

Evolution of support concept and design method of deep foundation pit

Lian-xiang Li*, Yingxue Hou*, Hongxia Xing*, Bin Jia* & Tian-yu Chen*
*School of Civil and Hydraulic Engineering, Shandong University, Jinan, China, Foundation Pit, Deep
Foundation Engineering Research Center, Shandong University, Jinan, China*

ABSTRACT: Based on more than 30 years of construction practice in China, this paper analyzes
the advantages and disadvantages of the existing deep foundation pit supporting concepts of "temporary" and "combination with the main structure", and put forward the "permanent" concept to
promote the high-quality development of foundation pit engineering. Combined with a foundation
pit project in Jinan, the paper demonstrated the progressive lifting of the three concepts and the
calculation content of the corresponding supporting structure, and defined the design methods
of "geotechnical method of structure " and "structural method of geotechnical" that adapted to
the concepts of "combination with the main structure" and "permanence"; this paper focused on
the analysis of composite soil nailing wall combined with anti-floating anchor, and the durability of anti-floating anchor in the permanent support structure composed of anti-floating anchor,
extended support and horizontal floor, and explained the evolution direction of the deep foundation
pit support concept and the corresponding content of design improvement. It had guiding value
for arousing the reflection and awakening of the geotechnical and structural design management
system, promoting the reform of foundation pit and foundation design theory and method, and
promoting the practice of "new development concept" in foundation pit engineering.

1 INTRODUCTION

The foundation pit is "the space excavated from the ground for the construction of the underground
part of the building (structure)". The foundation pit exists because of the underground structure
of the building (structure) and develops with the deepening of the underground structure and the
tension of the environment (Jiang 2022). The so-called "idea" refers to theories, concepts, or ideas,
and the foundation pit support concept is the foundation pit support function goal (positioning) and
its leading implementation behavior and is the fundamental direction and principle of foundation
pit support.

The evolution of the foundation pit support concept will inevitably be accompanied by the
improvement of the design method. Taking a specific project as an example, this paper explains
different design concepts and their design priorities, shows the three realms of foundation pit
engineering design behavior, and calls on the industry to actively promote the reform of foundation
pit engineering and promote the high-quality development of foundation pit engineering.

---

*Corresponding Authors: jk_doctor@163.com, hyx1130@126.com, 17865199982@163.com,
jiabincf@163.com and 371918616@qq.com

                                            DOI 10.1201/9781003344209-34

2 EVOLUTION OF DEEP FOUNDATION PIT SUPPORT CONCEPT

The evolution of the deep foundation pit support concept in China can be summed up in three stages, or embodied in three concepts.

The first is the temporary concept. The positioning of the foundation pit support structure is a temporary measure to ensure the safety of itself and the surrounding environment. However, the shortcomings are prominent, that is, the support and the main body respectively bear the earth's pressure due to the underground structure, and the short-term function of the support structure becomes an obvious waste.

The second is the concept of combining with the main structure, that is, the technology of combining foundation pit support with the main structure (Wang 2010, 2012; Xu 2005). The advantage is that the geotechnical engineering profession uses structural components or makes some supporting components part of the main structure in the construction stage of the underground structure, which reduces waste and promotes environmental protection. Increasing the use conditions of components is not widely accepted by the structural engineering profession. At present, it is mostly used in foundation pit inversion projects.

The third is the concept of permanence. This concept embodies the principle of " top-level design". Based on the same project reality as foundation pit support and the main underground structure, the support piles, and wall permanent structures are actively positioned before the underground structure design to overcome the disadvantages of waste, pollution, and carbon emissions in the temporary concept. The concept of combining improvement with the main structure is an important direction for implementing the new development concept and building a new development pattern in the new development stage of deep foundation pits. It hopes that the structural engineering major will cooperate with the subjective practice of foundation pit enclosure, and realize the division of labor and cooperation between the geotechnical and structural engineering majors.

An example of a foundation pit project is briefly introduced to fully illustrate the evolution of the above-mentioned foundation pit support concept and the improvement of the design method. Jinan Western Convention and Exhibition Center is located at the intersection of the central development axis of the core area of Jinan Western New City and West Second Ring Road, with Rizhao Road in the south, Weihai Road in the north, Binzhou Road in the west, and West Second Ring Road in the east. The basement elevation of the convention and exhibition center is -10.37m, and the basement elevation of the subway open-cut section is -18.92m. The convention and exhibition center adopts a raft foundation, and the base is provided with anti-floating anchors. The spacing is a 2m rectangular arrangement, the diameter is 250mm, and the length is 17m. The soil layer parameters are shown in Table 1:

Table 1. Soil parameters of soil layers.

Serial number	Soil layer name	Soil layer scope	Severe/ (kN/m^2)	Internal friction angle /(°)	Cohesion/ kPa
1	Silty Clay 1	−10.37−−11.7	19	38	twenty-three
2	Silty Clay 2	−11.7−−16.7	19.1	37	25
3	Silty Clay 3	−16.7−−31.15	18.6	36.5	13
4	Strongly Weathered Gabbro	−31.15−−51.2	18.8	55	30

3 TEMPORARY CONCEPT – DESIGN AND CONSTRUCTION OF SOIL NAIL WALL

According to the temporary design concept, the foundation pit of the subway station is supported by grading soil nailing walls (Figure 1). The concrete surface layer is made of $\phi 6.5@250\times250$ steel mesh, and C20 concrete is sprayed.

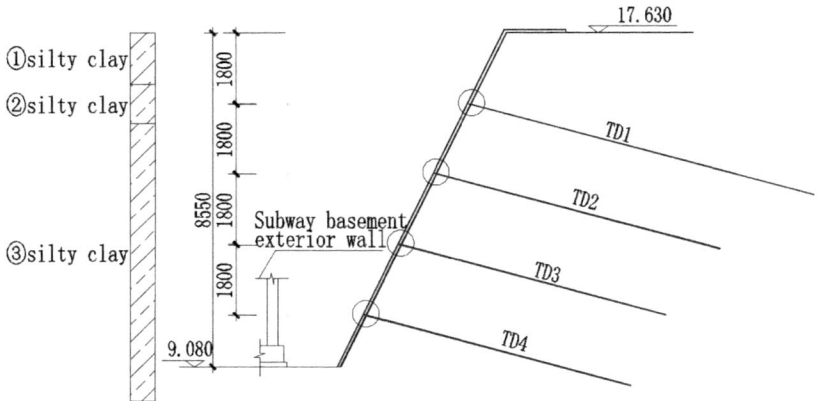

Figure 1. Schematic diagram of grading soil nailing wall support.

Table 2. Soil nailing parameters.

Soil nail number	Length (mm)	Horizontal spacing (mm)	Drilling diameter (mm)	Test value of pull-out bearing capacity (kN)
TD1	9000	2000	130	95.1
TD2	7500	1500	130	84.6
TD3	6000	1500	130	73.0
TD4	6000	1500	130	81.9

Specific construction steps:

(1) Excavate to 0.3m (-10.070) above the base of the exhibition center;
(2) Excavate in layers in sequence, and construct each layer of soil nails and concrete surface;
(3) Excavate to the bottom of the subway station (-18.920), construct the subway station structure upward, and reach the base level of the exhibition center (-10.370);
(4) Earthwork backfilling of fertilizer trough between the exterior wall and soil nail wall of subway station;
(5) Construction of anti-floating anchors, cushions, and floor structures at the bottom of the original convention and exhibition center.

4 THE CONCEPT OF COMBINING WITH THE MAIN STRUCTURE – THE "STRUCTURAL GEOTECHNICAL" DESIGN METHOD

4.1 Structural geotechnicalization

"Structural geotechnicalization" refers to the use of underground structural components to be supported by foundation pits. Because the industry generally believes that foundation pit support belongs to the category of geotechnical engineering, and the concept of combining with the main structure is dominated by the geotechnical engineering profession (Burland 1989). The technology of combining pit support with the main structure is called the "structural geotechnical" design method of foundation pit engineering. For this project, the composite soil-nail wall is formed by combining the anti-floating anchor rod (structural member) with the temporary soil-nail wall.

4.2 Design of composite soil nailing wall

According to the concept of combining support and main structure, the structure of the three rows of anchor rods adjacent to the subway station is densified to 1 m. The original design of the soil nail arrangement and the concrete surface remains unchanged.

4.3 Analysis of internal force and displacement of the anti-floating anchor

The combination of the temporary soil nailing wall and the main anti-floating anchors enables the anti-floating anchors adjacent to the subway station to bear the horizontal load during the excavation stage of the foundation pit. PLAXIS is used to numerically calculate the force and displacement of the anti-floating anchor (Hu 2019). The horizontal displacement, bending moment, and axial force of the anti-floating anchor at the end of excavation are obtained as shown in Figures 2 and 3.

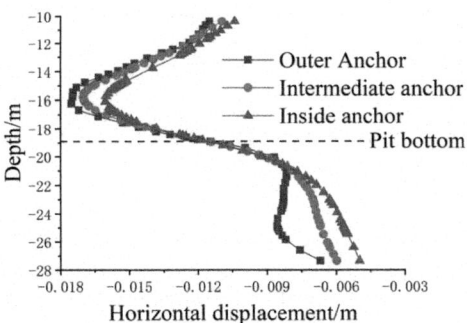

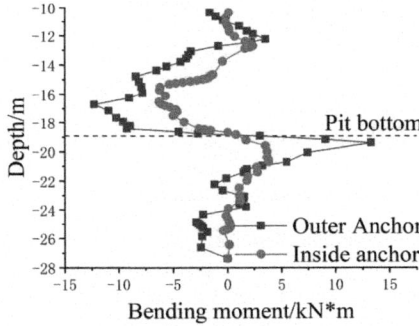

Figure 2.   Comparison of horizontal displacement of three rows of bolts from excavation to the bottom of the pit.

Figure 3.   Comparison of inner and outer anchor bolt bending moments when excavating to the bottom of the pit.

From Figure 2, the maximum displacement of the anchor rod is about 17mm in the lower half of the excavation depth, and the inclination is 0.10%, which does not affect the normal pull-out bearing capacity of the anchor rod (Yang 2008).

Figure 3 shows the bending moment diagram of the inner and outer anchor rods when excavating to the bottom of the pit. The bending moment of the outer anchor rod is larger than that of the inner anchor rod, and the maximum bending moment is taken $M = 13.30$ kN · m.

4.4 Durability analysis of anti-floating anchor

The anti-floating waterproof level of the project is -0.9m, the underground garage is planned to be built with 2 underground floors, the floor elevation of the convention and exhibition center is -10.37m, and the weight of groundwater is 10 kN/m^3. The stress diagram of the longitudinal tensile steel bar when the bending moment and the axial force are superimposed as shown in Figure 4 is obtained by calculation, and the stress level represents the contribution value to the crack of the bolt.

Analysis of Figure 4 shows that the contribution of the bolt bending moment to the crack plays a leading role, and the main influence section of the anti-buoyancy force on the bolted crack is

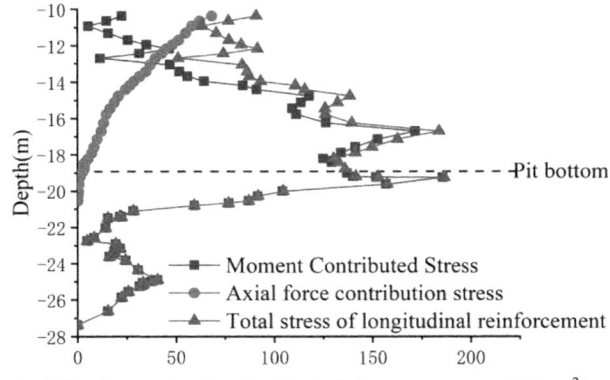

Figure 4. The stress diagram of the longitudinal reinforcement of the anti-floating anchor when excavating to the bottom of the pit.

in the upper half of the bolt. The final crack width of the bolt is 0.160mm, which is less than the requirement of the maximum crack width of 0.2mm. The anti-floating bolt support does not affect its durability.

5 PERMANENCE CONCEPT – "GEOTECHNICAL STRUCTURE" DESIGN METHOD

5.1 Geotechnical structuring

"Geotechnical structure" means that the foundation pit supporting components are used by the structural engineering profession and permanently becomes an integral part of the underground structure. Under the guidance of "top-level design", the geotechnical and structural majors each undertake their tasks, the foundation pit support is permanently designed, and the main structure is supported by the main structure (Li 2018; Zhou 2013).

5.2 Analysis of mechanical properties of anti-floating anchor

As the outrigger support gradually replaces the soil nail, the stress of the anti-floating anchor is rebalanced, the restraint position and restraint force change, and the anchor bending moment changes accordingly (Xia 2004). Figure 3 is the comparison diagrams of the bending moment and horizontal displacement of the inner and outer anchor rods of the permanent support.

From Figure 4, compared with the excavation to the base condition, the sudden change of the bending moment in the outer anchor bolt of the permanent support is transferred from the original soil nail position to the overhanging support position, and the peak value of the bending moment is −15.80kN m. The reason for this phenomenon is that the outer bolt is connected to the basement floor through the bottom support, and the bending moment is transferred from the bottom support to the basement floor, which provides the required restraint force.

5.3 Durability analysis of anti-floating anchor

The anti-floating anchor has been permanently converted, and its durability needs to be checked. When the bending moment and the axial force are caused by the anti-floating, the stress diagram of the longitudinal tensile steel bar is superimposed when the backfill is completed, as shown in Figure 6.

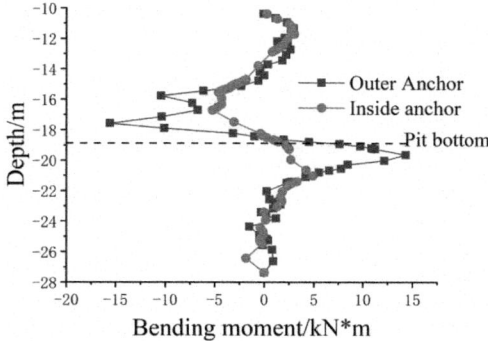

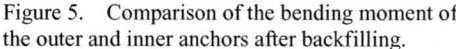

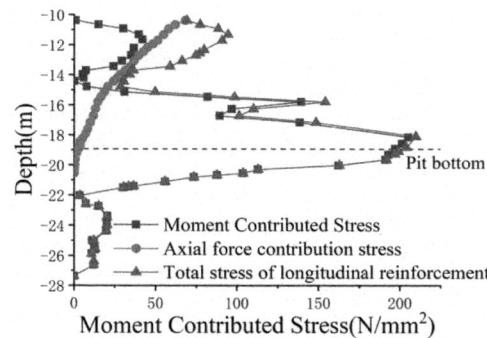

Figure 5. Comparison of the bending moment of the outer and inner anchors after backfilling.

Figure 6. Stresses of longitudinal tensile steel bars of anti-floating anchors under normal operating conditions.

Finally, the crack width of the bolt is obtained as 0.177mm, which is less than the requirement of the maximum crack width of 0.2mm. The anti-floating anchor rod plays a supporting role without affecting its durability.

5.4 *Analysis of mechanical properties of basement exterior walls*

According to the design and positioning of the permanent support structure, the earth pressure of the outer soil body received by the anti-floating anchor is transferred to the horizontal floor, and the basement exterior wall is only subjected to the earth pressure of the backfill part. It can be seen from this that the permanent foundation pit support not only greatly improves the economic, environmental, and social benefits but also promotes the reform of the basic structure theory and design method.

6 CONCLUSION

Based on the existing concepts of "temporary" and "combined with the main structure" of deep foundation pit support in China, the concept of "permanence" and its realization of high-quality development of foundation pit engineering are proposed and demonstrated.

Combined with examples, it shows the evolution of the three design concepts of deep foundation pit support. According to the mutual utilization of foundation pit support and underground structural components, the "structure" is defined as corresponding to the "integration with the main structure" and "permanent" concepts. "Geotechnical" and "Geotechnical structuring" design methods focus on the analysis of the working conditions of support members with different concepts and their durability control analysis.

REFERENCES

Hu Yong. (2019). *Deformation Mechanism and Control of Large-scale Foundation Pit with Cut-off Wall Inserting Rock Near the Yangtse River in Wuhan*. The China University of Geosciences.

Burland J.B. (1989). Ninth laurits bjerrum memorial lecture: "Small is beautiful"—the stiffness of soils at small strains. *Canadian Geotechnical Journal*, 26(4).

Jiang Weizhen, Tan Yong. (2022). Overview of failures of urban underground infrastructures in complex geological conditions due to heavy rainfall in China during 1994–2018. *Sustainable Cities and Society*, 76.

Li Lianxiang, Cheng Xiaoyang, Liu Bing. (2018). Analysis of the permanent intensive design of the supporting structure of the composite foundation. *Journal of Railway Science and Engineering* (08), 1971–1979. doi:10.19713/j.cnki.43-1423/u.2018.08.010.

Wang Wei-Dong, Sheng Jian. (2012). Design and analysis of unity of support piles and basement external walls. *Chinese Journal of Geotechnical Engineering*, (S1), 303–308.

Wang Wei-Dong, Xu Zhong-Hua. (2010). Design and construction of deep excavations supported by a permanent structure. *Chinese Journal of Geotechnical Engineering*, (S1), 191–199.

Xia Jiang, Xu Zhimin, Yan Ping, Zuo Renyu. (2004). Application of one pile with three uses technology in foundation pit work. *Architecture Technology*, (05), 340–341.

Xu Zhong-Hua, Deng Wen-Long, Wang Wei-Dong. (2005). Construction technique of a deep excavation supported with substructure. *Chinese Journal of Underground Space and Engineering*, (04), 607–610.

Yang Bao-Zhu, Wang Li, Zheng Gang. (2008). Analysis of bearing capacity of inclined piles due to adjacent excavation by use of finite element method. *Chinese Journal of Geotechnical Engineering*, (S1), 144–150.

Zhou Qiujuan, Chen Xiaoping, Xu Guangming. (2013). Centrifugal model test and numerical simulation of soft soil foundation pit. *Chinese Journal of Rock Mechanics and Engineering*, (11), 2342–2348.

Frontiers in Civil and Hydraulic Engineering – Mohamed A. Ismail and
Hazem Samih Mohamed (Eds)
© 2023 The Authors, ISBN 978-1-032-38247-0

Theory and design methods for geotechnically structured permanent support for deep foundation pits

Lian-xiang Li, Shengqun Li*, Hongxia Xing, Yingxue Hou & Zhixiao Han
School of Civil Engineering, Shandong University, Jinan, China

ABSTRACT: Based on an in-depth analysis of the current situation of the project's underground system design, reflecting on the disadvantages of waste, pollution, and carbon emission caused by the separation of geotechnical and structural professions for temporary foundation pit support, and adapting to the requirements of the new development concept, the theory and design method of deep foundation pit geotechnical structured permanent support system is proposed. It clarifies the composition and construction process of the components of the permanent support system and stipulates the design principles of the bearing capacity and normal use limits of the permanent support system. Moreover, the load combinations and calculation contents of the three stages of the foundation pit excavation, construction, and normal use are clarified, suggesting the finite element method and elastic support method applicable to the division boundary of deep and deeper foundation pits and their permanent support structures respectively. The durability analysis of the support components is emphasized, proving that the geotechnical structure is good. The deep foundation pit permanent support structure basement exterior wall is not subject to earth pressure, changing the current situation that the excavation stage support structure and the use stage basement exterior wall bear the earth pressure separately. Twice the building materials resist the earth's pressure. The theory of permanent support structures for deep foundation pits is the basis for the high-quality development of foundation pit engineering in the new development phase. The geotechnical structural design method is the technical support for the change from temporary to permanent foundation pit support.

1 INTRODUCTION

Currently, domestic standards (JGJ120-2012 2012, GB50007-2011 2011) lack a clear definition of deep and shallow pits. The engineering community has been using a depth of 10m or more to define the support structure of safety level one pit (GB50202-2002 2002, DG/TJ08-61-2018 2018), and the supporting structure is suitable for deeper pits of safety level one (JGJ120-2012 2012). Therefore, the depth of 10m or more is called a deep foundation pit, and the choice of the anchor-tensioned or internally braced pile or wall support (JGJ120-2012 2012) structure has general acceptance in the industry.

Pit support is the premise of underground space construction, and the main underground structure construction must support the structure's protection. The main body is the purpose, and the support is the means. The support and the main structure are interdependent and constitute the underground engineering system of the same project. Figure 1 shows a schematic diagram of the pit support and the main underground structure.

At this stage, the design of the underground engineering system is in a separate state. The main body of the project is designed by the construction institute, and the survey institute completes the

---

*Corresponding Author: 17350467648@qq.com

DOI 10.1201/9781003344209-35

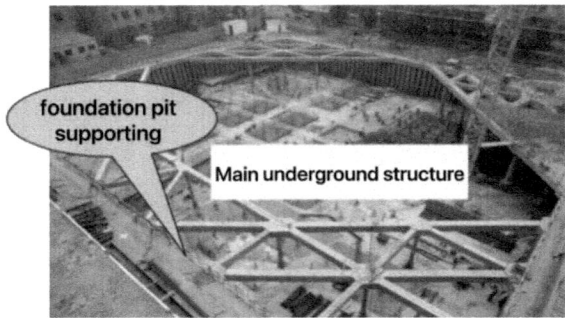

Figure 1. Schematic diagram of the foundation pit support and main underground structure.

foundation pit. The separation of different structures for the same objective, due to the division of labor, not only determines the continuation of the "temporary" concept of pit support but also leads to a second shift in the role of the horizontal load bearing of the underground structure during the construction process, namely the pit support during the construction phase and the basement facade in normal use. As a result, the two structures of the same project underground system bear all the external horizontal loads respectively. The existing management system and the temporary (JGJ120-2012 2012, GB50007-2011 2011, DG/TJ08-61-2018 2018) positioning of the foundation pit project allow twice the structural input of the underground project to bear only twice the load, namely the earth pressure.

Deep pit supports such as supporting piles and ground link walls, as a prerequisite for the underground structure and the necessary measures for its construction, will always exist despite the "temporary" construction phase, and the structural profession should design the underground space with the support measures as the environment. The permanent presence of temporary pit supports is considered a shield for the main underground structure (Li et al. 2018). As the underground structure precedes the design of the foundation support, the structural profession tends to ignore the presence of the support and carry out the design of the basement facade according to the soil environment. The current construction status of the underground engineering system has resulted in huge wastage, environmental pollution, and carbon emission of the foundation pit, especially the deep foundation pit support input. To be more precise, the interesting drive of "taking fees according to design cost" has overshadowed the overall light of "science and technology is the first productive force", resulting in the lack of scientific and technological innovation in the field of engineering construction to reduce the cost motivation.

Based on the new development stage and implementation of the new development concept, Li Lianxiang (Li 2018, 2019, 2020) calls for changing the current concept of "temporary" foundation pit support, and suggests the high-quality development direction of "full recovery of temporary support elements for shallow foundation pits and permanent support for deep foundation pits". A permanent support structure is constructed based on temporary support piles. The permanent support structure system based on temporary support piles was constructed (Li 2017). On this basis, the concept of "permanent" deep foundation pit support is further defined, and the "geotechnical structured" design method is explained (Li 2018, 2019, 2020). The permanent support and basement wall only bear the backfill load of the fat trench. The structural profession is used to construct the connecting beams and outreach supports to connect the foundation support completed by the geotechnical profession with the underground structure undertaken by the structural profession. The division of labor and close collaboration are clarified between the geotechnical and structural professions. The geotechnical scope of foundation support is transformed into a component of the main underground structure, inscribing the "geotechnical structuring". The process of "geotechnical structuring" is inscribed. Figures 2 and 3 show the schematic diagram of the permanent support structure system based on piles and walls and the definition of the components respectively.

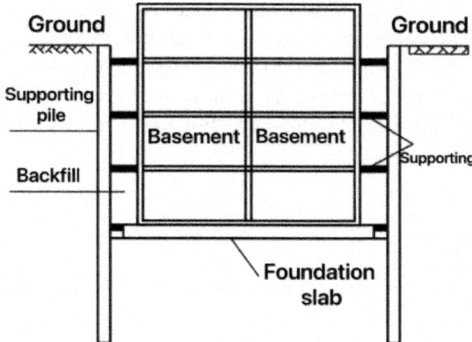

Figure 2. Permanent support system.

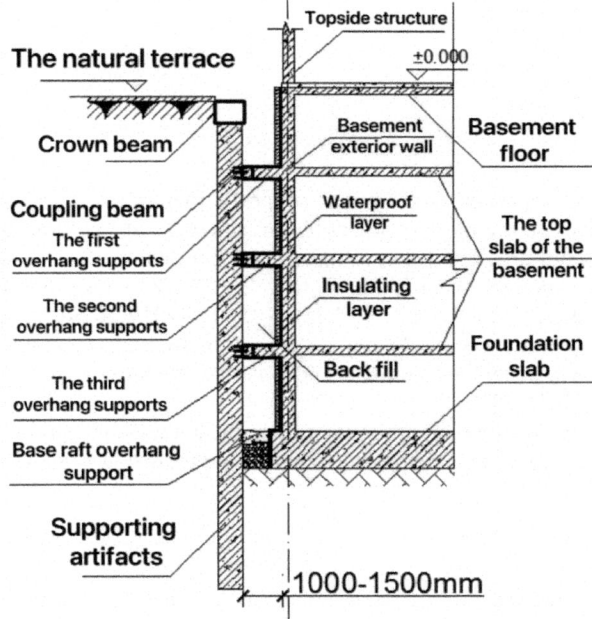

Figure 3. Definition of permanent support system components.

This paper provides a detailed analysis of the permanent support system shown in Figure 2, with examples of the general construction process, analysis theory, and design methods of "geotechnical structuring", showing personal efforts and dreams to promote the high-quality development of deep foundation pit projects.

2 DESIGN PRINCIPLES

2.1 *Load combinations for permanent support structures*

In contrast to temporary support which only includes the excavation phase, the theory of permanent support adds to the structural analysis of the construction and use phases. Permanent supports are mainly subject to horizontal loads from earth pressure and ground overload. The ground overloads consist of the gravity of the existing facilities in the surrounding environment and the loads from the construction process. The construction loads are variable. After normal use of the project, the construction loads are replaced by general road loads. The construction loads during the excavation phase of the support structure have been incorporated into the structural analysis and are generally considered to be around 20 kPa. The only load combination for the permanent support structure is the basic combination.

2.2 *Load capacity limit states*

(1) Supporting blocks, tie beams, supports, basement facades (Figure 2), and other components and their connecting nodes shall be of uniform concrete strength grade due to excess material strength or excessive deformation, which shall be required to meet.

$$\gamma_0 S_d \leq R_d \tag{1}$$

$$S_d = \gamma_F S_k \tag{2}$$

$$R_d = R(f_c, f_s, \dots) / \gamma_{Rd} \tag{3}$$

Where, $\gamma 0$ is the important factor of the support structure taken as 1.1; Sd is the design value of the effect on the basic combination of action (axial force, bending moment, etc.). In the seismic-proof area according to the seismic combination of action, Rd is the design value of the resistance of the structural members. γF is the comprehensive subfactor of the basic combination of action, and $\gamma F=1.35$ (GB50009-2019 2019). Sk is the effect of the basic combination of action. γRd is the resistance of structural members model uncertainty coefficient, which is generally taken as 1.0. γRE is taken for the design of seismically protected areas (GB50010-2010 2010).

(2) The calculation and verification of the stability in the permanent support structure, including overall, local, and seepage stability, shall be under the following formula.

$$\frac{R_k}{S_k} \geq K \tag{4}$$

Where Rk is the value of the slip resistance moment. Sk is the value of the slip moment. K is the stability safety factor. The construction phase is selected according to the literature (JGJ120-2012 2012 2012). The normal function at the end of construction is selected according to the literature (GB50007-2011 2011).

2.3 *Normal use limit state*

The permanent support structure encompasses the construction and post-completion use, with emphasis on the deformation of the support structure, the surroundings after excavation, and main construction to meet.

$$S_d \leq C \tag{5}$$

Where S_d is the design value of the effect of the standard or quasi-permanent combination of loads (displacement, settlement, etc.). C is the limit value of the surrounding environmental deformation, and the displacement of the support structure is determined by it. The environmental limits are the superposition of the existing displacements and the formation process of the pit excavation and permanent support, which are the basis for determining the permanent support displacements.

3 THEORY OF ANALYSIS OF PERMANENT SUPPORT STRUCTURES

3.1 *Excavation phase*

The current temporary support structure is designed by the geotechnical engineering profession to bear the horizontal loads outside the pit and to ensure the smooth excavation of the earth in the pit.

3.2 *Construction of the permanent support structure*

It is designed by structural engineering disciplines.

(1) Design and construction of tie beams

A tie beam is a member that connects the retaining elements and transmits the geotechnical pressure of the underground structure to the outreach support (Figure 3). The lowermost tie beam is formed by the outward extension of the foundation or mat, while the other tiers of tie beams are modified by retaining elements. The cross-sectional height of the tie beam is greater than the thickness of the floor slab. The width is not greater than 300 mm. The tie beam connection nodes are constructed as shown in Figures 4 and 5.

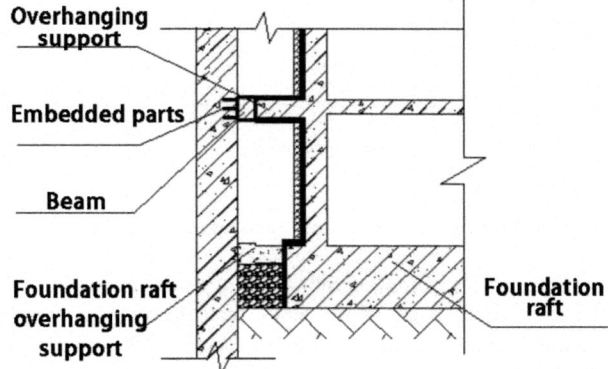

Figure 4.   Cross-sectional view of the node connecting the tie beam to the outreach support.

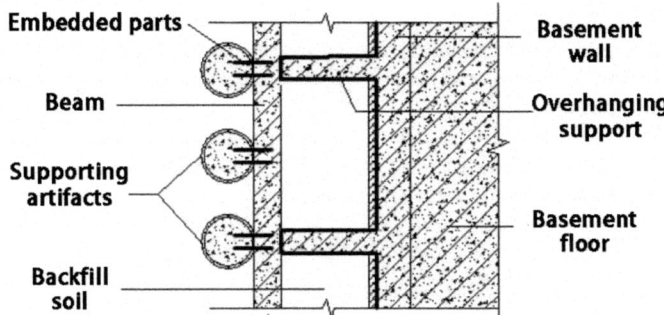

Figure 5.   Plan of the node connecting the tie beam to the outreach support.

(2) Design and construction of external extension bracing

The support extends from the external wall of the basement, with a section height flush of the floor slab which is within 300mm of the width. After the working conditions have

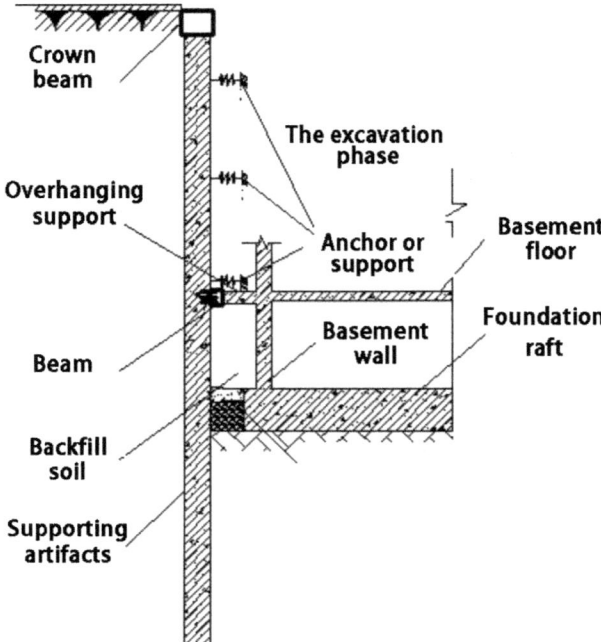

Figure 6.   Removal and replacement of bracing stage profiles.

been transformed, the outer extension support becomes a rigid support point for the retaining element.

(3) Replacement of temporary horizontal members by outreach supports

Figure 6 is a schematic diagram of a working condition in the process of changing and removing supports. The formation of the permanent support structure system is the process of replacing the temporary support horizontal members with outreach support. Each work condition change is the completion of the construction in one layer of outreach support as well as the removal of the temporary horizontal members similar to it above and below.

At the end of the excavation, along with the construction process of the underground main body, a connecting beam is set up on the retaining members, gradually replacing the temporary horizontal members of the support structure, such as anchor rods and internal bracing. By the outreach bracing, the retaining members are transformed into a permanent support structure, as shown in Figure 3. The process needs to be determined after calculation and selection.

(4) Basement wall design

The external wall of the basement only bears the limited soil pressure of the fat trench, and the literature (Li 2017) proves that the external wall of the basement can be designed only according to the minimum structural requirements.

3.3   *Theory of analysis of permanently supported structures*

3.3.1   *Content of structural analysis*

(1) Load calculation

Temporary deep foundation pit support structure mostly uses supporting structure support members and the basement wall between the need to reserve 1000–2000mm working space and backfill before completion. The loads on the supporting members and basement walls at different stages are shown in Table 1.

Table 1. Table of loads on the different stages of the support elements in the permanent support structure.

Stage	Components	Soil pressure*	Water pressure	Backfill pressure	Overloading q_0
		Load			
Excavating	support	✓	✓		✓
Constructing	support wall	✓	✓	✓	✓
Using	support wall	✓	✓ *	✓	✓

(*: 1 static earth pressure; 2 Class A buildings, curtain failure to be considered at 100-year life cycle.)

(2) Content of structural analysis

Table 2. Table of stages of permanent support theory formation and analysis.

Stage	Components			Control objectives
				Strength, deformation, and Stability
Excavating	Permanent support		Temporary horizontal elements	Load capacity and normal use limits
Constructing	support	beam	Anchor rods, internal bracing	
Using	The durability of support elements		Exterior wall durability	

3.3.2 *Structural analysis methods*

(1) Critical depth of " deeper" pits

There is a "depth effect" for deep foundation pits in soil bodies. Li Lianxiang (Li 2019) determined that the critical depth for the "depth effect" of foundation pits in Jinan alluvial strata is 15 m. In this paper, the foundation pits greater than 10 m and less than the critical depth are defined as "deeper foundation pits", and the depth of foundation pits exceeding the critical depth is defined as "Deep pits". The deeper pits are permanently supported by the elastic pivot method (JGJ120-2012 2012), and the deep pits are subjected to three-dimensional finite element analysis (Li 2019). The critical depth of foundation pits differs from different geological conditions. Therefore, the respective criteria for deep and deeper pits are established according to the geological characteristics of the city, corresponding to different analysis methods for the design of permanent support structures.

(2) Structural analysis methods

There are differences in the calculation of the internal forces in permanently used elements such as piles, walls, connecting beams, outreach supports, and temporary horizontal elements such as anchor rods and internal bracing.

A. *Permanently used elements*

The vertical retaining members are part of the main underground structure, which are calculated as positive section flexural members under the literature (GB50010-2010 2010). The maximum values of bending moment and shear force of retaining members Mkw, max, Vkw, and max and their locations are obtained in the excavation stage. The maximum values of bending moment and shear force of retaining members MGJ, max, VGJ, max, and their locations are clarified in the construction stage. The outreach support is appropriately set up with pre-stress. The size of which is based on the end of excavation working condition near the horizontal member tension (anchor rod) or pressure (internal support), which aims to control the deformation of the support member and ensure MGJ, max ≤ Mkw, max. The final working condition of the construction stage of the permanent support member represents the working condition for permanent use.

B. *Durability of permanent support elements*

Crack control is to achieve durability. Cracks in permanent support members are calculated according to the maximum bending moment of the positive section, and the crack control level is three, with the maximum crack control width $\omega lim = 0.2mm$ (GB50010-2010 2010). The maximum bending moment of the support member during the formation and use of the permanent support structure was selected, and the cracks of the rectangular section among the support member were tested according to the literature (GB50010-2010 2010) and the circular section according to the literature (JTS151-2011 2011). At the same time, the durability of the permanent support structure in a corrosive environment is verified according to the literature (GB50046-2018.2018).

C. *Temporary horizontal members*

Temporary horizontal members such as anchor rods and internal bracing only function in the excavation stage, and the design calculates the comprehensive load sub-factor $\gamma F = 1.25$, so there is no need to carry out crack control and durability test.

4 CONCLUSION

(1) To implement the new development concept, based on the same system view of the underground structure, a deep foundation pit permanent support system characterized by the "top-level design" of the project is proposed with the help of the existing results. It is also supported by a clear division of labor and close collaboration between the geotechnical and structural disciplines.

(2) The design principles of permanent support structures for deep foundation pits are clarified, and the specific performance and analysis of load-bearing capacity and normal use limits are explained. The uniformity of the strength grade of the concrete material is emphasized. The permanent support, connecting beam, supporting member, and basement outer wall clarify the comprehensive load. Subfactor $\gamma F = 1.35$ is for the permanent member and 1.25 for the temporary horizontal member. The content of structural analysis of excavation, construction, and use conditions of the permanent support structure is clarified, forming the theory of permanent support.

(3) Considering the "depth effect" of deep foundation pits, the deep foundation pits are divided into deeper and deeper pits. It is suggested that the elastic support point method and the finite element method be used to adapt to the permanent support system of deeper and deeper pits respectively.

(4) The deep foundation pit permanent support structure adapts to the requirements of the new development concept and overcomes the disadvantages of wasted construction materials, geological environment pollution, and increased carbon emissions. It is an important direction for foundation pit engineering workers to build a new development pattern based on the new development stage. The theory and design method of deep foundation pit geotechnical structured permanent support is solid support for the change and high-quality development of foundation pit engineering.

REFERENCES

DG/TJ08-61-2018. (2018). *Shanghai technical specification for foundation pit engineering* (DG/TJ08-61-2018). Shanghai: Shanghai General Station for Market Management of Construction and Building Materials Industry. (in Chinese)

GB50007-2011. (2011). *Code for the design of the building foundation. Beijing*: China Construction Industry Publishing House. (in Chinese)

GB50009-2019. (2019). *Code for structural loading of buildings*. Beijing: China Construction Industry Press. (in Chinese)

GB50010-2010. (2010). *Code for the design of concrete structures*. Beijing: China Construction Industry Publishing House. (in Chinese)

GB50046-2018. (2018). *Design code for corrosion protection of industrial buildings*. Beijing: China Planning Press. (in Chinese)

GB50202-2002. (2002). *Specification for acceptance of construction quality of building foundation works* (GB50202-2002). Beijing: China Construction Industry Press. (in Chinese)

JGJ120-2012. (2012). *Technical Specification for Retaining and Protection of Building Foundation Excavations. Beijing*: China Construction Industry Press. (in Chinese)

JTS151-2011. (2011). *Code for the design of concrete structures for water transport engineering*. Beijing: People's Traffic Publishing House. (in Chinese)

Li Lianxiang, Liu Bing, Cheng Xiaoyang. (2018). Influence of the permanent presence of foundation pit support piles on soil pressure distribution of basement exterior walls. *Journal of Shandong University*. 48(2): 30–39. (in Chinese)

Li Lianxiang, Liu Bing, Li Xianjun. (2017) Permanent support structure combining supporting piles and underground main structure. *Journal of Building Science and Engineering*. 34(2): 119–126. (in Chinese)

Li Lianxiang, Liu Jiadian, Li Kejin, et al. (2019) Selection and applicability of HSS parameters for typical strata in Jinan. *Geotechnics*. (10). (in Chinese)

Li Lianxiang. (2018). *Actively promoting change in foundation pit engineering-both permanent and recyclable support structures*: https://news.yantuchina.com/39312.html. (in Chinese)

Li Lianxiang. (2019). *Foundation pit engineering design concept leap and change thinking*: https://news.yantuchina.com/42070.html. (in Chinese)

Li Lianxiang. (2020). *Design examples and insights on the evolution of the foundation pit engineering concept*: https://news.yantuchina.com/46356.html. (in Chinese)

*Frontiers in Civil and Hydraulic Engineering – Mohamed A. Ismail and
Hazem Samih Mohamed (Eds)
© 2023 The Authors, ISBN 978-1-032-38247-0*

Stability analysis of double-layer reticulated shell based on member sensitivity

Di Liu* & Menghong Wang
School of Civil and Transportation Engineering, Beijing University of Civil Engineering and Architecture, Beijing, China

Jialiu Liu
Sinopec Guangzhou Engineering Co., Ltd., Guangzhou, China

ABSTRACT: Stability analysis and anti-collapse analysis are carried out on the clinker bank roof with a real engineering background. The large response area of the structure is obtained through the characteristic value buckling analysis, the concentrated area of the structure-sensitive rod is initially determined, and the structure is parameterized by changing the yield intensity. The cross-sectional size, initial defects, and support conditions of the structure are analyzed by using the double nonlinear analysis theory. By calculating the sensitivity of the members, the eigenvalue buckling analysis method verifies the accuracy of the sensitive members.

1 INTRODUCTION

In 2014, China issued a code for Collapse Design of Building Structures (CECS392:2014) to define the problems related to the continuous collapse of structures. Three methods are provided: pull structural part method, remove component method, and local strengthening method. At present, a large number of studies with practical engineering backgrounds are needed to provide the theoretical basis for the formulation of relevant standards.

Based on the method of buckling analysis, eigenvalue buckling analysis and nonlinear buckling analysis were used to study the stability of reticulated shells under specific loads. In the stage of eigenvalue buckling analysis, the internal forces of the bar are extracted and applied back to the structure by the principle of equivalent static force without considering the initial defects and double nonlinearity of the structure.

2 DEFINITION AND SIGNIFICANCE OF ROD SENSITIVITY

Structural redundancy can be considered as the ability of a structure to resist continuous collapse. If the structure is partially damaged, the ideal structure should still be able to bear the load without continuous collapse, but the ultimate bearing capacity will decrease (Tian 2016). The calculation formula of structural redundancy is as follows:

$$R = \frac{L_{\text{int}\,act}}{L_{\text{int}\,act} - L_{damage}} \tag{1}$$

Where is the ultimate load of the initial complete structure; Is the ultimate load of the residual structure after a component failure?

---

*Corresponding Author: 851014187@qq.com

DOI 10.1201/9781003344209-36

The sensitivity of a member is defined based on the bearing capacity of the whole structure. To simulate the local failure of the structure, the decline of the bearing capacity in the remaining structure after the removal of a single rod was analyzed and compared with the original structure. Moreover, whether the remaining structure could maintain certain integrity and avoid continuous collapse was analyzed.

Sensitivity index SI and redundancy R are reciprocal: SI=1/R. When the sensitivity index of the component is very small, the damage to the component will not affect the bearing performance of the structure. On the contrary, the damage to the highly sensitive rod will greatly affect the bearing capacity of the structure and even cause collapse.

The concept of rod sensitivity reflects the importance of rod to structure. Spatial structures are often composed of complex rod systems, and each rod contributes to the structure in different degrees. The greater the contribution of the rod is, the higher the sensitivity value will be, proving that the rod is more important. In the case of the failure of the rod, the risk of the overall collapse of the structure (Su 2006) is greater. In addition to locating the importance of the member itself, the sensitivity index of the member can also be used to evaluate the weak part of the structure. Through the key monitoring of the weak parts, people can maximize the efficiency of reinforcement and repair, which saves time and decreases economic costs.

3 STABILITY ANALYSIS OF DOUBLE RETICULATED SHELL BASED ON SENSITIVITY OF ROD

3.1 *Project overview*

The inner and outer layers of the clinker library mesh shell model are respectively composed of ribbed ring mesh shells. The rods are round steel tubes, with a total of 1557. The mesh shell is provided with 18 supports at the bottom, which are fixed in three directions, and 9 supports at the top hinged in three directions. There are 414 nodes, welded ball nodes, with a structural span of 50.5km. The structural form is shown in Figure 1.

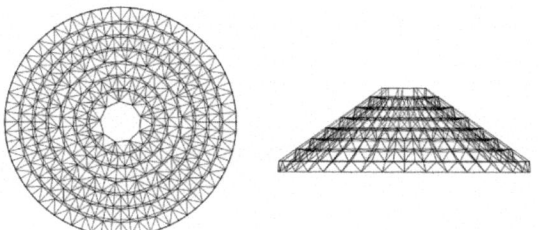

Figure 1.   Structural plan and elevation.

According to the stress characteristics of the double reticulated shell, the member is designated as the rod element with material Q235 steel, the elastic modulus of 206GPa, Poisson's ratio of 0.3, yield strength, and ultimate strength. The nonlinear buckling analysis adopts displacement control and selects 8 monitoring points in radial direction U3. According to "Technical Specification for Spatial grid Structure", the maximum deflection of a double reticulated shell is 1/250 of the span. The initial defect is adopted according to the lowest global buckling mode according to "Technical Standard for Steel Structures" 5.5.10. The maximum defect value is L/300, and L is the structural span. The initial defects of components shall be subject to clause 5.2.2. Axial P is adopted, and the relative length is 0.1. It is applied to the position where the relative length of the rod element is 0.5. The ideal buckling analysis and the whole process analysis considering the p-delta effect and large displacement are respectively used in the analysis, which is divided into eigenvalue buckling analysis and nonlinear stability analysis.

3.2 Eigenvalue buckling analysis

According to the study of single-layer reticulated shell in literature (Xu 2011), the large response area obtained by eigenvalue buckling analysis in reticulated shell structure coincides with the distribution area of sensitive rods of the reticulated shell. This conclusion serves as the basis for verifying the sensitivity standard of rods and selecting sensitive rods. By using the geometric symmetry of the structure, 1/9 member of the reticulated shell was taken as the representative to comprehensively judge the position of the large response area under each mode. The buckling modes of the first six-order eigenvalues of the reticulated shell are shown in Figure 2.

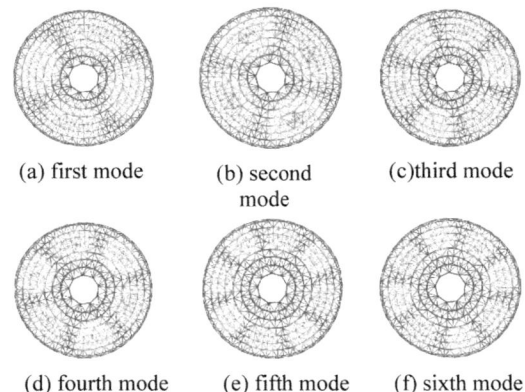

| (a) first mode | (b) second mode | (c)third mode |

| (d) fourth mode | (e) fifth mode | (f) sixth mode |

Figure 2.   Buckling modes of the first six eigenvalues.

Structure in each order buckling mode under large response is in the form of symmetrical distribution and regional focus. The shell deformation is close to the fifth and sixth rings and CHS, so the high sensitivity of rods can be preliminarily located in the center of the net shell. Combined with the "collapse resistance design code for the design of building structures" relevant provisions, bearing eight bars near sensitivity analysis are selected, with a total of 36 primary rods. It is shown in Figure 3.

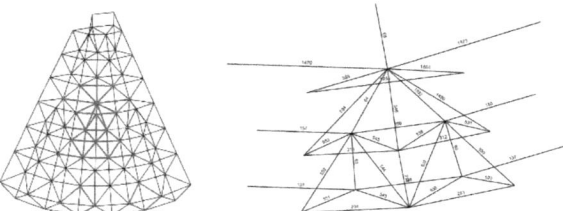

Figure 3.   Position and number of sensitive rod.

3.3 Nonlinear stability analysis of double-layer reticulated shell under different parameters

3.3.1 Influence of steel yield strength on rod sensitivity

By changing the yield strength of steel, the stability bearing capacity of double reticulated shell and the sensitivity of members are analyzed. In addition to the initial steel model Q235 of the double reticulated shell, the steel with yield strength Q345, Q390 and Q420 were selected for analysis and comparison. The load corresponding to the first critical point in the whole process curve was the ultimate load of the reticulated shell. According to Formula (1), the ultimate bearing capacity of the complete structure under different steel strengths and the ultimate bearing capacity of the remaining structure after the removal of the primary sensitive rod are calculated, and then the sensitivity index of the single rod is calculated. Due to space limitations, only the top 10 results in each case are listed, as shown in Tables 1–4.

Table 1. Sensitivity values of each rod when yield strength is Q235.

Bar code	$L_{\text{int act}}$/ kN	$L_{\text{int act}}$/ kN	SI	Bar code	$L_{\text{int act}}$/ kN	$L_{\text{int act}}$/ kN	SI
547	1202	998	0.1695	504	1202	1116	0.0709
512	1202	1035	0.1389	1321	1202	1119	0.0688
548	1202	1095	0.0884	1473	1202	1119	0.0685
539	1202	1107	0.0789	1328	1202	1121	0.0674
513	1202	1107	0.0786	505	1202	1128	0.0612

Table 2. Sensitivity values of each rod when yield strength is Q345.

Bar code	$L_{\text{int act}}$/ kN	$L_{\text{int act}}$/ kN	SI	Bar code	$L_{\text{int act}}$/ kN	$L_{\text{int act}}$/ kN	SI
547	1419	1212	0.1456	504	1419	1316	0.0721
512	1419	1215	0.1435	1321	1419	1317	0.0713
548	1419	1289	0.0915	1473	1419	1319	0.0705
539	1419	1303	0.0814	1328	1419	1320	0.0692
513	1419	1304	0.0810	505	1419	1330	0.0628

Table 3. Sensitivity values of each rod when yield strength is Q390.

Bar code	$L_{\text{int act}}$/ kN	$L_{\text{int act}}$/ kN	SI	Bar code	$L_{\text{int act}}$/ kN	$L_{\text{int act}}$/ kN	SI
547	1358	1174	0.1353	1321	1358	1260	0.0719
505	1358	1180	0.1312	513	1358	1265	0.0684
539	1358	1248	0.0807	1473	1358	1265	0.0684
504	1358	1259	0.0732	540	1358	1268	0.0662
548	1358	1259	0.0730	512	1358	1268	0.0662

Table 4. Sensitivity values of each rod when yield strength is Q420.

Bar code	$L_{\text{int act}}$/ kN	$L_{\text{int act}}$/ kN	SI	Bar code	$L_{\text{int act}}$/ kN	$L_{\text{int act}}$/ kN	SI
504	1495	1328	0.1120	505	1495	1386	0.0728
539	1495	1351	0.0964	1321	1495	1390	0.0699
512	1495	1361	0.0899	513	1495	1396	0.0662
547	1495	1380	0.0772	1473	1495	1402	0.0620
540	1495	1383	0.0750	548	1495	1404	0.0608

A comprehensive study on the sensitivity coefficient of the mesh shell rod with 1/9 of the structure can be found that no. 547 member is the most sensitive rod, which is located just above the support. At the support, the rod is usually subjected to a large reaction force, and the mesh shell deformation is extruded to a higher degree.

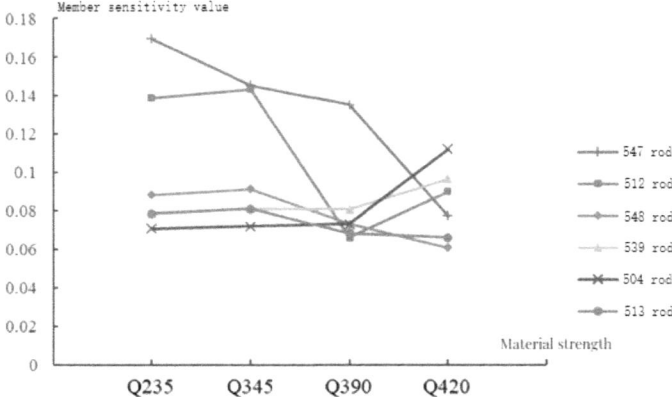

Figure 4.   Changes in rod sensitivity at each yield strength.

Figure 4 shows that the key link sensitivity values with the increase of the strength of materials are not the same trends, and bar sensitivity has a very large discreteness, proving that the change in the strength of materials and the influence of sensitivity on the bar is not monotonous and sensitive. There is a clear line between sensitive rods and the bar. For the convenience of the key areas of spatial structure in the project to strengthen the observation, the designer should pay special attention to these sensitive members.

3.3.2   Influence of rod section size on rod sensitivity

The influence of outer diameter and wall thickness on the stability performance of reticulated shell is investigated to comply with the requirements of section size of structural rod in "Seismic Theory and Design of Spatial Structures" and meet the checking conditions of structural stability. The cross-section size of the rod was set to ① φ 60mm×5mm, ② φ 60mm×3.5mm, ③ φ 70mm×5mm, ④ φ 70mm×4mm, and ⑤ φ 80mm×5mm (outer diameter × wall thickness) to obtain the ultimate bearing capacity curve. The influence of wall thickness and outer diameter on the sensitivity of the rod is analyzed. According to Formula (1), the sensitivity index of a single rod was calculated. Due to space limitations, only the sensitivity values of the top 7 rods were listed, as shown in Figure 5.

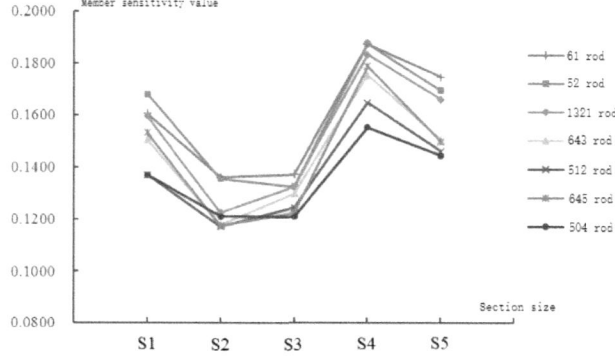

Figure 5.   Changes in rod sensitivity under different section sizes.

With the change in the cross-sectional area of the bar, the sensitivity of the bar shows a non-monotonic trend. Although the larger the outer diameter and wall thickness of the component is, the more stability and ultimate bearing capacity of the reticulated shell can be improved, but it also increases the project cost and steel consumption. Therefore, considering the reasonable size of the component section is the key to studying the stability of the reticulated shell.

No. 61 and No. 52 rods are located above the support. It can be seen from the deformation diagram analyzed by SAP2000 that the rod near the support has the largest deflection. In addition to the support area, other sensitive rods are concentrated in the middle of the reticulated shell, which also confirms the conclusion in Section 2.2. Sensitive rods of the double reticulated shell are concentrated in the large response area of eigenvalue buckling analysis.

3.3.3 Influence of initial defects on rod sensitivity

The initial geometric defects of spatial reticulated shell structure may lead to a decrease in carrying capacity, and even in the case of slight initial defects, the mechanical characteristics of the structure will be greatly changed (Cao 2006) (Zhang 2018). Therefore, the initial geometric defects of the structure must be introduced in the analysis of the defect-sensitive structure. For the method of simulating the initial geometric defects of spatial steel structure, the uniform defect method considers that the buckling mode of structure is the displacement trend when the structure is unstable, and the initial defects distributed according to the buckling mode will harm the structure stability (Cai 2015). The initial defect values of L/2000, L/1000, L/800, L/500, and L/400 were applied to the structure for stability analysis. Figure 6 shows the sensitivity change trend of the top six rods.

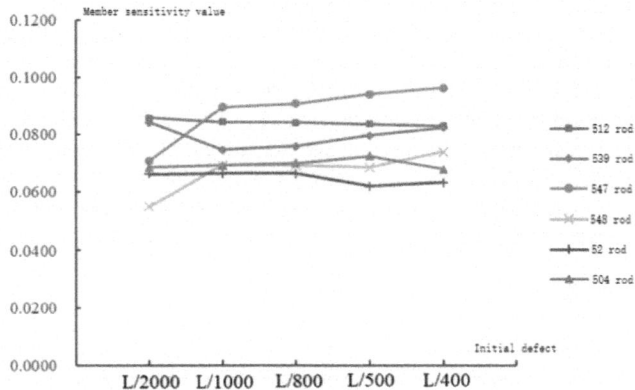

Figure 6.   Changes in rod sensitivity under each initial defect.

It can be seen that the sensitivity of the member is not greatly affected by the initial structural defects. Here, by comparing the results of single-layer rib-ring reticulated shell in literature (Ding 2012) and literature (He 2007), it can be seen that the in-plane stiffness of ordinary single-layer reticulated shell is much larger than that of out-of-plane, and the instability modes of the structure are mostly out-of-plane instability. The grid of the double-layer reticulated shell is denser, the overall stiffness of the structure is higher, and the joint constraint degree is richer, so the sensitivity of the rod is less affected by the initial defect.

3.3.4 Influence of supporting mode on rod sensitivity

To ensure the overall redundancy and stability of the structure, the lower support system is changed to three-hinged, and the "jump" mode is adopted. In other words, when the support is applied, the hinged support is arranged along the bottom node of the structure interval, and the whole process of the structure is analyzed.

The supporting form changes and the sensitivity variation trend of the top five rods are shown in Figure 7. It can be seen that the change of supporting conditions has little influence on it. A reasonable supporting mode should be adopted in structural selection and design.

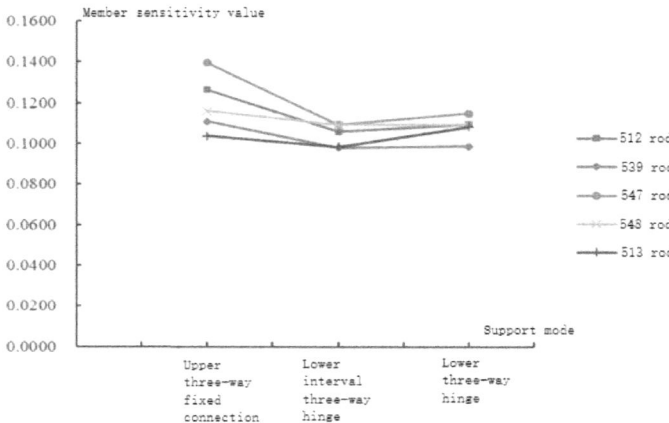

Figure 7. Changes in rod sensitivity under various supporting conditions.

4 CONCLUSION

The main work of this paper is to study the clinker warehouse roof with a double reticulated shell under a practical engineering background. SAP2000 finite element analysis software is used to analyze the stability bearing capacity of the structure and the following conclusions are drawn:

The large response region of the double-layer reticulated shell is obtained by eigenvalue buckling analysis, which coincides with the region of the highly sensitive rod in the reticulated shell. The yield strength and section size have a significant effect on the sensitivity of the rod, and there is a significant difference between the sensitive and non-sensitive parts. In practical engineering, safety and economy should be considered comprehensively to strengthen the observation of the key parts. The structure has low sensitivity to initial defects and support forms, so reasonable engineering conditions should be set up in engineering to avoid structural redundancy and material waste.

REFERENCES

Cai Jian, He Sheng, Jiang Zhengrong, et al. Study on the maximum initial geometric defect in stability analysis of single-layer reticulated shell structures [J]. *Journal of Building Structures*, 2015, 36(6): 86–92.

Cao Zhenggang, Fan Feng, Shen Shizhao. Elastic-plastic stability of single-layer spherical reticulated shell[J]. *Chinese Journal of Civil Engineering*, 2006, 39(10): 6–10.

Ding Leran. *Analysis of static performance and dynamic characteristics of rib-ring type giant grid chord-supported dome structure* [D]. Xi'an University of Architecture and Technology, 2012.

He Yongjun, Li Jia, Zhou Xuhong. The composition, support method, and mechanical model of spherical giant grid structure [J]. *Journal of Hunan University*, 2007, 34(3): 11–14.

Su Ci. *Research on the ultimate bearing capacity of large-span rigid spatial steel structures* [D]. Tongji University, 2006.

Tian Limin, Wei Jianpeng, Hao Jiping, Wang Xiantie. Evaluation method for important components of large-span spatial grid structures based on multiple responses [J]. *Journal of Hunan University* (Natural Science Edition), 2016, 43(11): 39–46.

Xu Gongyong. *Analysis of continuous collapse resistance of single-layer spherical reticulated shell* [D]. Southwest Jiaotong University, 2011.

Zhang Tianlong, Ding Yang, Li Zhongxian. Influence of initial geometric defects on the seismic bearing capacity of single-layer spherical reticulated shell structures [J]. *Space Structure*, 2018, 24(3): 3–9.

Frontiers in Civil and Hydraulic Engineering – Mohamed A. Ismail and
Hazem Samih Mohamed (Eds)
© 2023 The Authors, ISBN 978-1-032-38247-0

Failure mechanism on rigid pile composite foundation with pile damage

Chun Li
Jinan Rail Transit Group Construction Investment Co., Ltd. Jinan, China

Shengqun Li*
School of Civil Engineering, Shandong University, Jinan, China

Shuyin Feng
Department of Civil Engineering, University of BRISTOL, Queen's Building, University Walk, Bristol, UK

Xiangkai Ji & Lian-xiang Li
School of Civil Engineering, Shandong University, Jinan, China

ABSTRACT: A three-dimensional analytical model of an excavation adjacent to the existing rigid piles is established with a concrete damage plasticity model using ABAQUS software. The stress state and failure mechanism of piles and surrounding soil induced by the adjacent excavation are studied. The results show that during the process of excavation, when the support structure of the existing foundation is flexible, the soil between the piles will be flatly fractured before any damage occurs, and local bending failure will subsequently appear on the piles intersecting with the slip surface. When the support structure in the existing foundation is a raft foundation, the lateral movement of the composite foundation could be effectively restrained, which lessens the potential damage to piles and soil. When the support structure is experiencing convex or composite deformation, the near-pit piles will bend and fracture, and circular-shaped plastic development areas will occur in the soil among the piles, which will eventually develop into circular-shaped structural failure. Moreover, convex deformation in the retaining wall will bring more damage to the pile and the soil between the piles compared to the composite deformation.

1 INTRODUCTION

Due to the ongoing process of urbanization, more new buildings are planned to be built adjacent to an existing structures. However, the interaction effects between the foundation of the new building and the existing structure can influence the stress state of the existing composite foundation (e.g., Feng et al. 2016; Fu et al. 2017; Xing et al. 2016; Yue 2017), which also change the failure mode of the newly excavated foundation pit slope, especially when it is adjacent to the composite foundation. Thus, a detailed analysis of these effects is required to enable an effective and performance-based design.

The existing research on the interaction between foundation pit excavation and adjacent composite foundation mainly focuses on two aspects: lab testing (mostly centrifugal test) (Leung et al., 2006; Li et al., 2017; Li et al. 2018; Ong et al. 2006) and the finite element analysis (Fu et al. 2017; Goh et al., 2003; Zhang et al. 2013). Li et al. (2017) studied the lateral mechanical behavior

*Corresponding Author: 17350467648@qq.com

DOI 10.1201/9781003344209-37

of composite foundations and the influence of different replacement ratios during near pit excavation with a centrifugal test. Li et al. (2018) explained the deformation mechanism of excavated composite foundations and its influence on the safety of overlying structures based on centrifuge model tests. Leung et al. (2006) pointed out that the stability of the retaining wall affected the mechanical behavior of the pile behind the retaining wall with a centrifugal model test. Zhang et al. (2013) analyzed the horizontal displacement of piles and the soil of a composite foundation adjacent to an excavating pit by using a finite element model. Goh et al. (2003) used a simplified numerical procedure based on the finite-element method to analyze the pile response, which was in reasonable agreement with the measured results. These researches are mostly limited to the elastic behavior of rigid piles.

Studies have also been carried out on the composite foundation under road embankments. Broms (2004) discussed the possibility of progressive failure on piles using the concrete damage model. Yu et al. (2017) used the concrete damage plasticity model to simulate the stress migration in local damage and cracking of piles. By applying the "CUT-OFF" exit mechanism, Zheng et al. (2010) explained that the failure of the rigid pile-supported embankment was mostly caused by the bending damage of the piles under the embankment shoulder. The failure modes of piles at different locations may be different.

In this study, the ABAQUS software is used to establish a three-dimensional analysis model of a composite foundation adjacent to an excavating pit with a concrete damage plasticity model. The failure mode of the composite foundation with different support structure deformation is studied by modeling the damage and stress migration of rigid piles in the composite foundation adjacent to the excavating pit.

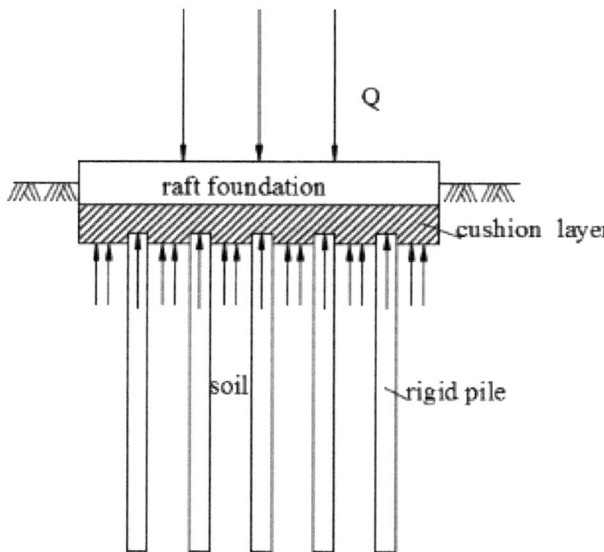

Figure 1. Structure and free-body diagram of rigid pile composite foundation.

The rigid pile composite foundation structure is designed to enable the most bearing capacity of both pile and foundation soil. It is mainly comprised of rigid piles, foundation soil, and a cushion layer (as shown in Figure 1). The load acts directly on a cushion layer which is supported by the rigid piles and foundation soils underneath. As the modulus of the pile is far greater than soil, the pile body will experience smaller settlement and spike upward into the cushion layer, which ensures the coordinated deformation within the system, thus further facilitating soil working accordingly with the rigid pile as a whole (Lv et al. 2012; Yang et al. 2008; Zheng 2007).

2 NUMERICAL MODELING

2.1 *Geometric dimensions*

The size of the model and the placement of piles are shown in Figure 2. Huang (2018) simulated the centrifuge test of the composite foundation by using ABAQUS. The results from the finite element simulation are consistent with the actual centrifuge test result. To reduce the number of units and facilitate the calculation process, a standard unit was selected for analysis through symmetry processing during the modeling process. Piles are labeled as piles 1~6 respectively with the increasing distance away from the pit.

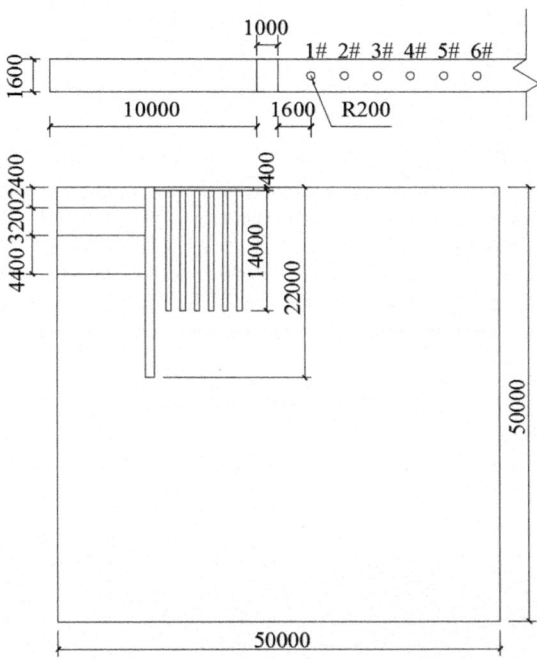

Figure 2.   Plane and elevation of composite foundation (mm).

Table 1.   Material parameters of concrete piles.

E/MPa	ν	$\rho/$ (kg·m^{-3})	Dilation angle	Initial compressive yield stress/MPa	Ultimate compressive yield stress/MPa	Tensile failure stress/MPa
26000	0.2	2400	36	18.1	24.4	2.41

2.2 *Damage plasticity model of pile concrete*

Fang et al. (2007) examined the effectiveness of the concrete damage plasticity model in the analysis of static problems with the help of ABAUQS by comparing the simulation results with the experimental results. The damage in the concrete damage plasticity model consists of two parts: tensile damage and compression damage. Both parts can be expressed as a function of plastic strain, temperature, and field quantity. The damage variable ranges from 0 (representing undamaged) to 1 (complete damage). All the concrete material parameters are listed in Table 1. Based on the maximum tensile strength of the material, the ultimate bending moment of the pile body should be 51 kN·m under pure bending.

2.3 Soil constitutive model and material parameters

The Mohr-Coulomb constitutive model is chosen in the numerical modeling for both the cushion layer and the sand layer. The adopted values for each parameter are listed in Table 2.

Table 2. Physical and mechanical parameters of soil layers.

Material	$\rho/(\mathrm{kg \cdot m^{-3}})$	E/MPa	c/kPa	$\varphi/(°)$	v
Cushion	2100	35	20	30	0.35
Sand	2200	56	0	35	0.30

2.4 Boundary conditions and loading steps

The mesh of the model is shown in Figure 3. The left and right sides of the model are fixed along the X direction, the front and back sides are fixed along the Y axis, and the bottom of the model is fixed in all three directions. The model element adopts the eight-node reduced integral element C3D8R in the space. The contact between pile and soil is set as surface-to-surface contact, the normal contact is set as hard contact which follows the default setting in ABAQUS, and the tangential contact is set to follow the penalty function model with the friction coefficient of 0.4.

The main steps for load applying during the calculation are listed as follows:

1) Balance ground stress;
2) Apply the gravity load of the building to 180 kPa step by step to simulate the construction process of the building;
3) Excavate the foundation pit adjacent to the composite foundation by layers to 10 m.

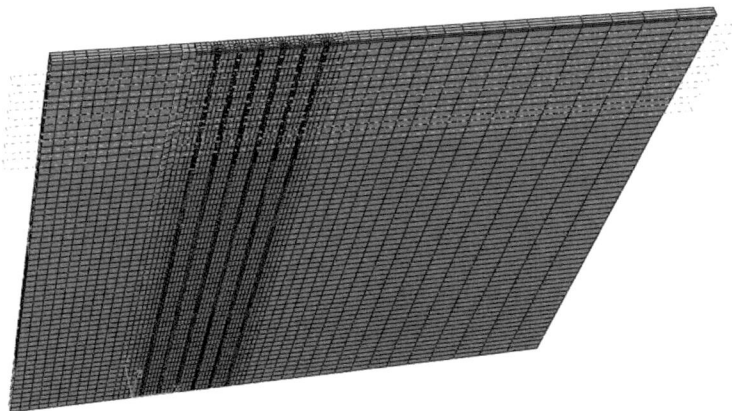

Figure 3. FEM mesh.

3 ANALYSIS OF FAILURE MODE OF COMPOSITE FOUNDATION

3.1 Analysis of failure mode of composite foundation under test conditions

This simulation is based on a centrifuge model test, and the composite foundation is loaded vertically by the air bag, which approximates a flexible foundation. The support structure exhibits cantilever deformation.

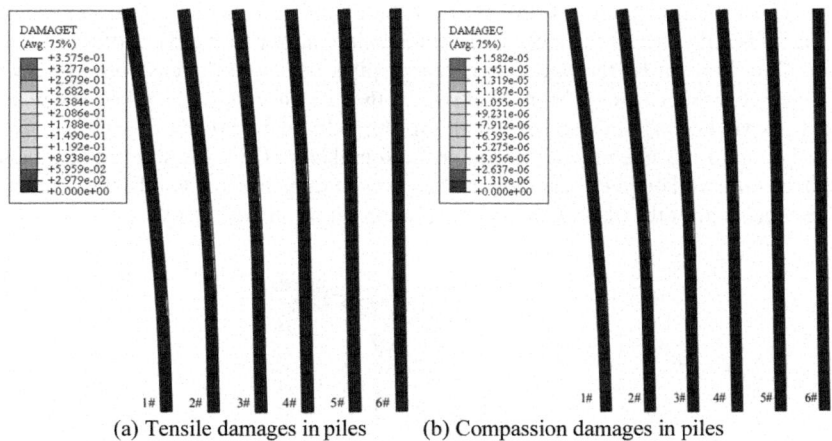

(a) Tensile damages in piles   (b) Compassion damages in piles

Figure 4.   Damages in piles under test conditions.

The distribution of tensile and compression damage of piles are shown in Figure 4. With an excavation depth of 10m, the tensile damage of piles 1 to 4 mostly occurs on the surface of the piles. The area of tensile damage locates higher on piles 3 and 4 than on piles 1 and 2, while no tensile damage can be observed on piles 5 and 6. Among all six piles, pile 3 experiences the most tensile damage which is up to 35.75%. The level of compression damage is still negligible. For composite foundations under the simulated test condition, only surface local damage appears on the rigid piles.

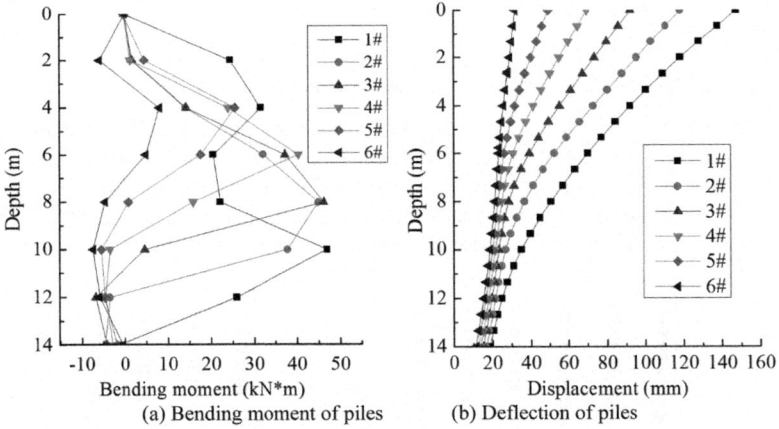

(a) Bending moment of piles   (b) Deflection of piles

Figure 5.   Moment and deflection of piles under test conditions.

Figure 5(a) shows the bending moment profile along piles behind the wall. The location of the maximum bending moment goes up as the pile is located farther away from the retaining wall. The maximum bending moment of the damaged pile is about 50 kN·m, which is close to the bending moment limit of the pile under the pure bending condition.

Figure 5(b) shows the deflection profile along the piles. The maximum deflection occurs at the top of the piles, and the magnitude of deflection decreases with the increase of the excavation depth and the distance to the retaining wall. As the depth of the reinforcement area and the pit are close, the piles are basically overturned. No sharp inflection point can be observed on the displacement curve, which indicates that whole section failure is unlikely to appear under the simulated test condition.

289

Figure 6 gives the distribution of the plastic zone of soil among piles at an excavation depth of 10m. The plasticity area of the soil by this time mainly appears on the top of piles. Due to the unloading effect brought by the lateral excavation, piles penetrate the cushion layer, where the soil has a lower bearing capacity. More load will be then transferred to the piles. The plastic zone continues to expand downwards and eventually forms a slip plane projecting from the excavation surface to the top of the side pile of the composite foundation. Only slight tensile damage on the surface can be observed on piles near the pit when the soil fails. In other words, the bending failure on the piles occurs after the plane failures in the surrounding soil of the pile.

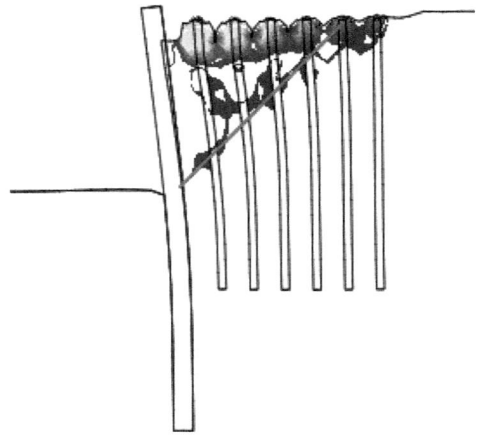

Figure 6.   Plastic zone of soil under test conditions.

3.2 *Analysis of failure mode of composite foundation under Raft Foundation*

As raft foundation is also one of the commonly used types of foundation, it is also added as part of the rigid pile composite foundation based on the centrifuge model. The raft foundation is simulated as a linear elastic material. Figure 7 shows the distribution of tensile and compression damage of piles under the raft foundation by the end of the excavation. Tensile damage is only observed on the upper part of pile 6. There is no tensile or compression damage appears on the other pile.

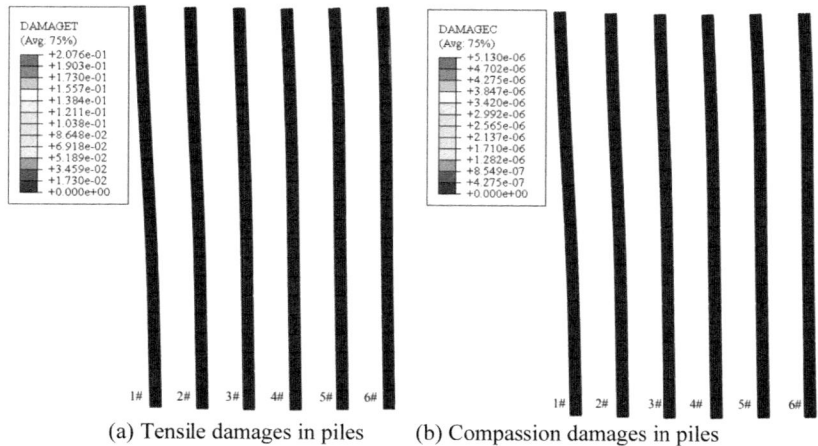

(a) Tensile damages in piles   (b) Compassion damages in piles

Figure 7.   Damages in piles under the raft foundation.

290

The raft foundation has successfully restrained the top of piles by friction between the raft and cushion layer, which further reduces the deflection and a bending moment of piles. The largest displacement of pile 1 is reduced by about 50%. That is why no damage occurs on piles 1 to 5.

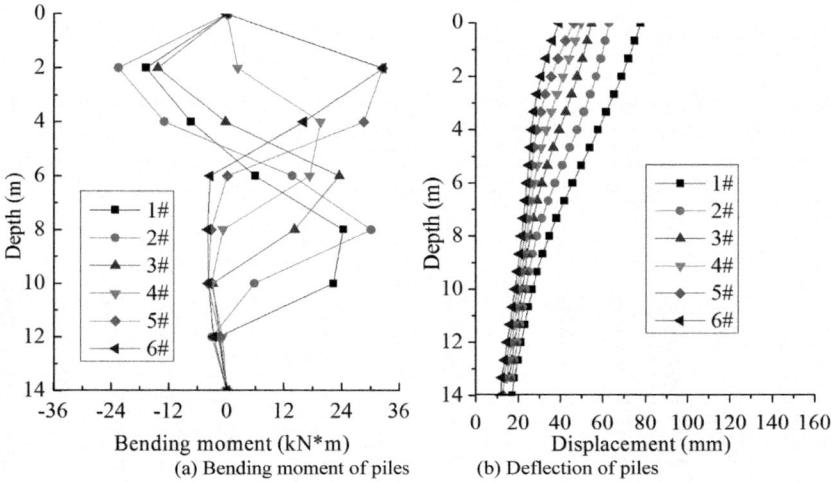

Figure 8.   Bending moment and deflection of piles under the raft foundation.

During the progress of excavation, a negative bending moment occurs in the upper part of the piles which are close to the excavation pit (piles s1 to 3), which partly conforms with the results in Li et al. (2017). Due to the restraint effect of the raft, the lateral displacement of the piles is smaller compared to the case without a raft (as shown in Figure 5(b)). The bending moment on the upper part of the side pile at the far-pit side (piles 4 to 6) keeps increasing until slight tensile damage occurs.

Figure 9 shows the distribution of plastic zone for foundation soils in the rigid pile's foundation with raft by the completion of excavation. Compared to Figure 6, it can be inferred that the raft reduces both the scope of the plastic zone and the degree of plastic development. The plastic zone mainly concentrates on the top of the composite foundation which is close to the excavated pit. There is no plastic zone in the deep soil between piles, which reduces the possibility of any global

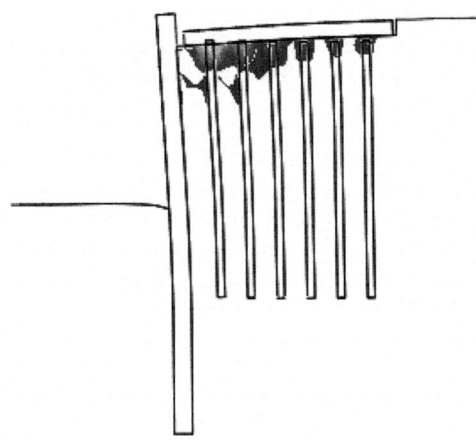

Figure 9.   Plastic zone of soil under the raft foundation.

instability. The soil around piles gradually loses its bearing capacity during the excavation and brings more load to the rigid piles which eventually leads to local compression damage on the piles.

3.3 *Analysis of failure mode of composite foundation with convex deformation in the retaining structure*

The deformation mode of the retaining structure has a great impact on both the deformation and the failure modes of the existing rigid piles and foundation soil. To control the deformation of the existing foundation, internal bracings are often needed during the excavation of a new foundation pit. Zheng et al. (2014) suggested that the deformation of the foundation pit slope with internal bracing is often convex.

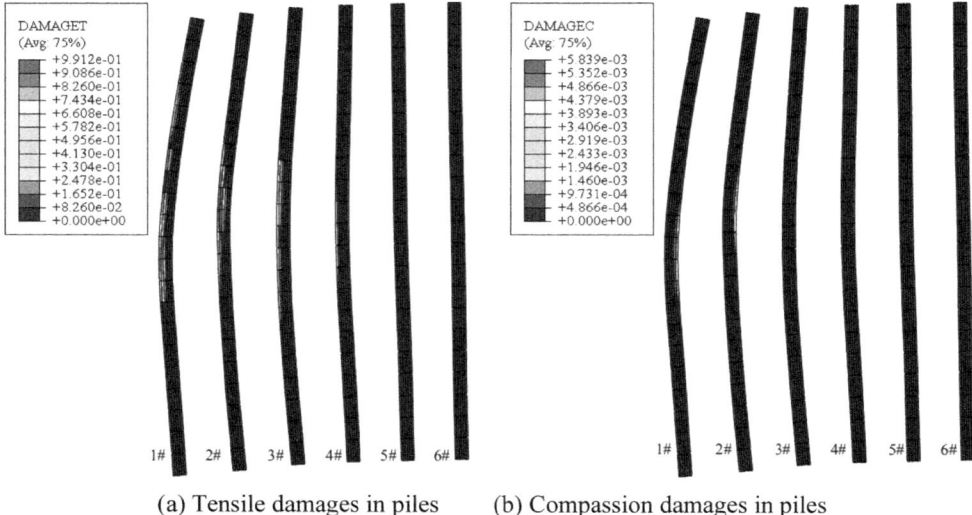

(a) Tensile damages in piles   (b) Compassion damages in piles

Figure 10.   Damages in piles when the wall is of convex type deformation.

Figure 10 shows the distribution of tensile and compression damage of piles when the retaining wall is of convex deformation. The degree of damage on piles was gradually mitigated with the increasing distance away from the excavation pit. The tensile damage which is up to 99% can be found in the middle section of piles 1 and 2. From the middle section, the damage spreads to most of the pile body. Pile 3 also experiences serious tensile damage which is up to about 50%, while little damage can be observed on piles 4 to 6.

Figure 11 shows that negative moment distributes along most piles near the excavated pit. As the deformation mode of the retaining wall is convex, soil near the pit may exhibit larger lateral deformation which will induce more displacement in the middle part of the rigid piles, while the soil at both ends of the rigid piles may have relatively small lateral displacement and brings more tensile and compression damages in the middle part of rigid piles near the foundation pit.

Figure 12 shows the distribution of plastic zone in soil among piles by the end of excavation when the retaining wall for the excavation pit is of convex deformation. Due to the convex deformation of the retaining wall, a large lateral displacement appears in the soil behind the retaining. Differential settlement of rigid piles and foundation soil form a large area of the plastic zone on the top of the rigid pile composite foundation.

The radius of the simulated circular slip surface in the foundation soil is 18.5m. It is noticed that within the failure zone, piles 1 to 4 all experience different levels of tensile damage. The simulation

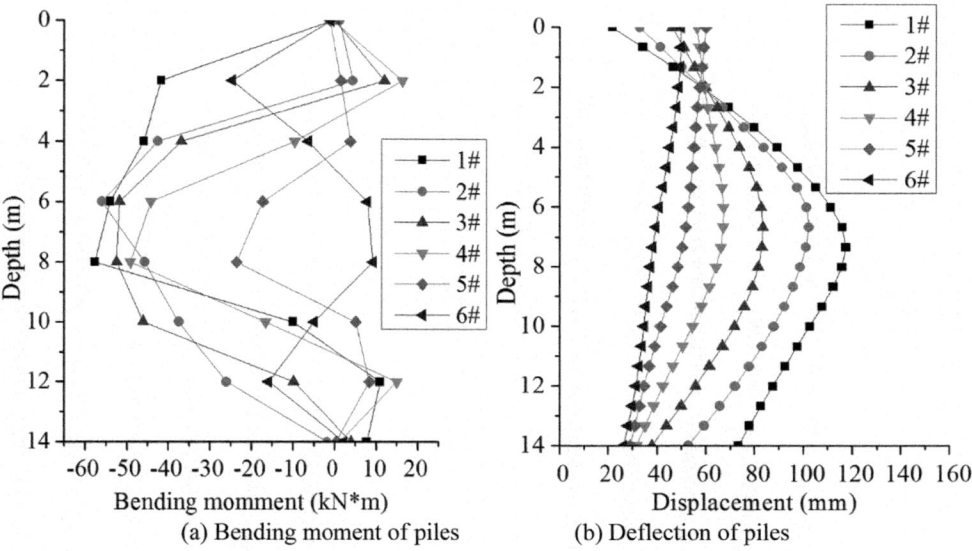

(a) Bending moment of piles   (b) Deflection of piles

Figure 11.   Bending moment and deflection of piles when the wall is of convex type deformation.

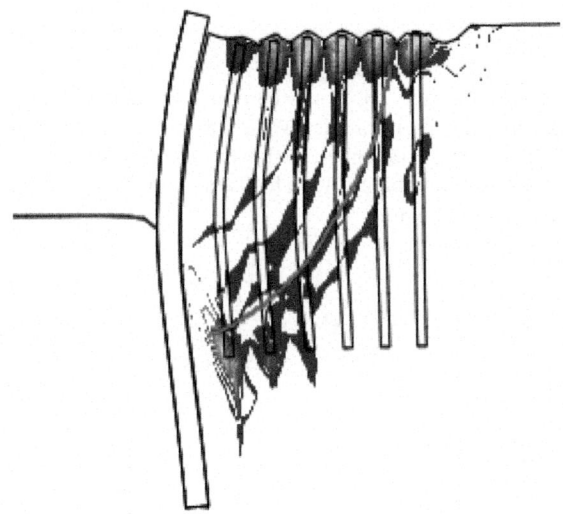

Figure 12.   Plastic zone of soil when the wall is of convex type deformed.

result shows that the excavation of the foundation pit will cause large lateral displacement and local plastic deformation in the foundation soil. Bending failure will appear in the middle part of the piles close to the foundation pit, after which a circular failure plane might develop in the foundation soil around the piles.

3.4   *Analysis of failure mode of composite foundation with combined deformation in retaining structure*

The pile-anchor support structure is also a commonly used retaining structure that is likely to experience combined deformation (Li et al. 2009). Therefore, it is particularly important to discuss the failure mode of rigid piles combined foundation with a combined and deformed retaining structure.

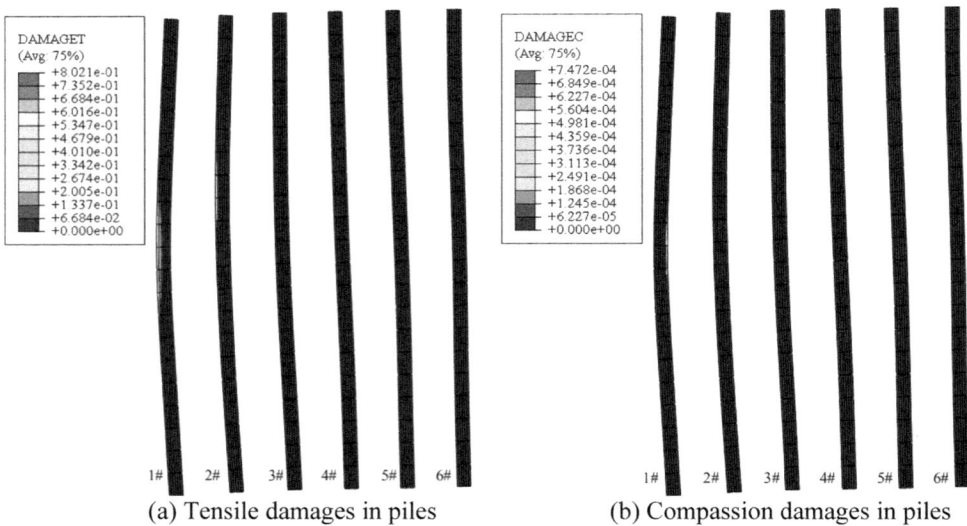

(a) Tensile damages in piles   (b) Compassion damages in piles

Figure 13.   Damages in piles when the retaining wall is of composite type deformation.

By comparing Figure 13 to Figure 10, it can be noticed that the influence of the excavation on both the piles and foundation soil is smaller. While the piles near the foundation pit still bend inward, there is also a large positive moment on the lower part of these piles. The deformation mode of piles gradually changes from "combined" type to "cantilever" type with the increasing distance away from the excavated foundation pit.

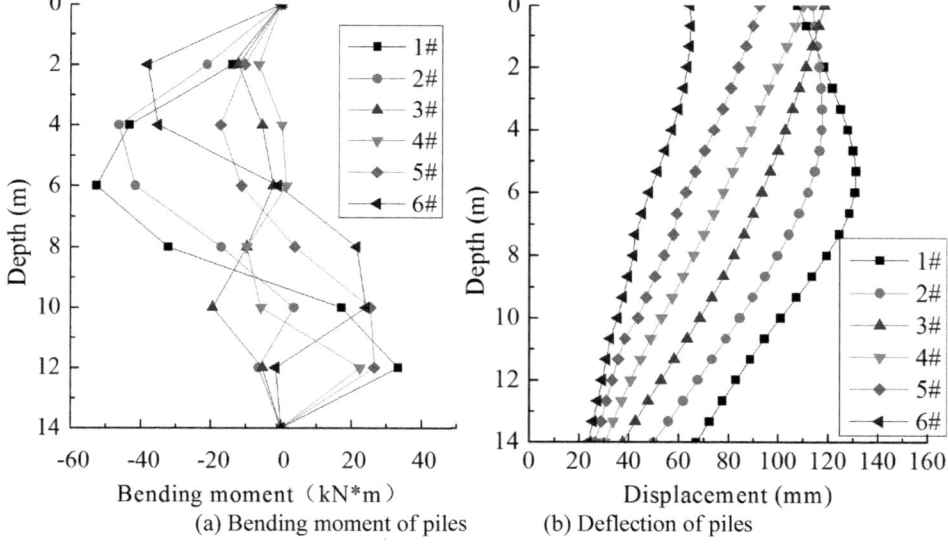

(a) Bending moment of piles   (b) Deflection of piles

Figure 14.   Bending moment and deflection of piles when the retaining wall is of composite type deformation.

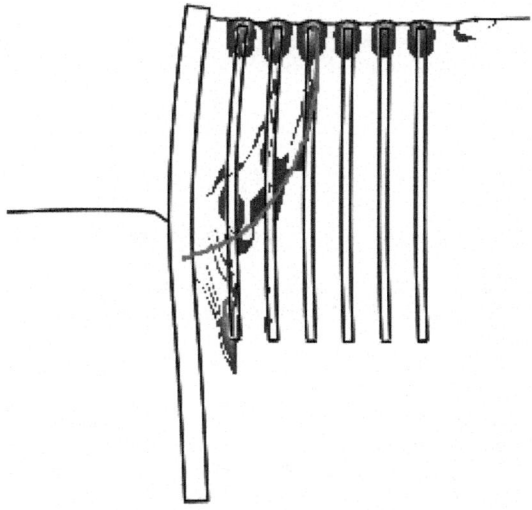

Figure 15.   Plastic zone of soil when the retaining wall is of composite type deformation.

Figure 15 presents the development of a plastic zone in the soil between piles when "combined" deformation on the retaining wall occurs. It can be noticed that compared to the convex deformation case, the plastic zone in the foundation soil is smaller when the retaining structure is under "combined" deformation.

The radius of the simulated circular slip surface is 12m. It can be observed in Figure 16 that within the failure zone, pile 1 and pile 2 both experience some level of damage. The simulated failure mode in the rigid pile combined foundation is generally similar to the case with a convex deformed retaining structure, with a smaller slip plane and less damaged piles.

4   CONCLUSIONS

This paper established a three-dimensional analysis model of an existing rigid pile combined foundation adjacent to a newly excavated pit with the help of the ABAQUS software under the concrete damage plasticity model. The failure modes of the rigid piles combined foundation under four different conditions are investigated by modeling the damage distribution and stress migration. The key findings from the numerical analysis are summarized as follows:

1) During the excavation process of the foundation pit adjacent to a flexible foundation, failure occurs firstly in the surrounding soil of piles, and then on the pile bodies. The local bending failure on the piles occurs only after the slip surface passes through the pile.
2) Raft can effectively reduce the influence of the pit excavation on the rigid piles by restraining the displacement and reducing the bending moment on the piles. The occurrence of global failure is less likely in the rigid pile foundation with a raft, but there is still a risk of local failure on piles.
3) The deformation type of the support structure for excavated pit can also affect the failure mode in the rigid pile combined foundation. Compared to the "convex" deformation, retaining structure with "combined" deformation can alleviate the degree of damage brought by the adjacent excavation to the existing rigid pile combined foundation.

ACKNOWLEDGMENTS

This work was supported by the Natural Science Foundation of Shandong Province (Grant No. ZR2018QEE008).

REFERENCES

Broms B.B. (2004) *Lime and lime/cement columns, ground improvement.* 2nd ed. London: Spon Press, 2004: 252–330.

Fang Q., Huan Y., Zhang Y.D., et al., (2007). "Investigation into static properties of a damaged plasticity model for concrete in ABAQUS." *Journal of PLA University of Science and Technology* (Natural Science Edition), 8(3): 254–260.

Feng S.J., Lu S.F. (2016). "Failure of a retaining structure in a metro station excavation in Nanchang City, China." *J. Perform. Constr. Facil.*, 30(4): 04015097.

Fu J.Y., Yang J.S., Zhu S.T., Shi Y.F. (2017). "Performance of jet-grouted partition walls in mitigating the effects of shield-tunnel construction on adjacent piled structures." *J. Perform. Constr. Facil.*, 31(2): 04016096.

Goh, A.T.C., K.S. Wong, C.I. Teh and Wen, D. (2003). "Pile response adjacent to the braced excavation." *Journal of Geotechnical and Geoenvironmental Engineering*, 129(4): 383–386.

Huang J.J. (2018). *Study on formation mechanism and mechanical properties of existing composite foundation with adjacent excavation.* Jinan: Shandong University. School of Civil Engineering, 2018.

Leung C.E., Ong D.E.L., Chow Y.K. (2006). "Pile behavior due to excavation-induced soil movement in Clay. II: Collapsed wall." *J. Geotech. Geoenviron. Eng.*, 132(1): 45–53.

Li L.X., Huang J.J., Han B., (2018). "Centrifugal investigation of excavation adjacent to existing composite foundation." *J. Perform. Constr. Facil.*, 32(4): 04018044

Li L.X., Huang J.J., Fu Q.H., Cheng X.Y., Hu F. (2017) ."A centrifugal test study on the influence of additional loads on mechanical properties of composite foundations with different displacement rates". *Rock and Soil Mechanics*, 38(S1), 131–139.

Li S. (2012). "Research on failure mechanisms and stability analysis method of embankment supported on composite ground reinforced with rigid piles". Tianjin: Tianjin University. *School of Civil Engineering*, 2012.

Ong D.E.L., Leung C.E., Chow Y.K. (2006). "Pile behavior due to excavation-induced soil movement in Clay. I: STable wall" *J. Geotech. Geoenviron. Eng.*, 132(1): 36–44.

Xing H.F., Xiong F., Jiemei Wu J.M. (2016). "Effects of pit excavation on an existing subway station and preventive measures." *J. Perform. Constr. Facil.*, 30(6): 04016063

Yang M., Zhou H.B., Yang H. (2005). "Analysis of interaction between foundation pit excavation and adjacent planting". *Journal of Civil Engineering*, 38 (4), 91–96.

Yu J.L., Wang C.W., Xie Y.M., Zhang J.L., Gong X.N. (2017). "Analysis of stress and failure mechanism on composite foundation improved by rigid piles under flexible foundation considering the damage of piles." *Journal of Central South University* (Science and Technology) DOI: 10.11817/j.issn.1672-7207.2017.09.023

Yue Choong Kog (2017). "Excavation-induced settlement and tilt of a 3-story building" *J. Perform. Constr. Facil.*, 31(1): 04016080.

Zhang J.Y., Qiao J.S., Liang L.J. (2013). "Impact of the soil lateral displacement on the performance of the adjacent composite ground." *Soil Eng. and Foundation*, 27(3), 89–92.

Zheng G., Deng X., Liu C., Liu Q.C. (2014). "Comparative analysis of the influence of deformation modes of different retaining structures on the displacement field of deep soil outside the pit." *Chinese Journal of Geotechnical Engineering*, 36(02): 273–285.

Zheng G., Liu L., Han J. (2010). "Stability of embankment on soft subgrade reinforced by rigid piles(1): Background and single pile analysis". *Chinese Journal of Geotechnical Engineering*, 32(11), 1648–1657.

*Frontiers in Civil and Hydraulic Engineering – Mohamed A. Ismail and
Hazem Samih Mohamed (Eds)*
© 2023 The Authors, ISBN 978-1-032-38247-0

Credibility of estimating the hysteretic energy demand of moment-resisting steel frames

Cuiling Ma*

College of Urban Construction and Transportation, Hefei University, China

ABSTRACT: Seismic design is of high importance. The energy-based seismic design method, which features a clear conception, is beneficial to the comprehensive evaluation of structural seismic performance. However, how to determine the energy demand is fundamental. In this paper, moment-resisting steel frames were designed as calculation examples in accordance with the current specifications and standards in China. Then, such structures were subjected to the elastic-plastic time history analysis. The analysis results revealed that the hysteretic energy (HE) demand obtained via the time history analysis was smaller than the estimated value. Under different seismic waves, the inter-layer hysteretic energy demand of different structures presented identical distribution laws, and the proportion of the hysteretic energy in each component in the total hysteretic energy was also relatively stable. In addition, the energy dissipated on the intermediate floor of the moment-resisting frames was large. Under rare earthquakes, all components designed by the small earthquake elasticity design methods entered the plastic stage, with an uncontrollable yield mode. Therefore, it is rather essential to investigate energy-based design methods.

1 INTRODUCTION

Earthquake is one of the natural disasters faced by human beings. Earthquakes, which occur frequently in China, have resulted in casualties and property losses to different degrees, making it important to perform the seismic design of buildings.

As one of the behavior-based structural aseismic design methods, the energy-based method simultaneously considers two factors of force and displacement, as well as the structural cumulative damage caused by ground motions, with a clear conception. However, this method stays at a developmental stage. How to calculate the ground motion energy demand of structures remains a primary problem. Hence, numerous scholars have carried out extensive research (Akiyama 1985; Chou & Uang 2000; Chou & Uang 2003; Danny & Mario 2007; Habibi et al. 2013; Housner 1956; Khashaee 2004; López-Almansa et al. 2013; Ma et al. 2022; Sun et al.2012; Sun et al.2017; Xiao 2004). As a critical link in the energy behavior-based design method, the energy demand estimation of MDOF system structures is of importance. In this research, the moment-resisting steel frames were designed as the calculation examples in accordance with the current specifications and standards in China, which aims to evaluate the reasonability of estimating the hysteretic energy consumption demands for such structures according to the equivalent velocity spectrum of SDOF elastic-plastic hysteretic energy consumption. It conforms to China's site classification standards in literature (Ma et al. 2019).

---

*Corresponding Author: 415997243@qq.com

DOI 10.1201/9781003344209-38

2 DESIGN OVERVIEW

The calculation examples were moment-resisting steel frames, with the seismic design parameters as follows. The seismic fortification intensity was 8 (0.3 g), and the site class was II. The designed earthquake belonged to the second group (Class C), and the designed service life of the building structure was 50 years. In the calculation examples, the building parameters were described below. Bidirectional spans were both 7.8 m, and each storey was 3.3 m in height (10 and 15 storeys in total, respectively). The spacing of secondary beams was 2.6 m. The height and thickness of parapet walls were 1.1 and 0.2 m, respectively. The exterior walls were 0.2 m in thickness.

The constant (active) loads of the floor and roof were 4. 5 (2.0) kN/m^2 and 5.0(2.0) kN/m^2, respectively. Wind loads were not considered, and Q235B steel was used. Welded I-section components were applied to beams and box-section components to steel columns, with joints subjected to rigid connection. Therein, the cross-sectional flange of the welded I-section components was a sheared edge. Moreover, the influences of the structural torsional effect and floor slabs on the structure were neglected. The plane form and elevation form of the calculated structure are respectively displayed in Figure 1.

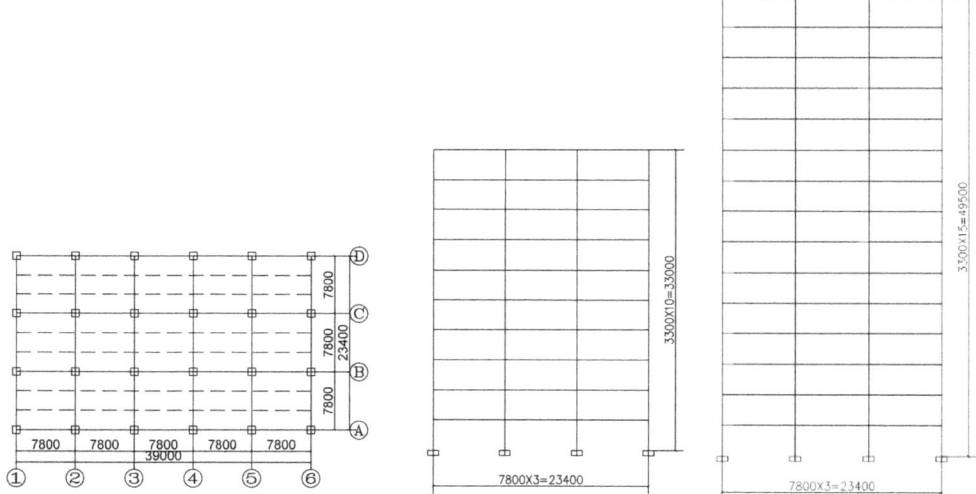

Figure 1.　Plan and elevation of moment-resisting steel frame.

3 DESIGN STANDARDS

The calculation examples were designed in accordance with the Standard for Design of Steel Structures (GB 50017-2017), Code for Seismic Design of Buildings (GB 50011-2010), Technical Specification for Steel Structure of Tall Building (JGJ 99-2015), and Load Code for the Design of Building Structures (GB 50009-2012). Among the calculation examples, the 10-storey structure possessed an anti-seismic grade III, while the anti-seismic grade II was conservatively considered for the 15-storey structure. The damping ratio of calculation examples was taken from Code for Seismic Design of Buildings during the elastic design under frequent earthquakes and taken as 0.05 during the energy analysis in case of rare earthquakes. The characteristic period T$_g$ was 0.40 s.

According to Technical Specification for Steel Structure of Tall Building, the wind loads and the vertical earthquake action are not considered. In this research, the beam and column sections of each calculation model were finally determined (Tables 1 and 2) through repeated modifications and trial calculations.

Table 1. Component sections of the 10-storey moment-resisting steel frame structure.

Floor	Column section	Beam section
1–3	600*20	H600*250*14*20
4–6	500*18	H600*250*14*20
7–9	400*16	H600*250*14*20
10	400*16	H600*250*14*20

Table 2. Component sections of the 15-storey moment-resisting steel frame structure.

Floor	Column section	Beam section
1–3	800*24	H600*300*14*24
4–6	700*20	H600*300*14*24
7–10	600*18	H600*250*12*20
11–14	450*16	H600*250*12*20
15	450*16	H600*250*12*20

4 ESTIMATED VALUE OF STRUCTURAL HYSTERETIC ENERGY CONSUMPTION DEMAND

The cumulative hysteretic energy demands of the MDOF system could be solved (Table 3) according to the mathematical expression for the equivalent velocity spectrum of SDOF elastic-plastic hysteretic energy consumption in literature (Ma et al. 2019). The conversion formula between the hysteretic energy consumption of the SDOF system and that of the MDOF system in literature (Sun et al. 2012) is considered as well. To obtain the relatively ideal structural cumulative hysteretic energy consumption demands, the sum of the mass participation coefficients $X_{mass,j}$ of participatory vibration modes are generally needed to be greater than 90%. For the structure influenced greatly by high-order vibration modes, the number of vibration modes should be appropriately increased.

Table 3. Hysteretic energy consumption demands of structures.

Calculation example	Cumulative hysteretic energy consumption of ESDOF $E_{h(ESDOF),j}/kN·m$			Hysteretic energy consumption of MDOF structure $E_{h(MDOF)}/kN·m$
	$E_{h(ESDOF),1}$	$E_{h(ESDOF),2}$	$E_{h(ESDOF),3}$	
10-storey moment-resisting frame	484.004	1107.132	1246.313	1300.537
15-storey moment-resisting frame	531.310	1300.236	1572.900	1638.285

5 TIME HISTORY ANALYSIS

In this research, the frame structures were subjected to the elastic-plastic time history analysis via finite element software ABAQUS. The beams and columns in both structures were simulated using B31 beam elements. A bilinear kinematic hardening model was selected, and the damping ratio of both structures was taken as 0.05 under rare earthquakes. The mesh size of the beam elements was determined as 0.2 m. A total of 10 eligible seismic waves were selected for each calculation example.

5.1 Comparison between analysis value and the estimated value of HE

In this section, the two designed calculation examples were subjected to the nonlinear time history analysis via ABAQUS, aiming to reasonably evaluate the credibility of estimating their hysteretic energy consumption demands. Next, the hysteretic energy consumptions of the two structures under the action of each seismic wave were extracted and averaged as the analysis value and were then compared with the estimated value (Table 3) obtained. They were specifically as shown in Figures 2–3.

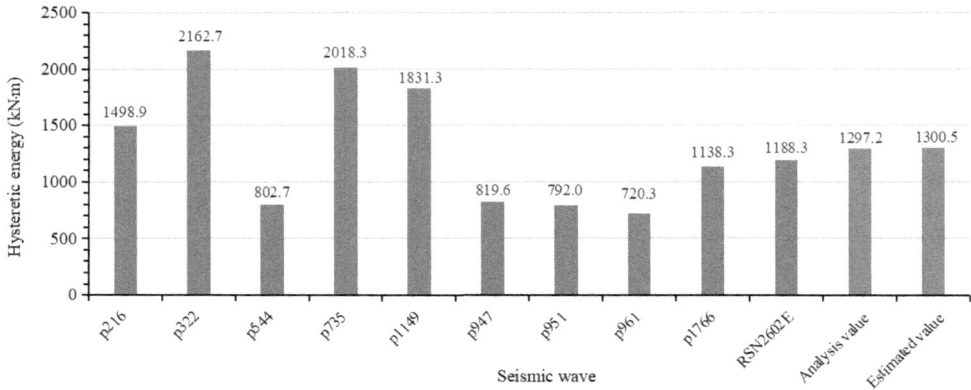

Figure 2. Comparison between analysis value and the estimated value of HE of 10-storey moment-resisting steel frame structures under different seismic waves.

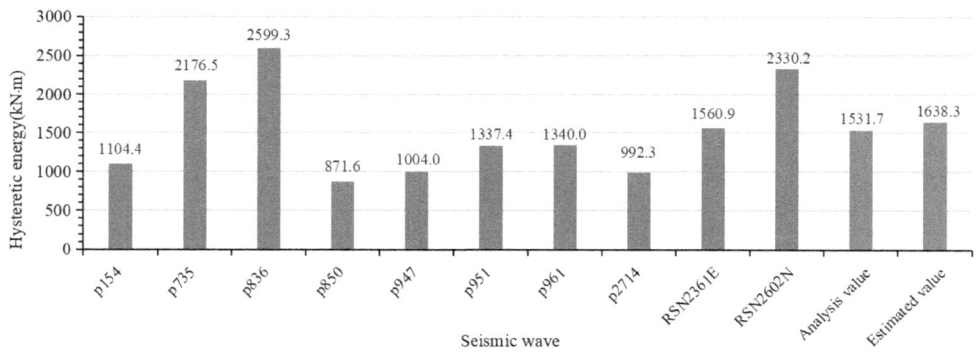

Figure 3. Comparison between analysis value and the estimated value of HE of 15-storey moment-resisting steel frame structures under different seismic waves.

It could be known from Figures 2 and 3 that the hysteretic energy of the same calculation example varied with seismic waves due to their discreteness and randomness though screened out according to a certain principle. For the same calculation example, the average hysteretic energy under multiple seismic waves basically coincided with the estimated value, and the analysis value was always slightly smaller than the estimated value, which was set with a certain safety margin.

5.2 Distribution laws of hysteretic energy consumption between components

The reliability of the equivalent velocity spectrum of hysteretic energy in this research could only be judged on the whole by comparing the analysis value and estimated value of each calculation example. Moreover, the proportion of the hysteretic energy consumption of differently structured components was varied, which resulted in the differences in their damage degree. Therefore, the distribution laws of hysteretic energy between different components remain to be analyzed.

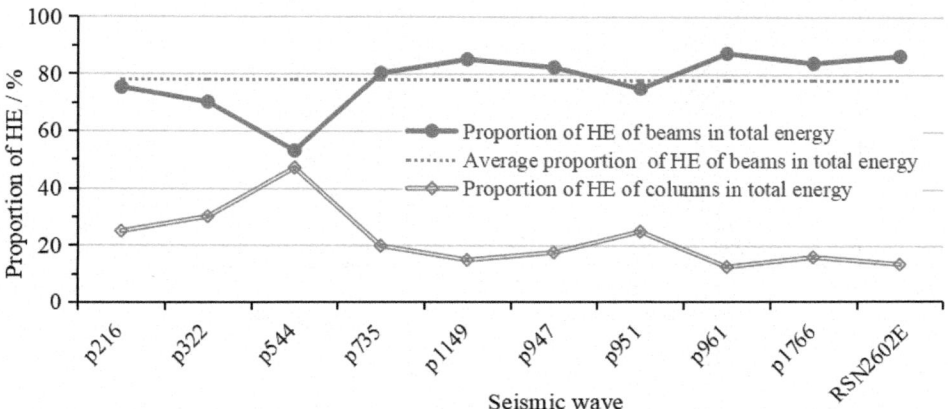

Figure 4. Hysteretic energy consumption distribution in 10-storey moment-resisting steel frame components under different seismic waves.

As observed in Figure 4, the components of the moment-resisting steel frames showed similar distribution laws of hysteretic energy under different seismic waves. All components were prone to plastic deformations and damaged to different degrees, among which the beams shared the most hysteretic energy (the proportion was about 78.1% for the 10-storey structure), while the side span columns consumed less energy.

In conclusion, the proportion of the hysteretic energy by different components in each calculation example did not change with the seismic wave input. Under rare earthquakes, all components were damaged to different degrees, and the post-yield state could not be controlled. Therefore, the methods of the strength-based small earthquake elasticity design, which were specified in the existing standards, should be further improved.

5.3 Interlayer distribution mode of hysteretic energy consumption

Whether there is any evident energy concentration phenomenon can be revealed by the distribution laws of hysteretic energy consumption along the structure height, thus providing a reference for comprehensively evaluating the structural seismic performance.

It could be known from Figure 5 that the hysteretic energy consumption of the same structure was varied and might be different several times, but the interlayer hysteretic energy consumption presented largely identical distribution laws under different seismic waves. For moment-resisting steel frames, the energy dissipated on floors within their height range of $\frac{H}{3} - \frac{2H}{3}$ was large. The cumulative hysteretic energy consumption on the 6th, 7th, 8th, 9th, and 10th floors of the 15-storey moment-resisting steel frame accounted for 54.7% of the total accumulative hysteretic energy consumption in this structure. The hysteretic energy consumption of different calculation examples

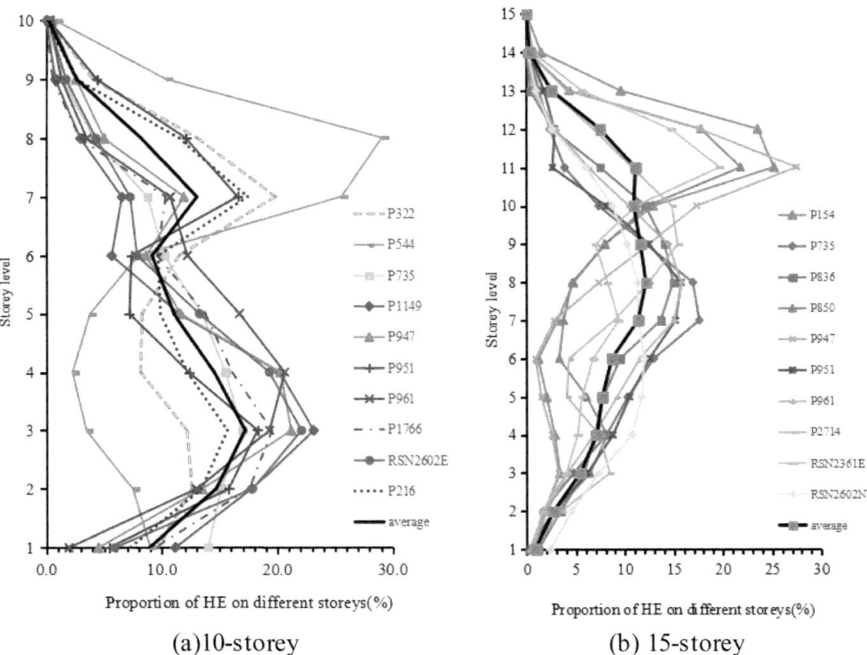

(a)10-storey   (b) 15-storey

Figure 5. The proportion of interlayer HE of moment-resisting steel frame structures in total HE under different seismic waves.

did not follow the single linear distribution along the direction of the structural height. Instead, the hysteretic energy consumption on adjacent floors with constant sectional dimensions of components in the same calculation example was nearly linearly distributed along its height direction. Variable sectional dimensions usually experienced sudden changes.

6 CONCLUSION

In this research, the credibility of the HE spectrum in estimating the energy demand of the moment-resisting steel frames was mainly evaluated. The research results reflected that nearly all the cumulative hysteretic energy consumption demands obtained through the time history analysis were smaller than the estimated value in this research. Under different seismic waves, the interlayer hysteretic energy of each structure presented largely identical distribution laws, and the proportion of the hysteretic energy consumption of each component in the total hysteretic energy consumption was also relatively stable. Besides, the energy dissipated on the intermediate floors in the moment-resisting steel frame was large. Under rare earthquakes, all components designed based on the small earthquake elasticity design method entered the plastic stage, and the yield mode was uncontrollable. Therefore, it is quite essential to investigate energy-based design methods.

ACKNOWLEDGEMENT

This research is funded by the 2020 Excellent and Top-notch University Talent Cultivation Program (No. gxyq2020068) and the 2021 Scientific Research Program for Talents in Hefei University (No. 21-22RC39).

REFERENCES

Akiyama H. (1985). *Earthquake-resistant limit state design for buildings*, University of Tokyo Press, Tokyo, Japan.

Chou C.C., Uang C.M. (2000). *An evaluation of seismic energy demand: An attenuation approach*[R]. PEER Report 2000/04 Pacific Earthquake Engineering Research Center, College of Engineering, University of California, Berkeley.

Chou C.C., Uang C.M. (2003). A procedure for evaluation of seismic energy demand of framed structures[J]. *Earthquake Engineering and Structural Dynamics*, (32): 229–244.

Danny A., Mario O. (2007). On the estimation of hysteretic energy demands for SDOF systems[J]. *Earthquake Engineering and Structural Dynamics*, 36: 2365–2382.

Habibi A., Chan R.W.K., Albermani F. (2013). Energy-based design method for seismic retrofitting with passive energy dissipation systems[J], *Engineering Structures*, 46(1):77–86.

Housner G.W. (1956). *Limit design of structures to resist earthquakes, Proceedings of the 1st World Conference Earthquake Engineering*, Berkeley, California, USA.

Khashaee P. (2004). *Energy-based seismic design and damage assessment for structures[D]. School of Engineering*, Southern Methodist University.

López-Almansa F., Yazgan A., Benavent-Climent, A. (2013). Design energy input spectra for high seismicity regions based on Turkish registers[J], *Bulletin of Earthquake Engineering*, 11(4): 885–912.

Ma C.L., Gu Q., Sun G.H. (2019). The mathematical expression of design hysteretic energy spectra based on Chinese soil type [J]. *Mathematical Problems in Engineering*, (11): 1–10. https://doi.org/10.1155/2019/3483516.

Ma H.W., Yang J.Y., Pan C.Q. (2022). Energy-based seismic design method for energy-dissipation steel frame [J]. *Building Structure*, 52(5): 55–62.

Sun G.H., Gu Q., He R.Q., Fang Y.Z. (2012). Evaluation of seismic energy of steel frames based on energy spectrums [J]. *China Civil Engineering Journal*, 45(5): 41–48.

Sun G.H., Gu Q., He R.Q., Fang Y.Z. (2017). A simplified normalized cumulative hysteretic energy spectrum[J]. *Earthquakes and Structures*, 12(2): 177–189.

Xiao M.K. (2004). *Research on performance-based analysis methods for the displacement and energy response of aseismic structures* [D]. Ph.D. dissertation of Chongqing University.

Frontiers in Civil and Hydraulic Engineering – Mohamed A. Ismail and
Hazem Samih Mohamed (Eds)
© 2023 The Authors, ISBN 978-1-032-38247-0

Numerical simulation of hydraulic characteristics of cylindrical roughness element on riverbed surface

Dingguo Jiang
China Three Gorges Corporation, Beijing, China

Haihua Wang*
Ningbo Hong Tai Water Conservancy Information Technology Co., Ltd, Zhejiang, China

Ji Yang
College of Water Conservancy and Hydropower Engineering, Hohai University Nanjing, China

Yiqing Gong
Institute of Water Science and Technology, Hohai University, Nanjing, China

Yanhong Chen
College of Water Conservancy and Hydropower Engineering, Hohai University Nanjing, China

ABSTRACT: In order to explore the hydrodynamic characteristics of roughness elements in a natural riverway and probe their influences on the habitat of aquatic organisms, a hydrodynamic mathematical model was constructed with a cylindrical roughness element as an example to carry out a three-dimensional (3D) numerical flow model research. Moreover, the reliability of this model was verified through a physical model experiment. Research results showed that when water flowed through the cylindrical roughness element, a low-velocity region was formed in front of this element due to the water-resistance of obstacles. This region developed towards two sides at the back of this cylindrical roughness element, thus forming a local high-velocity region. Finally, a low-velocity wake flow region was formed in the rear, presenting evident Karmen vortex street characteristics. The relevant research results will provide a reference for governing the ecological environment of riverways and improving the habitat of aquatic organisms.

1 INTRODUCTION

Large-scale riverbed roughness elements in riverways like block stones and gravels provide fish with upstream channels and rest areas by lowering the water flow velocity and enhancing the energy dissipation (Dm 2021). Meanwhile, the water-resisting effect of these gravels changes the water flow velocity of surrounding water bodies and affects the sediment transport process and the habitat of aquatic organisms. Most significantly, the interference of wake flow in the downstream wake flow region of obstacles exerts marked effects on the local average flow velocity field and fluctuating velocity field. Hence, it is necessary to deeply explore the mechanism relationship between the internal roughness elements of riverways and surrounding water flow and clarify the hydrodynamic characteristics of roughness elements.

At present, the water flow structures around large-scale roughness elements have been investigated both in China and abroad. For instance, this research (Moss 1980) explored the sediment transport rate considering the shade of large-scale particles. Some scholars (Lacey 2008) observed the prototype of the pebbled riverbed in a mountainous area, pointed out the areas where the water flow structure was intensely affected by large-scale roughness elements and analyzed the variation laws of the water flow structure around the cube beneath a rough riverbed with the relative water

---

*Corresponding Author: whh_hhu@163.com

DOI 10.1201/9781003344209-39

depth. The research (Zhong 2019) generalized roughness elements into three morphologies—cubes, spheres and tetrahedrons—to perform a flume experiment and acquired the relationship between the shape of large-scale roughness elements and the surrounding three-dimensional (3D) water flow structure. These researchers (Cao 2012) studied the flow field characteristics around large-scale particles by combining physical testing and numerical simulation and acquired the cross-sectional flow velocity distribution and the evolution characteristics of the vortex momentum. Meanwhile, the Reynolds stress nearby particles showed a 3D distribution state. The above results manifest that the flow-around pattern and structural characteristics can be more accurately reflected through 3D analyses. In addition, the combination of numerical simulation and physical model experiments is of very good applicability and reliability for solving around-flow and hydrodynamic problems.

However, the local water flow mechanism has been recognized in a limited way due to the 3D complexity of water flow around roughness elements in riverways. Most relevant research has been conducted from a macroscopic angle, while large-scale rough micro-flow characteristics have been less involved. Hence, the influence of a cylindrical roughness element on the surrounding water flow structure was investigated from the angle of single particles. With a physical model experiment as the reference basis, a 3D mathematical model was constructed to numerically simulate the water flow near the cylindrical roughness element, emphasizing changes in key parameters like riverbed shear force and flow velocity. Then, the hydrodynamic characteristics of the water body nearby this cylindrical roughness element were analyzed. Next, the relationship between the change in the fluid state and the riverway structure was preliminarily explored, thus providing technical support for the ecological development and protection of rivers and the engineering design of river regulation.

2 METHODOLOGY

2.1 Physical model experiment

This physical model mainly took the experimental flow around the cylindrical model on a fixed riverbed constructed by Roulund (Roulund 2005) as the reference. The cylinder with a smooth surface was 0.536 m in diameter (D), the water depth (h) was 0.54 m, the thickness of the boundary layer was δ, the average inflow velocity was 0.326 m/s, the water temperature was 20°C, and the corresponding coefficient of dynamic viscosity μ was 0.001 Pa · s (see detailed parameter settings in Table 1).

Table 1. Model parameters.

Cylinder diameter	Water depth	Average flow velocity	Roughness height of riverbed surface	Reynolds number	Friction velocity
D/m	h/m	$U/\,m{\cdot}s^{-1}$	k_s/cm	Re	$U_f/m{\cdot}s^{-1}$
0.536	0.54	0.326	0	1.75×10^5	0.013

The flow velocity was measured in the experiment using a rectangular flume and a point-type velocity meter. Then, the longitudinal and vertical flow velocity distributions of the fluid around the cylinder at 0.005, 0.01, 0.02, 0.05 and 0.1 m from the bed bottom were obtained.

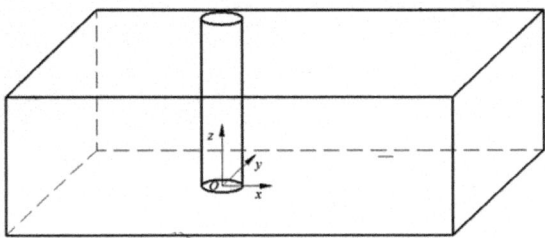

Figure 1. Schematic diagram of flume experiment of cylindrical roughness element.

2.2 Numerical simulation

2.2.1 Turbulence model

To improve the adaptability of calculation models, they put forward the *RNG k − ε* model based on the *Standard k − ε* model in 1986 (Yakhot 1986). This model shows favorable performance in considering the turbulent vortex and correcting the turbulent viscosity. Moreover, it can clearly display the action mechanism of small-scale vortices in turbulence. Hence, the *RNG k − ε* model was used to describe turbulence characteristics, with its governing equation as below:

$$\frac{\partial(\rho k)}{\partial t} + \frac{\partial(\rho k u_i)}{\partial x_i} = \frac{\partial}{\partial x_j}\left[\left(\mu + \frac{\mu_t}{\sigma_k}\right)\frac{\partial k}{\partial x_j}\right] + P_k - \rho\varepsilon \tag{1}$$

$$\frac{\partial(\rho\varepsilon)}{\partial t} + \frac{\partial(\rho\varepsilon u_i)}{\partial x_i} = \frac{\partial}{\partial x_j}\left[\left(\mu + \frac{\mu_t}{\sigma_\varepsilon}\right)\frac{\partial\varepsilon}{\partial x_j}\right] + C_{1\varepsilon}\frac{\varepsilon}{k}P_k - C_{2\varepsilon}\rho\varepsilon\frac{\varepsilon}{k} \tag{2}$$

Where:

$$\mu_t = \rho C_\mu \frac{k^2}{\varepsilon} \tag{3}$$

$$C_{2\varepsilon}^* = C_{2\varepsilon} + \frac{C_\mu \eta^2 \left(1 - \eta/\eta_0\right)}{1 + \beta\eta^2} \tag{4}$$

$$\eta = sk/\varepsilon \tag{5}$$

$$s = \left(2s_{ij}\right)^{1/2} \tag{6}$$

Where k, ε, μ, μ_t and C_μ stand for the turbulent kinetic energy, the energy dissipation efficiency, the kinematic viscosity coefficient, and the viscosity coefficient ($C_\mu = 0.0845$), respectively. σ_k and σ_ε represent the turbulent Prandtl numbers corresponding to the turbulent kinetic energy and dissipation rate, respectively, and $\sigma_k = 0.7194$ and $\sigma_\varepsilon = 0.7194$. The model constants include $C_{1\varepsilon} = 1.42$, $C_{2\varepsilon} = 1.68$, $\eta_0 = 4.38$ and $\beta_0 = 0.012$.

2.2.2 Calculation model and mesh generation

A cylindrical roughness element model was taken as the research object to mainly simulate the flow-around phenomenon on the surface of a smooth riverbed, while attention was paid to its flow field distribution characteristics. The computational domain (model diameter: D, water depth: h) was set as 60D (length), 10D (width) and 1.5h (height), and the model was placed at 40D of the inflow boundary. The computational domain and mesh generation are displayed in Figures 2 and 3, respectively. Given the presence of horseshoe vortices on the free surface and the bottom boundary, local meshes were densified to improve computational accuracy.

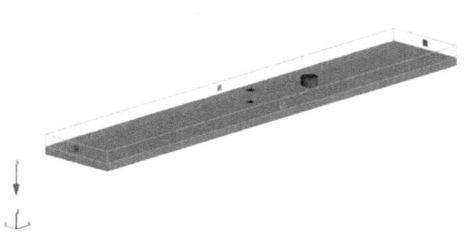

Figure 2.   Schematic diagram of computational domain.

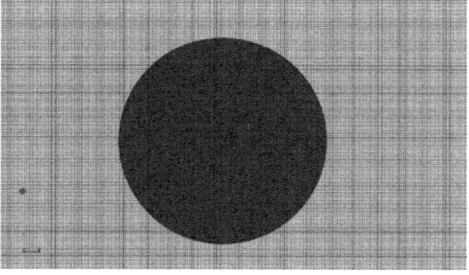

Figure 3.   Schematic diagram of mesh generation.

2.2.3 Boundary conditions

During the numerical calculation, the left side and right side of the computational domain were set as the inflow and free outflow boundaries, respectively. The riverbed surface and the model surface were set to be wall boundaries, which were regarded as non-slip boundaries with zero velocity. In the meanwhile, the bilateral boundaries were simplified into symmetrical boundaries. Then, the turbulence state of the river in the wall boundary region was simulated through the wall function method. For the inflow boundary, its flow velocity was determined according to the principle of logarithmic flow velocity profile, and the flow velocity distributions on the surface of the smooth riverbed and rough riverbed are expressed by the following formulas, respectively:

$$\frac{u}{u_*} = \frac{1}{\kappa} \ln\left(\frac{u_* y}{v}\right) + 5.5 \tag{7}$$

$$\frac{u}{u_*} = \frac{1}{\kappa} \ln\left(\frac{y}{k_s}\right) + 8.5 \tag{8}$$

Where κ represents the Karman constant, set as 0.4. k_s, v, u, u_* and y stands for the riverbed roughness, the kinematic viscosity coefficient, the flow velocity at different water depths, the friction velocity, and the water depth, respectively. The initial gas volume fraction of the simulated water body was taken as 0, and the volume fraction of the fluid was set to 1. The water flow velocity was set to be distributed on a logarithmic velocity profile, and the gas flow velocity was taken as 0.

3 RESULT ANALYSIS

3.1 Inlet velocity analysis

The simulated flow velocity gradually tended to be stable at 20 s. The vertical inlet velocity distributions obtained by the numerical simulation and physical experiment were compared, as shown in Figure 4. It could be observed that with the increase in the water level, the flow velocity was evidently accelerated. When the water depth ranged from 0 to 0.3, the model-calculated result coincided favorably with the physical model experimental result. When $Z > 0.3$, the simulated value was slightly greater than the experimental value, with the error falling within 10%, which was acceptable. In general, the data acquired through the numerical simulation was identical to the physical model experimental result.

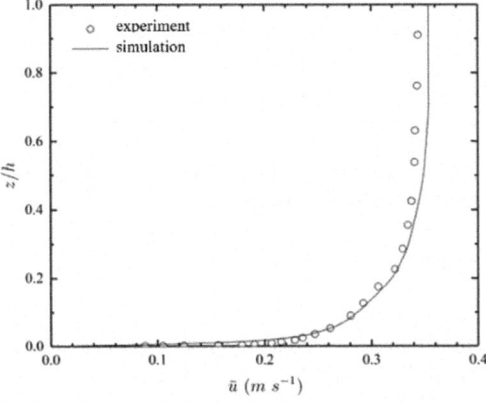

Figure 4.   Comparison of vertical inlet velocity distributions.

3.2 Flow velocity analysis

The flow velocity distributions at different water depths are exhibited in Figure 5. In two regions—incident flow side and dorsal flow side of the cylindrical roughness element, the simulated data and physical model experimental data could be comparatively analyzed. The figure showed that at the incident flow side of the cylinder, the numerically simulated values of horizontal flow velocity and distribution trend were basically consistent with the results obtained through the physical model experiment. However, slight differences were observed at the dorsal flow site, which was mainly attributed to the continuous time-dependent change in the vortex generated behind the cylinder and the horizontal flow velocity selected in the numerical simulation at a moment different from that selected in the physical model experiment. At the same time, the vertical flow velocity distributions at the incident flow side and dorsal flow side of the cylinder at different water depths were compared, as shown in Figure 5. The comparison results reflected that the two values were basically identical at the same water depth.

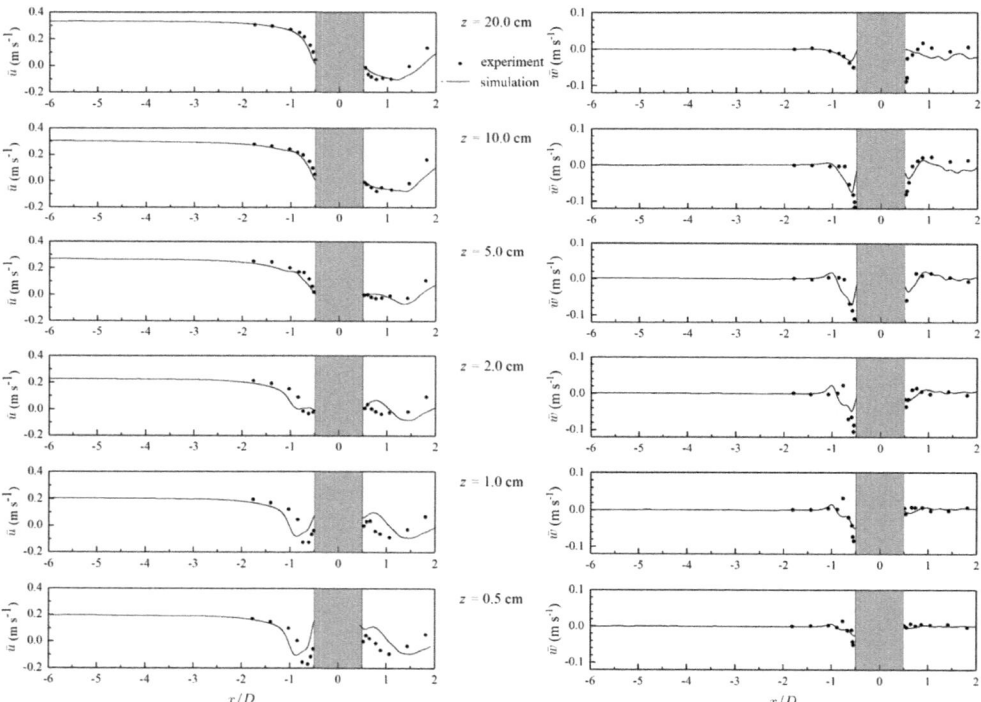

Figure 5. Horizontal flow velocity profiles (left) and vertical flow velocity profiles (right) at incident flow side and dorsal flow side of cylinder at different water depths.

3.3 Shear stress analysis

The shear force defining algorithm was used to simulate the shear stress of the bed surface around the cylinder. Meanwhile, the influence of the turbulent viscosity coefficient on the shear stress of the bed surface was also considered, as expressed by the following formula:

$$\tau = (\mu + \mu_T) \left[\nabla u + (\nabla u)^T \right] \tag{9}$$

Where μ, μ_T and ∇u represent the dynamic viscosity coefficient of water flow, the turbulent viscosity coefficient of the fluid, and the flow velocity parallel to the bottom bed surface, respectively

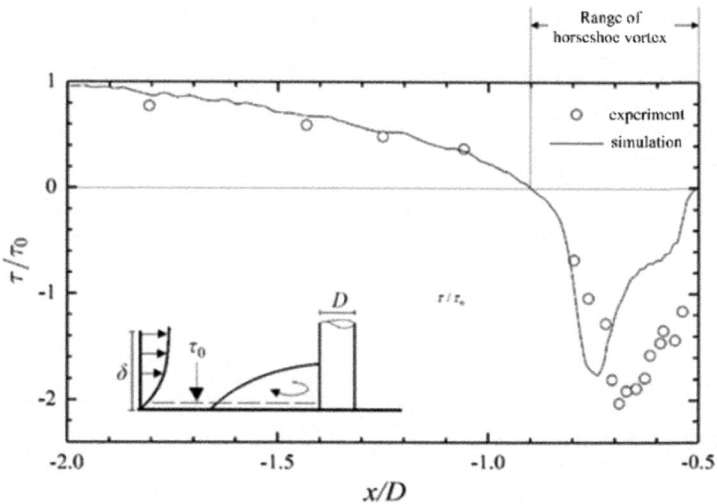

Figure 6.   Comparison of shear stress distributions.

As shown in Figure 6, the x-coordinate took the dot of the cylindrical roughness element as the datum point. D represents the diameter of the cylindrical roughness element. The outer wall at the incident flow side of the roughness element was located at an x-coordinate of -0.5. Therefore, the range generated by horseshoe vortices was from -0.1 to -0.5, and the comparative results between experimental values and simulated values showed that the vortex position rightly fell within this range.

It could be observed that the shear stress along the flow direction became greater and greater until the lowest point of the scour pit where horseshoe vortices were generated reached the maximum value. Before the generation of horseshoe vortices, simulated values coincided with experimental values, and the horseshoe vortex was generated more forward by the simulated value.

3.4   Flow field analysis

3.4.1   Time-averaged flow field analysis
In the flow field displayed in Figure 7, the flow velocity around the cylindrical roughness element was disturbed. A low-velocity region was formed before the pier, a local high-velocity region was generated at the shoulder before the bridge pier, and a low-velocity wake flow region was formed behind the pier, all of which manifested some characteristics of Karman vortex street (Hu 2009). Moreover, a transverse vortex rotating clockwise was formed, which moved along the bed surface towards two sides and finally interacted with the spiral flow around the bridge pier to form a new vortex system. This system was shaped like a horseshow, thus being referred to as a horseshoe vortex system (Ling 2007).

3.4.2   Turbulent flow field analysis
As shown in Figure 8, water flowed through the roughness element to form symmetrically shaped, oppositely rotating, and regularly arranged bilateral vortices in the rear. The vortex intensities at two sides reached extreme values at a rear position 40 cm from the pier center. Subsequently, the vortex intensity was gradually weakened and converged at about 100 cm from the center of this roughness element. Then, it was distributed downstream along the central axis (Zhang 2002).

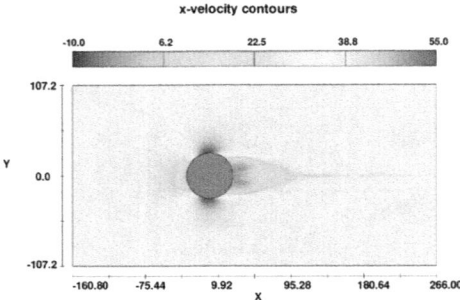

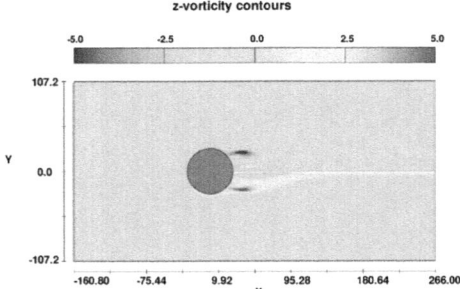

Figure 7. Middle-layer flow velocity distribution.   Figure 8. Middle-layer vorticity distribution.

4 CONCLUSIONS

In this research, two factors exerting the most prominent influences on the hydrodynamic characteristics of the cylindrical roughness element were selected according to the actual situation. Then, different operating conditions were set, and a mathematical model was established for the numerical simulation. On this basis, the hydrodynamic characteristics of the flow around the roughness element were acquired. Meanwhile, the main causes for the changes in the riverbed structure were analyzed by comparing the physical experiment with the numerical simulation.

(1) The hydrodynamic force of the riverbed in local areas exceeded the critical value, thus leading to the local scour of the riverbed around the roughness element. Due to the obstructing effect of this roughness element, the flow space around was reduced, and the water level at the incident flow side of the roughness element was elevated, which gave rise to the non-uniform pressure distribution of the surrounding water body and the change in the vertical flow velocity. Consequently, the downward flow and horizontal flow at the incident flow side were tangled at the riverbed to form x-axis vortices. Meanwhile, the lateral flow around the roughness element would result in the bottom flow concentration and generate a high-velocity flow, which formed a horseshoe vortex together with x-axis vortices, thus causing the local scour of the riverbed. The physical experiment and numerical simulation results revealed that horseshoe vortices existed at the bottom and water surface at the incident flow side and dorsal side of the roughness element.

(2) Because of the fluctuating turbulent flow velocity, the shear stress of the riverbed would increase instantaneously, which further induced the local scour of the riverbed around the roughness element. Water flowed through the roughness element to form symmetrically shaped, oppositely rotating, and regularly arranged bilateral vortices in the rear. Moreover, a local high-velocity region was generated, and finally, a low-velocity wake flow region was formed in the rear, manifesting the characteristics of Karman vortex street, which resulted in the local scour of the riverbed. The numerical simulation results showed that the lateral vortex intensity reached an extreme value at 40 cm from the rear part of the roughness element center.

ACKNOWLEDGMENTS

This research is funded by the Research Funding of China Three Gorges Corporation (202003251).

REFERENCES

Cao Y.G. (2012). *Flow dynamics characteristics around sediment particles with consideration of scale effect D*. Tianjin University.

Dm, A., Jm, B., Pf, C., Cb, C., & A. Hf (2021). Spoiler baffle patch design for improved upstream passage of small-bodied fish – sciencedirect. *Ecological Engineering*. 169.

Hu, X.Y., Zu, X.Y., Cheng, Y.Z., Li, Y.F., & Wang, Q.Y. (2009). Experimental study of lateral turbulent flow width at round-ended pier. *J. Hydro-Science and Engineering*, 8–13.

Lacey, R. & Roy, A.G. (2008). The spatial characterization of turbulence around large roughness elements in a gravel-bed river. *J. Geomorphology*. 102(3–4), 542–553.

Ling, J.M., Lin, X.P., & Zhao, H.D. (2007). Analysis of the three-dimensional flow field and local scour of riverbed around cylindrical pier. *J. Journal of Tongji University* (Natural Science), 582–586.

Moss, A.J., Walker, P.H., & Hutka, J. (1980). Movement of loose, sandy detritus by shallow water flows: An experimental study. *J. Sedimentary Geology*. 25(1–2), 43–66.

Roulund, A., Sumer, B.M., Freds, E.J., & Michelsen, J. (2005). Numerical and experimental investigation of flow and scour around a circular pile. *J. Fluid Mech*. 534, 351–401.

Yakhot, V., & Orszag, S.A. (1986). Renormalization group analysis of turbulence. I. basic theory. *J. Scientific Computing*. 1(1), 3–51.

Zhang, W., Wang, Y., Xu, Z., & Jin, W. (2002). Experimental investigation of vortex evolvement around a circular cylinder by digital particle image velocimetry (DPIV). *J. Acta Aerodynamica Sinica*, 379–387.

Zhong, L., Jiang, T., & Han, Z.G. (2019). Flow structures around large-scale artificial roughness elements. *J. Hydro-Science and Engineering*, 67–75.

Frontiers in Civil and Hydraulic Engineering – Mohamed A. Ismail and
Hazem Samih Mohamed (Eds)
© 2023 The Authors, ISBN 978-1-032-38247-0

Numerical simulation of deformation influence of foundation pit excavation on shield tunnel under support structures

Senyu Li
Faculty of Civil and Transportation Engineering, Guangdong University of Technology, China

Wenzhao Liu, Zhitang Huang & Genshen Liu
China Construction First Group Corporation Limited, China

Sheng Xie
Geological Construction Engineering Group Corporation of Guangdong Province, China

Shihua Liang*
Faculty of Civil and Transportation Engineering, Guangdong University of Technology, China

ABSTRACT: With the rapid development of the times, urban underground space engineering is more widely used, and the construction of deep foundation pit excavation will inevitably affect the shield tunnel under the support structure. Based on a large deep foundation pit support project in Zhuhai, this paper uses Midas GTS three-dimensional numerical simulation calculation software to analyze different width and thickness arrangements of foundation pit reinforcement area and the regularity changes of influence of foundation pit excavation on the deformation of the support structure and shield tunnel. The main conclusions are as follows: the thickness of the foundation pit reinforcement area increases compared with the width, which has a more obvious protection effect on the foundation pit support structure and shield tunnel. And when the width of the reinforcement area is increased to a certain value, the protection effect for the shield tunnel will decrease after 39m of this project. The comparison between the corresponding simulation data and actual monitoring data shows that the research results of this paper can provide a reference for other similar projects in deep soft soil areas.

1 INTRODUCTION

With the expansion of urban space driven by social and economic development, the construction of urban underground space projects is also increasing. The continuous emergence of subway tunnels and deep and large foundation pit projects makes the construction of underground space projects more and more difficult, and its safety has also been paid more and more attention by people. The construction of many deep foundation pit projects will have an impact on the existing subway tunnels nearby or below (Hu et al. 2003; Wang 2011). In practical engineering, the excavation of the foundation pit will cause the redistribution of site stress, and then break the stress balance of the underlying existing tunnel, causing the deformation of the tunnel. If the deformation is too large, it will cause uneven settlement of the shield tunnel, leakage and even the destruction of the main structure, leading to major engineering accidents, especially in the deep foundation pit excavation under the condition of deep and weak stratum, this phenomenon is more obvious (Lin et al. 2015; Qiu et al. 2021). The existing relevant specifications have extremely strict requirements on the deformation of the ground fall track that has been put into operation.

---

*Corresponding Author: shihua_l@gdut.edu.cn

DOI 10.1201/9781003344209-40

Scholars have used a variety of methods to study the impact of foundation pit excavation on adjacent or underpass shield tunnels. In terms of theoretical analysis, Zhang Zhiguo et al. used the two-stage analysis method to study the influence of foundation pit excavation on the longitudinal deformation of adjacent subway tunnels (Zhang et al. 2011). In terms of the model test, Huang et al. used centrifuge tests to verify that under soft soil geology, foundation pit excavation will cause uplift and deformation of existing tunnels (Huang et al. 2015). In the field monitoring research, Chang et al. analyzed the segment falling off accident of adjacent subway tunnel caused by foundation pit excavation (Chang et al. 2011). In terms of numerical simulation, Zhang Zhiguo et al. used Midas finite element software combined with field monitoring data to study the change law of existing tunnels and diaphragm walls under the divisional excavation of foundation pits (Zhang et al. 2018).

To sum up, at present, the methods to control the influence of foundation pit excavation on the deformation of existing tunnels are mainly divisional excavation, grouting protection technology, setting up barrier walls and foundation treatment (Zhang et al. 2018). In recent years, the foundation pit foundation reinforcement technology has been widely studied and applied. For example, Yu Jin et al., through the analysis of ANSYS finite element and field monitoring data, compared the tunnel uplift deformation under the two working conditions of foundation pit reinforcement and unreinforced. The reinforcement of the foundation pit can improve the uplift deformation of the tunnel (Yu et al. 2007); Huang Hongwei et al. compared and studied the impact of foundation pit excavation on the underlying highway tunnel through the finite element software plaxis-gid and the field monitoring results, obtained the scope of the impact of foundation pit excavation on the lower soil mass, and evaluated the protective measures for soil reinforcement (Huang et al. 2012).

Therefore, deep foundation pit excavation in soft soil areas is essential for foundation pit foundation treatment. The research content of this paper goes further on the research of foundation pit reinforcement on the existing tunnel. Based on a large deep foundation pit project in Zhuhai, based on the above research foundation, Midas GTS finite element software is used to establish a three-dimensional model to explore different foundation pit foundation treatment scopes, the influence law of foundation pit excavation on the deformation of the tunnel, which is then combined with the monitoring data to further compare and summarize the relevant engineering experience. It will provide a reference for related engineering construction in the future.

2 PROJECT OVERVIEW

2.1 *Foundation pit overview*

A large-scale deep foundation pit support project in Zhuhai is located in Hengqin new area, Zhuhai, and is located in the land area under the jurisdiction of Huandao East Road, Lianao Road, metropolis road and Guanao road. The construction area of the project is about 20000 square meters, and the underground building is set as two floors. Guangzhou Zhuzhou metro shield tunnel passes under the northeast corner of the foundation pit, and the vertical distance from the bottom of the foundation pit to the top of the tunnel is only 13.5 m. The excavation depth of the foundation pit is about 11.0 m. Considering that the excavation of the foundation pit is deep, the area is huge, and the geological conditions are poor, the soil displacement caused by the excavation of the foundation pit must be strictly controlled to avoid the impact on the surrounding environment and the rail transit tunnel. The safety level of the side wall of the foundation pit is level I. It is proposed to use cast-in-place piles + internal supports for support. The diameter of the support piles is 1200 mm, the spacing is 1400 mm, and two reinforced concrete supports are set; considering the abundant groundwater, it is arranged around the foundation pit $\phi750@450$ The mixing pile is used as the water stop curtain; at the same time, the sludge layer around the foundation pit is more than 17 m thick, and the upper half of the shield tunnel is also in the sludge layer, so it is proposed to adopt $\phi700@400$ The cement mixing pile pit is reinforced to protect the underpass tunnel, and the reinforcement range in the pit is 15 m wide and 13 m deep. The layout plan of the foundation pit support structure is shown in Figure 1, and the typical section of the tunnel and foundation pit support structure is shown in Figure 2.

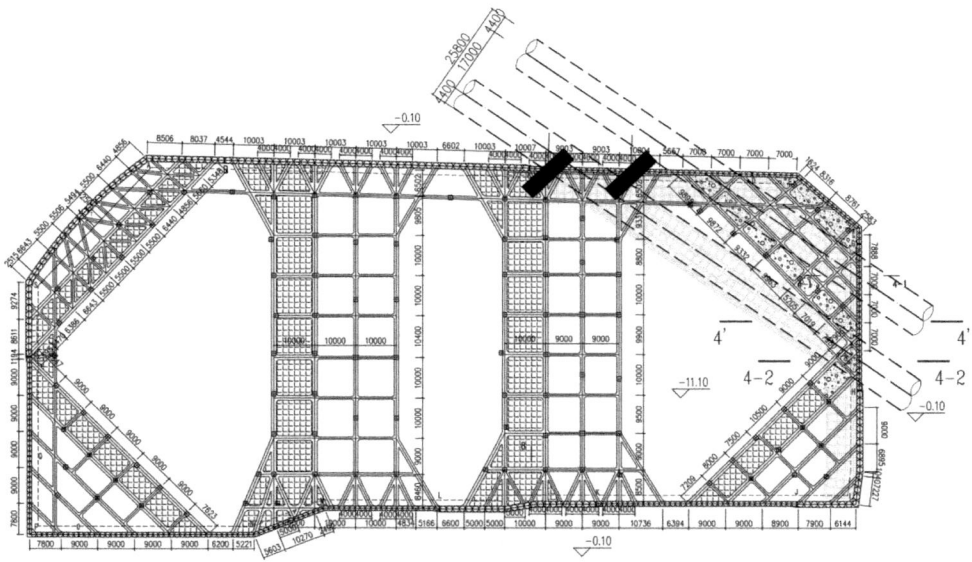

Figure 1.   Layout plan of foundation pit support structure.

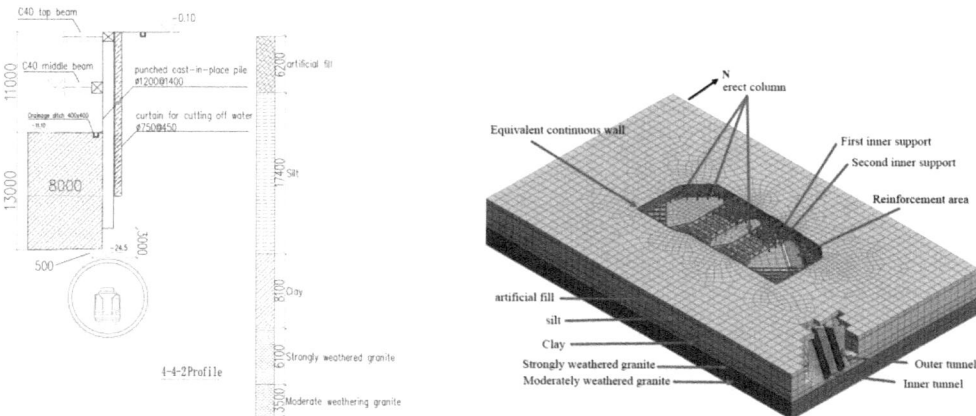

Figure 2.   Profile of foundation pit support structure (see Figure 1 for the specific location of profile 4-4-2).

Figure 3.   Midas/GTS finite element model.

2.2 Hydrogeological overview

The geological conditions of the project are relatively complex. According to the drilling results, the properties of the soil layer in the site range from top to bottom are mainly 2.2~22.3 m artificial fill, 0.4~20.6 m silt, 1.2~12.2m clay, 0.3~7.8 m strongly weathered granite, 0.6~6.8m completely weathered granite and 3.0~5.7 m moderately weathered granite, with the uneven stratigraphic distribution. The excavation depth of the foundation pit is 11 m, and its scope is mainly artificial fill and silt. It can be seen that the geological conditions of the excavation area are poor and the groundwater level is rich.

3 THREE DIMENSIONAL FINITE ELEMENT CALCULATION MODEL

3.1 Model establishment

In order to simulate the impact of deep foundation pit excavation on the lower span shield tunnel, a certain simplification is made according to the engineering practice, and the overall three-dimensional finite element model as shown in Figure 3 is established, with the model taken as 460 m × 280 m × 60 m (length × wide × High). Referring to the size and shape of the foundation pit in the actual project and the location of the shield tunnel, the foundation pit + tunnel model as shown in Figure 4 is established, in which the supporting pile is equivalent to the continuous wall according to the principle of equal bending stiffness, and the equivalent continuous wall width is 0.954m. From top to bottom, the soil layers of the model are artificial fill, silt, clay, strongly weathered granite and moderately weathered granite, and the thickness of each layer is 6.2, 17.4, 8.1, 6.1 and 3.5 m respectively. It can be seen from Figure 1 and Figure 2 that the location distribution of the foundation pit and the shield tunnel underneath makes the pile bottom of the supporting pile arranged above the shield tunnel in the sludge layer. Insufficient embedded depth will affect the stability of the foundation pit supporting structure. Therefore, it is necessary to carry out foundation treatment before foundation pit excavation, but the impact of foundation pit excavation on the foundation pit supporting structure, surrounding environment and shield tunnel will be different in different reinforcement ranges. Therefore, the three-dimensional finite element numerical simulation analyzes the impact of foundation pit excavation on the support structure, surrounding environment and shield tunnel under different reinforcement depths and different reinforcement widths, determines the reasonable scope of the reinforcement area, and simplifies the location of the reinforcement area and the tunnel accordingly, as shown in the reinforcement area + tunnel model in Figure 5.

Figure 4.   Foundation Pit + tunnel model.   Figure 5.   Reinforcement area + tunnel model.

3.2 Basic assumptions

The basic assumptions of the model are as follows: ① The constitutive model of rock and soil mass adopts Mohr Coulomb constitutive model, the constitutive model of support structure system and tunnel adopts the linear elastic model, and the physical and mechanical parameters of various materials in the finite element model are shown in Table 1; ② It is assumed that all soil layers are

Table 1.   Physical and mechanical parameters of materials in the finite element model.

Rock soil layer/support structure	Weight/ kN/m³	Cohesion/ kpa	Internal friction angle/°	SPT blow count/blow	Modulus of elasticity/mpa
artificial fill	17.2	8.0	10.0	4.5	12.47
silt	16.4	7.0	6.5	2.3	6.37
Clay	19.2	16.3	15.4	9.2	25.48
Strongly weathered granite	21.1	38.0	25.2	48.3	133.79
Moderately weathered granite	21.5	80.0	28.0	/	280.00
C40 supporting pile	25.0	/	/	/	32500
Mixing pile	20.5	/	/	/	150.00
shield tunnel	25.0	/	/	/	34500

stratified homogeneous and horizontally distributed; ③ The influence of groundwater in the process of foundation pit excavation is not considered; ④ Regardless of the time.

3.3 *Variables of finite element model*

In this paper, the influence of foundation pit excavation on the tunnel, the support structure and the surrounding environment in different reinforcement areas can be simulated by changing the properties of the reinforcement areas in different areas. The specific methods are as follows: 1) Different widths: parallel to the direction of the inner tunnel (as shown in Figure 6), parallel to the outer tunnel side for two meters, the line at this position and the range enclosed by the foundation pit are set as the first width variable condition, and then expand 2 meters in parallel in the opposite direction successively, expanding 4 times, that is, the width of the reinforcement area is set to 33 m, 35 m, 37 m, 39 m and 41 m, respectively, so five width variables are set; 2) different thickness: in the range of different widths, it extends 9 m, 11 m and 13 m from the bottom of the pit, that is, the thickness of the reinforcement area is set to 9 m, 11 m and 13 m, respectively, so three thickness variables are set. To sum up, according to the setting of reinforcement areas with different widths and thicknesses, a total of 15 working conditions of reinforcement areas with different ranges can be selected. See Figure 6 for the specific scope and layout of reinforcement areas.

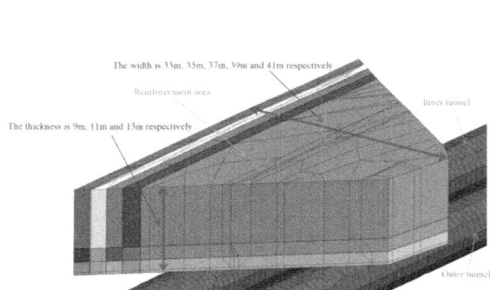
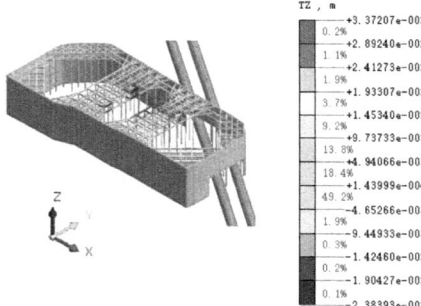

Figure 6.　Schematic diagram of reinforcement area.

Figure 7.　Simulation results of the influence of reinforcement area width on tunnel uplift.

4　NUMERICAL SIMULATION RESULTS AND ANALYSIS

Since there are 15 working conditions, only cloud images will be used to show the simulation results of some parts (a certain thickness or depth) below, and finally, all the results will be summarized with a broken line diagram to obtain the relevant laws.

4.1　*The influence of the thickness and width of the reinforcement area on the tunnel and foundation pit support structure (taking the thickness and width of the reinforcement area as 13m and 35m as an example)*

The simulation results of the reinforcement area thickness and width on the tunnel uplift are shown in Figure 7. The simulation results of the thickness and width of the reinforcement area on the horizontal displacement of the East and West supporting structures are shown in Figure 8. The simulation results of the horizontal displacement of the South and North support structures caused by the thickness and width of the reinforcement area are shown in Figure 9.

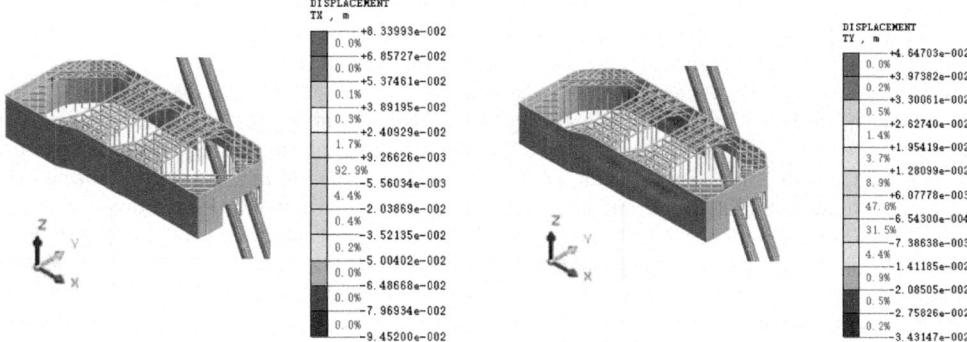

Figure 8. Simulation results of the influence of the reinforcement area width on the horizontal displacement of the East and West supporting structures.

Figure 9. Simulation results of the influence of the reinforcement area width on the horizontal displacement of the South and North supporting structures.

4.2 Result analysis

According to the simulation results in Figures 7–9, it can be concluded that the width and thickness of the reinforcement area have little effect on the south, west and north sides of the support structure, and the maximum horizontal displacement is concentrated at 35 mm, 27 mm and 25 mm. Therefore, the following is only to sort out the influence of the reinforcement area width on the tunnel uplift and the horizontal displacement of the support structure of the east and north sides of the tunnel. If the above simulation results are the same, the simulation results of other working conditions are sorted out in Figures 10–12.

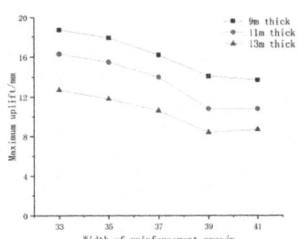
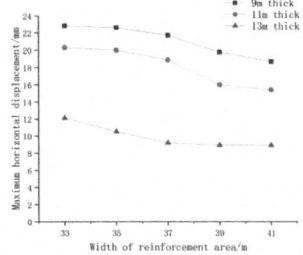
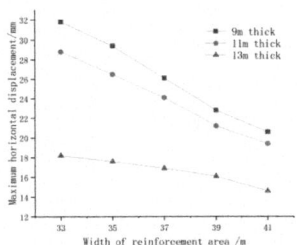

Figure 10. Variation law of width and thickness of reinforcement area on tunnel uplift.

Figure 11. Variation law of the width and thickness of the reinforcement area on the horizontal displacement of the support structure on the east side.

Figure 12. Variation law of the width and thickness of the reinforcement area on the horizontal displacement of some supporting structures of the tunnel under the north side.

Figure 10 shows that when the thickness of the reinforcement area (hereinafter referred to as H) is certain, the uplift of the shield tunnel will decrease with the increase of the width of the reinforcement area (hereinafter referred to as B), but when B ≥ 39 m, this phenomenon becomes not obvious and even has the opposite effect; Figure 12 shows that the variation law of the maximum horizontal displacement of the East support structure with B is similar to the uplift of the shield tunnel; Figure 12 shows that when H is certain, the maximum horizontal displacement of the supporting structure of the tunnel under the north side decreases with the increase of B. when H=13 m, the effect of the increase of B on the reduction of its value is obviously not as good as when H < 13 m.

At the same time, it can be seen from Figures 10 to 12 that the increase of H has a significant effect on reducing the uplift of the shield tunnel and the horizontal displacement of the support structure, and its effect is significantly greater than the increase of B.

To sum up, due to the obvious protective effect of H on foundation pit excavation, it is recommended to use H=13 m, and the actual project uses H of 13 m, that is, the lowest end of the reinforcement area is 0.5 m away from the top of the shield tunnel; the monitoring and early warning value of the deep horizontal displacement of the support structure is 36mm, the uplift deformation of the shield tunnel should be controlled below 15 mm, and the above working conditions of H=13 m and B ≥ 37 m are not exceeded. Considering the optimization effect, as shown in Figure 12, when H=13 m, the horizontal displacement of the support on the east side is significantly reduced with B from 33 m to 37 m. It is suggested that B should be 37 m, and considering economic factors, B selected in the actual project is 35 m, that is, the reinforcement area just covers the quadrangular area surrounded by the shield tunnel and foundation pit. If 33 m is selected, the short supporting pile arranged above the tunnel may cause the kick effect due to insufficient passive earth pressure. In the numerical simulation, because the supporting structure is equivalent to a continuous wall, the overall stiffness is improved, and the kick effect is not fully reflected, which will be different from the actual project, so it is not recommended that B be 33 m.

5 COMPARISON BETWEEN NUMERICAL SIMULATION RESULTS AND FIELD MONITORING RESULTS

According to the above analysis, the reinforced soil area arranged above the underpass tunnel of a large deep foundation pit support project in Zhuhai is 35 m wide and 13 m thick. The project monitors the deep horizontal displacement of the supporting structure. The specific monitoring scheme of the foundation pit is shown in Figure 13a. Based on S10 monitoring points of deep horizontal displacement of the foundation pit, the simulation results of the corresponding points are compared with the monitoring data, as shown in Figure 13b.

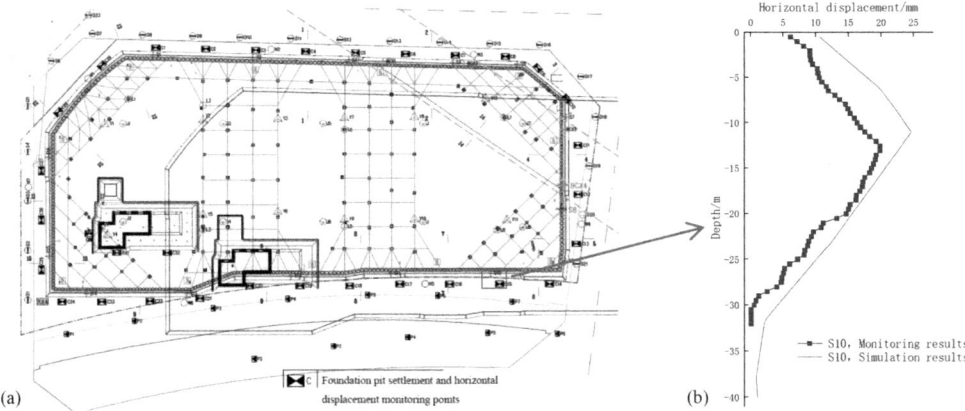

Figure 13.   (a) Layout plan of foundation pit monitoring. (b) Comparison between calculation results and monitoring results.

From the above analysis, it can be seen that the simulation results are in good agreement with the field monitoring data. Although the data obtained by numerical simulation will be larger than the field monitoring, the relevant change law is consistent. The reason may be that the unloading and loading modulus of Mohr-Coulomb constitutive model adopts the same modulus. To obtain a reasonable uplift value and a reasonable horizontal displacement of the support structure, the elastic modulus will be adjusted appropriately, but since the monitoring data does not give the actual value of pit bottom uplift, to obtain a reasonable value of pit bottom uplift, the horizontal displacement of the support structure may be too large, resulting in the aggravation of "belly bulge". At present, the project has been completed, and the application effect of the foundation pit support structure is good. At the same time, the shield tunnel under the foundation pit passing through Guangzhou Zhuzhou Metro has been completed and opened to traffic, and the use effect is good. The research results of this paper can provide a reference for other similar projects in deep soft soil areas.

6 CONCLUSION

The geological conditions of a large deep foundation pit support project in Zhuhai are poor, the groundwater is rich, and the excavation depth of the foundation pit is large. At the same time, the northeast corner passes through the shield tunnel involving the support structure, which puts forward higher requirements for the support structure of the foundation pit and the uplift control of the shield tunnel. This paper studies the influence law of the thickness and width of the reinforcement area on the foundation pit support structure and tunnel uplift, explore the reasonable layout of the reinforcement area, and the main conclusions are as follows:

(1) The excavation of foundation pit in deep soft soil layer will cause large uplift deformation of shield tunnel;
(2) Compared with the increase in the width, the increase in the thickness of the reinforced area has a more obvious protective effect on the horizontal displacement of the support structure and the uplift of the tunnel. Under similar engineering conditions and the premise that the construction of the reinforced soil does not affect the shield tunnel, the thickness of the reinforced soil area should be set as thick as possible.
(3) Compared with the increase of the thickness, the increase in the width of the reinforcement area does not play a significant role in protecting the horizontal displacement of the support structure and the uplift of the tunnel. Even when the width is set to a certain value, the protection effect is weakened after the project is 39 m. Therefore, under the conditions of similar projects, the thickness of the reinforcement area should completely cover the intersection area of the tunnel and the foundation pit, and it is not recommended to extend the reinforcement area too outward. From the perspective of economic and actual protection effects, extending 0~2 m outward is a reasonable range.

REFERENCES

Chang C.T., Sun C.W., Duann S.W., Hwang R.N. (2001). Response of a Taipei Rapid Transit System (TRTS) tunnel to adjacent excavation. *Tunnelling and Underground Space Technology*, 16(3), 151–158.
Hu Z.F., Yue Z.Q., Zhou J., Tham L.G. (2003). Design and construction of a deep excavation in soft soils adjacent to the Shanghai Metro tunnels. *Canadian Geotechnical Journal*, 40(5), 933–948.
Huang Hongwei, Huang Xu, Schweiger F. Helmut. (2012). Numerical analysis of the influence of deep excavation on underneath existing road tunnel. *China Civil Engineering Journal*, 45(3), 182–189.
Huang Hongwei, Huang Xu, Zhang Dongmei. (2015). Centrifuge modelling of deep excavation over existing tunnels. *Geotechnical Engineering*, 167(1), 3–18.
Lin Hang, Chen Jingyu, Guo Chun, Liu Qunyi. (2015). Numerical analysis on the influence of foundation pit excavation on deformation of adjacent existing tunnel. *Journal of Central South University* (Natural Science Edition), 46(11), 4240–4247.
Qiu, Jutao, Jiang Jie, Zhou Xiaojun, Zhang, Yuefeng, Pan Yingdong. (2021). Analytical solution for evaluating deformation response of existing metro tunnel due to excavation of adjacent foundation pit. *Journal of Central South University*, 28(6), 1888–1900.
Wang Qiang. (2011). Field and numerical analysis of a deep foundation in the sensitive environment. *China Civil Engineering Journal*, 44(S2), 98–101.

Yu Jin, Xu Qionghe, Xing Weiwei, Ding Yong, Cheng Wanzhao. (2007). Numerical analysis of up warping deformation of existing tunnels under a new tunnel's excavation. *Rock and Soil Mechanics*, 28, 653–657.

Zhang D.M., Huang Z.K., Wang R.L., Yan J.Y., Zhang J. (2018). Grouting-based treatment of tunnel settlement: Practice in Shanghai. *Tunnelling and Underground Space Technology*, 80, 181–196.

Zhang Zhiguo, Xi Xiaoguang, Wu Ling. (2018). Numerical simulation and site monitoring analysis of the influence of division excavation of foundation pit on adjacent large-diameter river-crossing tunnel. *Tunnel Construction*, 38(9), 1480.

Zhang Zhiguo, Zhang Mengxi, Wang Weidong. (2011). Two-stage method for analyzing effects on adjacent metro tunnels due to foundation pit excavation. *Rock and Soil Mechanics*, 32(7), 2085–2092.

Frontiers in Civil and Hydraulic Engineering – Mohamed A. Ismail and Hazem Samih Mohamed (Eds)

Spatial coupling study between waterfront settlements and water network in Henan Island of Guangzhou

Ying Pan, Yunlei Su & Ying Shi*

School of Architecture, South China University of Technology, Guangzhou, Guangdong, China

ABSTRACT: The system of the waterfront settlement is complicated and closely connected and chimeric with the water network. The ecology of the water network and the typical style of the waterfront settlement in Henan Island have been seriously destructed by rapid urbanization. This paper makes the spatial coupling analysis between the water network and the waterfront settlements in the historical and current periods of Henan Island.

1 INTRODUCTION

With the promotion of the "Three Old Reform" Policy, the protection and renewal of waterfront settlements in Henan Island have been paid more and more attention. This paper takes 24 typical waterfront settlements of Henan Island as the research objects, excluding areas in the northwest that had been urbanized during the Republic of China. The research explored the spatial coupling patterns of the waterfront settlements in the historical period and analyzed its current development challenges based on the spatial coupling theory. The term "coupling" in physics can vividly explain the dynamic linkage relationship. In recent years, coupling theory is often applied in the fields of economics and geography (Huang et al. 2015), and the theory of "spatial coupling" is gradually used in the sciences of human settlements (Liu et al. 2010). With the spatial coupling theory analysis of the waterfront settlements, the study hopes to enrich the perspective of waterfront settlements research and certain reference for future urban renewal.

2 SPATIAL COUPLING ANALYSIS IN THE HISTORICAL PERIOD

2.1 The water network environment of Henan Island in the historical period

Henan Island is located between the front and rear channels of the Pearl River, which is formed by the sediment of the Pearl River gradually connecting to the land around the hills in the northwest, east and south of Henan Island. The Henan Island was flat and low-lying, with main elevations below 5 m, and the highest point of the northwest hill reached 54.3 m in 1929.

The central, southern and eastern parts of Henan Island were densely covered with rivers in 1929, showing the reticulated river characteristics, most of which were tidal reach. The water network density in 1929 reached 3.71 km/km^2, and the water surface rate reached 8.50% on Henan Island, which was the highest in the urban area of Guangzhou (Liu 2020). The intricate linear rivers and punctiform ponds constituted the water network with a multi-level structure, which served as a natural basement to create diversified patterns of waterfront settlements.

---

*Corresponding Author: 246834@qq.com

2.2 Spatial coupling between water network and waterfront settlements distribution

This research implemented the kernel density estimation by simplifying the water network and ponds into polyline factors in ArcGIS (Figure 1). The result shows that all the waterfront settlements tend to be distributed in areas with a higher kernel density of water network, in which 21% of them were located in the central area with higher kernel density of rivers, and the remaining 79% were located in the transition zone between the high and low density of rivers, but the kernel density of ponds around such settlements was much higher than others.

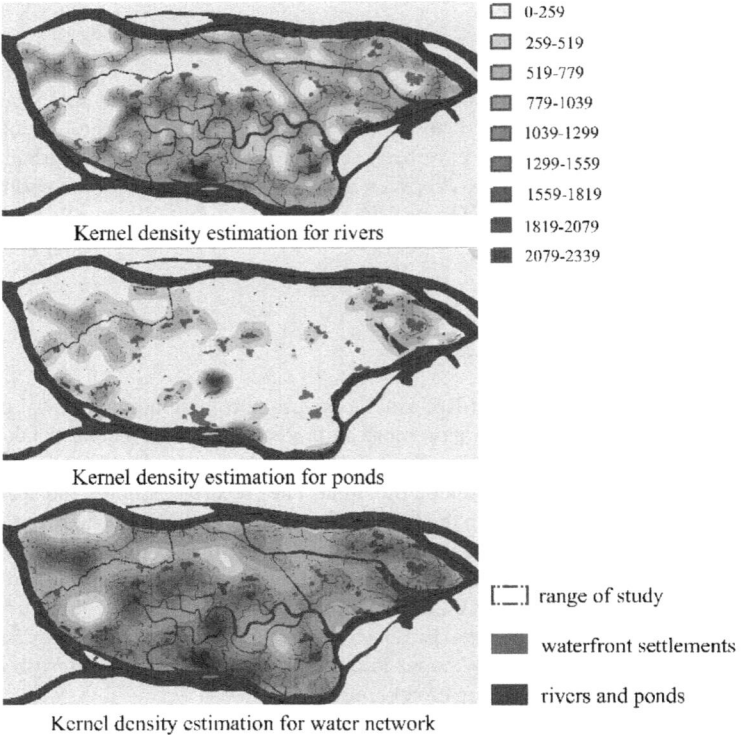

Kernel density estimation for rivers

Kernel density estimation for ponds

Kernel density estimation for water network

☐	0-259
☐	259-519
☐	519-779
☐	779-1039
■	1039-1299
■	1299-1559
■	1559-1819
■	1819-2079
■	2079-2339

[⎯] range of study

■ waterfront settlements

■ rivers and ponds

Figure 1. Kernel density estimation for the water network in 1929.

Regarding the analysis of settlement distribution, it is found that fewer waterfront settlements were distributed near the rivers with the width of 20 m and above in the central bottomland of Henan Island, and most of them are settlements with small areas and established later.

In general, waterfront settlements in Henan Island tended to choose sites that were covered by a high-density water network for daily life, agricultural production, and convenient transportation, while being located on the accumulation bank or land with high terrain for adequate expansion space and not being inundated by floods.

According to the positional relationship between settlements and the water network as well as the pattern and establishment time of settlements, three types of settlements can be divided: (1) blocky waterfront settlements surrounded by ponds; (2) linear waterfront settlements built along the river; (3) reticular waterfront settlements developed across the river.

These three types of waterfront settlements in Henan Island were characterized by a simplified three-layer-circular structure coupled with the water network structure, with linear waterfront settlements being concentrated in the central bottomland, blocky waterfront settlements far from the central bottomland being distributed at the foot of hills in the northwest and the northeast, and reticular waterfront settlements being distributed between linear and blocky waterfront settlements (Figure 2).

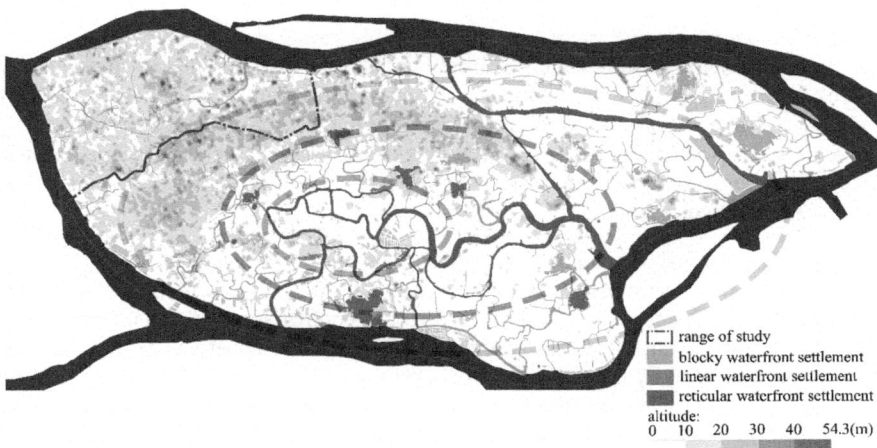

Figure 2.   The three-layer-circular structure of the waterfront settlements.

2.3   *Spatial coupling between water network and residential space*

The residential space of waterfront settlements and the waters in Henan Island have a closely coupling relationship. In addition to the direct use function of water resources and waterborne traffic, the ecological function of water storage and flood control, the water network also was the invisible enclosure of residential space and the extension of vision, as well as had geomantic aesthetics, which played a role of spatial division and connection. Rivers and ponds were often used as the boundary or skeleton in the construction of settlements (Figure 3).

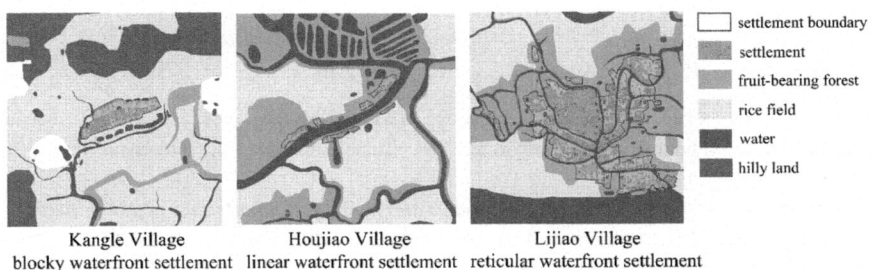

Figure 3.   Typical landscape pattern of three types of waterfront settlements.

2.3.1   *Blocky waterfront settlements*
This type of waterfront settlement was first established in the Song Dynasty, located at the foot of the hills in the northwest or northeast of Henan Island, relatively high in terrain. The settlement was formed into one or several concentrated clusters. Rivers were tangent to or far from the settlement, and ponds surrounded it. The landscape pattern of the settlement was "rice field – river – pond – settlement – hill" from front to back.

2.3.2   *Linear waterfront settlements*
This kind of waterfront settlement was mainly established in the Qing Dynasty, mainly distributed in the low-lying areas in the middle of Henan Island. The settlements were located on the dike

of the river on one or both sides, within a small area. They were the simplest type of waterfront settlements. The settlements presented a unilateral or symmetrical landscape pattern of "rice field fruit-bearing forest (dike of the river) – settlement – river".

2.3.3 *Reticular waterfront settlements*

These waterfront settlements were first established in the Song Dynasty, mainly distributed at the confluence of the river branches far away from the central bottomland. The settlements had a large area and were cut into several groups by rivers. The settlements were mostly built on an artificial platform to raise the foundation and reduce the damage of flood disasters. From the outside to the inside, the settlement presented a centripetal landscape pattern of "rice fields – fruit-bearing forest – river – settlement – river".

2.3.4 *Comparison analysis among three types of waterfront settlements*

The area of the linear waterfront settlements was the smallest, but the settlement's water surface rate and hydrophilic rate were the highest of the three. In comparison, the area of reticular waterfront settlements was the largest, with the highest water network density, water surface rate and hydrophilic rate, ranking only second to linear waterfront settlements. The coupling relationship between these two types of waterfront settlements and the water network was very close. The area of blocky waterfront settlements was only second to that of reticular waterfront settlements, but the water network density, the water surface rate and the hydrophilic were much lower than that of other types of settlements. The coupling relationship between blocky waterfront settlements and the water network was the weakest (Table 1).

Table 1. Spatial coupling of three types of waterfront settlements in 1929.

Three types of waterfront settlements	Settlement area* (unit: hectare)	Water network density within 100m around (unit: km/km^2)	Water surface rate within 100m around (unit: %)	Hydrophilic rate of the settlement (set 30m as the influence range of the water network, unit: %)
Blocky waterfront settlements	8.3	2.5	8.28	19.64
Linear waterfront settlements	2.4	5.1	17.09	57.30
Reticular waterfront settlements	12.5	7.4	12.76	55.59

*It divided the settlement boundary according to the 30m virtual boundary of the settlement (Pu 2012).

2.3.5 *Two patterns of spatial coupling between waterfront settlements and water network*

According to the above data analysis, it is found that the three types of waterfront settlements presented two spatial coupling patterns with the water network. One was pond coupling, such as the blocky waterfront settlements, which built a close connection with water by artificial ponds in front or around the settlement. The other was the river coupling model, such as the linear and the reticular waterfront settlements, in which the river flowed through the settlement and the development of the settlements matched the spatial structure of the rivers.

3 CURRENT SPATIAL COUPLING ANALYSIS

3.1 *The water network had decayed rapidly*

After the founding of new China, much of Henan Island's water network had been cut and straightened to improve agricultural production efficiency. Hills had been flattened to increase the arable land and construction land area. In order to build roads instead of waterways, rivers had been cut, filled and occupied. The development of these measures had achieved a specific purpose of construction and development. However, it also resulted in a chain reaction of continuous deterioration of the water ecological environment. The water network in Henan Island had declined rapidly, especially at the foot of the northwest hills, the north and the south. The density of the water network had decreased from 3.71 km/km^2 to 1.40 km/km^2, and the water surface rate was only 4.52%. Although attention has gradually been paid to ecological water protection, such as building dikes and sluices, connecting waterways (Chen 2010), demarcating ecological protection areas and building artificial lakes (Zhang 2015), it is not what it used to be.

3.2 *Spatial coupling between water network and waterfront settlements was weakened*

The decay of the complex water network weakened the spatial coupling relationship between most of the waterfront settlements and the water network (Table 2), and the landscape pattern of waterfront settlements had undergone qualitative changes (Figure 4).

The water network density, the water surface rate and the hydrophilic rate of these three types of waterfront settlements all decreased to different degrees. Reticular waterfront settlements had decreased most obviously, which was reflected in the sharp decline of water network density and water surface rate, mainly because a large number of rivers in the settlements had been buried for widening roads. The decrease in water surface and hydrophilic rate of linear waterfront settlements was noticeable, mainly because linear waterfront settlements expand far faster than any other settlements, and the connection between newly expanded settlements and the water network was

Table 2. Comparative analysis of 1929 and 2022.

Three types of waterfront settlements	Growth of area rate (unit: %)	Decrease of water network density (unit: %)	Decrease of water surface rate (unit: %)	Decrease of hydrophilic rate (unit: %)
Blocky waterfront settlements*	154.18	58.84	74.97	60.34
Linear waterfront settlements	476.15	48.42	64.79	74.75
Reticular waterfront settlements	138.66	71.71	76.19	65.26

*Pazhou village and Jiufenghuang Village, which had been completely renovated, were not included in the calculation.

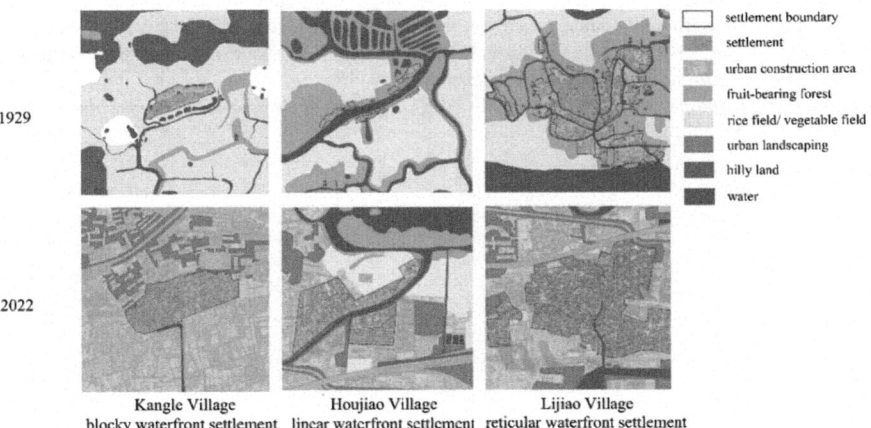

settlement boundary
settlement
urban construction area
fruit-bearing forest
rice field/ vegetable field
urban landscaping
hilly land
water

1929

2022

Kangle Village   Houjiao Village   Lijiao Village
blocky waterfront settlement   linear waterfront settlement   reticular waterfront settlement

Figure 4.   Comparative analysis on landscape pattern of waterfront settlements in 1929 and 2022.

poor. The water surface rate of the blocky waterfront settlements decreased most obviously, because the ponds had mainly been buried, and the hydrophilic characteristics of the settlements disappeared.

4 CONCLUSION

According to research, in the historical period, three kinds of waterfront settlement distribution were characterized by a three-layer-circular structure coupled with the water network. These three kinds of waterfront settlements could be classified as two spatial coupling patterns closely coupled to the water network. However, during the rapid development of urbanization, the water factor was ignored, which led to the gradual weakening of the spatial coupling relationship between the waterfront settlements and the water network.

In the historical period, the water network environment was a passive factor restricting the development of waterfront settlements on the one hand. On the other hand, it was a positive factor in settlement development. Villagers dug ponds and built dikes to live by the water to meet the needs of geomantic aesthetics, production and living, and actively promoted the continuous improvement of the coupling relationship between settlements and the water network in the development process of waterfront settlements. The spatial coupling model between the water network and the waterfront settlements contains the wisdom of the villagers in the traditional period to create a quality living environment and retains the city memory of suburban areas in Guangzhou, which is the intrinsic value we need to be aware of.

In the process of urban renewal, attention should be paid to the protection and restoration of water network ecology while the optimization of the human settlement environment. At the same time, it is necessary to promote the suitability and correlation between the settlement and the water network and consider the renewal strategy of the settlements based on the spatial coupling patterns in the historical period for sustaining the city's cultural vein.

REFERENCES

Bo Zhang (2015). The urban wetland restoration, utilization and sustainable development: With the protection and construction of Guangdong Haizhu National Wetlands Park of phase I and II. *J. Guangdong Landscape Architecture* (02), 49–53.

Chang Liu (2020). *The Study of Morphological Characteristics of Traditional Waterfront Settlements in Henan Island of Guangzhou*. D. South China University of Technology.

Mei Chen (2010). Case analysis of water diversion and replenishment project in the northern part of Shiliugang, Haizhu District, Guangzhou. *J. Guangdong Water Resources and Hydropower* (11), 5–8.

Qi Huang & Fanghua Tang (2015). The evaluation of coupling coordination development of ecology-economy society system in dongting lake area. *J. Economic Geography* (12), 161–167+202.

Song Liu & Binyi Liu (2010). Coupling analysis between urban green space and urban development: A case study of Wuxi. *J. Chinese Landscape Architecture* (03): 14–18.

Xincheng Pu (2012). *Quantitative research on the integrated form of the two-dimensional plan to traditional rural settlement*. D. Zhejiang University.

Frontiers in Civil and Hydraulic Engineering – Mohamed A. Ismail and
Hazem Samih Mohamed (Eds)

Research on pumped storage capacity demand for power grid peak regulation – A case study of a power grid in Liaoning

Xiaojun Li*, Hui Jin & Yang Liu
China Water Northeast Survey and Design Co Ltd, Changchun, China

ABSTRACT: The proportion of cogeneration units in Liaoning Power Grid is excessive, the electricity supply structure is not optimal and the peak regulation and accident reserve capacity for power system operation are insufficient. The optimal allocation of power is assessed by analyzing the major peak regulation paths available in the power grid in the province and the demand for peak regulation from pumped storage stations. The results indicate that the Liaoning Power Grid demands 11,200MW of pumped storage capacity by 2030 to achieve a suitable configuration of the system's power supply, which is economical and rational and can effectively ensure the safe and efficient operation of the power grid.

1 INTRODUCTION

The Liaoning Power Grid consists mainly of thermal power (Wang et al. 2015), with a large proportion of installed thermal power, accounting for 63.07% of the total installed capacity of the system, while the proportion of hydropower is only 5.27%; Moreover, Liaoning Province is in a cold region, influenced by climate and other factors, with a long heating cycle and more heating units, with thermal power units accounting for about 75.9% of the total thermal power capacity. The proportion of thermal power units is excessive, the power supply structure is not rational and during the winter heating and summer hydropower generation periods, the system operates with significantly insufficient peak regulation and accident reserve capacity (Ge et al. 2013). During the winter maximum capacity period, in particular, many thermal power units are forced to operate at extraordinarily high capacities. At present, peak regulation in the Liaoning Power Grid has become most complicated in winter, which is mainly attributed to the large proportion of thermal units in the power grid. In winter, the thermal units implement the principle of "power by heat", which makes it difficult for the units to be pressed down when the capacity is low, causing difficulties in peak regulation. The rapid development of wind power has also put greater pressure on the power grid to regulate peaks (Zhang et al. 2017). Due to the stochastic and intermittent characteristics of the wind, the power generated by wind turbines fluctuates significantly and the large-scale grid connection of wind power will cause impact and influence the power grid, which exacerbates the burden of grid regulation and increases the pressure on power grid peak regulation (Cao et al. 2019; Zhang & Sun 2018). Therefore, it is necessary to conduct thorough research on the demand for pumped storage capacity for power grid peak regulation to ensure the safe and efficient operation of the power grid.

2 ANALYSES OF THE MAJOR PEAK REGULATION METHODS FOR POWER GRIDS

2.1 Hydropower peak regulation

Hydropower in Liaoning Power Grid system accounts for a small proportion of installed hydropower, which is affected by the distribution of water resources. During the abundant water years,

---

*Corresponding Author: 14411300@qq.com

DOI 10.1201/9781003344209-42

hydropower mainly operates with base capacity and has to discard water for peak regulation. During the drought period, hydropower mainly operates with spikes, the system lacks waist load output, and there are many problems using conventional hydropower for peak regulation, such as the Shuifeng Power Station on the river between China and North Korea, a Chinese power station managed by the North Korean side on the Yalu River, which cannot be peak regulated following the intentions of China, and the Taipingwan Power Station has to carry the base load for the downstream power supply to Dandong City and Sinuiju City in North Korea. As a result, the conventional hydropower peak regulation of the Liaoning Power Grid is greatly restricted.

2.2 Thermal power peak regulation

The peak regulation mode of conventional thermal power units in Liaoning Power Grid consists of conventional peak regulation mode and non-conventional peak regulation mode of on-off peak regulation, in which the on-off peak regulation units are mainly 50 MW, 100 MW and 200MW (minority) units. Coal-fired thermal power peak regulation using load changes or on-off (200 MW and below conventional coal-fired thermal power) and other ways of peak regulation are the main measures to solve the peak regulation problem of thermal power-based power grids at present. Among the installed thermal power grid in Liaoning, thermal power accounts for nearly 75.9% of the installed coal power, and due to the special climatic conditions, thermal power units will still account for a significant proportion in the future, which basically cannot be used to regulate peaks in winter. Although the imported and domestic large-capacity coal-fired thermal power units can peak at about 50%, the coal-fired thermal power units cannot meet the requirements of rapid changes in system loads. Compared with hydropower, thermal power equipment accidents are more frequent and have a certain impact on the safe operation of the power grid. Moreover, the coal consumption of power generation rises during peak regulation operation, the power consumption rate of the plant is high, the equipment is seriously damaged, and the maintenance cost increases with the high power generation cost. With the accelerated shutdown of small thermal power units and the construction of "replacing small units with large ones" power projects, the capacity of small thermal power units will be gradually reduced.

2.3 Pumped storage power stations

Pumped storage power stations have the advantages of conventional hydropower stations, such as flexible start/stop, rapid load change, safety and reliability, long service life, simple operation and maintenance, low cost, low accident rate, and the ability to adapt to rapid changes in grid load. Pumped storage power stations also have a dual function of peak and valley regulation and can improve the operating conditions of thermal power units, which reduces the magnitude of their peak regulation, energy consumption of the system, accidents and wear and tear on thermal power units due to intensive peak regulation. In terms of natural conditions, Liaoning Province is a mountainous province, with mountains covering nearly 60% of the land area. There are many rivers and lakes, which have a certain amount of fall and sufficient water sources, providing the topographical conditions and water sources for the construction of pumped storage power stations. Through successive years of research and planning, there are many sites available for development. Using pumped storage power stations as the main peak regulation power source for the Liaoning Power Grid is an effective way to address peak regulation in the Liaoning Power Grid.

3 ANALYSES OF THE DEMAND FOR PUMPED STORAGE STATIONS IN THE POWER GRID

According to the load forecast and power supply construction planning of Liaoning Power Grid, a peak capacity surplus/loss analysis for 2030 was conducted, taking the system load reserve rate as 3% and the accident reserve rate as 10%. Based on the power supply construction plan, Liaoning Power Grid in 2030 will mainly consist of conventional hydropower, pumped storage power stations, coal power, nuclear power, wind power and solar power. By 2030, the planning hydropower capacity of Liaoning Power Grid is 1,846MW, of which 5 power stations are directly transferred from Liaoning Power Grid, namely Huanren, Huilong, Jinshan, Shuangling and Taiping, with a total installed capacity of 589.5 MW. The proportion of hydropower allocated to Liaoning Power Grid by the Northeast Power Grid is 2070 MW, and the available hydropower capacity of Liaoning Power Grid is 3916 MW with a peak capacity of 2095 MW and the actual peak capacity is used for system peak regulation. Pumped storage power stations can regulate peaks and fill valleys, with a peak regulation capacity of 200%. Nuclear power units, solar energy, biomass, etc. are considered as not participating in peak regulation Wind power units are considered as 40% counter-regulation to participate in system peak regulation. Considering the actual situation of thermal power flexibility transformation to increase system peak regulation capacity, the comprehensive peak regulation range of conventional thermal power units is considered 45%.

The balance of the peak capacity is based on the consideration of conventional hydropower, pumped storage units, nuclear power and wind power, with the insufficient or abundant part of the system capacity being regulated by conventional thermal power units. Based on the aforesaid principles, an analysis of the peak capacity gain/loss of the Liaoning Power Grid in 2030 was carried out. A peak capacity gain/loss analysis based on 45% of the combined peak capacity of conventional thermal power would indicate a lack of 9,119MW of peak capacity in Liaoning in 2030.

4 ANALYSES OF THE OPTIMAL CONFIGURATION OF POWER SUPPLIES

4.1 Calculation parameters and programming

Based on load forecasts and power supply planning for the Liaoning Power Grid, several scenarios were developed based on the 1200 MW of Pushi River storage already in operation, the 1800MW of Qingyuan storage and the 1000 MW of Zhuanghe storage, with an incremental increase of 1200 MW in the installed capacity of new pumped storage plants. We carried out power balance, production simulation and the present value of system cost calculations for the 2030 load level of the Liaoning Power Grid and analyzed the economically rational demand space for pumped storage capacity in the Liaoning Power Grid based on the principle of the minimum present value of system cost.

Among the calculation of cost values, concerning relevant engineering and power station results, the construction period of pumped storage power stations is taken to be 7 years, with a static investment of 5,100 RMB/kW for the starting unit, an increase of 60 RMB for each additional 1,200 MW of energy storage unit, and an operating cost rate of 2.4%. The investment in coal-fired thermal power units was investigated and taken to be RMB 4,000/kW, with a construction period of 3 years and an operating rate of 4% (excluding fuel costs). The price of coal is RMB800/t. The calculation period is taken as 47 years, of which the production period is 40 years. The social discount rate is 8% and the discount base year is the beginning of the first year of construction. Costs include investment, running costs and fuel costs.

4.2 Power sources participating in power balancing

(1) Pumped storage power stations

Based on the 1200 MW of Pushi River pumped storage power station, the 1800MW of Qingyuan energy storage and the 1000 MW of Zhuanghe energy storage, several schemes will be considered based on an incremental increase of 1200 MW in the installed capacity of new pumped storage power stations. The pumped storage capacity scheme for 2030 will range from 4000 to 12400 MW.

(2) Conventional hydropower stations

The hydropower stations participating in power balancing mainly include those participating in the unified scheduling of the Northeast Power Grid and the peak regulation of the Liaoning Power Grid. According to the Northeast Power Grid Passage Dispatching Rules, 46.67% of the Northeast Power Grid's regulated hydropower is allocated to the Liaoning Power Grid, with a capacity of approximately 1,577.5 MW participating in peak regulation, along with approximately 517.5 MW participating in peak regulation in Liaoning Province, for a total available hydropower capacity of approximately 2,095 MW. Details of the major hydropower stations participating in balancing are shown in Table 1.

Table 1. The major hydropower participating in the balancing of the Liaoning Power Grid (Unit: MW).

Scheduling relationships	Name	Capacity	Number of units * capacity	Notes
Northeast Power Grid Unified Control	1.Baishan	1500	300×5	The allocation to Liaoning was 46.67%
	2.Hongshi	200	50×4	
	3.Fengman	1480	140×2+200×6	
	4.Yunfeng	200	100×2	
	Subtotal	3380		
Liaoning Province Direct Transfer	1.Hengren	222.5	2×75+72.5	
	2.Jinshao	84	2×42	
	3.Shuangling	50	2×25	
	4.Taipingshao	161	4×40.25	
	Subtotal	517.5		

5 COMPARISON OF POWER BALANCE AND ECONOMY

Based on the indicators mentioned above, the present value of system costs was calculated based on power balance and production simulation review calculations (Program: P; Pumping Energy storage: PES; Hydroelectricity capacity: HC; Energy storage capacity: ESC; Thermal power capacity: TPC; Wind power capacity: WPC; Photovoltaic capacity: PC; Nuclear power capacity: NPC; Systems total installer capacity: STIC; System coal consumption: SCC; System cost present value: SCPV) for the Liaoning Power Grid in 2030, and the results of the calculations are detailed in Table 2.

Table 2. Comparison of pumped storage capacity allocation options for Liaoning Power Grid in 2030.

P	PES MW	HC MW	ESC MW	TPC MW	WPC MW	PC MW	NPC MW	STIC MW	SCC 10000 t	SCPV billion
1	4000	1846	1000	51722	27500	27496	14640	128204	10200.56	7775.2
2	5200	1846	1000	50402	27500	27496	14640	128084	10122.70	7735.6
3	6400	1846	1000	49082	27500	27496	14640	127964	10049.49	7699.1
4	7600	1846	1000	47761	27500	27496	14640	127843	9985.36	7667.8
5	8800	1846	1000	46442	27500	27496	14640	127724	9932.28	7642.7
6	10000	1846	1000	45122	27500	27496	14640	127604	9911.83	7635.9
7	11200	1846	1000	43802	27500	27496	14640	127484	9900.18	7634.0
8	12400	1846	1000	42482	27500	27496	14640	127364	9892.79	7634.6

Table 2 shows that as the capacity of the pumped storage plant increases, the present value of the system's cost gradually decreases, with the smallest present value of the system at 11,200 MW storage capacity, indicating that 11,200MW of pumped storage capacity is required in the Liaoning Power Grid by 2030 to achieve a rational configuration of the system's power supply. Excluding the 1200MW already built in Pushihe, the 1800 MW under construction in Qingyuan and the 1000MW in Zhuanghe, there is still space for 7200MW of pumped storage capacity in the system in 2030. At that time, the proportion of the installed capacity of pumped storage power stations in the Liaoning Power Grid to the total installed capacity of the system will be around 9%, and the proportion of the maximum load will be around 19%, which is similar to the proportion of pumped storage capacity in thermal-based power grids at a domestic and international level, indicating that it is economical and rational to allocate 11,200 MW of pumped storage capacity to the Liaoning Power Grid in 2030.

6 CONCLUSION

(1) According to the analysis of the current conditions, the conventional hydropower regulation of the Liaoning Power Grid is limited to a large extent.
(2) Through successive plans, there are many pumped storage sites available for development in Liaoning Province, and using pumped storage power stations as the main peak regulation power source for the Liaoning Power Grid is an effective solution for peak regulation in the Liaoning Power Grid.
(3) As the capacity of pumped storage stations increases, the present value of system costs gradually decreases, with the smallest present value of system costs at 11,200 MW of storage capacity, indicating that 11,200 MW of pumped storage capacity is required in Liaoning Power Grid by 2030 to achieve a rational configuration of system power sources and that the configuration results are more economical and rational.

REFERENCES

Cao, N., Man, L., Liu, A., Yang, S. & Wang, G. (2019). Research on wind power and photovoltaic acceptable ability of liaoning power grid. *Northeast Electric Power Technology*, 40, 6–11.
Ge, Y., Kong, X., Gao, K., Gao, L. & Ouyang, H. (2013). Research on the correlation between weather and load in liaoning grid. *Northeast Electric Power Technology*, 34, 6–9.
Wang, Y., Zhu, H. & Li, D. (2015). Study on nuclear power plants concerned with peak regulation of liaoning power grid. *Northeast Electric Power Technology*, 36, 28–33.
Zhang, K. & Sun, H. (2018). Research on improving absorptive capacity of wind and solar power in liaoning power system. *Northeast Electric Power Technology*, 39, 10–13.
Zhang, Z., Wang, S., Yu, J., Wang, D. & Xue, Q. (2017). Analysis of absorption capacity for offshore wind in liaoning province. *Northeast Electric Power Technology*, 38, 23–28.

Frontiers in Civil and Hydraulic Engineering – Mohamed A. Ismail and
Hazem Samih Mohamed (Eds)
© 2023 The Authors, ISBN 978-1-032-38247-0

Numerical simulation of flow field at upstream junction of stepped overflow dam face

Rui Yin* & Dong-yao Hong
Department of Water Engineering, Kunming University of Science and Technology Oxbridge College, Kunming, Yunnan, China

Jia-shi Cai & Ding Cui
School of Electric Power Engineering, Kunming University of Science and Technology, Kunming, Yunnan, China

ABSTRACT: We conducted a two-dimensional unsteady numerical simulation of the flow field at the upstream junction of the stepped overflow dam face of a gravity dam based on the VOF multiphase model. According to the analysis of the liquid phase, pressure, and flow line of the overflow dam face at different moments, we found that the linear lower edge of the lower drainage nappe didn't adhere to the overflow dam face at the initial stage, but formed a cavity at the concave corners of steps; the cavities on the second, third and fourth steps changed most dramatically, and reverse flows existed along the step surface; the first step was filled with the liquid phase quickly, while the second step slowly; the second, third, and fourth steps had wide negative pressure areas, which may lead to cavitation; the flow line formed a vortex structure on each step; the vortices on the second, third, and fourth steps were integrated, and there were many vortex structures on the second and third steps.

1   INTRODUCTION

The traditional overflow dam face is smooth, simple in structure, and easy to construct, but causes heavy fog and great fluctuations of tailwater during energy dissipation as stated by the State Key Laboratory of Hydraulics and Mountain River Development and Protection of Sichuan University 2016. As a new and efficient energy dissipater, the stepped overflow dam face can increase the energy dissipation rate by aerating and decelerating the water flow along the step surface step by step Lin 2009.

Wu (1998) carried out an experimental study on the flow pattern, energy dissipation effect, and the time-averaged pressure distribution on the steps in four dam face models with different step patterns on the right bank of Dongxiguan Hydropower Station on Jialing River, which greatly improved the energy dissipation effect of the dam face in a certain discharge range. To avoid cavitation, they are applicable in the case of small discharge per unit width. Wang et al. (2018) carried out a three-dimensional numerical simulation of the pressure distribution of transition steps with different combinations based on the model of the Ahai Hydropower Station. Chen et al. (2021) analyzed the test results of nine hydraulic models with "big step plus convex step" combinations by orthogonal experimental design, such as the aerated cavity length, air concentration, and negative pressure.

---

*Corresponding Author: 386072923@qq.com

DOI 10.1201/9781003344209-43

Most of the studies above were aimed at hydraulic models or a certain aspect of hydraulic characteristics. In this paper, we conducted an unsteady numerical simulation of the liquid phase, pressure, and flow line distribution at the upstream (the end of the weir crest curve section) junction of the stepped overflow dam face according to the actual engineering dimensions. This paper only considered the two-dimensional profile of the overflow dam face due to the large actual engineering dimensions and the small impact of transverse flow on the hydrodynamic characteristics.

2 NUMERICAL MODEL AND ALGORITHM

2.1 Governing equations

The VOF method Hirt 1981 and the RNG $k - \varepsilon$ turbulence model Launder 1974 were adopted herein. The governing equations include:
Continuity equation

$$\frac{\partial \rho}{\partial t} + \frac{\partial(\rho u_i)}{\partial x_i} = 0 \tag{1}$$

Momentum equation

$$\frac{\partial(\rho u_i)}{\partial t} + \frac{\partial(\rho u_i u_j)}{\partial x_j} = -\frac{\partial p}{\partial x_i} \\ + \frac{\partial}{\partial x_j}\left[(\mu + \mu_t)\left(\frac{\partial u_k}{\partial x_j} + \frac{\partial u_j}{\partial x_k}\right)\right] + \rho g_i \tag{2}$$

$$\frac{\partial(\rho k)}{\partial t} + \frac{\partial(\rho k u_i)}{\partial x_i} = \frac{\partial}{\partial x_i}\left[\left(\mu + \frac{\mu_t}{\sigma_k}\right)\frac{\partial k}{\partial x_i}\right] + G_k - \rho\varepsilon \tag{3}$$

$$\frac{\partial(\rho\varepsilon)}{\partial t} + \frac{\partial(\rho\varepsilon u_i)}{\partial x_i} = \frac{\partial}{\partial x_i}\left[\left(\mu + \frac{\mu_t}{\sigma_\varepsilon}\right)\frac{\partial \varepsilon}{\partial x_i}\right] + c_1 G_k \frac{\varepsilon}{k} - c_2\rho\frac{\varepsilon^2}{k} \tag{4}$$

where $\mu_t = \rho c_\mu \frac{k^2}{\varepsilon}$, $c_\mu = 0.085$, $c_2 = 1.68$, $\sigma_k = \sigma_\varepsilon = 0.7179$, $c_2 = 1.42 - \frac{\eta(1 - \frac{\eta}{\eta_0})}{1 + \beta\eta^3}$, $\eta = \frac{Sk}{\varepsilon}$, $S = \sqrt{2S_{ij}S_{ij}}$, $\eta_0 = 4.38$, $\beta = 0.015$.

2.2 Calculation method and engineering parameters

The PISO algorithm in Fluent 14.0 software was used for pressure-velocity coupling and to solve the unsteady incompressible RANS equation directly, in which the time step was 0.001s and the convergence criteria k and ε were both smaller than 1×10^{-5}.

The downstream weir of a gravity dam had a height of 53.715 m, a front length of 13.796 m, a crested head of 2.185 m, and a maximum discharge per unit width of 22.049 m^2/s. The overflow dam section had a WES profile. The downstream linear section had a slope coefficient of 0.75, with 38 steps 0.9 m high, 0.68 m wide, and 41.552 m along the dam face. The flip bucket was continuous with a bucket radius of 11 m, a deflection angle of 22°, and a height of 11.7 m.

2.3 Computational domain and grids

The computational domain started from the weir crest to 68 m downstream of the flip bucket via the downstream overflow dam face. The height of the overflow dam section was that of the guide wall, 3.994 m, and the height of the downstream area was 21.233 m above the riverbed. The domain was divided into structured grids. The closer to the dam face, the denser the grids are. The computational domain and grids are shown in Figure 1.

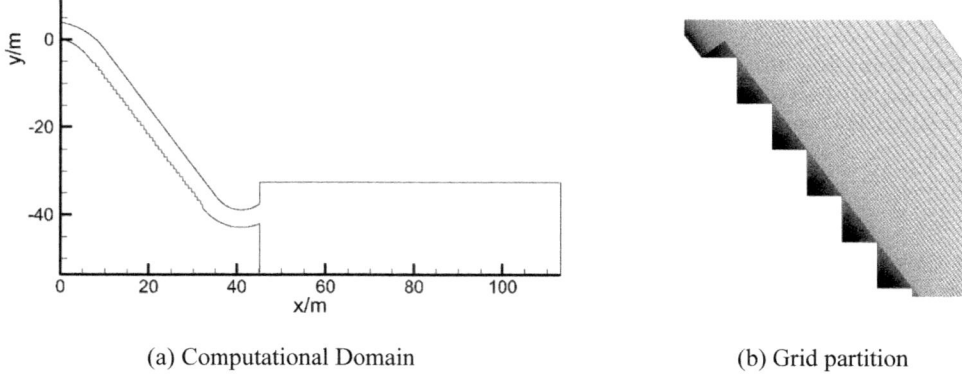

(a) Computational Domain   (b) Grid partition

Figure 1. Computational domain and grids.

3 ANALYSIS OF CALCULATION RESULTS

3.1 *Liquid phase on the overflow dam face*

Figure 2 shows the liquid phase distribution at the end junction of the weir crest curve at different moments.

According to Figure 2, the discharged water reached the end junction of the weir crest curve section for the first time when $t = 2.0$ s; the lower edge of the nappe did not adhere to the overflow

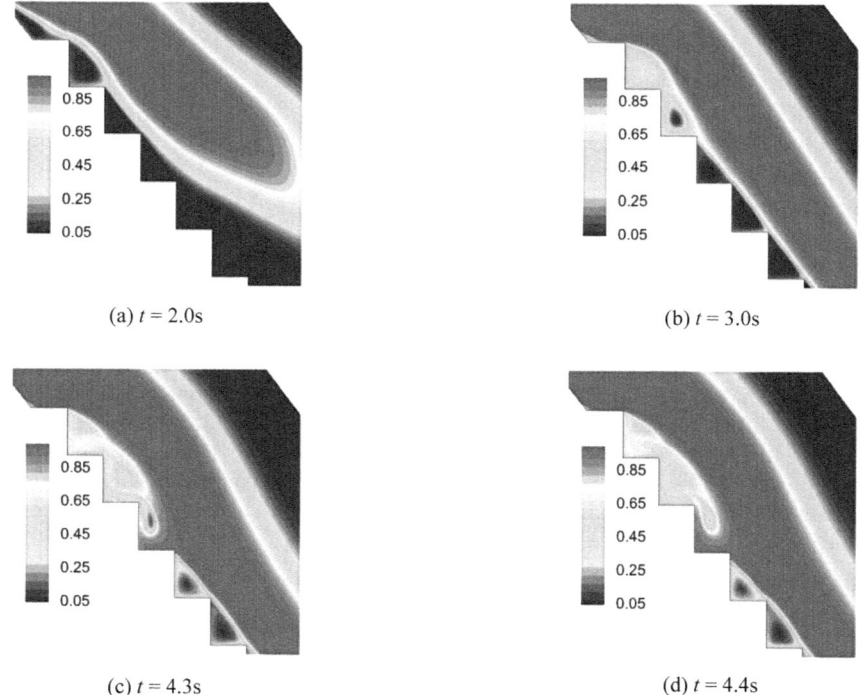

(a) $t = 2.0$s   (b) $t = 3.0$s

(c) $t = 4.3$s   (d) $t = 4.4$s

Figure 2. Liquid phase distribution on the overflow dam face.

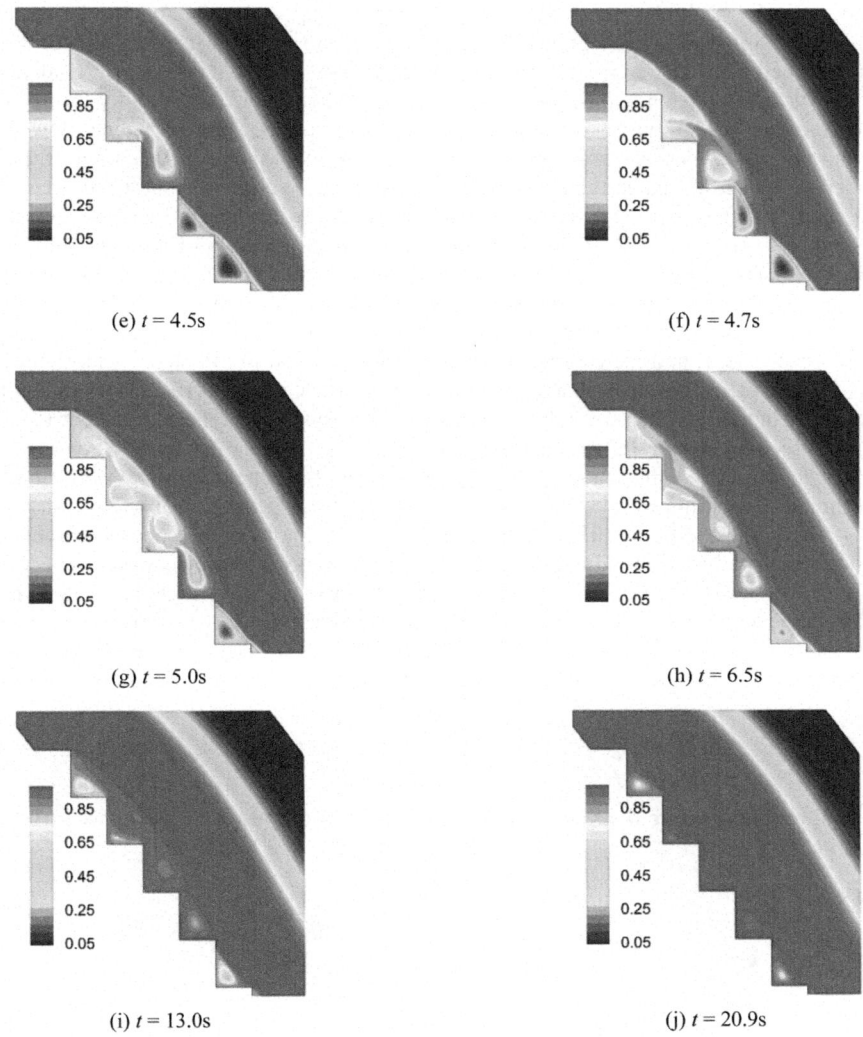

(e) $t = 4.5$s

(f) $t = 4.7$s

(g) $t = 5.0$s

(h) $t = 6.5$s

(i) $t = 13.0$s

(j) $t = 20.9$s

Figure 2. Continued.

dam face, but formed a cavity at the concave corners of steps, indicating that the water velocity was high and that the air between the nappe and the overflow dam face could not be discharged in time. When $t = 3.0$ s, the cavity at the concave corner of the first step became smaller rapidly and was almost filled with the liquid phase, while the proportion and area of the liquid phase in the cavities at the concave corners of the second and third steps both increased. At this time, the water depth on the steps increased, and an approximate triangular cavity was formed on the fourth and subsequent steps. Moreover, the bottom edge of the nappe was almost linear, indicating that the step surface did not fully exert its function of energy dissipation. The water flow pattern was similar to that in the case of a smooth overflow dam face. When $t = 4.3$ s, the cavity at the concave corner of the first step was very small, and the cavity areas at the concave corners of the second and third steps continued to increase; a cavity was formed above the concave corner of the fourth step, and the cavities on the second, third, and fourth steps were connected into a whole; an approximate circularly triangular cavity was formed at the concave corners of the fifth and subsequent steps. When $t = 4.4$ s, the liquid phase in the cavity formed above the concave corner of the fourth step increased gradually, and the water flow (reverse) below it dipped into the concave corner of the step clockwise so that the cavity gradually moved away from the step surface. When $t = 4.5$ s, the cavity formed above the concave corner of the fourth step showed a downward trend, and the reverse flow further dipped into the previous step surface. When $t = 4.7$ s, the cavity formed above the concave corner of the fourth step descended to the step surface, and the reverse flow further dipped into the concave corner of the previous step; the cavity at the concave corner of the fifth step tended to repeat the changing pattern of that at the concave corner of the fourth step. When $t = 5.0$ s, the first step was filled with the liquid phase, the cavity disappeared, and the reverse flow was divided into two again, one further dipping into the concave corner of the third step and the other into the second step; there were liquid phase residues in a small range at the concave corner of the fourth step, and the cavity at the concave corner of the fifth step was similar to that at the concave corner of the fourth step when $t = 4.5$ s. When $t = 6.5$ s, the reverse flow continuously rose to the second step, and the cavities on the second, third, and fourth steps were divided into two parts, which were concentrated at and far away from the concave corners. The cavity at the concave corner of the fifth step was relatively independent, and those at the concave corners of the sixth and subsequent steps were still approximate circularly triangular cavities. When $t = 13.0$ s, the cavities on the second, third, and fourth steps away from the concave corners were first filled with the liquid phase, and those at the concave corners of the fifth and subsequent steps were gradually filled with the liquid phase. When $t = 20.9$ s, the cavity at the concave corner of the fourth step was filled with the liquid phase, and then the cavities at the concave corners of the fifth ($t = 31.3$ s), third ($t = 31.3$ s), seventh ($t = 32.6$ s), ninth ($t = 38.4$ s), eighth ($t = 38.7$ s), tenth ($t = 42.6$ s), eleventh ($t = 43.3$ s), sixth ($t = 45.3$ s), twelfth ($t = 47.3$ s), thirteenth ($t = 49.8$s), fourteenth ($t = 50.1$ s), fifteenth ($t = 50.8$ s), sixteenth ($t = 52.2$ s), seventeenth ($t = 53.6$ s), eighteenth ($t = 54.9$ s), and second ($t = 55.9$ s) steps were successively filled with the liquid phase.

3.2 Pressure of overflow dam face

Figure 3 shows the pressure distribution at the end junction of the weir crest curve at different moments.

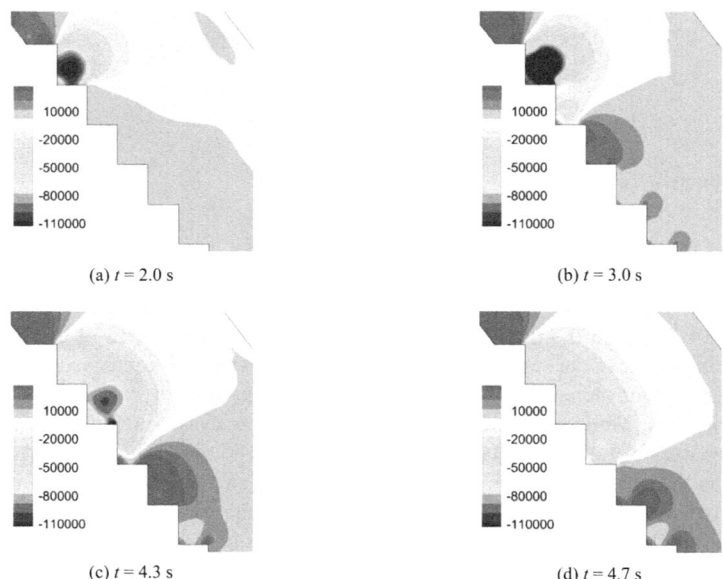

Figure 3.   Pressure distribution on overflow dam face (Pa).

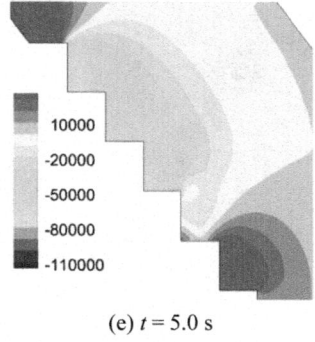
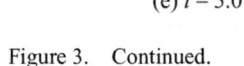
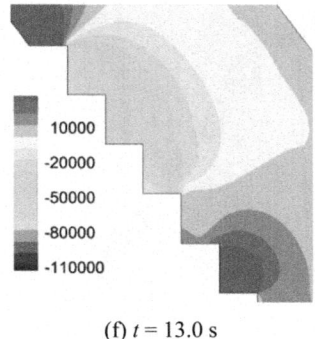

(e) $t = 5.0$ s   (f) $t = 13.0$ s

Figure 3.   Continued.

According to Figure 3, there was a strong high-pressure area on the first step and a strong negative-pressure area on the second step when $t = 2.0$ s. When $t = 3.0$ s, the strong high-pressure area on the first step and the strong negative-pressure area on the second step increased, and there was a small-scale weak high-pressure area at the concave corner of the third step, a weak negative-pressure area on the step surface, a strong high-pressure area on the fourth step, especially at the concave corners and the upstream and downstream convex corners, and a weak high-pressure area at both the concave and convex corners of the fifth and subsequent steps. When $t = 4.3$ s, the negative-pressure area on the second step decreased, and that on the third step increased; there were local weak high-pressure areas at the top and bottom of the fourth step surface and a strong high-pressure area on the fifth step; weak high-pressure areas at concave and convex corners of the sixth step increased. When $t = 4.7$ s, the second, third, and fourth steps were in a weak negative-pressure area, and the strong high-pressure area on the fifth step shifted downstream. When $t = 5.0$ s, the whole weak negative-pressure area on the second, third, and fourth steps extended to the fifth step, and there was a strong high-pressure area on the sixth step. When $t = 13.0$ s, the weak negative-pressure area extending to the fifth step was replaced by a weak high-pressure area, and the subsequent pressure distribution had no obvious change.

3.3   Flow line on the overflow dam face

Figure 4 shows the distribution of flow lines at the end junction of the weir crest curve section at different moments.

According to Figure 4, the vortex on the first step was larger, and there was a single vortex on the second and subsequent steps when $t = 2.0$ s. When $t = 3.0$ s, the vortex on the first step became very small, and that on the second step increased and tended to combine with the vortex on the third step. When $t = 4.4$ s, the vortices on the second, third, and fourth steps were combined into a whole and far away from the concave corners, and there were small-scale vortices at the concave corners of the second and third steps. When $t = 13.0$ s, the vortices at the concave corners of the second and third steps and on the fourth and fifth steps increased, and those on the second and third steps away from the concave corners decreased, but the flow line was dense and the flow velocity was high, and the subsequent flow line distribution did not change greatly.

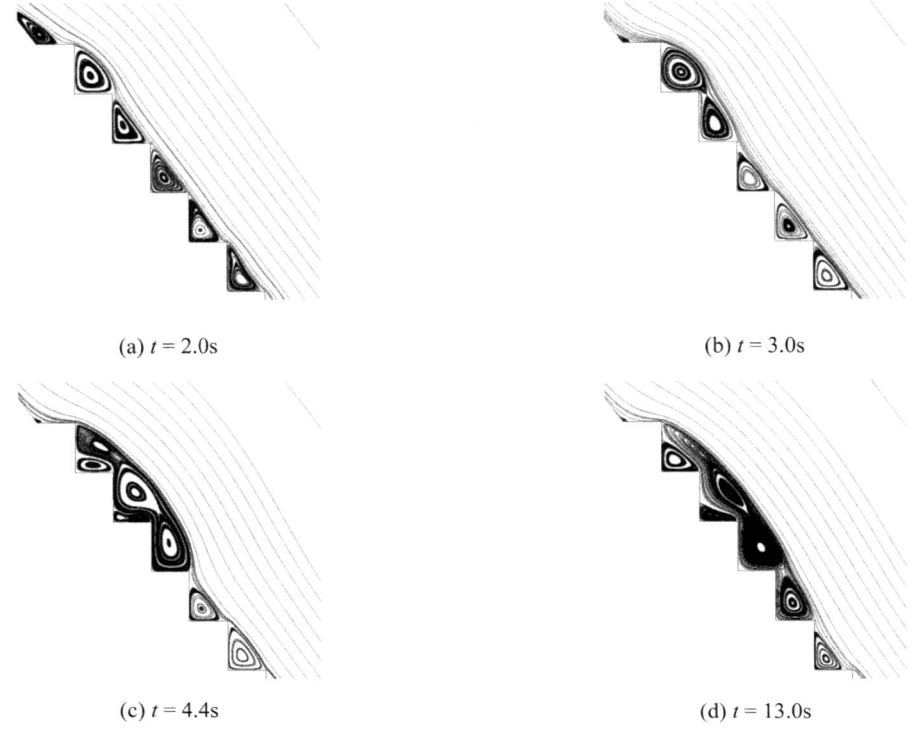

(a) $t = 2.0$s   (b) $t = 3.0$s

(c) $t = 4.4$s   (d) $t = 13.0$s

Figure 4. Distribution of flow lines on the overflow dam face.

4 CONCLUSION

According to the unsteady numerical simulation analysis of the flow field at the upstream junction of the stepped overflow dam face, we obtain the following conclusions:

(1) The lower linear edge of the lower drainage nappe did not adhere to the overflow dam face at the beginning, but formed a cavity at the concave corners of steps. The cavities on the second, third, and fourth steps changed most dramatically, and reverse flows existed along the step surface; the first step was filled with the liquid phase quickly, whereas the second step was slowly;

(2) The second, third, and fourth steps had wide negative pressure areas, which may lead to cavitation;

(3) Each step in the flow line produced a vortex structure; the vortices on the second, third, and fourth steps were integrated, and there were many vortex structures on the second and third steps.

ACKNOWLEDGEMENT

This research is supported by the Yunnan Provincial Department of Education Science Research Fund Project.

REFERENCES

Chen W.X., Yang J.R. Analysis on the aeration characteristics of combined steps based on orthogonal design [J]. *Journal of Wuhan University* (Engineering Edition), 2021, 54(11): 1008–1014. doi: 10.14188/j.1671-8844.2021-11-2014.

Hirt, Cyril W., and Billy D. Nichols. "Volume of fluid (VOF) method for the dynamics of free boundaries." *Journal of computational physics* 39.1 (1981): 201–225.

Launder B.E. Spalding D.B. The numerical computation of turbulent flows[J]. *Computer Methods in Applied Mechanics and Engineering*, 1974, 3: 269.

Lin J.Y., Wang G.L. *Hydraulic Structures* (5th edition) [M]. Beijing: China Water & Power Press, 2009.

State Key Laboratory of Hydraulics and Mountain River Development and Protection, *Sichuan University. Hydraulics Volume I* (5th edition) [M]. Beijing: Higher Education Press, 2016.

Wang Q., Yang J.R., Long Y.S., et al. Influence of transition steps with different combinations on pressure characteristics of stepped overflow dam face [J]. *Hydropower*, 2018, 44(03): 42–46.

Wu X.S. Preliminary study on hydraulic characteristics of stepped overflow dams [J]. *Sichuan Water Power*, 1998(01): 73–77+95.

*Frontiers in Civil and Hydraulic Engineering – Mohamed A. Ismail and
Hazem Samih Mohamed (Eds)*
© 2023 The Authors, ISBN 978-1-032-38247-0

Discussion on pipe jacking construction technology and safety management

Zhaoxiao Yong* & Ran Gang
*Chengdu Municipal Waterworks Co., Ltd., Chengdu Environment Investment Group Co., Ltd.,
Chengdu, China*

ABSTRACT: With the increasing construction of municipal projects, the new construction technology has a great development space. Pipe jacking construction technology is gradually formed on the basis of a complex environment. The article firstly introduces the overview of the water pipeline project of Chengdu city beltway, then elaborates on the basic principle of pipe jacking construction, related types and advantages and disadvantages, and finally discusses the causes of safety risks and countermeasures in the process of pipe jacking construction, in order to provide a reference basis for effective implementation of the pipe jacking construction technology and safety risk control in the construction process.

1   INTRODUCTION

The construction of the water pipeline of Chengdu city beltway needs to cross the Margo manor in Xindu district of Chengdu city, the cross section is distributed in the east-west direction, starting from the east side of the manor and ending at the southwest side of the parking lot, with a total length of about 174 meters. There are artificial landscapes, ponds, structures, and living facilities on the surface of the crossing section, and the surrounding environment is complicated. According to the site geological exploration and underground pipeline detection results, the pipe jacking section is mainly distributed at the silty clay layer, and shallowly buried and interspersed with all kinds of municipal pipelines, which, in essence, belongs to the typical large-diameter and large-buried-depth urban complex regional sand layer pipe jacking construction. Therefore, considering the construction requirements and pipe jacking conditions, DN2436mm steel pipe is used for pipe jacking, with an average burial depth of about 9.0 m. In addition, the pipe jacking working well is 11.0 m in inner diameter and 900 mm in wall thickness, and the receiving well is 5.5 m in inner diameter and 450 mm in wall thickness.

2   PRINCIPLE, TYPE, AND MODE SELECTION OF PIPE JACKING CONSTRUCTION

2.1   *Principle*

In the process of pipe jacking, the excavation of the soil at the front of the pipe is carried out manually or mechanically, and the cavity is excavated according to the actual pipe diameter and shape, and then the pipe is pushed into the cavity by the jacking device, which is a continuous cycle until the pipe is all pushed into the designed area. The specific jacking principle is shown in Figure 1.

---

*Corresponding Author: mrzhao@vip.126.com

DOI 10.1201/9781003344209-44

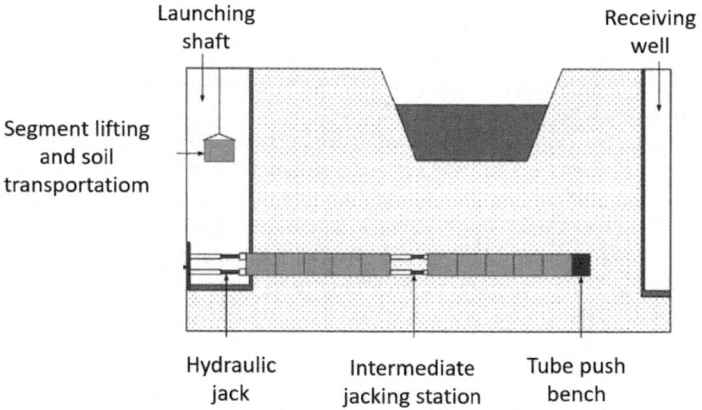

Figure 1. Principle of pipe jacking construction.

2.2 *Classification of pipe jacking method*

According to the different ways of excavation in front of the pipeline, the pipe jacking method can be broadly divided into two categories: manual pipe jacking and mechanical pipe jacking (refer to Figure 2). Manual pipe jacking refers to the manual excavation method for the front end of the pipeline soil excavation, and mechanical pipe jacking means mechanical excavation for the front end of the pipeline soil excavation. According to the different working principles of the jacking head, mechanical pipe jacking can be divided into soil pressure balance type, mud and water balance type, and pneumatic balance type. In the actual construction process, it is necessary to consider the construction site, engineering geological conditions, construction safety risks, construction costs, construction period, and other factors, and choose one or more ways to combine the pipe jacking construction method.

(a) Manual pipe jacking   (b) Mechanical pipe
                                   jacking

Figure 2. Type of pipe jacking method.

341

2.3 Selection of construction method for pipe jacking

This section is located at the silty clay layer, which is smooth with high humus content and high natural water content mostly under a soft plastic or semi-liquid plastic state, accompanied by low foundation strength and large deformation.

It's reported that in the engineering case of artificial pipe jacking through the silty clay layer, the extrusion method is selected to work in this geological layer (Yan 2017). Other researchers have discussed the range of soil suitability and engineering examples of mechanical pipe jacking. The results show that mechanical pipe jacking is suitable for a wide range of soils including clay layers (Meng 2016). The above studies have shown that both manual and mechanical pipe jacking are suitable for silt-clay soils, but multiple factors need to be considered to select the pipe jacking method. For this reason, the construction cost and duration are further discussed in the context of the actual project.

From the perspective of costs, under the same construction distance, the cost per meter of manual pipe jacking is 26609.86 yuan, the total price is 4630116 yuan; the cost per meter of mechanical pipe jacking is 36626 yuan, the total price is 6372924 yuan, which is 1876524 yuan higher than the total cost of manual pipe jacking, so the construction side prefers to use the manual pipe jacking method. In terms of the construction period, mechanical pipe jacking is featured with high efficiency, fast construction, and a short construction period (only 20 days to complete the construction, while manual pipe jacking is expected to last 30 days). However, since both construction methods meet the construction target set by the builder, the duration has little influence on the selection of the construction method.

In summary, the builder finally adopted the manual pipe jacking method.

Table 1. Cost comparison of different pipe jacking construction methods.

Construction method	Distance (m)	Comprehensive unit price (m/yuan)	Total (yuan)
Manual pipe jacking	174	26609.86	4630116
Mechanical pipe jacking	174	36626	6372924

3 PIPE JACKING CONSTRUCTION DESIGN

3.1 Calculation of pipe jacking force

In the process of pipe jacking construction, the calculation of the jacking force is a very important part, which needs to be calculated according to the actual situation.

According to the actual position of the designed water supply pipe and related data, the pipe jacking parameters are statistically sorted out. The top of the water pipe is covered with the soil of 6.18 m, the diameter of the steel jacking pipe is D=2436 mm, and the thickness of the pipe wall is 18 mm.

(1) Head-on resistance of pipe jacking machine

$$N_F = \pi/4 D_g^2 P = 3.14 \div 4 \times 2.6^2 \times 400 = 2122.64 \text{ kN} \tag{1}$$

where N_F = head-on resistance of pipe jacking machine; D_g = outer diameter of pipe jacking machine (m), being 2.6 m; P = soil pressure (kPa), being 400 kPa.

(2) Total pipe jacking force

$$F_0 = \pi D_1 L f_K + N_F = 3.14 \times 2.436 \times 174 \times 9 + 2122.64 = 14101.03 \text{ KN} \tag{2}$$

where F_0 = standard value of total pipe jacking force (kN); D_1 = outer diameter of water pipe (m), is 2.436 m; f_K = average frictional resistance of pipe wall to soil (kN/m^2), being 9 kN/m^2 in this project using the thixotropic mud drag reduction technology.

According to the maximum jacking force value of the pipe provided by the design, it can be seen that the DN2436 pipe jacking in this project cannot meet the maximum effective distance, and a relay room needs to be set up.

3.2 *Number of intermediate jacking stations*

$$N = \frac{\pi D_1 f_k(L+50)}{0.7 \times f_0} - 1 = \frac{\pi \times 2.436 \times 9 \times (174+50)}{0.7 \times 7762} - 1 = 1.83 \approx 2 \tag{3}$$

where f_0 = allowable pipe jacking force of each intermediate jacking station (kN); N = the number of intermediate jacking stations.

In order to prevent the steel pipe from being deformed, it is necessary to set up intermediate jacking station when the jacking force reaches 80% of the design value.

$$7762 \times 80\% = 2122.64 + 3.14 \times 2.436 \times 9 \times L \tag{4}$$

So, we can get that L is 59 m. The jacking force reserve value of the second intermediate jacking station should not be less than 30%. Similarly, the second intermediate jacking station should be arranged within 126 meters behind the pipe jacking machine.

The situation of the intermediate jacking station is shown blow: The first one is 67 m behind the pipe jacking machine, and the second one is 126 m behind the pipe jacking machine.

4 ADVANTAGES AND DISADVANTAGES OF PIPE JACKING CONSTRUCTION TECHNOLOGY

4.1 *Main advantages*

In the construction of urban municipal engineering projects, the pipe jacking method is often used to achieve the crossing of public (railway) roads, buildings, and rivers, reduce the cost of bypass construction and prevent the waste of land resources. Moreover, the application of pipe jacking technology does not cause traffic breaks, which is conducive to improving the overall construction effect of municipal engineering. In addition, the use of pipe jacking technology in municipal engineering construction will not produce noise, dust, or other environmental pollutants.

In this project, a full-scale garden landscape and pavilions have been built on the surface of the cross section. If the open excavation construction method is adopted, although the construction cost is lower than the pipe jacking, the overall investment costs will be significantly increased considering tens of millions of compensation costs. Although pipe jacking technology per meter cost is higher than the open excavation, underground construction avoids the destruction of surface buildings and landscapes, saving a huge amount of compensation costs. This reflects the advantages of pipe jacking technology in cost control.

In addition, there are famous tourist attractions near this pipe jacking construction section, and the flow of people is large. The pipe jacking construction can avoid the impacts of dust, noise, and other environmental pollutants on the surrounding area, and create a construction image of municipal engineering.

4.2 *Main disadvantages*

In the operation of pipe jacking construction technology, the use of multi-curve combination construction will increase the difficulty in the process of pipe jacking construction, and may also lead to the failure of pipe jacking construction progress control. If a soft soil layer is encountered during the pipe jacking construction, this will lead to the occurrence of deviation in the pipe jacking construction, and the pipelines laid in the project will also settle, and the accuracy of the operation of pipe jacking construction technology will be affected, so it is not suitable for soft soil.

The progress and quality of construction of this pipe jacking section are constrained by the geology of the operating section and underground obstacles. On the one hand, the silt-clay soil layer is generally adaptable to the pipe jacking technology, and an engineering case of the collapse of a pipe jacking project in the process of digging in the clay layer has been reported (Chen 2019). On the other hand, in the development process of urbanization construction in recent years, the soil used in some landscape improvement areas comes from construction slags, which contain large hard objects difficult to destroy and remove if the pipe jacking method is used.

5 SECURITY RISK ANALYSIS

The pipe jacking construction method is widely used because of various advantages such as less land occupation, less damage, less impact on traffic, a short construction period, low comprehensive cost, etc. However, while we choose the pipe jacking method, we are also bound to bear the potential risks attached to the pipe jacking project such as large burial depth, complex geology, deep pit, under-passing structures, and crossing complex pipelines. In the construction process, if good risk analysis and management are not carried out in advance, it may lead to safety accidents. In order to avoid accidents, the main safety risks in this pipe jacking project are now analyzed as follows.

5.1 *Hydrogeological conditions*

According to the drilling data, the groundwater within the construction scope of the project is mainly divided into two categories: pore water and bedrock fissure water. Groundwater is mainly stored in the Quaternary unconsolidated sedimentary rocks, usually recharged by atmospheric precipitation, surrounding surface water, and part of the groundwater. The buried depth of the water level is stable at 2.60–5.60 m below the ground all year round, the variation of the buried depth of the water level is 3.0 m, and the annual variation of the groundwater level is about 1.0 m. The pipeline is mainly located in the bedrock, the lithology is mudstone and sandstone, being mainly moderately weathered, some sections are strongly weathered, and the water richness is weak to moderate, there will be large water leakage locally during the construction, so it is necessary to strengthen the water stopping and drainage prevention measures.

5.2 *Deep foundation pit*

According to the "Measures for the Safety Management of Sub-item Projects with Greater Risk", the excavation of earthwork exceeding 5 meters is a deep foundation pit project. The depth of the project pipe jacking construction reaches 9 meters. Therefore, the pipe jacking work well belongs to the deep foundation pit, which requires the timely and effective implementation of support measures during the construction of the work well to avoid the collapse of the foundation pit. After the completion of the work well construction, the staffing requirements of the wellhead materials and earth and the quality of the wellhead edge protection facilities should be strictly controlled, and the ladder should be stable for people. Once the relevant protective facilities are not set up in a standard way, there is a possibility of safety accidents such as falling objects from height.

5.3 *Underground pipelines and surrounding structures*

There are obstacles on the surface and underground of the project operation area. Before the construction of pipe jacking, the distribution of existing buildings and pipelines within the construction area should be investigated in detail, and a targeted reinforcement plan should be formulated according to the structure form, location, and other relevant parameters of the building facilities. If the actual situation of existing structures or pipelines in the surrounding area cannot be ascertained, and the construction is carried out blindly, there is a possibility of cracking, the collapse of structures, or destruction of pipelines, and other problems.

5.4 *Harmful gases*

There are often different strata in the pipe jacking construction area, so the pipe jacking construction process may cross the humus layer, silt layer, etc., and these humus layers that exist in the underground structure all year round are very likely to contain a large amount of toxic gas (Wang 2016). If the detection of toxic and harmful gases is not carried out according to the actual situation, and the ventilation effect of the working environment is not good, the problem of poisoning or suffocation of construction workers is very likely to occur.

5.5 *Subjective factors*

The subjective factors are mainly caused by the operating personnel. The level and quality of the construction team are uneven, and some construction personnel may not pay attention to the safety in production, and the management staff fails to carry out safety awareness training and construction technology disclosure in a timely manner, which can easily lead to the illegal operation or poor implementation of safety measures in the actual construction, and then lead to the occurrence of production safety accidents.

6 SECURITY CONTROL

6.1 *Organizational measures*

To set up a safety management organization at the construction site, the project manager is responsible for the project safety, according to the scale of the project with sufficient professional and technical management personnel and professional safety management personnel (He 2020; Zhu 2020). Each post should undertake clear safety responsibilities, and professional and technical personnel and safety management personnel should both divide the work and cooperate with each other, forming a clear safety management structure, from the construction management organization structure to the construction safety.

6.2 *Technical measures*

(1) The investigation of hydrogeology in the area of pipe jacking path should be strengthened, which serves as the basis to develop the design of pipe jacking.
(2) By reference to the investigation and exploration report, the excavation and support plan of the foundation pit should be made in strict accordance with the specification.
(3) Before construction, a detailed pipeline survey should be conducted along the pipe jacking path to clarify the relative position of each pipeline and the new pipe jacking pipeline.
(4) Protective measures should be taken for limited space operations during construction, and the gas in the pipeline should be detected on a daily basis as required.

Before the construction of the pipe jacking project, a detailed survey of the surface structures and underground pipelines was conducted, and a special program for the protection of structures and pipelines was formulated to minimize the impacts of construction on the surrounding obstacles.

The project management targeted to strengthen safety management, especially when limited space operations are involved, and to maintain 24-hour mandatory continuous ventilation, with air quality tested at 15-minute intervals, accompanied by first-aid drugs and rescue supplies. In the end, no poisoning or suffocation accident occurred during the project implementation stage.

6.3 *Inspection measures*

As the main body of construction project management, on the one hand, the training of managers and operators should be strengthened, and the awareness of personnel should be enhanced to ensure fine management of supervisors and safety operations and further effectively weaken the safety risks subjectively caused (Ma 2007). On the other hand, a safety inspection system should be established with regular inspections and checks in order to eliminate safety hazards in the bud. The focus should be on clear safety risk control in the whole construction, and the inspection patrol should be strengthened.

7   CONCLUSION

Although the pipe jacking construction technology has many advantages, there are still many significant safety risks in the implementation process of the pipe jacking project. All staff must constantly identify and analyze the safety risks and control them effectively so that the pipe jacking construction method can give full play to its advantages on the premise of ensuring safety.

ACKNOWLEDGMENTS

The authors gratefully acknowledge the support from Chengdu Environment Investment Group Co., Ltd. (0426KY020010).

REFERENCES

Chen, R. (2019). Example of pipe jacking machine selection and accident handling under unknown geological conditions. *J. China Standardization*, 12, 76–77+80.

He, Z.J. (2020). Safety risks and management measures of pipe jacking construction. *J. Engineering Technology Research*, 5, 178–179.

Ma, B. & Wang, W.H (2007). Safety management points in pipe jacking construction. *J. Construction Safety*, 10, 7–9.

Measures for the Safety Management of Sub-item Projects with Greater Risk. S. Decree No. 37 of the Ministry of Housing and Urban-Rural Development, PRC, 2018, 6.

Meng, X.H. & Pang, D.G. (2016). Selection and application of construction technology of sewer pipe jacking under complex geological conditions. *J. China Water Supply and Drainage*, 32, 134–138.

Wang, P. (2016). Discussion on construction management technology of long-distance large diameter pipe jacking. *J. China Water Supply and Drainage*, 32, 119–122.

Yan, Z.Q. & Zhong, X. (2017). Problems of pipe jacking method construction in complex strata and the solution measures taken. *J. Sichuan Hydropower*. 36, 85–87.

Zhu, R. (2020). Analysis of safety risks and management measures of pipe jacking construction. *J. Electrical and Mechanical Information*, 5, 68–69+71.

*Frontiers in Civil and Hydraulic Engineering – Mohamed A. Ismail and
Hazem Samih Mohamed (Eds)*
© 2023 The Authors, ISBN 978-1-032-38247-0

Analysis of the impact of declining rate on the internal force of the unstable bank slope

Yanqin Dai*
Yunnan Water Resources and Hydropower Vocational College, Kunming, China

Wenchao Xu*
Zhejiang Water Group, Jiaxing, Zhejiang, China

Feng Pan* & Yuehong Sha*
Yunnan Water Resources and Hydropower Vocational College, Kunming, China

ABSTRACT: Influenced by the water level fluctuation, rockfall and landslides have taken place many times on the lining highway slope of the Jinsha River Hydropower Station Reservoir area, so it is necessary to analyze and calculate the stability of the reservoir bank. Through the GEO-SLOPE software, a two-dimensional highway slope seepage model is established to simulate the water level fluctuation of the reservoir. The results show that as the water level drops more rapidly, the bank becomes more stable, with a low pore water pressure and increasing normal force and shear resistance, and the safety coefficient of soil will be improved. In this study, related slope protection measures are proposed, expecting to provide a basis for slope treatment and the construction of new highways.

1   INTRODUCTION

The slumped area of the reconstruction and expansion project from Lijiang Ancient Town of LiNing Road to NINGLANG Kaiyuanqiao section is located in the V-shaped valley terrace topography of the Jinsha River, which is deeply cut into the middle and high mountain canyon and is an ancient collapse-type landform area. The collapsed section of K34+899.191–K35+860 is located in the reservoir area 12 kilometers upstream of the Dam of Jin'anqiao Power Station of Jinsha River Cascade Power Station. The subgrade elevation of this section is low, the elevation of the section subgrade is 1423.741 meters, the normal water level of the reservoir is 1418 meters, and the height difference is only 5.741 meters. The center line of the roadbed is close to the shore, which is only 8.32 meters. The rising water level changes the water content of the reservoir bank soil, which increases the bulk density and decreases the shear strength, and causes the soil to soften, collapse and even collapse. When the water surface expands, the wind and waves increase, and the reservoir bank is eroded, impacted, and washed by waves, the lower soil is washed out and wave erosion niches are formed, the upper soil loses support and collapses, and the shoreline moves backward. The slippage is transported and sorted by water flow to form underwater shoals. Under the continuous action of water and wave, the collapse of the reservoir bank continues to happen, and the shoreline moves backward continuously and has a trend of continuous development and extension. Part of the roadbed of the original highway has collapsed, as shown in Figure 1. The previous treatment plan is to take the slope gradient of the upper slope as 59° according to the

---

*Corresponding Authors: 276851849@qq.com; 244227332@qq.com; 513527033@qq.com and 596130471@qq.com

slope ratio of 1:0.75 to the slope (53°) of the subgrade for the step excavation, the edge of the subgrade is provided with a drainage ditch, and the upper slope foot is set with a retaining wall for protection (Xu 2016). However, the bank slope angle of this reservoir is steep, the slope lithology is relatively soft, the soil water content in the lower part of the subgrade is relatively high, the soil layer lithology is relatively soft, and the strength is low. Under the action of reservoir water, rainstorm or earthquake, some sections of roadbed still collapse, and the operation of highway is seriously affected, so it is necessary to calculate the stability of the reservoir bank (Liu 2005; Song 2004).

Figure 1. Collapse diagram of Kuan-bank highway (Xu 2015).

2 NUMERICAL SIMULATION AND ANALYSIS

2.1 *Model establishment*

In the actual situation, the water level of the reservoir plummets, and the slope drainage has a lag effect, so slope drainage is considered. In this paper, the SLOPE drainage seepage model is established by geo-slope software and the Morgenstem-price method of limit equilibrium. The seepage boundary of the original river level of Jinsha River is the initial water head, the fixed water head boundary is below the initial water level, the zero flow boundary is above it, and the impermeable boundary is the side and bottom of the mountain. The normal and declining conditions of reservoir water level can be reflected by changing the boundary conditions. The total water head is taken as a variable and a function of time. Using triangular and quadrilateral elements, the two-dimensional seepage model generates 10,317 grid elements with 10,417 nodes. In this paper, the fifth exploration section of mileage pile number K35+460.000 is taken as the calculation section to simulate the variation relationship of normal force, pore water pressure, and shear force under the condition that the reservoir water level drops to the same water level at different rates.

Reservoir bank materials are divided into three zones. Based on the borehole data, a geotechnical test of engineering investigation, relevant literature, and the inversion method, the values of reservoir bank materials are shown in Table 1, in which the unsaturated permeability coefficient function is predicted by the Van Genuchten model (Xu 2006).

Table 1. Values of reservoir bank materials.

Bank material	d_{60} (mm)	γsat (kN/m³)	Φ (°)	C (kPa)	c' (kPa)	Drop value after immersion Φ	Drop value after immersion C	K (cm/s)	k_s (cm/s)	Saturated water content (m³/m³)	Liquid limit (%)
Gravel 1	77	20.4	30.7	13.3	13.3	3	7	1.16×10^{-4}	9.26×10^{-4}	0.2	26
Gravel 2	75	21.9	30.7	15.3	15.3	3	7	9.26×10^{-5}	4.32×10^{-4}	0.18	27
Basalt	–	27.1	45.3	1420	1420	5	100	6.95×10^{-4}	5.26×10^{-6}	–	–

In this paper, the water level of the reservoir is simulated to drop from the normal storage level of 1418.0 meters to 20 meters from the dead level of 1398.0 meters, the minimum safety factor is obtained by simulating four working conditions: 5 days down 20 m, 10 days down 20 m, 15 days down 20 m, and 20 days down 20 m. We find out the most easily landslide failure surface of slope body through the automatic search function for analysis.

2.2 Numerical analysis

In this paper, the SLOPE drainage seepage model is established by geo-slope software and the Morgenstem-price method of limit equilibrium. The minimum safety factor of the bank under different descent rates is obtained, as shown in Table 2.

Table 2. Minimum safety factors under different working conditions.

Working condition	It dropped by 20 m in 5 days	It dropped by 20 m in 10 days	It dropped by 20 m in 15 days	It dropped by 20 m in 20 days
Minimum safety factor	0.448	0.486	0.499	0.502

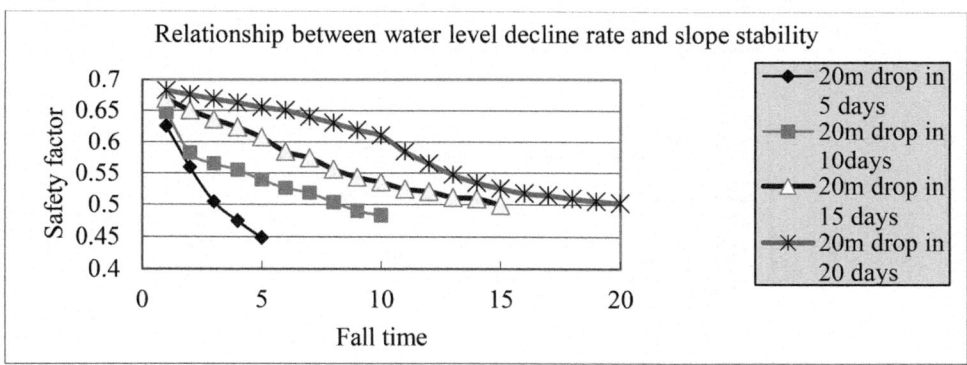

Figure 2. Relationship between water level decline rate and slope stability.

It can be seen from Figure 2 that in the process of reservoir drainage, the water level drops, and the support of water to the reservoir bank disappears, causing the safety factor corresponding to different precipitation rates to decrease monotonically. In addition, it takes a shorter time to lower the same water level, and the safety factor of the reservoir bank tends to decrease.

It can be seen from Table 2 that the minimum safety factor under various working conditions is far less than 1, and the slope has begun sliding or is prone to landslide. This situation is consistent with the actual situation. The length of drainage time of the bank slope depends on the influence of the precipitation rate. When the water level drops to the same level, the longer time it takes, the lower the decline rate, the longer the drainage time of the bank slope, and the safety factor tends to increase. Therefore, slope stability is inversely proportional to the precipitation rate.

2.3 Internal force analysis

The internal force variation of the soil strip with the minimum safety factor is analyzed when the precipitation rate of the reservoir decreases by 20 m in 5 days, 20 m in 10 days, 20 m in 15 days, and 20 m in 20 days, respectively. The results are shown in Table 3 and Figures 3–5.

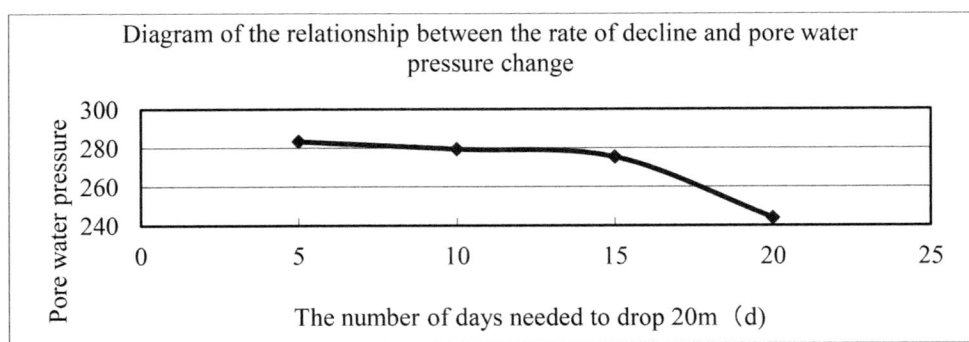

Figure 3.   Relationship between decline rate and pore water pressure change.

Table 3.   Relationship between decline rate and internal force change.

Drop rate	Pore water pressure (KPa)	The normal force (KN)	Shear (KN)	Normal force/ shear force
It dropped by 20 m in 5 days	283.27	350.06	19.943	17.55303
It dropped by 20 m in 10 days	279.36	387.09	23.63	16.38129
It dropped by 20 m in 15 days	275.22	417.26	28.932	14.42209
It dropped by 20 m in 20 days	243.84	424.46	45.066	9.41863

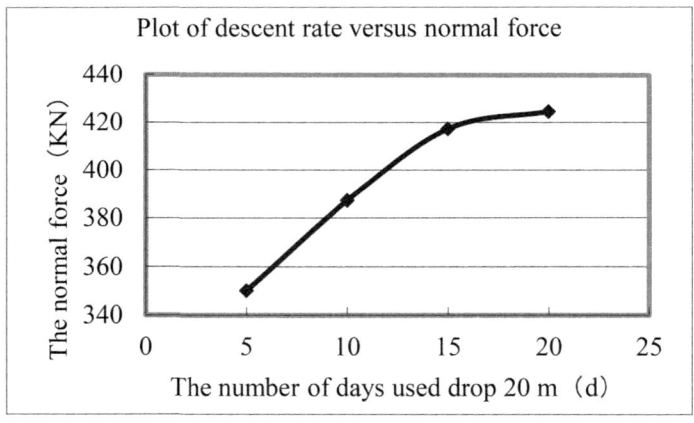

Figure 4.   Relationship between descent rate and normal force.

From Table 3 and Figures 3–5, it can be known that when the water level declines from the normal storage level down to the dead level by up to 20 days, the internal pore water pressure is constantly decreasing within five days, so are the normal force and the shearing resistance, only the normal force rendering on the growth of the convex and concave under shearing resistance present growth. As the water level drop rate constantly decreases, the matrix suction is bigger, the pore water pressure decreases, the saturation line falls slowly, the stability of the soil bank slope in the normal storage level still can maintain, and the internal normal force is constantly increasing, the soil shear resistance also increases unceasingly, and the sliding resistance of soil increases relatively. The safety factor of soil is increasing.

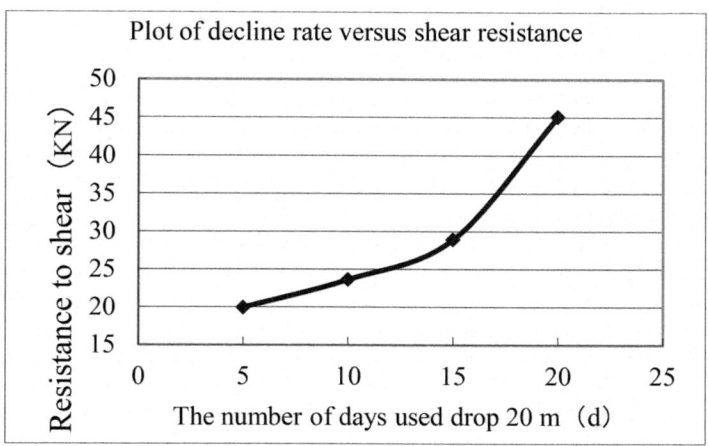

Figure 5.   Relationship between decline rate and shear resistance.

3   IT IS SUITABLE FOR SLOPE PROTECTION MEASURES OF LINING HIGHWAY IN JIN'ANQIAO RESERVOIR AREA

There have been dozens of landslides on the slope of Lining Highway in Jin'anqiao reservoir area. The width of the landslide is 2–15 meters, the vertical depth is 8–30 meters, and the slope is 37°–53° . All kinds of deformation forms are widely distributed and threaten the stability of the proposed highway. The slope collapse under the subgrade in Jin'anqiao reservoir area is a typical gravel-soil collapse type, while the slope collapse on the subgrade is a traction gravel-soil slip type. The following protective measures are proposed for the bank collapse in the reservoir area:

(1) Because of the high groundwater level and strong seepage effect, the soil of the lower slope is seriously softened, the shear strength is low, and it is easy to slide. Therefore, reinforced concrete row anti-slide piles should be used at the base of the subgrade and the lower slope. And considering the high river level and wave erosion effect, appropriate masonry protection should be done. However, the lower slope is steep, the canyon is deep, the river is winding, the water flow is rapid, and the effect of riprap may not be good, so it is advisable to adopt the method of soil nail, anchor pipe, and hinge sink (Zhu 2002).
(2) Due to the high underground water level and low bearing capacity of the foundation, a small dead weight is selected. The supporting wall of abandoned earthwork can also be used. Slope cutting, step excavation, and retaining wall should be used together on the slope of the roadbed. The retaining wall should be equipped with drainage holes to facilitate drainage. The upper slope is a high and steep slope, and slope protection should be carried out to prevent heavy rain from causing debris flow and damaging road facilities. Reinforced concrete lattice should be used in combination with grass protection by spraying. In addition, an anti-slide pile can also be combined to prevent shallow sliding failure and enhance the anti-sliding stability of the slope body (Zhang 2006).

4   CONCLUSION

In the Jinshajiang Power Station area, many landslides have occurred on the Lining highway, which affects the normal operation of the Lining expressway and threatens the safety of people's life and property. In this paper, the software is used to simulate the stable deformation of the gravel soil slope along the reservoir bank, and the following conclusions are drawn:

(1) In the process of reservoir water level decline, the support of water to the reservoir bank disappears, causing the safety factor corresponding to different precipitation rates to decrease monotonically. In addition, it takes a shorter time to lower the same water level, accompanied by a greater decline rate, and a smaller safety factor of the reservoir bank. Therefore, slope stability is inversely proportional to the precipitation rate.

(2) The reservoir water level drops from the normal storage water level to the dead water level, and the longer time it takes to drop to the same water level, the lower the decline rate will be. The pore water pressure inside the reservoir bank keeps decreasing, the matrix suction is relatively large, the normal force increases, and the shear resistance keeps increasing. The safety coefficient of soil also increases with the relative increase of anti-sliding force, but it is still less than 1 and in an unsafe state (Chen 1996).

(3) When the water level of the reservoir falls at the same height and at different rates, the safety coefficient of the bank slope of the reservoir is less than 1. Because the bank slope of this section is formed by the accumulation of ancient landslides, it is consistent with the actual situation that landslides have occurred or continue to occur. Relevant slope prevention and control measures adapted to this section are proposed to ensure the stability of this section before the completion of the construction of the new road, provide relevant reference materials for the stability and safety of the roadbed of the target road, and provide geological information for the treatment of collapse.

A two-dimensional model is established by the software in this paper, while the slope stability problem has a significant three-dimensional effect, and the study is relatively inadequate. Hence, a better and more realistic model should be established in the future to study geotechnical slope engineering.

REFERENCES

Chen C.H., Deng B.Q. Journal of Chongqing Jiaotong University, Vol.15 Suppl. 1996:50–58.
Liu X.X., Xia Y.Y., Zhang X.S., et al. *Chinese Journal of Rock Mechanics and Engineering*, 2005, 24(8): 1439–1444.
Lu G.L., Tao Y.X. Cause analysis of bank collapse in sanmenxia reservoir area, henan province [J]. *Journal of Yellow River Conservancy Technical Institute*, 2002, Vol.14 No.1,14–5.
Song Y., Duan S.W., Chen S.W. Analysis on influencing factors of bank collapse of guanting reservoir [J]. *Design of Water Resources and Hydropower Engineering*, 2004(1):34–37.
Xu J.C., Shang Y.Q. *Chinese Journal of Rock Mechanics and Engineering*, 2006, 25(11):2264–2271.
Xu W.C. *Study on safety and prevention of bank collapse of canyon reservoir in mountainous area, Master dissertation*, Kunming University of Science and Technology, 2015.4
Xu W.C., Zhang L.H. et al., Study on anti-sliding stability of gravel soil slope in jin anqiao hydropower station reservoir area, *Journal of Hydraulic Power Generation*, 2016,
Zhang W.J., Zhan L.T., Ling D.S. *Journal of Zhejiang University* (Engineering Science), 2006, 40(8):1365–14278.
Zhu D.L., Ren G.M., Nie D.X. et al. Prediction of influence of reservoir water level change on landslide stability [J]. *Hydrogeology and Engineering Geology*, 2002, 45(3):6–9.

Frontiers in Civil and Hydraulic Engineering – Mohamed A. Ismail and
Hazem Samih Mohamed (Eds)
© 2023 The Authors, ISBN 978-1-032-38247-0

Research on seepage safety review of dam in Yuankou Reservoir

Lin-feng Luo*
Zhejiang Water Conservancy Science and Technology Promotion Service Center, Hangzhou, China

Xiao-cheng Xu
Zhejiang Guangchuan Engineering Consulting Co., Ltd, Hangzhou, China

ABSTRACT: Yuankou Reservoir is a medium-sized reservoir, and its dam type is a homogeneous earth dam. Due to the leakage problem of the dam, it was reinforced in 2005 to eliminate dangers, mainly using the dam foundation and bank composed of dam foundation curtain grouting and seepage prevention walls. In order to fully evaluate the effectiveness of the seepage prevention measures in the Yuankou Reservoir and the current seepage safety issues, this paper analyzes the dam seepage safety measurement data of the Yuankou Reservoir for more than 10 years. With the dam finite element simulation calculation results combined, the effectiveness of the seepage prevention measures and the seepage safety of the dam of the Yuankou Reservoir are comprehensively evaluated, and a relatively accurate and objective evaluation conclusion is obtained, which can provide a useful reference for the seepage prevention treatment measures and seepage safety evaluation of large earth dams.

1 INTRODUCTION

The Yuankou Reservoir in Wuyi County is a medium-sized reservoir that is mainly used for flood control and irrigation, combined with water supply, power generation, and aquaculture. The height is 43.30 m. The construction of the Yuankou Reservoir started in December 1968, the dam was completed in August 1978 and the preliminary acceptance was completed, and the project was completed and accepted in August 1980. During the operation, there was serious leakage of the water reservoir at the source mouth. In October 2005, reinforcement was carried out. In November 2006, the reinforcement of the dam was basically completed. A reverse filter area of 0.5 m in thickness and a transition area of 0.5 m in thickness outside the reverse filter area are set up. The rockfill area is between the external part of the transition area and the upstream and downstream dam slopes. To prevent the loss of clay particles, the clay core wall and the interval are set up. A layer of anti-filter geotextile is arranged. In the middle of the original clay core wall, a low-elastic modulus concrete anti-seepage core wall 80 cm in thickness is set. The bottom of the anti-seepage wall is about 0.5 m deep into the weakly weathered bedrock, the maximum depth of the wall is about 55 m, and the elevation of the top of the wall is 208.0 m. A C20 concrete pressure beam with a width of 0.8–1.6 m and a height of 3.0 m is set on the top of the anti-seepage wall, and a C20 concrete wave break wall with a height of 3.5 m is placed on the pressure beam, and the elevation of the dam crest is 214.50 m. In order to evaluate the effectiveness of the anti-seepage structure of the dam of the Yuankou Reservoir, by burying the piezometer and combining the analysis of the measured seepage data with the finite element calculation and analysis of the dam, as well as the on-site safety inspection, Zhan (2018) explored the seepage safety of the dam in the Yuankou Reservoir And the dam safety monitoring data is regularly compiled and analyzed (Jiang et al 2018;

---

*Corresponding Authors: 425700459@qq.com and 187369867@qq.com

Zhu & Ma 2021), which has guiding significance for the future seepage safety review of large and medium-sized reservoirs.

2 DAM ANTI-SEEPAGE TREATMENT

The anti-seepage structure of the dam of the Yuankou Reservoir mainly includes the original core wall of the dam body, the curtain grouting of the dam foundation, and the anti-seepage wall, so as to form a closed anti-seepage system with the dam foundation and bank slope.

(1) Geological conditions are good. The upper part of the upper and lower valleys of the dam is distributed with silty clay or sandy silt, with a thickness of about 1–2 m; the lower part is distributed with sand and pebbles, partly containing boulders, with a thickness of 0.5–9.7 m, being slightly dense to dense, and engineering geological conditions are general.
(2) During the construction of the reservoir, the cobblestone layer of the dam foundation shall be removed from the tooth grooves of the core wall and placed on the weakly to strongly weathered bedrock. There are many broken zones or faults in the bedrock, and only local deep excavation is carried out to seal with concrete or mortar. After the dam is reinforced, the anti-seepage wall can bear most of the seepage pressure, which reduces the possibility of seepage damage at the outlet of the core wall, which is beneficial to the seepage safety of the dam.
(3) The measured bedrock surface of the anti-seepage wall is basically the same as the predicted bedrock surface in the initial design stage, and the depth of the anti-seepage wall into the weakly weathered bedrock meets the design requirements. According to the test results of the concrete mechanical properties of the anti-seepage wall, the compressive strength, tensile strength, elastic modulus, permeability coefficient, and hydraulic gradient of the anti-seepage wall concrete all meet the design requirements.
(4) A total of 28 holes are completed for the foundation curtain grouting of the dam body near the two abutments, and the hole depth is 5.0 m below the relatively impermeable layer (q ≤5 Lu). In the curtain grouting of the dam foundation, the average unit cement injection amount in each sequence shows a decreasing trend, and the final hole water permeability is also less than the design control standard 5 Lu, which meets the specification requirements. After the curtain grouting is completed, three inspection holes are arranged in the grouting area. After the pressure water test, the water permeability of each section is q=0.28–2.1 Lu, all being less than 5 Lu, which meets the design and specification requirements for the seepage effect.

3 ANALYSIS OF SEEPAGE MONITORING DATA

3.1 *Layout of seepage monitoring facilities*

Seepage observation mainly refers to the monitoring of the seepage field generated by the dam under the action of the upstream and downstream water level difference, including the seepage pressure and seepage flow observation of the dam body and dam foundation (Mi & Xiong 2018; Zhang 2022). This seepage monitoring part mainly observes the contact seepage between the dam body and the dam foundation and the seepage around the dam.

(1) Seepage observation of dam body and dam foundation
   According to the seepage characteristics of the dam body and the dam foundation, four seepage observation sections are arranged along the dam axis, with pile numbers of dam 0+091.0 m, dam 0+193.0 m, dam 0+289.0 m, and dam 0+385.0 m, totaling 29 osmometers only.
(2) Observation of seepage around the dam
   According to the requirements of the specification and combining the characteristics of the Yuankou Reservoir dam, a total of 8 piezometers are arranged on both sides of the dam for seepage monitoring around the dam, including 3 on the left bank and 5 on the right bank; each piezometer has a built-in piezometer.

3.2 *Hydrograph analysis of dam seepage monitoring data*

The reservoir water level, rainfall, and piezometer water level monitoring data after calibration and adjustment from the end of July 2006 to the end of December 2017 are drawn into a process line, as shown in Figures 1 to 4.

(1) Section 0+091

The water level of the piezometer on the upstream side of the anti-seepage wall at the same elevation is significantly higher than the water level on the downstream side. It is higher than P1-4 on the downstream side, indicating that the anti-seepage effect of the anti-seepage wall is obvious.

(2) Section 0+193

The water level of the piezometer on the upstream side of the anti-seepage wall at the same elevation is significantly higher than the water level on the downstream side. It is higher than P2-4 on the downstream side, indicating that the anti-seepage effect of the anti-seepage wall is obvious.

Since the osmometers P2-4 (1) and P2-4 (2) have been in the process of continuous decline in water level since the reinforcement and reinforcement, the average water level has dropped by about 20 m, and there is no convergence trend. It shows that the seepage medium on the downstream side of this measuring point has changed significantly, and the permeability of the sand and pebble layer is increasing, which may be related to the lack of back-checking on the downstream side of the anti-seepage core wall. The wall bears most of the water head, but the permeability gradient is relatively large, which has reached about 30. Close attention should be paid to the piezometer of this section, especially the change of P2-3.

(3) Section 0+289

During the normal water storage period, the water level difference between the upstream and downstream sides of the foundation of the anti-seepage wall is large, and the water head decreases significantly. The head reduction of the downstream measuring point on the seepage prevention wall inside the core wall of the dam body is also obvious. Before April 2011, the upstream side P3-3 and P3-4 measuring points on the upstream side of the seepage prevention wall with an elevation of 175.0 m were higher than the upstream side. The water level on the downstream side was in line with the general law, but then the water level on the downstream side was higher than the water level on the upstream side, and the P3-3 measuring point was lower than the water level of the upper P3-1 and lower P3-5 measuring points on the same survey line, which is not in line with the earth-rock dam. The general law of seepage indicates that there is a seepage channel connected downstream of the concrete anti-seepage wall at this position. After April 2011, the anti-seepage effect began to gradually lose, and monitoring should be strengthened to analyze the data in a timely manner.

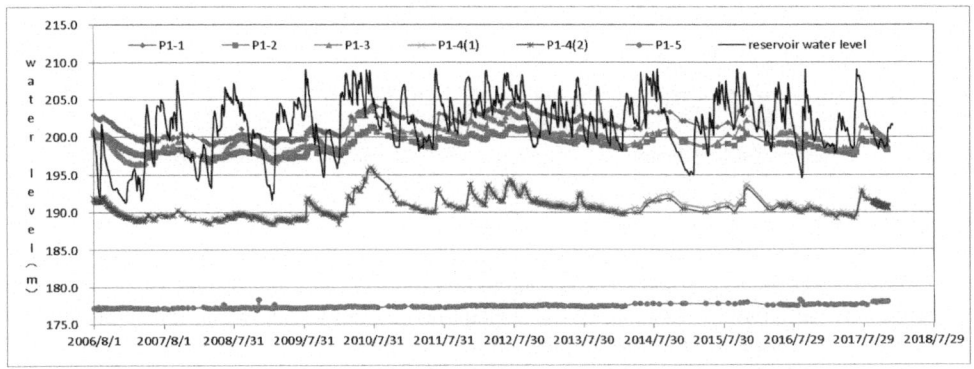

Figure 1.   Change process line diagram of seepage pressure and water level at section 0+091.

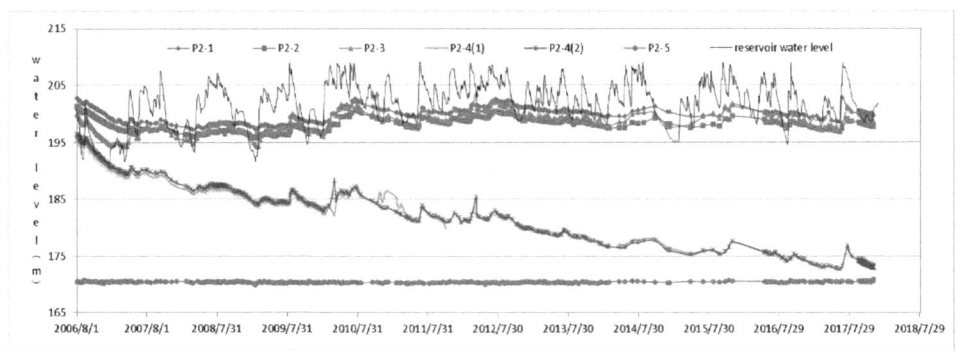

Figure 2.   Change process line diagram of seepage pressure and water level at section 0+193.

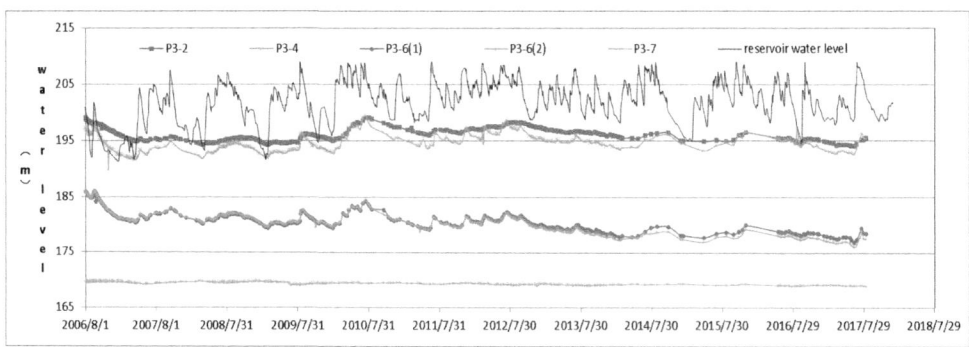

Figure 3.   Change process line diagram of seepage pressure and water level at section 0+289.

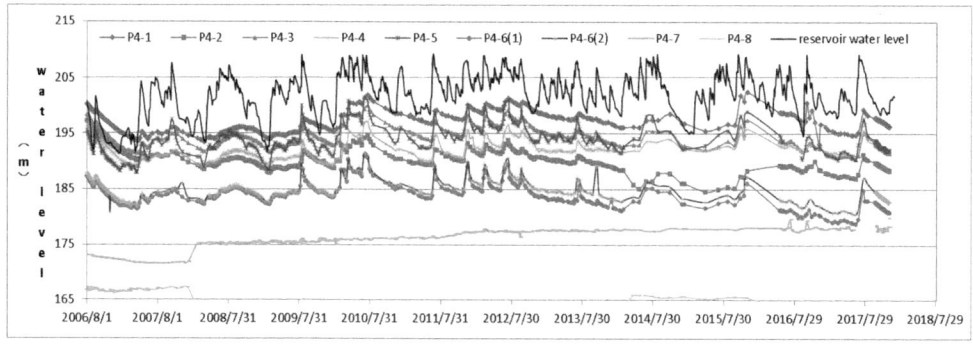

Figure 4.   Change process line diagram of seepage pressure and water level at section 0+385.

(4) Section 0+385

The water level of P4-1 at the same elevation is higher than that of P4-2, but the water level difference between P4-3 and P4-4 was large before 2015, and then tended to be consistent, which did not conform to the general seepage law of earth-rock dams, indicating that the anti-seepage wall loses the anti-seepage effect. The water level of P4-7 on the downstream side has a slow upward trend. Therefore, attention should be paid to its changes and the data analysis should be strengthened.

3.3 *Analysis of seepage monitoring data around the dam*

In order to monitor the seepage around the dam on both banks, a total of 8 pore water pressure gauges were buried. The observation began in July 2006. By arranging and calculating the measured data, the water level and time course of each measuring point are drawn as shown in Figure 5 to Figure 6. As can be seen from the figure:

(1) The measuring points L1, L2, and L3 are located on the left bank slope, and the buried elevation is high, so the water level of the piezometric pipe is high, which is higher than the reservoir water level, which is mainly affected by rainfall and the diving level of the mountain.

(2) There is a good correlation between the water level of the R1 measuring point on the right bank of the seepage around the dam and the reservoir water level, which is in line with the actual situation of the project. The measuring points R2 and R3 are located on the right bank slope, being mainly affected by rainfall and the phreatic level of the mountain, and the water level is higher than the reservoir water level in most periods; the measuring points R4 and R5 are located on the first- and second-level horse road slopes on the downstream side of the right bank of the dam, respectively. The piezometer water level is lower than the reservoir water level, but the correlation with the reservoir water level is not obvious.

In summary, the measured value is relatively stable, and there is no obvious upward or downward trend. There is less possibility of seepage around the dam on the left and right banks.

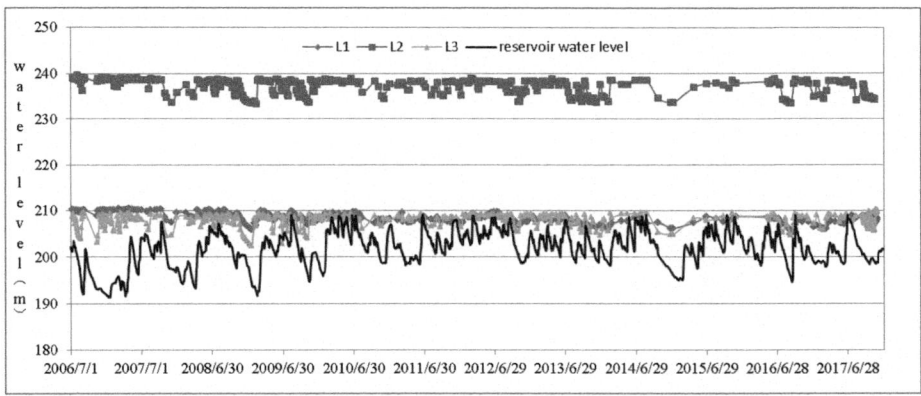

Figure 5.   Line diagram of the change process of seepage pressure and water level around the dam on the left bank.

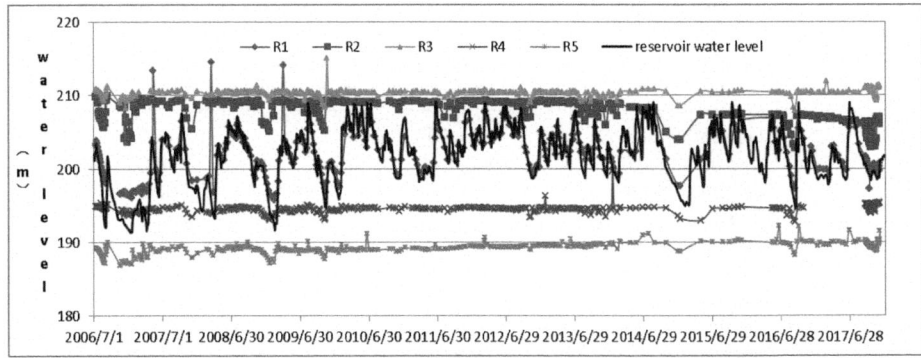

Figure 6.   Line diagram of the change process of seepage pressure and water level around the dam on the right bank.

4 SIMULATION CALCULATION AND ANALYSIS OF DAM SEEPAGE

4.1 *Calculated section and calculation parameters*

With reference to the geological survey data and the characteristics of the dam's own layout, the 0+385 section is selected as the typical section for the two-dimensional seepage finite element calculation and analysis. The section material partition is shown in Figure 7, and the seepage plane finite element mesh is shown in Figure 8. See Table 1 for the permeability coefficient of each material.

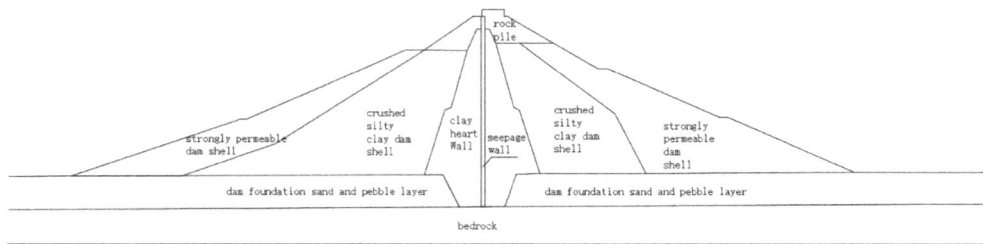

Figure 7.   Sectional material zoning diagram.

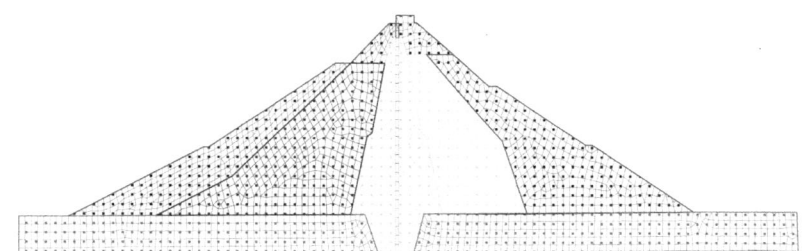

Figure 8.   Finite element mesh of seepage.

Table 1.   Values of initial permeability coefficient of dam rock and soil layer.

Serial number	Material	Horizontal permeability coefficient k (cm/s)	Vertical permeability coefficient k (cm/s)
1	Clay heart wall	5.5×10^{-5}	6.1×10^{-5}
2	Crushed silty clay	8.1×10^{-5}	8.6×10^{-5}
3	Mud-gravel sand dam shell	1.1×10^{-4}	1.1×10^{-4}
4	Dam foundation gravel pebble layer	1.5×10^{-3}	1.5×10^{-3}
5	Dam foundation permeable bedrock	1.0×10^{-5}	1.0×10^{-5}
6	Concrete anti-seepage wall	1.0×10^{-8}	1.0×10^{-8}

4.2 *Calculation condition*

The calculation conditions are selected according to the stable seepage operating conditions, as shown in Table 2.

Table 2. Field calculation conditions of seepage.

Working condition	Name	Upstream water level (m)	Downstream water level (m)
1	normal water level	208.20	166
2	design flood level	212.12	166
3	check flood level	212.50	166

4.3 *Analysis of calculation results*

(1) From the analysis of the flow network morphology of the typical section (maximum section), the infiltration line behind the anti-seepage wall of the dam is relatively high, the water level drop between the anti-seepage wall and the core wall is small, and the anti-seepage wall and the core wall are combined to prevent osmosis. The water potential between the clay core wall and the dam shell soil is obviously cut off. The downstream infiltration surface of the clay core wall is located in the sand and pebble layer, and the water level is low, indicating that the combined anti-seepage effect of the clay core wall and the anti-seepage wall is obvious. The downstream water level is low and does not escape from the dam slope, and the escape point is lower than the dam drainage prism.

(2) After calculation, under the conditions of check flood level, design flood level, and normal water storage level, the seepage ratio through the bottom of the intercepting trough is 1.8, 2.05, and 2.15, respectively. The maximum permeability gradients of bedrock, dam shell, and seepage prevention wall are 9.87, 0.42, and 20.6, respectively, which are less than the allowable permeability gradients of the material and meet the requirements of the specification.

(3) After calculation, the maximum seepage flow is 2.3 $m^3/d/m$, and the overall seepage flow is small, which meets the requirements of the specification.

5 CONCLUSIONS

(1) During the normal water storage period, the water level difference between the upstream and downstream sides of the foundation of the anti-seepage wall is relatively large, and the water head is significantly reduced. The head reduction of the downstream measuring point on the seepage wall inside the dam core wall is also obvious, but it was obvious at the measuring point P3-3 on the upstream side of the seepage wall and P3-4 on the downstream side of the seepage wall at the 0+289m section with an elevation of 175.0 m in April 2011. The difference in the back head is very small, and the water level on the upstream side is lower than the water level on the downstream side. The head difference of P4-3 on the upstream side of the anti-seepage wall and P4-4 on the downstream side of the anti-seepage wall at an elevation of 175 m at the 0+385 section has become smaller and smaller since 2017, indicating that the concrete anti-seepage wall at this part gradually loses its anti-seepage effect, and monitoring should be strengthened to analyze the data in a timely manner.

(2) Section 0+193 piezometer P2-4(1) and piezometer P2-4(2) have been in the process of continuous decline of water level since the reinforcement and reinforcement. The average water level has dropped by about 20 m, and there is no convergence trend. It shows that the seepage medium on the downstream side of this measuring point has changed significantly, and the permeability of the sand and pebble layer is increasing, which may be related to the lack of

back-checking on the downstream side of the anti-seepage core wall. The wall bears most of the water head, and the penetration phenomenon has not yet occurred, but the permeability gradient is relatively large and has reached about 30. Section 0+289 is similar to section 0+193, and close attention should be paid to the change in the piezometer water level in this section, especially the changes in the water level at P2-3.

(3) The water level at each measuring point of seepage around the dam is mainly affected by rainfall and fissure water, and the observation data show that there is no obvious seepage around the dam on both sides of the dam body.

(4) According to the simulation calculation, the seepage prevention wall of the dam body and the clay core wall play a role in preventing seepage together, the seepage gradient is less than the allowable seepage gradient of the material, and the seepage flow is generally small, meeting specification requirements.

(5) According to the decreasing trend of the water level of some osmometers, it shows that the seepage medium is changing and the seepage is unstable. It is recommended to strengthen the observation and analyze the monitoring data in time to ensure the safe operation of the reservoir.

REFERENCES

Jiang C., Liu Y.S., Fan L.R., etc. Analysis of seepage monitoring data in the dam of zhuanjiaolou reservoir [J]. *Dams and Safety*, 2018, (1): 43–48.

Mi C.Y., Xiong G.W. Analysis of dam safety monitoring data in meixi reservoir [J]. *Zhejiang Water Conservancy Science and Technology*, 2018, 46(6): 50–54.

Zhu F.J., Ma Ha. Monitoring and analysis of seepage in loam core sand-gravel shell dams [J]. *Agricultural Science and Technology and Information*, 2021, (4):103–105.

Zhan W.Q. Analysis of seepage monitoring data of dam in tiegang reservoir [J]. *Water Conservancy Construction and Management*, 2018, 38(3):41–43.

Zhang Y. Analysis and evaluation of safety monitoring data of yanwangbi reservoir [J]. *Northeast Water Resources and Hydropower*, 2022, 40(1):63–65.

*Frontiers in Civil and Hydraulic Engineering – Mohamed A. Ismail and
Hazem Samih Mohamed (Eds)
© 2023 The Authors, ISBN 978-1-032-38247-0*

Numerical simulation of seawater intrusion in typical area based on FEFLOW

Hao Liang
Shandong Provincial Territorial Spatial Ecological Restoration Center, Jinan, China

Linxian Huang*
School of Water Conservancy and Environment, University of Jinan, Jinan, China

Yongwei Zhang
Shandong Provincial Territorial Spatial Ecological Restoration Center, Jinan, China

Zhong Han
Institution of Geology and Mineral Resources Exploration of Shandong Province, Weihai, China

Jinxiao Hou
School of Water Conservancy and Environment, University of Jinan, Jinan, China

ABSTRACT: Seawater intrusion is a kind of environmental geological deterioration phenomenon caused by seawater immersion in the underground fresh water layer due to the decline of the land's fresh water level. The groundwater numerical model can effectively simulate the migration path, migration range, and pollution concentration of seawater intrusion, and can provide quantitative data for seawater intrusion control and groundwater resource management. Taking the typical area of Weihai City in Shandong Province as the research goal, this paper simulates and predicts the pollution characteristics of seawater intrusion under the original conditions through FEFLOW software. This research has certain practical significance for effectively preventing environmental geological disasters caused by seawater intrusion in Weihai City, and has a certain positive effect on the concept of sustainable development of coastal zones.

1 INSTRUCTION

The coastal area around Weihai is an area with a highly concentrated population and rapid economic development. The excessive demand for fresh water resources in the area leads to the overexploitation of groundwater resources, and the continuous and substantial decline of groundwater level, resulting in changes in the interface between salty seawater intruding into fresh water aquifers and fresh water, and the salinity of groundwater increases; the water quality gradually turns into salty water and then evolves into 'seawater intrusion' (Jiang et al. 2021, Sun et al. 2022).

Over the past decades, many Chinese scholars and experts have done a lot of work on the prediction of seawater intrusion. Among them, the more successful prediction methods are as follows: Yang Y X et al. established the Verhulst prediction model by using the gray system method (Yang et al. 1994). Li X Y et al. used the gray system model GM (1, 1) to predict the development trend of seawater intrusion along the southeast coast of Laizhou Bay (Li et al. 1994). Zhang J D et al. established a prediction model of two-dimensional unstable flow and convection-dispersion water quality for seawater intrusion simulation and prediction (Zhang et al. 1995). Yuan Y R et al.

---

*Corresponding Author: stu_xinglt@ujn.edu.cn

DOI 10.1201/9781003344209-47

proposed an upwind operator splitting algorithm for seawater intrusion (Yuan et al. 2001). Pan Shulan et al. studied the causes and changing trends of seawater intrusion in Laizhou Bay by using isotope technology (Pan et al. 1997).

In this study, FEFLOW software was used to establish a two-dimensional variable density solute transport model to simulate and predict the pollution characteristics of seawater intrusion under the original conditions (Gao et al. 2012, Lu et al. 2010). This study has certain practical significance for effectively preventing environmental geological disasters caused by seawater intrusion in Weihai City.

2 OVERVIEW OF STUDY AREA

This paper selects the typical seawater intrusion harbor area in Weihai City as the research object. The administrative district belongs to Rushan District (Figure 1), and the hydrogeological unit belongs to Muzhu River Basin region. It starts from Yuanli reservoir in Rushan City in the west, extends to Baishatan Town in the East, Rushan urban area in the north and the seaside in the south. The extreme geographic coordinates are 121°19′52″–121°37′36″E and 36°41′09″–36°59′31″N. The numerical model range is shown in Figure 1 (red shading).

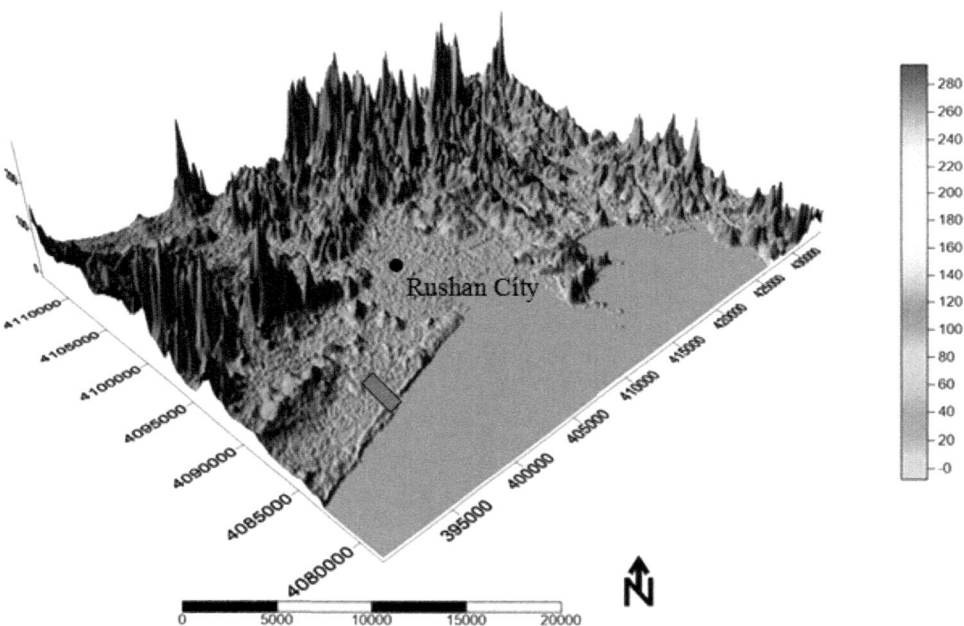

Figure 1.   Three-dimensional elevation structure model of the study area.

3 NUMERICAL MODEL OF SEAWATER INTRUSION

At present, the majority of seawater intrusion models take the coastline and waterline in the plan as the boundary of the whole water-bearing system. In fact, coastal aquifers will extend to the seafloor, with their boundaries distant, to some extent, from the shoreline. This problem clearly and decisively affects the distribution of brackish and fresh water. This treatment may lead to incorrect simulation results of the numerical models. In order to solve this kind of problem, the usual method is to establish a numerical profile model of seawater intrusion in the study area.

362

Therefore, in order to solve the problem of boundary generalization in coastal areas, a numerical model of seawater intrusion profile flow is established in this study to simulate the changes in the seawater intrusion seepage field under the original conditions.

3.1 Numerical model of seawater intrusion

The numerical model of the seawater intrusion profile displays a rectangular area (Figure 2). Among them, the eastern and western long edges are perpendicular to the water groundwater level contour, so they can be generalized as a non-flow boundary with a length of 5000 m; The northern boundary receives groundwater recharge and is generalized as a discharge boundary. The south boundary is the sea, which is generalized as the constant head boundary. The upper boundary has vertical water exchange with the outside world, such as receiving atmospheric precipitation infiltration supply, evaporation, and discharge.

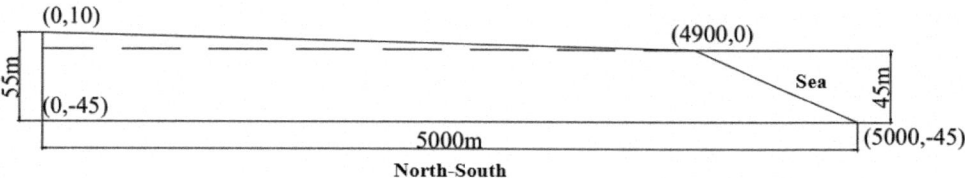

Figure 2. Schematic diagram of three-dimensional numerical model profile of seawater intrusion.

There is a height difference of 10 m between the north and south sides of the model according to the topographic changes and the flow model in the study area. The thickness of the model is 55 m on the north side and 45 m on the south side. The relevant parameters in the model are shown in Table 1 below.

Table 1. Hydrogeological parameters.

Parameter	Numerical value	Parameter	Numerical value
Left boundary supply strength	0.02 m/d	Upper boundary supply strength	0.0005 m/d
Seawater salinity	35 g/L	osmotic coefficient	8.64 m/d
Density coefficient	0.0245		

3.2 Subdivision of model gridding

First, the basic map and relevant graphics of the research area are imported into the model and digitized into super units. The important regional sidelines and points in the area, such as coastline, hydrogeological unit line, and mining well points, are also digitized. The quadrilateral meshing method is used for grid dissection, which is divided into 11 slices, each of which contains 20,000 nodes.

3.3 Numerical simulation results

For the simulation of water quality, only one observation point was selected for comparison due to the lack of water quality observation data. It can be seen from Figure 3 that the simulation results of the numerical model of seawater intrusion can well fit the observation data at observation points.

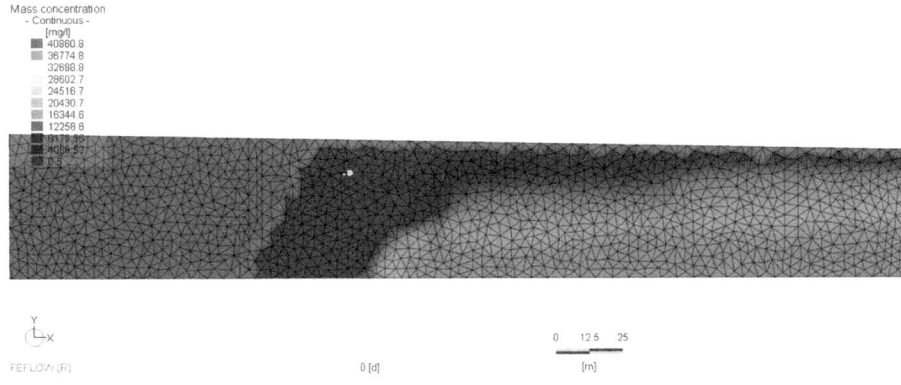

Figure 3.   Concentration of fitting.

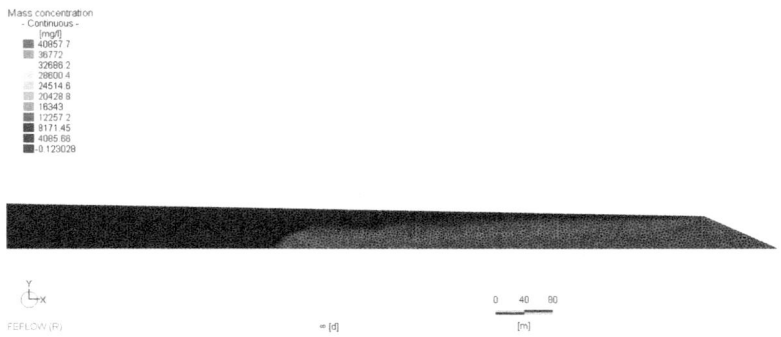

Figure 4.   Distribution diagram of slice concentration contour.

The simulation results of Cl^- concentration distribution range of seawater intrusion are shown in Figure 4. It can be seen from the figure that the toe position of the 0.5 mg/L concentration contour of Cl^- is exactly where the flow velocity changes direction. Seawater enters the aquifer at the lower right boundary, mixes with fresh water, and flows out at the upper right. Taking the migration range of Cl^- 0.5 mg/L as the standard, the distance of seawater intrusion is 726 m. After entering the groundwater aquifer, the concentration of seawater decreases due to mixing with fresh water.

It can also be seen from the Darcy velocity distribution map (Figure 5) that seawater enters the groundwater aquifer from the lower part on the right, and the interface between brackish and fresh

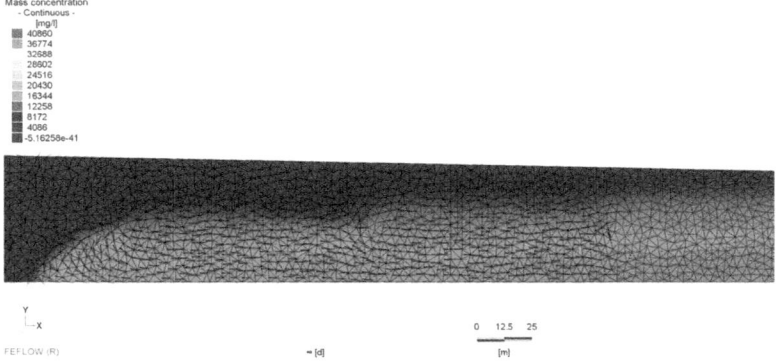

Figure 5.   Schematic diagram of darcy velocity of seawater intrusion.

water appears at the concentration contour of 0.5 mg/L. Salty water flows out at a relatively large flow rate at the upper right side.

4 CONCLUSION

In this study, a numerical model of seawater intrusion profile flow was established by FEFLOW numerical software, and the changes in the seawater intrusion seepage field under different scenarios were simulated numerically, and the following conclusions were drawn:

Under the original conditions, seawater enters the aquifer at the lower part of the right boundary, mixes with fresh water, and then flows out at the upper part of the right boundary. Taking the migration range of Cl^- 0.5 mg/L as the standard, the distance of seawater intrusion is 726 m. After seawater enters the groundwater aquifer, the concentration of seawater decreases due to mixing with fresh water.

The above seawater intrusion phenomenon is very common in coastal areas. When groundwater is extracted near the boundary between brackish and fresh water in sensitive coastal areas, the mining wells gradually move inland as the well water becomes saltier, and the groundwater funnel also moves inland, leading to further invasion of seawater. It can be seen that an important reason for seawater intrusion is the shortage of regional water resources and the overexploitation of groundwater. Therefore, an effective way to prevent seawater intrusion lies in the reasonable allocation of water resources and reduction of groundwater exploitation.

ACKNOWLEDGMENTS

This study was supported by the Natural Science Foundation of Rizhao ([2021]40-31), the Shandong Provincial Natural Science Foundation (ZR2019MD029), the Key Laboratory of Groundwater Remediation of Hebei Province, and the China Geological Survey (SK202103KF01), Shandong Province Innovation Team Project of Colleges and Universities (Major Scientific and Technological Innovation Project) (2021GXRC070), and the Science and Technology Planning Project of University of Jinan (XKY2015).

REFERENCES

Gao H.Q., Yang M.M., Hei L., et al. 2012. Application of MODFLOW and FEFLOW in groundwater numerical simulation in China [J]. *Groundwater*, (04): 13–15.

Jiang H.J., Wu Y., Tian X.Y., et al. 2021. *The role of water Environment Information System in Water resources management and protection in Weihai City* [C]//. China Water Conservancy Society 2021 Academic Conference Proceedings of the Third Volume, 385–391.

Li X.Y., Jiang W.M., Zhang N.X. 1994. Correlation analysis and trend prediction of seawater intrusion on the southeast coast of Laizhou Bay [J]. *Journal of Geological Hazard and Prevention of China*, (04):33–39.

Lu W., Zhu Z.Y., Liu W.P. 2010. Numerical simulation of seawater intrusion based on FEFLOW [J]. *Ground Water*, 32(03):19–21,129.

Pan S.L. & Ma F.S. 1997. *Isotopic study of seawater intrusion* [C]//. Proceedings of the 6th National Symposium on Isotopic Geochronology and Isotope Geochemistry. 321–323.

Sun N.B., Chen X.Q., Li Q., et al. 2022. Hydrochemical characteristics and genesis analysis of groundwater in Weihai city [J]. *Water-saving Irrigation*, (07):85–90.

Yang Y.X., He P.Q., Xie Y.Q., et al. 1994. Grey model prediction of seawater intrusion in Qinhuangdao [J]. *Chinese Journal of Geological Hazard and Prevention*, (S1):181–183.

Yuan Y.R., Liang D., Rui H.X. 2001. Prediction of aftereffect of seawater intrusion and control engineering [J]. *Applied Mathematics and Mechanics*, (11):1163–1171.

Zhang J.D. & Liu W.L. 1995. Dynamic simulation and prediction of seawater intrusion along the coast of Funing [J]. *Hebei Water Science and Technology*, (04):31–35.

*Frontiers in Civil and Hydraulic Engineering – Mohamed A. Ismail and
Hazem Samih Mohamed (Eds)
© 2023 The Authors, ISBN 978-1-032-38247-0*

A groundwater optimization model for identification of groundwater discharge characteristics under unknown locations

Peng Yang
*Institute of Geology and Mineral Resources Exploration of Shandong Province, Rizhao, China
Key Laboratory of Nonferrous Metal Ore Exploration and Resource Evaluation of Shandong
Provincial Bureau of Geology and Mineral Resources, Rizhao, China
Rizhao Big Data Research Institute of Geology and Geographic Information, Rizhao, China
Rizhao Key Laboratory of land quality evaluation and pollution remediation, China*

Linxian Huang*
School of Resources and Environment, University of Jinan, Jinan, China

Zhong Han
Institution of Geology and Mineral Resources Exploration of Shandong Province, Weihai, China

Hao Liang
Shandong Provincial Territorial Spatial Ecological Restoration Center, China

Jinxiao Hou
School of Resources and Environment, University of Jinan, Jinan, China

ABSTRACT: Accurate determination of the discharge characteristics of groundwater pollution sources is an important prerequisite for rational disposal of groundwater pollution accidents. Using pollutant monitoring data to deduce the emission characteristics of groundwater pollution sources belongs to groundwater inversion, and the inversion of pollution sources under the condition of unknown location is the difficulty of this kind of problem. A simulation optimization model was designed by coupling location search algorithm, compound evolutionary algorithm (SCE-UA), groundwater flow numerical model (MODFLOW), and groundwater solute transport numerical model (MT3DMS). Through the example verification under different situations, the results show that: (1) Compared with the traditional optimization algorithm, the simulation optimization model can accurately identify the pollution source location and emission characteristics under the condition of unknown pollution source location; (2) The simulation optimization model has high inversion accuracy under multiple scenarios such as single source and multiple sources, stable flow and unsteady flow, fixed position and variable position.

1 INSTRUCTION

Groundwater inversion identification methods mainly include a geochemical method, geophysical prospecting method, isotope tracing method, and optimization algorithm, among which the optimization algorithm is the most effective and common method for groundwater inversion identification. At present, scholars have carried out a lot of research on the optimization algorithm of groundwater pollution inversion. Skaggs et al. used a method based on Tikhonov regularization to invert the release intensity of groundwater pollution (Skaggs et al. 1994). Woodbury uses the method of minimum relative entropy to inverse and identifies the release history of pollutants (Woodbury et al. 1996). Alapati et al. used a nonlinear least squares method to identify the release

---

*Corresponding Author: stu_huanglx@ujn.edu.cn

DOI 10.1201/9781003344209-48

characteristics of pollution sources (Sidauruk et al. 1998). Hou Zeyu et al. coupled a simulated annealing algorithm and particle swarm optimization algorithm to invert the transport process of pollutants in groundwater (Hou et al. 2019). However, all the above optimization algorithms have some disadvantages, such as the least square method, pattern search method, and other methods. The accuracy of the inversion results of these methods depends on the selection of the initial value. In particular, most of the above inversion optimization methods need to know the location of the pollution source to perform the inversion identification of the emission characteristics of the pollution source. Nevertheless, in practical applications, the location of pollution sources is often uncertain, or can only be roughly delineated within a certain range, which limits the scope of application of such methods. A simulation-optimization model for groundwater pollution identification at an unknown pollution source location is designed through coupled location search algorithm, compound evolutionary algorithm SCE-UA, numerical model MODFLOW of groundwater flow and numerical model MT3DMS of groundwater solute transport, and case studies are carried out to verify the model.

2 SIMULATION OPTIMIZATION MODEL

2.1 Model structure

The simulation optimization model of groundwater pollution sources can be divided into three parts: a location search algorithm, a numerical model of groundwater flow, and a solute transport and optimization algorithm model. The location search algorithm is mainly used to generate all possible combinations of pollution source locations in the predefined subareas. The numerical model MODFLOW and the solute transport numerical model MT3DMS are mainly used to simulate and predict the spatial and temporal migration and transformation characteristics of pollutants in aqueous media; the optimization algorithm model mainly generates population sampling points through the SCE-UA optimization algorithm, and continuously evolves the generated sample points to search for possible pollutant emission characteristics in the global scope.

2.2 Objective function

In the process of groundwater pollution inversion and identification, the objective function E can be defined as:

$$E^t = \frac{1}{n_{obs}} \sum_{i=1}^{n_{obs}} |C_{i,sim}^t - C_{i,obs}^t| \tag{1}$$

where: E^t is the objective function value at the time t; $C_{i,sim}^t$ is the simulated value of pollutant concentration of the i (th) monitoring well at the time t (mg/L); $C_{i,obs}^t$ is the monitoring value of pollutant concentration of the i (th) monitoring well at the time t (mg/L); n_{obs} is the number of monitoring wells.

2.3 Working principle

The working principle of the simulation optimization model can be summarized as the following six steps:

(1) The location search algorithm is used to generate a possible combination of pollution source locations;

(2) The SCE-UA optimization algorithm is adopted to generate a series of population sampling points (i.e., estimated pollutant concentration) at possible pollution source locations;

(3) The numerical models MODFLOW and MT3DMS are utilized to simulate and predict the pollutant concentration value $C_{i,sim}^t$ at the monitoring well;

(4) The objective function value E^t is calculated according to the simulated value $C_{i,sim}^t$ of pollutant concentration and the observed value $C_{i,obs}^t$ of concentration;

(5) According to the objective function value E^t, the optimization algorithm SCE-UA mutates, reflects and evolves the population sample points, and searches for possible pollutant emission concentration values in the global range;

(6) When the objective function value E^t is small enough, it can be approximately considered that the estimated value of the concentration is equal to the true value of the concentration.

3 RESULTS AND ANALYSIS

3.1 Case study

The case refers to a pollution source inversion optimization model proposed by Singh et al. (Raj et al. 2004). The scope of the model is a 100 m*100 m square area, which is a homogeneous isotropic two-dimensional aquifer (as shown in Figure 1). The thickness of the aquifer is 10 m, and the permeability coefficient is 1 m/d; the whole study area is divided into 10 rows and 10 columns; the North-South boundary of the aquifer is the non-flow boundary; the East-West boundary is the hydraulic head boundary (the water head value of the east boundary is 7 m; the water head value of the west boundary changes linearly from north to south, the water head value of the north endpoint is 9.7 m, and the water head value of the south end point is 0.4 m); the dispersion coefficient of the aquifer is 20 m.

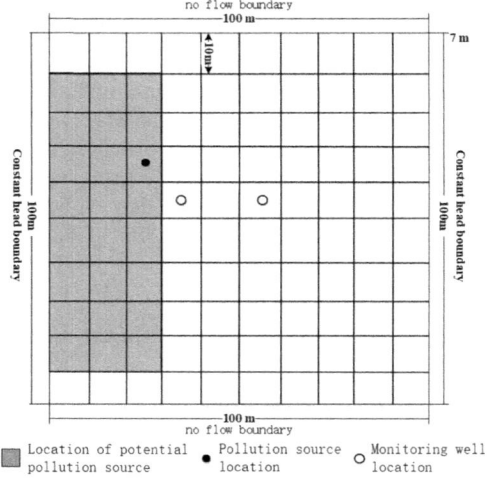

Figure 1. Aquifer model domain of case study.

3.1.1 Case 1

In the aquifer shown in Figure 1, there is a pollution source A (row=4, column=3), and its real pollutant emission concentration is $C_{act1} = 80$ mg/L. In the inversion process, its location is assumed to be unknown, and its potential location can be determined in the shaded area of Figure 1. The shaded area contains a total of 24 cells, and the pollution source may be located in any of them, so 24 locations need to be searched. Downstream, there are two monitoring wells located at O_1 (row=5, column=4) and O_2 (row=5, column=6), and the monitoring concentrations of pollutants in the monitoring wells are $C_{obs1} = 26.95$ mg/L and $C_{obs2} = 6.19$ mg/L respectively. It is set that each population contains 20 composite types, and each composite type contains 5 vertex numbers. When the evolutionary algebra is 10 generations, the inversion results are shown in Table 1.

It can be seen from Table 1 (Table 1 only shows part of the inversion results) that the location of the pollution source and the emission concentration of pollutants can be accurately inverted. The NE value (NE value=|identified value − actual value|/actual value*100) of the optimal inversion result is 0.106, which can meet the requirements for inversion accuracy. In addition, it can be seen

Table 1.   Identified results for generation=10.

| Sort | Pollution source | Location of pollution sources | | Emission concentration/mg·L^{-1} | | NE/% |
		Actual value	Identified value	Actual value	Identified value	
1	A	(4, 3)	(4, 3)	80	79.915	0.106
2	A	(4, 3)	(4, 3)	80	79.69	0.388
3	A	(4, 3)	(3, 3)	80	99.785	24.731
4	A	(3, 2)	(5, 3)	80	58.218	27.228

from Table 1 that when the inversion of the location of the pollution source is wrong, the retrieved pollutant emission concentration is very different from the real value. The influence of evolutionary algebra on the inversion results is further studied, and the evolutionary algebra is increased to 300 generations.

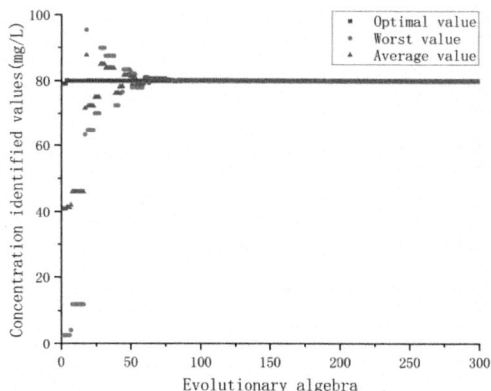

Figure 2.   Effect of variation in generation number.

When the evolutionary algebra is 300 generations, the identified value of pollutant concentration is 80.00006 mg/L, and the inversion accuracy is further improved compared with that when the evolutionary algebra is 10 generations. The variation trends of inversion concentration are shown in Figure 2. It can be seen from Figure 2 that the optimal concentration inversion value of the inversion model has basically been stable around the 5th generation. The average value of concentration inversion fluctuates greatly at the beginning of inversion and deviates from the real value, and basically becomes stable when it evolves to about 75 generations. The worst value of concentration inversion also fluctuates greatly in the initial stage, but it can also reach a stable state when it evolves to about 75 generations, indicating that the optimization model has achieved the optimal solution.

3.1.2   Case 2

Case 2 is closer to the actual situation than case 1, which includes two pollution sources A_1 and A_2, and their real location is assumed to be unknown, the possible location is in the shaded part of Figure 3. At the same time, in order to verify the inversion effect of the inversion model on unsteady flow conditions, case 2 is set as an unsteady flow model containing four stress periods (stress period=1 year). Moreover, the positions of A_1 and A_2 are not fixed. In the first two stress periods, they are located at row=3, column=2, and row=3, and column=3, respectively, and in the last two stress periods, they are located at row=3, column=3, and row=4, column=3, respectively. There are four monitoring wells O_1 (row=2, column=4), O_2 (row=3, column=6), O_3 (row=4, column=7) and

O_4 (row=5, column=8) in the downstream industries of the aquifer. The emission concentration of pollutants from the two pollution sources will also change in different stress periods. The monitored concentration values and emission concentration values of pollution sources in each stress period are shown in Table 2 below.

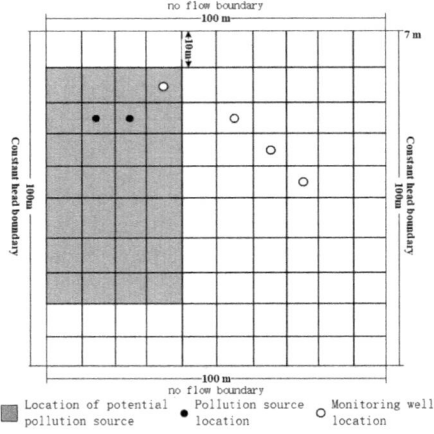

Figure 3.   Aquifer model domain of case 2.

Table 2.   Actual and identified concentration for each stress period.

Number	Stress period 1	Stress period 2	Stress period 3	Stress period 4
A_1	30	0	20	20
A_2	60	30	40	40
O_1	20.61	16.79	16.81	18.44
O_2	8.31	12.54	14.64	17.17
O_3	1.4	6.69	10.01	12.98
O_4	0.1	2.08	5.07	8.05

When the evolution algebra is set to 100 generations, the simulation optimization model can accurately identify the location of the two pollution sources and the emission concentration of pollutants in different stress periods. The inversion results are shown in Figure 4. As can be seen

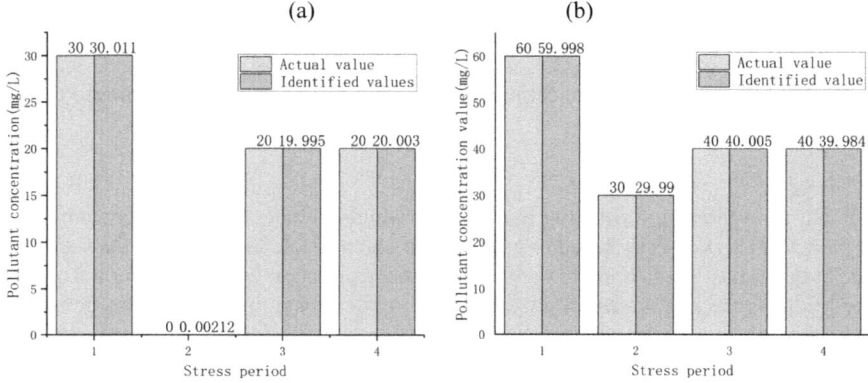

Figure 4.   Identified results for generation=100 in A_1 (a) and A_2 (b).

from Figure 4, the identified values of pollutant emission concentrations in all four stress periods are very close to the actual values, indicating that the inversion results are highly accurate, and the designed simulation inversion model can obtain satisfactory results. In the inversion process, the inversion of the current stress period depends on the inversion result of the previous stress period, so the inversion error will continue to accumulate. The solution is to increase the number of evolutionary algebras or population individuals to make each stress period more accurate.

4 CONCLUSION

In this paper, a simulation optimization model that can identify the emission characteristics of pollution sources under the unknown location of groundwater pollution sources is proposed and verified by an example:

(1) By compiling FORTRAN interfaces program, the optimization algorithm SCE-UA is coupled with the numerical models MODFLOW and MT3DMS, which greatly improves the calculation efficiency of the inversion optimization model;
(2) Aiming at the disadvantage that most of the traditional groundwater inversion models cannot optimize the inversion under the unknown pollution source location, a location search algorithm is developed, so that the inversion model can automatically search all possible combinations of pollution source locations in the predefined subareas;
(3) Through the multi-scenario case test, the inversion model can efficiently and uniformly converge to the global optimal solution for stable flow, unstable flow, variable pollution source location, and variable emission concentration, accompanied by good stability and fast convergence speed;
(4) Because the location searches algorithm needs to search all possible combinations of pollution source locations in the predefined subareas, the calculation burden of the inversion optimization model is large and the calculation time is increased. The next step is to focus on how to improve the calculation efficiency of the location search algorithm and introduce parallel algorithms to make the inversion model more practical.

ACKNOWLEDGMENTS

This study was supported by the Natural Science Foundation of Rizhao ([2021]40-31), the Shandong Provincial Natural Science Foundation (ZR2019MD029), the Key Laboratory of Groundwater Remediation of Hebei Province, and the China Geological Survey (SK202103KF01), Shandong Province Innovation Team Project of Colleges and Universities (Major Scientific and Technological Innovation Project) (2021GXRC070), and the Science and Technology Planning Project of University of Jinan (XKY2015).

REFERENCES

Hou Z.Y., Lu W.X., Wang Y. 2019. Identification of groundwater pollution sources by DNAPLs inversion based on the alternative model [J]. *China Environmental Science*, 39(1): 188–195.
RAJ Mohan Singh & BITHIN Datta. 2004. Groundwater pollution source identification and simultaneous parameter estimation using pattern matching by artificial neural network [J]. *Environmental Forensics*, 5(3): 143–153.
Sidauruk P., Chang A.H.D., Quazar D. 1998. Ground water contaminant source and transport parameter identification by correlation coefficient optimization[J]. *Ground Water*, 36(2): 208–214.
Skaggs T.H. & Kabala Z.J. 1994. Recovering the release history of a groundwater contaminant [J]. *Water Resources Research*, 30(1): 71–79.
Woodbury A.D. & Ulrych T.J. 1996. Minimum relative entropy inversion: Theory and application to recovering the release history of a groundwater contaminant[J]. *Water Resources Research*, 32(9): 2671–2681.

*Frontiers in Civil and Hydraulic Engineering – Mohamed A. Ismail and
Hazem Samih Mohamed (Eds)
© 2023 The Authors, ISBN 978-1-032-38247-0*

Research on the construction of a BIM-based model for cross-floor indoor navigation maps

Jia-huan Mao*, Jing Yang*, Ru-ping Shao* & Wen-zhang Wang*

College of Electrical Engineering & Control Science, Nanjing Tech University, Jiangsu Nanjing, China

ABSTRACT: With the increasing maturity of indoor positioning and indoor data acquisition technologies, indoor road network extraction has become an important part of indoor navigation work. A method of constructing a cross-floor indoor navigation map model is designed for existing public buildings with high floors and complex internal spatial structures. Based on the method of BIM technology and graph theory, the IfcOpenShell library is applied to extract the required component information for indoor navigation, staircases are considered as transformation nodes, and the extraction method and intersections in the corridor are optimized. Combined with path planning algorithms, a visual operating system is designed to realize the automatic construction of indoor road network models and indoor path planning guidance. Practice shows that the proposed model, when combined with the pathfinding algorithm, is able to efficiently find reliable paths for indoor navigation.

1 INSTRUCTION

In recent years, with the acceleration of urbanization and modernization, people spend more time indoors than outdoors on average every day, and the indoor environment has become the main activity area for people to live and work, where public buildings such as airports, shopping malls, hospitals, museums, and railway stations have more complex internal structures, more diverse functional partitions and a higher degree of freedom to walk, making it more difficult for people to find their way around large and complex public buildings (Schougaard 2012).

Many indoor navigation models for indoor path planning have been proposed for different application areas, mainly being divided into grid-based road network models (Sturtevant 2012) and topological relationship-based road network models (Cheng 2015). However, the above-mentioned navigation models are based on a single-floor two-dimensional road network model, which lacks information on indoor components and connectivity between floors, making it impossible to navigate across floors, thus creating obstacles for users. In order to solve the above problems, this paper proposes a method to build a cross-floor indoor navigation map model based on Building Information Modeling (BIM) technology and graph theory modeling methods, considering stairs and escalators as conversion nodes, and using node optimization and edge extraction algorithms. The experimental results show that by applying the Dijkstra pathfinding algorithm to the improved model, the nodes and distances of the optimal path can be observed visually and concretely, thus improving the efficiency of indoor navigation.

*Corresponding Authors: orangecat211@163.com; muyi_qing@njtech.edu.cn; njsrplw@163.com and 1635035395@qq.com

DOI 10.1201/9781003344209-49

2 BIM TECHNOLOGY AND DIAGRAM THEORY

2.1 *BIM and component information*

The core of BIM technology relies on a three-dimensional information model created by relevant software to achieve efficient and intelligent whole building process management, linking data, operations, processes, and resources at all stages and providing a complete description of the building for use by all parties involved in the construction. The IFC (Industry Foundation Classes) standard is a standardized digital description developed by the building SMART organization to help the engineering and construction industry achieve an object-oriented file format for data inter-operability. Currently, there are many products and systems that provide data exchange interfaces oriented to the IFC standard to achieve data information transfer and sharing between different software and BIM systems in different stages. Therefore, BIM models built based on Revit can achieve interaction with IFC standard data information.

2.2 *Concepts and expressions of graph theory*

A graph is a mathematical abstraction consisting of objects and relations between objects in the objective world, where concrete objects in the objective world are represented by points and relations between objects are represented by edges (Xue 2020). A graph is defined as the set of nodes V and the set of edges E connecting these nodes together, denoted as G= (V, E). Graphs generally include undirected graphs, directed graphs, and weighted graphs.

The storage of graphs is usually expressed in terms of adjacency matrices, as shown in Figure 1 below, where the weighted directed graph of graph (a) is stored by means of the adjacency matrix of the graph (b), where there are connected edges from nodes n_2 to n_1 with edge weights of 5 and non-connected edges between n_1 and n_4 with edge weights of ∞.

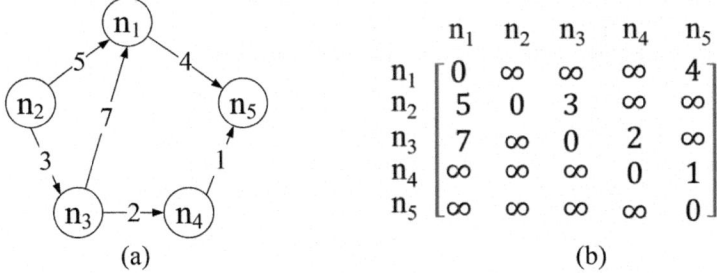

Figure 1.   Weighted directed graphs and their adjacency matrix storage forms.

3   CONSTRUCTION OF AN INDOOR NAVIGATION MAP MODEL

The indoor navigation map model proposed in this paper is based on the concept of graph theory and describes the indoor road network model through undirected graphs. By extracting information about the building components from the IFC file, the indoor components are abstracted as nodes, and the connecting edges between the nodes are abstracted as edges and given weights using the Manhattan distance between the node coordinates as a reference. Each element in the model can refer to its corresponding IFC class. And by transforming the connections between the nodes, the cross-floor navigation model is constructed.

3.1 *Types of IFC interior architectural elements*

IFC 2×3 is the current common IFC file specification, which contains 653 entities and over 300 supplementary data types and extensible attribute sets (Wang 2021). The key step is to extract the elements for indoor navigation from the IFC file, which is generated based on a hierarchical construct in which all elements are inherited from the entity ifcProduct. This paper summarizes the IFC entities and their corresponding building space element types for indoor navigation, as shown in Table 1, from the perspective of indoor navigation requirements.

Table 1.   Selected IFC Entities and their corresponding building space component types.

IFC entity type	Building space elements
IfcDoor	Doors
IfcStair	Stairs
IfcWall	Walls
IfcStairfilght	Elevators
IfcSpace	Rooms and other areas
IfcColumn	Columns
IfcBuildingStorey	Storey

3.2   IFC Indoor component information extraction

For the semantic information of the indoor components to be extracted, the IFC file defines the attribute information of each solid component element, but as the IFC data file is described using the EXPRESS language, the data in it cannot be processed and analyzed directly.

IfcOpenShell is a Python-based IFC file parsing tool that can parse the names, materials, topology, and other information of buildings and components in IFC files, and it is easy to operate and has flexible function modules. In this paper, we use IfcOpenShell to parse the IFC source file, which is read by ifcopenshell.open() method and the components of the specified type are obtained by calling ifc_file.by_type() method; the geometric representation of the components is obtained by using IfcProduct.Representation method. The core code for the parsing operation is shown in Figure 2.

```
import ifcopenshell                        #Importing the ifcopenshell library
ifc_file = ifcopenshell.open('/path/to/your/file.ifc')#Loading IFC files
slabs = ifc_file.by_type('IfcSlab')        #Get the slabs
walls = ifc_file.by_type('IfcWall')        #Get the walls
doors = ifc_file.by_type('IfcDoor')        #Get the doors
columns = ifc_file.by_type('IfcColumn')    #Get the columns

geometry = wall.Representation. Representations[0].Items[0] #Take the example of a stretched solid wall
print(geometry.Position.Location[0])       #Reading the relative coordinates of the centre of mass
```

Figure 2.   IFC File information parsing core code.

3.3   Spatial node selection

The generation of interior spatial nodes requires the extraction of interior spatial elements, their simplification, and the acquisition of spatial coordinates. For the navigation-oriented needs of the interior diagram model, this paper divides the interior spatial nodes into three main categories: room nodes, door nodes, and stair nodes. The room nodes and door nodes are extracted from the mass of the interior elements, and the stair nodes are abstracted as nodes depending on their form, in addition to the entrances and exits of the stairs. In order to observe and analyze the distribution of each spatial node in each floor more intuitively, the height of each node in the z-direction is reduced to the floor height of the floor where it is located in this paper.

The door and room nodes are calculated by extracting geometric information from the ifcDoor and ifcSpace classes, respectively, and the stair nodes are calculated from the ifcStair and ifcStair-Filght classes. As each component in IFC uses a relative coordinate system rather than a mapped coordinate system, it is necessary to obtain the positioning information of the component from IfcObjectPlacement and the geometric representation of the component from IfcProductRepresentation. Finally, by means of the transformation matrix, the relative coordinates of the mass of the components are converted to absolute coordinates using the IfcSite site coordinate system as a reference. The node-specific extraction process is shown in Figure 3.

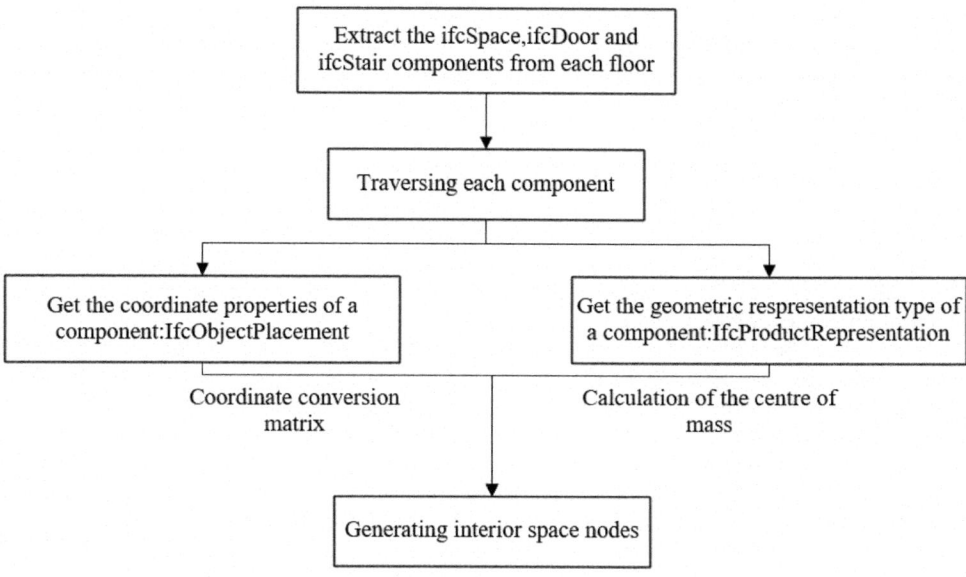

Figure 3.   Spatial node extraction process.

3.4   *Generation of connected edges*

The edges of the interior navigation diagram model include horizontal and vertical edges. Horizontal edges connect rooms, corridors, and doors, and vertical edges connect stairs. Horizontal paths include the following 3 types: room to the door, door to the corridor, corridor to stairs, and vertical paths consider stairs between floors.

3.4.1   *Corridor median extraction*

The corridor is the connecting channel between rooms in an indoor environment and is a key point for the extraction of information about the indoor space. If the corridor space is simply abstracted as a node, it cannot truly represent the topological relationship of indoor space and cannot meet the application requirements of indoor navigation. In this paper, we improve the medial transformation method (Blum 1967; Taneja 2011) based on the Delaunay algorithm (Lu 2014; Li 2014), which takes the boundary of the corridor as input data and outputs the corridor path, and the specific process of generating the corridor skeleton is as follows:

(1) The geometric profile of the interior corridor in the BIM model in X-Y plane projection is extracted.

(2) The boundaries of the corridor element outlines are used as constraining edges, and the outline nodes are encrypted so that more closely spaced points participate in the construction of the Delaunay constrained triangular network.

(3) Triangles are classified into three categories according to the number of adjacent triangles of each triangle. Triangles with only one adjacent element are class I triangles; triangles with two adjacent elements are class II triangles, and triangles with three adjacent elements are class III triangles.

(4) For Type I triangles, which often occur at the start or end points, the midpoints of their non-boundary sides are extracted; for Type II triangles, the midpoints of their two non-boundary sides are extracted and connected; for Type III triangles, the center of gravity is extracted and connected to the midpoints of the cross-boundary sides respectively.

The corridor skeleton generated by the above methods may have some closed areas or irregular shapes at the corridor boundaries, which leads to many redundant line segments. In this paper, we propose a method to extract the center of mass of the closed areas to achieve the purpose of removing redundant nodes from the corridor boundary buffer and maintaining the symmetry and topology of the corridor.

4 APPLICATION OF INDOOR NAVIGATION MAP MODELS

This paper selects a primary school in Nanjing to build a BIM model and design a 3D indoor path planning system based on BIM technology. The system is not only able to interact with the BIM model but also enables the automatic construction of indoor road network models and the visualization of indoor path planning guidance.

Based on the above method of extracting and optimizing the components of the indoor road network, this paper selects the Python-based software package Network X. The network topology map is generated by inputting the original node and edge weights and coordinates into Network X. In the road network data visualization interface, the "Load Road Network Model" button in the interface is clicked to extract the corresponding indoor road network map in layers. Once the extraction is complete, the network model will be automatically generated and displayed in the building model. The final extracted road network model is shown in Figure 4.

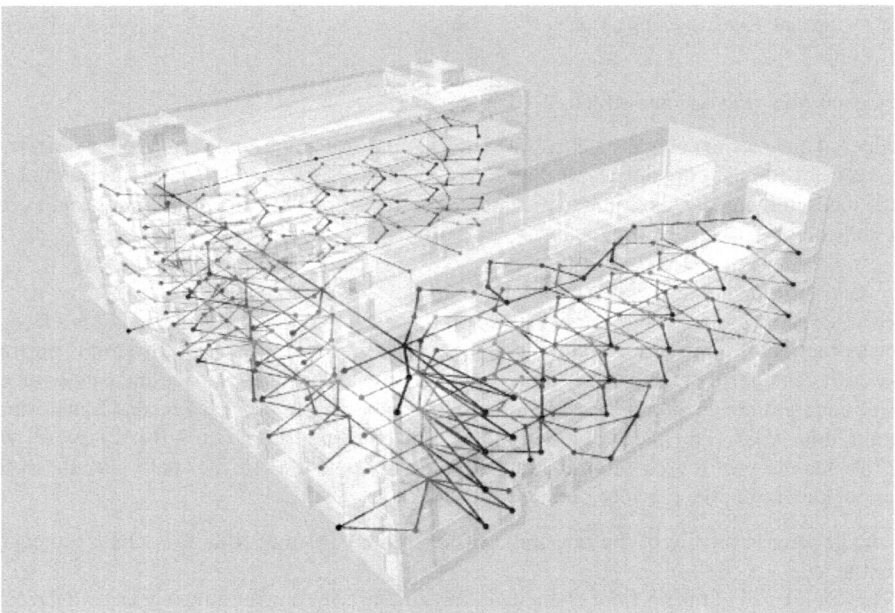

Figure 4.   Extracted road network model.

In this paper, a lounge on the fourth floor is selected as the starting point and a laboratory on the first floor is selected as the endpoint to simulate the indoor pathfinding process. After selecting the start and end points in the system, the corresponding nodes are marked in green and the optimal path is generated by clicking the "Show optimal path" button and applying the above pathfinding algorithm to obtain the nodes in turn and display the direction of passage between them. The final path generation diagram is shown in Figure 5.

Figure 5 shows the optimal path selected in an indoor environment after determining the start and end points, mainly including the optimal path. This includes information such as the nodes that the path passes through in sequence and the final length of the path. The visualization of the optimal path allows the indoor personnel to observe their position more intuitively and realistically to reach their destination quickly, which is of some practical importance.

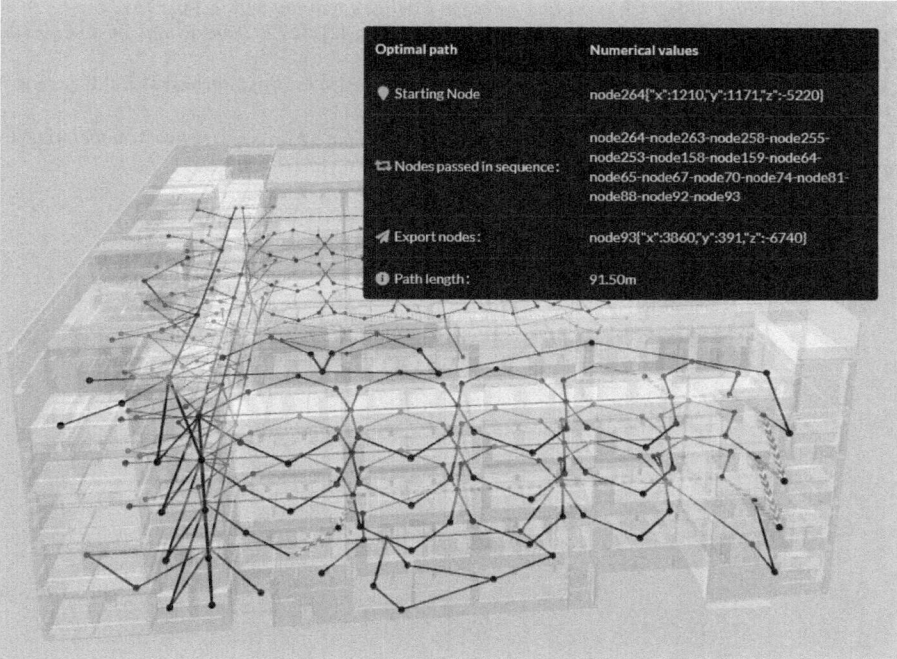

Figure 5.   Results of indoor navigation.

5   CONCLUSION

This paper investigates the construction method of the indoor road network model using the BIM model as the data source, proposes considering staircase nodes as conversion nodes, optimizes the extraction method of corridor medians and intersection processing and puts forward a navigation-oriented method to construct a cross-floor indoor road network map model. The IfcOpenShell library is used to extract the IFC entities and the adjacency relationships between the entities in the indoor road network model, the abstract diagram is used to store the structure of the road network model, and NetworkX is used to visualize the network topology data.

The experimental results show that the combination of the indoor road network model and the path planning algorithm proposed in this paper can find reliable navigation paths in public buildings with high floors and complex internal space structures, which is of positive significance for practical indoor path planning in buildings.

REFERENCES

Blum H. (1967). *A transformation for extracting new descriptors of shape, Models for the perception of speech and visual form*, 19, pp. 362–380.

Cheng H., Chen H., Liu Y. (2015). Topological indoor localization and navigation for autonomous mobile robot [J]. *IEEE Transactions on Automation Science & Engineering*, 12(2):729–738.

Schougaard, K.R., Grønbæk, K. & Scharling, T. (2012). *Indoor pedestrian navigation based on hybrid route planning and location modeling* [M]. Berlin Heidelberg: Springer, 289–306.

Li M., Li W., Qian L.T. (2014). Research on personalized map service model based on text mining [J]. *Mapping and Spatial Geographic Information*. 37(5):39–41.

Lu W., Wei F.Y., Zhang S., et al. (2014). Research on indoor location service applications based on Zigbee [J]. *Mapping and Spatial Geographic Information*. 37(10):75–77.

Taneja, S., Akinci, B., Garrett, J.H., Soibelman, L., East, B. (2011). *Transforming an IFC-based building layout information into a geometric topology network for indoor navigation assistance, Proceedings of the 2011 ASCE International Workshop on Computing in Civil Engineering*, vol. 1, pp. 315–322.

Sturtevant N.R. (2012). Benchmarks for grid-based pathfinding [J]. *IEEE Transactions on Computational Intelligence & Ai in Games*, 4(2):144–148.

Wang R.X., Tang Y.J. (2021). Research on parsing and storage of BIM information based on IFC Standard [J]. *IOP Conference Series: Earth and Environmental Science*, 643(1).

Xue L. (2020). *Research on indoor emergency path planning based on supernetwork model* [D]. China University of Mining and Technology. DOI:10.27623/d.cnki.gzkyu.2020.000305.

*Frontiers in Civil and Hydraulic Engineering – Mohamed A. Ismail and
Hazem Samih Mohamed (Eds)
© 2023 The Authors, ISBN 978-1-032-38247-0*

Soil erosion forecasting of substations based on ELM and environmental factor

Naiyong Wang* & Lei Lei*
*State Grid Shaanxi Electric Power Co Ltd., Electric Power Research Institute, Xi'an, Shaanxi Province,
China*

Mengying Hu*
School of Electrical Engineering, Xi'an University of Technology, Xi'an, Shaanxi Province, China

Huaiwei Yang* & Kai Wu*
State Grid UHV Engineering Construction Company, Xi'an, Shaanxi Province, China

Miao Wei*
State Grid Shaanxi Electric Power Co., Ltd, Xi'an, Shaanxi Province, China

ABSTRACT: The intelligent forecasting and early warning of soil erosion with high precision is
necessary for the safe construction and operation of substations. However, there are few researches
on the intelligent forecasting of soil erosion based on machine learning algorithms and environmental factors. A soil erosion forecasting method based on extreme learning machine (ELM) and
environmental factor was proposed in this paper. Firstly, the correlation between the soil erosion
and the rainfall was analyzed comprehensively. Then, the forecasting model was built based on historical soil erosion and rainfall, using ELM algorithm. Finally, the forecasting model was validated
with the online monitoring data of soil and water conservation in the substation. The accuracy of
the soil erosion forecasting method is verified to be high, and the mean absolute percentage error
is 4.07%. It is proved that the intelligent forecasting model of soil erosion based on ELM and
environmental factor is effective.

1 INSTRUCTION

During the construction of substation projects, the construction activities, such as digging and
filling, cause the damage to the soil and vegetations. With the action of wind, water and gravity, the
soil and water are eroded. Usually, the erosion of soil and water damages the ecological environment
of the construction areas, and poses a threat to substation projects (Bai 2015; Li 2020). Therefore,
soil & water conversation is a research emphasis in substation projects. It is necessary to forecast
the future soil erosion amount of substation projects, because the forecasting results can be used
for the early warning, and for setting up engineering measures of water and soil conservation
reasonably (Pan 2022).

Soil erosion is a complex nonlinear dynamic process, which is influenced by the soil and vegetation conditions of the site and environmental factors. Soil erosion forecasting methods include
mathematical model method, analogy method and field measurement method. The revised universal soil loss equation is used to estimate the soil loss of Tiemiersu River cascade hydro-power
stations. The basic data includes vegetation and land use information extracted from multi-source
remote sensing, rainfall data from weather station and soil data from soil database. The result has

---

*Corresponding Authors: shuibaozu@163.com; ll2494590@163.com and hmy@xaut.edu.cn a certain reliability, which can provide a scientific basis for the soil erosion control in Tiemiersu River cascade hydro-power stations (Li 2015). Several common soil erosion forecasting methods, including analogy analysis method, empirical formula method and field measurement method, are introduced. And the applicability of each method is discussed (Xu 2004). In these researches, the universal soil loss equation depends on the several empirical parameters, and is mainly applicable to the soil erosion forecasting of a year. Using this method, it is difficult to forecast the soil erosion amount of the next several months with high precision according to the changes of influence factors. The forecasting accuracy of soil erosion by analogy method is low, because of regional difference and influence factor difference. There are few researches on the forecasting of soil erosion, considering environmental factors. The intelligent algorithms such as machine learning are not used in soil erosion forecasting of substation projects.

In this paper, a forecasting method of soil erosion based on ELM and environmental factor was proposed. The correlation between the soil erosion and rainfall was analyzed using the monitoring data of soil and water conservation. Then, the soil erosion forecasting model, which comprehensively considers the influence of rainfall factor and involves intelligent algorithm of ELM, was proposed. The soil erosion and environmental factor of S substation in China was collected to verify the effectiveness of this forecasting method.

2 METHODOLOGY

2.1 Data collection and soil erosion calculation

Data are obtained from the online monitoring system of soil and water conservation in S substation. The online monitoring system of soil and water conservation consists of several survey pins with ultrasonic sensors, an environmental factor sensor, a power supply and a control system. The soil erosion thickness is measured by the ultrasonic sensor. The rainfall, wind speed, wind direction and temperature are measured by the environmental factor sensor.

The soil erosion thickness from the ultrasonic sensor is needed to be calculated to the soil erosion amount. According to the ultrasonic ranging theory, the distance L from the ultrasonic probe to the ground is as follows:

$$L = C \times t \tag{1}$$

In the above equation, C is the speed at which ultrasonic waves travel through the air, t is the time from launching to reflecting back from the ground.

The distance from the ultrasonic probe to the ground of each day in a week is l_i ($i = 1, 2, \ldots, 7$), i is the number of a day in a week. The soil erosion thickness of the first day in a week is subtracted from the soil erosion thickness of the seventh day in a week to obtain the weekly soil erosion thickness L_s.

$$L_s = l_7 - l_1 \tag{2}$$

Using the soil bulk density, the slope area, the slope gradient and the calculated soil erosion thickness of a week, the weekly soil erosion amount S_r is obtained.

$$S_r = \gamma_s S L_s \cos \theta \times 10^3 \tag{3}$$

In the above equation, γ_s is the soil bulk density, S is the slope area of the observation area, θ is the slope gradient of the observation area.

2.2 Correlation analysis between the soil erosion and environmental factor

There are many factors affecting soil erosion, such as slope gradient, slope area and rainfall (Nie 2022; Zhang 2013). In water erosion areas, rainfall is one of the most important influencing factors for soil erosion. After data collection and soil erosion calculation from S substation, the weekly rainfall from daily accumulations and the weekly soil erosion amount are shown in Figure 1, from Jul.1, 2021 to Oct.20, 2021.

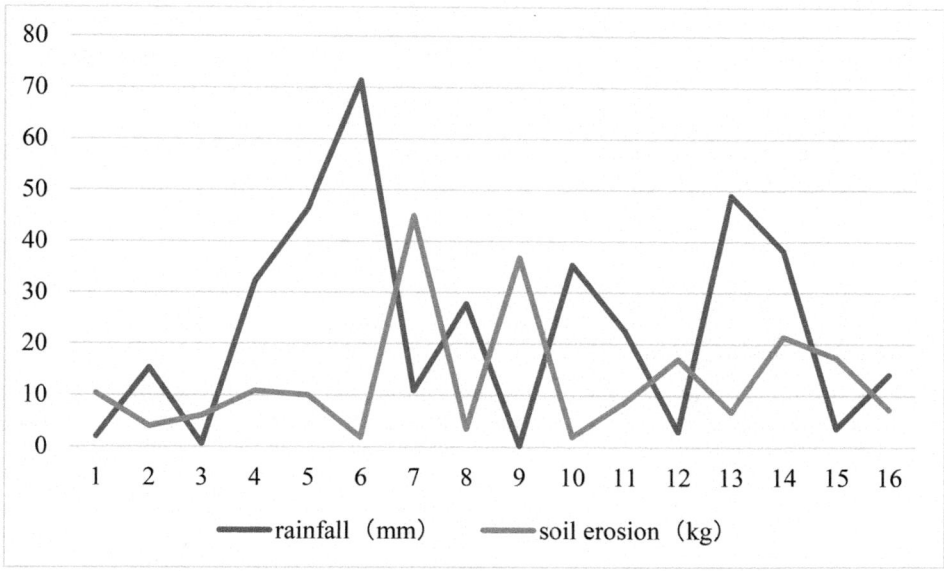

Figure 1. Weekly soil erosion and rainfall.

The weekly soil erosion is constantly changing, and is highly correlated with the cumulative effect of the rainfall in the last or the previous weeks. Rainfall usually intensifies the soil erosion, and the soil erosion lags behind the rainfall. It can be seen that rainfall has a great impact on the soil erosion of S substation. Therefore, as an environmental factor, rainfall is introduced into the forecasting model.

2.3 ELM

ELM is an intelligent machine learning method. Since ELM is suitable for nonlinear characteristic relation mapping, it has been widely used in time series analysis, intelligent forecasting and other fields (Hong 2022).

There are N different samples (X_i, t_i), $X_i = [x_{i1}, x_{i2}, \ldots, x_{in}]^\mathrm{T} \in \mathrm{R}^\mathrm{n}$, and $t = [t_{i1}, t_{i2}, \ldots, t_{im}]^\mathrm{T} \in \mathrm{R}^\mathrm{m}$. The equation of ELM with L nodes at the hidden layer is as follows.

$$\sum_{i=1}^{L} \beta_i g(W_i \cdot X_j + b_i) = o_j, \quad j = 1, 2, \ldots, N \tag{4}$$

In the above equation, $g(x)$ is the excitation function, W_i is the input weight from the input layer to the hidden layer, b_i is the offset of the hidden layer node, β_i is the output weight from the hidden layer to the output layer, and o_j is the forecasting output.

The sample training by ELM aims at minimizing the error between the forecasting values and the actual values.

$$\sum_{i=1}^{N} \left\| o_j - t_j \right\| = 0 \tag{5}$$

There are the matrices β_i, W_i and b_i to gain the following equations.

$$\sum_{i=1}^{L} \beta_i g(W_i \cdot X_j + b_i) = t_j, \quad j = 1, 2, \ldots, N \tag{6}$$

$$H\beta = T \tag{7}$$

In the above equations, $H = \begin{bmatrix} g(w_1 \cdot x_1 + b_1) & \cdots & g(w_L \cdot x_1 + b_L) \\ \vdots & \ddots & \vdots \\ g(w_1 \cdot x_N + b_1) & \cdots & g(w_L \cdot x_N + b_L) \end{bmatrix}$ is the output of the hidden layer, β is the output weight matrix and T is the desired output of ELM.

When w_i and b_i have been produced in ELM, training the model equals calculating β. The calculation equation of β is as follows.

$$\beta = H^+ T = (H^T H)^{-1} H^T T \tag{8}$$

2.4 Soil erosion forecasting model based on ELM and environmental factor

Considering the influence of the rainfall on the soil erosion, the forecasting model based on ELM and rainfall is established to reflect the nonlinear mapping relationship between the soil erosion and rainfall.

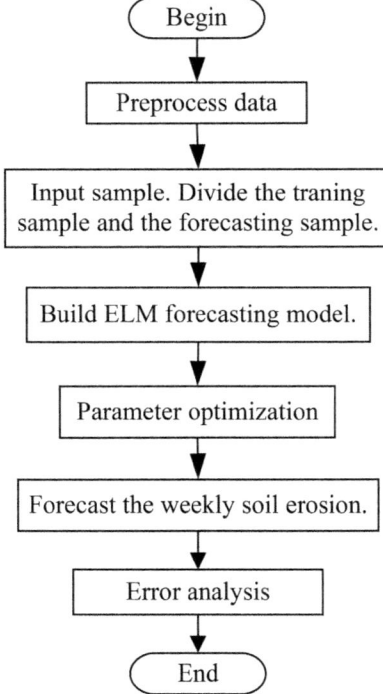

Figure 2.　Forecasting flow.

The forecasting flow is shown in Figure 2. The forecasting steps are as follows:

The data is preprocessed and the weekly rainfall and soil erosion are calculated.

The training sample and the forecasting sample are divided. The historical weekly rainfall is selected as the input of the training sample, and historical weekly soil erosion is selected as the output of the training sample.

An ELM forecasting model is built and the parameters are optimized.

The trained ELM forecasting model is used to forecast the future weekly soil erosion, and analyze the error and the forecasting results.

3 CASE STUDY

Since weekly soil erosion lags one week behind the rainfall, the weekly rainfall from Jun.24, 2021 to Sep.15, 2021 is selected as the input of the training sample, and the weekly soil erosion from Jul.1, 2021 to Sep.22, 2021 is selected as the output of the training sample. The weekly rainfall from Sep.16, 2021 to Oct.13, 2021 is selected as the input of the forecasting sample, and the weekly soil erosion in the future 4 weeks (Sep.23, 2021 to Oct.20, 2021) can be forecasted.

The precision of the soil erosion forecasting is described by the absolute percentage error (APE) and mean absolute percentage error (MAPE) (Zhang 2021). APE and MAPE are defined as follows.

$$APE = 100 \left| \frac{y(t) - y'(t)}{y(t)} \right| \tag{9}$$

$$MAPE = \frac{100}{n} \sum_{t=1}^{n} \left| \frac{y(t) - y'(t)}{y(t)} \right| \tag{10}$$

In the above equation, $y(t)$ is the actual value, $y'(t)$ is the forecasting value, and n is the quantity of the output of the forecasting sample.

The actual soil erosion, the forecasted soil erosion, APE and MAPE are shown in Table 1. The weekly soil erosion forecasting result is shown in Figure 3.

Table 1. Actual soil erosion, forecasted soil erosion and errors.

Week	Actual soil erosion(kg)	Forecasted soil erosion (kg)	APE (%)
1	6.80	6.93	1.85
2	21.40	22.65	5.86
3	17.28	16.63	3.78
4	7.30	6.95	4.78
MAPE (%)			4.07

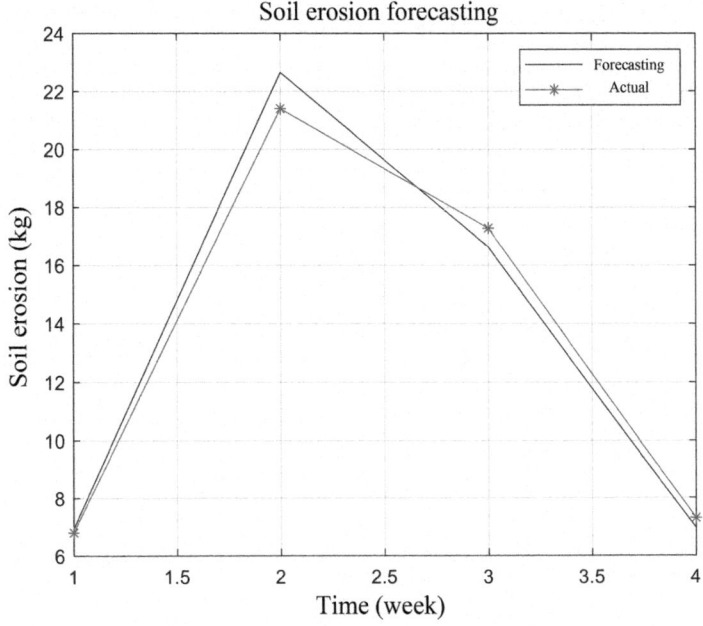

Figure 3. Forecasting result of soil erosion.

From Figure 3 and Table 1, the weekly soil erosion varies. According to the error analysis, the maximum APE of the weekly soil erosion forecasting is 5.86% and the MAPE is 4.07%. It is proved that the weekly soil erosion is forecasted well. High precision forecasting can be achieved by using the forecasting model based on ELM and rainfall factor.

4 CONCLUSION

In this paper, the influence of environmental factor on the soil erosion was analyzed. The soil erosion forecasting method based on ELM and the rainfall factor was proposed. The online monitoring data of soil and water conservation in S substation was collected to validate the proposed forecasting model. The forecasting result shows that the MAPE of the weekly soil erosion forecasting is 4.07%. It is proved that the proposed forecasting method is effective and the precision of the forecasting model is high. The intelligent forecasting of the future soil erosion with high precision is significant for the safe construction and operation of substations.

REFERENCES

Bai Z.W. (2015). Dynamic monitoring of the soil erosion during the construction of Xiaowan hydropower stations. *Journal of Southwest Forestry University* 5: 58–63.
Hong M. (2022). Mid-long term runoff forecasting based on FPA – ELM model in Yalong River Basin. *Yangtze River* 6: 119–125.
Li J.X. (2020). Research on the influencing factors of soil loss in dumping sites of Nuozhadu hydropower station based on principal component analysis. *Hydroelectric Power* 4:1–5+34.
Li X.H. (2015). Soil erosion estimation within Tiemiersu river cascade hydropower stations in Xinjiang using revised universal soil loss equation and remote sensing technology. *China Rural Water and Hydropower* 10: 184–189.
Nie H.Y. (2022). Characteristics of erosion environment and soil and water losses induced by power transmission and transformation project in Hilly area. *Research of Soil and Water Conversation* 2: 50–56.
Pan M.J. (2022). Characteristics and prevention measures of soil erosion in power transmission and transformation projects in Xinjiang region. *Applied Technology of Soil and Water Conservation* 2: 28–31.
Xu Y.N. (2004). *The prediction of soil loss on disturbed surface of development and construction project.* SWCC 3: 29–31.
Zhang H.Y. (2021). *Research on prediction model of deformation behavior of concrete face rockfill dams based on the improved support vector machine.* Xi'an University of Technology.
Zhang W.T. (2013). Spatial-temporal evolution of soil water erosion in Donghu watershed based on cellular automata. *Journal of Hydroelectric Engineering* 1: 96–100.

Frontiers in Civil and Hydraulic Engineering – Mohamed A. Ismail and
Hazem Samih Mohamed (Eds)
© 2023 The Authors, ISBN 978-1-032-38247-0

Analysis of the variation law of negative moment impact factor of continuous T-girder bridge

Jun Tian
Ordos Institute of Technology, Inner Mongolia, China

Yelu Wang*
Chang'an University, Shanxi, China

Pei Jin
Ordos Institute of Technology, Inner Mongolia, China

ABSTRACT: The development of computers and the development and application of finite element analysis software make it possible to solve the complex vehicle-bridge coupled vibration system. The dominant modes of the negative moment section and the positive moment section of the continuous girder bridge are different, and the dynamic impact mechanism of the vehicle load on them is also different. The existing research results do not provide reliable data support for the negative moment impact factor of the fulcrum section of continuous girder bridges. In this paper, the variation law of the impact factor at the negative bending moment of the continuous T-girder bridge will be studied through the coupled vibration of the vehicle bridge, and the relevant calculation results will be analyzed accordingly. Finally, in the case of deploying 3 vehicles, for the multi-piece girder continuous T girder, the impact factor of the partial load is greater than the impact factor of medium load, and the maximum deviation can reach 60%; the live load effect of the fulcrum 2# section is large, while the impact factor is relatively small. The study of its change law on the impact factor lays a foundation for further exploring the influence of the impact factor in the negative bending moment region.

1 INTRODUCTION

According to the current research status of the vehicle-bridge coupled vibration problem, it can be found that the difficulty of the vehicle-bridge coupled vibration problem is that the stiffness matrix, damping matrix, and load matrix of the system all change instantaneously with time (Deng 2010; Moghim 2008; Zhou 2015), which makes the solution of the finite element equation of the system more complicated. At present, Chinese and foreign scholars have achieved some results in the research on the analytical solutions of simple vehicle models and simple bridge vehicle-bridge coupled vibrations. However, through the study of its derivation process, it can be found that it is very unrealistic to extend the analytical solution to the complex vehicle model and the complex vehicle-bridge coupled vibration of the bridge. The road vehicle-bridge coupled vibration has been investigated for more than 100 years. In recent years, the rapid development of computer and finite element technology has made it possible to numerically solve complex vehicle-bridge coupled vibration problems. In this paper, the variation law of the impact factor at the negative bending moment of the continuous T-girder bridge will be studied by means of the vehicle-bridge coupled vibration.

---

*Corresponding Author: 767905358@qq.com

DOI 10.1201/9781003344209-51

This paper takes the continuous T-girder bridge as the research object and combines theoretical analysis, numerical simulation, and real bridge verification to study the variation law of the negative bending moment impact factor of the continuous T-girder bridge. The research results reveal the vehicle dynamic behavior and mechanism of continuous girder bridges, which can provide a reference for dynamic evaluation of the bridge and promote its wide application and development.

2 THEORETICAL ANALYSIS

2.1 Bridge calculation model

In the study of vehicle-bridge coupled vibration, the bridge structure is usually discretized and a spatial finite element model is established. As a system with infinite degrees of freedom, the vibration equation of the bridge is:

$$[M]\{\ddot{Y}\} + [C]\{\dot{Y}\} + [K]\{Y\} = \{F\} \tag{1}$$

where: $[M]$, $[C]$ and $[K]$ are the mass matrix, damping matrix, and stiffness matrix, respectively;
$\{\ddot{Y}\}$, $\{\dot{Y}\}$ and $\{Y\}$ represent the nodal acceleration, velocity, and vertical displacement vector, respectively;
$\{F\}$ stands for the force column vector acting on the node.
The bridge structure is usually solved by the mode superposition method because of its relatively large number of degrees of freedom. For a structural system with more degrees of freedom, a relatively accurate dynamic response can be obtained by considering only a few mode shapes, thus reducing the redundant workload.

2.2 Vehicle calculation model

In order to study the coupled vibration effect between the vehicle and the bridge structure, it is necessary to establish a vehicle calculation model. Combined with the mainstream vehicles used in the actual highway bridge load test, this paper selects a 1/2 five-degree-of-freedom three-axis plane vehicle as the vehicle calculation model (Caprani 2013; Jung 2013; Szurgott 2011; Song 2012) as shown in Figure 1.

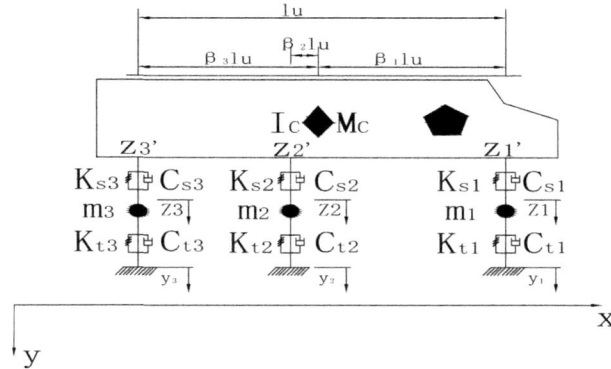

Figure 1.   Five-degree-of-freedom three-axis plane vehicle model.

Assuming that the wheels do not leave the bridge deck when the vehicle is running and keep in contact all the time, the spatial dynamic balance equation of the vehicle can be established and expressed as a matrix:

$$[M]\{\ddot{Z}\} + [C]\{\dot{Z}\} + [K]\{Z\} = \{P\} \tag{2}$$

where:

$$[M] = \begin{bmatrix} M_G & & & & & & & & \\ & m_1 & & & & & & & \\ & & m_2 & & & & & & \\ & & & m_3 & & & & & \\ & & & & J_{xg} & & & & \\ & & & & & J_1 & & & \\ & & & & & & J_2 & & \\ & & & & & & & J_3 & \\ & & & & & & & & J_{yg} \end{bmatrix}$$

$$[K] = \begin{bmatrix} K^1 & K^2 \\ K^3 & K^4 \end{bmatrix}$$

$$[K^1] = \begin{bmatrix} \sum_{m=1}^{3}\sum_{n=1}^{2} K_{smn} & -\sum_{n=1}^{2} K_{s1n} & -\sum_{n=1}^{2} K_{s2n} & -\sum_{n=1}^{2} K_{s3n} & -\sum_{m=1}^{3}\sum_{n=1}^{2} K_{smn}(-1)^n b \\ & \sum_{n=1}^{2} K_{s1n} + \sum_{n=1}^{2} K_{t1n} & 0 & 0 & \sum_{n=1}^{2} K_{s1n}(-1)^n b \\ & & \sum_{n=1}^{2} K_{s2n} + \sum_{n=1}^{2} K_{t2n} & 0 & \sum_{n=1}^{2} K_{s2n}(-1)^n b \\ & & & \sum_{n=1}^{2} K_{s3n} + \sum_{n=1}^{2} K_{t3n} & \sum_{n=1}^{2} K_{s3n}(-1)^n b \\ \text{Symm} & & & & \sum_{m=1}^{3}\sum_{n=1}^{2} K_{smn} b^2 \end{bmatrix}$$

$$[K^2] = \begin{bmatrix} \sum_{n=1}^{2} K_{s1n}(-1)^n b & \sum_{n=1}^{2} K_{s1n}(-1)^n b & \sum_{n=1}^{2} K_{s1n}(-1)^n b & \sum_{n=1}^{2} K_{s1n} l_1 - \sum_{n=1}^{2} K_{s2n} l_2 - \sum_{n=1}^{2} K_{s3n} l_2 \\ -\sum_{n=1}^{2} K_{s1n}(-1)^n b - \sum_{n=1}^{2} K_{t1n}(-1)^n b & 0 & 0 & -\sum_{n=1}^{2} K_{s1n} l_1 \\ 0 & -\sum_{n=1}^{2} K_{s1n}(-1)^n b - \sum_{n=1}^{2} K_{t1n}(-1)^n b & 0 & -\sum_{n=1}^{2} K_{s2n} l_2 \\ 0 & 0 & -\sum_{n=1}^{2} K_{s1n}(-1)^n b - \sum_{n=1}^{2} K_{t1n}(-1)^n b & -\sum_{n=1}^{2} K_{s3n} l_2 \\ -\sum_{n=1}^{2} K_{s1n} b^2 & -\sum_{n=1}^{2} K_{s2n} b^2 & -\sum_{n=1}^{2} K_{s3n} b^2 & \begin{matrix} -\sum_{n=1}^{2} K_{s1n}(-1)^n b l_1 \\ +\sum_{n=1}^{2} K_{s2n}(-1)^n b l_2 \\ +\sum_{n=1}^{2} K_{s3n}(-1)^n b l_2 \end{matrix} \end{bmatrix}$$

$$[K^3] = [K^2]^T$$

$$[K^4] = \begin{bmatrix} [\sum_{n=1}^{2} K_{s1n} + \sum_{n=1}^{2} K_{t1n}] b^2 & 0 & 0 & \sum_{n=1}^{2} K_{s1n}(-1)^n b l_1 \\ & [\sum_{n=1}^{2} K_{s2n} + \sum_{n=1}^{2} K_{t2n}] b^2 & 0 & -\sum_{n=1}^{2} K_{s2n}(-1)^n b l_2 \\ & & [\sum_{n=1}^{2} K_{s3n} + \sum_{n=1}^{2} K_{t3n}] b^2 & -\sum_{n=1}^{2} K_{s3n}(-1)^n b l_2 \\ \text{Symm} & & & \sum_{n=1}^{2} K_{s1n} l_1^2 + \sum_{n=1}^{2} K_{t2n} l_1^2 + \sum_{n=1}^{2} K_{s3n} l_2^2 \end{bmatrix}$$

$$[C] = \begin{bmatrix} C^1 & C^2 \\ C^3 & C^4 \end{bmatrix}$$

$$[C^1] = \begin{bmatrix} \sum_{m=1}^{3}\sum_{n=1}^{2} C_{smn} & -\sum_{n=1}^{2} C_{s1n} & -\sum_{n=1}^{2} C_{s2n} & -\sum_{n=1}^{2} C_{s3n} & -\sum_{m=1}^{3}\sum_{m=1}^{2} C_{smn}(-1)^n b \\ & \sum_{n=1}^{2} C_{s1n} + \sum_{n=1}^{2} C_{t1n} & 0 & 0 & \sum_{n=1}^{2} C_{s1n}(-1)^n b \\ & & \sum_{n=1}^{2} C_{s2n} + \sum_{n=1}^{2} C_{t2n} & 0 & \sum_{n=1}^{2} C_{s2n}(-1)^n b \\ & & & \sum_{n=1}^{2} C_{s3n} + \sum_{n=1}^{2} C_{t3n} & \sum_{n=1}^{2} C_{s3n}(-1)^n b \\ \text{Symm} & & & \sum_{m=1}^{3} C_{t3n} & \sum_{m=1}^{3}\sum_{n=1}^{2} C_{smn} b^2 \end{bmatrix}$$

$$
[C^2] = \begin{bmatrix}
\sum_{n=1}^{2} C_{s1n}(-1)^n b & \sum_{n=1}^{2} C_{s1n}(-1)^n b & \sum_{n=1}^{2} C_{s1n}(-1)^n b & \sum_{n=1}^{2} C_{s1n}l_1 - \sum_{n=1}^{2} C_{s2n}l_2 - \sum_{n=1}^{2} C_{s3n}l_2 \\
-\sum_{n=1}^{2} C_{s1n}(-1)^n b - \sum_{n=1}^{2} C_{t1n}(-1)^n b & 0 & 0 & -\sum_{n=1}^{2} C_{s1n}l_1 \\
0 & -\sum_{n=1}^{2} C_{s1n}(-1)^n b - \sum_{n=1}^{2} C_{t1n}(-1)^n b & 0 & -\sum_{n=1}^{2} C_{s2n}l_2 \\
0 & 0 & -\sum_{n=1}^{2} C_{s1n}(-1)^n b - \sum_{n=1}^{2} C_{t1n}(-1)^n b & -\sum_{n=1}^{2} C_{s3n}l_2 \\
& & & -\sum_{n=1}^{2} C_{s1n}(-1)^n bl_1 \\
-\sum_{n=1}^{2} C_{s1n}b^2 & -\sum_{n=1}^{2} C_{s2n}b^2 & -\sum_{n=1}^{2} C_{s3n}b^2 & +\sum_{n=1}^{2} C_{s2n}(-1)^n bl_2 \\
& & & +\sum_{n=1}^{2} C_{s3n}(-1)^n bl_2
\end{bmatrix}
$$

$$[C^3] = [C^2]^T$$

$$
[C^4] = \begin{bmatrix}
[\sum_{n=1}^{2} C_{s1n} + \sum_{n=1}^{2} C_{t1n}]b^2 & 0 & 0 & \sum_{n=1}^{2} C_{s1n}(-1)^n bl_1 \\
& [\sum_{n=1}^{2} C_{s2n} + \sum_{n=1}^{2} C_{t2n}]b^2 & 0 & -\sum_{n=1}^{2} C_{s2n}(-1)^n bl_2 \\
& & [\sum_{n=1}^{2} C_{s3n} + \sum_{n=1}^{2} C_{t3n}]b^2 & -\sum_{n=1}^{2} C_{s3n}(-1)^n bl_2 \\
\text{Symm} & & & \sum_{n=1}^{2} C_{s1n}l_1^2 + \sum_{n=1}^{2} C_{t2n}l_1^2 + \sum_{n=1}^{2} C_{s3n}l_2^2
\end{bmatrix}
$$

$$
[P] = \begin{bmatrix}
0 \\
-\sum_{n=1}^{2} K_{t1n}Z_{01n} + C_{t1n}\dot{Z}_{01n} \\
-\sum_{n=1}^{2} K_{t2n}Z_{02n} + C_{t2n}\dot{Z}_{02n} \\
-\sum_{n=1}^{2} K_{t3n}Z_{03n} + C_{t3n}\dot{Z}_{03n} \\
0 \\
-\sum_{n=1}^{2} (-1)^n b(K_{t1n}Z_{01n} + C_{t1n}\dot{Z}_{01n}) \\
-\sum_{n=1}^{2} (-1)^n b(K_{t2n}Z_{02n} + C_{t2n}\dot{Z}_{02n}) \\
-\sum_{n=1}^{2} (-1)^n b(K_{t3n}Z_{03n} + C_{t3n}\dot{Z}_{03n}) \\
0
\end{bmatrix}
$$

$$[Z] = [Z_G, Z_1, Z_2, Z_3, \theta_{xg}, \theta_1, \theta_2, \theta_3, \theta_{yg}]$$

where: M, m_1, m_2 m_3 —mass of the body and each axle;

$Z_G, \dot{Z}_G, \ddot{Z}_G$—vertical displacement, velocity and acceleration of the body, respectively;

$Z_m, \dot{Z}_m, \ddot{Z}_m$—the vertical displacement, velocity and acceleration of the m-th axis ($m = 1, 2, 3$), respectively;

$Z_{rmn}, \dot{Z}_{rmn}$—the function and derivative of the bridge deck concave and convex value at the position of each wheel with respect to time;

C_{smn}, C_{tmn}—the suspension spring of the m-th axis and the wheel damping coefficient, respectively ($m = 1, 2, 3; n = 1, 2$);

K_{smn}, K_{tmn}—the suspension spring and wheel stiffness of the mth axis, respectively. ($m = 1, 2, 3; n = 1, 2$);

$J_{xg}, J_{yg}, J_1, J_2, J_3$—the moment of inertia of each axle;

$\theta_{xm}, \dot{\theta}_{xm}, \ddot{\theta}_{xm}$—the rotation angle, angular velocity, and angular acceleration of each axis of lateral turbulence ($m = 1, 2, 3$);

$\theta_{yg}, \dot{\theta}_{yg}, \ddot{\theta}_{yg}$—the longitudinal bump body angle, angular velocity and angular acceleration, respectively;

$y(t,x_{mu})$, $\dot{y}(t,x_{mu})$—the deflection and velocity at the location of each wheel on the beam, respectively;

b, l_1, l_2—the half of the left and right wheel bases, the distance between the front and rear axles, and the distance between the middle and rear axles, respectively.

From the above theoretical analysis, it can be seen that under the action of uniform moving loads, the solution process of the forced vibration equations of the bridge structure in each order mode is very complicated and needs to be solved by a computer processor.

3 PROJECT EXAMPLES

In order to study the change law of the impact factor at the negative bending moment of the continuous girder bridge, this paper takes the continuous T-girder bridge as an example to carry out the numerical simulation analysis, and the numerical simulation process of the continuous girder bridge is not described in detail.

3.1 Bridge model and parameters

Taking a continuous T-girder bridge with a calculated span of 30 m as the research object, the finite element spatial beam lattice model was established by ANSYS, and the dynamic response and impact effect of the transition region of positive and negative bending moments were analyzed.

(1) Elevation layout

In order to ensure that the width-span ratio of the bridge is as close as possible, this paper selects a 3×30 m continuous T-girder bridge with a width-span ratio of 3/8–4/8.

(2) Cross-sectional layout

The cross-sectional form of the 3×30 m continuous T-girder bridge is shown in Figure 2.

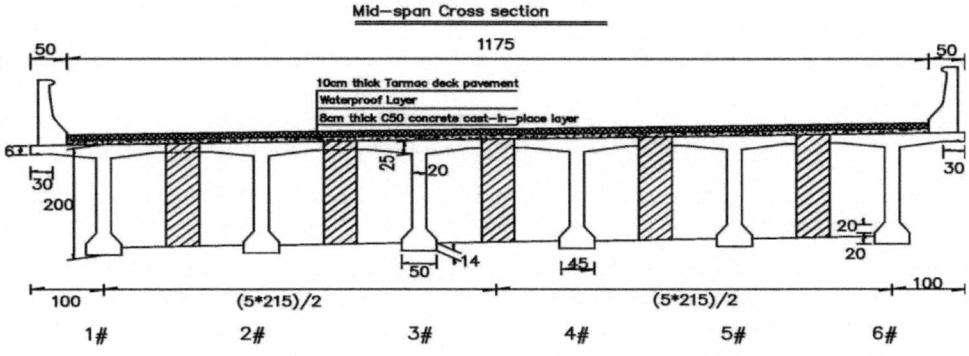

Figure 2.   Cross-sectional view of prestressed concrete continuous T-girder bridge (Unit: cm).

(3) Materials

Concrete: C50 is used for prefabricated T girders and bridge decks, $E_c = 3.45 \times 10^4$MPa and $\rho = 2500$kg/m^3. The structural damping ratio is taken as 0.05.

(4) Computational model

Ignoring the influence of the substructure, the finite element full superstructure model is established by ANSYS, as shown in Figure 3

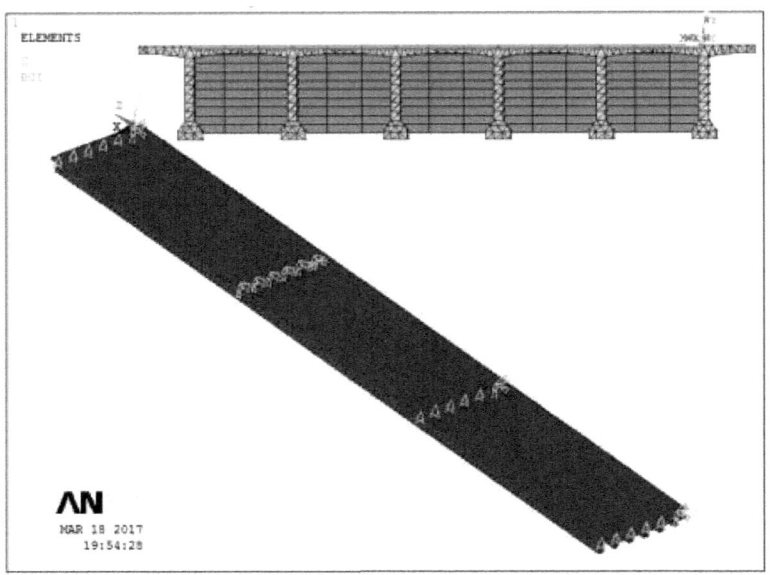

Figure 3. Schematic diagram of continuous T box girder model.

3.2 *Vehicle model and parameters*

In order to analyze the spatial dynamic response of the axle more realistically, this paper adopts a five-degree-of-freedom three-axis plane vehicle model, and makes the following assumptions.

(1) During the vibration of the vehicle, the body is regarded as an absolutely rigid inertial body, and the entire mass of the vehicle is concentrated on the axle.
(2) Both the wheel and the suspension system are considered ideal elastic bodies, and their damping is proportional to the speed of the vehicle.
(3) During the driving process of the vehicle, the vehicle tires maintain close contact with the bridge deck.

3.3 *Time history curve and impact factor analysis of positive and negative bending moment transition region*

In order to study the influences of the transverse driving position and the number of vehicles on the impact factor of negative bending moment and the change law of the impact coefficient of the longitudinal negative bending moment of bridges, this paper takes the three-span continuous T girder introduced above as an example to study it in detail.

3.3.1 *Design conditions*

Five typical cross-sections in the transition area of positive and negative bending moments are selected, and the specific locations are shown in Figure 4 below.

In order to study the influences of the driving position and the number of vehicles on the impact factor of different main beam negative bending moment regions, two working conditions are designed in this paper, as shown in Table 1 below. This paper gives the running position of the continuous T girder (Figures 5 and 6).

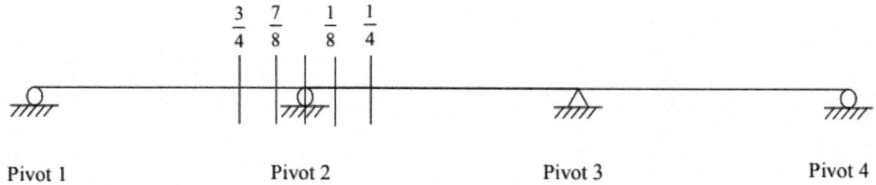

| | $\frac{3}{4}$ | $\frac{7}{8}$ | $\frac{1}{8}$ | $\frac{1}{4}$ |

Pivot 1   Pivot 2   Pivot 3   Pivot 4

Figure 4.   Schematic diagram of key section of three-span continuous girder.

Table 1.   Parameters of each working condition.

Operating condition number	Operating condition name
Operating condition 1	Unbalanced load - 3 cars
Operating condition 2	Medium load - 3 cars

3.3.2   *Calculation results of impact factor*

The APDL module in ANSYS software is used to solve the vehicle-bridge coupled dynamic response under the corresponding parameters, and the time-history curves of the bending moment of the prestressed concrete continuous T-girder bridge are obtained respectively.

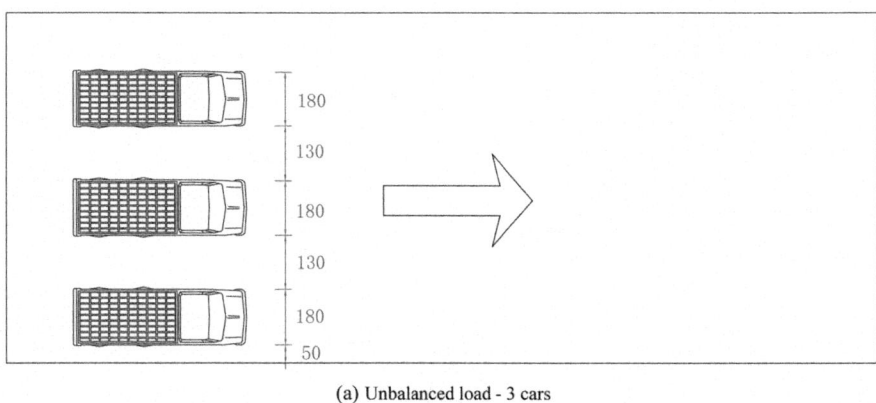

(a) Unbalanced load - 3 cars

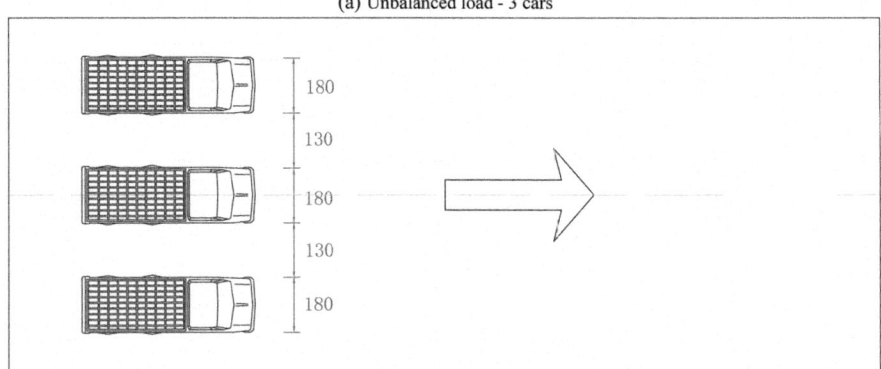

(b)Medium load - 3 cars

Figure 5.   Schematic diagram of the layout of the longitudinal section of the bridge under various working conditions.

391

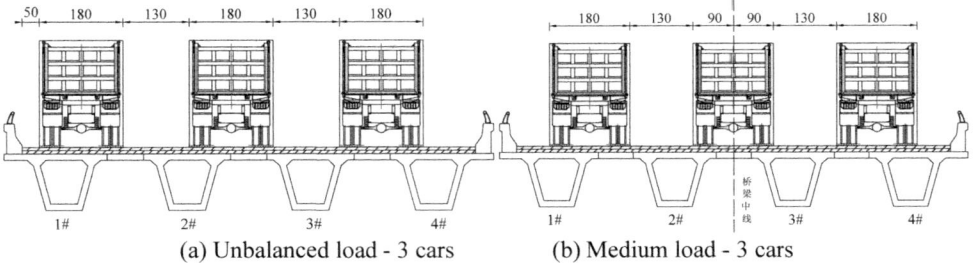

50	180	130	180	130	180

1#   2#   3#   4#

(a) Unbalanced load - 3 cars

180	130	90	90	130	180

1#   2#   3#   4#

(b) Medium load - 3 cars

Figure 6.   Schematic diagram of cross-sectional layout of bridges under various working conditions.

In the time history curve, different impact factors will be obtained due to the different selection of the left or right trough values, and the difference cannot be ignored (Han 2013; Han et al. 2011; Song & He 2001). In this paper, the left wave peak and the right wave peak corresponding to the maximum effect are selected to calculate the impact factor μ_1 and μ_2, and the mean μ_{mean} value of the two respectively, so as to correct the difference in the selection of left and right wave peaks in the traditional test method, which is regarded as the corrected test method. The calculation formula is expressed as the following Equation (5), and the calculation process is shown in Figure 7.

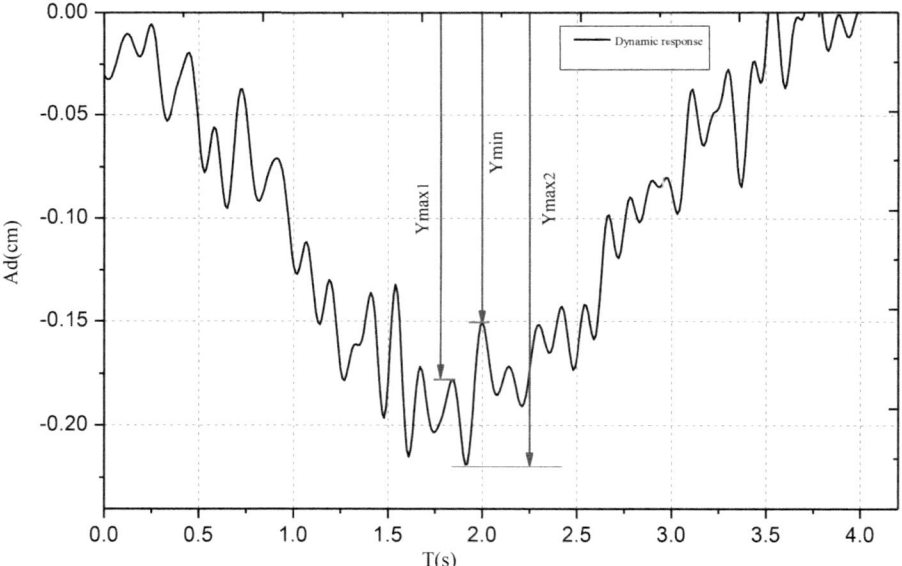

Figure 7.   Schematic Diagram of the Modified Test Method for Calculating the Impact Factor.

$$\mu_1 = \frac{2Y_{max}}{(Y_{min\,1} + Y_{max})} - 1 \qquad (3)$$

$$\mu_2 = \frac{2Y_{max}}{(Y_{min\,2} + Y_{max})} - 1 \qquad (4)$$

$$\mu_{mean} = \frac{(\mu_1 + \mu_2)}{2} \qquad (5)$$

$Y_{min\,1}$—the left-side peak (deflection or strain) value corresponding to Y_{max};

$Y_{\min 2}$—the right-side peak (deflection or strain) value corresponding to $Y_{\max}$;

μ_1—left impact factor calculated using the left crest and trough;

μ_2—right impact factor calculated using the right crest and trough;

μ_{mean}—This measuring point uses the test method impact factor calculated from the left wave crest and the right wave crest.

The calculation results of the impact factor for each working condition are shown in Figure 8.

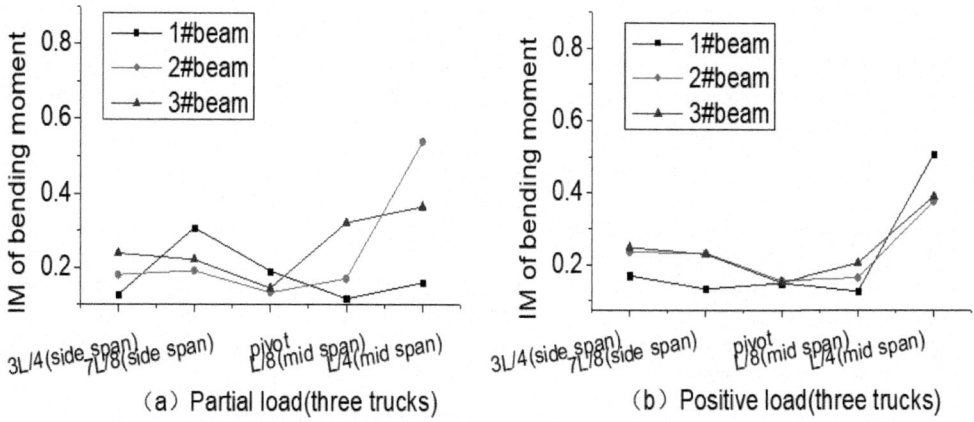

(a) Partial load(three trucks)   (b) Positive load(three trucks)

Figure 8. Bending Moment Impact Factor of Continuous T Girder.

It can be seen from Figure 8 that:

Under the eccentric load, the 1# girder of the continuous T girder has a larger bending moment effect than other girders, and the impact factor is relatively small, which is the working beam under the design load.

Under the medium load, the 3# girder of the continuous T girder has larger bending moment effect than other girders, and the impact factor is relatively small, which is the working girder under the design load.

Therefore, this paper only takes the working girder under the corresponding design load as the research object, and further studies the influence of the number of vehicles on the impact factor of the negative bending moment.

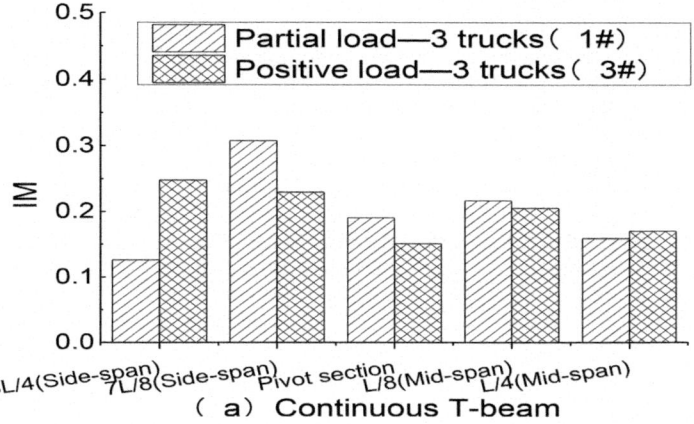

(a) Continuous T-beam

Figure 9. Impact Factor of Design Conditions.

It can be seen from Figure 9 that in the case of deploying 3 vehicles, for the multi-piece girder type continuous T girder, the impact factor of partial load is greater than the impact factor of medium load, and the maximum deviation can reach 60%. For the multi-girder continuous girder bridge, the negative bending moment value of the fulcrum 2# section is obviously larger than that of the side span 7/8 section and the mid-span 1/8 section.

4 CONCLUSION

In this paper, a continuous T girder is taken as an example to study the variation law of the impact factor of multiple main girders at different loading positions in the alternating region of positive and negative bending moments.

(1) Under the eccentric load, the 1# girder of the continuous T girder has a larger bending moment effect than other girders, and the impact factor is relatively small. Under the medium load, the 3# girder of the continuous T girder has a larger bending moment effect than other girders, and the impact factor is relatively small;

(2) In the case of deploying 3 vehicles, for the multi-piece girder type continuous T girder, the impact factor of the partial load is greater than the impact factor of medium load, and the maximum deviation can reach 60%. The live load effect of the fulcrum 2# section is large, and the impact factor is relatively small;

In this paper, the continuous T girder is taken as an example, and three vehicles are loaded to study the change law of the impact factor, which lays a foundation for further exploring the impact of the impact factor in the negative bending moment region.

REFERENCES

Caprani C.C. Lifetime highway bridge traffic load effect from a combination of traffic states allowing for dynamic amplification [J]. *Journal of Bridge Engineering*, 2013, 18(9): 901–909.

Deng L. and Cai C.S. Development of dynamic impact factor for performance evaluation of existing muti-girder concrete bridges [J]. *Engineering Structures*, 2010, 32: 21–31.

Jung H., Kim G., and Park C., Impact factors of bridges based on natural frequency for various superstructure types [J]. *KSCE Journal of Civil Engineering*, 2013, 17(2): 458–464.

Moghimi H. and Ronagh H.R. Impact factors for a composite steel bridge using non-linear dynamic simulation [J]. *International Journal of Impact Engineering*, 2008, 35: 1228–1243.

Song Du. *Research on Performance Test and Evaluation Method of Bridge Structure Based on Vibration Time-history Curve* [D]. Chongqing: Chongqing Jiaotong University, 2012.

Szurgott P., Wekezer J.K., Wasniewski L., et al. Experimental assessment of dynamic responses induced in concrete bridge by permit vehicles [J]. *Journal of Bridge Engineering*, 2011, 16(1): 108–116.

Han W.S., Wang T., Li Y.Q., Li Y.W., Huang P.M. Vehicle-bridge coupling vibration analysis system based on model modified grillage method [J]. *China Journal of Highway and Transport*, 2011, 24(5): 47–55.

Song Y.F., He S.H. Analysis of impact factors of highway bridges [J]. *Journal of Xi'an Highway Traffic University*. 2001, 21(2): 47–48.

Han Z.Q. *Study on impact coefficient of long span curved continuous rigid frame bridge with high piers* [D]. Xi'an: Chang'an University, 2013.

Zhou Y.J., Ma Z.J., Zhao Y., Shi X.W., and He S.H. Improved definition of dynamic load allowance factor for highway bridges. *Structure Engineering and Mechanics*, 2015, 54(3): 561–57.

Frontiers in Civil and Hydraulic Engineering – Mohamed A. Ismail and
Hazem Samih Mohamed (Eds)
© 2023 The Authors, ISBN 978-1-032-38247-0

The use of multi-source information fusion in dam safety evaluation

Cheng Fang*
Nanjing Hydraulic Research Institute, Nanjing, China
College of Water Resources and Hydropower, Hohai University, Nanjing, Jiangsu, China

Haipeng Wang
Nanjing Hydraulic Research Institute, Nanjing, China

ABSTRACT: With the improvement of dam safety and the diversification of monitoring methods, the demand for theoretical methods of a comprehensive evaluation of dam safety is getting higher and higher. As a newly emerging frontier and promising research field, the information fusion theory has been widely used in the process of multi-decision information processing. In this paper, the application of the multi-source information fusion method in dam safety evaluation is briefly introduced, and the theory of multi-source fusion optimization of VIKOR is proposed, which makes the final evaluation results more reasonable and reliable.

1 INTRODUCTION

Among the 98,566 reservoir dams in China, the number of dangerous reservoirs occupies a large proportion. Considering the funding and reinforcement design and construction conditions, the reservoir dam removal and reinforcement is a long-term and arduous system project. At the same time, it is difficult to rank the severity of dangerous reservoirs only by forming a group of experts. But with the progress and development of technology and the advent of the digital era, most reservoir dams are digitally supervised, and monitoring data is stored in the cloud (Zhang 2008). The data analysis of dam safety status, in combination with expert evaluation, makes the dam safety evaluation results more reliable.

The monitoring data information and expert evaluation information, which are two information sources, can provide evidence support for different states of evaluation indicators. Due to the complexity of dam projects and the variability of the external environment of operation, their multi-source information is also uncertain and incomplete (Xu et al. 2007). When conducting dam safety evaluation, it is necessary to organically fuse the evidence from different information sources. The theory of multi-source information fusion can well handle and integrate the evidence of multiple uncertainty information and determine the nature of the observed object.

Because of the special geological structure of the landslide body in the plateau reservoir area, the traditional landslide safety evaluation method of a single information source has been difficult to meet the needs of engineering safety. Yang J used the multi-source information fusion technology, entropy weight method, and hierarchical analysis method to achieve the fusion of data information, constructed a dynamic weight function based on monitoring data and established a landslide safety evaluation index system through the hierarchical analysis method (Yang et al. 2022).

Tao C C (Tao et al. 2010) found that the traditional dam safety evaluation methods could not solve the contradiction between the accuracy of safety evaluation and the uncertainty of dam safety. By introducing the D-S evidence fusion theory method in dam safety evaluation, it has been found

---

*Corresponding Authors: fangcheng9804@foxmail.com and 30666329@qq.com

DOI 10.1201/9781003344209-52

that the D-S evidence theory has better distinguishability and also reduces uncertainty compared with the traditional evaluation methods.

The specific application of multi-source information fusion theory in dam safety evaluation is more detailed, but there is a single information source, which may lead to excessive uncertainty coefficients, and the credibility assignment of each information source in multi-source decision information fusion is not perfect. This paper summarizes the application of multi-source information fusion theory in dam safety evaluation and proposes to solve the information conflict between different information sources by introducing the VIKOR method, screening out the key information sources, and giving the credibility of the evidence sources to correct their basic probability assignments to make the final evaluation results more reasonable.

2 THEORETICAL APPROACH TO MULTI-SOURCE INFORMATION FUSION

(1) Bayesian inference methods

Bayesian methods for multi-source information fusion require that the possible decisions of the system should be independent of each other. This allows us to consider these decisions as a division of the sample space and to solve the decision problem of the system using the formula of Bayesian conditional probability.

The possible decisions of the system are set as $A_1, A_2, \cdots, A_m$, and the observation B is provided by a certain information source. If the prior knowledge of the system and the characteristics of this information source can be used to obtain each prior probability $p(A_i)$ and conditional probability $p(B|A_i)$, then the Bayesian probability formula is used for calculation.

$$p(A_i|B) = \frac{p(A_iB)}{p(B)} = \frac{p(B|A_i)p(A_i)}{\sum_{j=1}^{m} p(B|A_j)p(A_j)} \tag{1}$$

The prior probability $p(A_i)$ is updated to the posterior probability $p(A_i|B)$ based on the observations of the information sources. When there are two information sources observing the system, in addition to the information source observation information result B introduced above, the other information source observes the system results in C. Its conditional probability about each decision A_i is $p(C|A_i) > 0$ $(i = 1, 2, \cdots, m)$, then the conditional probability formula can be expressed as

$$p(A_i|B \cap C) = \frac{p(B \cap C|A_i)p(A_i)}{\sum_{j=1}^{m} p(B \cap C|A_j)p(A_j)} \tag{2}$$

It is often difficult to calculate the prior conditional probability $p(B \cap C|A_i)$ $(i = 1, 2, \cdots, m)$ of calculating the simultaneous occurrence of B and C. To simplify the calculation, a further independence assumption is proposed. AB and C are assumed to be independent of each other, so $p(B \cap C|A_i) = p(B|A_i)p(C|A_i)$, then the above equation can be rewritten as:

$$p(A_i|B \cap C) = \frac{p(B|A_i)p(C|A_i)p(A_i)}{\sum_{j=1}^{m} p(B|A_j)p(C|A_j)p(A_j)} \tag{3}$$

This result is also available to the case of multiple sources. When there are n sources and the observations are $B_1, B_2, \cdots, B_n$, respectively, which are assumed to be independent of each other and conditionally independent of the observed object, the total posterior probability of each decision when the system has n sources can be obtained as:

$$p(A_i|B_1 \cap B_2 \cap \cdots B_n) = \frac{\prod_{k=1}^{n} p(B_k|A_i)p(A_i)}{\sum_{j=1}^{m} \prod_{k=1}^{n} p(B_k|A_j)p(A_j)} i = 1, 2, \cdots, m \tag{4}$$

Finally, the decision of the system can be given by certain rules, such as taking the one with the maximum posterior probability as the final decision of the system.

Bayesian methods were the first methods used to deal with uncertainty inference, and Bayesian formulas are simple, basic and easy to understand (Luo, et al. 2015). Bayesian methods only require moderate computational time, but Bayesian methods require all probabilities to be independent (Song et al. 2021), which poses great difficulties to practical systems, some of which are even impractical.

(2) D-S theory of evidence

The evidence theory was proposed by Dempster in 1967 and later expanded and developed by Shafer, so it is also known as the D-S theory. The evidence theory can deal with uncertainties. It uses trust functions rather than probabilities as metrics, and establishes trust functions by imposing constraints on the probabilities of some events without having to state precise unobtainable probabilities, and it proceeds to become probabilistic when the constraints are restricted to strict probabilities (Yan et al. 2003).

(1) Basic concept of evidence theory

U denotes a set of all possible values of X in a theoretical domain and all elements within U are mutually incompatible, and then U is said to be the identification framework of X. The function $m: 2^U \rightarrow [0, 1]$ (2^U is all subsets of U) satisfies the following conditions.

1) $m(\phi) = 0$

2) $\sum_{A \subset U} m(A) = 1$

Then $m(A)$ is said to be the basic probability assignment of A

The trust function $BEL(A) = \sum_{B \subset A} m(B)$ ($\forall A \subset U$) denotes the sum of the probability measures of all subsets of A. Then, it represents the total trust in A. Thus, it is known that $BEL(\emptyset) = 0, BEL(U) = 1$

(2) Combinatorial rules of evidence theory

$m_1, m_2, \ldots m_n$ are set as n mutually independent basic probability assignments on 2^U. The problem now is how to determine the combined basic probability assignments: $m = m_1 + \ldots + m_m$

$[BEL_1, \ldots, BEL_n]$ stans for n trust functions on the same identification frame U, and $m_1, m_2, \ldots, m_n$ are their corresponding basic probability assignments, then:

$$K_1 = \sum_{A_i \cap B_i \cap \ldots \cap Z_k = \emptyset} m_1(A_1) \cdot m_2(B_2) \cdot \ldots \cdot m_n(Z_k) < 1 \tag{5}$$

$$m(C) \begin{cases} \frac{\sum_{A_i \cap B_i \cap \ldots \cap Z_k = \emptyset} m_1(A_1) \cdot m_2(B_2) \cdot \ldots \cdot m_n(Z_k)}{1 - K_1} \forall C \subset U_C \\ 0 \end{cases} \tag{6}$$

K is a measure of the degree of evidence conflict, and when $K \neq 1$, evidence fusion can be performed; when $K \leq 1$, the evidence is considered to be in conflict with each other and cannot be fused.

Compared with Bayesian theory, the evidence theory can be free of prior and conditional probabilities and has a stronger theoretical foundation, i.e., it can deal with uncertainties due to randomness and ambiguity. Relying on the accumulation of evidence, the hypothesis set is continuously reduced, i.e., the evidence theory has the ability to model the restricted hypothesis set when the evidence increases. However, Dempster's combination rule is combinatorially sensitive, and sometimes a small change in the underlying probability can lead to a large change in the outcome. In addition, applying evidence theory requires that the evidence is independent, which is inconvenient in practice.

3 THE USE OF MULTI-SOURCE INFORMATION FUSION IN DAM SAFETY EVALUATION

According to the Implementing Rules for Safety Inspection of Hydropower Station, the identification framework of the safety evaluation level of the dam is made $V = \{v_1, v_2, v_3, v_4, \} =$

{safe, safer, less safe, unsafe}, where V is the identification framework of the decision state of the indicator, and V_i is the safety evaluation state of a layer of the indicator. Its function m satisfies $\sum_i m(v_i) = 1$, which is said to be the basic probability assignment of v_i, that is, the trust degree of the evaluation indicator in v_i

(1) Monitoring data information fusion

The various types of data obtained through the dam monitoring equipment itself has a certain degree of uncertainty, so there may be anomalous values that are significantly different from other monitoring data. The abnormal values may be caused by the failure of monitoring instruments or changes in the monitoring environment, and may also be a sign of a sudden change in the nature of monitoring indicators leading to safety accidents. In order to improve the credibility of the information and reduce the ambiguity of the information, monitoring data should be identified and extracted when pre-processing the abnormal values. The monitoring data is used to analyze the safety status of the dam and combine with mathematical operations to derive the probability value of the safety status of the dam.

(2) Expert evaluation information fusion

Expert evaluation is also a multi-source information fusion step, where each expert scores the safety evaluation index of the dam based on the preliminary engineering design, inspection and safety evaluation reports of previous years, and other information based on experience. This section applies the NGT (Nominal Group Technique) method of group decision making for expert evaluation. In the face of all information data, n experts make a judgment independent of the information reflected in the degree of safety of the indicators, and give a specific degree of safety value p_i. On this basis, the average value of the evaluation of the group of experts $\bar{p}$ can be obtained as the next input, that is:

$$\bar{p} = \frac{1}{n}\sum_{i=1}^{n} p_i \tag{7}$$

(3) Multi-source information fusion

The expert system is used as the information source to expand the number of underlying indicators, and the multi-source information fusion theory is applied to fuse the monitoring system with the expert system for decision making. The monitoring system and the expert system as the final evidence source may be highly conflicting, when the credibility of the fusion results using the traditional multi-source information theory is not high. Therefore, this paper introduces the VIKOR Method (Liu 2021) to assign the credibility of the final evidence source, gives the basic probability assignment of the two information sources to be corrected, and uses the multi-source information fusion theory to fuse with more credibility.

There are m alternatives $S_i(i = 1, 2, k, m)$, and n attributes $C_j(j = 1, 2, k, n)$ are used to evaluate each alternative. There are decision makers to collect objective data and subjective empirical judgments for each decision option to obtain the decision matrix $A = [a_{ij}]_{m \times n}$ matrix. Generally, the weight of attribute C_j is set as $\omega_j(j = 1, 2, k, n)$ satisfying $\omega_j \geq 0$ and $\sum_{j=1}^{n} \omega_j = 1$.

The original matrix $A = [a_{ij}]_{m \times n}$ is normalized using the polar variation method. If C_j is a benefit type attribute,

$$b_{ij} = \frac{a_{ij} - a_j^{\min}}{a_j^{\max} - a_j^{\min}}, i = 1, 2, \ldots, m, j = 1, 2, \ldots, n \tag{8}$$

If C_j is a cost attribute,

$$b_{ij} = \frac{a_j^{\max} - a_{ij}}{a_j^{\max} - a_j^{\min}}, i = 1, 2, \ldots, m, j = 1, 2, \ldots, n \tag{9}$$

The original matrix $A = \left[a_{ij}\right]_{m \times n}$ is normalized and noted as $B = \left[b_{ij}\right]_{m \times n}$.
The group utility value Ed_i, individual regret value EH_i and VIKOR value Q_i are calculated for each data as:

$$Ed_i = \sum_{j=1}^{n} \omega_j b_{ij}, \quad i = 1, 2, \ldots, m \tag{10}$$

$$EH_i = \max_{1 \leq j \leq n} \left\{\omega_j b_{ij}\right\}, i = 1, 2, \ldots, m \tag{11}$$

$$Q_i = \nu \frac{Ed_i - Ed^+}{Ed^- - Ed^+} + (1 - \nu) \frac{EH_i - EH^+}{EH^- - EH^+} \tag{12}$$

Where $Ed^+ = min\{Ed_i\}$, $Ed^- = max\{Ed_i\}$, $EH^+ = min\{EH_i\}$ and $EH^- = mAX\{EH_i\}$. Both Ed_i and EH_i are cost-based variables, i.e., the smaller the outcome of the variables, the better. Q_i is the combined evaluation value of the i (th) option, and again the smaller the value of Q_i, the better the option. $\nu \in [0,1]$ is the decision coefficient, which is used to indicate the decision maker's preference. The smaller the value of Q_i, the better the decision is, and the decision corresponding to the value of Q_i is the optimal decision and the key evidence. The credibility of each evidence can be calculated based on the Q_i value to correct the basic probability assignment.

$$\boldsymbol{Sup}_i = \frac{Q_i^{-1}}{min(Q_i)} \quad i = 1, 2, 3 \tag{13}$$

The basic probability assignment correction formula under each evidence is as follows:

$$m\left(v\right)' = \boldsymbol{Sup}_i m\left(v\right) \tag{14}$$

4 CONCLUSION

Dam safety evaluation is a complex task, although many reservoir dams are now set up with monitoring equipment, but at present only local monitoring can be realized without all-round full coverage, so the dam safety evaluation should be implemented by combining monitoring data analysis and expert information-based evaluation. The two different sources of information are integrated through the VIKOR method to filter the important information sources as the main source of evidence reference. Such a multi-source information fusion method makes the evaluation results more reliable and the theoretical system of dam safety evaluation more complete.

ACKNOWLEDGMENTS

This work was supported by National Natural Science Foundation (51909172) and Fundamental Research Funds of Nanjing Hydraulic Research Institute (Y721001).

REFERENCES

Hsiao-Hui Luo (2015). *A multi-source information fusion grid fault diagnosis method based on improved Bayesian network and Hibert-Huang transform* [D]. Southwest Jiaotong University.

Liu Dal, Zhang Q. (2021) Research on multi-attribute decision problem based on VIKOR method, *J. Computer Knowledge and Technology*, 17(23): 132–133.

Roy Yan. (2003) *A study of several algorithms for information fusion D.* Jiangsu: Nanjing University of Technology.

Song Z.L., Jia X., Guo B., Cheng Z.J. (2021) Bayesian fusion and simulation-based system remaining life prediction, *J. Systems Engineering and Electronics Technology*, 43(06): 1706–1713.

Tao C.C., Xu B., Wang Z.L. (2010). Application of D-S evidence theory in dam safety monitoring, *J. Red Water River*, 29(02): 34–37.

Xu, H.C. & Wu, S. (2007). An introduction to multi-sensor information fusion technology *J. Journal of Tung Wah University of Science and Technology* (Natural Science Edition), 30(2): 189–192.

Yang J., Wang J., Lv Q.P., Tian Z.H. (2022) Research on dynamic safety evaluation of landslides in highland reservoir area based on multi-source information fusion *J. Hydropower Technology* (in Chinese and English), 53(04):165–171.

Zhang G., Li L., Peng X.H. (2008). Techniques for evaluating the degree of disease risk based on dam safety identification and expert experience. *Chinese Journal of Safety Science*, (09):158–166.

Frontiers in Civil and Hydraulic Engineering – Mohamed A. Ismail and
Hazem Samih Mohamed (Eds)
© 2023 The Authors, ISBN 978-1-032-38247-0

Research and application of digital technology in the construction period for Foshan Fulong Xijiang Bridge

Yu Liang, Qinghan Bu*, Zhaohui Tang, Kaibing He, Xin Li & Daqing Zhang
China-Road Transportation Verification and Inspection Hi-Tech Co. Ltd., Beijing, China

ABSTRACT: Foshan Fulong Xijiang Bridge is one of the first batches of centennial quality demonstration projects announced by the General Office of the Ministry of Transport. The large scale of construction, great technical difficulties, complex construction environment, and difficult organization and management of the project have brought significant challenges to the construction management of the project. Relying on BIM, GIS, cloud computing, big data, Internet of Things, and other technologies, the project focuses on the application of BIM technology, project management system development and application, intelligent site construction, data mining, and analysis, and other digital technology research and application, to comprehensively improve the level of lean construction and fine management of the project.

1 INTRODUCTION

With the development of cloud computing, the Internet of Things, big data, artificial intelligence, and other new technologies, the state, and industry authorities have issued a series of policy documents requiring the development of digitalization and promoting the digital transformation of traditional infrastructure such as transportation. In order to explore the theoretical system, technical route, and development direction of transportation infrastructure digitization, many scholars and engineers have carried out research and application. Wen Y F et al. took the bridge which crossed the Baotou-Maoming Expressway as the project carrier and applied BIM to the design and construction of the project to improve the communication efficiency of the various shareholders (Wen et al. 2020). Zhang G Z developed a BIM construction management platform for Hutong Yangtze River Bridge, which built a bridge construction site perception system based on the Internet, solved the lightweight problem of the BIM model, and raised the overall level of management and control (Zhang 2018). Zhang Y et al. have carried out application and research on a BIM-based railway engineering project management system, which has been successfully applied to the Dali-Ruili Railway construction project, greatly enhancing the ability of model data sharing and effectively improving the level of engineering construction control (Zhang et al. 2019). Wang T J carried out the innovation and practice of railway engineering construction management based on BIM technology and proposed a new railway engineering construction management mode based on engineering data objects (Wang 2018). By analyzing the problems existing in the traditional engineering management process and the advantages and values of BIM-aided construction management, Guo H L et al. proposed the BIM application mode in the construction stage and developed a construction management platform architecture composed of a simulated and real-time BIM system (Guo et al. 2017). Ou H L et al. designed a smart site management platform based on BIM technology and developed application modules such as the BIM model, real-name management system, and video surveillance system to create an all-around and intelligent construction management mode for highway projects (Ou et al. 2020). Relying on the waterway regulation project of the Yangtze River Estuary of the

---

*Corresponding Author: buqinghan@foxmail.com

DOI 10.1201/9781003344209-53

Beijing-Hangzhou Grand Canal, Zhao J H et al. carried out the comprehensive application of the BIM technology and smart site, which solved typical problems such as mass concrete pouring, long-line construction process control, process inspection of complex concealed works, as well as construction and monitoring data management (Zhao et al. 2022).

The above research discusses the innovative application of digital technologies such as BIM, project management systems, and smart sites in the construction of transportation infrastructures. However, there are still some common problems that need to be solved. For example, the engineering volume is large, the structure is complex, and some design problems exist in the construction drawings, which affect the engineering quality; during the construction process, the project progress, quality, cost, archives, and other business data are relatively independent of each other, which is not conducive to enhancing the efficiency of project management; engineering archives are still mostly paper files, which is challenging to guarantee the availability, integrity, authenticity, and security of archives; the hidden dangers on the construction site are not dealt with in time, which may cause great safety risks. In order to resolve these problems, we rely on Foshan Fulong Xijiang Bridge to conduct the research and application of digital technology, focusing on BIM technology application, project management system research and application, smart site construction, data mining, analysis, etc. This article will expand on the technical content and application effects.

2 INTRODUCTION OF FOSHAN FULONG XIJIANG BRIDGE

Foshan Fulong Xijiang Bridge, with a total length of about 5.81 km, starts from Fuwan Town, Gaoming District, and then joins Yangxi Avenue, passes Gaoming Hecheng Street and Baini Town, Sanshui, crosses Xijiang River from west to east, and ends at Jihua Road West extension line. The construction of the project adopts the first-class highway standard. The design speed of the main line is 80 km/h. The main bridge and the approach bridge use eight lanes, and the rest of the road uses six lanes. The entire line is equipped with 4 large bridges and 2 interchanges. The main bridge is a 580 m cable-stayed bridge.

Foshan Fulong Xijiang Bridge is a major project of Guangdong province during the 13th Five-Year Plan period and a demonstration project for Foshan city to promote the construction of the Guangdong-Hong Kong-Macao Greater Bay Area. The completion of the project is conducive to the integration of the Guangzhou-Foshan western area into the Guangzhou-Foshan fast road network. It accelerates the traffic integration, industrial integration, and urban integration of the areas along the line.

Figure 1. Overview of main bridge of Foshan Fulong Xijiang Bridge.

3 RESEARCH AND APPLICATION OF DIGITAL TECHNOLOGY

The digital technology research and application of Foshan Fulong Xijiang Bridge are mainly divided into BIM technology application, project management system development and application, smart site construction, and data mining and analysis. The technical route and application will be introduced in these four parts.

3.1 *Application of BIM technology*

In this project, the BIM models of the main structures during the construction period, terrain BIM models, three-dimensional real scene models, and geological BIM models of the karst development area are established. Using Catia software and parametric technologies, the BIM models of the main structures during the construction period of the project are established, and the model accuracy is not less than LOD300. First, the design parameters in the construction drawing are sorted in Excel to form a parameter element table. Second, the BIM model skeleton and parametric engineering templates are established in Catia software. The data resource tables are built through the EKL language. The mapping relationships among the data resource tables, parameter element tables, and parameter engineering templates are established. Finally, the automation program in Catia software is run to generate the refined vector BIM models quickly.

Multi-source data such as terrain CAD drawings, measurement points, contour lines, and feature lines are input into Catia software. The software automatically extracts spatial coordinate data and forms 3D point cloud data. After fitting, vector terrain BIM models are generated. The main bridge is located in the karst area, which is prone to quality problems during pile foundation construction. Therefore, based on drilling data, three-dimensional geological models are established to reflect the spatial distribution of underground karst caves and provide reliable technical support for pile foundation construction. Uav is used for oblique photography, and high-precision 3D real scene models with an image resolution of 5 cm are constructed to visually display the structures on the engineering site, which is used to assist in land expropriation and demolition, scheme demonstration, and other work.

Figure 2.   BIM +GIS platform.

3.2 *Development and application of project management system*

The project management system includes schedule management, quality management, cost management, safety management, supervision management, archives management, and more than ten business modules covering all aspects of project management. According to the requirements of standardized management guidelines for highway construction in Guangdong province, the sub-item bill of quantities is decomposed to the minimum measurement and management unit. The estimated and budget item code of the sub-item bill of quantities is taken as the basic code of the project. The estimate and budget project code are associated and mapped with the bill of quantities, construction procedures, sub-item coding, drawings, BIM models, etc., to form the fundamental information base of the project. Based on the uniform project code of estimate and budget, the construction personnel apply the APP terminal for construction procedures inspection on the project site. The supervision personnel confirm it to timely collect the project's actual progress. The staff fill in the quality inspection form online and imprint the digital signature to form the quality inspection electronic file in the quality management module. The cost management module automatically obtains the actual project progress and quality inspection electronic files as the basis of measurement approval. After online process approval and digital signature, electronic files of measurement and payment are formed. Lastly, the electronic files formed by each business are automatically shoved to the archive management system to realize the organic correlation and closed-loop management of schedule, quality inspection, measurement, archives, and other business data, which effectively improves the efficiency of the project management.

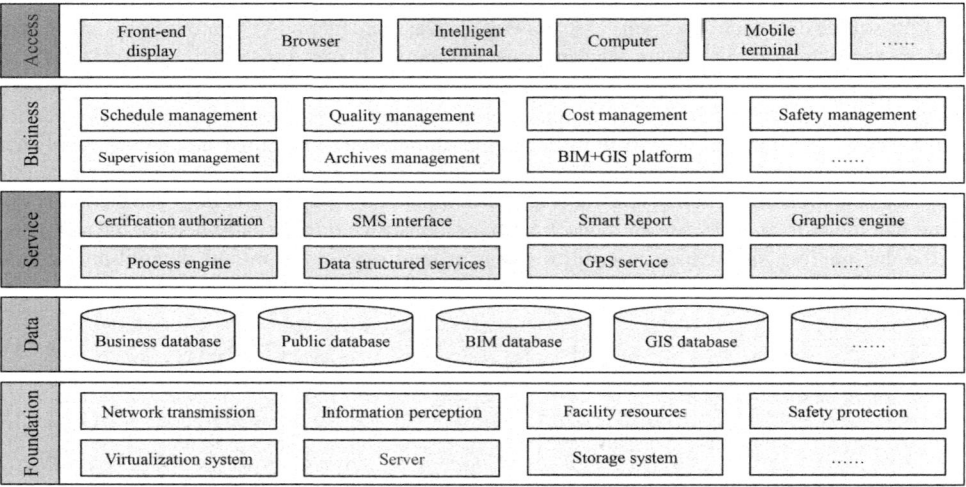

Figure 3.　Project management system architecture.

3.3　Smart site construction

Smart site construction mainly includes information management of special mechanical equipment, data acquisition and monitoring of laboratory and mixing stations, video monitoring, construction monitoring measurement, etc. The information technology is used to record the entry, exit, maintenance, and other information of special mechanical equipment, and GPS locators are installed to track the running track of special mechanical equipment in real-time. The system automatically collects the test data of universal machines, press machines, bending machines, and other equipment in the laboratory for statistics, analysis, and early warning. The system automatically collects the production data of the mixing station and monitors the production mix ratio of concrete. Cameras are installed in the main bridge, beam yard, laboratory, mixing station, and other important construction sites to monitor the scene in real-time. A cable-stayed bridge safety detection system integrating a series of hardware and software facilities such as sensors, acquisition equipment, software, data transmission, and processing has been developed to monitor more than 20 items such as vehicle load, wind load, structure temperature, ship collision, cable force and support reaction force of the main bridge to ensure the stability and safety of the main bridge structure.

3.4 *Data mining and analysis*

The BI technology is used to develop the project data management system, which collects statistical data and analyzes the project management data and construction production data. In terms of progress management, the system automatically updates the progress of milestone nodes, makes statistics on the completion progress of pile foundation, pier, prefabricated beam, and other engineering structures, and compares and analyzes the planned output value and the actual output value of the project. In terms of hidden dangers management, the number of hidden dangers of each participant in a closed loop is compared and analyzed. The key construction sites where hidden dangers occur are reflected on the two-dimensional map through thermal maps. Regarding land expropriation and demolition management, the system automatically calculates the land transfer rate, land expropriation rate, the completed amount of housing expropriation and demolition, and the completed amount of pipeline relocation. In mechanical equipment management, the statistical monthly variation trends in the numbers of equipment in and out as well as the proportion of the number of various types of equipment on the construction site are sorted. The data can be more intuitive and flat through statistics and analysis of various data types and better support the management and decision-making.

System Application Statistics (From July 18, 2022 to July 24, 2022)											
Contract Section	Management approval module		Quality hazard troubleshooting module		Process inspection module		Pre-class education module		Supervision side station module		Supervision inspection module	
	This week	Subtotal	This week	Subtotal	This week	Subtotal	This week	Subtotal	This week	Subtotal	This week	Subtotal
FLSG-01	18	21814	0	57	0	8	0	19	0	0	0	0
LXSG-01	2	9141	0	10	0	5	0	42	0	0	0	0
LXSG-02	40	23765	0	32	0	1	0	31	0	0	0	0
NHJL-03	0	0	0	0	0	134	0	0	1	2065	0	470
NHJL-02	0	0	0	0	1	243	0	0	0	931	0	121
Total	60	54720	0	99	1	391	0	92	1	2996	0	591

Figure 4. System application statistics generated by the system.

4 CONCLUSION

Through the application of BIM technology, the BIM models of the main structure, topography, and karst development area during the construction period were established, which enhanced the design quality. The project management system has been developed, covering all aspects of project management. The collaborative management mechanism of schedule, quality, measurement, archives, and other business data has been established, and the refined management of construction quality and safety has been enhanced. The smart construction site has been built to comprehensively manage special machinery and equipment, laboratory and mixing station data, video monitoring, construction monitoring, etc., and improve the visualization level of the construction site. Based on the project data management system, data mining and analysis are carried out to improve the management value of data. Through the research and application of digital technology, the level of lean construction and fine management of the project has been improved comprehensively.

ACKNOWLEDGMENT

The authors gratefully acknowledge the funding provided by China-Road Transportation Verification & Inspection Hi-Tech Co., Ltd. (No. 0293E620107100).

REFERENCES

Zhang G.Z. (2018). Research and application of BIM construction management platform for hutong changjiang river bridge. *Bridge Construction*. 48(5), 6–10.

Ou H.L., Wu E.M. (2020). Application of intelligent construction site based on BIM technology in highway engineering projects. *China Water Transport*. 20(4), 48–49.

Guo H.L, Pan Z.Y. (2017). The mode and process of BIM-aided construction management. *J Tsinghua Univ (Sci & Technol)*. 57(10), 1076–1082.

Zhao J.H, Qin B., Wu D.Y. (2022). Comprehensive application of BIM technology and intelligent construction sites in inland waterway regulation works. *Port & Waterway Engineering*. 3, 139–145.

Wang T.J (2018). Innovation and practice of railway engineering construction management based on BIM technology. *Journal of the China Railway Society*. 41(1), 1–9.

Wen Y.F., Chen J.P., Yang Y.M. (2020). Research on lean project management application of bridge spanning baotou-maoming expressway based on BIM. *Highway*. 10, 245–252.

Zhang Y., Huang C.Z, Zhu C. (2019). Research on the application of railway engineering project management system based on BIM. *Journal of Railway Engineering Society*. 9, 98–103.

Frontiers in Civil and Hydraulic Engineering – Mohamed A. Ismail and Hazem Samih Mohamed (Eds)
© 2023 The Authors, ISBN 978-1-032-38247-0

Numerical simulation of oil spill accident during submarine cable laying near the Lvhua Island of Zhoushan sea area

Wenlong Liu*
School of Naval Architecture and Maritime, Zhejiang Ocean University, Zhoushan, China

Yifan Gu
Zhejiang Qiming Ocean Electric Power Engineering Co., Ltd., Zhoushan, China

Wei Chen
School of Marine Engineering Equipment, Zhejiang Ocean University, Zhoushan, China

ABSTRACT: Specific sea areas, such as Lvhua Island, are characterized by economic and social benefits associated with a threat affected by oil particles that spread under the influence of tidal currents and wind. In order to properly address the issue of oil spill accidents and study the hazard, a 2D hydrodynamic model in Lvhua Island and its surrounding area is built with MIKE 21. A particle tracking model, MIKE 21 Particle Tracking (PT), is used to establish the prediction model for oil spill accidents. The sweeping area, the drift trajectory of oil film, and its impact on the environment are analyzed when an oil spill occurs under SE wind at different time. In the case of high tide, after 6 hours of the oil spill, sensitive areas of West Lvhua Island begin to be affected. In the event of the ebb tide, the sensitive areas near Huaniao Island are relatively less affected, and the oil particles do not significantly affect the sensitive areas 12 hours after the oil spill.

1 INTRODUCTION

Both the Lvhua Island and the Huaniao Island are located in the northern part of the Zhoushan Islands and the northwest corner of the Ma'an Archipelago, belonging to Shengsi County in Zhe-jiang Province (Figure 1). There are quite a number of environmentally sensitive points in Lvhua Island and its surrounding waters. The main points are the marine ranch demonstration area of the Ma'an Archipelago in Shensi County (see the red markings in Figure 2), the culture zone of Lvhua in Shengsi County (see those numbered 1–7 in Figure 2), and the marine aquaculture area of Huaniao Island (see those numbered A–F in Figure 2). The marine oil spill has a serious impact on the marine ecological environment (Zhang & Wang 2006). It is necessary to analyze the diffusion trajectory of the leaked crude oil after the accident so as to take appropriate measures to reduce the impact of the accident on the environment in time.

In order to better study the trajectory of crude oil in water, the oil spill model is established to analyze the oil spill accident. As early as the last century, there was a prototype of numerical simulation. In the 1990s, Cuesta et al. (1990) compiled a two-dimensional oil spill model with FORTRAN language and simulated and verified the oil spill situation of the Amoco Cadiz accident, which verified the reliability of the model for the long-term prediction of oil spill drift in oceans. Sugioka et al. (1999) established a 3D oil spill model and analyzed the oil particle's trajectory in horizontal and vertical space during the oil spill accident in Tokyo Bay. With technological development, oil spill models are developed and used as software in production and construction,

---

*Corresponding Authors: z20086100048@zjou.edu.cn; 603020387@qq.com and wchen@zjou.edu.com

such as the GNOME model (Xu et al. 2013), OILMAP model (Spaulding et al. 2017), MS4 model (Xu 2006), Navy model (Wu et al. 2015), Delft3D-PART module (Kuang et al. 2016), MIKE 21 SA model (Vethamony et al. 2007), etc.

This paper is based on the Lvhua Island submarine cable laying project. An oil spill accident is assumed to occur on a construction vessel in the project area. The drift trajectory of oil particles, sweeping area, and their effects on engineering are analyzed.

Figure 1. Location of the project point.

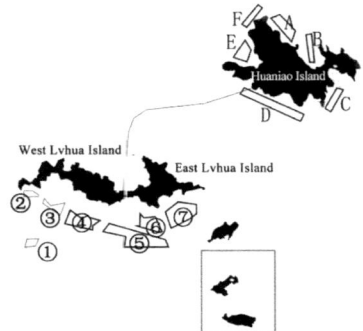

Figure 2. Distribution of environmental sensitive points in engineering sea area.

2 NUMERICAL MODEL

2.1 Model introduction

The MIKE model is a numerical model developed by Danish Hydraulic Research (DHI). It contains a variety of modules, which can simulate different situations with different modules. It has a very strong pre and post-processing capability (Dhi 2011). The particle tracking module (PT) is one of the modules for water quality and environmental assessment included in MIKE21. It is mainly used to simulate the diffusion trajectory and the state of suspended substances, and can be used to predict oil spill problems and analytically evaluate accident oil spill plans.

2.2 Model area and meshes

The calculation region of the model covers Cangqian in the west of the area, which is to the north of Luchao Port, to the south of Xiangshan Mountain, and 124 ° E on the east boundary. It includes the sea areas of Hangzhou Bay and Zhoushan Islands. The transverse width of the calculation domain is about 378 km, the longitudinal length is 216 km, and the calculated area is about 81 648 km^2. In order to improve the computational efficiency of the model, the meshes far from the model engineering area are relatively sparse, and the resolution of the mesh is between 200 and 4000 m. There are many islands in the sea area near the project area, the topography and tidal current structure are complex, and the tidal power has a strong impact on the environment. In order to obtain enough calculation accuracy, the meshes of the project area are densified and the minimum mesh size is 10 m, which can describe the underwater topography and shoreline in the front of the project. The mesh generation is shown in Figure 3.

2.3 Setting of relevant parameters

The time step of the model is 0.000 1-30 s. The model is calculated by cold start with a Colon Number of 0.8, and the setting parameter of the bed resistance applied to the Manning coefficient is within the range of 0.012–0.014. The river boundary is controlled by flow, and the water boundary condition is controlled by the tide time series file which is calculated by the East China Sea model.

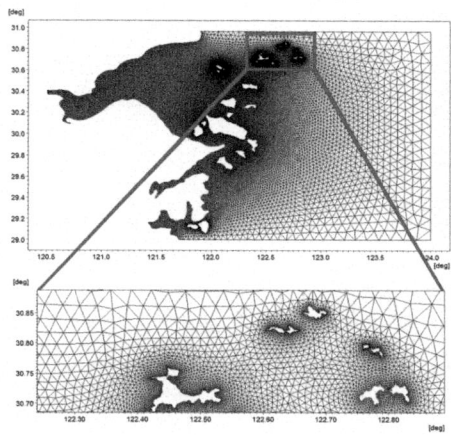

Figure 3. Computational domain and meshes.

There are shoals in the coastal area with tide fluctuations, and the phenomenon of flooding and the alternation of dew will appear. So, the dry-wet grid method is used to deal with the problem. The critical water depth of the dry point is 0.005 m and that of the wet point is 0.05 m. The open boundary of the East China Sea model only takes the open boundary of the open sea into account. The tidal boundary data extracted from the global tide model that comes from the MIKE 21 software package is used for the open boundary of the open sea. The constituents included in the model are M_2, S_2, K_2, N_2, S_1, K_1, O_1, P_1, Q_1, and M_4.

2.4 Model verification

In the Lvhua Island submarine cable project, a hydrological survey was carried out on March 2, 2018. The fieldwork was completed on March 14. A temporary tide gauge station (L1 temporary station, location shown in Figure 1) was set up for tide level observation, which lasted for 12 days. The temporary leveling elevation was not lower than the requirement of the map root level accuracy. The height of the system is National Vertical Datum 1985, aiming at the sea location and tidal current characteristics of the project. The reliability of the model is verified by the measured data of 1 temporary tidal observatory and 6 tide gauges, and the results are verified well. The specific locations of tidal current stations and tidal level stations used in the verification data are shown in Figure 1, and the specific measuring time is as follows: from March 2, 2018, to March 3, 2018, for spring tidal current measurement; from March 10, 2018, to March 11, 2018, for neap tidal current measurement.

This project selected L1 temporary tidal observatory data as tide level verification data. The observation time started at 14:00 on March 2, 2008, and ended at 13:00 on March 11, 2008. The continuous tide level observation lasted for 12 days. The verification results are shown in Figures 4 and 5. The above tidal level data cover the duration of hydrological tests in spring, meddle, and neap tidal floods.

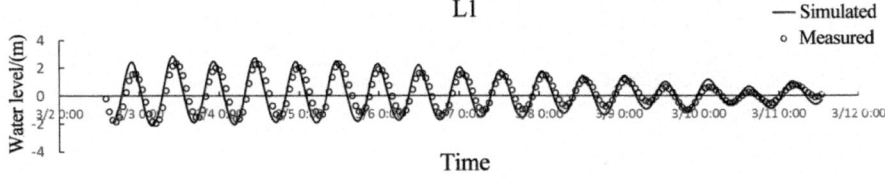

Figure 4. Tide level verification of Lvhua station.

409

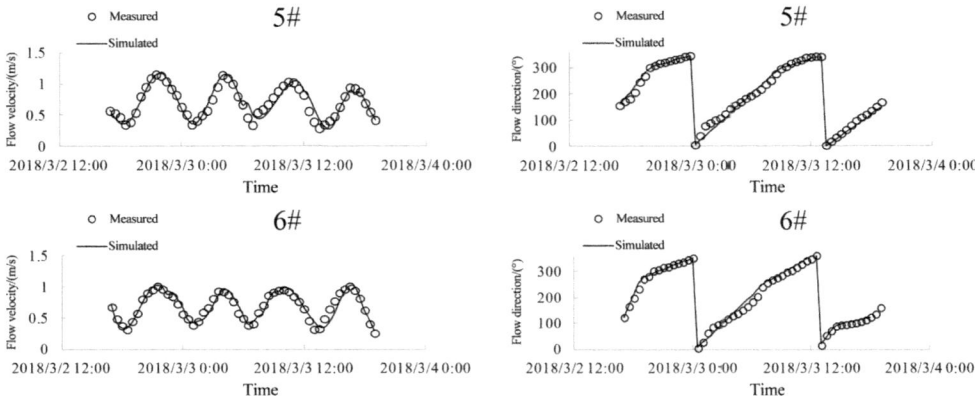

Figure 5. Verification of depth-averaged velocity and direction of the current during spring tide.

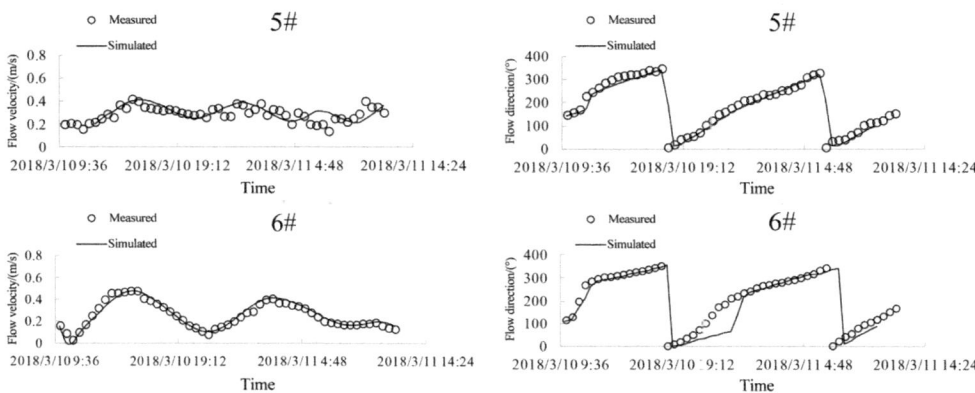

Figure 6. Verification of depth-averaged velocity and direction of the current during neap tide.

The validation results of tidal level and tidal current shown in Figures 5 and 6 show that the established hydrodynamic model can reflect the real situation of the sea area near the project area, and the validation results of flow field calculation are reasonable and reliable.

2.5 Current situation simulation and analysis of power flow

The simulation results show that the main motive force to maintain the flow field of the large regional model is the tidal waves coming from the East China Sea and the southeast and east directions. The velocity of the regional model increases gradually from the east to the west. Figure 7 shows the local area of the flood and ebb rapids of the small-scale model. It can be seen from the map that in the engineering area between Lvhua and Huaniao Islands, the east and west sides of the project are basically open sea areas, constrained by the narrow terrain, the fluctuation tide basically moves back and forth, and the flow direction of the fluctuation tide in the nearshore area is roughly parallel to the shoreline, and the velocity of the main channel is obviously greater than that at the beachside; the path of the fluctuation tide is basically the same.

2.6 Oil spill model

In the "Oil Particle" model, the motion and deformation of oil films are reflected by oil particles. In the process of moving, the Deterministic Method—Lagrange Method—is used to simulate the

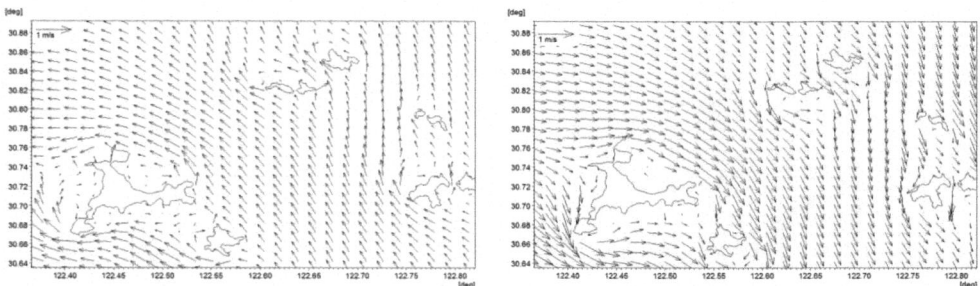

Figure 7.   Flow field in local area during flood and Ebb.

translation process, and the diffusion process is simulated by an Uncertain Method—the random walk method (Zhuang et al. 2007). Moreover, oil particles can predict the process of oil film edge expansion and the elliptical expansion of oil films under the influence of flow fields and wind fields by their own dispersion.

According to the total capacity (100 m³) of the four tanks on the construction ship, under the most disadvantageous circumstances, namely, when the four tanks are completely broken, 80 t of diesel oil spilled can be calculated by the specific gravity. The leakage occurs after a collision at the waterway in the period of flood and ebb tide. The annual average wind speed of the dominant wind direction SE is predicted to be 7.1 m/s.

3   ANALYSIS OF NUMERICAL RESULTS

According to the calculation results of the oil spill, the sweeping area of oil spill particles in the period of the flood tide and ebb tide under SE wind is calculated respectively. The detailed analysis is as follows:

3.1   *Oil spill at flood maximum*

When oil spills in the flood tide, oil particles are affected by the flood current and spread northward. Because the main body of the tidal current is consistent with the direction of the wind as shown in Figure 7 (a), oil particles accelerate northwest drift under the superposition of wind and tidal current as shown in Figure 8. After 1 hour of the oil spill, the westward diffusion width of oil particles at the far end of the oil spill trajectory is about 1 km, and oil particles continue to diffuse eastward and westward. After 6 hours, oil particles begin to migrate northwest. In the West Lvhua Island aquaculture area, the 2 parts of the aquaculture area begin to be slightly affected by oil. After 12 hours, oil particles have spread to most of the sea areas near Lvhua Island and Huaniao Island. Except for aquaculture area C, other sensitive areas are partly or wholly affected by oil. After 48 hours, all environmentally- sensitive areas are affected by oil.

3.2   *Oil spill at ebb maximum*

After oil spill at the ebb maximum, oil particles are affected by the tidal current and diffuse to the southeast direction. As the wind direction is opposite to the main direction of the tidal current as shown in Figure 7 (b), oil particles drift southeast under the combined influence of wind and tidal current as shown in Figure 9. After 1 hour, oil particles drift away from the source point by about 2 km. Because of the tidal current, oil particles drift southeast and diffuse eastward and westward, which makes the oil film coverage area form a fan-shaped distribution with oil spill points as the axis. After 6 hours, oil particles drift to the south of East Lvhua Island and Huaniao Island under the influence of the tidal current. Since then, due to the weakening of the southeast ebb tide, oil

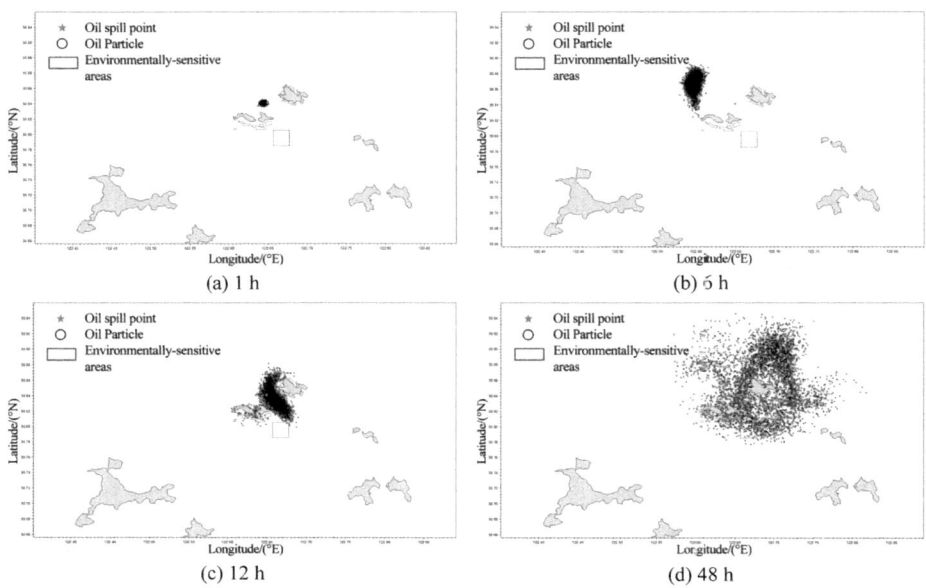

Figure 8.　Spread range of oil particles during flood tide.

particles begin to drift northwest, and the marine pasture demonstration area begin to be affected by oil particles. After 12 hours, oil particles appear near most of the shoreline of West Lvhua Island and East Lvhua Island. Most sensitive areas in breeding regions 1–7 and the marine pasture demonstration area are covered by oil particles. After 48 hours, oil particles diffuse to almost all visible areas in the study area under the action of flood and ebb tidal currents, and all sensitive points are affected by oil.

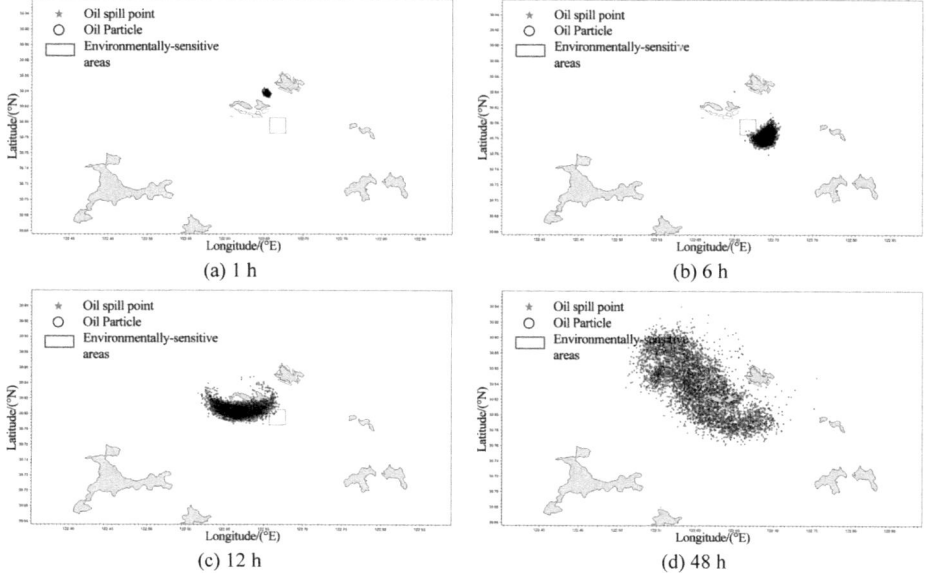

Figure 9.　Spread range of oil particles during Ebb tide.

4 CONCLUSION

After oil spill at the flood stage, oil particles accelerate northward drift under the superposition of wind and tidal currents due to the consistency of the current direction with the wind direction. In the process of northwest drift, oil particles continue to diffuse eastward and westward. After 6 hours, oil particles begin to migrate northwest. Oil spills begin to affect the 2 parts of the farms on West Lvhua Island. After 12 hours, the sensitive areas are affected by oil particles partly or wholly, except for C culture areas. After 48 hours, oil particles basically cover the whole study area. After an oil spill at the ebb tide stage, oil particles drift southeast under the action of ebb tide and wind. After 1 hour, oil particles drift away from the source point by about 2 km. After 6 hours, the marine pasture demonstration area begins to be affected by oil. After 48 hours, oil particles diffuse to almost all visible areas in the study area under the repeated action of fluctuating tidal currents.

After the oil spill, oil particle migration is affected by wind and tidal currents. In the case of high tide, after 6 hours of the oil spill, the sensitive areas of West Lvhua Island begin to be affected. In the event of the ebb tide, the sensitive areas near Huaniao Island are relatively less affected, and the oil particles do not significantly affect the sensitive areas 12 hours after the oil spill.

REFERENCES

Cuesta, I., Grau, F. X. & Giralt, F. (1990). Numerical simulation of oil spills in a generalized domain. *Oil and Chemical Pollution*, 7(2), 143–159.

DHI (2011). MIKE21 & MIKE3 Flow Model FM-Hydrodynamic and Transport Module Scientific Documentation. Demark: *DHI Water & Environment*.

Kuang, C., Xie, H., Su, P. & Chen, K. (2016). Numerical simulation and analysis of transport a fate of penglai 19-3 oil spill. *Journal of Tongji University* (Natural Science), 10, 1585–1594.

Spaulding, M., Li, Z., Mendelsohn, D., Crowley, D., French-McCay, D. & Bird, A. (2017). Application of an integrated blowout model system, OILMAP DEEP, to the Deepwater Horizon (DWH) spill. *Marine Pollution Bulletin*, 120(1–2), 37–50.

Sugioka, S.I., Kojima, T., Nakata, K. & Horiguchi, F. (1999). A numerical simulation of an oil spill in Tokyo Bay. *Spill Science & Technology Bulletin*, 5(1), 51–61.

Vethamony, P., Sudheesh, K., Babu, M.T., Jayakumar, S., Manimurali, R., Saran, A.K. & Srivastava, M. (2007). Trajectory of an oil spill off Goa, eastern Arabian Sea: Field observations and simulations. *Environmental Pollution*, 148(2), 438–444.

Wu, Y., Dong, S., Gao, J. & Wang, Z. (2015). Numerical simulation study of oil spilling diffusion of haimiao port. *Transactions of Oceanology and Limnology*, 02, 177–184.

Xu, Q., Li, X., Wei, Y., Tang, Z., Cheng, Y. & Pichel, W. G. (2013). Satellite observations and modeling of oil spill trajectories in the Bohai Sea. *Marine Pollution Bulletin*, 71(1-2), 107–116.

Xu, Y. (2006). *A study on the weathering and predicting model of spilled oils at sea*. Qingdao: China Ocean University.

Zhang, L. & Wang, H. (2006). Optimization design of countermeasures to clean up the marine oil pollution in different sea surface conditions. *Ocean Technology*, 25(3), 1–6.

Zhuang, X., Chen, J. & Sun, Q. (2007). Numerical simulation and visualization technology of marine spilled oil. *Navigation of China*, 01, 97–100.

Frontiers in Civil and Hydraulic Engineering – Mohamed A. Ismail and
Hazem Samih Mohamed (Eds)
© 2023 The Authors, ISBN 978-1-032-38247-0

Study on safety evaluation of Earth-Rock Dam based on Fuzzy AHP

Liang Wang*, Peijiang Cong, Zhigang Yin & Bo Zhang
Changchun Institute of Technology, Changchun, China

ABSTRACT: The number of small and medium-sized earth-rock dams that have been built in China has been increasing year by year, and it is imperative to solve the problem of their safe operation. Now, through the investigation of the reinforcement of an earth-rock dam and the safety appraisal report, the hierarchical analysis method and fuzzy mathematical theory are used to establish a reasonable fuzzy comprehensive evaluation model for the safety of the earth-rock dam in the province, which combines the quantitative factors and qualitative factors affecting the earth-rock dam, and the safety evaluation indicator system constructed is not only in line with the safety appraisal content but also in line with the current situation of the reservoir in Jilin Province. The proposed safety evaluation model is applied to five small and medium-sized earth-rock dam cases to compare feasibility and applicability. The results show that the evaluation results obtained by the model are similar to the actual safety situation of the dam, and the comprehensive score of the safety evaluation is finally quantified by the evaluation model, which can distinguish the safety status of the disease-risk dam. Therefore, this research method can provide a reasonable evaluation basis for the safety identification of earth-rock dams and the reinforcement of risk removal.

1 INTRODUCTION

China has nearly 100,000 reservoir dams, and the earth-rock dam type accounts for more than 95% of the total number of reservoir dams in China, and small reservoirs account for about 95%. Most of China's reservoir projects were built in the 1950s–1970s, and 79,500 reservoirs of various types were built in 30 years, accounting for about 81.7% of the total number of reservoirs (Su 2017; Zhang 2013). Affected by the economic and technological restrictions and a series of realities at that time, there were many "trilateral" projects and "half-pulled" projects, low dam construction standards and poor quality, coupled with insufficient long-term investment in management funds, inadequate maintenance, serious project aging and disrepair, and "congenital deficiency, acquired disorders", which led to a large number of disease-risk reservoirs in China (Zhang 2017). At present, most of the existing earth-rock dams reinforced in China were built in the middle and late period of the last century, and the "trilateral" project has remarkable characteristics, especially for small and medium-sized reservoirs, such as poor engineering quality and low flood control standards, and the reservoirs built after 1985 are mainly small and medium-sized reservoirs (Hou 2017).

Earth-rock dams have been studied a lot at home and abroad. Salazar et al. (Salazar 2017) proposed a predictive model based on enhanced regression trees to detect anomalous behaviors in the early days of dams. Spross et al. (2019) proposed a general calculation algorithm based on the reliability of civil infrastructure's alarm threshold. Li M C et al. (2020) used the TOPSIS method to measure the safety performance of dams through numerical safety scores. Mta (2011) combined the artificial neural network model with the integrated evaluation system and proposed a corresponding method for evaluating the working characteristics of a dam. Shu B (2006) used the fuzzy synthesis method to make a hierarchical fuzzy judgment on an earth-rock dam, and the

*Corresponding Author: 1427292631@qq.com

DOI 10.1201/9781003344209-55

comprehensive assessment yielded the final assessment result. Li Z K et al. (2003) established a multi-level fuzzy safety evaluation system for earth-rock dams to assess the operating status of earth-rock dams. Wei W B (1996) introduced the fuzzy theory into the evaluation of the measured nature of the dam and established the dam safety evaluation system. It can be seen that at this stage, studies have been carried out regarding the safety evaluation of earth-rock dams for many years and a large number of results have been achieved. However, the current research on dangerous reservoirs is still insufficient, and analysis and evaluation systems still rely on the experience of experts, which requires high data accuracy or is highly subjective. Therefore, it is necessary to carry out research on the safety evaluation of earth-rock dams. Through scientific evaluation methods for the safe operation of earth-rock dams, we can ensure that water resources are reasonably utilized under safe operation conditions, maximize the social and economic benefits of the reservoir, and ensure downstream public safety.

The article collects the safety appraisal reports of typical earth-rock dams for statistical analysis, selects reasonable evaluation indicators, constructs a safety evaluation system, calculates the comprehensive value of dam safety, and quantitatively analyzes the safety status of the dam so that the safety appraisal of the dam can more intuitively present the safety form of the reservoir and refer to the reinforcement of risk removal. Finally, the case verification shows that the proposed method can better assess the safe operation status of the earth-rock dam.

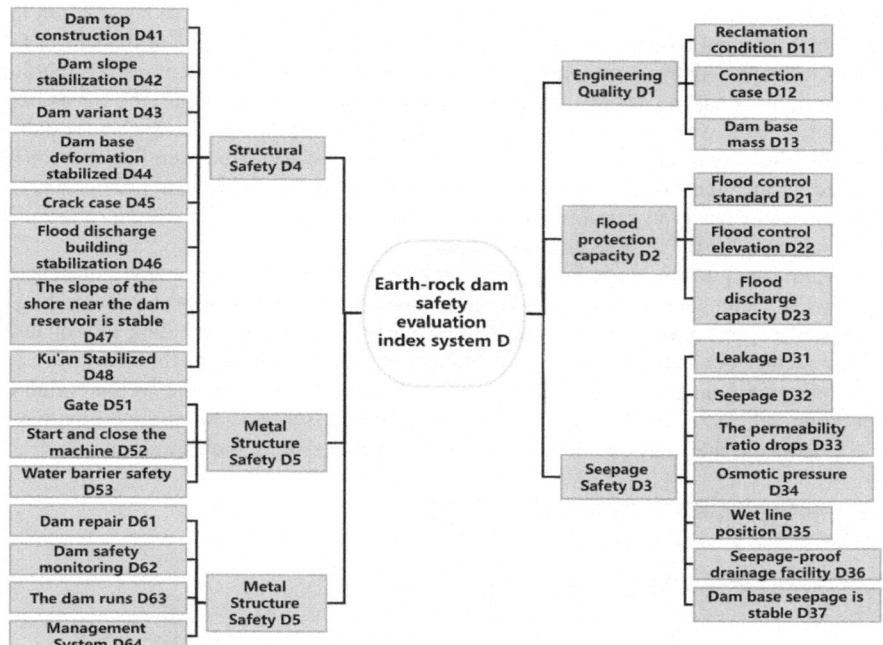

Figure 1. Evaluation indicator system for the safe operation of Earth-rock dams.

2 FUZZY COMPREHENSIVE EVALUATION BASED ON ANALYTIC HIERARCHY PROCESS

2.1 *Establishment of an indicator system*

The evaluation indicator is not only related to whether the dam is normal but also related to the accuracy of the evaluation results, so this paper uses the guidelines for the safety evaluation of reservoir dams and the relevant norms involved and combines the research of relevant scholars to select the indicators, and on this basis, according to the data analysis of 10 reservoirs, the evaluation indicator system for the evaluation of the safe operation status of the earth-rock dam is established.

The indicator system constructed for the safe operation of earth-rock dams is shown in Table 1, in which the criterion layer is consistent with the report of the safety identification requirements, and the specificity and meaning are shown in Table 2.

Table 1.   The indicator system is explained at the standard level.

Symbol	Meaning	Paraphrase
D1	Engineering quality	Dam soil mass, dam base mass
D2	Flood protection capabilities	The current flood resistance of the dam
D3	Seepage safety	Dam seepage properties
D4	Structural safety	About the durability and safety of the dam structure
D5	Metal construction	Status of the existing metal structure of the dam
D6	Run management	Dam operation safety is required

2.2   *Fuzzy analytic hierarchy process (AHP)*

It is mainly to establish a multi-level hierarchical structural model to form a safety evaluation system for earth-rock dams, and then, on the basis of analyzing the determined evaluation factors, the fuzzy mathematical theory is used to determine the relative importance of inter-level factors, construct a judgment matrix between factors, perform compound operations on the relationship between multiple levels, and determine whether each judgment matrix can meet the consistency test so as to determine the weight of each indicator. Finally, an appropriate evaluation set is established, a fuzzy comprehensive evaluation is carried out at different levels, and the overall evaluation results are obtained through the calculation of comprehensive importance, which can provide a reference for the sequence of reservoir reinforcement for the evaluation results of multiple reservoirs, and provide a reference for the reinforcement of single reservoirs.

At the same time, at this stage, in order to fully consider the rich experience of experts and engineering and technical personnel, it is necessary to score qualitative indicators according to the relevant specifications and the actual situation of the project and convert them into quantitative indicators, and then obtain the corresponding evaluation matrix according to the structural equation of the quantitative indicators, and multiply the evaluation matrix with the weight value of the corresponding indicators, and the result is the comprehensive score of the safety evaluation of the earth-rock dam, and the analysis flow chart of the earth-rock dam is shown in Figure 2.

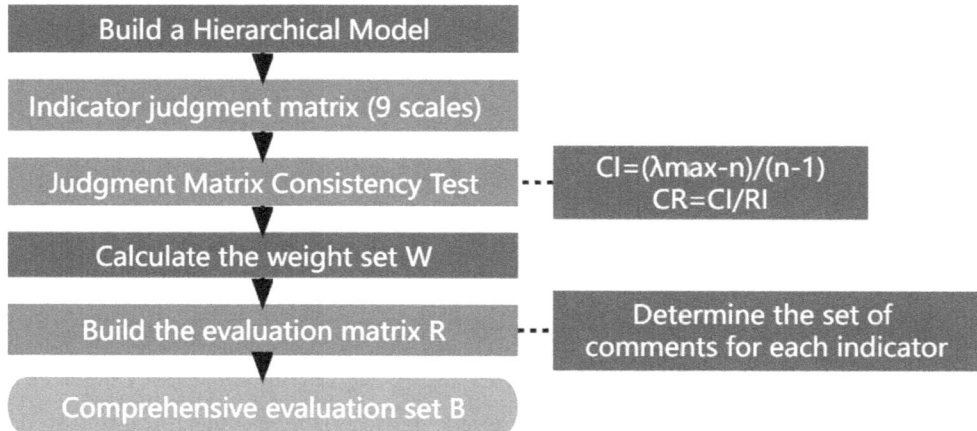

Figure 2.   Flowchart of Fuzzy AHP of Earth-rock Dam.

In the comprehensive evaluation in this paper, five levels of comments are set for each indicator, that is, V = [V1, V2, V3, V4, V5] = [normal, basic normal, mild abnormality, severe abnormality, malignant abnormality], and the value is V = [100,90,80,70,60]. The indicator value system is evaluated by a number of experienced personnel, and each expert scores each qualitative indicator of the indicator layer individually, and at the same time, the standard and convenient experts have a detailed perception of the indicators. Comprehensive safety evaluation guidelines and survey analysis, a descriptive introduction of the selected evaluation indicators, for example, are seen in Table 2 (Liu 2016).

Table 2. Scoring criteria for indicator D48.

Stable and safe state of the reservoir shore	Judging Criteria	Score
Severe abnormalities	The soil on both sides of the river bank collapses severely	60
Moderate abnormalities	The soil on both sides of the river bank collapses in a small area	70
Mild abnormalities	A small area of soil erosion on both sides of the riverbank	80
Basically normal	Partial peeling of soil on the shore of the reservoir	90
Normal	The treasury is stable	100

3 CASE STUDY

This evaluation model is now a feasibility analysis of five reservoirs for which the safety appraisal has been completed, and the reservoir data is shown in Table 3.

Table 3. Basic information about the reservoir.

Reservoir name	Reservoir details
Reservoir 1	Medium-sized reservoirs mainly for flood control and irrigation, and comprehensive utilization
Reservoir 2	Medium-sized reservoirs mainly for flood control and irrigation, and comprehensive utilization
Reservoir 3	Flood control and irrigation are the mainstay and comprehensive utilization of small (2) type reservoirs
Reservoir 4	Small reservoirs, raising flood control standards, and building food bases
Reservoir 5	Flood control and water supply are the mainstays, and the medium-sized reservoirs are comprehensively utilized

3.1 AHP weight calculation

According to the specific conditions of the safety evaluation of each sub-factor indicator, the construction matrix of the level 1 indicators in the safe operation evaluation indicator system of the earth-rock dam in Table 1 is shown in Table 4.

Through the calculation of the judgment matrix, the maximum characteristic root is 6.3296.

A consistency test is conducted for the judgment matrix.

$$CI = \frac{\lambda_{max} - n}{n - 1} = \frac{6.3296 - 6}{6 - 1} = 0.0659$$

Table 4. Level 1 indicator judgment matrix.

Level 1 indicator	D_1	D_2	D_3	D_4	D_5	D_6
D_1	1	1/4	1/2	1/3	1/3	2
D_2	4	1	3	2	2	5
D_3	2	1/3	1	1/2	1/2	3
D_4	3	1/2	2	1	1	4
D_5	3	1/2	2	1	1	4
D_6	1/2	1/5	1/3	1/4	1/4	1

The stochastic consistency ratio is calculated.

$$CR = \frac{CI}{RI} = \frac{0.0659}{1.26} = 0.0523 < 0.10$$

According to the calculation results, it is believed that the AHP results have satisfactory consistency, that is, the allocation of weight coefficients is very reasonable. The weights of the calculated criteria layers are shown in Table 5.

Table 5. Weights of level indicators.

Indicator layer	D_1	D_2	D_3	D_4	D_5	D_6
Weight	0.1454	0.2148	0.1574	0.1787	0.1787	0.1249

In the same way, the basic index discriminant matrix is constructed, and the comprehensive subjective weights of the indicators are calculated as shown in Table 6.

Table 6. Weights of level 2 indicators.

Indicator	D11	D12	D13	D21	D22	D23	D31
Weight	0.0312	0.0052	0.0128	0.0417	0.0139	0.0417	0.0056
Indicator	D32	D33	D34	D35	D36	D37	D41
Weight	0.0202	0.0275	0.0275	0.0275	0.0089	0.0777	0.0122
Indicator	D42	D43	D44	D45	D46	D47	D48
Weight	0.0763	0.1186	0.1486	0.0441	0.0502	0.0502	0.0468
Indicator	D51	D52	D53	D61	D62	D63	D64
Weight	0.0379	0.0126	0.0379	0.0181	0.0181	0.0042	0.0027

3.2 Fuzzy synthetic judgment

The weights corresponding to the metrics included in the project quality are as follows.

$$W_1 = [0.3928 \quad 0.1744 \quad 0.2883]$$

Experts score according to the actual situation of reservoir 1 dam and the survey report, and can get the fuzzy evaluation matrix corresponding to the indicators.

$$R_1 = \begin{bmatrix} 1 & 0 & 0 & 0 & 0 \\ 0 & 1 & 0 & 0 & 0 \\ 1 & 0 & 0 & 0 & 0 \end{bmatrix}$$

418

The engineering quality evaluation vector is as follows.

$$B_1 = W_1R_1 = [0.6813 \quad 0.1744 \quad 0 \quad 0 \quad 0]$$

The same can be done for B2, B3, B4, B5 and B6.

According to the overall evaluation matrix of the reservoir dam and the sub-report, the weight can be calculated, and the comprehensive evaluation of dam B can be calculated.

$$B = [0.5989 \quad 0.0227 \quad 0.1405 \quad 0.0525 \quad 0.1767]$$

In the process of conversion to quantitative assessment, $F = VB^T$ is used to express the comprehensive safety evaluation value of the reservoir dam.

$$F = VB^T = [100 \quad 90 \quad 80 \quad 70 \quad 60] \begin{bmatrix} 0.5989 \\ 0.0227 \\ 0.1405 \\ 0.0525 \\ 0.1767 \end{bmatrix} = 87.45$$

The overall score of the reservoir is calculated to be 87.45, and the overall state of the reservoir is between unsafety and basic safety.

In the same way, a comprehensive evaluation is made, and the safety identification score report scores are listed in Table 7.

Table 7. Level 1 indicator scoring.

Indicator layer	D_1	D_2	D_3	D_4	D_5	D_6
Score	83.8225	85.344	97.475	91.0596	60.0000	66.8416

After calculation, in the evaluation of the criterion layer, the dam operation management safety and metal structure safety review evaluation value obtained are small, the corresponding safety degree is low, the seepage safety evaluation value is larger, the corresponding safety degree is higher, the dam should be replaced and maintained for metal structured gates and opening and closing machines, the monitoring equipment and management of operation management should be strengthened, and the dam seepage monitoring and structure testing should be continued. The situation reflected in the evaluation available from the table above is consistent with the actual situation and has practical significance.

On this basis, the safety evaluation of five reservoirs is carried out, and the scores are listed in Table 8.

Table 8. Reservoir scoring.

Reservoir	1	2	3	4	5
Score	87.608	92.590	86.416	80.256	95.434

The situation reflected in the evaluation is consistent with the actual situation, which is of practical significance and can provide a reference for the safety appraisal of small and medium-sized earth and rock dams in the future.

At this stage, the evaluation guidelines highlight the advantages that there are norms to follow, but in the evaluation of the severity of disease risk, the level of disease risk cannot be quantified, and the intensity of the need to strengthen the dam cannot be determined or prioritized. The method used in this study can complete the overall comprehensive evaluation of the project, but the degree of disease risk of the dam is ambiguously classified. This point is supplemented and distinguished using scores, with the evaluation of the severity of the disease risk of similar dams. At the same time, a series of evaluation indicators are all based on the investigation and analysis of the reservoir and combined with the guidelines and the opinions of the engineers, and the consideration of the evaluation indicators is more comprehensive and practical, which can be used to solve the priority reinforcement problem of similar dams.

4 CONCLUSIONS

The safety of the dam body affects the life, and economic and social development downstream of the reservoir. In this paper, an evaluation index system is constructed through AHP, and finally, the fuzzy comprehensive evaluation method is used to evaluate the safe operation status, and the conclusions are as follows.

(1) In this paper, in view of the safe operation of the earth-rock dam combined with the safety evaluation of the earth-rock dam at the current stage, through the investigation and analysis of the earth-rock dam in Jilin Province, the evaluation system suitable for the local area is constructed, including six sub-reports of the complete evaluation report, and 28 basic indicators are established.

(2) The safety evaluation of the earth-rock dam using this research method is consistent with the safety appraisal results obtained by the reservoir, indicating that the research method in this paper is effective and feasible, and finally transformed into a quantitative assessment, which can compare and sort the dam body of the same level safety state, and facilitate the distinction between the priority of dam de-risking and reinforcement. It can also provide a reference for reinforcing different parts of the same dam. It is a simple method suitable for dam safety assessment.

To sum up, the fuzzy comprehensive method based on AHP is proved to be able to distinguish the safety degree of the dam body very well, thus solving the problem of the pros and cons of dams with similar defects; the pros and cons of dams with similar consequences; dams with similar defects and consequences, the pros and cons of dams that are not easy to report to the relevant departments; and the problem of ranking the pros and cons of dam defects. Through the scientific ranking of the pros and cons of the safe operation status of dams, those dams that should be firstly reinforced and strengthened in time can be reinforced in time, and the dam safety work can be carried out quickly and accurately.

REFERENCES

Hou J.B., Zheng X.Y., Wang M., et al. Remote sensing extraction and pattern analysis of reservoir information in Northeast China [J]. *Wetland Science*, 2017, 15(04): 582–587.

Li Z.K., Zhou J., Zheng J.X. Research on multi-level fuzzy pattern recognition method for measured properties of earth-rock dam[J]. *Journal of Hydraulic Engineering*, 2003, 34(9): 0083–0087.

Liu X. *Analysis on the diagnosis and analysis of safety state affiliation of in-service earth-rock dams* [D]. Nanchang University, 2016.

Li M.C., Si W., Ren Q.B., et al. An integrated method for evaluating and predicting long-term operation safety of concrete dams considering lag effect [J]. *Engineering With Computers*. 2020.

Mat J., Interpretation of concrete-dam behavior with artificial neural network and multiple linear regression models, *Engineering Structures*, vol. 33. no. 3. pp. 903–910, 2011.

Ministry of Water Resources of the People's Republic of China, *National Bureau of Statistics of the People's Republic of China. Communiqué of the First National Water Census*[M]. Beijing: China Water Resources and Hydropower Publishing House, 2013.

Ministry of Water Resources of the People's Republic of China. 2017 *National Water Conservancy Development Statistical Communiqué*[M]. Beijing: China Water Resources and Hydropower Press, 2018.

Su N. Reservoir dam safety risks attract attention[N]. *China Energy News*, November 13, 2017(011).

Salazar F., Toledo MÁ, González J.M., O-ñate E. (2017) Early detection of anomalies in dam performance: A methodology based on boosted regression trees. *Struct. Control Health Monit.*, 24(11): e2012.

Shu B. Fuzzy comprehensive evaluation of the safety of earth-rock dams in small and medium-sized reservoirs[D]. *Hefei University of Technology*, 2006.

SL258-2017, *Guidelines for Safety Evaluation of Reservoir Dams*[s]. China: Ministry of Water Resources of the People's Republic of China, 2017.

Spross J., Gasch T. (2019) Reliability-Based alarm thresholds for structures analyzed with the finite element method. *Struct. Saf.*, 76:174–183.

Wei W.B., Li Z.Z. Research on Fuzzy Comprehensive Judgment decision-making method of dam safety[J]. *Hydropower Station Design*, 1996(01):1–8+26.

Zhang J.Y., Yang Z.H., Jiang J.P. Analysis on the disease risk and rupture law of reservoir dams in China[J]. *Science in China: Technical Sciences*, 2017, 47(12):1313–1320.

Zhang J.Y., Yang Z.H., Jiang J.P. *Research and warning of reservoir dam disease risk and dam collapse*[M]. Beijing: Science Press, 2013.

Frontiers in Civil and Hydraulic Engineering – Mohamed A. Ismail and Hazem Samih Mohamed (Eds)
© 2023 The Authors, ISBN 978-1-032-38247-0

Numerical simulation of the flow field at stepped spillway dam faces

Rui Yin* & Dong-yao Hong
Department of Water Engineering, Kunming University of Science and Technology Oxbridge College, Kunming, Yunnan, China

Jia-shi Cai & Ding Cui
School of Electric Power Engineering, Kunming University of Science and Technology, Kunming, Yunnan, China

ABSTRACT: A two-dimensional steady numerical simulation of the flow at the smooth spillway face and the stepped spillway face of a gravity dam has been carried out in this study using the VOF multiphase model, with a comparative analysis of energy dissipation characteristics, pressure, and flow lines at a spillway dam face. The study reveals that for stepped spillway dam faces, there is an overall greater water depth and higher energy dissipation efficiency, with a relative energy dissipation rate of approximately 24.45%. Besides, for such dam faces, alternating higher pressure and negative pressure zones alternatively occur at the junction of weir-top curve ends and the junction of reverse-arc beginning ends, with a large number of vortex structures in the flow lines, so the above two types of junctions need to be designed meticulously.

1 INTRODUCTION

Jetting dissipation of energy is a common way to dissipate energy in gravity dams, in which, however, the fog is heavy and the tailwater fluctuates greatly (State Key Laboratory of Hydraulics and Mountain River Development and Protection of Sichuan University 2016). Therefore, a new type of efficient energy dissipaters is the key to optimizing the energy dissipation effect. The stepped spillway face allows the water flow to be aerated, decelerated, and energy-dissipated step by step along the dam, with an energy dissipation rate 40%–80% higher than that of a conventional smooth spillway dam face (Lin 2009).

Wu X S (1996) conducted an experimental study on the flow pattern, energy dissipation effect, and time-averaged pressure distribution on the steps in a cross-sectional model of the stepped spillway dam on the right bank of Wangwei East-West Hydropower Station. The stepped spillway dam could significantly improve the energy dissipation effect within a certain flow range, and no cavitation damage would occur on the dam face under reasonable operating conditions. Tang S C et al. (2008) studied and discussed the starting aeration position, water surface line, pressure, and energy dissipation rate on the stepped spillway dam face, combined with a hydropower station engineering hydraulic model test, concluding that the main factors affecting the energy dissipation rate of a stepped dam face were stepped dam's height, stepped protrusion's height, stepped dam's slope and flow rate of unit width, and meanwhile the calculation formula of energy dissipation rate was deduced based on the test data. Dong L Y et al (2018) carried out hydraulic model tests on the integrated energy dissipators of Y-shaped flaring gate pier + stepped spillway dam + energy dissipation pond under a total of four scenarios for three transitional steps and the original

---

*Corresponding Author: 386072923@qq.com

                    DOI 10.1201/9781003344209-56

Ahai working conditions, and compared and analyzed the differences in maximum hourly average pressure, maximum negative pressure on the stepped surface and energy dissipation rate. Wang Q et al. (2020) employed three-dimensional numerical simulation to explore the effects of different step sizes of the transition steps on the aeration and negative pressure characteristics of the integrated joint energy dissipator based on the integrated joint energy dissipator of the Ahai hydropower station.

Most of the above studies are aimed at three-dimensional hydraulic models. In this paper, numerical simulations are carried out for the smooth spillway dam face and stepped spillway dam face with respect to the actual project dimensions. Due to the large size of actual projects and the small effect of lateral water flow on the hydrodynamic characteristics, only two-dimensional profiles of the spillway dam faces are considered in the study.

2 NUMERICAL MODELS AND ALGORITHMS

2.1 Control equations

This paper adopts the VOF method (Hirt 1981) and the RNG k-ε turbulence model (Launder 1974), and the control equations are:

$$\frac{\partial \rho}{\partial t} + \frac{\partial (\rho u_i)}{\partial x_i} = 0 \tag{1}$$

$$\frac{\partial (\rho u_i)}{\partial t} + \frac{\partial (\rho u_i u_j)}{\partial x_j} = -\frac{\partial p}{\partial x_i} + \frac{\partial}{\partial x_j}\left[(\mu + \mu_t)\left(\frac{\partial u_k}{\partial x_j} + \frac{\partial u_j}{\partial x_k}\right)\right] + \rho g_i \tag{2}$$

$$\frac{\partial (\rho k)}{\partial t} + \frac{\partial (\rho k u_i)}{\partial x_i} = \frac{\partial}{\partial x_i}\left[\left(\mu + \frac{\mu_t}{\sigma_k}\right)\frac{\partial k}{\partial x_i}\right] + G_k - \rho\varepsilon \tag{3}$$

$$\frac{\partial (\rho\varepsilon)}{\partial t} + \frac{\partial (\rho\varepsilon u_i)}{\partial x_i} = \frac{\partial}{\partial x_i}\left[\left(\mu + \frac{\mu_t}{\sigma_\varepsilon}\right)\frac{\partial \varepsilon}{\partial x_i}\right] + c_1 G_k \frac{\varepsilon}{k} - c_2 \rho \frac{\varepsilon^2}{k} \tag{4}$$

Among which, $\mu_t = \rho c_\mu \frac{k^2}{\varepsilon}$, $c_\mu = 0.085$, $c_2 = 1.68$, $\sigma_k = \sigma_\varepsilon = 0.7179$, $c_2 = 1.42 - \frac{\eta(1-\frac{\eta}{\eta_0})}{1+\beta\eta^3}$, $\eta = \frac{Sk}{\varepsilon}$, $S = \sqrt{2S_{ij}S_{ij}}$, $\eta_0 = 4.38$, $\beta = 0.015$.

2.2 Computational methods and engineering parameters

The pressure-velocity coupling has been carried out using the PISO algorithm in Fluent 14.0 software, and the steady incompressible RANS equation has been solved directly with a convergence criterion of k and ε below 1×10^{-5}.

The downstream weir height of a gravity dam is 53.715 m, the front-edge length is 13.796 m, the head at the top of the weir is 2.185 m, and the maximum flow rate of unit width is $q = 22.049$ m²/s. The spillway dam section adopts the WES profile, the downstream straight-section side slope is 0.75, the step height is 0.9 m, and the width is 0.68 m, with a total of 38 steps; the length along the dam face is 41.552 m. The flip bucket is continuous-type, with a radius of 11 m, a flipping angle of 22°, and a height of 11.7 m.

2.3 Computational domain and grid

The computational domain starts from the top of the weir, passes through the downstream spillway dam face, and ends 68 m downstream of the flip bucket. The height of the spillway dam section is 3.994 m above the guide wall and the height of the downstream area is 21.233 m above the river bed. The paper uses the structural grid for mesh division, indicating that the closer to the dam face, the denser the meshes, as shown in Figure 1 for the computational domain and meshes of a stepped spillway dam face, which are similar to those for a smooth spillway dam face.

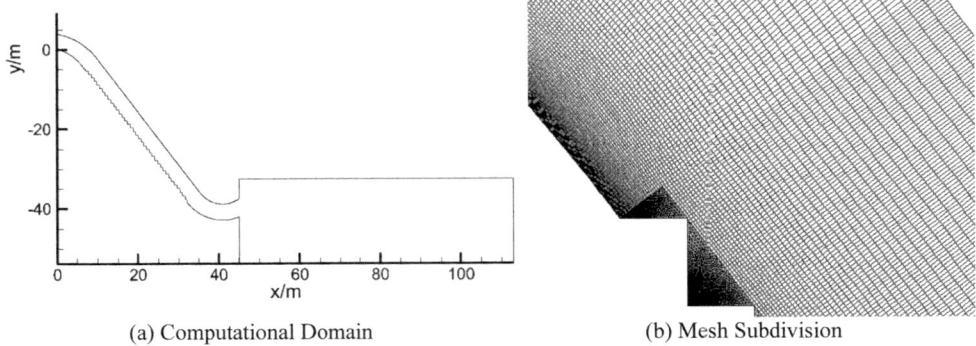

(a) Computational Domain   (b) Mesh Subdivision

Figure 1. Calculation domain and meshes of a stepped spillway dam face.

3 ANALYSIS OF CALCULATION RESULTS

The calculation results are worked out all for the same working conditions and when the flow has been stabilized, via comparing the smooth spillway dam face with the stepped spillway dam face.

3.1 *Energy dissipation characteristics of the spillway dam face*

Figure 2 shows the overall situation of the water surface line at the spillway dam face.

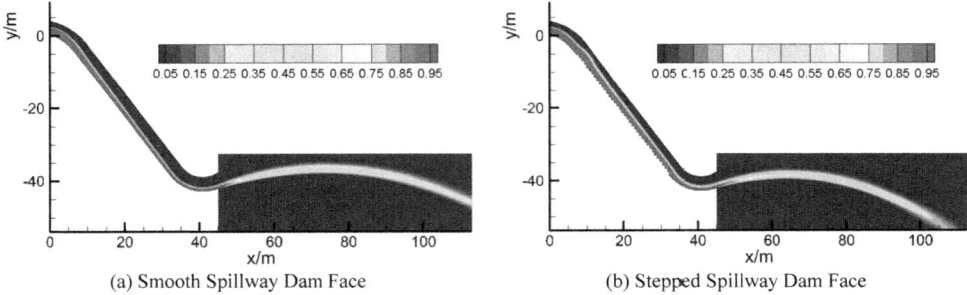

(a) Smooth Spillway Dam Face   (b) Stepped Spillway Dam Face

Figure 2. Overall situation of the water surface line at the spillway dam face.

As can be seen from Figure 2(a), for smooth spillway dam faces, the water depth gradually decreases from the top of the weir to the flip bucket, indicating a gradual increase in flow velocity.

From Figure 2(b), for the stepped spillway dam surface, there is an obvious rising water surface line at the beginning of the step surface, indicating that the flow velocity here gradually decreases; the water depth does not change significantly throughout the step surface, indicating that the flow velocity is more uniform; there is a water surface line at the end of the stepped surface, indicating that the flow velocity here gradually increases.

The overall greater depth of water in a stepped spillway dam face results in lower flow velocities and smaller trajectory distances, indicating that the energy dissipation efficiency of the stepped spillway dam face is higher. In order to characterize the energy dissipation efficiency of a stepped spillway dam face, the relative energy dissipation rate of a stepped spillway dam face compared to a smooth spillway dam face is introduced as ζ:

$$\zeta = \frac{E_1 - E_2}{E_1} \tag{5}$$

where E_1 and E_2 are the cross-sectional specific energy at the lowest point of the reverse arc for smooth and stepped spillway dam faces, respectively, while the cross-sectional specific energy at the lowest point of the reverse arc is:

$$E = h_c + \frac{q^2}{2g\varphi^2 h_c^2} \tag{6}$$

where h_c is the water depth at the lowest point of the reverse arc, φ is the flow coefficient, taken as 0.95, and g is the acceleration of gravity. The measured water depths at the lowest point of the reverse arc for the smooth spillway dam face and the stepped spillway dam face are $h_{c1} = 0.771$ m and $h_{c2} = 0.891$ m, respectively, and $\zeta = 24.45\%$ from Equations (5) and (6), which is not a large relative energy dissipation rate.

3.2 Pressure at the spillway dam face

Figure 3 shows the distribution of pressure at the face of the spillway dam.

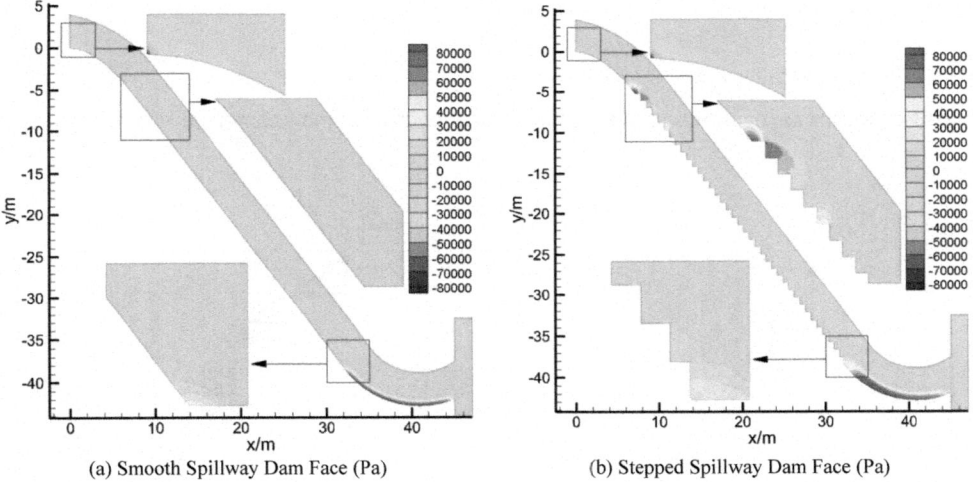

(a) Smooth Spillway Dam Face (Pa)   (b) Stepped Spillway Dam Face (Pa)

Figure 3.   Distribution of pressure at the spillway dam face.

From Figure 3(a), for the smooth spillway dam surface, the pressure is distributed more evenly from the top of the weir to the flip bucket. There is a negative pressure zone at the top of the weir, with the negative pressure at the top of the weir being the highest, and the negative pressure value gradually decreases along the flow direction and vertical direction, demonstrating that cavitation may occur at the top of the weir, especially at the top of the weir. The pressure does not change significantly at the beginning of the linear section, manifesting that the flow is smoothly connected; but the pressure does not change significantly throughout the linear section, presenting that the flow velocity does not change significantly. There is a high-pressure zone at the end of the linear section, which is distributed more evenly from the beginning of the reverse-arc section to the flip bucket, suggesting that the pressure in the reverse-arc section is higher and more even.

From Figure 3(b), it can be seen that for the stepped spillway dam face, the pressure is not uniformly distributed from the top of the weir to the flip bucket, with a negative pressure zone at the top of the weir similar to that of the smooth spillway dam face, reflecting that the stepped face does not have an obvious effect on the upstream pressure. The pressure is clearly more variable at the beginning of the stepped face, with a high-pressure zone at the first step, and a negative pressure zone at the second to the fourth steps, which gradually decreases in degree. At the fifth step, the pressure rises to normal values again. In the sixth step, a secondary high-pressure zone arises again, and at each subsequent step there is only one local high-pressure zone at the convex corner with the pressures in the other areas being normal; at the last step, a secondary negative pressure zone occurs again. The above phenomenon indicates that the pressure is relatively high at the interface between the curved section at the top of the weir and the step surface. Cavitation is prone to appear in the second to the fourth steps, whose degree gradually decreases; and the pressure gradually increases at the fifth to the sixth step, which, at each subsequent step, is relatively high at the convex corner, with the pressure values at the other areas being normal; cavitation may appear at the convex corners of the last step. At the end of the step face, there is a high-pressure zone that is similar to but more extensive than that of the smooth spillway dam face. The extent of the high-pressure zone drops along the flow direction, indicating that the pressure in the reverse-arc section is higher but decreases along the flow direction.

Although the stepped spillway dam face has higher energy dissipation efficiency, its pressure distribution is more complex, so there is both a high-pressure area with higher pressure and a negative pressure area that may cause cavitation. Especially at the junction of weir-top curve ends and the junction of reverse-arc beginning ends, the high-pressure area and the negative pressure area alternatively develop, and therefore, the design should be highlighted for the junction between the curve section of the top of the weir and the step face.

3.3 Flow lines at the spillway dam face

Figure 4 exhibits the distribution of flow lines at the junction of the weir-top curve ends.

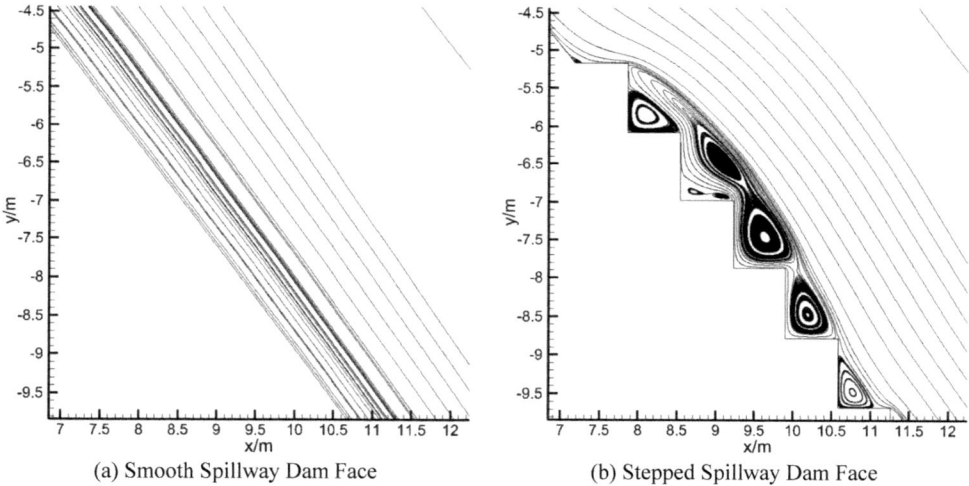

(a) Smooth Spillway Dam Face   (b) Stepped Spillway Dam Face

Figure 4.   Distribution of flow lines at the junction of weir-top curve ends.

As can be seen from Figure 4(a), for a smooth spillway dam face, the flow lines are nearly parallel to each other in a straight line, implying that the flow is approximately uniform.

From Figure 4(b), it can be seen that for the stepped spillway dam, the vortex always occurs at the concave corner of the steps, with the vortex of the first step being not obvious and the vortex of the second step increasing without occupying the whole step. In addition, the vortex of the third step is divided into two parts—one part is larger and far from the concave corner, and the other part consists of two smaller vortices and approaches to the concave corner, with the flow lines between the two parts being relatively smooth. From the fourth step, the vortices appear in all subsequent steps, which may be the main reason for the improvement of the energy dissipation efficiency. The water flow at the junction of the weir-top curve ends is smooth, but the change of the flow lines at the beginning of the step surface is more complicated, which may be the reason for the complicated pressure distribution.

Figure 5 shows the distribution of flow lines at the beginning junction of the reverse-arc sections.

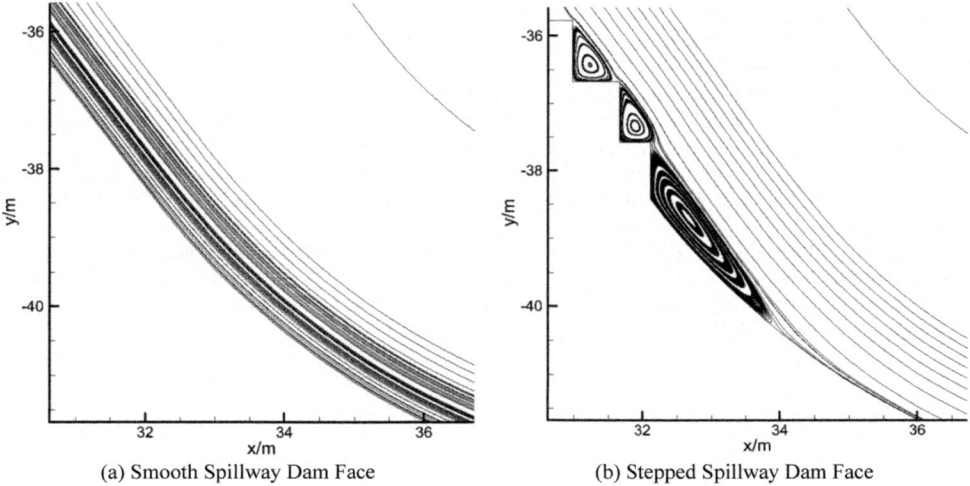

(a) Smooth Spillway Dam Face   (b) Stepped Spillway Dam Face

Figure 5.   Distribution of flow lines at the junction of reverse-arc beginning ends.

From Figure 5(a), the flow lines are nearly parallel to each other in arcs for the smooth spillway dam face, indicating that the flow is approximately non-uniform and asymptotic.

From Figure 5(b), it can be seen that for the stepped spillway dam face, there is a large banded vortex at the beginning junction of the reverse-arc sections, showing that the flow lines change more dramatically here, which may be the cause of the complex pressure distribution.

4   CONCLUSION

The following conclusions can be reached through the comparative analysis of numerical simulations of the water flow at the two spillway dam faces:

(1) The stepped spillway dam face has an overall greater water depth and higher energy dissipation efficiency, with a relative energy dissipation rate of approximately 24.45%.
(2) For the smooth spillway dam face, except for the negative pressure zone at the top of the weir and the high-pressure zone at the reverse-arc section, there is no significant change in the pressure at the rest of the locations, with the flow lines nearly parallel to each other. While for the stepped spillway dam face, there are alternating high-pressure zones and negative pressure zones; especially at the junction of weir-top curve ends and the junction of reverse-arc beginning ends, there are a large number of vortex structures in the flow lines near the steps, and the changes at the above two junctions are significant.

ACKNOWLEDGEMENT

This research is supported by the Yunnan Provincial Department of Education Science Research Fund Project.

427

REFERENCES

Dong L.Y., Yang G.R., Li S.J. Influence of transition stepped types on the hydraulic characteristics of integrated energy dissipators[J]. *Journal of Hydroelectric Engineering*, 2018, 37(03): 50–58.

Hirt, Cyril W., and Billy D. Nichols. Volume of fluid (VOF) method for the dynamics of free boundaries. *Journal of Computational Physics*, 39.1 (1981): 201–225.

Launder B.E. Spalding D.B. The numerical computation of turbulent flows[J]. *Computer Methods in Applied Mechanics and Engineering*, 1974, 3: 269.

Lin J.Y., Wang G.L. *Hydraulic Buildings* (5th ed.) [M]. Beijing: China Water & Power Press, 2009.

State Key Laboratory of Hydraulics and Mountain River Development and Protection of Sichuan University. *Hydraulics 1st Volume* (5th ed.) [M]. Beijing: Higher Education Press, 2016.

Tang S.C., Jin F., Shi J.H. Experimental study and energy-dissipation rate calculation of stepped spillway dams[J]. *People's Yangtze River*, 2008(12): 43–45+114. DOI:10.16232/j.cnki.1001–4179. December 2, 2008.

Wang Q., Yang G.R., Yang Z.L., et al. Study on the effect of transition-step size on the aeration and negative-pressure characteristics of the integrated joint energy dissipation dam[J]. *Water Power*, 2020, 46(10): 56–62.

Wu X.S. Experimental study of spillway dam section on the right bank of the dongxiangguan hydropower station[J]. *Design of Hydroelectric Station*, 1996(03): 82–86.

Frontiers in Civil and Hydraulic Engineering – Mohamed A. Ismail and Hazem Samih Mohamed (Eds)

Application of 3D printing technology in prefabricated buildings

Dongmei Yan
Guangxi Industry Research Institute Intelligent Agriculture Research Institute Co., Ltd., Nanning, Guangxi, China

Mingyuan Dou*, Jing Yang, Shuai Zou & Qing Feng
College of Mechanical Engineering, Guangxi University, Nanning, Guangxi, China

ABSTRACT: The emergence of prefabricated buildings is of great significance to the transformation and upgrading of traditional buildings in China. 3D printing technology and BIM Technology have played an important role in promoting the development of prefabricated buildings. This paper introduces the related concepts and development history of an assembly building and 3D printing. On this basis, this paper introduces the relevant applications of 3D printing in building engineering, expounds on the situation and difficulties in integrating the 3D printing technology into prefabricated buildings at present, and puts forward some solutions, in order to provide some reference for the development of prefabricated buildings in China.

1   INTRODUCTION

Prefabricated building refers to that some or all components of the building are transported to the construction site by corresponding transportation methods after production in the component prefabrication factory, and the components are assembled by reliable installation methods or installation machinery to become a building with use functions (Zhang 2017). Prefabricated buildings are characterized by high efficiency, energy saving, and environmental protection, and have been widely used in Europe, the United States, Japan, and other countries. Most components of prefabricated buildings are prefabricated in the factory and then transported to the construction site. These components are assembled through reliable connections to build a new building. Construction industrialization is the only way for China's construction industry to get rid of labor shortage, resource waste, environmental pollution, and frequent safety accidents. It is of great significance to the transformation and upgrading of China's traditional construction industry and the realization of green and sustainable development.

3D printing technology is also known as additive manufacturing technology. This technology is mainly based on a digital model. Through the control system, the materials are stacked layer by layer from the bottom to the top by spraying and extrusion, and finally, a complete structural entity is formed (Zhang 2021). Under the background of building materials manufacturing automation and intelligence, 3D printing technology has been rapidly developed and applied. In recent years, the 3D printing technology has received more and more attention, among which it has been more mature in the metal processing field (Liu 2020; Wang 2016), medical field (Gan 2021; Li 2017), and aerospace field (Lv 2021; Tan 2016; Wu 2013).

In order to combine the 3D printing technology combined with prefabricated buildings more closely, the application of the BIM technology is also required. BIM technology (building information modeling) simulates the real information of buildings through the establishment of 3D models and digital information simulation to build a complete, systematic, and global information

---

*Corresponding Author: doumingyuan@st.gxu.edu.cn

DOI 10.1201/9781003344209-57

building (Huang 2021). It is a geometric information structure or parameter components such as point, line, face, and volume in traditional architectural drawing technology, such as a wall, beam, and slab. In addition to geometric information, these parameter components can also carry information such as material, cost, use, and upgrade. This is also a variety of possibilities for BIM technology in architecture. The biggest feature of BIM technology is that the building model carries a lot of information. BIM technology integrates the engineering information, material requirements, and construction of buildings into one model and provides this information to 3D printing and assembly buildings, and provides relevant coordination information for building design, building material manufacturing, and building construction, so as to ensure the consistency of all phases of building engineering and make the building engineering develop towards intelligent, automation and low-carbon directions. With the introduction of the BIM technology, the information generated in the design and construction stage can be effectively integrated and recorded, and this information can be continued to the building operation stage, even to the demolition and recovery of the building after the final destruction, and such information can also be used as the basis for decision-making.

The development of the construction industry cannot rely on the extensive and quantitative growth mode. Instead, it should seize the opportunity of the development of the construction industry to improve the comprehensive technical level, explore the leapfrog application of multi-domain technologies, combine the development of 3D printing technology, BIM technology, and prefabricated buildings, and give full play to the leading role of science and technology, promote the green and low-carbon industrial construction mode, upgrade and rebuild the construction industry chain, and implement the sustainable development strategy.

2 DEVELOPMENT AND CURRENT SITUATION OF PREFABRICATED BUILDINGS IN CHINA

2.1 *Development of prefabricated buildings in China*

In order to solve a series of major technical problems such as quality, performance, safety, efficiency, energy saving, environmental protection, and low carbon in the process of traditional housing construction, it is inseparable from the development of prefabricated buildings. The earliest prefabricated buildings in China began after the completion of the first Five-Year Plan of new China. At that time, China established a preliminary industrial foundation and began large-scale infrastructure construction, which promoted the large-scale application and development of prefabricated buildings.

By the 1970s, China had initially established a prefabricated building technology system, such as a large plate housing system, a large template housing system, and a frame light plate housing system (Liu 2019). After the reform and opening up in 1978, with the influx of foreign capitals, prefabricated buildings developed rapidly, and a corresponding and perfect standard system of prefabricated buildings was initially established. By the 1990s and the beginning of the 21st century, the state, local governments, enterprises, and joint scientific research institutions have continuously issued policies and standards related to prefabricated buildings and introduced the third-party supervision mechanism. The manufacturing, construction, and acceptance management system of prefabricated buildings has become increasingly perfect.

In 2016, the *Opinions of the CPC Central Committee and the State Council on Further Strengthening the Management of Urban Planning and Construction* and the *Guiding Opinions on Vigorously Developing Prefabricated Buildings* (hereinafter referred to as the Guiding Opinions) were issued. The Guiding Opinions clearly pointed out that the construction goal of prefabricated buildings is to strive to make the ratio of prefabricated buildings to new buildings reach 30% in about 10 years. Since then, the state and the Ministry of Housing and Urban-Rural Development have issued a series of policy documents to promote the healthy development of prefabricated buildings, making the development of prefabricated buildings in China enter a new period.

With the development of the social economy, the requirements for durability, earthquake resistance, and low-carbon environmental protection of buildings are constantly improving. Public buildings and various industrial buildings will also be built in an industrialized manner. On the other hand, China's infrastructure construction and transformation will also start the new-type industrial upgrading and progress toward industrialization. At present, the subway tunnel, urban water supply, drainage, highway, railway, and new energy construction projects in Chinese cities have basically adopted prefabricated assembly structures and formed large-scale industrialization. From 2016 to 2020, the newly started prefabricated building area in China increased year by year, showing a good development trend, as shown in Table 1. Therefore, China's construction industrialization market is very broad and the development potential is very huge.

Table 1.   Newly started prefabricated building area in China from 2016 to 2020.

Particular year	2016	2017	2018	2019	2020
Built-up area (km^2)	114000	160000	289000	418000	630000

2.3 *Significance of developing prefabricated buildings*

(1) The development of prefabricated buildings is the inevitable requirement of the country's sustainable double carbon economy

The construction method of cast-in-situ or masonry structures in the traditional construction industry requires a large number of templates and scaffolds. In the process of wet operation on site, the consumption of wood, steel, cement, and water is huge and wasteful, and the energy consumption is large. The industrialized building components are produced in the factory. The production water and formwork can be recycled, which can reduce the large number of wet operations on the construction site and reduce the consumption of resources and energy. The prefabricated building construction technology can effectively reduce the strong noise at the construction site, save power consumption during the construction process, and greatly reduce the generation of construction wastes in traditional construction. The energy-saving rate of power consumption of prefabricated buildings is 36.4%, and the energy-saving rate of waste discharge is 27.6%, as shown in Table 2.

Table 2.   Power consumption and waste discharge of prefabricated buildings and traditional cast-in-situ buildings.

Project	Prefabricated buildings	Traditional cast-in-situ buildings
Power consumption (kWh/m^2)	10.86	17.10
Waste discharge (m^3/m^2)	0.0110	0.0152

(2) The development of prefabricated buildings is an inevitable trend to deal with the shortage of human resources

Construction workers will be gradually reduced, and labor costs will increase significantly. The lack of construction workers and the increase in labor costs will narrow the difference between the cost of construction industrialization and the cost of traditional construction production. If most of the on-site operations of the building can be transferred to the prefabrication factory, and some migrant workers can be transformed into industrial workers, so that they can enjoy a better working environment and social welfare conditions, this will certainly stimulate the enthusiasm of workers, improve the utilization rate of human resources, significantly reduce the labor intensity of workers and the incidence of safety accidents, and also reduce the number of on-site workers, It reduces the dependence of the production process on skilled workers, thus greatly improving labor productivity.

(3) The development of prefabricated buildings is an inevitable choice for the construction industry to improve efficiency and quality

The industrial construction mode can provide high-quality, durable, energy-saving, and environment-friendly building products, and solve various quality problems existing in the construction industry for a long time. Since most of the components are manufactured in the factory, the production quality of the products can be controlled according to certain operating procedures and strict process standards, which can easily meet the requirements of quality standards. The on-site lifting and a few node connection operations can greatly reduce the workload and labor intensity of the on-site workers, and create good conditions for ensuring the construction quality.

The construction period of industrial construction can generally be shortened by about 30%. A shorter construction period can improve the developer's ability to resist risks during the construction period, elevate the turnover rate of investment funds, improve the financial situation and enhance profitability.

3 COMBINATION OF 3D PRINTING TECHNOLOGY AND PREFABRICATED BUILDING WITH BIM TECHNOLOGY

With the intelligent development of science and technology and the integration and innovation of BIM technology and 3D technology, the construction industry is gradually developing in the direction of intelligence, security, and greenness. In the process of combining the BIM technology, 3D printing technology, and assembly building, the 3D model of the assembly building is established through the BIM Technology, and the model is further optimized, and then the building structures are disassembled, optimized, and debugged. Designers import the 3D model established and optimized by BIM into the 3D printing-related software, and then use the 3D printing equipment to print the model. The 3D printing model can be used for the selection of prefabricated building schemes to adapt to different construction conditions and user needs. In addition, the construction scheme and progress of prefabricated buildings can also be displayed to users and investors through 3D-printed solid models. At the same time, some heterosexual components can be printed to judge the feasibility and safety.

3.1 *Development and application of 3D printing technology in construction engineering*

Originating in 1997, building 3D printing was proposed by Joseph Pegna (Pegna 1997), an American scholar, as a construction method of free-form components suitable for cement materials to be accumulated layer by layer and selectively solidified. At present, 3D printing technology is mainly applied to three processes in the construction field: D-shape (Soar 2012), contour craft (Khoshnevis 2006), and concrete printing (Ma 2014).

(1) D-shape process

The D-shape technology was first proposed by Italian engineers. The D-type process is a method to realize stacking molding by spraying and extruding adhesives to selective gels and hardening layers by layered gravel powders. There are hundreds of nozzles at the bottom of the D-type process printer, and sand is sprayed on the magnesium adhesive ejected from the nozzles to cast into the stone solid. Such repetition forms a combination of layers of cement and sand and finally forms a substantial building. In the working state of the 3D printing equipment, the printer will move back and forth along the horizontal axis beam and four vertical columns, and the printer head will only form a thickness of 5 mm–10 mm when printing one layer. The printer can be operated by electronic equipment such as computers. The finished main building material after printing is similar to marble. The building built by this method has higher material strength than concrete and does not need to be reinforced with additional steel pipes. The D-type process is similar to the selective powder deposition process, but the selection of materials is different. The material commonly used for building printing is magnesium oxychloride cement (Wang 2015). In addition, the unbonded powder material will remain in place to support the printing of subsequent layers. At this stage, the D-type process printer has been successfully applied to building ducts, hollow columns, and other components, as well as internal curves, partitions, and other building structures. When the printing task is finished, the printed products can be removed from the loose powder platform. Unbound materials can be recycled by a vacuum cleaner for other printing materials.

For the application of D-type process, Cesaretti et al. (Cesaretti 2014) used lunar soil (weathered layer) as the printing material to print residential buildings suitable for the lunar environment; the mechanical properties were tested in the air and vacuum environment for future space development. In addition, in coastal areas or saline-alkali lands, magnesium oxychloride cement can be used as 3D printing material to print the substrate structure, which not only reduces the manufacturing cost and improves the efficiency, but also lengthens the service life and enhances the durability and safety of this building in the local area.

(2) Contour crafting

Contour craft was first proposed by Professor Khoshnevis (Khoshnevis 2006, 2004) at the University of Southern California, and then developed rapidly. It has become one of the commonly used 3D printing processes. The contour technology is proposed to realize the automatic construction of the whole components and auxiliary components of complex buildings.

Compared with the additive construction method of layered bonding and stacking of sand and stone powder in the D-type process, the materials in the contour process are all extruded from the nozzle. The nozzle will spray concrete materials at the designated place according to the instructions of the design drawing. This kind of method belongs to the additive construction method of layered concrete spraying and extrusion. The advantage of contour molding is that it does not need to use a mold. The building contour printed by the printing equipment will become a part of the building, which can improve construction efficiency. At present, contour technology also has some disadvantages. For example, certain technical requirements are put forward for printing materials; the size, height, and freedom of the printing structure are restricted by the adhesive strength of the layers (Zhang 2013); the surface accuracy of the contour printing process also depends on the post-processing of the auxiliary parts such as the trowel.

At present, the contour printing technology based on 3D printing has made new progress in the field of architecture and has been applied to a certain extent. Paul (2007) and others improved the 3D contour printing process, added the cable system, and used the rigid frame as the mechanical skeleton. The three-dimensional movement of the terminal nozzle is controlled by 12 cables, and the light steel frame is used to replace the heavy gantry, which improves the portability and flexibility of the whole printing equipment, makes the equipment transfer more convenient, and makes on-site 3D printing possible. In addition, domestic companies have successfully printed residential buildings that can be used for human habitation by using the of 3D contour printing process (Panda 2018).

(3) Concrete printing

The concrete printing technology was first proposed by Professor Lim (Lim 2012) from the Innovation and Architecture Research Center of Loughborough University in the UK. The concrete printing process is similar to the contour process. The nozzle in the printing equipment needs to be installed on the sedan car. The nozzle is used to extrude the concrete, and the components are constructed by the stacking method. The difference is that concrete printing technology needs to ensure the high accuracy of material ejection.

A concrete printing system includes a computer digital control system and the printing machinery. Its equipment is a gantry robot of a certain size (Zhu 2018). The whole concrete printing process includes three stages: data collection, concrete preparation, and component printing. In the data acquisition stage, after inputting the high-precision 3D model, the technicians will optimize the printing path, like modeling using reverse scanning technology, in order to improve the working efficiency of the printer and ensure the working performance of the printing materials. In the concrete preparation stage, in order to ensure the fluidity of the concrete in the pipeline and prevent the concrete from solidifying during transportation, it is necessary to control the transportation time of the concrete. In the component printing stage, according to the different geometry of the printing components, the printing head needs to be adjusted to the appropriate geometry and angle. In addition, in the printing process, it is also necessary to pay attention to the consistency between the printing speed and the concrete transmission speed to ensure that the printing materials have good uniformity, good adhesion between layers, and good accuracy of the printed products.

In terms of construction technology, 3D concrete printing has been applied to the printing of prefabricated building walls. Before printing, the size and space position of pipes and windows shall be reserved in advance from the BIM model; meanwhile, the positions of tie bars and embedded parts shall be reserved; after the components are transported to the site, the secondary concrete filling is required to complete the connection of the wall. In addition, the 3D printing of the internal structure of the prefabricated building wall can also be optimized acoustically and mechanically according to special environmental requirements.

3.2 Application difficulties of 3D printing technology and prefabricated buildings under BIM technology

(1) Low efficiency of 3D printing and assembly building printing
At present, prefabricated components of prefabricated buildings need to be manufactured by the manufacturer in advance and then transported to the site for installation. However, the 3D printing efficiency cannot meet the construction schedule, and its production speed and mode will affect the progress of the assembly installation of each project and thus affect the construction period.

(2) BIM technology has not received due attention
BIM Technology has broad technical coverage, which can effectively reduce the rework rate and error rate in the construction process, with a great commercial value. However, due to the huge investment in the BIM technology in the early stage, the investment in BIM projects often accounts for 2%–5% of the project funds, which brings great capital pressure to the early stage of the project; it is easy to cause the purpose of the construction unit and the construction unit to be inconsistent, thus affecting the progress of the project.

(3) 3D printing does not match with prefabricated buildings
At present, 3D printing is demanding more for material selection. The printed buildings are also not concrete buildings in the traditional sense, as they lack thermal insulation, waterproofing, and decoration. Moreover, the 100% house printing fails. Especially when the printing problem occurs, if the building is defective, there is no relevant technical quality standard to solve it. There are no relevant quality standards and particularly no sufficient experimental data to support the safety, seismic resistance, and fire resistance of the house.

(4) The adaptability of 3D printing and BIM needs to be improved
There is no software that integrates 3D printing and BIM Technology. The adaptability of different software can easily lead to errors in converting BIM model information into 3D printing model information and easily lead to data losses or translation errors. Such data errors will greatly affect the progress and quality of 3D printing and waste engineering materials.

3.3 Description of specimens

(1) Formulate and improve relevant technical quality standards
BIM-related national standards have been issued, and all regions, industries, and enterprises are not aware of the basic direction and strategy of BIM development; the industry and leading

enterprises should be actively encouraged and guided to participate in the establishment and implementation of local standards, which can promote the informatization development of the construction industry, and thus accelerate the release and implementation of the 3D printing technology and related standards for prefabricated buildings.

(2) Establish big data and cloud sharing platforms

New vitality should be injected into the construction industry in the big data era. Using big data and cloud contribution, we can refer to the previous cases of 3D printing combined with architecture, so as to get the inspiration of combining with prefabricated architectures. Problems generated in the project can be uploaded to the cloud in the form of data so that practitioners can share experiences.

(3) Actively innovate 3D printing materials

The material of 3D printing is one of the main reasons that restrict its combination with prefabricated buildings. At present, the main material for housing construction is reinforced concrete, which cannot be used as a 3D printing material. The research and development of new materials to meet the 3D printing and manufacturing needs of prefabricated buildings, and material innovation have become the key to development. It is one of the mainstream directions of the current development to continuously develop 3D printing materials to meet building needs.

4 CONCLUSION

Since 2016, the state has issued a series of policy documents on prefabricated buildings, proposing to vigorously develop prefabricated buildings from the policy level. Under the policy support and facing the competitive pressure of the traditional construction market, how to expand the influence, give play to the advantages of technical products and resources, cope with competition, increase income, realize large-scale market application and supporting services, and create an all-round 3D printing-prefabricated building and housing industry is the inevitable development trend of the national construction industry. Construction industrialization is the only way for China to get rid of labor shortage, resource waste, environmental pollution, and frequent safety accidents, which is of great significance to the transformation and upgrading of China's traditional construction industry and the realization of green and sustainable development.

REFERENCES

Bosscher, P., Williams, R.L., Bryson, L.S. et al. (2007). Cable-suspended robotic contour crafting system. *J. Automation in Construction*. 17(1): 45–55.

Cesaretti, G., Dini, E., De, K.X. et al. (2014). Building components for an outpost on the Lunar soil by means of a novel 3D printing technology. *J. Acta Astronaut*. 93: 430–450.

Gan, J.H., Men, H.R., Zhang Z.M. et al. (2021). Application status and prospect of 3D Printing technology in medical related areas. *J. Journal of Modern Clinical Medicine*. 47(05): 385–387.

Huang, H.B. (2021). Application value analysis of BIM Technology in prefabricated buildings. *J. Cities and Towns Construction in Guangxi*. (09): 84–85.

Khoshnevis, B. (2004). Automated construction by contour crafting—related robotics and information technologies. *J. Automation in Construction*. 13(1): 5–19.

Khoshnevis, B., Hwang, D., Yao K.T. et al. (2006). Mega-scale fabrication by Contour Crafting. *J. International Journal of Industrial and Systems Engineering*. 1(3): 301–320.

Li, Y.K. (2017). *Research on 3D printing dental implant technology*. D. Changchun: Changchun University of Technology. 56–57.

Lim, T.T., Austin, S.A., Lim, S. et al. (2012). Hardened properties of high-performance printing concrete. *J. Cement and Concrete Research*. 42(3): 558–566.

Liu, C.Y., Zhao, B.B., Li, L.J. et al. (2020). Research progress of 3D printing technology for metallic materials. *J. Powder Metallurgy Industry*. 30(02): 83–89.

Liu, R.N., Zhang, J., Wang Y. et al. (2019). Development background and current situation of prefabricated buildings in China. *J. Housing and Real Estate*. (32): 32–47.

Lv, J.K. & Cao, G.M. (2021). Research progress of metal parts surface finishing technology applied to additive manufacturing in aviation industry. *J. New Technology & New Process*. (08): 1–7.

Ma, J.W., Jiang, Z.W. & Su, Y.F. (2014). Development and prospect of 3D printing concrete technology. *J. China Concrete*. (07): 41–46.

Panda, B., Tay, Y.W.D., Paul, S.C. et al. (2018). Current challenges and future potential of 3D concrete printing. *J. Material Wissenschaft and Werkstofftechnik*. 49(5): 666–673.

Pegna, J. (1997). Exploratory investigation of solid freeform construction. *J. Automation in Construction*. 5(5): 427–437.

Soar, R. & Andreen, D. (2012). The role of additive manufacturing and physiomimetic computational design for digital construction. *J. Architectural Design*. 82(2): 126–135.

Tan, L.Z. & Fang, F. (2016). 3D printing technology and its application in aerospace industry. *J. Tactical Missile Technology*. (04): 1–7.

Wang, M. (2016). Discussion on the application of 3D printing technology in metal processing. *J. Modern Manufacturing Technology and Equipment*. (01): 110–111.

Wang, Z.M. & Liu, W. (2015). 3D printing technology and its application in architecture. *J. China Concrete*. (01): 50–57.

Wu, F.X., Liu, L.M., Xu, Y. et al. (2013). Development trends of 3D printing technology in foreign aerospace field. *J. Aerodynamic Missile Journal*. (12): 10–15.

Zhang, C., Deng, Z.C., Ma, L. et al. (2021). Research progress and application of 3D printing concrete. *J. Bulletin of the Chinese Ceramic Society*. 40(06): 1769–1795.

Zhang, H.S. (2017). Application of 3D printing technology in prefabricated buildings. *J. Project Management*. (08): 79–81.

Zhang, J. & Khoshnevis, B. (2013). Optimal machine operation planning for construction by contour crafting. *J. Automation in Construction*. 29: 50–67.

Zhu, B.R., Pan, J.L., Zhou, Z.X. et al. (2018). Advances in large-scale three-dimensional printing technology applied in construction industry. *J. Materials Reports*. 32(23): 4150–4159.

Frontiers in Civil and Hydraulic Engineering – Mohamed A. Ismail and Hazem Samih Mohamed (Eds)
© 2023 The Authors, ISBN 978-1-032-38247-0

Research and application of auxiliary control system linkage technology in pumping storage power station

Congying Wu* & Han Liu
State Grid Economic and Technological Research Institute Co., LTD., Beijing, China

Siming Zhuo & Yonghong Sun
Guodian Nanjing Automation Co., LTD., Nanjing, Jiangsu, China

ABSTRACT: In this paper, the auxiliary control system linkage ability of a pumping storage power station was explored, and the reason for the poor linkage technology for the auxiliary control system was studied based on the communication protocol and information model of auxiliary control equipment and the plant model. The linkage was built, and real-time monitoring, fire alarm, condition monitoring, video surveillance, and other typical linkage strategies were put forward. The possible problems in the linkage process were analyzed, including hierarchical linkage alarm, the priority of auxiliary control policy, monitoring and evaluation of auxiliary control equipment, and countermeasures were proposed. The practice has proved that the application of the linkage technology of auxiliary control systems plays a positive role in improving the safety and comprehensive benefits of the power station.

1 INTRODUCTION

Under the requirement of a "dual carbon" goal, accelerating the development of pumping storage power stations (hereinafter referred to as "pumping storage power stations") is the only way to solve the problem of multi-energy complementarity. As new energy sources such as wind power photovoltaic are characterized by intermittent, random, and unstable power (Li 2021), if there are no good means of peak regulation, the power grid may be subjected to major risks. Pumping and storage power stations have better functions of peak regulation and energy storage, which is of great significance to increasing the absorption capacity of the new energy power system and ensuring the safe and stable operation of the power grid.

Pumping and storage power station equipment includes main equipment and auxiliary control system. The auxiliary control system provides the main equipment with such functions as system temperature control and drainage, equipment lubrication protection, and power station fire protection (Li 2013), which is an important guarantee for the safe operation of the power station.

At present, the monitoring and linkage ability of auxiliary control system equipment of the power station is poor, and the linkage cannot be timely and effectively targeted at abnormal phenomena in the system operation process or safety-sensitive factors in the environment, which is very unfavourable to improve the safety protection and comprehensive benefits of the power station. With the development of pumping storage power stations from digital to intelligent directions, it is very important to realize automatic linkage by the auxiliary control system.

In this paper, the linkage technology of the auxiliary control system of a power station is deeply studied, and the key technologies and difficulties in linkage are discussed, so as to give full play to the role of the auxiliary control system and improve the comprehensive management level and benefits of the power station.

---

*Corresponding Author: wcycongying@163.com

DOI 10.1201/9781003344209-58

2 STATUS OF THE AUXILIARY CONTROL SYSTEM OF THE POWER STATION

The auxiliary control system of the pumping and storage power station mainly includes oil, gas, and water systems and the fire control system.

From the current situation of the pump storage power station, the reason for the auxiliary control system part of the equipment to realize linkage is mainly embodied in several aspects:

(1) Auxiliary control equipment varies with the structure of the power station, and there is no unified communication interface standard; there are various auxiliary control devices, a wide range of layouts, and scattered data sources. Especially in the case of sudden security accidents, managers cannot make correct decisions quickly when faced with a large number of alarm information, and manual operation will lead to delay, which seriously affects the protection efficiency of the auxiliary control system.

(2) Some auxiliary control equipment adopts traditional equipment with low intelligence and difficult data acquisition, transmission, and processing, which does not meet the requirements of intelligent power stations; for example, for the ventilation system, the traditional exhaust fan needs to be measured by a hand-held anemometer at regular or irregular times (Liu 2013). The low ventilation efficiency cannot be found in time and can only be found when the iron core temperature is too high to cause other problems. When the ambient temperature rises and the humidity is high, it cannot be automatically dehumidified and cooled down, leading to the load reduction operation of the unit (Ma 2015). Most of the water pipes in the fire protection system are buried, and the flowmeter and pressure gauges with communication functions are not installed, so water leakage or rust cannot be found in time; the valve adopts the traditional valve, which cannot be controlled automatically (Ma 2016).

(3) Subsystems of auxiliary equipment are self-contained. Although they also belong to automatic control, they are in a self-circulation mode and are not associated with the main equipment, such as technical water supply systems and oil pressure device systems.

3 RESEARCH ON LINKAGE TECHNOLOGY OF THE AUXILIARY CONTROL SYSTEM OF THE POWER STATION

To link the auxiliary control system in the power plant, you first need to realize the communication between the various auxiliary equipment, on this basis, the related equipment and subsystems of linkage can give full play to the function of the auxiliary control system in case of any emergency so that management personnel can provide technical support to make quick decisions.

3.1 *Architectural design of auxiliary control system linkage hardware*

According to safety requirements, pumping and storage power stations are generally divided into safety zones I, II, and III (GB/T 36572 2018). The auxiliary control system is arranged as follows: a ventilation system and five microcomputer-aided defense systems are deployed in zone I; the status monitoring system is deployed in zone II; the industrial TV system, the access control security system and the production management system, are deployed in zone III. Firefighting systems are deployed in all three zones.

3.2 *Communication protocol and information model of auxiliary control equipment*

3.2.1 *IEC61850 overview*

The IEC61850 standard is the only global standard in the field of power system automation. It is based on the concept of object-oriented modeling, and establishes a unified modeling standard for various types of equipment. It provides a foundation for interoperability between equipment. Its expansibility and maintainability meet the technical requirements for equipment linkage of the

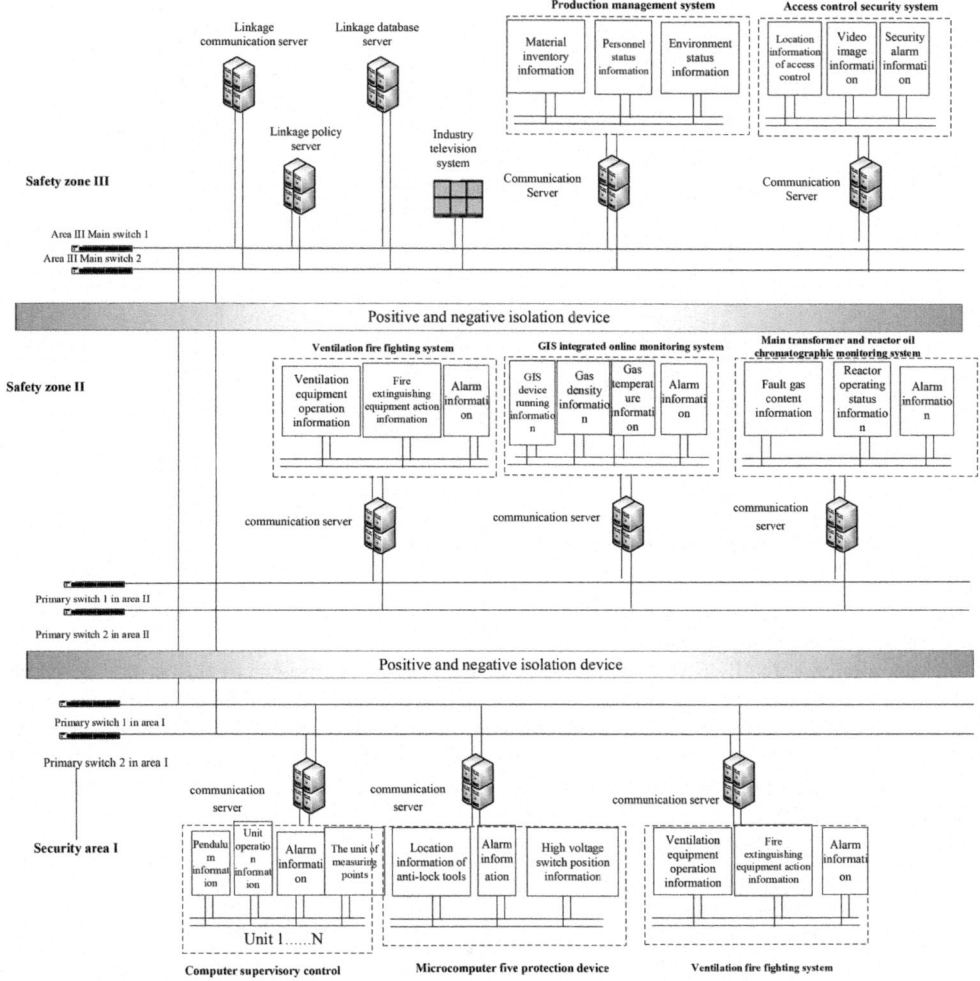

Figure 1. Distribution diagram of the auxiliary control system.

auxiliary control system in pumping and storage power stations. Therefore, this communication protocol standard is adopted as the modeling standard of the auxiliary control system.

3.2.2 *Information model hierarchy based on IEC61850*

The auxiliary control system in the pumping and storage power station adopts a tree-like hierarchical structure model, from the root to branches and leaves, being the server, the logical device (LDevice), the logical node (LNode), and the data object (Do) (Min 2011). An IEC61850 Intelligent Electronic Device (IED) consists of one or several servers. Each level contains one or more of the contents of the next level, and so on (Wang & Zhang & Wen 2012).

3.2.3 *Information modeling method based on IEC61850*

The SCL language is used to describe the power station information model, which is based on IEC61850.

The information model consists of Header, Communication, IED, and DataTypeTemplates. Header and Communication are about version information and communication information, which

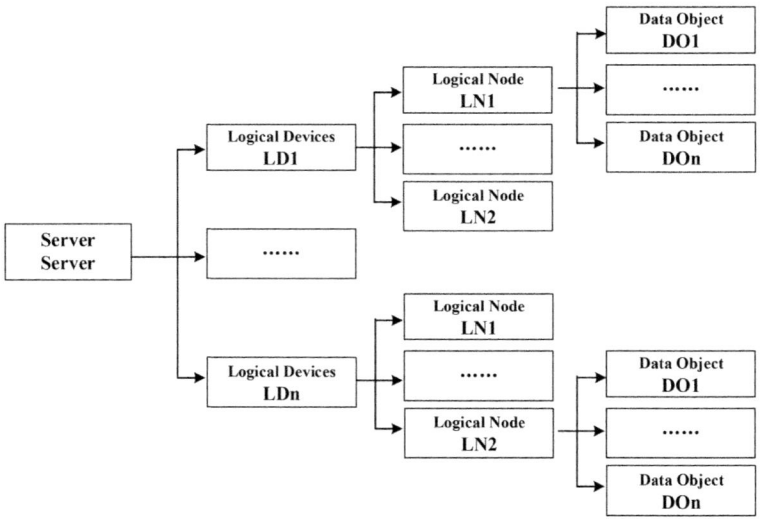

Figure 2. Schematic diagram of IED layered information model.

are relatively fixed for a pumping and storage power station. The IED and DataTypeTemplates sections vary from plant to plant.

In accordance with IEC61850 standards, the IED logical equipment of this project is firstly determined, and then, its logical node is determined according to the logical equipment. Next, the data object is determined based on the logical node; the data objects have multiple attributes that describe them; finally, the above information is combined with common information such as communication and system configuration to synthesize the IED information model (Wang & Zhang & Wen 2012).

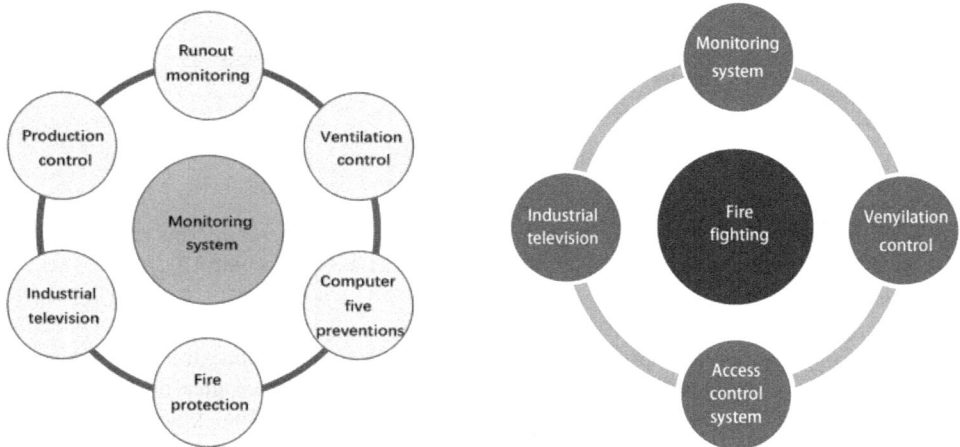

Figure 3. Real-time monitoring linkage.   Figure 4. Fire fighting linkage.

3.3 Linkage of the auxiliary control system

Due to the large quantity of equipment in the auxiliary control system of this power station, equipment and subsystems are composed of different linkage bodies in order to achieve different

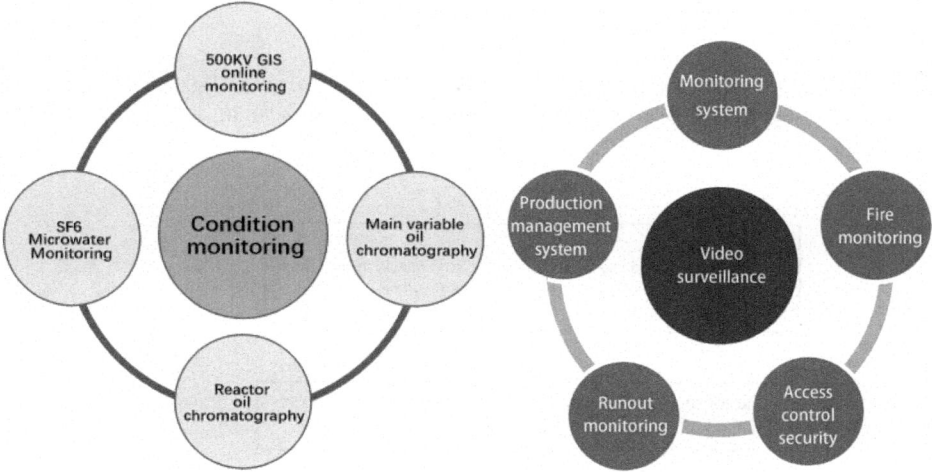

Figure 5. Condition monitoring linkage.　　　　Figure 6. Video monitoring linkage.

linkage purposes. It can be mainly divided into real-time monitoring linkage, fire fighting linkage, five anti-locking linkages, condition monitoring linkage, video surveillance linkage and so on. The central subject of each linkage acts as the active part, and the related part acts as the follower. When the active part discovers or produces an event, the action of the follower is triggered.

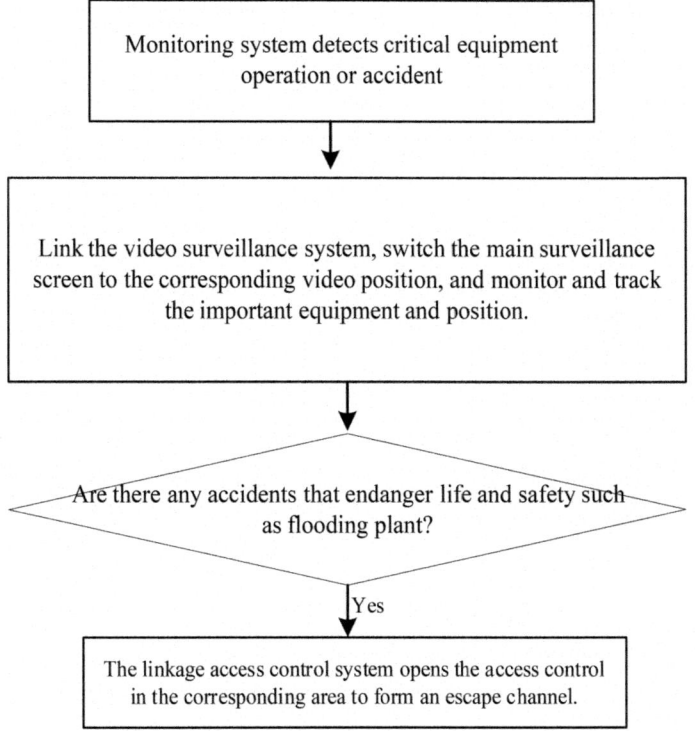

Figure 7.　Schematic diagram of real-time monitoring linkage strategy.

3.4 *Linkage policies for the auxiliary control system*

3.4.1 *Overview of linkage policies*
A linkage policy describes the master/slave action relationship between linkage objects in the auxiliary control system, including the description, verification, search, and execution of the linkage policy.

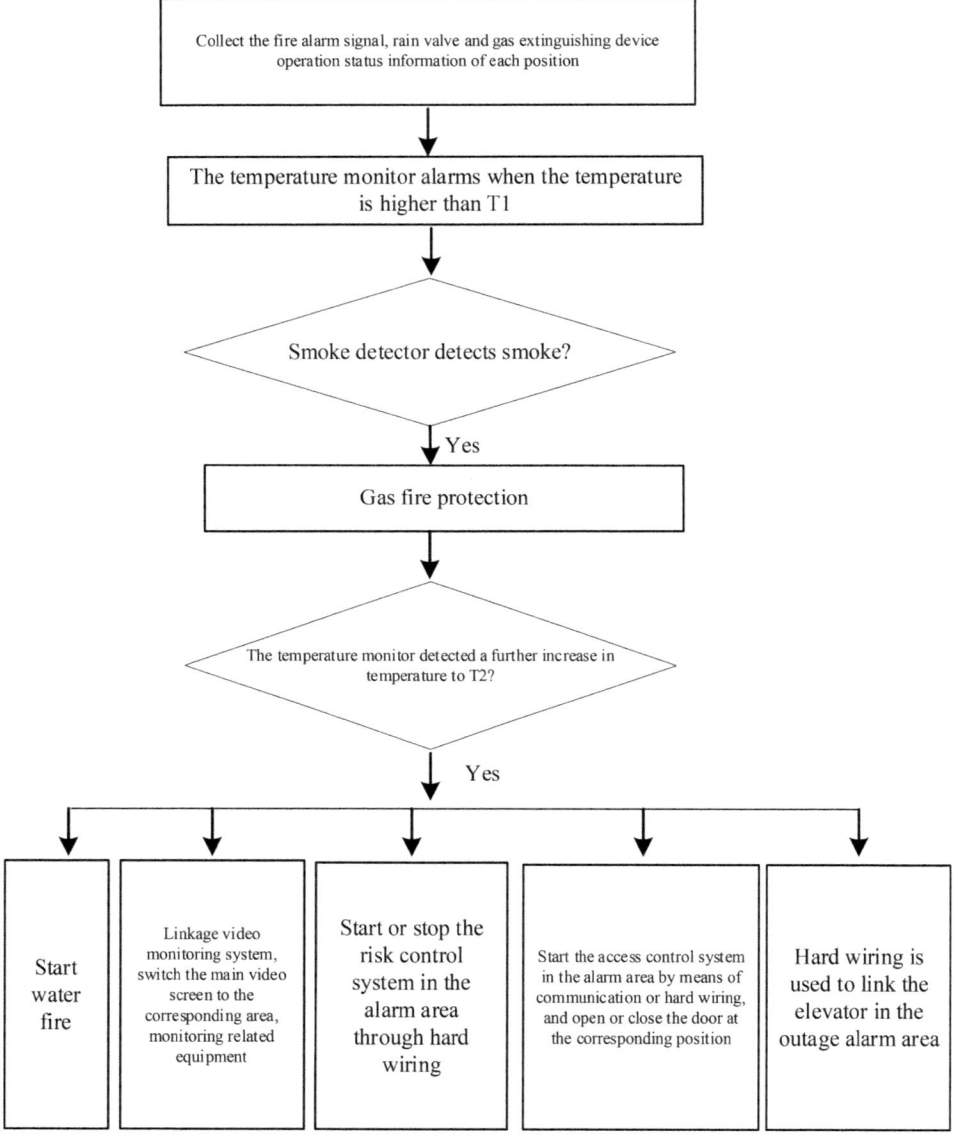

Figure 8.   Schematic diagram of fire fighting progressive linkage strategy.

3.4.2 *Typical linkage policy*
(1) Real-time monitoring of system linkage policies

The real-time monitoring system monitors the normal production process, focuses on the operation of important equipment and accident-prone points, and timely links to the access control system to form escape channels in case of life-threatening accidents.

(2) Fire fighting linkage strategy

The fire protection system and industrial TV system play an extremely important role in ensuring fire safety for the plant area and the construction of the hydropower station (Xiao 2020).

Fire detectors mainly include temperature sensing, smoke sensing, light sensing, combustible gas, and compound fire (Yin 2014).

When the fire control system detects a fire signal, it adopts dual fire control measures of gas fire and water fire. According to different fire situations, it carries out progressive fire control, and simultaneously links video surveillance, exhaust system, access control, elevator, etc., to ensure personal and equipment safety:

(3) State monitoring linkage strategy

Parameter values can be obtained in real-time by means of online parameter monitoring. If abnormal variable values are found, corresponding operations such as starting the exhaust fan are generated and an alarm is generated.

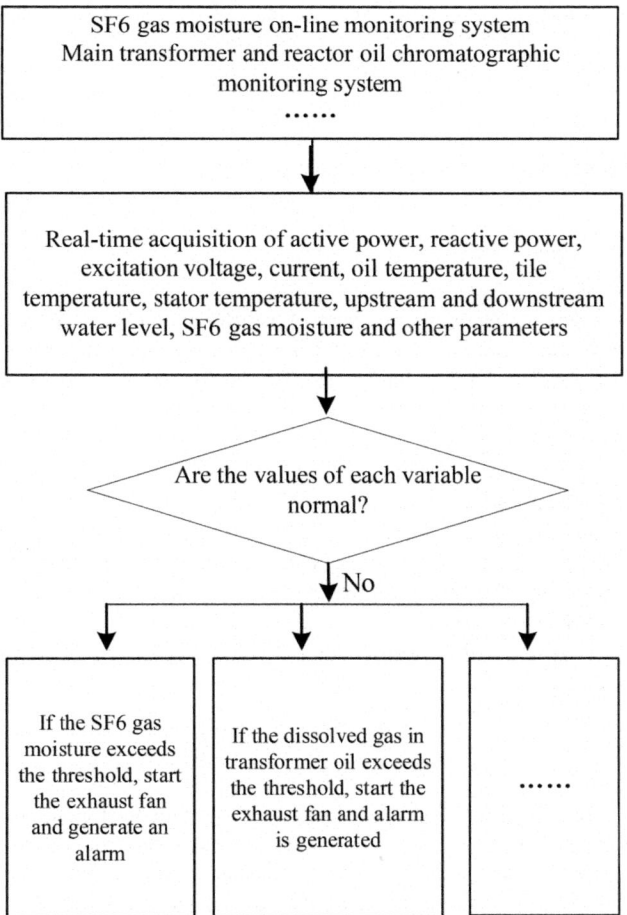

Figure 9. Schematic diagram of condition monitoring linkage strategy.

Video surveillance linkage strategy

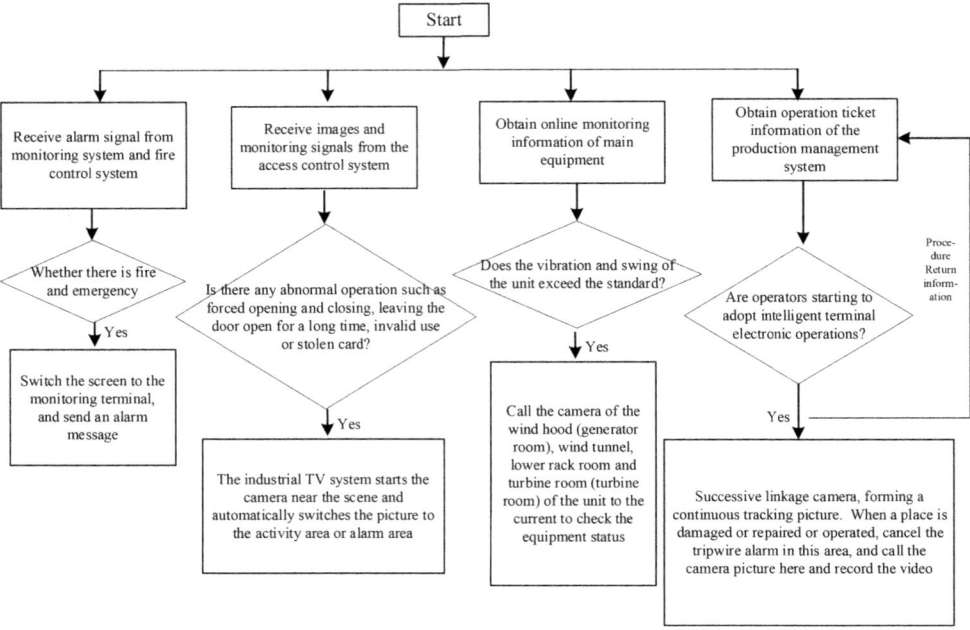

Figure 10.   Schematic diagram of video monitoring linkage strategy.

3.5   *Analysis of key technologies*

3.5.1   *Hierarchical alarm linkage*

The auxiliary control device may generate alarms for different events. In order to distinguish the causes of alarms, different modes are used to level alarms. Hierarchical alarms can be pushed in different frequencies, durations, and volumes, and can be notified by voice or SMS. For example, in the most critical case, the alarm is pushed for 5 s and stopped for 5 s until the emergency is cleared. The lower-level alarm is pushed for 10 s and stopped for 10 minutes, and so on.

3.5.2   *Priority of linkage policies*

As auxiliary control devices are in different linkages, conflicts may occur among linkage policies. For example, the general risk control system is in the monitoring system and the fire control system at the same time, and the access control system is in the fire control system and the video surveillance system at the same time.

The priority of each linkage policy is calculated based on the dynamic priority algorithm:

$$E = \sum_{i=1}^{n} A_i \cdot \eta_i - B_i \cdot \theta_i$$

Among them, E is the priority of the auxiliary control equipment. A_i is the role of the auxiliary control equipment in the linkage strategy i, being irreplaceable or replaceable. η_i is the auxiliary control equipment in the linkage strategy i in the importance of the policy, being very important, important, general or not important. B_i is the injury caused, which is divided into personal safety, equipment safety, environmental safety, etc. θ_i is the degree of the injury that may be caused, being very serious, serious, or not serious.

Based on the formula, the overall policy priorities are as follows: personal safety > device security > environmental safety.

3.5.3 *Health assessment of auxiliary control equipment*

To ensure the normal operation of the linkage system, the auxiliary controller should be in a healthy state to avoid device failure when a linkage event occurs. Based on the daily operation data and health trend analysis of the equipment, a two-level health assessment model is established through big data analysis to carry out health assessment of the auxiliary control equipment.

The first level is to study the health status assessment of individual equipment. The second layer studies the health status assessment of the complex auxiliary control equipment system of the pumping and storage power station composed of single equipment, thus forming the framework for the health status assessment system of the auxiliary control equipment. The diagram of the two-layer model framework for the health assessment is shown as follows:

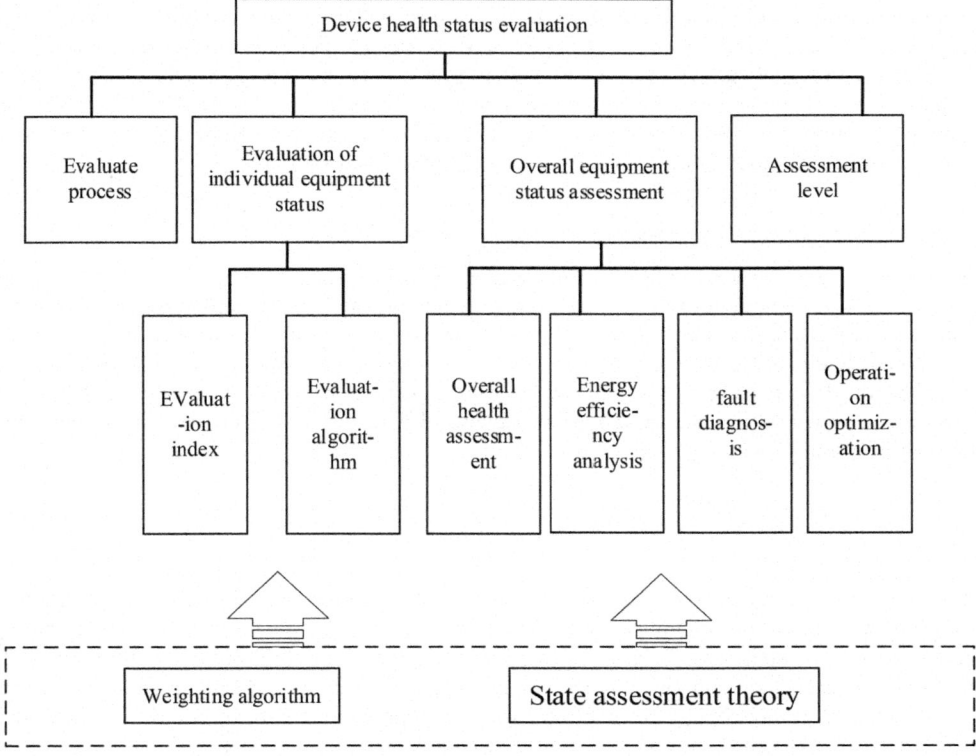

Figure 11.   Schematic diagram of the two-layer health assessment model.

4   ANALYSIS OF APPLICATION BENEFITS OF THE LINKAGE TECHNOLOGY

After the application of the above linkage system in a pumping storage power station in Xiamen, Fujian Province, good effects have been achieved, which are mainly reflected in the following aspects.

4.1 The incidence of safety accidents is greatly reduced

The number of safety accidents in the pumping storage power station is reduced by about 70%, from an average three more years before to below the average, mainly manifested in the following aspects: (1) According to different levels during a fire in fire control, the door of the corresponding position is opened or closed through the communication or the hard wiring linkage to start the alarm linkage area of the entrance guard system; the alarm area of the elevator is stopped using the hard wiring linkage to reduce casualties; (2) a remote linkage alarm is given when the unit is repaired and drained and the power station leakage drainage system is tested and drained; pumping and power generation are implemented and a remote linkage alarm is given to reduce water drowning accidents; (3) when the plant lifting equipment starts the equipment, the remote linkage alarm prevents personnel injury accidents.

4.2 The power station has a significant effect on energy saving and consumption reduction

The effects of saving energy and reducing consumption in pumping storage power stations are remarkable. With the ventilation system as an example, air leakage in the generator ventilation system will lead to low cooling efficiency of the core (Liu 2020), and the traditional ventilation system cannot be detected in time, resulting in energy waste. The auxiliary control system is linked and problems are timely found through the automatic detection of air leakage to avoid energy wastes caused by air leakage or undercooling; the wind speed is automatically adjusted, and the effective cooling air volume is increased or decreased. The electricity consumption of the ventilation system decreases from 35% to 25% of the plant's annual electricity consumption.

4.3 The service life of the generator set is significantly prolonged

The service life of the generating set in the pumping storage station is significantly prolonged. Because the auxiliary control system can automatically adjust the flow and pressure of cooling water according to the oil temperature and water temperature, avoiding the overheating damage of the machine and insufficient pressure or flow of the cooling water system of the unit, the service life of the machine is increased by 30% on average. Calculated with the installed capacity (2000 MW) of the power station, the annual maintenance and purchase cost of the machine is saved by about one million yuan.

5 CONCLUSION

Due to the functions of fast start-up and fast load tracking, the conversion of power generation and pumping conditions is complicated. With the change in working conditions, the corresponding auxiliary control system also needs changes, and the linkage effect of the auxiliary control system becomes obvious. Since pumping storage power stations started late in China, more data need to be accumulated to realize the all-around linkage, and training and learning should be combined with artificial intelligence knowledge to provide a technical basis for further realizing smart power stations.

REFERENCES

GB/T 36572, *Guidelines for Network Security Protection of Power Monitoring System* [S], Beijing: Standardization Administration of China, 2018.

Li Y.X., *Auxiliary Equipment of Hydraulic Generating Set* [M], Beijing, China Water and Power Press, 2013, 1, 10–24.

Li C.J., Study on carbon reduction benefit of pumping storage power station [J], *Hydropower and Pumping storage*, 2021, Vol.7, No. 6 (total No.40), 45–48,

Liu B.K., Improvement of generator ventilation system of daxia hydropower station [J], *China Machinery*, 2020 (2), 34–35.

Ma X.M., Technical transformation of water fire control system of tianshengqiao hydropower station [J]. *Yunnan Hydropower*, 2016, Vol. 32, No. 6, 177–181.

Min Z., Xu.J., Wang J.L., Intelligent hydropower plant modeling technology based on IEC 61850 [J], *Hydropower Automation and Dam Monitoring*, 2011, Vol. 35, No. 4, 1–6.

Ma Z.J., Analysis on the renovation technology of the turbine generator ventilation system of Longshou-1 hydropower station [J], *East China Science and Technology: Academic Edition*, 2015, No. 8 180.

Wang C., Zhang Y., Wen Z.G., Information modeling of hydropower plant monitoring System based on IEC 61850 standard [J], *Hydropower Automation and Dam Monitoring*, 2012, Vol. 36, No. 6, 1–4.

Xiao L, Design and implementation of fire control system and video linkage in hydropower station [J], *Hydropower Plant Automation*, 2020, Vol. 40, No. 2, 21–23.

Yin Q, Application of new fire alarm system in Xiangjiaba hydropower station [J], *Hydropower*, 2014, Vol. 40, No. 1, 57–59.

*Frontiers in Civil and Hydraulic Engineering – Mohamed A. Ismail and
Hazem Samih Mohamed (Eds)*
© 2023 The Authors, ISBN 978-1-032-38247-0

Numerical simulation of flow field at the downstream joint of a stepped overflow dam

Rui Yin* & Dong-yao Hong

Department of Water Engineering, Kunming University of Science and Technology Oxbridge College, Kunming, Yunnan, China

Jia-shi Cai & Ding Cui

School of Electric Power Engineering, Kunming University of Science and Technology, Kunming, Yunnan, China

ABSTRACT: The VOF multiphase model is used to simulate the two-dimensional unsteady flow field at the junction downstream of the stepped overflow surface of a gravity dam (beginning of the ogee section). By analyzing the liquid phase, pressure, and streamline of the overflow dam surface at different time, it is found that the lower edge of the discharge nappe is not attached to the overflow dam surface at the initial stage but forms a cavity at the concave corner of the step, with a linear lower edge, the most-violent one of which is the cavity change of the last step, with a countercurrent along the step surface. The penultimate step is filled with faster liquid, while the last step is filled with slower liquid. Before the discharge flow is stable, there is a strong negative pressure area at the concave corner of the last step, and after the discharge flow is stable, there is a weak negative pressure area at the concave corner of the penultimate step. The streamline forms a vortex structure on each step. The closer the streamline is to the downstream, the denser and faster the flow velocity.

1 INTRODUCTION

In order to overcome the defects of large fog and tailwater fluctuation when the traditional overflow dam surface dissipates energy (State Key Laboratory of Hydraulics and Mountain River Development and Protection of Sichuan University 2016), the stepped overflow dam surface is widely used as a new-type high-efficiency energy dissipator. The stepped surface aerates and decelerates the water step by step, which is able to improve the energy dissipation rate. The deceleration can improve the energy dissipation rate (Lin 2009).

Wang et al. (2017) studied the influence of four different sizes of transition steps on hydraulic characteristics. And the energy dissipation rate gradually increases with the increase in the size of the transition steps.

Chen et al. (2019) conducted hydraulic model tests on four different numbers of stepped bulges under the condition of a 1 m high aerator and 10° angle, and studied many hydraulic characteristics such as flow pattern, water surface profile of the stilling basin, near-bottom velocity, negative pressure, time average pressure, energy dissipation rate, and other hydraulic characteristics.

---

*Corresponding Author: 386072923@qq.com

DOI 10.1201/9781003344209-59

Tang et al. (2020) studied the influences of the aerator height of 11.67 mm, and 16.67 mm, the angle of 8°, and 10°, and the combined action of the four transition steps on the aerating characteristics of the stepped surface.

Most of the above studies are based on hydraulic model experiments, and more attention has been paid to the upstream (end of the weir crest's curved section) junction of the stepped overflow dam face. In this paper, for the actual project size, the unsteady numerical simulation is conducted on the liquid phase, pressure, and streamline distribution at the downstream (beginning of the ogee section) junction of the stepped overflow dam face. Due to the large actual project size, the transverse flow has little impact on the hydrodynamic characteristics. Therefore, this paper only studies the two-dimensional profile of the overflow dam surface.

2 NUMERICAL MODEL AND ALGORITHM

2.1 Governing equations

In this paper, the VOF method (Hirt 1981) and the RNG k-ε turbulence model (Launder 1974) are used, and the governing equations are:

Continuity equation:

$$\frac{\partial \rho}{\partial t} + \frac{\partial (\rho u_i)}{\partial x_i} = 0 \tag{1}$$

Momentum equation:

$$\frac{\partial (\rho u_i)}{\partial t} + \frac{\partial (\rho u_i u_j)}{\partial x_j} = -\frac{\partial p}{\partial x_i} + \frac{\partial}{\partial x_j} \left[(\mu + \mu_t) \left(\frac{\partial u_k}{\partial x_j} + \frac{\partial u_j}{\partial x_k} \right) \right] + \rho g_i \tag{2}$$

$$\frac{\partial (\rho k)}{\partial t} + \frac{\partial (\rho k u_i)}{\partial x_i} = \frac{\partial}{\partial x_i} \left[\left(\mu + \frac{\mu_t}{\sigma_k} \right) \frac{\partial k}{\partial x_i} \right] + G_k - \rho \varepsilon \tag{3}$$

$$\frac{\partial (\rho \varepsilon)}{\partial t} + \frac{\partial (\rho \varepsilon u_i)}{\partial x_i} = \frac{\partial}{\partial x_i} \left[\left(\mu + \frac{\mu_t}{\sigma_\varepsilon} \right) \frac{\partial \varepsilon}{\partial x_i} \right] + c_1 G_k \frac{\varepsilon}{k} - c_2 \rho \frac{\varepsilon^2}{k} \tag{4}$$

In which, $\mu_t = \rho c_\mu \frac{k^2}{\varepsilon}$, $c_\mu = 0.085$, $c_2 = 1.68$, $\sigma_k = \sigma_\varepsilon = 0.7179$, $c_2 = 1.42 - \frac{\eta(1-\frac{\eta}{\eta_0})}{1+\beta\eta^3}$, $\eta = \frac{Sk}{\varepsilon}$, $S = \sqrt{2S_{ij}S_{ij}}$, $\eta_0 = 4.38$ and $\beta = 0.015$.

2.2 Calculation method and engineering parameters

Using the PISO algorithm in Fluent14.0 software for pressure-velocity coupling, the unsteady incompressible RANS equation is directly solved. The time step is 0.001 s, and the convergence criteria are k and ε, both of which are lower than 1×10^{-5}.

The downstream weir height of a gravity dam is 53.715 m, the front length is 13.796 m, the weir crest head is 2.185 m, and the maximum unit width flow is 22.049 m²/s. The overflow dam section adopts the WES profile. The downstream straight section slope is 0.75, the step height is 0.9 m, and the width is 0.68 m. With 38 stages in total, the length along the dam surface is 41.552 m. The flip bucket is continuous, the ogee radius is 11 m, the projection angle is 22°, and the flip bucket height is 11.7 m.

2.3 Computational domain and grids

The calculation area starts from the weir crest, passes through the downstream overflow dam surface, and reaches 68 m downstream of the flip bucket. The height of the overflow dam section is 3.994 m, which is the same as that of the guide wall, and the height of the downstream area is 21.233 m above the riverbed. The structural grid is used for grid division. The closer the dam surface is, the denser the grids are. The calculation domain and grids are shown in Figure 1.

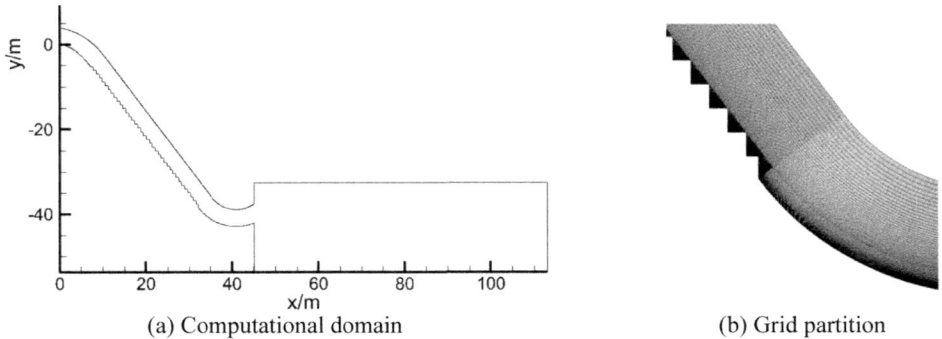

(a) Computational domain   (b) Grid partition

Figure 1.   Computational domain and grids.

3   ANALYSIS OF CALCULATION RESULTS

3.1   *Liquid phase on the overflow dam face*

Figure 2 shows the distribution of the liquid phase at the junction of the beginning of the ogee section at different time.

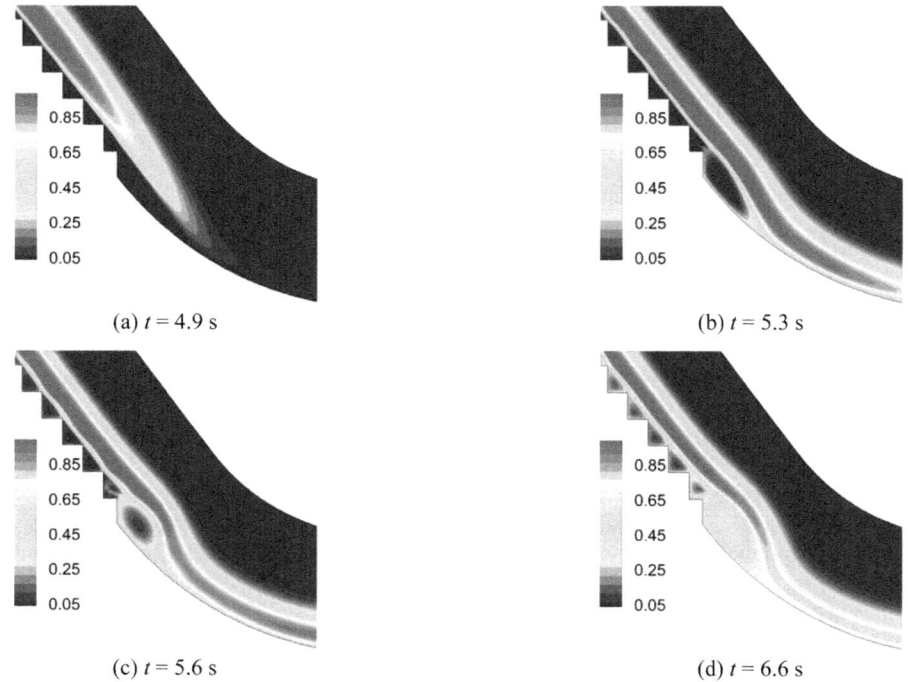

(a) $t = 4.9$ s   (b) $t = 5.3$ s

(c) $t = 5.6$ s   (d) $t = 6.6$ s

Figure 2.   Distribution of liquid phase on overflow dam surface.

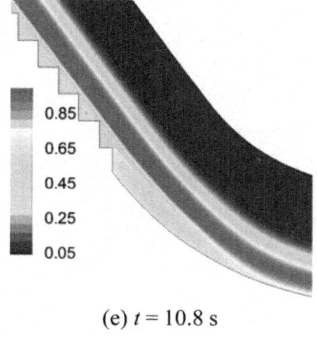

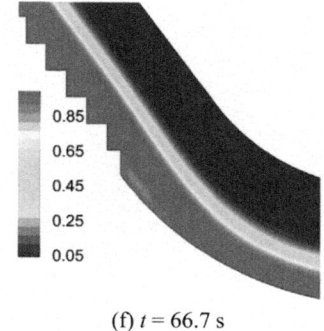

(e) $t = 10.8$ s
(f) $t = 66.7$ s

Figure 2.　Continued.

　It can be seen from Figure 2 that when $t = 4.9$ s, the discharged water reaches the beginning of the ogee section for the first time. The lower edge of the water nappe is not attached to the overflow dam surface, which is almost linear, indicating that the stepped surface does not give full play to its energy dissipation function at this time, and the flow pattern is similar to that of the smooth overflow dam surface; when $t = 5.3$ s, the end of the water nappe is led to the downstream by the ogee section, and a cavity is formed at the concave corner of the step, indicating that the water flow speed is fast at this time. The air between the water nappe and the overflow dam surface cannot be discharged in time. The cavity formed at the concave corner of the last step is large, and some water vapor flows upstream to the penultimate step in a counterclockwise direction, dividing the concave corner of the penultimate step into two cavities; when $t = 5.6$ s, the cavity formed at the concave corner of the last step continues to increase, and the liquid phase increases. At this time, the water depth increases, and the countercurrent water vapor continues to flow upstream in a clockwise direction; when $t = 6.6$ s, the height of the cavity formed at the concave corner of the last step increases to the maximum and begins to decrease gradually, but the length begins to increase, from "fat" to "thin". The two cavities at the concave corner of the penultimate step are basically replaced by the main cavity far away from the concave corner, and the rest of the steps form a cavity at the concave corner of the step, and the liquid phase in it is increasing; when $t = 10.8$ s, the length of the cavity formed at the concave corner of the last step increases to the maximum and begins to decrease gradually. And the liquid phase in the remaining cavities of the other steps continues to increase; when $t = 66.7$ s, the cavity at the concave corner of the penultimate step is filled with the liquid phase, and the cavities at the concave corner of steps 19 ($t = 56.2$s), 20 ($t = 57.3$s), 21 ($t = 58.5$s), 22 ($t = 59.6$s), 23 ($t = 60.7$s), 24 ($t = 61.7$s), 25 ($t = 62.7$s), 26 ($t = 63.7$s), 27 ($t = 64.6$s), 28 ($t = 65.5$s), and 29 ($t = 66.4$s) are filled with the liquid phase successively. Then. the cavities at the concave corners of steps 30 ($t = 67.3$s), 31 ($t = 68.1$s), 32 ($t = 68.9$s), 33 ($t = 70.0$s), 36 ($t = 70.4$s), 34 ($t = 70.6$s), and 35($t = 71.1$s) are successively filled with the liquid phase, and the cavities at the concave corners of the last step are filled with the liquid phase at $t = 99.1$s.

3.2　Pressure on the overflow dam face

Figure 3 shows the pressure distribution at the beginning of the ogee section at different time.
　It can be seen from Figure 3 that when $t = 4.9$ s, there is only a weak high pressure area at the convex corner of the step surface and the first contact point between the water nappe and the ogee section, and other areas are basically atmospheric pressure areas; when $t = 5.3$ s, there is a strong negative pressure area at the concave corner of the last step, a strong high pressure area at the downstream ogee section, and a weak high pressure area at the concave corner of the penultimate step; when $t = 5.6$ s, the height of the strong negative pressure area at the concave corner of the last step increases, with its strong high pressure area of the downstream ogee section extending upstream, and the pressure value of the weak high pressure area at the concave corner

451

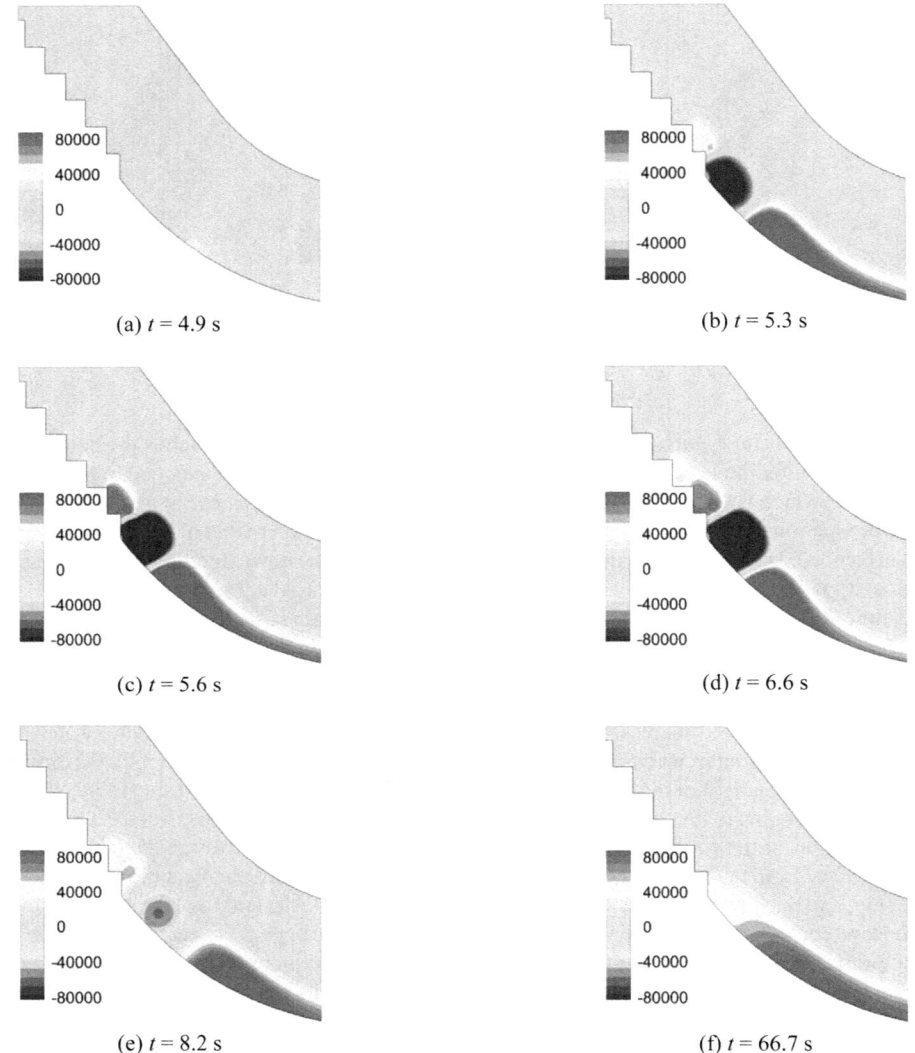

(a) $t = 4.9$ s (b) $t = 5.3$ s

(c) $t = 5.6$ s (d) $t = 6.6$ s

(e) $t = 8.2$ s (f) $t = 66.7$ s

Figure 3.   Pressure distribution on the overflow dam surface (Pa).

of the penultimate step increases; when $t = 6.6$ s, the height of the strong negative pressure area at the concave corner of the last step continues to increase to the maximum and begins to decrease gradually, and the pressure value of the weak high pressure area at the concave corner of the penultimate step begins to decrease and transfers to the upper step, but its value begins to decrease; when $t = 8.2$ s, the pressure value of the strong negative pressure area at the concave corner of the last step decreases continuously, the pressure value of the weak high pressure area at the concave corner of the penultimate step decreases continuously and transfers to the downstream, and the strong high pressure area in the downstream ogee section extends upstream; when $t = 66.7$ s, there is a weak high-pressure area at the concave corner of the last step, a weak negative pressure area at the concave corner of the penultimate step, and a small range of weak high-pressure areas at the convex corners of the other steps. After that, there is no obvious change in the pressure distribution.

3.3 Streamline of the overflow dam face

The streamline distribution at the beginning of the ogee section at different time is displayed in Figure 4.

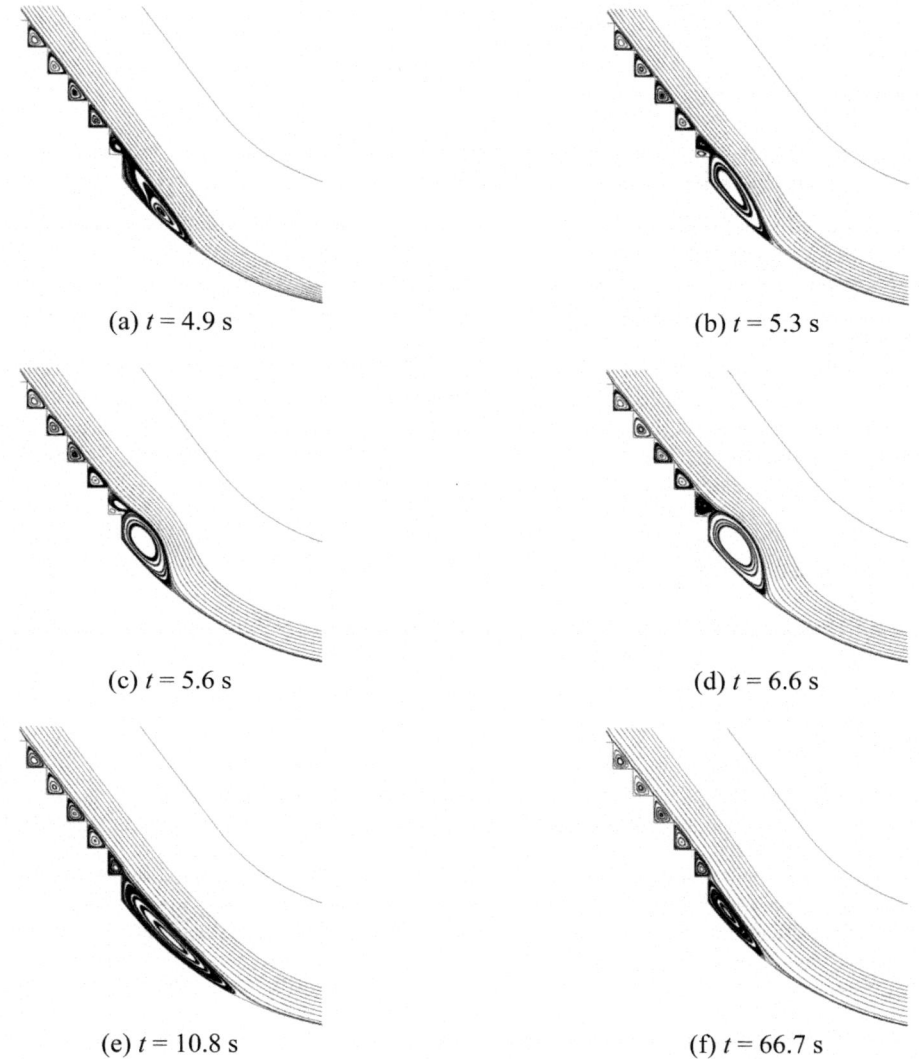

(a) $t = 4.9$ s (b) $t = 5.3$ s

(c) $t = 5.6$ s (d) $t = 6.6$ s

(e) $t = 10.8$ s (f) $t = 66.7$ s

Figure 4.   Distribution of flow lines on the overflow dam surface.

It can be seen from Figure 4 that when $t = 4.9$ s, the vortex on the last step is large, there is a certain space between the vortex on the penultimate step and the concave corner, and there is a single vortex on the other steps; when $t = 5.3$ s, the vortex height on the last step increases, and the vortex on the penultimate step is divided into two parts; when $t = 5.6$ s, the height of the vortex on the last step continues to increase, the length decreases, the vortex streamlines near the concave corner on the penultimate step become sparse, and the flow velocity slows down;

When $t = 6.6$ s, the height of the vortex on the last step increases to the maximum and begins to decrease gradually, but the length begins to increase, from "tall and fat" to "short and thin", and the vortex away from the concave corner on the penultimate step basically occupies the step surface;

when $t = 10.8$ s, the vortex length on the last step increases to the maximum and begins to decrease gradually, the vortex streamline on the penultimate step becomes sparse, and the flow velocity slows down; when $t = 66.7$ s, the length and height of the vortex on the last step are reduced to the minimum. The closer the streamline is to the downstream, the denser it is, and the faster the velocity is. After that, the pressure distribution has no obvious change.

4 CONCLUSION

Through the unsteady numerical simulation analysis of the flow field at the downstream junction of the stepped overflow dam face, it is concluded that:

(1) The lower edge of the discharge nappe is not attached to the overflow dam surface at the initial stage but forms a cavity at the concave corner of the step, with a linear lower edge, the most-violent one of which is the cavity change of the last step, with a countercurrent along the step surface. The penultimate step is filled with faster liquid, while the last step is filled with slower liquid.
(2) Before the discharge flow is stable, there is a strong negative pressure area at the concave corner of the last step. After the discharge flow is stable, there is a weak negative pressure area at the concave corner of the penultimate step.
(3) The streamline forms a vortex structure on each step. The closer the streamline is to the downstream, the denser it is, and the faster the flow velocity is.

ACKNOWLEDGEMENT

This research is supported by the Yunnan Provincial Department of Education Science Research Fund Project.

REFERENCES

Chen W.X., Yang J.R., Dong L.Y., et al. Effect of the number of raised steps on the hydraulic characteristics of integrated energy dissipater[J]. *Journal of Drainage and Irrigation Machinery Engineering* (JDIME), 2019, 37(8): 686–691, 736.
Hirt, Cyril W., and Billy D. Nichols. "Volume of Fluid (VOF) method for the dynamics of free boundaries." *Journal of Computational Physics* 39.1 (1981): 201–225.
Launder B.E. Spalding D.B. The numerical computation of turbulent flows[J]. *Computer Methods in Applied Mechanics and Engineering*, 1974, 3: 269.
Lin J.Y., Wang G.L. *Hydraulic Structures* (5th Edition) [M] Beijing: China Water Resources and Hydropower Press, 2009.
State Key Laboratory of Hydraulics and Mountain River Development and Protection, *Sichuan University Hydraulics Volume I* (5th Edition) [M] Beijing: Higher Education Press, 2016.
Tang J.Q., Yang J.R. The influence of the combined action of aerator and transition steps on the aeration characteristics of stepped surface[J]. *China Rural Water and Hydropower*, 2020(03): 98–104.
Wang Q., Yang J.R., Zhang Z.Y., et al. Impact of transition ladder size on energy dissipation of integrated energy dissipater[J]. *Water Power*, 2017, 43(07): 46–52.

*Frontiers in Civil and Hydraulic Engineering – Mohamed A. Ismail and
Hazem Samih Mohamed (Eds)*
© 2023 The Authors, ISBN 978-1-032-38247-0

Sensitivity analysis of Duncan Chang *E-B* model parameters to the dam deformation in different construction periods based on the single-factor analysis method

Lingyun Yang*, Yupeng Chang* & Weiwei Li*
Power China Northwest Engineering Corporation Limited, China

ABSTRACT: Based on the single-factor analysis method, the change rate of indexes can be positive or negative, and the change range of each parameter of the model is also different, which makes its quantitative evaluation less intuitive and limited in the traditional process of evaluating the sensitivity of Duncan-Chang *E-B* model parameters. For the above reasons, a relative change rate that allows the calculation results of different parameters and different ranges of changes to be directly compared is proposed, and this index is used to analyze and study the sensitivity of the dam's material parameters of the Duncan-Chang *E-B* model during the filling and impounding period. Research shows that during filling and impounding periods, the sensitivity of each parameter of the Duncan-Chang *E-B* model to the deformation is different, and the sensitivity of Duncan-Chang *E-B* model parameters is not only related to the index but also closely related to the reference value, change range, and direction of parameters.

1   INTRODUCTION

Compared with the elastic-plastic model, the nonlinear elastic model—Duncan-Chang E-B model—has the advantages of a clear concept and simple parameter measurement, so it is widely used in the stress and deformation analysis of concrete-faced rockfill dams. In order to correctly grasp the influence of the seven parameters of the Duncan Chang E-B model (the coarse-grained soil friction C is 0) on the calculated value of the dam stress and deformation, many scholars have done a lot of research on the parameter sensitivity of the Duncan Chang E-B model. He (2002) and Ngoko (1999) studied the corresponding change rate of calculation indexes (stress or deformation) when the eight parameters of the Duncan-Chang E-V model changed within a certain range, and evaluated the sensitivity of the parameters of the Duncan-Chang E-V model through the magnitude of the change rate. Yang (2013) studied the influence of the sample space size and sample point location on the sensitivity of Duncan Chang E-B model parameters. Zhang (2008) analyzed the sensitivity of Duncan Chang E-V model parameters by comparing the change rate of corresponding indexes (stress or deformation) caused by the same parameter change rate. Yin (2004), Li (2013), Yan (2010), and Li (2019) studied the sensitivity of Duncan Chang E-B model parameters based on the orthogonal experimental method. Although many scholars have conducted many studies on the sensitivity of Duncan Chang model parameters, the single factor method is still one of the commonly used methods in the research process. The single-factor analysis method usually selects the index value and changes one of the parameters while other parameters remain unchanged at the same time. The sensitivity of each parameter is reflected by comparing the relation curve between the benchmark index value and the parameter change, that is, the sensitivity can be evaluated through the change rate. However, the change rate can be positive or negative, and the change

---

*Corresponding Authors: 478666138@qq.com; 2018126092@chd.edu.cn and 81081462@qq.com

range of each parameter is also different, which makes the quantitative evaluation of the change rate less intuitive and have great limitations during the sensitivity analysis of Duncan Chang E-B model parameters.

Therefore, based on the change rate of traditional indexes, this paper standardizes the parameter change, puts forward the relative change rate that enables the calculation results of different parameters and different change ranges to be directly compared, and applies this index to study the sensitivity of Duncan-Chang E-B model parameters. Generally speaking, the higher the relative change rate is, the stronger the sensitivity of the parameters is; the lower the relative change rate is, the weaker the parameter sensitivity is. The specific calculation process is as follows:

$$\triangle \delta = \frac{\vartheta_i - \vartheta_0}{\vartheta_0} \times 100\% \tag{1}$$

where $\triangle \delta$ represents the index change rate, ϑ_i represents the index value of the current scheme, and ϑ_0 represents the index value of the benchmark scheme.

$$\triangle \delta R = |\frac{\triangle \delta}{\omega_i - \omega_0} \times \omega_0| \times 100\% \tag{2}$$

where $\triangle \delta_R$ represents the relative change rate of the index, ω_i represents the current parameter value, and ω_0 represents the benchmark value of the parameter.

2 ESTANLISHMENT OF SENSITIVITY ANALYSIS AND CALCULATION MODEL FOR DUNCAN CHANG E-B MODEL PARAMENTS

2.1 *Project profile*

The concrete-faced rockfill dam of Liushugou Hydropower Station has a crest elevation of 1499.00 m, a toe slab foundation elevation of 1397.00 m in the riverbed section, and a dam height of 102 m. The two-dimensional finite element model is established with Autobank software, as shown in Figure 1. The x-axis is horizontally along the river, with the dam axis as the origin and pointing to the downstream as positive; the z-axis is vertical, with the elevation as the coordinate and pointing to the upstream as positive. The x-direction boundary is a unidirectional constraint, and the bottom is a bidirectional constraint. The model adopts graded loading to simulate the construction process and water storage process of the dam body, such as layered rolling and filling, pouring of slabs, etc., which is divided into 28 levels, and each load level is completed at one time. The number of elements in this model is 861, the number of nodes is 898, and the number of boundaries is 147.

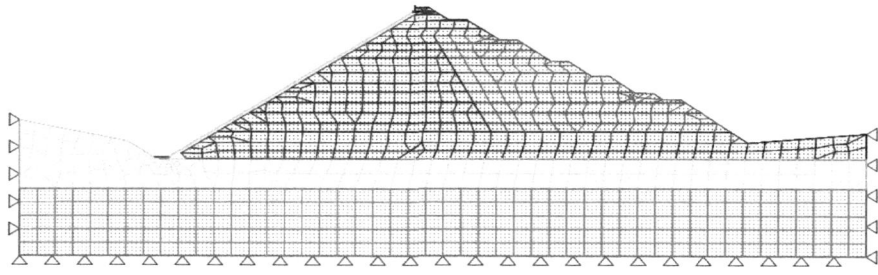

Figure 1. Two-dimensional finite element model of concrete-faced Rockfill Dam of Liushugou hydropower station (Dam Transverse 0+76.550).

2.2 Model parameters

The material zoning of the dam body mainly includes the cushion area, the upstream main rockfill area, and the downstream secondary rockfill area. The material zoning of the dam foundation mainly includes the overburden and weakly weathered rock mass. The Duncan-Chang E-B model is used for the rockfill material of the dam body and the overburden of the dam foundation. The model parameters are shown in the following table:

Table 1. Duncan Chang E-B model parameters of dam materials.

Material	γ (kN·m^{-3})	$\varphi_0(°)$	K	n	R_f	K_b	m	$\Delta\varphi$ (°)
Bedding	23.4	49.6	1000	0.44	0.87	583	0.34	8.3
Main rockfill	22.5	55.2	1100	0.257	0.68	709	0.32	9.7
Secondary rockfill	22.2	53.4	1050	0.22	0.64	668	0.33	8.8
Overburden	23.7	48.5	800	0.44	0.84	414	0.37	7.2

2.3 Calculation scheme design

In order to simplify the experiment process and ensure the safety of the dam body, this paper assumes that the construction material of the concrete-faced rockfill dam of Liushugou Hydropower Station is a new type of material.

The overburden with the lowest model parameter value is selected as the benchmark, and the 7 parameters are changed by −10% or −2°, 5% or 1°, and 10% or 2°, respectively, based on the benchmark value. The single-factor analysis method is adopted, and only one parameter is changed in each experiment, and the other parameters remain unchanged. The parameter values and specific plans are shown in Table 2.

The dam's vertical displacement V, horizontal upstream displacement H_u, and horizontal downstream displacement H_d are selected as test indexes for the sensitivity analysis.

Table 2. Calculation scheme.

Scheme	$\varphi_0(°)$	$\Delta\varphi(°)$	R_f	K	n	K_b	m
1	48.5	7.2	0.84	800	0.44	414	0.37
2	46.5	7.2	0.84	800	0.44	414	0.37
3	49.5	7.2	0.84	800	0.44	414	0.37
4	50.5	7.2	0.84	800	0.44	414	0.37
5	48.5	6.2	0.84	800	0.44	414	0.37
6	48.5	8.2	0.84	800	0.44	414	0.37
7	48.5	9.2	0.84	800	0.44	414	0.37
8	48.5	7.2	0.756	800	0.44	414	0.37
9	48.5	7.2	0.882	800	0.44	414	0.37
10	48.5	7.2	0.924	800	0.44	414	0.37
11	48.5	7.2	0.84	720	0.44	414	0.37
12	48.5	7.2	0.84	840	0.44	414	0.37
13	48.5	7.2	0.84	880	0.44	414	0.37
14	48.5	7.2	0.84	800	0.396	414	0.37
15	48.5	7.2	0.84	800	0.462	414	0.37
16	48.5	7.2	0.84	800	0.484	414	0.37
17	48.5	7.2	0.84	800	0.44	372.6	0.37
18	48.5	7.2	0.84	800	0.44	434.7	0.37
19	48.5	7.2	0.84	800	0.44	455.4	0.37
20	48.5	7.2	0.84	800	0.44	414	0.333
21	48.5	7.2	0.84	800	0.44	414	0.3885
22	48.5	7.2	0.84	800	0.44	414	0.407

3 SESITIVITY ANALYSIS OF DUNCAN CHANG E-B MODEL PARAMETERS TO DAM DEFORMATION DURING FILLING AND IMPOUNDING PERIODS

3.1 *Sensitivity analysis of parameters to vertical displacement V*

The sensitivity analysis results of each parameter to THE vertical displacement are shown in Table 3. It can be seen from this table that the relative change rate of the parameter φ_0 to the vertical displacement V during the filling period is the largest, reaching 90.35–112.03%, while the relative change rate of the parameter m to the vertical displacement is the smallest, only being 8.94–11.92%, so the sensitivity of the parameter φ_0 to the vertical displacement V is the strongest. The sensitivity of the parameter m is the lowest. According to the evaluation of the relative change rate, the sensitivity of each parameter to the vertical displacement V during the filling period is in the descending order of φ_0, K_b, R_f, n, K, $\triangle\varphi$, and m. During the impounding period, the relative change rate of the parameter φ_0 to the vertical displacement V is the largest, reaching 94.48–111.98%, while the relative change rate of the parameter $\triangle\varphi$ to the vertical displacement V is the smallest, only being 9.87–10.91%. Therefore, the sensitivity of the parameter f_0 to the vertical displacement V is the strongest, while the sensitivity of the parameter $\triangle\varphi$ is the lowest. According to the evaluation of the relative change rate, the sensitivity of each parameter to the vertical displacement V in the impounding period is in the descending order of φ_0, K_b, R_f, n, K, m, and $\triangle\varphi$. By comparing the sensitivity of each parameter to THE vertical displacement V during the filling period and impounding period, it can be seen that the regularity of the sensitivity of each parameter to the vertical displacement in the two periods are basically the same. Although the sensitivity of $\triangle\varphi$ and m is slightly different, the difference in the relative change rate of the two parameters to the vertical displacement V in the two periods is almost negligible. At the same time, it can be seen from the relative change rate of the two periods that the sensitivity of each parameter to the vertical displacement V differs little.

Table 3. Calculation results of vertical displacement V.

Scheme			Filling period			Impounding period		
			Calculated value	Change rate	Relative change rate	Calculated value	Change rate	Relative change rate
Original value		1	−0.671	0.00%	0.00%	−0.693	0.00%	0.00%
φ_0	−2°	2	−0.702	4.62%	112.03%	−0.725	4.62%	111.98%
	+1°	3	−0.658	−1.94%	93.96%	−0.679	−2.02%	97.98%
	+2°	4	−0.646	−3.73%	90.35%	−0.666	−3.90%	94.48%
$\triangle\varphi$	−2°	5	−0.653	−2.68%	9.66%	−0.674	−2.74%	9.87%
	+1°	6	−0.682	1.64%	11.80%	−0.703	1.44%	10.39%
	+2°	7	−0.693	3.28%	11.80%	−0.714	3.03%	10.91%
R_f	−10%	8	−0.649	−3.28%	32.79%	−0.669	−3.46%	34.63%
	+5%	9	−0.683	1.79%	35.77%	−0.705	1.73%	34.63%
	+10%	10	−0.696	3.73%	37.26%	−0.718	3.61%	36.08%
K	−10%	11	−0.682	1.64%	16.39%	−0.703	1.44%	14.43%
	+5%	12	−0.667	−0.60%	11.92%	−0.688	−0.72%	14.43%
	+10%	13	−0.664	−1.04%	10.43%	−0.685	−1.15%	11.54%

(continued)

Table 3. Continued.

	Scheme		Filling period			Impounding period		
			Calculated value	Change rate	Relative change rate	Calculated value	Change rate	Relative change rate
N	−10%	14	−0.688	2.53%	25.34%	−0.709	2.31%	23.09%
	+5%	15	−0.663	−1.19%	23.85%	−0.685	−1.15%	23.09%
	+10%	16	−0.656	−2.24%	22.35%	−0.678	−2.16%	21.65%
K_b	−10%	17	−0.735	9.54%	95.38%	−0.759	9.52%	95.24%
	+5%	18	−0.644	−4.02%	80.48%	−0.665	−4.04%	80.81%
	+10%	19	−0.620	−7.60%	76.01%	−0.639	−7.79%	77.92%
M	−10%	20	−0.679	1.19%	11.92%	−0.701	1.15%	11.54%
	+5%	21	−0.668	−0.45%	8.94%	−0.689	−0.58%	11.54%
	+10%	22	−0.665	−0.89%	8.94%	−0.686	−1.01%	10.10%

3.2 *Sensitivity analysis of parameters to horizontal upstream displacement H_u*

The sensitivity analysis results of each parameter to the horizontal upstream displacement H_u are shown in Table 4. From this table, it can be seen that during the filling period, the relative change rate of the parameter φ_0 to the horizontal upstream displacement H_u is the largest, reaching 362.07–505.21%, while the relative change rate of the parameter K_b to the horizontal upstream displacement H_u is the smallest, reaching 6.94–17.36%. Therefore, the parameter φ_0 has the strongest sensitivity to the horizontal upstream displacement H_u, while the parameter K_b has the lowest sensitivity. According to the evaluation of the relative change rate, it is concluded that the sensitivity of each parameter in the filling period to the horizontal upstream displacement H_u from large to small is φ_0, R_f, K, n, $\triangle\varphi$, m, and K_b. During the impounding period, the relative change rate of the sensitivity of the parameter φ_0 to the horizontal upstream displacement H_u is the largest, reaching 606.25 is the largest 894.22%, while the relative change rate of the sensitivity of the parameter K_b to the horizontal upstream displacement H_u is the smallest, reaching 25.00–37.50%. Therefore, the parameter φ_0 has the strongest sensitivity to the horizontal upstream displacement H_u, while the parameter K_b has the lowest sensitivity. According to the evaluation of the relative change rate, it is concluded that the sensitivity of each parameter in the impounding period to the horizontal upstream displacement H_u from large to small is φ_0, R_f, K, m, $\triangle\varphi$, n, and K_b.Comparing the sensitivity of each parameter to the horizontal upstream displacement H_u during the filling period and impounding period, it can be seen that φ_0, R_f, K, and K_b have almost the same regularity of sensitivity to the horizontal upstream displacement H_u in two periods. while the sensitivity of n, $\triangle\varphi$, and m is slightly different, but the relative change rate of the three parameters to the horizontal upstream displacement H_u during the two periods is not significantly different. At the same time, the conclusions that f_0, R_f, and K have the greatest impact on the index H_u are consistent with those in literature. The difference is that the sensitivity of K to the index H_u is higher than that of R_f in literature. It is inferred that the reason for this difference is that the reference value and variation range of the parameter K are different. In this paper, the reference value of the parameter K is high ($K = 1250$) and the variation range is large (33%). The influence of the parameter K is more obvious at this level. It is noteworthy that the relative change rate of each parameter to the horizontal upstream displacement H_u during the impounding period is higher than that during the filling period, indicating that parameters are more sensitive to the horizontal upstream displacement H_u during the impounding period. This may be related to the action of the water load during the impounding process. After impounding, the displacement of the dam body decreases upstream and increases downstream. The decrease in the datum value makes the change in parameters more sensitive to the index.

Table 4. Calculation results of horizontal upstream displacement H_u.

Scheme			Filling period			Impounding period		
			Calculated value	Change rate	Relative change rate	Calculated value	Change rate	Relative change rate
Original value		1	−0.288	0.00%	0.00%	−0.160	0.00%	0.00%
φ_0	−2°	2	−0.348	20.83%	505.21%	−0.219	36.88%	894.22%
	+1°	3	−0.265	−7.99%	387.33%	−0.138	−13.75%	666.88%
	+°	4	−0.245	−14.93%	362.07%	−0.12	−25.00%	606.25%
$\triangle\varphi$	−2°	5	−0.266	−7.64%	27.50%	−0.138	−13.75%	49.50%
	+1°	6	−0.301	4.51%	32.50%	−0.172	7.50%	54.00%
	+2°	7	−0.315	9.38%	33.75%	−0.185	15.63%	56.25%
R_f	−10%	8	−0.25	−13.19%	131.94%	−0.124	−22.50%	225.00%
	+5%	9	−0.31	7.64%	152.78%	−0.182	13.75%	275.00%
	+10%	10	−0.334	15.97%	159.72%	−0.206	28.75%	287.50%
K	−10%	11	−0.317	10.07%	100.69%	−0.189	18.13%	181.25%
	+5%	12	−0.276	−4.17%	83.33%	−0.148	−7.50%	150.00%
	+10%	13	−0.265	−7.99%	79.86%	−0.139	−13.13%	131.25%
N	−10%	14	−0.299	3.82%	38.19%	−0.167	4.38%	43.75%
	+5%	15	−0.283	−1.74%	34.72%	−0.156	−2.50%	50.00%
	+10%	16	−0.279	−3.12%	31.25%	−0.153	−4.38%	43.75%
K_b	−10%	17	−0.293	1.74%	17.36%	−0.156	−2.50%	25.00%
	+5%	18	−0.287	−0.35%	6.94%	−0.163	1.88%	37.50%
	+10%	19	−0.285	−1.04%	10.42%	−0.165	3.13%	31.25%
M	−10%	20	−0.284	−1.39%	13.89%	−0.151	−5.63%	56.25%
	+5%	21	−0.291	1.04%	20.83%	−0.164	2.50%	50.00%
	+10%	22	−0.293	1.74%	17.36%	−0.168	5.00%	50.00%

3.3 Sensitivity analysis of parameters to horizontal downstream displacement H_d

The relative change rate of each parameter to the horizontal downstream displacement H_d is shown in Table 5. It can be seen from this table that during the filling period, the relative change rate of the parameter φ_0 to the horizontal downstream displacement H_d is the largest, reaching 269.44–307.94%, while the relative change rate of parameters K_b and m to the horizontal downstream displacement H_d is the smallest, reaching 15.87%. So, the parameter φ_0 is the most sensitive to the horizontal downstream displacement H_d. According to the evaluation of the relative change rate, it is concluded that the sensitivity of each parameter in the filling period to the horizontal downstream displacement H_d from large to small is φ_0, R_f, K, n, $\triangle\varphi$, m, and K_b. During the impounding period, the relative change rate of the parameter φ_0 to the horizontal downstream displacement H_d is the largest, reaching 199.71–228.24%, while the relative change rate of parameters K_b and m to the horizontal downstream displacement H_d is the smallest, reaching 0–5.88%. Therefore, the parameter φ_0 has the strongest sensitivity to the horizontal downstream displacement H_d, while parameters K_b and m have the lowest sensitivity. According to the evaluation of the relative change rate, it is concluded that the sensitivity of each parameter in the impounding period to the horizontal downstream displacement H_d from large to small is φ_0, R_f, K, n, $\triangle\varphi$, m, and K_b. Comparing the sensitivity of each parameter to THE horizontal downstream displacement H_d during the filling period and impounding period, it can be seen that the regularity of sensitivity of parameters to the horizontal downstream displacement H_d in the two periods is consistent. The difference is that the relative variation rate of the variation of parameters to the horizontal downstream displacement H_d during the impounding period is small.

Table 5. Calculation results of horizontal downstream displacement H_d.

Scheme			Filling period			Impounding period		
			Calculated value	Change rate	Relative change rate	Calculated value	Change rate	Relative change rate
Original value		1	0.126	0.00%	0.00%	0.170	0.00%	0.00%
φ_0	$-2°$	2	0.142	12.70%	307.94%	0.186	9.41%	228.24%
	$+1°$	3	0.119	−5.56%	269.44%	0.163	−4.12%	199.71%
	$+2°$	4	0.112	−11.11%	269.44%	0.155	−8.82%	213.97%
$\Delta\varphi$	$-2°$	5	0.118	−6.35%	22.86%	0.161	−5.29%	19.06%
	$+1°$	6	0.13	3.17%	22.86%	0.175	2.94%	21.18%
	$+2°$	7	0.135	7.14%	25.71%	0.18	5.88%	21.18%
R_f	-10%	8	0.114	−9.52%	95.24%	0.156	−8.24%	82.35%
	$+5\%$	9	0.132	4.76%	95.24%	0.177	4.12%	82.35%
	$+10\%$	10	0.139	10.32%	103.17%	0.183	7.65%	76.47%
K	-10%	11	0.14	11.11%	111.11%	0.186	9.41%	94.12%
	$+5\%$	12	0.12	−4.76%	95.24%	0.164	−3.53%	70.59%
	$+10\%$	13	0.115	−8.73%	87.30%	0.158	−7.06%	70.59%
n	-10%	14	0.133	5.56%	55.56%	0.178	4.71%	47.06%
	$+5\%$	15	0.123	−2.38%	47.62%	0.166	−2.35%	47.06%
	$+10\%$	16	0.12	−4.76%	47.62%	0.163	−4.12%	41.18%
K_b	-10%	17	0.124	−1.59%	15.87%	0.171	0.59%	5.88%
	$+5\%$	18	0.127	0.79%	15.87%	0.170	0.00%	0.00%
	$+10\%$	19	0.128	1.59%	15.87%	0.171	0.59%	5.88%
m	-10%	20	0.124	−1.59%	15.87%	0.169	−0.59%	5.88%
	$+5\%$	21	0.127	0.79%	15.87%	0.170	0.00%	0.00%
	$+10\%$	22	0.128	1.59%	15.87%	0.171	0.59%	5.88%

4 CONCLUSION

In this paper, a method to evaluate the parameter sensitivity of the Duncan Chang E-B model by using the relative change rate index is proposed. Based on this method, the parameter sensitivity of the Duncan Chang E-B model during the filling period and impounding period of the concrete face rockfill dam of Liushugou Hydropower Station is studied, and the following main conclusions are obtained:

(1) During the filling period, the sensitivity analysis results of Duncan Chang E-B model parameters to the dam's vertical displacement V, upstream horizontal displacement H_u, and downstream horizontal displacement H_d, are basically consistent with the previous relevant research results, which verifies the reliability of the method in studying the parameter sensitivity of the Duncan Chang E-B model.

(2) During the filling period and impounding period, the relative change rate of φ_0 to the vertical displacement V, horizontal upstream displacement H_u, and horizontal downstream displacement

H_d, is the highest, meaning the greatest sensitivity; the relative change rate of parameters R_f, K, and K_b to the index is moderate, meaning a relatively high sensitivity level; the relative change rate of parameters n, m, and $\triangle\varphi$ to the index is least, meaning a relatively low sensitivity level.

(3) When the main sensitive parameter φ_0, K, and K_b decreases and R_f increases, each index becomes more sensitive. When the main sensitive parameters f_0, K, and K_b increases and R_f decreases, each index has the less sensitivity. Therefore, it is very important to select the change range of each parameter when analyzing the sensitivity of Duncan Chang E-B model parameters.

REFERENCES

He, C.G. & Yang, G.F. (2002). Effects of parameters of Duncan-Chang model on calculated results. *J. Chinese Journal of Geotechnical Engineering*, 24(2), 170–174. (in Chinese)

Li, Y.L., Li, S.Y. & Ding, Z.F. (2013). TU Xing. The sensitivity analysis of Duncan-Chang E-B model parameters based on the orthogonal test method. *J. Journal of Hydraulic Engineering*, 44(7), 873–879. (in Chinese)

Li, Y.L., Zhang, J.H., Zhang Z.W. & Liu, Y.H. (2019). Sensitivity analysis of E-B model parameters in high modulus zone based on orthogonal test method. *J. Advances in Science and Technology of Water Resources*, 39(1), 34–38. (in Chinese)

Ngoko. (1999). Sensitivity analysis of parameters for rockfill dams. *J. Journal of Hohai University*, 27(5), 94–99. (in Chinese)

Wu, C.B. & Yan, Q. (2010) Sensitivity analysis of Duncan E-B model parameters for rockfill. *J. Water Resources and Power*, 28(8), 94–96. (in Chinese)

Yan, Q., Wu, C.B. & Zang, Y. (2010). Analysis of parameter sensitivity of Duncan E-B model based on uniform design. *J. China Rural Water and Hydropower*, (7), 82–85. (in Chinese)

Yang, Y.S., Liu, X.S., Zhao, J.M. & Wang, X.G. (2013). Parameter sensitivity analysis of Duncan E-B model. *J. Journal of China Institute of Water Resources and Hydropower Research*, 11(2), 81–86. (in Chinese)

Yin, R.R. & Zhu, H.H. (2004). Analysis on parameters Sensitivity of Duncan-Chang model. *J. Underground Space*, 24(4), 434–437. (in Chinese)

Zhang, J.Z. & Miao, L.C. (2008). Analysis of parameters sensitivity of Duncan-Chang model and study on the controlling deformation. *J. Industrial Construction*, 38(3), 75–79. (in Chinese)

*Frontiers in Civil and Hydraulic Engineering – Mohamed A. Ismail and
Hazem Samih Mohamed (Eds)*
© 2023 The Authors, ISBN 978-1-032-38247-0

Optimization of geometric parameters of the three-step method for four-lane tunnels with large sections

Min Liu & Jiangling He
Guangdong Province Freeway Co. LTD, Tianhe, Guangdong, China

Tian He*
School of Civil Engineering, Chongqing Jiaotong University, Chongqing, China

Xinghong Jiang
*China Merchants Chongqing Communications Technology Research and Design Institute Co., Ltd.,
Chongqing, China*

ABSTRACT: In this paper, the stability and safety of the surrounding rock can be improved and the advantages of the construction techniques can be brought into full play by optimizing the geometrical parameters of the step method under the condition of the soft surrounding rock. In the case of large section tunnel engineering with single hole excavation width of 22.6 m and excavation height of 14.5 m in the soft surrounding rock, the FLAC 3D simulation step method was adopted for construction, and simulation analysis was carried out for 8 different working conditions of step height and step length, and the stability of tunnel was determined by displacement criterion. The settlement of the arch face, the horizontal displacement of the two sides and the stability of the arch face are used as the basis for optimization selection. According to the research results, the three-step excavation method is adopted for the tunnel under the condition of the soft surrounding rock. Considering the balance of construction space and production efficiency, the recommended step height should be controlled at 1/4 h~1/3 h (H is the height of the tunnel), and the excavation height should be controlled at 3.5 m~4.5 m. It is recommended that the length of the step should be controlled within 4m, and the initial supporting structure of the tunnel should have reasonable stress and convenient operation of construction machinery.

1 INSTRUCTIONS

A soft rock tunnel with a large section is an important type of modern traffic engineering. Because soft rock has the characteristics of weak lithology, low strength and poor self-stability, it often leads to large deformation around the tunnel and poor stability of the face during the construction of soft rock tunnel. Wang Weifu (2017) et al. used MIDAS finite element software to establish the stratigraphic structure model to conduct the tunnel – strain – stress simulation analysis on the stages of the construction system of the surrounding rock to determine the structure of the excavation process of risk, and drawn the step method which satisfies the Shi Yanling large cross-section of shallow bias tunnel construction safety requirements and the applied technology is easy to operate, efficient and quick. According to engineering practice and theoretical analysis, some scholars (Cheng 2020; Kwon 1999; Zhao 2015, 2019) put forward that the three-step method has the best effect on surrounding rock deformation control in the construction of large sections of soft

---

*Corresponding Authors: 623365453@qq.com; hejl0620@126.com; 1647636435@qq.com and jiangxinghong@cmhk.com

surrounding rock tunnels. Therefore, to better control the deformation of soft rock tunnel when the step method is adopted in the construction of soft rock tunnel, rapid and timely support (Liu 2020) and advanced support measures are needed to ensure the stability of the excavation working face.

It is of practical significance to improve the stability and safety of surrounding rock and give full play to the advantages of the construction technology by optimizing the geometrical parameters of the step method. Zhang Yuanyuan (2021), Deng Tao (2017) et al. studied the height and length optimization of excavation steps using the three-step method based on the safety of the initial supporting structure, stability of face and convenience of construction by using numerical simulation combined with on-site monitoring data. Zou Chenglu (2017) and Song Shuguang (2011) used the method of numerical simulation to compare the vault settlement, horizontal convergence and extrusion deformation of the face of the tunnel under different step heights, revealing the evolution law of the mechanical effect of surrounding rock in the construction process of step method, and finally obtained the optimal step height. Jiang Liang (2019), Wang Haijun (2016), Wu Jianlin (2019) et al. used a numerical simulation method to simulate step construction and analyzed the similarities and differences between vault settlement and surface settlement caused by tunnel excavation under conditions of different step lengths. The surface settlement caused by tunnel excavation can be effectively reduced by reasonably reducing the length of step excavation. The optimization analysis of step height and length should also consider the stability analysis of surrounding rock in the process of three-step excavation. Ju Sensen (2020) conducted a numerical simulation of the tunnel excavation process through ANSYS finite element software, and analyzed and evaluated the safety of tunnel construction. Hu Shouyun (2016) and others used the elastoplastic theory and the basic principle of the discrete element method to establish a discrete element program UDEC numerical calculation model which is verified by the measured data of the numerical model. The results show that the three-step method can effectively restrain the surrounding rock of the soft rock tunnels during the construction of large deformation. The vertical displacement is mainly concentrated in the tunnel vaults and arches and the horizontal displacement is mainly concentrated in the tunnel arch feet on both sides.

At present, researchers have carried out optimization analysis on the geometrical parameters of large section tunnel construction by step method from the perspective of controlling structural stress and tunnel deformation, but there is still a lack of research on the optimization analysis of geometrical parameters of large section tunnel construction by the three-step method in the soft surrounding rock. This article mainly aims at single hole excavation width of 22.6 m, the excavation height is about 14.5 m large cross-section tunnel in weak rock engineering, construction of the FLAC 3D simulation steps, and the length of the step height, step 8 different working condition, simulation analysis and with the construction process, through the analysis of the corresponding were arranged to tunnel displacement as the criterion of tunnel stability, The settlement of tunnel vault, horizontal displacement of both sides of the arch waist and stability of the face can provide a reference for the application of step method in the construction of super-large section tunnel in the future.

2 MODEL ESTABLISHMENT AND CALCULATION CONDITIONS

2.1 *Establishment of computational model*

In this study, FLAC3D was used to establish a three-dimensional numerical analysis model according to the actual situation of the tunnel. According to Saint Venant's principle, to reduce the influence of boundary conditions on the model, the distance between the left and right boundaries of the model and the surrounding rock of the tunnel is 5 times the tunnel span. The constraint conditions are as follows: the surface is a free boundary, and the buried depth is 60 m, without any constraint; The left and right boundaries are constrained by the displacement along the X axis, and the bottom boundary of the model is constrained by the displacement along the Y axis. The calculation model is shown in Figure 1.

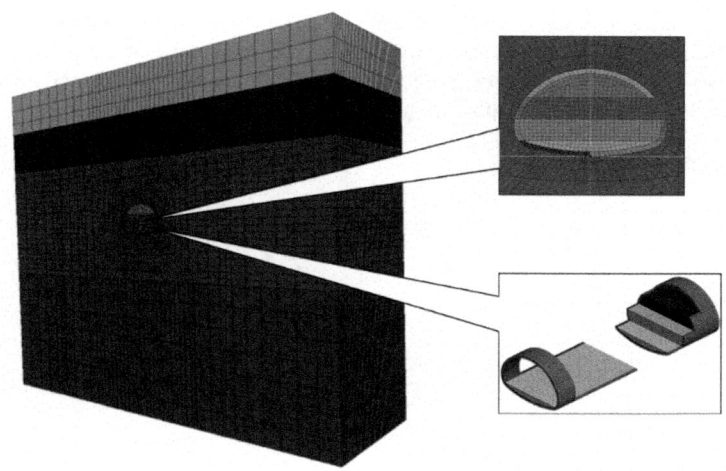

Figure 1.   Calculation model.

2.2 *Calculation parameter determination*

Based on the principle of "adjusting measures to local conditions and selecting rationally", this study conducted a standardized study on the step height and step length of the super-large span four-lane step method of the Shenzhou-Shantou West Tunnel. The research method is still based on finite difference software, using the plane strain numerical calculation method. Solid element is used in surrounding rock, shell element is used in initial support, and the Mocoulomb constitutive model is used in soil. Relevant mechanical parameters are shown in Table 1 and Table 2.

Table 1.   Physical and mechanical indexes of rock mass.

Level	Gravity γ (kN/m³)	Elastic modulus E (GPa)	Poisson's ratio μ	Friction angle $\psi(°)$	force of cohesion c (MPa)
IV	23.5	3.65	0.325	33	0.45

Table 2.   Mechanical parameters of initial support.

Material	Density γ (kN/m³)	Elastic modulus E(GPa)	Poisson's ratio μ
Initial support	22.50	26.94	0.20

3   RESULTS AND ANALYSIS

3.1 *Mechanical response analysis of step height*

The excavation simulation of the tunnel is carried out under four working conditions: 1/5 h, 1/4 h, 1/3 h and 2/5 h (H is the height of the tunnel). The surrounding rock deformation analysis, primary tensile stress analysis and plastic classification are monitored in the simulation process.

(1) Calculated monitoring point

In this study, the displacement criterion (Zhu 2001) was adopted, and stability was judged by vault settlement and horizontal convergence combined with domestic and foreign experience of long-span rock tunnels. The settlement point is set in the vault, and the level measuring point is set on the upper, middle and lower steps respectively. Three settlement monitoring points are set up in the vault, and six horizontal monitoring points are set up in the upper, middle and lower steps. The monitoring points of the tunnel calculation model are shown in Figure 2 below.

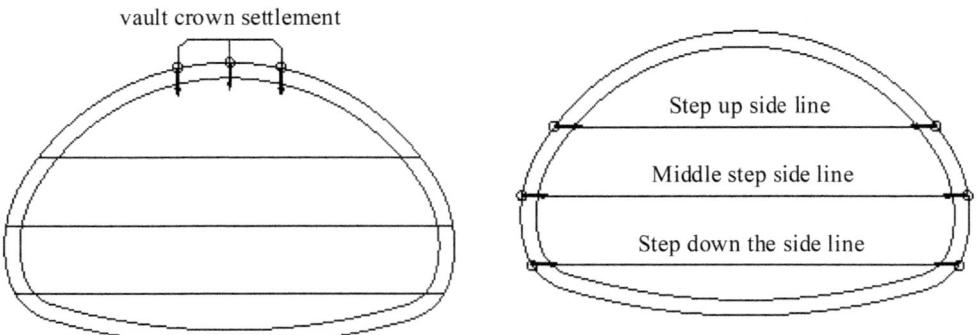

Figure 2. Layout of monitoring points.

(2) The deformation of surrounding rock

At each step height, the cumulative results of tunnel vault settlement and maximum horizontal convergence are shown in Figure 3 and Figure 4. It can be seen from Figure 3 and Figure 4 that the deformation of the tunnel will increase with the increase of step height, and the increasing trend of vault settlement is more obvious than the horizontal convergence.

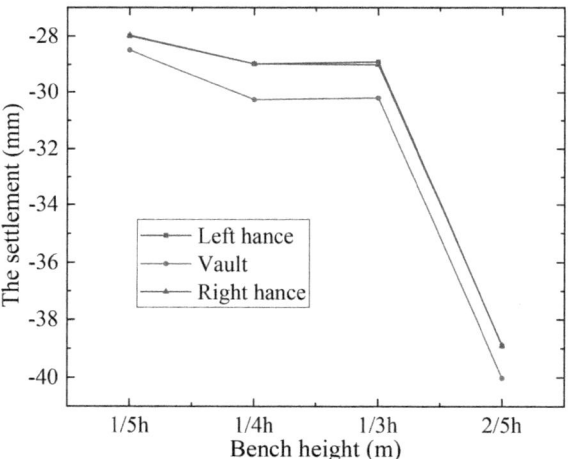

Figure 3. Dome settlement of surrounding rock varies with step.

Figure 3 shows the variation curves of tunnel vault settlement at different step heights. On the whole, the variation of vault settlement is greater with the increase of step height during tunnel excavation. After the completion of tunnel excavation, the subsidence of the vault is 28.50 mm, 30.26 mm, 30.20 mm and 40.02 mm respectively. According to the data, if the step height is 1/3 h as the benchmark and the step height is 2/5 h, the excavation area is too large, and the step height accounts for about half of the excavation height of the tunnel. The settlement of the vault is increased by 32.50% relative to the step height of 1/3 h. The step height has an obvious influence on the settlement of the tunnel. The smaller the step height is, the smaller the settlement of the vault is.

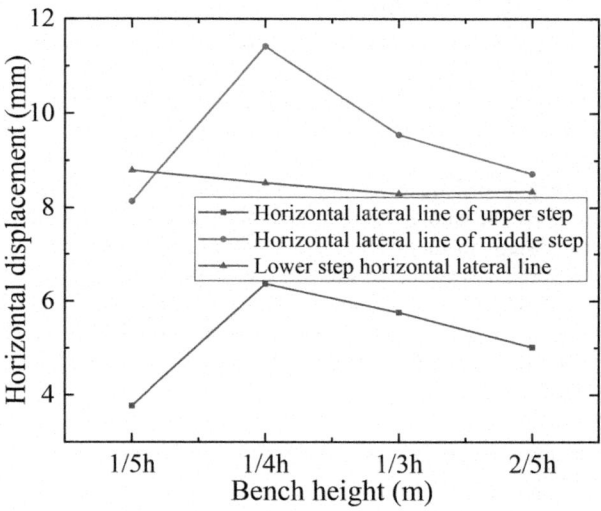

Figure 4.   Change of horizontal displacement of surrounding rock with a step height.

Figure 4 shows the variation curves of horizontal displacement of surrounding rock at different step heights. As can be seen from the figure, the horizontal displacement of the upper step decreases by 12.8%, and that of the lower step increases by 0.48% when the step height is 1/3 h and 2/5 h respectively. When the step height is 1/5 h, the horizontal displacement of the upper step decreases by 34.5%, the horizontal displacement of the middle step increases by 14.8%, and the horizontal displacement of the lower step is about the same. When the step height is 1/4 h, the horizontal displacement of the upper step increases by 19.6%, the horizontal displacement of the middle step increases by 10.6%, and the horizontal displacement of the lower step is about the same. On the whole, the horizontal convergence value at the arch produces little deformation of the surrounding rock, which meets the engineering requirements.

3.2   Mechanical response analysis of step length

(1) Calculated monitoring point

In this study, the displacement criterion was adopted to judge the stability through vault settlement, invert uplift and horizontal convergence combined with domestic and foreign experience of large-span rock tunnels. Settlement points were set at the vault, vertical measuring points were set at the bottom invert, and four horizontal monitoring points were set at the upper and lower steps. The monitoring points of the tunnel calculation model are shown in Figure 5 below.

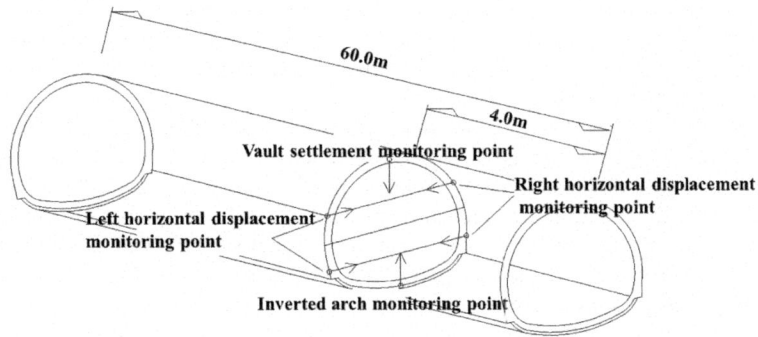

Figure 5.   Layout of monitoring points.

(2) The Deformation of Surrounding Rock

In the process of excavation simulation, the monitoring point was not moved and the face of the arch was advanced. The settlement and maximum horizontal convergence of the vault were recorded once at the end of each cycle of excavation. The cumulative results of vault settlement and maximum horizontal convergence are shown in Figure 6 and Figure 7 respectively in the process of palm surface advancement.

Now select the step lengths of 2 m, 4 m, 6 m, and 8 m for the four kinds of conditions of tunnel excavation simulation. From the whole, in Figure 6, in the process of tunnel excavation, excavation of the constraints on the steps to monitoring cross-section, near the arch sedimentation rate of change is bigger, as the constraints on the steps to the longitudinal excavation, vault subsidence rate gradually flatten out. The final settlement tends to be stable. And with the increase of step length, the greater the variation of vault settlement, the greater the final cumulative settlement. The final cumulative settlement of steps with lengths of 2 m, 4 m, 6 m and 8 m is 17.71 mm, 18.84 mm, 19.82 mm and 20.53 mm, respectively.

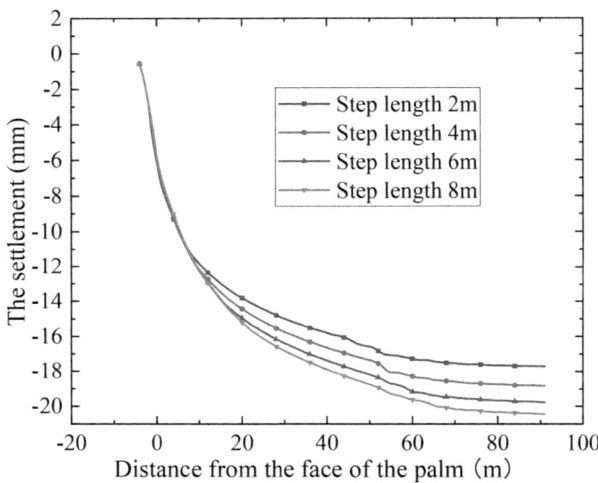

Figure 6.   Settlement curve of the vault under different step lengths.

As can be seen from Figure 7 as a whole, during tunnel excavation, when the distance from the face of the inverted arch is less than 20 m, the change rate of the inverted arch uplift value decreases with the increase of the step length. When the distance from the face of the inverted arch is greater than 20 m, the change rate of the inverted arch uplift value tends to be gentle, and the bottom uplift tends to be stable. With the increase of step length, the final cumulative uplift value of the invert increases. The final cumulative uplift value of 2 m, 4 m, 6 m and 8 m steps is 18.96 mm, 19.19 mm, 19.82 mm and 20.44 mm, respectively. Therefore, in order to ensure the safety of construction, the change rate of uplift and the final cumulative uplift value should be considered comprehensively when setting the reasonable step length.

As can be seen from Figure 8 as a whole, during the excavation of the upper step, the horizontal displacement of the upper step first squeezes outwards, then the excavation of the upper step reaches the monitoring section, and the horizontal displacement of the upper step begins to deform inwards, and the variation of the horizontal displacement increases rapidly to gradually moderate. As the step length increases, the horizontal displacement of the upper step tends to be stable for a longer time (the greater the distance between the palm surface). The cumulative horizontal displacement of the upper step with the step length of 2 m, 4 m, 6 m and 8 m is 2.94 mm, 2.80 mm, 2.89 mm and 2.96 mm, respectively. As can be seen from the figure, the final cumulative horizontal displacement of the upper step is in descending order: 8 m>2 m>6 m>4 m.

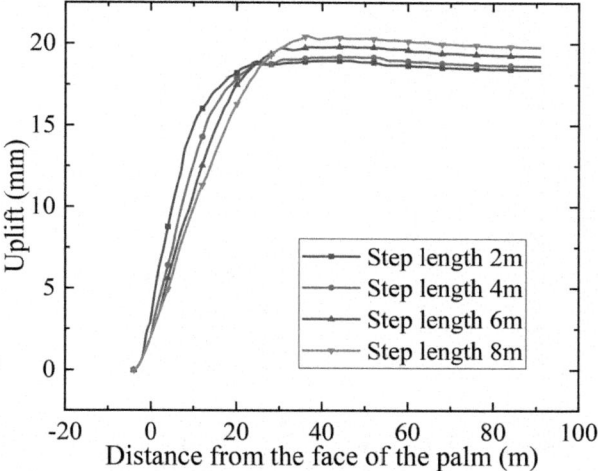

Figure 7. Inverted arch uplift curves with different step lengths.

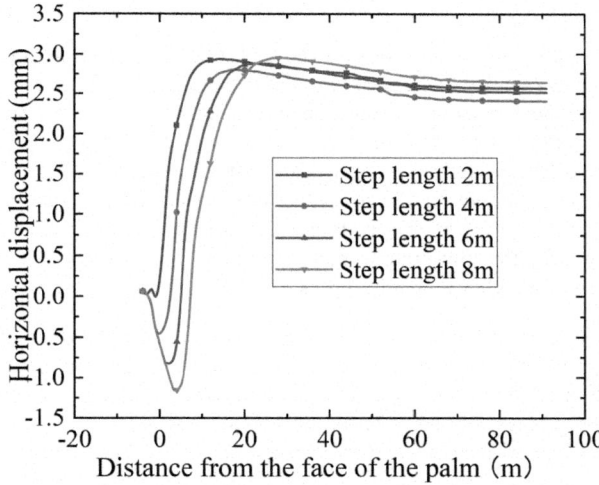

Figure 8. Horizontal displacement curves of lower and upper steps with different steps lengths.

As can be seen from Figure 9, the horizontal displacement of the lower step changes relatively little when the upper step is excavated. Then, when the middle step is excavated, the horizontal displacement deformation of the lower step begins to rise rapidly, and the horizontal displacement changes rapidly. With the increase of the distance from the face of the face, the horizontal displacement gradually tends to moderate. As the step length increases, the horizontal displacement of the lower step tends to be stable for a longer time (the greater the distance between the palm surface). The final cumulative horizontal displacement of the lower step with a step length of 2 m, 4 m, 6 m, and 8 m is 5.58 mm, 4.83 mm, 6.19 mm and 7.04 mm, respectively. As can be seen from the figure, the final cumulative horizontal displacement of the lower step is in descending order: 8 m>6 m>2 m>4 m.

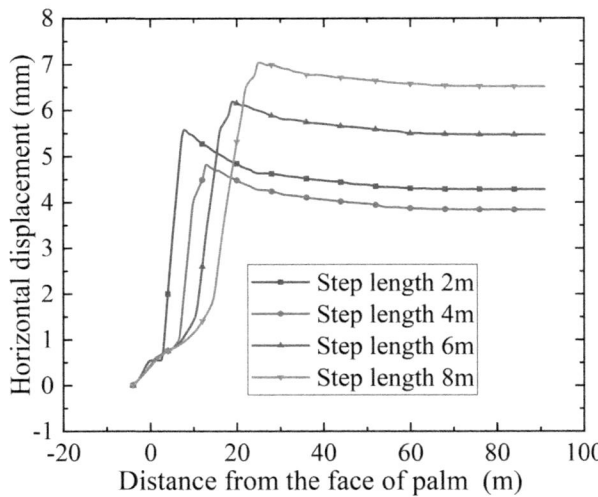

Figure 9. Horizontal displacement curves of lower and upper steps with different steps lengths.

4 CONCLUSION

After calculation, the following conclusions are drawn for the optimization of construction parameters of the three-step excavation method: From the calculation analysis, can vault settlement as an index, from the perspective of vault subsidence, etc, according to the displacement criterion, calculation and contrast the effect of bench height on three, on the bench height as small as possible, to control the settlement and horizontal convergence, the better comprehensive calculation of construction space and production efficiency balanced consideration, recommend bench height takes the bench height 1/4 h∼1/3 h, That is, the excavation height is 3.5 m∼4.5 m. The following conclusions are drawn for the longitudinal key parameters of the three-step excavation method under the condition of grade IV surrounding rock:

(1) With the increase in step length, the final cumulative settlement of the vault and the final cumulative uplift of the inverted arch are larger, and the longer the step length is, the more adverse the vertical deformation of the tunnel is. From the perspective of vertical displacement, the smaller the step length is, the safer the tunnel excavation is.
(2) From the perspective of horizontal displacement, when the step length is 2 m or 8 m, the horizontal displacement of the upper step is relatively large, indicating that too large or too small step length is unfavorable to the deformation of tunnel shoulder; When the step length is 8 m or 6 m, the horizontal displacement of the horizontal survey line of the lower step is relatively large, indicating that the long step is unfavorable to the deformation of the arch foot position of the tunnel. In comprehensive consideration, the recommended step length should be controlled within 4 m.

Overall consideration, under the condition of IV grade surrounding rock, the step height should be recommended to be increased by 1/4 h∼1/3 h, that is, the excavation height should be 3.5 m∼4.5 m, and the step length should be controlled within 4 m.

REFERENCES

Cheng Guangfu. Application of the three-step method in construction of railway double-track soft rock tunnel [J]. *Equipment Management and Maintenance*, 2020(20): 139–140.

Deng Tao, Liu Dagang, CAI Minjin, Zhao Sicguang, He Wei. Optimization analysis of geometric parameters based on the step method for the zhongtiaoshan tunnel [J]. *Tunnel Construction*, 2017, 37(12): 1550–1556.

Hu Shouyun, Wu Tingyao. A Study on deformation characteristics of surrounding rock during excavation of sanming railway tunnel by the three-step method [J]. *Construction Technology*, 2016, 45(19): 86–90.

Jensen. A Safety analysis of ANSYS simulation tunnel excavation with the three-step method [J]. *Urban Construction Theory Research* (electronic version), 2020(06): 47.

Jiang Liang, Xiong Chengyu. Comparative analysis of numerical simulation on the excavation of different steps in tunnel construction [J]. *Highway Engineering*, 2019, 44(06): 161–165.

Kwon Swilsonjw. Deformation mechanism of the underground at the WIPP site [J]. *Rock Mechanics and Rock Engineering*, 1999, 32(2): 101.

Liu Zhenggang. Application of the step method in large section tunnel engineering in soft surrounding rock [J]. *China High and New Technology*, 2020(15): 127–128.

Song Shuguang, Li Shucai, Li Liping, Liu Qin, Yuan Xiaoshuai, Shi Shaoshuai. A study on mechanical effect law in the construction process of super large section tunnel with soft broken surrounding rock by the step method [J]. *Tunnel Construction*, 2011, 31(S1): 170–175.

Wang Haijun. *Journal of Lanzhou Institute of Technology*, 2016, 23(01): 67–71.

Wang Weifu, Mei Zhu. Application of the step method in ultra-large section shallow buried bias tunnel [J]. *Tunnel Construction*, 2017, 37(12): 1578–1584.

Wu Jianlin, Wu Xun, Jiang Renguo. *Journal of Nanjing Institute of Technology* (Natural Science Edition), 2019(02): 54–57.

Zhang Yuanyuan. Research on optimization of excavation parameters of clastic sedimentary strata section of diversion tunnel by the three-step method [J]. *Technical Supervision of Hydraulic Engineering*, 2021(04): 172–174.

Zhao Xiangping. Research on reasonable excavation method of large section tunnel in soft surrounding rock [J]. *Railway Construction Technology*, 2015(07): 16–19+33.

Zhou Yu. Research on rational excavation method of large section tunnel in soft surrounding rock [J]. *China Standardization*, 2019(16): 122–123.

Zhu Yongquan. Stability displacement criterion of tunnel [J]. *China Railway Science*, 2001(06): 81–84.

Zou Chenglu, Shen Yusheng, Jin Zongzhen. Optimization of geometric parameters for the construction of large section tunnel by step method in soft broken surrounding rock [J]. *Highway Engineering*, 2013, 38(02): 27–31+35.

Frontiers in Civil and Hydraulic Engineering – Mohamed A. Ismail and
Hazem Samih Mohamed (Eds)
© 2023 The Authors, ISBN 978-1-032-38247-0

Evaluation and research on the development potential of prefabricated construction industry in Chongqing

Qili Gan* & Wanqing Chen*
Chongqing College of Architecture and Technology, China

Qiming Huang*
Chongqing Construction Residential Engineering CO. LTD., China

Langhua Li* & Junxia Sun*
Chongqing College of Architecture and Technology, China

ABSTRACT: In recent years, prefabricated buildings have entered a period of rapid development with the strong support of the Chinese government. The development of prefabricated buildings can greatly save resources and energy, and reduce construction waste, dust, noise, and other environmental problems. It can promote high integration of the construction industry with industrialization and informatization, and realize the transformation and upgrading of the development of the construction industry and the reconstruction of the traditional construction industry chain. It can significantly shorten the construction period and improve the quality of construction products. Its role in all aspects has been widely recognized in China's domestic construction and real estate fields. As the second batch of prefabricated building demonstration cities in China, the development of prefabricated buildings has also ushered in great opportunities and challenges. Therefore, an accurate understanding of the development potential of prefabricated buildings in Chongqing is very critical to promote the development of prefabricated buildings in Chongqing, and also provides a reference for other cities to evaluate the development potential of prefabricated buildings.

1   INTRODUCTION

Prefabricated buildings in foreign countries have mainly attracted people's attention since the beginning of the 20th century and were finally realized under the first attempt in Britain, France, the Soviet Union, and other countries in the 1960s (Chen 2018). Scholars (Buncio 2019) believe that the current construction industry is facing forces of mutual collision, which makes our current trajectory become wasteful, inefficient, and unsustainable, which requires us to enter the era of prefabricated and modular architecture. The development of prefabricated buildings in China has entered a period of rapid development since 2016 (Shen 2022). Because of their outstanding characteristics in energy conservation and environmental protection, fast construction, controllable cost, safety and durability, and prefabricated buildings have become the future development direction of China's construction industry. Since 2013, China's prefabricated buildings have been mentioned in national policy documents many times in recent years and incorporated into the government's development plan. Since 2015, the newly-built prefabricated building area, the scale of the construction industry, and the output value of prefabricated related industries have continued to rise, and urbanization is proceeding at an ultra-high speed. By 2019, the scale of China's prefabricated construction market has climbed to 42.75% compared to the world's construction

---

*Corresponding Authors: 850425529@qq.com and 380022735@qq.com

DOI 10.1201/9781003344209-62

market (Liu 2021). Many scholars have also actively engaged in research on the development of prefabricated buildings. For example, Zheng (2020) believes that prefabricated buildings in China are in a stage of steady progress, continuous enhancement of endogenous power in the industry, the gradual establishment of a policy support system, and preliminary establishment of technical support system. More researchers (Gu 2018; Guo 2020; Li 2018; Xiong 2020) believe that high production costs, incomplete industrial chains, imperfect policies and norms, low technical levels, and the impact of the market environment will restrict the further development of prefabricated buildings; however, breaking through these constraints will also become a factor to promote the development of prefabricated buildings (long 2017). At present, most of the domestic research is aimed at the development potential of prefabricated buildings, and few scholars study the development potential of prefabricated buildings in a city. The economic, technological, and social levels of each region are different, and they are also affected by the development of strategic objectives and relevant supporting policies of each city. The development potential of prefabricated buildings must be very different. Therefore, the paper establishes the urban development potential evaluation model of prefabricated buildings from the perspectives of economy, technology, policy, and market. This paper also takes Chongqing as an example to make an empirical analysis of the development potential of prefabricated buildings, and obtains its development potential evaluation, in order to provide more objective suggestions for promoting the development of prefabricated buildings in Chongqing.

2 DEVELOPMENT STATUS OF PREFABRICATED BUILDINGS IN CHONGQING

2.1 Analysis of the current situation of prefabricated building policy in Chongqing

After the general office of the State Council issued the guiding opinions on vigorously developing prefabricated buildings, the policy and standard system of prefabricated buildings in Chongqing have also been basically established and gradually improved after several years of efforts. In 2017, the Chongqing Municipal People's government issued the implementation opinions on vigorously developing prefabricated buildings, clarifying the objectives, tasks, and incentive measures for the development of prefabricated buildings. The government implemented them from the aspects of planning guidance, strengthening land security, strengthening fiscal and tax support, optimizing the pre-sale of commercial housing, supporting scientific and technological innovation, and supporting transportation and logistics, creating a good policy environment (Luo 2018). From 2018 to 2022, the general office of the Chongqing Municipal People's government and the Chongqing urban and Rural Development Commission issued more than 20 documents gradually, which provided policy guarantees in terms of prefabricated building evaluation management, industrial base management, pricing basis, incentive policies, and constantly emphasized the development of prefabricated construction to accelerate the modernization of the construction industry and form a relatively perfect construction industry chain, Promote the improvement of construction level and quality, drive the high-quality development of the construction industry, further improve the policies and systems of prefabricated buildings, and escort the sustainable development of prefabricated buildings.

2.2 Analysis of the current situation of the prefabricated building market in Chongqing

At present, according to the principle of zoning promotion and gradual promotion of prefabricated buildings in Chongqing, the key development areas, active development areas, and encouraged development areas of all districts and counties in Chongqing are defined. The development of key development areas first drives the development of the encouraging development areas and then promotes the overall development of prefabricated buildings in Chongqing. The development areas of prefabricated buildings in Chongqing are shown in Figure 1. At present, the annual production capacity of concrete components is 430000 M^3, 1.55 million tons of steel structural components, 12.2 million square meters of light wall panels, with an annual output value of 20 billion yuan, and the production capacity of prefabricated components meets the construction demand of 20 million square meters of prefabricated buildings every year. The output value of prefabricated components and industrial decoration in the city has exceeded 100 billion, and the total amount and scale of the prefabricated construction market have been expanding.

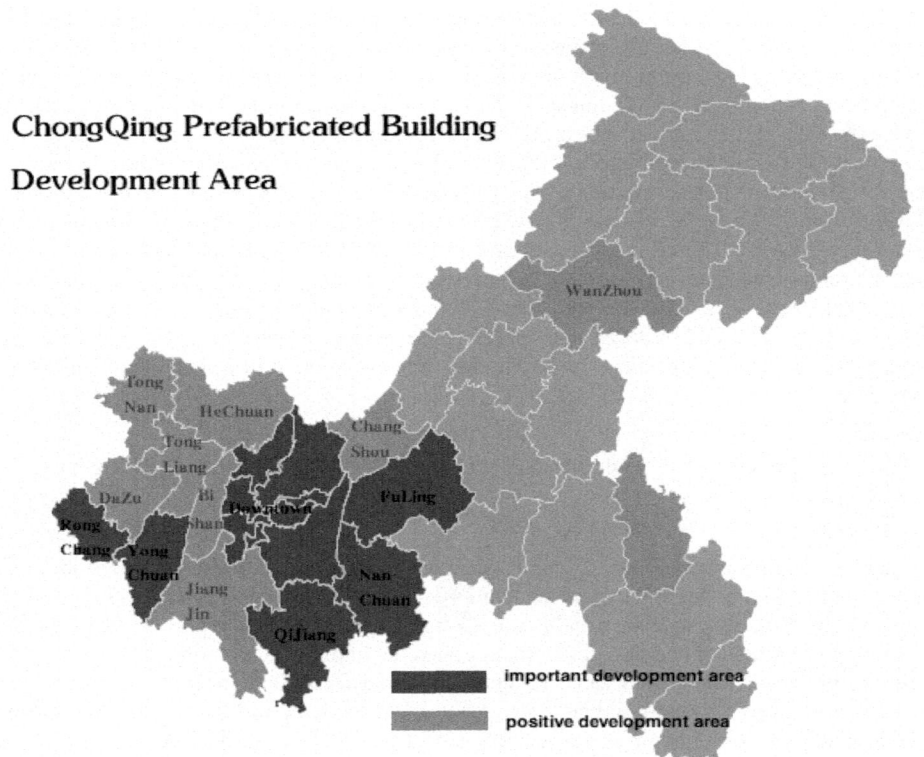

Figure 1.   Chongqing prefabricated building development area map.

2.3   *Analysis of the current situation of prefabricated building technology in Chongqing*

According to statistics, Chongqing has initially established a standard system for prefabricated buildings, with a total of 69 standards, including 31 currently in force and 38 under preparation. These standards are divided into basic standards, general standards, and special standards, covering the whole process of design, production, construction and acceptance, including the evaluation, review and informatization of prefabricated buildings, prefabricated concrete buildings, steel structure buildings, composite structure buildings, prefabricated building components, prefabricated municipal engineering, prefabricated decoration, equipment pipelines, etc. Chongqing municipal government, relevant enterprises, and universities have been committed to the research of technical standards. Atlas and technological innovation are related to prefabricated buildings, and the technical system has gradually matured.

3   PROBLEMS EXISTING IN THE DEVELOPMENT OF PREFABRICATED BUILDINGS IN CHONGQING

It can be seen from the current situation of the development of prefabricated buildings in Chongqing that in recent years, the relevant policy system of prefabricated buildings in Chongqing has been continuously improved, the total scale has been increasing, and the strength of technological innovation has been continuously enhanced, with better development. However, although the state and the government strongly support the development of prefabricated buildings, the development process has been relatively slow so far. It can be seen from the "Chongqing prefabricated building industry development plan (2018-2025)" issued by the Chongqing Municipal Commission of urban and rural development that the main reason is still the weak support of industrial policies, low project demand, and incomplete industrial categories. The implementation ability of enterprises is not strong and the actual implementation is difficult.

4 EVALUATION OF THE DEVELOPMENT POTENTIAL OF PREFABRICATED BUILDINGS IN CHONGQING

4.1 Selection of evaluation indicators for the development potential of prefabricated buildings

This chapter carries out the preliminary selection of indicators through literature research and then carries out the final selection of indicators according to the principles of comprehensiveness and comparability, combined with the availability of data, and finally determines the index system to comprehensively evaluate the development potential of prefabricated buildings in Chongqing. The comprehensive evaluation index system is shown in Table 1.

Table 1. Comprehensive evaluation index system for the development potential of prefabricated buildings.

Indicator class (level 1)	Indicator subclass (level ii)	Index item (level iii)
Economics	Economic development level	Per capita GDP (x1)
		Per capita local fiscal revenue (x2)
	Residents living standards	Urban per capita disposable income (x3)
		Consumption level of urban residents (x4)
	Economic structure	Proportion of fiscal expenditure in GDP (x5)
		Proportion of tertiary industry in GDP (x6)
Technology	Technological innovation fund input	Research and experimental development expenditure per capita (x7)
		Proportion of science and technology expenditure in local GDP (x8)
		Proportion of local fiscal expenditure on science and technology in local fiscal expenditure (x9)
	Technical talents cultivation of	Number of engineering technology and scientific research professionals in local enterprises and institutions (x10)
		Research and experimental development equivalent number of full-time personnel (x11)
		Number of research and experimental development equivalent personnel per 10000 people (x12)
		Number of engineering students in colleges and universities (x13)
	Technology research and development	Number of research and development institutions (x14)
		Number of invention patents (x15)
		Number of prefabricated construction industrial bases (x16)
	Relevant standards	Number of relevant standards, specifications, and atlas (x17)

Table 1. Continued.

Policy		Whether there are government promotion measures (x18)
		Whether to propose construction objectives and tasks (x19)
		Whether there are government support and incentive policies (x20)
		Whether there is a government supervision mechanism (x21)
		Is there a mandatory implementation policy of the government (x22)
Market	Construction supply and demand	Per capita investment in real estate development (x23)
		Per capita real estate construction housing area (x24)
		Per capita real estate transaction area (x25)
		Annual growth of urban population (x26)
	Related enterprises	Number of real estate development enterprises (x27)
		Production base of assembled parts (x28)

4.2 Construction of evaluation model for the development potential of prefabricated buildings

According to the basic theories of principal component analysis and the TOPSIS method, the article believes that the combination of them has great advantages for the evaluation process of the development potential of prefabricated buildings in a certain area in a certain year or a certain area in a certain year. The weighted principal component TOPSIS evaluation model overcomes the problem that a single TOPSIS method cannot be weighted, and the principal component analysis is also used to simplify and analyze the evaluation indicators. According to the model construction process, we can also see that the final assembled development potential can be judged scientifically and accurately.

4.3 Evaluation of the development potential of prefabricated buildings in Chongqing

The evaluation of the development potential of prefabricated buildings in Chongqing is to summarize the collected data on the economy, technology, policy, and market, to get a total evaluation of the original data, and to use SPSS and Excel software for analysis. Finally, the analysis results of the comprehensive evaluation index data of the development potential of prefabricated buildings in Chongqing from 2014 to 2021 are shown in Table 2.

Table 2. Evaluation results of the overall potential for the development of Chongqing prefabricated buildings from 2014 to 2021.

Year	Distance D of positive ideal solution+	Negative ideal solution distance D-	Relative proximity C	Sort results
2021	0.372	2.337	0.863	1
2020	0.402	2.103	0.840	2
2019	0.587	1.801	0.754	3
2018	1.069	1.329	0.554	4
2017	1.309	1.108	0.458	5
2016	1.504	0.933	0.383	6
2015	1.568	0.836	0.588	7
2014	1.897	0.502	0.209	8

It can be seen from Table 2 that the overall development potential of prefabricated buildings in Chongqing is on the rise, which shows that Chongqing has made continuous progress in promoting the development of prefabricated buildings in recent years. Every year, it is making continuous efforts to the development of prefabricated buildings, providing good economic, technical and other conditions, and providing a strong backing for the future development of prefabricated buildings.

5 CONCLUSION

According to the empirical analysis results, we can see that in terms of time, Chongqing's economic, technological, policy, and market development potential are gradually improving, which shows that Chongqing is making continuous efforts in the development of prefabricated buildings every year. If prefabricated buildings have great development potential in the city, they must have a solid economic foundation, strong technical strength, and broad market prospects. Therefore, in order to stimulate the development potential of prefabricated buildings in Chongqing, the government needs to pay close attention to its deficiencies in the economy, technology, policy and market, especially in economy and technology, supplemented by corresponding policy support, and fully release its development potential of prefabricated buildings. The suggestions for Chongqing to improve the development potential of prefabricated buildings are as follows:

1) We will strengthen government policy support for prefabricated buildings. The government has an important guiding role in the development of prefabricated buildings in this area, guiding the healthy development of prefabricated buildings. The government should formulate a medium and long-term plan for the development of prefabricated buildings, and guide the benign development of prefabricated buildings, supplemented by effective incentives to promote the healthy development of prefabricated buildings.
2) Create a good environment for the development of prefabricated buildings. At present, Chongqing's overall economic strength is good, but it still can't catch up with the first-tier cities, and the talent training and capital investment in technology are far less than those in other regions. Therefore, in order to promote the development of prefabricated buildings, the government should economically adjust its industrial development structure according to local conditions, increase the proportion of the tertiary industry and per capita local GDP, and provide sufficient financial support for the development of prefabricated buildings. In addition, it can increase residents' income through multiple channels, improve the wage level of employees in enterprises and institutions, implement policies and measures conducive to expanding consumption, comprehensively improve the consumption level of residents, and provide protection for the development and consumption end of prefabricated buildings.
3) Optimize the whole industrial chain of prefabricated buildings. The industrial chain of prefabricated buildings in Chongqing is incomplete. Most of the industrial bases are manufacturers of prefabricated concrete components, steel structures, and other components. There is still a blank in the industrial fields of logistics and transportation, machinery and equipment manufacturing, interior components and components, and there are few component R&D enterprises, and the training of technical personnel cannot keep up. Therefore, in the process of local industrial development, the government should do a good job in policy guidance, optimize the industrial structure, increase the construction of demonstration production bases, and guide enterprises to invest more funds and land for the development and production of prefabricated construction products.
4) Strengthen the publicity of prefabricated buildings. Insufficient awareness of prefabricated buildings by enterprises and consumers is also an important factor that hinders the development of prefabricated buildings. The government should take multichannel publicity of prefabricated buildings, deepen the comparative understanding of traditional buildings and prefabricated buildings between enterprises and consumers, enhance consumers' recognition of prefabricated building products, enhance the social influence of prefabricated buildings, and bring a social atmosphere in which enterprises actively develop and consumers voluntarily purchase prefabricated construction products.

REFERENCES

Buncio, Founder. (2019). The epic rise of industrialized construction [J]. *Building Design & Construction*, 60(4): 18–19.

Chen B. (2018). A preliminary study on bidding mode and scheme of prefabricated buildings [J]. *Bidding and Tendering*, 6(04): 57–58.

Liao L.P. (2019). Development status and strategies of green prefabricated buildings [J]. *Enterprise Economy*, 38(12): 139–146.

Liu K., Wang X. & Xue J.B. (2021). Comparison of the development status of prefabricated buildings at home and abroad and research on countermeasures [J]. *Engineering Construction*, 53(07): 19–24.

Li L.B. (2018). Zhang Xiaoming. Analysis of restrictive factors and countermeasures for the development of prefabricated buildings [J]. *Liaoning Economy*, (4): 80–81.

Long Y.F. (2017). Break through bottlenecks and promote the development of prefabricated buildings [J]. *Housing and Real Estate*, (02): 76–77.

Luo G. & Liao H.P. (2018) Village level multidimensional poverty measurement and poverty type classification from the perspective of geographical capital – based on the survey data of 1919 Municipal Poverty Villages in Chongqing [J]. *Agricultural Resources and Regionalization in China*, 39(08): 244–254.

Gu X.J. (2018). Analysis of restrictive factors and countermeasures for the development of prefabricated buildings [J]. *Jiangxi Building Materials*, (13): 3–4.

Guo X.X. (2020). Development status and prospects of prefabricated buildings [J]. *Sichuan Cement*, (01): 283.

Shen F. (2022). Comparison and selection of K15 school prefabricated building design schemes based on the whole life cycle [D]. Hejiang University, 144.

Xiong L. & Zhu L.L. (2020). Analysis of the development status and influencing factors of prefabricated buildings [J]. *Sichuan Architecture*, 40(03): 14–16.

Zheng M.L. (2020) Development status and existing problems of concrete structure prefabricated buildings [J]. *Jiangxi Building Materials*, (03): 5–6.

Frontiers in Civil and Hydraulic Engineering – Mohamed A. Ismail and
Hazem Samih Mohamed (Eds)
© 2023 The Authors, ISBN 978-1-032-38247-0

Research on heritage value evaluation of irrigation engineering based on AHP method

Xinru Mao* & Jianfeng Yao*
College of Civil Engineering and Architecture, Zhejiang University of Water Resources and Electric Power, Hangzhou, China

ABSTRACT: Under the guidance of water control policy and the identification of water conservancy heritage, various water conservancy departments have jointly carried out the comprehensive excavation, protection, and inheritance of ancient irrigation engineering heritages. Aiming at all kinds of irrigation project heritage, establishing a set of accurate and appropriate value evaluation systems is an effective method to identify and protect water conservancy heritages. Based on the characteristics of ancient irrigation projects, this paper extracts corresponding indexes around the composition of their values, studies the weights by an analytic hierarchy process, and finally constructs a comprehensive value evaluation system of irrigation project heritages from all aspects including science, economy, society, history, and ecology by using a fuzzy evaluation method. Taking Gaoyou Irrigation District as an example, the effectiveness and scientificity of the established evaluation system of irrigation canal engineering heritage values are verified.

1 INTRODUCTION

China has five thousand years of civilization. As an agricultural country, China has a long history of irrigation development. By 2021, China has successfully excavated a total of 23 world irrigation engineering heritages (Quan 2019; Zhang 2019), all of which are class A and are still exerting irrigation benefits. Each irrigation heritage reflects the situation of water conservancy construction in different periods and regions and its relationship with politics, economy, society, culture, environment, and ecology to different degrees. Among them, the Gaoyou Irrigation Area of Li Canal is the first selected world irrigation engineering heritage in Jiangsu Province, and effective protection and utilization will be of great significance to improve the high-quality development of water conservancy reform.

Referring to the interpretation of the value of antiquities heritage in the Guidelines for the Protection of Cultural Relics and Monuments in China (China National Committee 2000), the values of world irrigation project heritages are summarized into four aspects: scientific value, artistic value, cultural value, and social value. Among them, irrigation engineering heritages focus on aesthetic design and aesthetic experience, economic benefits, and agricultural income brought by irrigation, so this paper adopts economic value instead of artistic value.

At the same time, for a new type of heritages like irrigation heritages, a functional model needs to be established to evaluate the value through the ratio of multiple value indicators and appropriate weight coefficients, and the heritage value of irrigation channel projects need to be analyzed comprehensively. Through literature collection and on-the-spot investigation, the evaluation index mainly includes the following five aspects: scientific value, economic value, social value, historical value, and ecological value.

---

*Corresponding Authors: 1656443086@qq.com and yaojf@zjweu.edu.cn

DOI 10.1201/9781003344209-63

2 THE CONSTRUCTION OF A HIERARCHY

At present, there are many evaluation methods for engineering heritage in China, such as principal component analysis, data envelopment analysis, and so on, but these methods are not suitable for irrigation engineering heritage evaluation. Therefore, this paper puts forward the AHP method to quantify the value and calculate the weight of irrigation and canal engineering heritage. The key lies in reasonably decomposing the relevant factors into the target layer, criterion layer, scheme layer, etc., making qualitative and quantitative analysis of the decision-making methods, and finally evaluating the comprehensive value of irrigation and canal engineering heritage through the calculation of comprehensive scores (Tan 2013; Zhang 2013), as shown in Figure 1.

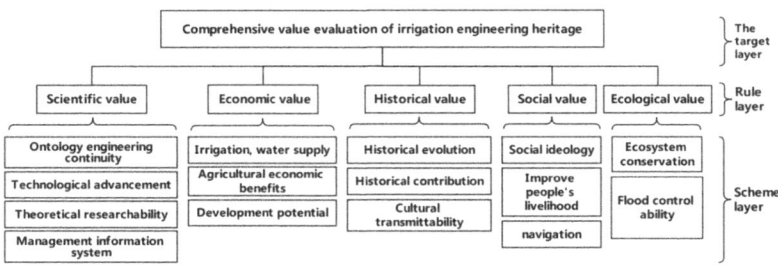

Figure 1.   Quantitative stratification of the comprehensive value evaluation system for irrigation engineering heritages.

3 METRIC QUANTIFICATION

To quantify the performance indexes of each scheme layer, the important performance indexes are extracted using a fuzzy evaluation method (Lai 2013; Li 2013), and then each quantitative factor is divided into four grades according to the evaluation system, and each grade is assigned points by the principle of the percentile system. The evaluation grades are divided into 25 points, 50 points, 75 points, and 100 points. When quantifying the score, the principle of a high score is adopted, and the corresponding score can be taken as the value evaluation score of the performance index according to the satisfaction degree. For example, the quantitative score of the performance evaluation index of the scientific value in the system is shown in Table 1.

Table 1.   Evaluation index of scientific value of irrigation channel engineering channel heritage.

Evaluation indicators	Performance evaluation metrics	Evaluation quantification factor	Evaluation level assignment			
			100	75	50	25
Scientific value	Ontology engineering continuity	Whether the ontology project exists in its entirety, has perfect functions, and is complete in the system. Whether it can be used continuously	Sufficiently reflect	More fully reflected	Partly reflect	Not
	Technological advancement	Whether the process technology is ahead of the times, whether the structural concept of its engineering design, material technology, construction technology, etc. is representative of science and technology	Sufficiently reflect	More fully reflected	Partly reflect	Not

(continued)

480

Table 1. Continued.

Evaluation indicators	Performance evaluation metrics	Evaluation quantification factor	Evaluation level assignment			
			100	75	50	25
	Theoretical feasibility	Whether it has scientific research value and whether it has made scientific and technological contributions to contemporary engineering theories and development methods	Sufficiently reflect	More fully reflected	Partly reflect	Not
	Manage information systematically	Whether it has a sound management system, a modern information management system, and can be used as a model for sustainable operation management	Sufficiently reflect	More fully reflected	Partly reflect	Not

4 WEIGHT AND COMPREHENSIVE TOTAL SCORE CALCULATION

For the comprehensive value of quantitative assessment of the irrigation project, the first is to quantify natural endowments of engineering value evaluation indexes, then to construct a judgment matrix to determine the weight of each evaluation factor and ratio at every levels. Using the level of a single weight and consistency check of total sequencing guarantees the rationality of the allocation, and each index of quantitative score according to certain weight addition. Thus, the comprehensive value quantization value of irrigation engineering heritages can be obtained, and the specific formula can be expressed as:

$$V = \sum_{i=1}^{n} w_i \times v_i \tag{1}$$

$$w_i = w_i \times v_{bi}$$

Where: V is the quantified value of the comprehensive value of the project, v_i is the quantified value of evaluation indexes at each scheme level, w_i is the weight of the comprehensive value of the plan layer index to the target layer, w_{ai} is the weight of the criterion layer index to the comprehensive value of the target layer, and w_{bi} is the weight of scheme layer index to the criterion layer index. Then, a pairwise comparison is made between the importance of indicators at each level and the importance of indicators at the next level, and the assigned values constitute a judgment matrix, as shown in Table 2.

Table 2. Index relative importance degree assignment value.

a_i/a_j	Equally important	Slightly more important	Obviously important	Strongly important	Extremely important	Median value of two adjacent judgments	Reciprocal
Assigned points	1	3	5	7	9	2, 4, 6, 8	a_j/a_i

Finally, the correlation calculation is verified by a consistency test. When **CR ≤ 0.1**, the consistency of the judgment matrix is considered acceptable.

5 APPLICATION CASE OF IRRIGATION PROJECT VALUE EVALUATION IN GAOYOU IRRIGATION AREA

Gaoyou Irrigation District, located in Lixiahe area of Jiangsu Province, is the source of the East Route of the South-to-North Water Transfer Project and the first large-scale irrigation area with an area of over 500,000 mu. The irrigation area is irrigated by water from Beijing-Hangzhou Grand Canal, with a total water diversion capacity of 150 flows. The total area of the irrigation area is 649 square kilometers, the cultivated land area is 632,200 mu, and the effective irrigation area is 588,900 mu. The current agricultural water consumption accounts for more than 80% of the total domestic water consumption.

5.1 Brief introduction to the irrigation area

Gaoyou Irrigation Area, located in Lixiahe Area of Jiangsu Province, is the source of the eastern route of the South-to-North Water Diversion Project and the first large irrigation area with an area of more than 333.34 square kilometers. The irrigated area is diverted from the Beijing-Hangzhou Grand Canal for self-flow irrigation, with a total water diversion capacity of 150 flows. The total area of the irrigated area is 649 square kilometers, the cultivated area is 421.47 square kilometers, and the effective irrigated area is 392.6 square kilometers. The current agricultural water consumption accounts for more than 80% of the total amount in China.

5.2 Index weight and weighting calculation

The calculation of the value index weight of the irrigation area mainly includes three aspects. One is the relative weight of criterion level index to the target level, the other is the weight of scheme level to criterion level, and the last is the comprehensive weight of scheme level performance index to the overall goal. Taking the comprehensive value evaluation of Gaoyou Water Transmission and Distribution Project as an example, a judgment matrix is constructed for the relative importance of the criterion layer to the target layer, namely, scientific value, economic value, social value, historical value, and ecological value to the comprehensive value, and a value of is calculated by using MATLAB software, as shown in Table 3.

Table 3. Weight of the comprehensive value assessment criteria layer*.

Combined value	Scientific value	Economic value	Social value	Historical value	Ecological value	Weight w_{ai}
Scientific value	1	2	4	6	5	0.4325
Economic value	1/2	1	3	7	5	0.3115
Social value	1/4	1/3	1	4	3	0.1428
Historical value	1/6	1/7	1/4	1	1/2	0.0447
Ecological value	1/5	1/5	1/3	2	1	0.0685

*The data obtained by the deviation test are in line with the requirements (**CR = 0.0387**, the weight of total target = **1.0000**).

The second is the relative importance of the scheme layer to the criterion layer. The weight of the scheme layer of the scientific value assessment index to the criterion layer is shown in Table 4. Finally, the weights of each index are calculated and weighted, and the comprehensive scores obtained are shown in Table 5.

Table 4. Weight of the scientific value evaluation index scheme layer*.

Scientific value	Ontology engineering continuity	Technological advancement	Theoretical feasibility	Manage information systematically	Weight W_{bi}
Ontology engineering continuity	1	3	6	4	0.5587
Technological advancement	1/3	1	3	2	0.2280
Theoretical feasibility	1/6	1/3	1	1/2	0.0781
Manage information systematically	1/4	1/2	2	1	0.1352

*The data obtained by the deviation test are in line with the requirements (CR = 0.0115, the weight of total target = **0.4325**).

Table 5. Comprehensive value assessment of channel projects in Gaoyou Irrigation District.

Comprehensive indicators	Evaluation indicators	Index weight	Assigned points	Sub-score	Value score
Scientific value	Ontology engineering continuity	0.2416	100	24.16	37.6325
	Technological advancement	0.0986	75	7.395	
	Theoretical feasibility	0.0338	50	1.69	
	Manage information systematically	0.0585	75	4.3875	
Economic value	Feature usability	0.1812	100	18.12	30.3075
	Agricultural economic benefits	0.0963	100	9.63	
	Development potential	0.0341	75	2.5575	
Historical value	Historical evolution	0.0143	75	1.0725	3.2225
	Historical contribution	0.0055	50	0.275	
	Cultural dissemination	0.0250	75	1.875	
Social value	Social ideology	0.0312	75	2.34	13.5
	Improve people's livelihoods	0.0900	100	9	
	Navigation	0.0216	100	2.16	
Ecological value	Ecosystem conservation	0.0220	50	1.1	4.5875
	Flood protection capabilities	0.0465	75	3.4875	
Total items	Overall score	1.0002	1175	**89.25**	89.25

6 CONCLUSION

The comprehensive value evaluation system established in this study can fully reflect the values of the irrigation canal engineering heritage, and provide an index basis for qualitative evaluation of irrigation engineering heritage values. Through the quantitative analysis and weight calculation of the evaluation index, the comprehensive values of the irrigation project can be quantitatively evaluated, and the single value of the project can be sorted. It can be seen from the above table that the scientific value > economic value > social value > ecological value > historical value of Gaoyou Irrigation District, which can be used as the basis for grading and focusing on the protection of irrigation engineering heritage, giving full play to the resource advantages of water conservancy heritage. In addition, this comprehensive evaluation index system can not only reduce the cost of protecting the irrigation project heritage but also provide a reference for the value evaluation of other water conservancy project heritages.

REFERENCES

Guidelines for the protection of cultural relics and historic sites in China (2000). Z. *China National Committee of the International Council on Monuments and Sites*.

Lai, Z.X. & Z. Li. (2013). Evaluation of flood prevention risk in irrigation division of water conservation project based on hierarchical analysis-fuzzy comprehensive evaluation method. *J. Agricultural Engineering* 8 (01): 71–77.

Zhang, N.Q., X. Tan & T.H. Wang (2013). Study on comprehensive value evaluation of ancient water conservancy projects of Beijing-Hangzhou Canal. A. Water Conservancy History Research Association of China Water Conservancy Society, Zhejiang Provincial Water Resources Department and Shaoxing Water Conservancy Bureau. *Proceedings of 2013 Annual Meeting of Water Conservancy History Research Association of China Water Conservancy Society and Strategic Forum on Protection and Utilization of Water Conservancy Heritage of Grand Canal of China*.

Zhang, X. & C.Y. Quan (2019). A review of research on the cognition and value of the world's irrigation engineering heritage. *A. Urban Planning Society of China*, Chongqing Municipal People's government. Vibrant Urban and Rural Living-Proceedings of 2019 China Urban Planning Annual Meeting (09 Urban Cultural Heritage Protection). C. Urban Planning Society of China, Chongqing Municipal People's Government: Urban Planning Society of China, 89–103.

Frontiers in Civil and Hydraulic Engineering – Mohamed A. Ismail and Hazem Samih Mohamed (Eds)
© 2023 The Authors, ISBN 978-1-032-38247-0

Research on multi-dimensional life cycle management system for Island projects based on BIM-GIS integration

Jiayi Tian
School of Earth Sciences and Engineering, Sun Yat-sen University, Zhuhai, Guangdong, China

Yan Gao*
School of Earth Sciences and Engineering, Sun Yat-sen University, Zhuhai, Guangdong, China, China Southern Marine Science and Engineering Guangdong Laboratory (Zhuhai), Zhuhai, Guangdong, China

ABSTRACT: The report of the 19th National Congress of the Communist Party of China proposed a strategy to accelerate the development of China into a strong maritime country. It is imperative for China to develop and build islands and reefs in the South China Sea, in the long run. However, the development of islands and reefs in the South China Sea also faces many problems, such as complex geological conditions, the corrosive effect of seawater on engineering structures, and vulnerability to earthquake and typhoon disasters. This paper proposes a BIM-GIS integrated application framework, which can build an interactive platform for visualization of the island and reef projects, enabling the integration of BIM single-project microscopic and GIS multi-project macroscopic scales, the multi-dimensional fusion of multi-source heterogeneous data, and the life cycle management of island and reef projects.

1 INTRODUCTION

The South China Sea is a vast area, rich in fishery and mineral resources, and has important geographical advantages. With the concept of the "21st Century Maritime Silk Road", China has carried out large-scale infrastructure construction on the islands and reefs in the South China Sea.

The integration of BIM-GIS has become a major research hotspot in the field of construction engineering in recent years and is currently mainly used in the management phase of construction projects. With the maturity and diversification of technology application, this technology is also gradually used in the pre-project site selection, cost estimation, feasibility analysis, as well as the operation and maintenance phase of the building (Sun et al. 2022).

In particular, the operation and maintenance phase of a building is the most-costly and longest-lasting phase in the life cycle of a building, which has a huge amount of information but often lacks proper storage and management. This is particularly true for island and reef projects. Compared with other types of projects, the site selection and feasibility analysis in the early stage of island and reef projects, and the operation and maintenance and risk management in the later stage are more important. Because island and reef projects are closely related to their geographical locations in terms of site selection, material transportation, and risk prediction, BIM-GIS technology has great potential for application in the field of island and reef projects, but no researcher has yet applied this technology to island and reef projects.

---

*Corresponding Author: gaoyan25@mail.sysu.edu.cn

DOI 10.1201/9781003344209-64

2 BIM-GIS DATA INTEGRATION AND ITS SIGNIFICANCE TO THE LIFE CYCLE MANAGEMENT OF ENGINEERING

2.1 *Characteristics of BIM and GIS and their data integration*

BIM (Building Information Modeling) can digitally express the geometric and functional information of a building. With the development of the BIM technology, the range of building-related information carried by BIM has become wider. BIM provides an advanced digital tool and information sharing platform for project life cycle management and interaction between stakeholders, effectively reducing the costs and risks of building planning, design, construction, operation, and maintenance (Fernández-Mora et al. 2022). In the field of building information modeling, to achieve the purpose of information exchange and data exchange between various engineering disciplines, the basic standards for BIM data have been based on three aspects, namely terminology, storage, and process. The three corresponding IBM standards are the International Semantic Dictionary Framework (IFD), the Industry Foundation Classes (IFC), and the Information Delivery Manual (IDM). The IFC standard was published by buildingSMART and became an international standard in 2013 (ISO 16739). It is a complex data standard covering nine domains such as architecture, structure, HVAC, and electrical, specifying a data structure based on a 3D geometric model and object-oriented language representation. It is a prerequisite for the interdisciplinary and inter-technical application of BIM (Söbke et al. 2021).

Geographic Information Systems (GIS) can acquire, store, analyze, manage, and display many types of geographic data (Gluck & Yu 1999), and it is becoming an important part of civil engineering, urban planning, and land surveying and management operations. The Open Geographic Information Systems Consortium (OGC) has established the international standard Geographic Markup Language GML for modeling, transmitting and storing geographic and geographically related information. GML is an XML encoding language that defines the data that can be used for data exchange. CityGML is one of the data formats dedicated to urban modeling. CityGML model consists of two levels: semantic and geographic features. The model consists of buildings, transportation networks, hydrology, vegetation, etc.

In the existing BIM-GIS application, the BIM system is usually exported to the IFC data format, converted to the CityGML format, and then imported into the GIS system. That is because the GIS system mostly uses surface models such as B-rep (Boundary representation) to express model geometric relationships, and they are more difficult to be converted to the solid model in the BIM system. In terms of practical engineering applications, Wang et al. (2019) established a data integration framework for urban underground pipelines, which includes geometric models for each stage of the project, database management system (DBMS), and management platform, solving the problems of lack of a digital information model of a comprehensive database, lack of unified platform management, and difficulties in condition assessment and planning maintenance for urban underground pipeline management. Tan et al. (2018) developed a network system based on BIM and GIS to solve the optimization of ship transportation plan of offshore platforms, including the layout on the ship and the transportation plan between each platform, BIM is used to provide the information of ship layout optimization, GIS is used to provide the spatial information of ship transportation, and the optimal plan is obtained by comparing the results of genetic algorithm, particle swarm algorithm, and fuzzy algorithm in the system. The results indicate that this system can significantly improve the efficiency of offshore platform hoisting.

2.2 *The significance of BIM-GIS integration for the life cycle management of engineering*

The theory of total life cycle management first originated in the aerospace manufacturing industry and is referred to as PLM (Product Lifecycle Management). It improves the use of resources and helps in decision making by sharing and managing data, which improves project and enterprise performance (Aram et al. 2013). In engineering projects, the design planning, construction, operation and maintenance aspects involve many industries and are highly fragmented, especially for the disconnection of information between operation and maintenance and other parts of the project after completion is particularly serious. These problems may lead to project cost overruns, schedule delays, operational inefficiencies, and maintenance difficulties.

BIM enables the exchange of information among multidisciplinary teams during the life cycle of an engineering project. The National Building Information Modeling Standard (NBIMS-US™) defines BIM as "a digital representation of the geometric and functional characteristics of engineering structures". BIM is a shared information resource about engineering structural information and helps make reliable decisions during the life cycle of an engineering project. The combination of the BIM and product life cycle management theory has led to the recent hot research direction of Building Lifecycle Management (BLM), which is a holistic approach developed to achieve the integration of engineering "design-construction-operation" management process. However, to realize the life cycle management of engineering projects, the influence of the surroundings of the project is also an important part, especially for island projects, the project site selection, material selection and transportation, risk prediction, etc. They are closely related to the geographical location, and the application of BIM-GIS technology can better solve these problems.

Both BIM and GIS are associated with information repositories and data management. BIM features detail-oriented information inclusion, while GIS focuses on analysis and management of geospatial data. Combining the life cycle information management capabilities provided by BIM with the location clarity provided by GIS is conducive to achieving better project life cycle management (Huang et al. 2021).

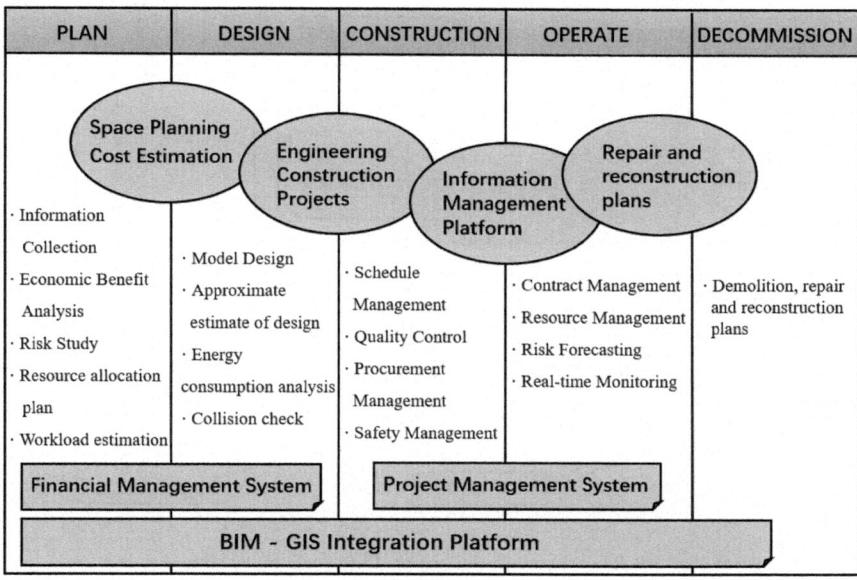

Figure 1.   Content and technology of project life cycle management.

3   ISLAND ENGINEERING LIFE CYCLE MANAGEMENT SYSTEM ARCHITECTURE

3.1   *Information flow*

Good information flow allows for more efficient communication between individuals and teams, delivery of results, indication of requirements, specification inquiries, design solutions, etc. In today's engineering construction, the amount of data generated during the engineering process has increased significantly with the changes in construction apparatus, the application of new sensors, and the overall trend toward smarter engineering projects. Complex processes and large amounts of data can jeopardize proper workflow between teams. Poor process design can lead to various types of waste, such as excessive rework and revision cycles, design errors, quality degradation cost overruns, and schedule delays (Ballard 2002). To convert the huge amount of data into actual information quickly and to exchange information among all stakeholders more efficiently, a more rational information flow and a data management solution need to be designed.

In a general BIM-GIS application, the information flow model is relatively fixed. Ma and Ren (Ma & Ren 2017) identified three information flow models, i.e., from GIS to BIM, from BIM to GIS, and from BIM/GIS to third-party systems. The information flow model proposed in this paper is shown in Figure 2. Its overall model belongs to "BIM/GIS to third-party system", and the overall information flow of the management system is designed based on the previous model for the needs of the life cycle management of the island and reef project, where the arrow indicates the direction of the information flow.

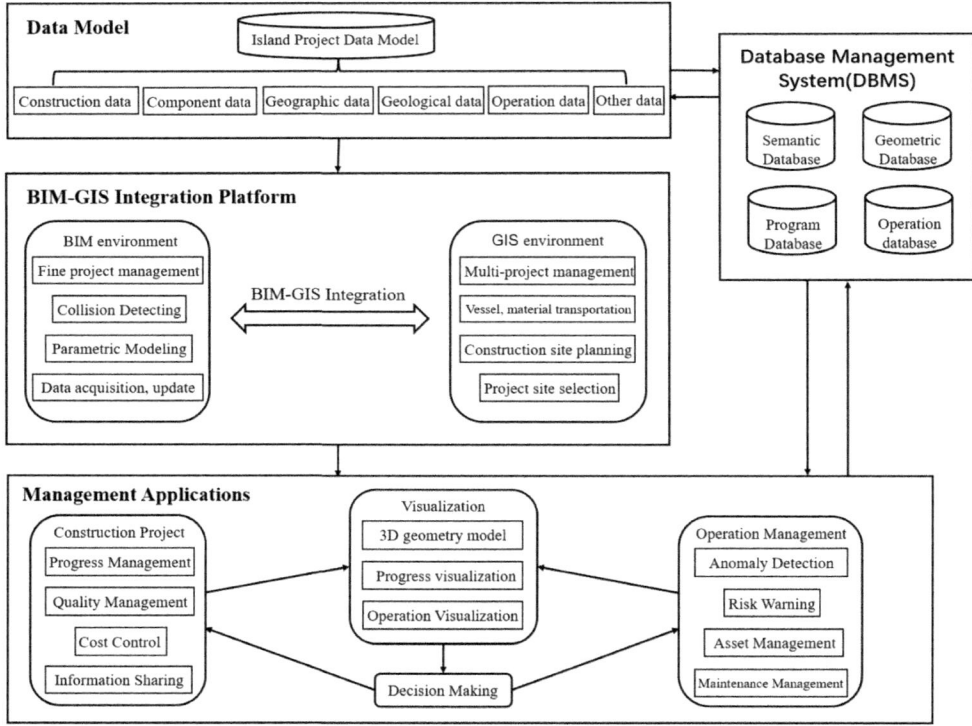

Figure 2.   Information flow model of the multi-dimensional life cycle management system.

3.2 *Management system contents*

The multi-dimensional island engineering life cycle management system proposed in this paper covers from the site selection and feasibility analysis in the pre-development stage of the island to the engineering construction of transportation facilities and buildings on the island and the operation stage of the island. The overall content is shown in Figure 3.

In the early stage of the island and reef project, through geological data collection, combined with the GIS system related to meteorology, disaster analysis, and other information, the system can compare and choose the project address, simulate the project process, and fully analyze the possible problems and risks of the project. Through the life cycle management system of the island and reef

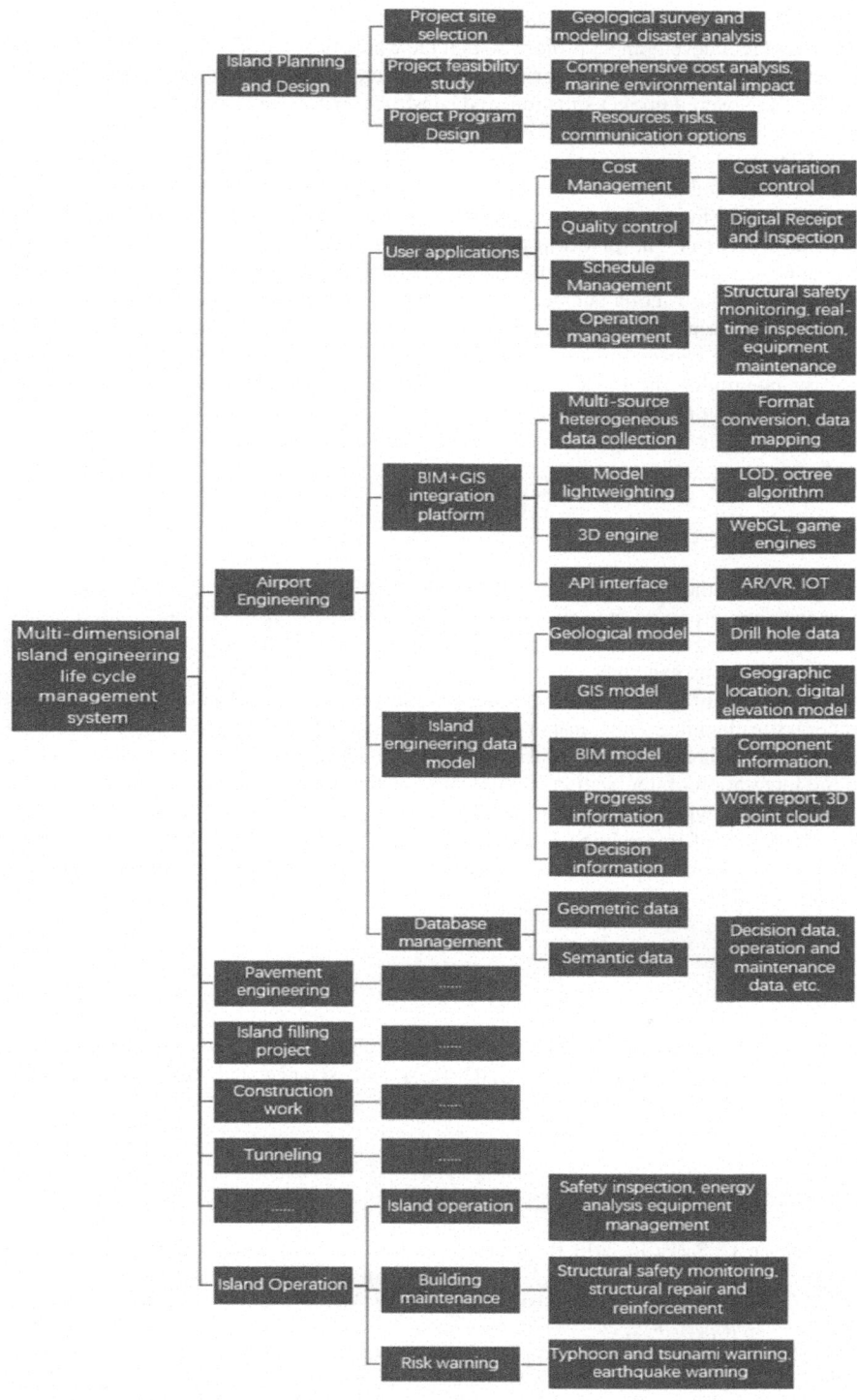

Figure 3.　Functions and contents of the multi-dimensional life cycle management system.

project, stakeholders can effectively view, manage, and communicate the information related to each island and reef development project, so that the whole process of island reef development can be informatized and visualized. This enables the decision-making, planning, and design personnel to understand the overall project situation, and construction personnel and management personnel to understand the specific information of the project.

BIM-GIS technology is applied to convert the format of BIM and GIS files. The model is then meshed, rendered, and displayed on the user terminal through the 3D engine, covering all types of projects involved in island development, such as island reclamation projects, airport projects, pavement projects, etc. The platform collects multi-source heterogeneous data for multi-scale planning and management of the project. The 3D engine enables the visualization of data and content related to construction engineering, operation, and maintenance. Different personnel can access relevant information through mobile devices, personal computers, and AR/VR devices, which facilitates the management of engineering projects, information interaction, and decision making. By integrating the modern monitoring sensor communication technology, the system can realize the functions of building safety dynamic assessment, operation management, decision management, and risk warning for island operation and maintenance.

4 CONCLUSION

This paper analyzes the significance of BIM-GIS technology for the life cycle management of the island project, and proposes the realization method of applying BIM-GIS integration technology to the life cycle management of island and reef projects. The BIM module can realize the functions of component level management, cost control, collision check, proposed geological model, and effect display of the project. The GIS module can carry out multi-project integrated management, construction site planning, ship transportation planning, disaster warning and analysis, and analysis of the degree of influences of regional seawater on the project structure. The whole system can realize the integration of applications in all stages of the project, the integration of management of different project types, the integration of fine single-project and macro multi-project scales, and the integration of information in the whole-time dimensions from project design planning to construction operation, proposing an efficient and comprehensive information management solution for island and reef projects.

ACKNOWLEDGEMENT

The research was supported by National Natural Science Foundation of China (42072295, 41807244), Guangdong project (2017ZT07Z066) and Science and Technology Program of Guangzhou, China (201904010415).

REFERENCES

Aram, S., Eastman, C., Sacks, R., 2013. Requirements for BIM platforms in the concrete reinforcement supply chain. *Automation in Construction*. 35, 1–17. https://doi.org/10.1016/j.autcon.2013.01.013

Ballard, G., 2002. Managing work flow on design projects: A case study. Engineering, *Construction and Architectural Management*. 9, 284–291. https://doi.org/10.1108/eb021223

Fernández-Mora, V., Navarro, I.J., Yepes, V., 2022. Integration of the structural project into the BIM paradigm: A literature review. *Journal of Building Engineering*. 53, 104318. https://doi.org/10.1016/j.jobe.2022.104318

Gluck, M., Yu, L., 1999. Geographic information systems, in: Chapman, E.A. (Ed.), *Advances in Librarianship, Advances in Librarianship*. Emerald Group Publishing Limited, pp. 1–38. https://doi.org/10.1108/ S0065-2830(1999)0000023003

Huang, M.Q., Ninić, J., Zhang, Q.B., 2021. BIM, *Machine learning and computer vision techniques in underground construction: Current status and future perspectives*. Tunnelling and Underground Space Technology 108, 103677. https://doi.org/10.1016/j.tust.2020.103677

Ma, Z., Ren, Y., 2017. Integrated Application of BIM and GIS: An Overview. *Procedia Engineering*. 196, 1072–1079. https://doi.org/10.1016/j.proeng.2017.08.064

Söbke, H., Peralta, P., Smarsly, K., Armbruster, M., 2021. An IFC schema extension for BIM-based description of wastewater treatment plants. *Automation in Construction*. 129, 103777. https://doi.org/10.1016/j.autcon.2021.103777

Sun, S., Song, Y., 2022. Research on construction management of water conservancy project life cycle based on BIM+GIS. *China Rural Water and Hydropower*. 4, 1–13.

Tan, Y., Song, Y., Zhu, J., Long, Q., Wang, X., Cheng, J.C.P., 2018. Optimizing lift operations and vessel transport schedules for disassembly of multiple offshore platforms using BIM and GIS. *Automation in Construction*. 94, 328–339. https://doi.org/10.1016/j.autcon.2018.07.012

Wang, M., Deng, Y., Won, J., Cheng, J.C.P., 2019. An integrated underground utility management and decision support based on BIM and GIS. *Automation in Construction*. 107, 102931. https://doi.org/10.1016/j.autcon.2019.102931

*Frontiers in Civil and Hydraulic Engineering – Mohamed A. Ismail and
Hazem Samih Mohamed (Eds)
© 2023 The Authors, ISBN 978-1-032-38247-0*

Construction of landscape ecological tea garden based on homeostasis theory: A case study of Wuyishan National Forest Park

Feifan Weng, Yali Wang, Xufang Li & Zheng Ding*
College of Landscape Architecture and Art, Fujian Agriculture and Forestry University, China

ABSTRACT: Since ancient times, China has been a big country in tea cultivation. Tea culture and tea farming civilization are very profound. As one of the agricultural resources, an ecological tea garden not only has the form of a garden landscape but also has the value of tourism development. Due to its unique geographical and climatic conditions, Wuyishan has been an important source of tea in China since the Tang Dynasty. Wuyishan National Forest Park, as an important producing area of high-quality tea in Wuyi Mountain, not only has a good tea planting environment, and rich tea resources, but also has accumulated a long history of tea planting. It has unique human and natural advantages in the construction of landscape ecological tea gardens. Taking Wuyishan National Forest Park as the research object, this paper analyzes the ecological elements and landscape elements of tea planting in this area, which receives the construction strategy of landscape ecological tea garden under the harmonious state of "internal balance" between the ecological elements and landscape elements.

1 INTRODUCTION

The homeostasis paradigm of ecosystem balance originated from Eugene, a famous ecologist. In the ecological balance theory proposed by Odum, the concept of integrity is placed at the core of ecology, making the principle of integrity the first principle in contemporary ecological theories (Shi 2012). In Odum's view, ecology is a discipline specializing in the study of "harmony", carrying out relevant studies on the structure of ecological systems. In his theory of "homeostasis", the ecosystem, as its basic unit, focuses on the interconnection and interaction of various elements in the ecosystem, and believes that when various ecological elements in the system reach the state of "homeostasis". The ecosystem can achieve its ultimate goal of health and order. In addition, Utrom integrates the human world with the natural ecological environment to create a new systematic whole. In the harmonious development process of "homeostasis" in the ecological system, human society and human behavior become an indispensable part. It is similar to the natural ecological theory of "unity of nature and man" in ancient China, which undoubtedly has great historical value and reference significance for constructing a more harmonious community of nature-human life in contemporary times.

As the tea name card of China and even the world, Mount Wuyi has a profound tea culture and rich tea resources. Zhengshan souchong and Wuyi rock tea were born in Wuyi Mountain area in history (Liao 2018). Relying on its unique natural conditions, Wuyishan National Forest Park has been an important node of tea production since ancient times. However, in the context of the booming development of contemporary tea production and the continuous rise of tea demand, the disorderly development of tea gardens appeared in Wuyishan National Forest Park. In addition, with the increase in the number of tourists, the external interference of tea gardens in the National

---

*Corresponding Author: 2678097376@qq.com

DOI 10.1201/9781003344209-65

Forest Park was gradually strengthened, and the ecology of tea gardens suffered a certain degree of imbalance. It is urgent to construct a landscape ecological tea garden that balances the ecological requirements of tea planting and the landscape requirements of tourists. In this study, from the homeostasis of the ecological theory, internal tea growing in the Wuyishan National Forest Park was analyzed. Combined with the status quo of internal human factors in the park, analysis is carried on, thus drawing the construction of the landscape with an ecological tea garden strategy. In the Wuyi mountain area of the landscape, ecological tea plantation and tea garden ecological protection dedicate a meager force. Details can be seen in Equation 1 below:

Based on the slope calculation formula, field-related data can be obtained:

$$\tan\angle\alpha = h/l$$
$$i = h/l \times 100\% \tag{1}$$

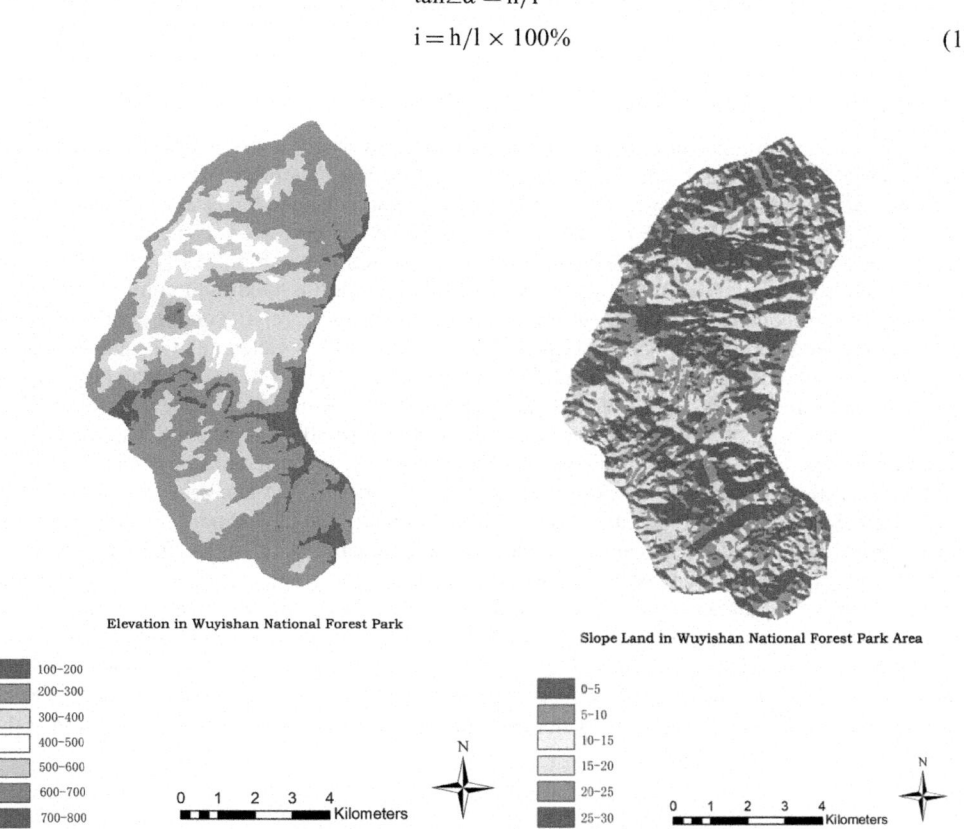

Figure 1.   Elevation in Wuyishan National Park.   Figure 2.   Slope in Wuyishan National Park.

Table 1.   The proportion of field slope.

Slope (degrees)	0–5	5–10	10–15	15–20	20–25	25–30
Elevation (meter)	100–400	200–400	400–500	400–500	500–700	500–800
Proportion (percentage)	27%	25%	8%	16%	7%	17%

2 ECOLOGICAL TEA GARDEN AND ITS LANDSCAPE PLANNING AND DESIGN

2.1 *Overview of landscape ecological tea garden*

Ecological tea gardens are the sustainable development of the tea industry and tea culture. The traditional tea garden and the special connotation of tea culture are served as the basis of landscape construction. At the same time, relying on science, tea production patterns of green ecological development will be in the form of tourism and the organic combination of the tea industry, realizing the sustainable development of the tea industry and building the ecological tea garden. It is very important to choose an environment with superior natural conditions. To realize the organic combination of the dual properties in ecological tea garden production and tourism, it is crucial to combine the ecological and scientific industrial model with the characteristics of art appreciation, which can not only satisfy the visual and sensory needs of tourists but also meet their material needs (Yang 2017).

2.2 *Analysis of key elements in the construction of landscape ecological tea garden*

The traditional tea garden construction mode in mountainous areas has disadvantages of water and soil conservation, low yield, and low resource utilization rate. Therefore, the construction of a landscape ecological tea garden with multiple ecologies and balanced landscape elements should grasp the tight coupling balance between elements. At the ecological level, the body balance is mainly for landscape tea plantations in the altitude, slope, light, moisture, heat, temperature, soil fertility, and other ecological elements. The coupling between the maximum combination takes advantage of the ecological environment and characteristic industry, relying on the good ecological system of the landscape ecological tea garden inside to achieve high yield and high-quality results. Location selection of ecological elements should be in the fresh air, water, soil, clean clear, trying to stay away from the noisy place of national forest park. As for the main hub, it should be less than 1900 m above sea level, and the slope should be less than 30° (Liu 2019). Using the GIS to the elevation and slope of Wuyishan National Forest Park for the detection, it is concluded that most areas in the region meet the above conditions. In addition, the site selection should have good irrigation conditions, good land structure, and soil type. The land should be with shade and sunrise, more cloud and fog, high humidity, and scattered light intensity. At the landscape level, the body balance is mainly for ecological tea garden landscape, geographical features, irrigation and drainage system and layout, road network, and connection. Doers of the word woods and tie-in, green landscape, and the coupling among the landscape elements such as layout only give full play to the process of developing ecological tea garden element homeostasis effect on the ecological landscape. Only in this way can we effectively regulate and control the dominant superposition relationship between various ecological elements and landscape elements in the construction of the whole landscape ecological tea garden.

3 THE STRATEGY SUGGESTION OF LANDSCAPE ECOLOGICAL TEA GARDEN CONSTRUCTION

3.1 *Adapt measures to Local conditions: scientific site selection of landscape ecological tea garden*

Landscape ecological tea garden is based on tea tree species and landscape for production. As a supplement to other landscape elements, scientific location can make regional ecological elements a reasonable combination and landscape elements, achieving homeostasis of the ecological and landscape system, and giving full play to the tea tree production. Therefore, the function complementarity of each living organism can be achieved in the landscape ecological tea garden and the advantage of superposition of environmental factors in the biosphere. On the landscape ecological tea garden's location, one should choose the fresh air, water, soil, clean clear, trying to stay away from the noisy place national forest park. The main hub should be in places that are less than 1900 m above sea level. The slope is less than 30m. The site selection should be close to the water at the same time, which is advantageous to the irrigation activities. Secondly, the soil in the region should have moderate acidity, good soil structure, tight but unsolid subsoil, and core soil, which is convenient for air permeability and water circulation, and also conducive to water storage and fertilizer preservation. Finally, in the problem of site selection, the shade and sunrise should be preferred with cloudy, humid, and scattered light.

3.2 Protecting mountains and forests: environmental construction of landscape ecological tea garden

Under the situation that the forest environment is destroyed due to the wanton development of tea gardens, it is very important to carry out planned afforestation in the landscape ecological tea garden. First, soil erosion in the tea garden should be prevented, and trees and grass should be planted beside roads and channels to prevent rain erosion. Secondly, the shelter belt should be built according to the terrain's hierarchical progressive structure to play a dual role of protection and landscape. At the same time, tree species should be planted reasonably in the shelter belt, not affecting the growth of the tea and tea industry and considering the appreciation of the landscape. Fourth, the grass cover should be replanted reasonably. Reclamation of tea gardens will inevitably damage the grass cover plants in the region, which is easy to cause soil erosion and other problems. Therefore, vegetation restoration and soil fertility improvement should be implemented in accordance with the overall planning and design of the tea garden.

3.3 The way of principle: the construction and management of landscape ecological tea garden

For the construction of a landscape ecological tea garden, scientific methods should be taken as the construction guideline. Scientific planning should be made, and reasonable areas of the tea garden should be controlled. Tea quality should be guaranteed, and the overall construction idea of sustainable development should be followed. The second is to implement project driving and pay attention to integrated demonstration. The principle of priority of ecological protection should be adhered to, highlighting green garden construction and promoting the coordination between industry and ecology. Various elements should be coordinated in the landscape ecosystem, improving the coordination mechanism, strengthening the green guarantee, and guaranteeing the internal balance of the ecosystem.

3.4 Unity of nature and man: landscape construction of landscape ecological tea garden

In the configuration of the plant landscape in the landscape ecological tea garden, an effective protection system of "tree-shrub-vegetation" should be formed in the front layout, and a linked ecological niche of "tree-tea tree-green manure" should be formed in the plane, which plays a positive role in regulating the ecological factors of crops in the lower layer by trees in the upper layer (Liu 2019). The landscape construction of a landscape ecological tea garden should be arranged according to local conditions and related topography. At the same time, a tea garden branch road can be built for sightseeing and a fast and slow road for sports and fitness. Sightseeing buildings can be set in a higher position to give full play to the functional diversity of the landscape ecological tea garden.

4 CONCLUSION

The construction of a landscape ecological tea garden cannot only create more abundant economic benefits for the tea garden but also increase the appreciation and carry forward the excellent tea culture. When planning and constructing landscape ecological tea gardens, it is necessary to follow the principle of harmonious coexistence between landscape elements and ecological elements, so that the ecosystem and landscape system can reach a state of internal balance, and the construction of landscape ecological tea gardens in the new era is undoubted of great significance.

ACKNOWLEDGMENT

The work was financially supported by Research on the integration and innovation development mode of Fujian rural Courtyard and economic industry (FJKX-A2008).

REFERENCES

Lia Lingyun, Hou Shuyu, Yang Rui. (2018). Comparative study on the construction and management of tea garden in Wuyi mountain from the perspective of ecological wisdom. *J. Chinese Landscape Architecture*, 34(07), 53–58.

Liu Penghu, Luo Xuhui, Wang Yixiang, Zhang Wenjin, Wang Dingfeng, Weng Boqi. (2019). The coupling of "production-ecology-life" in the mountain tea garden and its technical countermeasures. *J. Acta Ecologica Sinica*, 39(19), 7047–7056.

Liu Penghu, Wang Yixiang, Huang Ying, Luo Xuhui, Gao Chengfang, Weng Boqi. (2019). Research on the tea plantation model of "production-production-production" coupling in the mountainous area. *J. Chinese Journal of Eco-Agriculture* (in Chinese), 27(05), 785–792.

Shi Lina, Kingpin. (2012). Eugene Odum's ecological balance theory and its practical significance. *J. Scientific Theory*, 12, 42–43.

Yang Chao, Li Hai-Yi, Wang Shuang, Zhang Ying-yue, Chen Shu-ying. (2017). Study on landscape ecological planning of tea garden. *J. Fujian Tea*, 39(04), 102–103.

Frontiers in Civil and Hydraulic Engineering – Mohamed A. Ismail and
Hazem Samih Mohamed (Eds)
© 2023 The Authors, ISBN 978-1-032-38247-0

Energy-saving renovation of residential building envelope based on BIM

Lun Gao*

Tianjin University Research Institute of Architectural Design and Urban Planning Co., Ltd, Tianjin, China

ABSTRACT: At present, China's residential buildings consume a large amount of natural energy, which seriously affects the healthy development of the ecosystem. To improve the building energy efficiency, this paper proposes research on the energy-saving renovation of residential building envelopes based on building information modeling (BIM). BIM technology is used to establish a residential model, and the energy consumption of residential buildings is simulated and analyzed. Combined with the analysis results, the external walls and roofs of residential buildings are retrofitted for energy-saving. Finally, the energy consumption simulation software is used to set up comparative experiments, and the energy consumption of the proposed building energy-saving renovation scheme is compared with that of the conventional scheme. The experimental results show that the energy-saving retrofit scheme of building envelope based on the BIM technology has lower energy consumption and a better energy-saving effect. It is hoped that this study can provide theoretical and practical guidance for the promotion of energy-saving renovation of residential buildings.

1   INTRODUCTION

With the improvement of people's environmental protection awareness, most construction units will incorporate the concept of green energy saving and environmental protection in the design and construction process. The application of BIM technology can simulate the energy-saving design and construction scheme, calculate its building energy consumption and building material consumption, and achieve the purpose of energy saving and environmental protection through this process. BIM technology can further optimize and analyze the building according to the environmental climate characteristics of the construction area and the construction site conditions. For example, using BIM technology to simulate the lighting around the building can optimize the design of solar energy resources to achieve building energy efficiency (Moortel 2022). Currently, the construction industry is one of the four pillar industries in my country, and many buildings have been built in our country, of which residential buildings are the main buildings. Due to the lack of good high-tech for a long time, its energy consumption is relatively large. Although my country has advocated paying attention to its greenness, energy-saving, and environmental protection in the process of architectural design and construction, its application effect is still not good, and the current residential buildings still have problems such as high energy consumption and poor comfort. With the increasing harm of residential buildings to the environment, it is urgent to solve the problem of building energy consumption in my country (Jia 2022). BIM technology can continuously improve the level of construction technology and the rational use of resources. At the same time, it also improves the scientificity of the construction scheme. Simply speaking, a green building is an energy-saving building (Vb 2022).

---

*Corresponding Author: aatu02@126.com

The three-dimensional visual simulation of the BIM technology is to use computer software to accurately estimate and calculate building energy consumption and building material consumption. In addition, BIM technology also optimizes the design of solar energy resources through the lighting around the building to achieve the purpose of building skills, and ultimately realize the reuse of energy in people's real life (Huang 2021). The total number of buildings in China is large. The proportion of residential buildings is high, and the energy consumption generated by residential buildings is huge. Discordant energy consumption patterns and housing construction methods have led to high energy consumption and poor comfort in current residential buildings. So it is urgent to solve the problem of building energy consumption in China (Las-Heras-Casas 2021). Residential energy-saving renovation design is an important part of achieving the goal of energy-saving and emission reduction in China. At the same time, with the development and application of BIM technology, the quality of energy-saving design has been greatly improved. So it will be of great practical significance to combine BIM technology with the energy-saving renovation of old residential buildings (Wu 2021).

The conventional evaluation system usually has the problems of low accuracy of energy consumption simulation and incomplete evaluation dimensions, which leads to the fact that the energy-saving renovation of the residential building envelope in China can not achieve the prescribed energy-saving goals, which is very unfavorable to the sustainable development of ecological environment (Aydn 2020). Therefore, it is necessary to develop a set of scientific renovation schemes to promote the development of building energy-saving renovation in China (Xu 2022).

2 ENERGY-SAVING RECONSTRUCTION OF THE RESIDENTIAL BUILDING ENVELOPE

2.1 *Energy consumption analysis of the residential building envelope*

Before the energy-saving renovation, BIM technology is used to establish the building model and analyze the energy consumption to provide data support for the subsequent renovation work (Mi 2021). Because the Revit Architecture software in the BIM technology is superior to other building simulation software in function, characteristics, data exchange, and other aspects, its powerful computing power and simulation performance are very suitable for building simulation analysis, so Revit Architecture is selected as the modeling software for this building model. The specific modeling steps are shown in Figure 1 below (Long 2021).

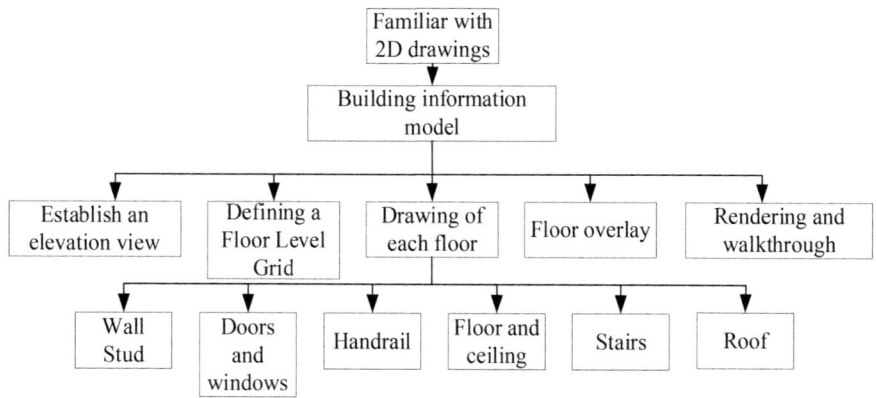

Figure 1.   Flow chart of Revit construction steps.

The BIM technology is used to understand and master the drawings, establish the building information model, and input the material, thickness, height, and other information of the building's internal components into the software. Then we establish elevation views, define grid widths and heights for floor levels, and detail drawings for each floor, including components such as wall columns, doors and windows, handrails, floor ceilings, stairs, and roofs. Finally, the information model of the whole building is obtained by superimposing and integrating the drawn floors and rendering them (Dai).

After obtaining the building model established by BIM technology, the energy consumption is calculated and analyzed by professional BIM energy consumption analysis software, mainly analyzing the thermal insulation performance of the building. The better the thermal insulation performance of the building envelope, the lower the energy consumption effect, and the more energy-saving effect can be achieved. Before analyzing the energy consumption of the envelope, the model parameters are debugged first, including meteorological information, material information, and environmental properties. After entering the basic information, the energy consumption can be analyzed. Through the energy consumption analysis results, the specific values of building heating and cooling energy consumption in different months can be obtained, which provides data support for material selection and scheme determination of energy-saving renovation. In addition, the requirement table of the building envelope simulation object can also be derived, as shown in Table 1 below (Lu 2022).

Table 1. Simulation object parameter requirements.

Parameter properties	Demand value
Ambient temperature	5–40°C
Ambient humidity	≤70%
Ventilation flow	≥0.15 m³/s
Atmospheric pressure	80–106kpa
Heating energy consumption	≤100000 Wh
Refrigeration energy consumption	≤100000 Wh

As shown in Table 1, the derived table specifies the requirements for the parameters of the simulation object, including the requirements for heating and cooling energy consumption. The energy consumption of each month and the average energy consumption of each month can help architects better understand and control the energy consumption of building internal components, and accurately evaluate the direction of energy-saving renovation.

2.2 Renovation of external walls and roofs

With the residential model constructed by the BIM technology in the above steps and the simulated energy consumption of the building as the data support, the building envelope is specifically transformed for energy-saving, mainly divided into the transformation of the external wall and roof envelope.

Generally, there are three types of external wall insulation: internal insulation, intermediate insulation, and external insulation. Since external insulation has better energy consumption and high durability, external insulation is selected for the energy-saving transformation of the external wall. The materials with less energy consumption are selected for construction based on energy-saving requirements. The specific structure is shown in Table 2 below.

The cement vermiculite insulation layer is used for the roof enclosure structure, and the original protective layer needs to be removed. In order to ensure compatibility between the insulation layer and the base layer of the roof, the waterproof layer shall be treated and removed according to the actual situation of the waterproof layer. If necessary, the leveling layer shall be repaired after removal to ensure the stability of the roof. The specific construction and transformation process is shown in Figure 2 below.

Table 2. Basic structure of rubber powder polystyrene particle sandwich polystyrene board.

Bonding layer	Insulation layer	Leveling course	Anti-crack protective layer	Veneer layer
Base course interface mucilage + glue powder polyphenyl granule bonding evelling material	Polyphenyl plate + surface with a trapezoidal groove	Adhesive powder polyphenyl granule bonding evelling slurry	Anti-crack mortar composite alkali-resistant glass fiber mesh cloth	Flexible waterproof putty + coating

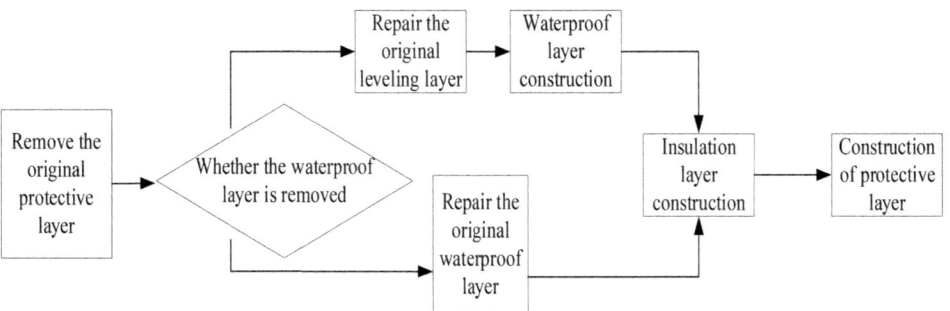

Figure 2. Construction process of roof envelope structure reconstruction.

The construction method from top to bottom is adopted for the reconstruction of the roof enclosure structure. The 25 mm anti-cracking mixed mortar protective layer is selected for the protective layer; a layer of non-woven polyester chemical fiber cloth is laid for heat insulation; the thickness of the original mixed mortar leveling layer is 25 mm; the thickness of the cement vermiculite insulation layer is 60 mm, and the thinnest part of the original cement slag sloping layer is kept at 25 mm.

3 EXPERIMENTAL ANALYSIS

To better illustrate that the proposed energy-saving renovation scheme of residential envelope based on BIM technology is superior to other renovation schemes in terms of energy consumption, after the completion of the theoretical part of the design, the experimental test link is constructed, and the actual energy consumption effect of the energy-saving renovation scheme of the residential envelope is compared and analyzed.

3.1 Description of the experimental environment

In this experiment, the HONTECT energy consumption simulation software is used to analyze the energy consumption of the designed transformation scheme. First of all, the model maps of the two transformation schemes are input into the software, and then the meteorological data are input. The average monthly temperature, air humidity, and radiation temperature in northern China are selected as specific parameters, and finally, the comparison parameters are set. The index used in this comparative experiment is the monthly energy consumption value of the transformation scheme, and the smaller the value, the more significant the energy-saving effect.

3.2 Calculation of building heat consumption index

After adjusting the test parameters, it is necessary to confirm the energy consumption index. The most intuitive indicator of the level of building energy consumption is the heat consumed by the building, specifically the heat consumed by the residential building at a certain time and provided by the construction of internal heating equipment, in the unit W/m². The specific calculation formula is as follows:

$$q_H = q_{HT} + q_{INF} - q_{IH} \tag{1}$$

In the above formula, q_H represents the heat consumption index of the building; q_{HT} represents the heat conducted by the building envelope within a certain period of time on the unit building area; q_{INF} represents the heat conducted by the air within a certain period of time on the unit building area; q_{IH} represent the heat consumed by the component itself within a certain time on the unit building area. The formula for the heat consumption of unit building area through the heat conduction of the building envelope and the heat consumption of air conduction is as follows:

$$q_{HT} = q_{Hq} + q_{Hw} + q_{Hd} + q_{Hmc} + q_{Hy} \tag{2}$$

$$q_{INT} = (t_n - t_e)(C_p \cdot \rho \cdot N \cdot V)/A_O \tag{3}$$

In the above formula, q_{Hq}, q_{Hw}, q_{Hmc} and q_{Hy} respectively represent the heat consumed by the unit residential area through the internal and external walls, upper and lower floors, strata, and balconies within a certain period of time. t_n represents the average temperature inside the room; t_e represents the average temperature outside the room; C_p represents the heat capacity per unit mass of air; ρ represents the mass per unit volume of air; N represents the frequency of ventilation; A represents the area of the dwelling.

Substitute the simulated values into the above formula. The specific index of building heat consumption can be obtained, which provides a comparison standard for experimental comparison.

3.3 Analysis of experimental results

According to the above experiment, the experimental results are shown in Figure 3 below.

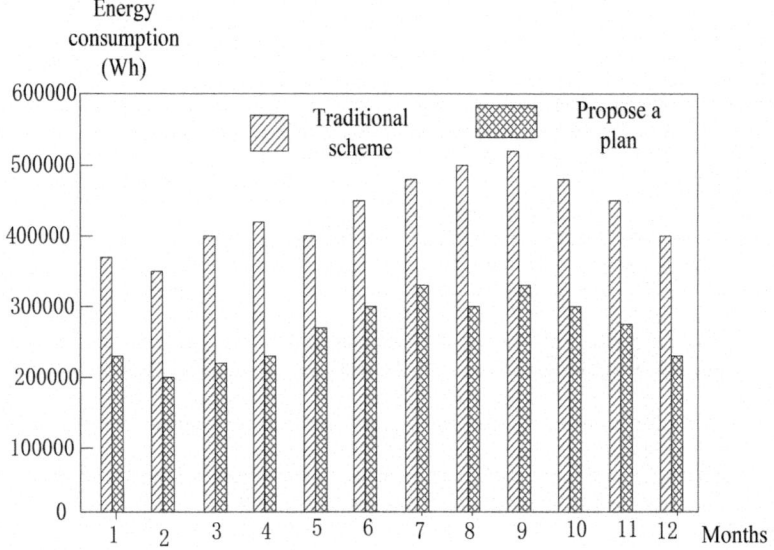

Figure 3. Analysis of experimental results.

The experimental results show that the building energy consumption is generally high in summer and relatively low in winter, which is caused by the large energy consumption of air conditioning and other refrigeration in summer. By comparing the specific energy consumption value, it can be seen that the energy consumption value of the traditional transformation scheme is higher. The highest month can reach 500000Wh, and it fluctuates greatly under the influence of the season.

The energy consumption value of the proposed energy-saving scheme of the residential building envelope based on the BIM technology is significantly lower than that of the traditional renovation scheme, which is maintained below 300000Wh and is not easily affected by seasonal changes. The fluctuation of energy consumption is more stable, which shows that the proposed renovation scheme has a better energy-saving effect and can effectively reduce the energy consumption of the building envelope.

4 CONCLUSION

The energy-saving renovation scheme of the residential building envelope proposed in this paper is effectively combined with BIM building information model technology, which greatly improves the efficiency of energy-saving renovations of the residential building envelopes. It can scientifically and reasonably analyze the specific situation of building energy consumption, and effectively improves the accuracy of energy consumption simulation.

The research method of this paper is helpful in providing scientific guidance for architects to carry out building energy-saving renovations and promote the intelligent development of residential building energy-saving renovation in China. It can be used to guide the energy-saving renovation of the residential building envelopes, and has a positive role in promoting energy conservation and improving the living environment of Chinese residents.

REFERENCES

Aydn D., Mhlayanlar E. (2020) A case study on the impact of building envelope on energy efficiency in high-rise residential buildings. *Social Science Electronic Publishing*, 2020(1).

Dai Aibing, Energy-saving analysis of low carbon building based on BIM, *Low Carbon World*, pp. 140–141.

Huang J., Wang S., Teng F. (2021) Thermal performance optimization of envelope in the energy-saving renovation of existing residential buildings. *Energy and Buildings*, 247(8):111103.

Jia J., Huang Z., Deng J. (2022) Government performance evaluation in the context of carbon neutrality: Energy-saving of new residential building projects. *Sustainability*, pp. 14.

Las-Heras-Casas J., LM López-Ochoa, L.M., López-González, et al. (2021) Energy renovation of residential buildings in hot and temperate mediterranean zones using optimized thermal envelope insulation thicknesses: The case of Spain. *Applied Sciences*, 11(1): 370.

Long R., Li Y. (2021) Research on energy-efficiency building design based on BIM and artificial intelligence. IOP Conference Series: *Earth and Environmental Science*, 825(1):6.

Lu Jingjing, Xu Chuanchao. (2022) Structural design and construction of bucket foundation breakwater BIM applications. *Water Transport Engineering*, pp. 1–7.

Moortel E., Allacker K., Troyer F.D. (2022) *Dynamic versus static life cycle assessment of energy renovation for residential buildings.*

Mi P., Zhang J., Han Y. (2021) Study on energy efficiency and economic performance of district heating system of energy-saving reconstruction with photovoltaic thermal heat pump. *Energy Conversion and Management*, 247:114677.

Vb A., Pms B., Pgr B. (2022) Business models for supporting energy renovation in residential buildings. *The case of the on-bill programs.*

Xu J. (2022) Research on energy consumption control method of the green building based on BIM technology. *International Journal of Industrial and Systems Engineering*, pp. 40.

Wu W., Meng Q., Lv Y., et al. (2021). An optimization and evaluation method construction of zero-energy residential building renovation. *IOP Conference Series: Earth and Environmental Science*, 675(1):8.

Frontiers in Civil and Hydraulic Engineering – Mohamed A. Ismail and
Hazem Samih Mohamed (Eds)

Digital reconstruction and control technology of railway tunnel construction equipment in high altitude area – Based on drilling and blasting method

Ling Dai*
China State Construction Railway Investment Engineering Group Co., Ltd., Beijing, China
China State Construction Corporation Limited, Beijing, China

Pengpeng Zhao & Feng Xu
China State Construction Railway Investment Engineering Group Co., Ltd., Beijing, China

Shuguo Zhang
China State Construction Railway Investment Engineering Group Co., Ltd., Beijing, China
China State Construction Corporation Limited, Beijing, China

Xian Cui, Wei Wang & Kun Jia*
China State Construction Railway Investment Engineering Group Co., Ltd., Beijing, China

ABSTRACT: Aiming at complicated geological conditions, large buried depth, excessive toxic or harmful gases, and deep water wave height in the Sichuan-Tibet region, this paper investigates and analyzes the status of large mechanical construction equipment in a tunnel section of the Sichuan-Tibet Railway, and proposes conducting digital transformation measures for large-scale mechanical equipment, so that it can collect and transmit relevant key construction parameters in real-time. Then, the tunnel construction process is analyzed, predicted, and controlled based on the interactive control system of visual IoT for tunnel construction. Promoting digital tunnel construction through advanced means can reduce personnel investment, lower the probability of safety accidents, improve construction efficiency, ensure construction quality and reduce construction cost.

1 INTRODUCTION

As a major strategic project in the implementation of the "14th Five-Year Plan", the Sichuan-Tibet Railway runs from Chengdu in the east to Lhasa in the west, with a total length of 1,567 kilometers. The construction of the Sichuan-Tibet Railway would greatly impact the economic and social development of Sichuan and Tibet and has special significance in enhancing national unity and maintaining border stability (Chen 2022). The Sichuan-Tibet Railway spans the Hengduan Mountain region, with complex and changeable terrain and strong geological processes along the route. Under various coupled environments, adverse geological environments such as high stress, toxic and harmful gases, high-pressure water, high cold, high altitude, and frequent earthquakes are formed (Guo 2021). Under the influence of a complex geological environment and harsh construction conditions, the Sichuan-Tibet Railway tunnel is faced with engineering disasters such as large deformation of soft rock, mud and water bursting, rock burst, and a roof collapse in the construction process, which increases the difficulty of construction (Xue 2020).

At present, digital tunnel construction mainly focuses on TBM tunneling and operation monitoring and evaluation after the completion of construction. The content of data collection, analysis,

---

*Corresponding Authors: dailing@cscec.com and 2992334530@qq.com

and control in the construction process is less involved and partial to theoretical research (Zhang 2022). Although the railway department has implemented the information management platform and accumulated massive data in the process, these data have not been developed and cannot give feedback on the problems existing on the construction site (Zhu 2022). The tunnel of the Sichuan-Tibet Railway accounts for as high as 84.4% (Xie 2022), most of which are constructed by drilling and blasting methods. Unfortunately, most of the related equipment constructed by these methods does not have the conditions for data collection, storage, and transmission, so it is impossible to make a timely and effective judgment and control on-site construction.

Based on the field research and the actual situation of the tunnel construction site, this paper puts forward the digital transformation scheme of the equipment, such as the rock drilling rig, dust removal rig, lining rig, and wet spray machine. Through hardware transformation and software upgrades, the equipment can obtain various real-time data and key parameters. The tunnel scene integration interactive control system based on BIM+GIS technology would connect the acquired data to the system and improve the system functions, so as to cover the entire cycle of tunnel construction, integrate geological prospecting information, design information, comprehensive monitoring, advanced geological prediction, personnel and equipment positioning, electronic fence, monitoring and measurement, and other information. The organic and heterogeneous fusion between relevant information and physical environment hardware can be realized, and real-time control can be carried out on the site to realize digital tunnel construction.

2 INVESTIGATION AND ANALYSIS OF TUNNEL CONSTRUCTION ENVIRONMENT

Through the field survey along the Sichuan-Tibet line and a bidding section, we learn that the Sichuan-Tibet railway has the characteristics of complex terrain, geology, climate conditions, fragile ecological environment, strong plate movement, frequent geological disasters, etc., making the construction difficulty difficult. Consequently, the project adopts three-arm rock drilling rigs, wet spray rigs, lining rigs, and other large tooling construction.

The model of the three-arm rock drilling rig is ZYS113G, which can realize the functions of automatic loading and unloading rod, automatic lubrication, precise automatic positioning, and automatic contour scanning to protect the safety of construction personnel, and has the function of remote control operation in a complex and risky environment. It can realize pipe shed, advanced drilling, blasting, bolt hole, and other drilling operations. The type of wet spray rig is GHP3017E. The pumping system adopts a large cylinder diameter and long-stroke concrete pump, which has good wear resistance and strong material absorption ability. The concrete conveying pipe adopts a large diameter wear-resistant pipe with strong resistance to blocking. There is no blind area in the construction of the working arm, and the mobile and flexible transition is convenient. It can take into account multiple working faces simultaneously, and the construction efficiency is high. The operator can operate from a distance, avoid dust pollution, and have fewer personnel needs. The lining rig is a CQZ1309AG-01 intelligent formwork rig, which adopts a 360° rotary concrete distributor, and the pipeline is automatically telescopic to reduce labor intensity. A pipeline layering layout can achieve layering pouring and overcome the shortage of pressure transport from a hole to the bottom and hopper chute caused by concrete segregation and miter slope cold joint defects. The insertion and attachment combination vibration are that the vault adopts four insertion vibrators, and 24 pneumatic vibrators are configured on the arch shoulder and arch wall to realize automatic walking, automatic concrete cloth, and a window into the mold.

From the above analysis, we can summarize that although the construction of the Sichuan-Tibet Railway project with large machinery and equipment has many advantages, such as improving efficiency and speeding up progress, the system and correlation between the equipment are not strong during the construction process. Moreover, the equipment does not have the ability for data collection and storage, so it is impossible to judge the key parameters of the equipment during work effectively and carry out real-time monitoring and control of the construction process.

3 REFORM MEASURES FOR MAIN TUNNEL CONSTRUCTION MACHINERY AND EQUIPMENT

In order to collect the real-time parameters of the construction process, the equipment transformation plan and measures are formulated according to the actual needs of the construction site and the investigation of the relevant equipment manufacturers.

3.1 *Reform measures of three-arm rock drilling rig*

Adding accessories such as an intelligent gateway and onboard webcam to the existing equipment can meet the functional requirements of video monitoring, real-time data acquisition, online transmission of post-construction logs, etc. The specifically required accessories are shown in Table 1.

Table 1.   Retrofitting of arm rock drilling rig.

No	Accessory Name	Number	Unit
1	Intelligent Gateway	1	Piece
2	2.4G High Gain Antenna	1	Piece
3	High-frequency Shunt-type Antenna	1	Piece
4	Webcam /2.8mm	1	Piece
6	Cable	20	Meter
7	Cable Plug	6	Piece
8	Two Socket Jacket /DT06-2S	4	Piece
9	Two-plug Jacket /DT04-2P	4	Piece
10	Double Plug Female Lock /W2S	4	Piece
11	Two Insert Male Latch /W2P	4	Piece
12	Male Needle / 1060-16-0122	8	Piece
13	Mother Needle / 1062-16-0122	8	Piece
14	Controlling Cable	20	Meter
15	Low Smoke Halogen Free Wire and Cable	2	Meter
16	Steel Rail	100	Millimeter
17	M5 Guide Rail Mounting Screw M5	4	Piece

Through the above transformation, the three-arm rock drilling rig can achieve the following functions:

(1) Realize the quantity and depth record storage and transmission of the leading small catheter/tube shed.
(2) Realize the excavation drilling log recording and transmission (including time, positioning mileage, position coordinates, drilling number, drilling position, drilling depth, borehole layout, and other information).
(3) Realize the vehicle video monitoring and image transmission.
(4) Realize the data obtained in real-time 24 hours during operation and transmit relevant data to the tunnel network platform through the existing network in the engineering site.

3.2 *Reform measures of wet spray rig*

Adding hardware, software, and other accessories to the existing equipment can meet the functional requirements of real-time data collection, storage, and online transmission of wet spray rig operation. The specifically required accessories are shown in Table 2.

Table 2. Accessories needed for wet spray rig transformation.

No	Accessory Name	Description	Number	Unit
1	Remote/Local Communication Integration	Integrated communication functions	1	Set
2	Data Analysis Function	Data analysis, recording, etc	1	Set
3	Hardware Design	Schematic diagram, PCB design, structure	1	Set
4	Software Design	Operational software	1	Set
6	Intelligent Detection	Working parameters, angle, distance, etc.	1	Set
7	Raw Material	Components, boards, etc	1	Set

Through the above transformation, the wet spray rig can achieve the following functions:

(1) Realize the function of automatic recording, reading, and uploading of wet shotcrete construction parameters (including concrete injection quantity, injection speed, accelerating agent consumption, accelerating agent mixing ratio, wind pressure, etc.). The relevant data is transmitted to the tunnel network platform through the existing network at the project site.

(2) By installing the intelligent space sensing system, the angle and vertical distance between the nozzle and the spray surface are monitored timely, and the relevant data is transmitted to the tunnel network platform through the existing network at the project site.

(3) Complete the statistics of spraying parameters in the to-be-sprayed area of grid tunnel space and output to the digital visual terminal.

(4) The CAN communication interface is extended to information on the parameters of the vehicle body, and the sensor communication of the vehicle body is triggered in a directional and real-time manner to obtain the real-time ratio parameters of shotcrete amount, accelerating agent amount, wind pressure, soil amount and accelerating agent amount. At the same time, the obtained parameters and the current grid area are synchronized and serialized to the Pickle local file for later data recording, analysis, and query.

3.3 Reform measures of lining rig

By adding the protocol conversion module to the existing lining rig, the lining rig in the tunnel converts the internal data of the rig into MQTT protocol through TCP or 485 through the protocol conversion module and sends it to the specified information acquisition system. The site needs to provide a network interface and can use the wired network or 4G network. The specifically required accessories are shown in Table 3.

Table 3. Accessories needed for wet spray rig transformation.

No	Material Name	Number	Unit
1	Protocol Conversion Module	1	Piece
2	Install Attachment	1	Piece

Through the transformation aforementioned, the lining rig can achieve the following functions:

(1) The function of automatic recording, reading, and uploading of construction parameters (including concrete pouring quantity, pouring pressure, vault pressure, pouring temperature, working efficiency and working time, etc.) of the lining rig can be realized, which can be transmitted to the developed tunnel network platform through the existing network in the project site.

(2) Through the protocol conversion module, the relevant construction parameters are transmitted to the tunnel network platform through the existing network of the project site.

(3) It can obtain data in real-time 24 hours during operation and transmit relevant data to the tunnel network platform through the existing network in the engineering site.

3.4 *Reform measures of dust removal rig*

The intelligent gateway and dust concentration sensor are added to the existing plateau common dust removal rig to meet the functional requirements of dust removal information data collection and transmission. The specific accessories are shown in Table 4.

Table 4. Accessories required for the transformation of dust removal rig.

No	Accessory Name	Number	Unit
1	Intelligent Gateway	1	Piece
2	2.4G High Gain Antenna	1	Piece
3	High-frequency Shunt-type Antenna	1	Piece
4	Cable	20	Meter
5	Cable Plug	6	Piece
6	Four-socket Sheath DT06-4S	2	Piece
7	Four-insert Female Lock W4S	2	Piece
8	Four-pin Male Latch W4P	2	Piece
9	Four-plug Sheath DT04-4P	2	Piece
10	Mother Needle 770520-1	2	Piece
11	Mother Needle W2S	2	Piece
12	Male Needle 1060-16-0122	2	Meter
13	Dust Concentration Sensor	1	Piece
14	Control Cable	30	Meter

Through the above transformation, the wet spray rig can achieve the following functions:

(1) Realize the automatic recording, reading, and uploading functions of the construction parameters of the dust removal truck (including dust concentration, dust removal air volume, operation time, etc.), and transmit them to the developed tunnel network platform through the existing network of the project site.

(2) It can obtain data in real-time 24 hours during operation and transmit relevant data to the tunnel network platform through the existing network in the engineering site.

4 DIGITAL TUNNEL CONSTRUCTION CONTROL TECHNOLOGY

4.1 *Intra-tunnel network laying*

To meet the requirements of device data transmission during tunnel construction, a good network environment arranged in the tunnel is important.

4.1.1 *Network equipment building standards*
According to the maximum network load on the site and the tunnel network configuration, the tunnel Intranet is configured as follows:

(1) The minimum bandwidth carrying capacity is 30Mbps (equipment only).

(2) The maximum uplink rate is 30Mbps inside and outside the tunnel.

(3) The maximum downlink rate is 4 Mbps outside and inside the tunnel.

4.1.2 *Typical supply network in tunnel*

The network is configured according to the requirements in Section 4.1.1, and the network in the tunnel is constructed according to the typical equipment supply network as follows:

(1) The service provider's network is holed in optical fiber mode.
(2) The information cabinet is arranged on the lining platform car behind the waterproof board rig, and the network devices such as optical cat, router, and switch can be integrated into it.
(3) The WiFi transmission points are set up near the waterproof pallet rig.
(4) The double WiFi channel is set up in the tunnel to realize the man-machine shunting and avoid interference.
(5) The signal between the wireless transmitter point and the palm surface is free from occlusion, and the distance is less than or equal to 100m.

4.1.3 *Provide network for equipment*

(1) Supply Network to Structural Equipment
    The lining rig directly connects the operator network to the rig router through the wired connection mode to realize the equipment networking.
(2) Supply Network to Mobile Devices
    By configuring the intelligent gateway carried by the equipment, such as rock drilling and dust removal, the network channel between the tunnel and wireless WiFi is established. When the equipment is working, the wireless network connection is established independently to realize the network of mobile equipment.

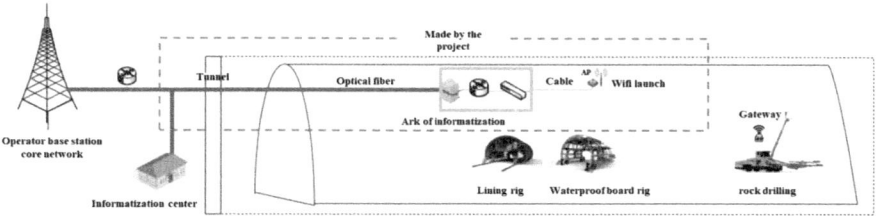

Figure 1. Tunnel network layout.

4.2 *Interactive control technology of visual iot in tunnel construction*

4.2.1 *Visual object interactive control system for tunnel construction*

The tunnel construction visual IoT interactive control system takes BIM+GIS and IoT heterogeneous fusion technology as the starting point. It takes "total factor, data sensing, heterogeneous fusion, and interaction" as the core during construction and builds a "BIM+GIS technology-based tunnel scene integration interactive control system platform." The platform can cover the whole cycle of tunnel construction, integrate geological prospecting information, design information, comprehensive monitoring, advanced geological prediction, personnel and equipment positioning, electronic fence, monitoring and measurement, and all kinds of key construction data in the construction process, and realize the organic heterogeneous fusion between virtual information and physical environment hardware.

The system is also a project implementation level management. The information interaction of the production data of personnel, machine, material, method, rings, and measurements in the tunnel through the Internet of things can achieve three-dimensional management. Besides, the system can also integrate the personnel service information, tooling production data, harmful gas monitoring, adverse geological information, monitoring data, and so on. Combined with the engineering big data technology and in-depth analysis of key data operation, it can optimize the key working procedures such as excavation and primary support parameters and decision management in the stages of safety, quality, progress, material, realizing intelligent identification, positioning, tracking, and monitoring.

4.2.2 *Control technology of tunnel construction and reconstruction equipment*

In this paper, through the digital transformation of large mechanical equipment such as the three-arm rock drilling rig and lining rig, the equipment can automatically collect and store the key parameters in the construction process and then transmit the data to the tunnel construction visual IoT interactive control system platform through the network.

The platform can carry on the integration and analysis of the data and make early warnings of unsafe behaviors and factors. For example, it can identify and determine the drilling quantity insufficiency of three-arm drilling jumbo, drilling position errors, small drilling depth, excessive casting stress of lining shutter, unqualified wet spray jumbo jet velocity, angle, and distance, undesirable dust removal effects of pallet, dust exceeding the limits, etc. Then the system can issue instructions according to the predetermined threshold, remind the relevant management personnel to take timely measures to rectify mistakes and control the process, and prevent quality and safety problems. At the same time, based on the tunnel construction visualization platform, real-time construction dynamics can be understood, and various construction machinery on the site can be reasonably dispatched to interweave the working procedures, to ensure the efficiency and fluency of all kinds of machinery combined operation.

5 CONCLUSION

This paper is based on the digital transformation of the main large-scale construction equipment on the site, so that relevant equipment can collect, store and transmit the key construction data in real time. Then the relevant construction parameters are transmitted to the tunnel construction visual IoT interactive control system platform through the existing network of the engineering site. Based on the tunnel construction visual IoT interactive control system, the site dynamics can be understood in real time, the compliance of various works can be judged, and the existing problems can be fed back automatically in time. In this way, personnel input can be reduced, construction progress can be accelerated, safety and quality risks can be reduced, and digital construction level can be improved, to better control the site, which has important reference significance for guiding site construction and digital construction.

REFERENCES

Chen S.K., Zhou H., Liao X., Chen X.Q., Zhao X.Y., Wang Z.M. Route selection for disaster reduction of high ground stress tunnel of Sichuan-Tibet Railway [J]. *Geoscience* 2022, 47(03): 803–817.

Guo C.B., Wu R.A., Jiang L.W., Zhong N., Wang Y., Wang D., Zhang Y.S., Yang Z.H., Meng W., Li X., Liu G. Typical geological disasters and geological engineering problems in Ya'an – Nyingchi section of Sichuan-Tibet Railway[J]. *Geoscience* 2021, 35(01):1–17. DOI:10.19657/j.geoscience.1000-8527.2021.023.

Xue X.G., Kong F.M., Yang W.M., Qiu D.H., Su M.X., Fu K., Ma X.M. Main adverse geological conditions and engineering geological problems along Sichuan-Tibet Railway [J]. *Journal of Rock Mechanics and Engineering* 2020, 39(03): 445–468. DOI:10.13722/j.cnki.jrme.2019.0737.

Xie H.P., Zhang R., Ren L., Zhang A.L., Zhang Z.L., Deng J.H., Xu Z.X., Zhang G.Z., Feng T., Wang D., Wang Z.W., Yi X.J., Lin Z.H., Li J.Y., Zhang Z.T., Yuan D., Jia Z.Q. Disaster analysis and consideration on the surrounding rock of deep tunnel of Sichuan-Tibet Railway [J]. *Engineering Science and Technology* 2022, 54(02): 1–20. DOI: 10.15961/j.jsuese.202101178.

Zhang H.P., Chen K. Research on data extraction technology of tunnel group equipment for multi-source heterogeneous information fusion [J]. *Tunnel Construction* (English and Chinese), 2022, 42(01): 41–47.

Zhu Q., Wang S.Z., Ding Y.L., Zeng H.W., Zhang G.L., Guo Y.X., Li H.K., Wang W.Q., Song S.B., Hao R., Cheng Z.B. *Construction method of safety quality progress knowledge graph for intelligent management of railway tunnel drilling and blasting construction*[J/OL]. Geomatics of Wuhan University (Information Science): 1–16 [2022-03-28]. http://kns.cnki.net/kcms/detail/42.1676.TN.20211028.1223.004.html

Frontiers in Civil and Hydraulic Engineering – Mohamed A. Ismail and
Hazem Samih Mohamed (Eds)
© 2023 The Authors, ISBN 978-1-032-38247-0

Resilience development model of green infrastructure in urban rivers and lakes in middle and lower reaches of Yangtze River

Yuan Xiong, Nan Wen*, Yuan Yao, Daqing Ma & Ying Zhang
Changjiang Survey, Planning, Design and Research Co., Ltd., Wuhan, Hubei, China

ABSTRACT: The traditional infrastructure of rivers and lakes in cities and towns often causes various problems, such as insufficient flood control function, weak ecological function, and fragmented shore-city connection while performing flood control functions. Based on the concept of resilient landscape, these rivers and lakes with a single function have great potential for improvement in terms of flood control efficiency, ecological services, landscape appearance, and public vitality, and have become key fields for modern urban renewal and development. The resilient development of green infrastructure of rivers and lakes in towns and cities in the middle and lower reaches of the Yangtze River is summarized in three main models (engineering resilience, ecological resilience, and social resilience) to achieve a balance between flood control and landscaping. Taking the Yichang Xiaoting section shoreline enhancement project as an example, ten resilience development strategies of 3 modes are applied comprehensively according to the site characteristics, current problems, and design objectives. Finally, the key issues that need to be solved for resilient landscape construction in project implementation are summarized with a view to proposing a resilient development model for green infrastructure suitable for riverbank projects in the middle and lower reaches of the Yangtze River.

1 GENERAL INSTRUCTIONS

1.1 *Research background*

With the intensification of global climate change, urban water security problems occur frequently, and the traditional grey infrastructure can no longer meet the needs of urban development and ecological construction. Compared with the traditional concept of "defending rivers," the attitude towards rivers has gradually changed from "defending" to "adapting." Flood prevention and control measures no longer rely on confrontation through hard engineering techniques such as building dikes and dams, but advocate resilient green infrastructure to absorb and adapt (Chen 2020). In recent years, many related project practices have been implemented at home and abroad, such as the Houtan Park in Shanghai, where ecological barging, riverbank re-greening, and sponge infrastructure are used to build a resilient landscape (Yu et al. 2015), and the "10,000-mileage Greenway" project in Guangdong, which demonstrates the concept of "treating the waterfront together." The "Wan Li Bi Dao" project in Guangdong demonstrates the concept of "water and bank governance," which extends river management from traditional waters to water and land complex areas. Although there are many excellent cases like them, most of the waterfront landscapes of rivers and lakes are restricted by land constraints and flood control functions, and there are still problems such as difficulty in expanding the flooding space of rivers, a single mode of embankment utilization, weak ecological functions, and poor landscape appearance. In terms of technology, there is a lack of systematic green infrastructure resilience development strategies between macro theory

---

*Corresponding Author: 980302512@qq.com

DOI 10.1201/9781003344209-68

and specific technology implementation (Manyena 2006). In this paper, the author proposes to study the resilient development model of green infrastructure in the riverside and lakefront zones in the middle and lower reaches of the Yangtze River, to improve the risk resistance of the riparian zones and to protect water ecology, water security, water landscape, and water resources. This is a positive response to the strategic objective of "ecological priority and green development" of the great protection of the Yangtze River and provides a reference for the resilient development of green infrastructure in the riverside and lakefront basins under the great protection of the Yangtze River.

1.2 *Concept analysis*

Green infrastructure is a resilient organic ecological network system that integrates natural areas and artificial environments by linking various landscape elements, covering ecological functions (Liu et al. 2010). It has a certain ability of adjustment and accommodation and can effectively cope with environmental impacts through mitigation, absorption, adjustment, and adaptation, and even develop the ability to adapt to future dynamic changes. In terms of spatial composition, it mainly includes three element categories: hubs, links, and sites (Liu et al. 2012). Hubs are often represented by forests, lakes, wetlands, parks, and other areas with large land areas and rich natural ecological resources; links are linear green spaces connecting various hubs, including ecological corridors, greenways, scenic corridors, etc.; sites are landscape nodes scattered around linear links, and often represented by pocket parks, urban squares, small green areas, etc.

The term "resilience" originated in the industrial field and was first applied to the ecological discipline in 1973. The connotation of a resilient landscape mainly focuses on three dimensions: engineering resilience, ecological resilience, and social resilience (Shao et al. 2015). In this evolutionary process, the pursuit of restoration to the original state is no longer the only objective; rather, continuous adaptation is the objective, and the equilibrium state restored after disturbance may not be consistent with the original equilibrium state.

2 PROBLEMS IN THE CURRENT INFRASTRUCTURE CONSTRUCTION OF RIVERS AND LAKES IN TOWNS AND CITIES IN THE MIDDLE AND LOWER REACHES OF THE YANGTZE RIVER

2.1 *Insufficient flood control capacity*

Most of the current infrastructure is based on the rigid demand for urban flood control and water transportation, and the centralized "hard" engineering mode with a single function is often adopted, such as curve cut-off, stone slope protection, and high embankment construction (Zhang 2020). Although this approach can increase the flooding speed, it restricts the horizontal development of the water area in space. When the water of upstream rivers and lakes increases dramatically, the rivers and lakes can neither be drained in time nor temporarily relieved by lateral spreading, which will lead to the water level rising or even exceeding the predetermined flood control level, threatening urban safety (Li 2020). In addition, from the perspective of the development of the whole basin, with the continuous expansion of urban construction and the construction of an increasing number of rivers and lakes embankments, water accumulated in the upper reaches of the Yangtze River during the flood season continues to increase, which in turn leads to an increase in flooding pressure in the downstream cities, resulting in difficulties and risks in the flood discharge of the rivers during the flood season, and the flood control capacity needs to be improved.

2.2 *Weak ecological function*

As a dynamic and continuous ecological link, rivers can be connected by water systems to make them an organic whole, providing rich habitats and biological migration links, which is of great significance for maintaining regional ecological balance (Jing 2014). However, at present, most of the river and lake revetments use cement mortar to replace the original river and lake floodplain, which greatly reduces the living space of flora and fauna in the riparian zone of rivers and lakes. The grey infrastructure such as culverts and pipes, which are mainly for rapid discharge, are directly connected to the river and lake water system, making the polluted rainwater runoff be discharged into the receiving water body without purification treatment, which further harms the water quality, habitats, flora and fauna of the rivers and lakes, and finally making the ecological environment within the rivers and lakes present a very fragile state.

2.3 *Fragmented shore-city connection*

Living and playing by water have been a human preference since ancient times. With the progress and development of society, the traditional gray infrastructure can no longer meet the needs of citizens' water-friendly activities and urban ecological construction. In the early days, the function of rivers and lakes was mainly focused on flood control and flood discharge, which failed to take into account the urban hydrological cycle, and often adopted the single form of high river embankment or stone slope protection, making the current riverside road and waterfront shoreline often have a large height difference, weakening the accessibility of riverside belt in the river and lake basin, and severely cutting off the spatial connection between waterfront and city. Much hard concrete blocks the continuity of the rivers and lakes with the surrounding natural landscape, weakening the visual aesthetics.

3 RESILIENT DEVELOPMENT MODE OF GREEN INFRASTRUCTURE IN URBAN RIVERS AND LAKES IN THE MIDDLE AND LOWER REACHES OF THE YANGTZE RIVER

With the promotion of ecological civilization, the construction of waterfront space in the middle and lower reaches of the Yangtze river is no longer limited to the traditional design of flood control engineering but gradually transformed into a green infrastructure integrating composite functions of water conservation and flood control, landscaping, and social benefits. To this end, through

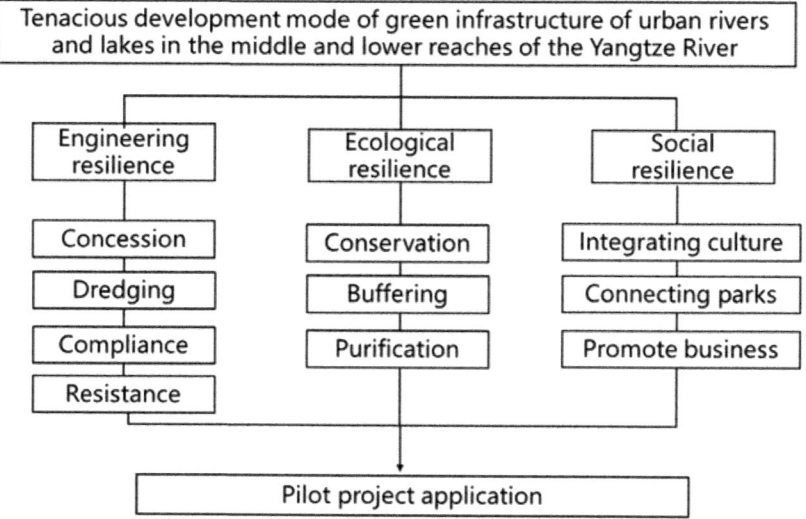

Figure 1. The resilient development system of green infrastructure figure.

the study of many excellent cases and related theories, a scientific and orderly model of resilient development is summarized. The idea of resilient landscape development and construction of green infrastructure is summarized into three strategies of "water conservancy protection, ecological conservation, and public vitality." The resilient development system of green infrastructure in the middle and lower reaches of the Yangtze River is shaped by three dimensions: engineering resilience, ecological resilience, and social resilience, which can be used in combination according to the current characteristics of the site in the actual engineering construction.

3.1 Engineering resilience model

3.1.1 Concession
Concession is a resilient development strategy to increase the floodable space of rivers and lakes through engineering technical means. This strategy is to advance by retreating, which can be combined with step expansion, or locally retreat to widen the dike, forming a transition zone to increase the flooding space and flood buffer zone and enhance the resilient flood control capacity of the river.

3.1.2 Dredging
Dredging mainly includes dredging and guidance in practical application. Dredging refers to enhancing the flood discharge capacity of the river using dredging, such as clearing the obstacles in the river and deeply dredging to increase the vertical gradient of the river. Diversion is to divert excess flood water to other areas for retention or discharge by engineering means, such as diversion to reduce the flood control pressure in the main river area. The famous Dujiangyan fish mouth project in China is a typical water conservancy project with a dredging strategy as its core.

3.1.3 Compliance
Compliance aims to incorporate natural hazards as one of the site constraints into the design application to create a flood adaptability landscape. It often includes two types, floodable landscapes, and floating landscapes. Floodable landscapes refer to that with the rise of water level, and some sites within the design range can be flooded to form a new landscape effect, which is often accompanied by fewer artificial facilities and water-resistant plant species. A floating landscape is a floating space on the water surface created by some floating structures, which can move up and down with the change of water level, not only does not affect the river flood control but also can reduce disaster losses.

3.1.4 Resistance
To a certain extent, the resistance strategy can be regarded as an extension of the traditional flood control concept, which is based on the original engineering facilities, by combining with public facilities to raise the dam or expansive reinforcement of the dam. With the functional use of the landscape, it presents a flood control infrastructure with rich types and complex functions, which is often applied in areas with restricted river width or weak flood control.

3.2 Ecological resilience model

From the ecological perspective, we can link the riparian zone of rivers and lakes with the natural environment, and restore and improve the habitat conditions by creating rich habitat ecological environment. In combination with the application of green and sustainable technologies, the ecological flood control and flood management mode of source dissipation and storage, process deceleration and energy dissipation, and end elastic adaptation is formed. According to the ecological functions and technical characteristics of green space, the ecological resilience model can be summarized into three core strategies (Zhao 2018).

3.2.1 Conservation

Conservation is an infiltration storage facility that collects surface runoff in a concave form. It often has good permeability and certain flood detention and storage capacity, which can undertake runoff, store, and utilize it, and has the function of regulation and storage. The accumulated water can replenish the underground water source through the infiltration of soil, forming a reduced planar structure. Common ecological technologies include a rain garden and a wet pond.

3.2.2 Buffering

Based on making full use of the original topography of the riparian zone of rivers and lakes, buffering is to "soften" and "ecology" the revetment combined with appropriate plant design, effectively detaining, and slowing down the flow velocity of runoff by using the stems and leaves of plants, thus forming a transmission structure that slows down energy dissipation in the middle to improve the flood control capacity and benefit the creation of animal and plant habitats. Common ecological technologies include stepped ecological revetment and grass planting ditch.

3.2.3 Purification

Purification is to absorb and degrade nitrogen and phosphorus molecules carried by runoff through the absorption and enrichment of plant roots to reduce the water pollution load. At the same time, the content of dissolved oxygen in water is increased to provide a good living environment for aerobic microorganisms. Many aquatic plants have ecological and landscaping value, such as Typha, Lythrum salicaria, Calamus, Scirpus tabernaemontani, and reed. Among the most common ecological technologies are ecological floating beds and purification ponds.

3.3 Social resilience model

According to the needs of the society, we should strengthen the social function resilience of water-front green space with three strategies: integrating culture to shape characteristics, connecting "parks" to conserve ecology, and promoting "business" for green production. Through landscape construction, we should shape the city's characteristic image, drive regional development, convert ecological benefits into economic benefits, and increase revenue to realize the value connotation of "lucid waters and lush mountains are invaluable assets" (Bao et al. 2018).

3.3.1 Integrating culture

Cultural integration is to condense and integrate urban cultural resources and shape urban spatial characteristics so that the materialized environmental forms of the city are different from other cities (Yang et al. 2013). Combining and integrating the urban context will allow us to explore the characteristics of historical and cultural blocks, dock cultural blocks, as well as refine the urban cultural symbols to achieve the purpose of shaping the characteristics of the city (Xiao 2006).

3.3.2 Connecting parks

Connecting parks refers to building a slow-moving system based on greenways and connecting it with the green space system around the city in series to enhance accessibility. Integrate greenways with rivers, air ducts, ecological corridors, and cultural tourism routes, connect ecological networks, increase the connectivity of ecological sites, and build ecological green corridors along the Yangtze River to connect the multiple spatial elements such as rivers, mountains, forests, cities to build a multifunctional urban green space network.

3.3.3 Promote business

It is a new format of cross-border integration using facility embedding, function integration, and scene introduction, innovating park consumption formats and consumption patterns, and realizing green production. With a variety of landscape nodes and industrial forms, various full-time and all-age activities are introduced, and the riverside activity system with varied styles will be shaped to realize the sustainable management of the site.

4 CASE APPLICATION

4.1 *Case status analysis*

The case study area is located from the entrance of the Bailin River in Wujiagang District, Yichang City, Hubei Province, to the ancient battlefield of Xiaoting District, with a total length of 7.2 kilometers. According to the current situation investigation and data collection, the problems of "poor road traffic continuity, disordered landscape style and single functional type of waterfront space" are generally presented.

4.2 *Study on remediation and resilience development model*

According to the basic construction requirements of the upper-level planning, guided by the three planning and design strategies of "water conservancy protection, ecological conservation, and public vitality," the three objectives of "highlighting landscape resources, upgrading riverside ecology, and promoting public services" are proposed for the remediation and resilience development of Yichang Yangtze River shoreline, and the detailed design of resilience development and rehabilitation was carried out from three aspects: engineering resilience, ecological resilience, and social resilience.

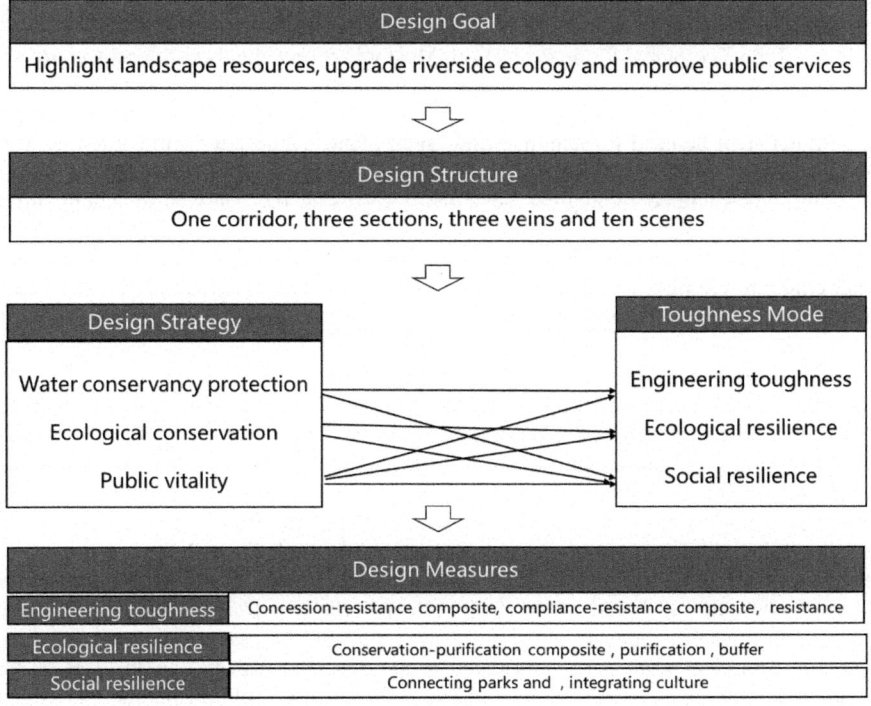

Figure 2.   Yichang Changjiang riverside space resilience development technical framework system.

4.2.1 *Engineering resilience design*

A. Concession-Resistance Composite. This section is close to the entrance of the Bailin River, and the surrounding residential areas are densely populated. Under the condition of ensuring flood discharge safety, it needs to meet the needs of citizens for daily leisure and diversified activities. Following the principle of earthwork balance in design, the revetment is provided with gentle slope concession, to increase the flood discharge space and flood buffer zone.

B. Compliance-Resistance Composite. This section is close to the ancient battlefield of Xiaoting, connecting people with the Yangtze River through the dock, aiming to create an interactive experience of dock culture. In the design, the revetment excavation will form a sinking landscape area to meet the hydrophilic needs of different spatial scales. Two 4m wide slow-moving greenways will be added to connect the riverside open space with the vitality green belt to create a rich interactive experience of wharf culture.

C. Resistance Type. This section is close to Jingmen Landscape Poetry Corridor, which mainly reflects the poetry culture of Jingmen Mountain through intention expression. Design undulating art mounds to create a rich spatial experience. At the same time, the embankment is skillfully expanded and reinforced in the form of a grass mound, becoming a green infrastructure integrating flood control function and landscape art.

4.2.2 *Ecological resilience design*

A. Conservation and Purification Type-Reclaimed Water Garden. The reclaimed water garden is in the space under the Wujiagang Yangtze River Bridge. It aims to design a reclaimed water park by using the tailwater of a sewage treatment plant, and at the same time, meet the requirements of sponge city design, and play the ecological functions of a rainwater garden in the rainy season, such as water collection, water retention, water collection, and water purification. The reclaimed water garden can supplement groundwater by reducing the proportion of hard pavement, increasing the soft space of vegetation and soil, and strengthening water storage and infiltration.

B. Buffering Type-Ecological Slope Protection. Most wharf revetments in the project are vertical retaining wall revetments, and the revetment of the bank slope and the ramp of the launching quay considered in the design. The existing bank slope shall be reserved, the oil hemp rattan hanging net shall be used for hanging, and green plants shall be used to form a good buffer zone.

C. Buffering Type-Stepped Ecological Revetment. By using the nearly 5 m height difference between the highway and the revetment to design the terrace, and by weakening it step by step step by step with step concessions, the surface runoff can be effectively retained. At the same time, it can combine the plants to create a beautiful pastoral scenery of fields, flowers, and fruit forests.

4.2.3 *Social resilience design*

A. Connecting Parks-Traffic Connection. In the design, firstly, the connection of greenways in the site shall be met. Secondly, the rationality of the opening from the highway into the site and disaster prevention requirements should be considered. At the same time, sufficient parking spaces and emergency shelters should be reserved for the future flow of people. Within site, many sightseeing routes follow the terrain, such as the main channel for viewing the river by the railing, the hiking trails on the grassy slopes, and the trails through the woods.

B. Integrating Culture-Cultural Revival. Taking ecological culture, poetry culture, and wharf culture as the cultural thread of the project, to reconnect the city, people, and nature, we will bring into play the educational significance of ecological restoration, return to the green mountains and clear waters that originated in the city, and reshape the spirit of destroying the old and building the new. We will abandon the form of cultural preaching, expand the display carrier and expression form of culture, convey cultural connotation and realize cultural resonance in the process of interesting participation and environmental interaction.

5 CONCLUSIONS

The focus of this study is to transform the river and lake embankment projects in the middle and lower reaches of the Yangtze River into a landscape infrastructure integrating water conservancy protection, ecological protection, and public vitality based on the traditional flood control infrastructure, integrating green infrastructure and resilient landscape concept. The purpose is to guide the change of the flood control concept and advocate the transformation from simple flood prevention to flood adaptation and even making friends with floods.

At present, the resilience development of green infrastructure mainly includes three modes: engineering resilience, ecological resilience, and social resilience, and ten transformation measures, which are also compatible with the element categories of green infrastructure construction in terms of spatial composition. The hub resilience development is mainly based on the engineering resilience mode, which mainly plays the functions of flood control and flood discharge and has the functions of ecological leisure. The resilience development of green sites is mainly based on the ecological resilience model, which mainly plays the functions of dissipation, storage, and ecological purification. The resilience development of green links is mainly based on social resilience, mainly increasing the connectivity of blue-green sites and constructing a multi-dynamic resilience network system through linear green space. These measures are often used in combination in engineering construction rather than in a single application. In practical application, it is necessary to make full use of the current conditions of the site, select low-cost and applicable renovation measures, and have better performance and identification characteristics.

At the same time, in view of the limitations of data collection, there are still many shortcomings. The resilient development of green infrastructure requires multi-disciplinary communication and cooperation, as well as evidence-based feedback from practice, to better adapt to future challenges and development.

ACKNOWLEDGMENT

Hubei Key Laboratory of Basin Water Security Support Project (CX2021Z03) 2021 Independent Innovation Projects of Changjiang Survey, Planning, Design and Research Co., Ltd.

REFERENCES

Bao L.L., Wei L. Research on layout and function of green space in beijing parks under current urban development stage [J]. *Forestry Science and Technology*, 2018 (06): 69–73.

Chen J.S. Research on the integration strategy of blue and green river space under the theory of resilient city [J]. *Planner*, 2020, 36 (14): 5–10.

Jing Y.L. *Research on urban river ecological landscape planning based on green infrastructure theory* [D]. Nanjing. Nanjing Forestry University, 2014.

Li S.B. *Study on design of hydro-elastic landscape in restricted area of small city waterfront* [D]. Suzhou. Suzhou University, 2020.

Liu B.Y., Wang P. Development history of green space ecological network planning and the frontier of chinese research [J]. *Chinese Garden*, 2010, 26 (03): 1–5.

Liu J.J., Li B.F., Nancy Ruo, Ning Y.F. Constructing city's life support system-case study of urban green infrastructure in seattle [J]. *Chinese Garden*, 2012, 28 (03): 116–120.

Manyena S.B. The concept of resilience revisited[J]. Disasters, 2006(4): 433–450.

Shao Y.W., Xu J. Urban resilience: Conceptual analysis based on international literature review [J]. *International Urban Planning*, 2015, 30 (02): 48–54.

Xiao M. Urban spatial context and urban characteristics [J]. *Jiangsu Urban Planning*, 2006(5): 4.

Yang J.Y, Hu X.Y. Approaches and methods of urban spatial characteristic planning [J]. *Urban Planning*, 2013 (6): 8.

Yu K.J., Xu T., Li D.H., Wang C.L. Research progress of urban water system elasticity [J]. *Journal of Urban Planning*, 2015 (01): 75–83.

Zhang P. *Study on the construction strategy of suburban river landscape in northern cities under the concept of flexible design* [D]. Xi'an. Xi'an University of Architecture and Technology, 2020.

Zhao W. *Application and research of flexible landscape design concept in waterfront environment* [D]. Jinan. Shandong Jianzhu University, 2018.

Frontiers in Civil and Hydraulic Engineering – Mohamed A. Ismail and
Hazem Samih Mohamed (Eds)
© 2023 The Authors, ISBN 978-1-032-38247-0

Comparative experimental study on different soil proportion and rain type of green infrastructure in Zhejiang

Shaopeng Qiu*
Zhejiang Institute of Hydraulics & Estuary (Zhejiang Surveying Institute of Estuary and Coast), China

Haoan Zheng*
Jinhua Water Resources Bureau, China

Yiheng Liu* & Long Ye*
Zhejiang Institute of Hydraulics & Estuary (Zhejiang Surveying Institute of Estuary and Coast), China

ABSTRACT: According to the study of rainfall and underlying surface in Zhejiang area, planting soil and park soil were mixed with fine sand and coarse sand respectively, and rainfalls of two-year return period and five-year return period in Hangzhou Xiasha area were selected to simulate the rainfall garden experiment by artificial rainfall. The results show that: under the same rainfall pattern, the maximum infiltration flow of planting soil is greater than that of park soil; under the same soil condition, the maximum infiltration flow under the condition of five-year return rainfall is greater than that of two-year return rainfall; after mixing the two kinds of soil with sands, the maximum infiltration flow containing park soil is greater than that of containing planting soil.

1 INTRODUCTION

In recent years, China's urbanization process has been accelerating, which to a large extent has changed the original hydrological conditions of runoff and confluence, and the overall trend shows increased confluence and frequent flooding. In order to solve this problem, the sponge city gathers the advantages of "stagnation, storage, infiltration, purification, use and discharge" to effectively realize multiple runoff rainwater control, reconstruct the ecological structure of the city, enrich the urban function, and reduce the adverse impact of urban development (Endreny 2009; Roy-poiriera 2010; Trowsdale 2011). As a new type of ecological rainwater control and utilization facility, the rainwater garden is the basic element and supplement for the construction of a sponge city, which has the advantages of reducing urban rainwater runoff, purifying rainwater water quality and replenishing groundwater. The form was first used in Prince George's County, Maryland, in the early 1990s (Luo 2008). After that, this design concept was recognized by all countries around the world, and the measures were gradually implemented and further developed in Germany, the United States, Australia and other places. The EMERSON study found that over time, vegetation roots and microbes in the soil reduced the impact of fine particle deposition in the planting soil layer, resulting in an insignificant decrease in the permeability of rain gardens over time (Emerson 2008). Studies by Shuangcheng and others found that loess had good infiltration ability, and there was basically no overflow occurring in the relatively humid 2011 (Tang 2012). Pengcheng et al. found that the vegetation selection and the particle size of the artificial packing layer had the most significant influence on the storage and detention effect of rainwater garden, and the selection of planting soil layer ingredients and artificial packing layer had the weakest influence (Li 2012).

---

*Corresponding Authors: 635933990@qq.com; zhenhaoan123@163.com; lyh123456@163.com and 996384338@qq.com

DOI 10.1201/9781003344209-69

At present, research on our country's urban rainwater resource utilization and urban rainwater runoff non-point source pollution control is relatively backward, rainwater garden rainwater storage engineering design and planning basis is insufficient, and there is an urgent need to put forward a set of suitable for different areas of rainwater garden technical specifications, and determining different parameters of rainwater garden storage is the key to guide the rainwater garden design. In this paper, the rainwater garden box model is designed according to two local soil types in Hangzhou. By changing the rainfall recurrence period and the mixed ratio of rainstorm intensity and soil types, the results can provide technical reference for determining the parameters of the design parameters of rainwater garden in similar areas.

2 RESEARCH MATERIALS AND METHODS

2.1 Rainwater garden box model

The typical general model selected in the study is a rectangular box, whose section is shown in Figure 1. It consists of a covering layer, a growth medium layer, a geotextile layer and a filter layer in which different types of typical plants are planted. The covering layer, i.e. vegetation layer, is selected as common vegetation winter grass. Hangzhou local soil is selected as a growth medium layer, which is planting soil and Park Soil respectively. Six different soil types with 50 cm thickness are obtained by mixing them with coarse sand and fine sand respectively. The filtering water storage layer is paved with 30 cm block gravel. Rainfall is carried out by means of artificial rainfall using rainfall data of 2-A and 5-A recurrence periods in Hangzhou.

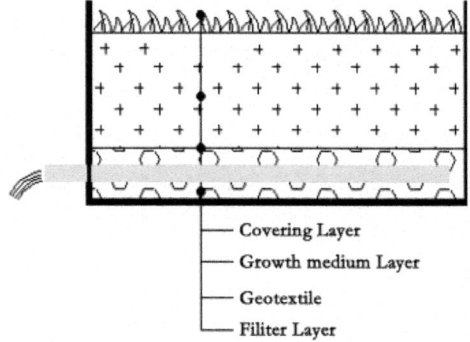

- Covering Layer
- Growth medium Layer
- Geotextile
- Filiter Layer

Figure 1.   Schematic diagram of typical model test configuration of rain garden.

2.2 Selection of rainfall data

According to the notice on publicizing the rainstorm intensity formula of cities in Zhejiang Province (JKF [2008] No. 89) issued by the Construction Department of Zhejiang Province, the rainstorm intensity formula adopts the annual maximum sampling method and the exponential distribution curve fitting. The rainfall duration is 5, 10, 15, 20, 30, 45, 60, 90 and 120 min, with a total of 9 durations. The rainfall return period is counted as 2, 3, 5, 10, 20 a. the sampling data is up to 2006. The applicable range of the rainstorm intensity formula is that the rainfall duration is not more than 120 mins and the return period is not more than 20 a. According to this topic, the rainstorm intensity formula in Hangzhou adopts the annual maximum method for sampling. The specific formula is:

$$i = \frac{57.694 + 53.476 \lg P}{(t + 31.546)^{1.008}} \tag{1}$$

Formula: i is the design rainstorm intensity [mm/min]; t is the rainfall duration (min); and P is the design recurrence period (a).

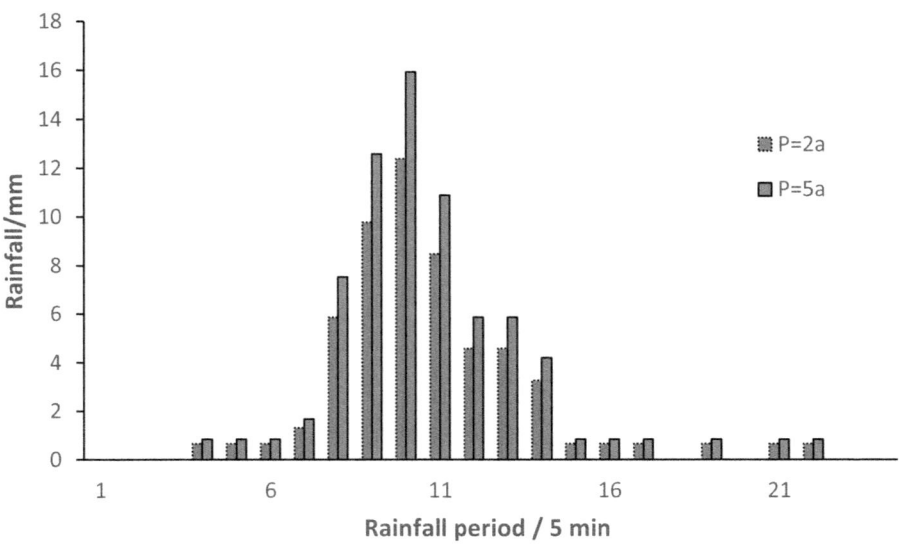

Figure 2.   Short-duration rainstorm process in the the xiasha area.

According to the formula of rainstorm intensity in Hangzhou, the short duration rainfall in the Xiasha area is calculated. The rainfall in the return period of 2 years is 56 mm; The rainfall in the return period of 5 years is 72 mm. The short-term measured rainfall in Xiasha area on July 21, 2015 is used as a typical rainfall pattern to distribute the rainfall under the return period of 2-year return period and 5-year return period.

Table 1.   Distribution of 2-h rainfall under different rainfall conditions in Hangzhou.

Rainfall period/5 min	Once-in-2-a/(mm/5min)	Once-in-5-a/(mm/5min)
1	0.00	0.00
2	0.00	0.00
3	0.00	0.00
4	0.65	0.84
5	0.65	0.84
6	0.65	0.84
7	1.30	1.67
8	5.86	7.53
9	9.77	12.56
10	12.37	15.91
11	8.47	10.88
12	4.56	5.86
13	4.56	5.86
14	3.26	4.19
15	0.65	0.84
16	0.65	0.84
17	0.65	0.84
18	0.00	0.00
19	0.65	0.84
20	0.00	0.00
21	0.65	0.84
22	0.65	0.84
23	0.00	0.00
24	0.00	0.00

Test group times

Two test groups were used for this study, including the rain frequency test group and the soil ratio test group. The rain frequency experiment group is tested together by changing the rainfall return period; soil ratio and sand grains are mixed together, as shown in Table 2 and Table 3.

Table 2.   Rain frequency test groups

Test number	Rainfall reproduction period	Somatomedin
A1	the once-in-2-a rainfall condition	Planting soil
A2	the once-in-5-a rainfall condition	

Table 3.   Test groups of soil mass proportioning

Test number	Rainfall reproduction period	Somatomedin
B1		Planting soil
B2		Park soil
B3	the once-in-2-a rainfall condition	Planting soil, fine sand (1:1)
B4		Soil and fine sand in the park (1:1)
B5		Planting soil, coarse sand (1:1)
B6		Soil and coarse sand in the park (1:1)

3   ANALYSIS OF THE TEST RESULTS

3.1   *Geotechnical test results*

Geotechnical tests were carried out on 4 soil samples and 2 original soil samples. The results show that the proportion of clay and silt in planting soil in Hangzhou area reaches 53.4%, which belongs to fine-grained soil. The proportion of clay and silt in the soil of the park reaches 95.1%, 40.0% more than that of planting soil, and it is also fine-grained soil. The proportion of silt and clay in planting soil mixed with fine sand was 26.8% and that in coarse sand was 22.6%. The proportion of silt and clay after being mixed with fine sand is 45.5% and that of coarse sand is 34.6%. In conclusion, it can be concluded that the proportion of clay and silt decreases with the mixing of fine sand and coarse sand.

Table 4.   Results of particle analysis for soil samples with different ratios %.

Sample number	gravel		sand			silt	cosmid
	10~2mm	2~0.500 mm	0.500~0.250 mm	0.250~0.075 mm	0.075~0.005 mm	<0.005 mm	
1 # Planting soil	9.3	8.9	10.9	17.5	41.4	12.0	
2 # Park soil	0.2	0.1	0.5	4.1	88.0	7.1	
3 # Planting soil + fine sand	5.9	6.2	7.5	53.6	19.7	7.1	
4 # Planting soil + coarse sand	3.4	10.6	46.5	16.9	15.5	7.1	
5 # Park soil + fine sand	0.6	0.2	1.5	52.2	41.7	3.8	
6 # Park soil + coarse sand	2.3	7.3	44.8	11.0	30.8	3.8	

3.2 Rain Frequency test results

Figure 3 shows the process line of seepage flow in the rainwater garden tank test in different rainfall return periods. It can be seen from Figure 3 that the trend of the seepage flow curve is basically the same in the two rainfall return periods, which can be divided into four stages. First, the seepage flow increases rapidly from 0.0 cm³/s to over 2.0 cm³/s within the period from 20 min to 60 mins at the beginning of the experiment, and then slowly decreases from the maximum to about 1.5 cm³/s within the period of 1 h to 5 h at the beginning of the experiment. Then it is rapidly down to 0.3 cm³/s in the next 5 to 10 minutes, and finally slowly down to 0.0 cm³/s. There are two stages that have changed in the once-in-5-a rainfall condition compared with the once-in-2-a rainfall condition. In the first stage, the maximum seepage flow rate of 2.3 cm³/s in the once-in-5-a rainfall condition is larger than that of 2.1 cm³/s in the once-in-2-a rainfall condition. During the second stage, the once-in-5-a rainfall condition is 30 minutes,which is longer than the once-in-2-a rainfall condition. The once-in-5-a rainfall increased by more than 25% compared with the once-in-2-a rainfall, while the maximum seepage flow only increased by about 10%. The once-in-5-a condition has reached the maximum seepage capacity of planting soil.

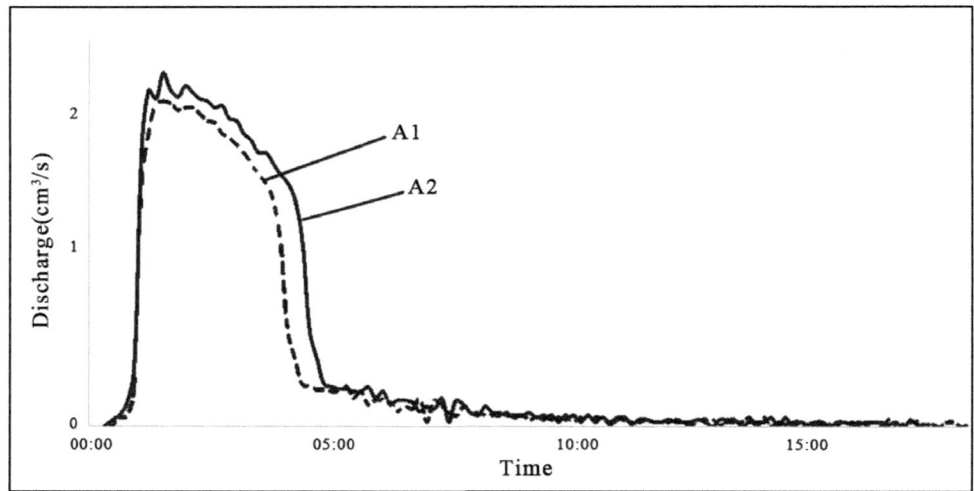

Figure 3.   Infiltration flow process diagram of rainwater garden box test in different rainfall reproducibility periods.

3.3 Soil body ratio test results

Figure 4 shows the seepage flow process line of the soil-sample rainwater garden tank test with different ratios. It also shows that the maximum seepage flow rate of B1 (planting soil) is 2.1 cm³/s under the condition of 2 a precipitation. The maximum seepage flow of B2 (park soil) is 0.7 cm³/s; The maximum seepage flow of B3 (planting soil + fine sand) is 4.8 cm3/s. The maximum seepage flow of B4 (park soil + fine sand) is 7.7 cm³/s. The maximum seepage flow of B5 (planting soil + coarse sand) is 5.6 cm³/s. The maximum seepage flow of B6 (park soil + coarse sand) is 9.2 cm³/s. From the above results, the maximum seepage flow of B1 is greater than that of B2, that of B4 and B6 is greater than that of B3 and B5, and that of B5 and B6 is greater than that of B3 and B4 respectively. It is concluded that both planting soil and park soil are fine-grained soil, which belongs to clayey soil and has more sand content, so the seepage capacity is higher. When the two soil samples are mixed with coarse sand or fine sand, the sand grains account for a large proportion, while the silt and clay grains account for a small proportion and become sand. The same is sandy soil. After being mixed with coarse sand or fine sand, the seepage ability of sandy soil mixed by planting soil is smaller than that of sandy soil mixed into the park because of the large proportion of clay grains in the planting soil. When the same soil sample is mixed with sand grains to become sand, the diameter of coarse sand is larger than that of fine sand, so the infiltration capacity of sand mixed with coarse sand is higher than that of sand mixed with fine sand.

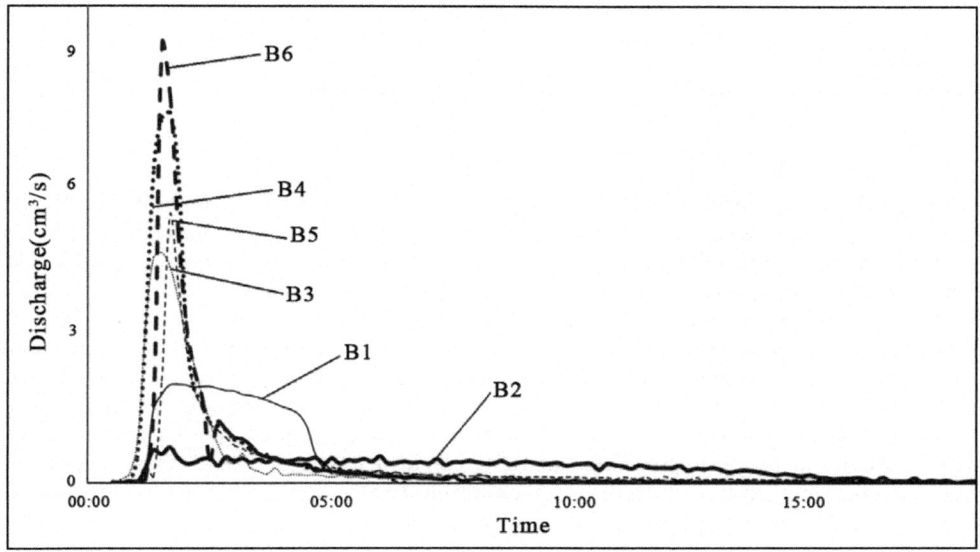

Figure 4.   Infiltration flow process diagram of soil sample rainwater garden box test with different ratios.

4   CONCLUSION

Based on the rainfall and soil conditions in Zhejiang Province, this paper establishes a physical model of sponge city, obtains experimental data under different conditions by changing the ratio of return period to the soil, and elaborates and analyses the causes of relevant test results, which provides ideas for subsequent tests. Relevant experimental results can provide a reference for rainwater garden design in Hangzhou and similar hydrological and geological areas.

REFERENCES

Emerson C.H., Traver R.G. Multiyear and seasonal variation of infiltration from storm-water best management practices [J]. *Journal of Irrigation and Drainage Engineering*, 2008, 134(5): 598–605.
Endreny T., Collins V. Implications of bioretention basin spatial arrangements on stormwater recharge and groundwater mounding [J]. *Ecological Engineering*, 2009, 35(5): 670–677.
Li Pengcheng, Lu Changxing, Gong Xueliang. Research on the structural parameters based on Orthogonal Test [J]. *Water-saving irrigation*, 2018(8): 14–18,29.
Luo Hongmei, Che Wu, Li Junqi, et al. Application of rainwater garden in the control and utilization of Rainflood [J]. *China Water supply and drainage*, 2008, 24(6): 48–52.
Roy-poiriera, Champagne, Filiony. Review of bioretention system research and design: Past, present, and future [J]. *Journal of Environmental Engineering*, 2010, 136(9): 878–889.
Tang Shuangcheng, Luo Wan, Jia Zhonghua, Yuan Huangchun. Experimental study on rainwater runoff in Xi'an [J]. *Journal of Soil and Water Conservation*, 2012, 26(6): 75–79,84.
Trowsdale S.A., Simcock R. Urban stormwater treatment using bioretention [J]. *Journal of Hydrology*, 2011, 397(3): 167–174.

*Frontiers in Civil and Hydraulic Engineering – Mohamed A. Ismail and
Hazem Samih Mohamed (Eds)
© 2023 The Authors, ISBN 978-1-032-38247-0*

Sustainable transformation design of Lingnan vernacular architecture and landscape

Feifeng Zhong*
Guangzhou Nanyang Polytechnic College, Guangzhou, Guangdong, China

ABSTRACT: Given the lack of sustainable planning, the destruction of the original ecological environment, and the loss of traditional regional cultural elements in the transformation of Lingnan vernacular architecture and landscape, a systematic and sustainable design methodology for the transformation of vernacular architecture and landscape is studied. Guided by the idea of "ecological network system" in the theory of vernacular geography and architectural ecology, this paper proposes approaches to sustainable transformation design of architectural ecology, architectural culture and architectural cultural tourism industry, as well as approaches to sustainable transformation design of landscape design ecosystem, renewable Lingnan vernacular landscape resources and Lingnan vernacular culture so as to address the existing challenges faced by Lingnan vernacular architecture and landscape, improve Lingnan vernacular habitats, carry forward the vernacular culture, and contributing to achieve ecologically sustainable development and accomplish the Chinese government's goal of carbon peaking and carbon neutrality.

1 INTRODUCTION

The term "vernacular" is used to describe the environmental complex involving settlement, production and ecology formed by the long-term interaction between human beings and the natural environment. The vernacular habitat is largely composed of vernacular architecture and landscape, the most crucial physical expressions of the vernacular habitat. To revitalize the rural areas, a systematic and sustainable development path should be pursued in the design of vernacular architecture and landscape transformation so that the vernacular architecture and the ambient environment can not only satisfy contemporary residents' needs for habitats while providing sustainable habitats for future generations of residents but also satisfy the needs of vernacular residents while accommodating urban tourists for sightseeing tours, leisure vacations, cultural experiences, among other activities (He 2020). The design for sustainable transformation of Lingnan vernacular architecture and landscape is not only in line with the national development program, but also plays a significant role and is of great significance in achieving the nation's carbon peaking and carbon neutrality goal and the development strategy of ecological progress and sustainable development.

---

*Corresponding Author: 6574030@qq.com

DOI 10.1201/9781003344209-70

2   NECESSITY AND SIGNIFICANCE OF SUSTAINABLE TRANSFORMATION DESIGN FOR LINGNAN VERNACULAR ARCHITECTURE AND LANDSCAPE

2.1   *National policy requirements*

The Report of the 19th National Congress of the Communist Party of China pointed out that "building an ecological civilization is vital to sustain the Chinese nation's development, and we should, acting on the principles of prioritizing resource conservation and environmental protection and letting nature restore itself, develop spatial layouts that help conserve resources and protect the environment." (Wang 2017). To make overall arrangements for carbon peaking and carbon neutrality, China released two documents in 2021. They are Working Guidance of the CPC Central Committee and State Council for Carbon Dioxide Peaking and Carbon Neutrality in Full and Faithful Implementation of the New Development Philosophy and Notice by the State Council of the Action Plan for Carbon Dioxide Peaking Before 2030. These policies have brought China's sustainable development requirements to an unprecedented level.

According to the China Building Energy Consumption Report (2021) released by the Committee of Energy Consumption under China Association of Building Energy Efficiency at the end of 2021, the building sector plays a key role in achieving the carbon neutrality target and is expected to reduce emissions by 72% by 2060. Given the carbon peaking target, the total building energy consumption should be controlled at 1.2 billion tce and carbon emissions should be controlled at 2.5 billion tons at the end of the 14th Five-Year Plan period. See Figure 1 below for the analysis of building energy consumption scenarios.

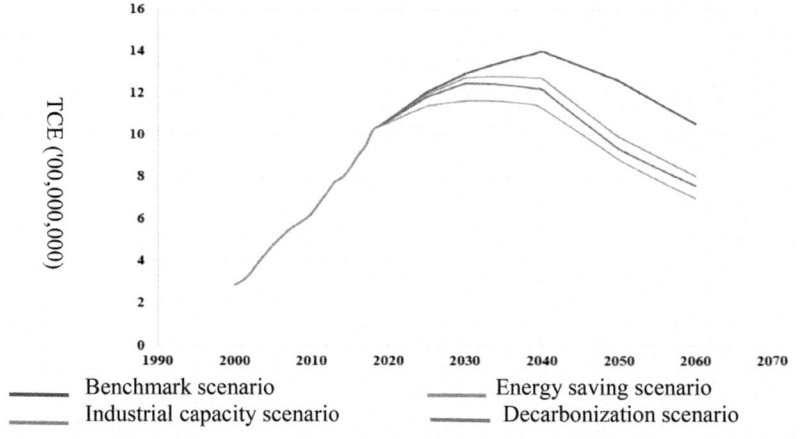

	2060 ('00,000,000 tce)	Peak (tce)	Peak time	End of 14th Five-Year Plan period
Benchmark scenario	10.58	14.03	2040	12.08
Energy saving scenario	8.08	12.82	2035	11.97
Industrial capacity scenario	7.62	12.48	2031	11.83
Decarbonization scenario	7.03	11.66	2031	11.40

Figure 1.   Building energy consumption scenario analysis.

To revitalize China's rural areas, it is therefore urgent to explore a sustainable path, which needs the support of vernacular architecture and landscape sustainability science before achieving the national goal of carbon peaking and carbon neutrality. In a bid to meet the requirements of rural revitalization and achieve ecologically sustainable development, the integration of sustainability philosophy in the design of vernacular architecture and landscape transformation is an important prerequisite for improving the vernacular environment, obtaining effective ecological and environmental benefits and carrying forward the vernacular culture.

2.2 Necessity of sustainable transformation of lingnan vernacular architecture and landscape

As urbanization progresses, mass destruction of vernacular architecture, landscape and ecological environment has aroused people's attachment to vernacular architecture, landscape and culture with traditional characteristics and their concern for the vernacular environment. The sustainable transformation design of vernacular architecture and landscape is conducive to improving the habitat of China's vernacular areas and the appearance of the national territory and plays a significant role in the inheritance, protection and sustainable development of the regional vernacular culture as well as the improvement of the vernacular ecological environment and the realization of the national "dual carbon" goals.

2.3 Significance to national economic and social development

Building a beautiful rural area has always been the expectation of the people, and living in a pleasant architectural and landscape environment is the indispensable ground for villagers to live in harmony and work contentedly. With this ground, it is more likely to promote rural prosperity and farmers' income, which will undoubtedly facilitate rural revitalization and even the economic and social development of the country. The sustainable transformation design of Lingnan vernacular architecture and landscape is not only an important means to make full scientific and reasonable use of natural landscape resources, effectively protect the ecological environment, regulate and embellish the vernacular habitat, and truly enable a beautiful vernacular habitat that is ecologically livable (Chen 2021), but also an important means to save social energy and costs, realize resource recycling and achieve carbon peaking and carbon neutrality. Moreover, it is of great importance to drive China's economic and social development.

3 OVERVIEW OF RESEARCH ON VERNACULAR ARCHITECTURE AND LANDSCAPE AT HOME AND ABROAD

3.1 Overview of foreign research on vernacular architecture and landscape

In terms of theoretical research on vernacular architecture, scholars in Europe and the United States, with a focus on the special relationship between architecture and culture, region and climate under the context of vernacular landscape planning and land utilization, have proposed new ideas such as "ecological network system" and "spatial concept" based on the theory of vernacular geography and landscape ecology, as well as the methodology of describing multi-objective vernacular land utilization planning and landscape ecology design by using geographic information systems. The objectives of village renewal in foreign countries and regions are mostly to ensure the sustainable development of agricultural production and the rational layout of living and production spaces; to secure the coordinated development of villages in the architectural, economic and social fields; to protect the intrinsic values and autonomy of villages; to maintain the historical and cultural "fundamentals" and "matrix" on which human beings depend and to strengthen the cohesiveness of villagers (Deng 2015)

In terms of practical research, scholars in Europe and the United States, with a focus on the application of building technology and local materials, are committed to the study of economic and energy-saving and environment-friendly building technology; They emphasize the integration of man and nature, and design for agricultural production, farming landscape, sightseeing and recreation while taking into account both vernacular development and landscape protection. They also place importance on the inheritance, protection and development of vernacular culture, integrating cultural elements and cultural symbols of traditional folk activities in design activities, highlighting the characteristics of regional tangible and intangible cultural landscape and cultivating unique vernacular landscape and folk culture.

3.2 *Overview of chinese research on vernacular architecture and landscape*

In terms of theoretical research on vernacular architecture, Chinese scholars focus on the historical value and practical significance of ancient architectural relics of traditional villages, as well as the in-depth investigation of the principles and methods of creating ancient wooden and stone structures and regional folklore architecture. With foreign theories of landscape ecology as their basis for the research on vernacular landscape design theory, Chinese scholars elaborated the fundamentals, principles, measures, methods and effectiveness evaluation of vernacular landscape design by focusing on the application of the "ecological design" theory and interdisciplinary integration, while a few Chinese scholars put forward the corresponding planning principles and approaches in combination with specific cases they have worked on. However, it takes a long time to form a complete research system because of the lack of depth and breadth of research.

In terms of practical research on vernacular architecture, Chinese scholars are dedicated to the research on new technologies and new materials, structural design and exterior renovation of sustainable buildings, as well as innovative exploration of B&B architecture in the new era. In terms of practical research on the vernacular landscape, Chinese scholars, in the context of national strategies of building new socialist countryside and beautiful vernacular environment, and achieving rural revitalization, are interested in inheriting historical heritage, restoring architecture and landscape "as it was", or combining traditional culture with modern aesthetics. They use modern design techniques for landscape renovation and design innovation, and actively explore for "making it possible for people to enjoy the natural landscape and retain their love of nature".

3.3 *Summary of research at home and abroad*

Foreign scholars have formed a fairly complete theoretical and methodological system in terms of the research on vernacular architecture and vernacular landscape. Although Chinese scholars have conducted some theoretical and practical exploration on the historical heritage of rural architecture and landscape, it is still necessary to vigorously promote the research on the theoretical and methodological aspects of systematic planning for the sustainability of vernacular architecture and landscape, especially given the very few theoretical research and practical cases for the sustainable transformation design of Lingnan vernacular architecture and landscape.

4 EXISTING CHALLENGES IN THE TRANSFORMATION OF LINGNAN VERNACULAR ARCHITECTURE AND LANDSCAPE

4.1 *Absence of sustainability planning for construction familiar to*

Various kinds of unexpected and unrealistic buildings and landscapes appear in the vernacular environment in large numbers as the local vernacular governments blindly apply urban construction templates and the relevant planning and design officials are not familiar with the term vernacular and fail to conduct enough surveys and studies. Due to the lack of sustainable planning in vernacular architecture and landscape construction, the massive use of various modern new materials and technologies such as reinforced concrete instead of recycling vernacular recycled resources, renewable resources and recyclable resources and developing vernacular traditional techniques and sustainable construction methods, adverse results include the waste of resources and the loss of local characteristics, incompatibility of reconstructed vernacular architecture and landscape environment with the original vernacular landscape and the destruction of the original vernacular ecosystem.

4.2 *Continuous loss of traditional architecture with regional culture characteristics*

Urbanization and modernization have not only brought advancement and prosperity to the vernacular regions but also greatly transformed and deconstructed the regional cultural characteristics of the vernacular society. New ideas, new materials and new methods are integrated into the design of vernacular architecture, but there is no systematic planning for these emerging stuff, whereby the local architectural styles that are originally unique to the region are replaced by mixed Chinese and Western, retro-modern and other styles. The local unique historical buildings such as earthen buildings and stone street landscapes are replaced by concrete buildings and modern road pavements, and the traditional buildings with regional cultural characteristics are continuously dislocated.

4.3 *Lack of vernacular landscape culture*

As time progresses, the vernacular landscape is increasingly exposed to the influence of foreign ideas and culture, resulting in the crowding of space by modern elements and foreign cultural elements in the ecological local culture of the vernacular landscape. The vernacular regions are packed with European-style promenades and pavilions, modern flower ponds, straightened rivers, hardened riverbanks, neat and uniform vegetation, and sparse flowers and weeds. Moreover, traditional cultural landscape elements are replaced by foreign landscape elements, the indigenous features of the vernacular landscapes are gradually disappearing, the vernacular taste is gone, and the "nostalgic" landscape is absent.

5 METHODOLOGY OF SUSTAINABLE TRANSFORMATION DESIGN OF LINGNAN VERNACULAR ARCHITECTURE AND LANDSCAPE

5.1 *Approaches to sustainable transformation design of lingnan vernacular architecture*

5.1.1 *Sustainable transformation design of lingnan vernacular architecture from the perspective of ecology*

Guided by the idea of "ecological network system" in the theory of vernacular geography and architectural ecology, this paper, by studying the relationship between architecture and the region, climate and ecology, proposes systematic and sustainable renovation design methodologies for Lingnan vernacular architecture. Specifically, they include creating open and ventilated building spaces, introducing different building openings, adopting passive building energy-saving design to satisfy the requirements of environmental-friendly buildings with comfortable indoor environment, improving the natural ventilation and natural lighting of the buildings, setting reasonable building windowing area, optimizing building thermal insulation conditions to minimize overall building energy consumption, reducing solar heat radiation and area ratio of window to wall, applying auxiliary facilities, louvers, vertical greening and other methods to address the problem of building overheating, improving the air tightness of building components, reducing window gaps, changing the placement of thermal insulation layer, minimizing the loss of building energy consumption to reduce carbon emissions, employing various green building materials such as thermal insulation and energy-saving materials, circular and recycled materials, vernacular ecological materials, ecological cement, new eco-friendly materials made of straw, etc. to reduce construction costs and realize building energy conservation, adopting vernacular traditional construction technology, incorporating vernacular traditional construction technology with modern building construction technology, and developing new energy-saving and environment-friendly construction technology to achieve sustainable development of construction technology, integrating traditional vernacular materials with new ecological materials, incorporating traditional vernacular materials with new construction technology, and combining new ecological materials with traditional vernacular construction technology to achieve ecological sustainable development of construction materials and construction technology.

5.1.2 *Sustainable transformation design of lingnan vernacular architecture from the perspective of culture*

From the perspective of culture, the sustainable transformation design of Lingnan vernacular architecture should conform to the development rules and evolution trends of Lingnan villages and promote the sustainable integration of Lingnan vernacular architecture with the overall environment and vernacular culture. Moreover, it should follow the ideas of gathering and upgrading, refining characteristics and protecting cultural heritage depending on the history and culture, folk customs, geographical environment, regional characteristics and resource endowment of different villages. It should also conduct sustainable inheritance and transformation design of Lingnan vernacular architectural culture in terms of vernacular architectural exterior, architectural materials, architectural colors and architectural texture, focus on the application of Lingnan vernacular traditional architectural techniques, traditional artisan techniques and local materials to reshape the facade of vernacular architecture, improve vernacular recognition and restore indigenous vernacular architectural features, strengthen the construction of vernacular infrastructure facilities, unify planning, harmonize the overall style of vernacular architecture, and study the methods of planning and improving the layout and construction of village architecture so that Lingnan vernacular architectural style can be inherited and developed in a sustainable manner.

5.1.3 *Sustainable design of lingnan vernacular architecture from the perspective of the cultural tourism industry*

Due to the high volume of young immigrants working in urban areas for better lives, many rural houses are left idle and hollowed out in rural areas. To enrich the amateur life of villagers and improve their quality of life, it is suggested that idle houses, warehouses, factories and other buildings in villages be transformed into public activity spaces such as book houses (libraries), villagers' activity rooms, cultural rooms, cultural halls, service centers, post stations and multifunctional rooms shared by villagers. To accommodate foreign tourists to travel and consume in the countryside, offer more jobs and increase the income of villagers, it is suggested that idle buildings be transformed into restaurants, B&Bs, resort hotels, agricultural science base display centers and rural culture exhibition halls while developing such businesses as catering, accommodation, agricultural products and rural culture display with vernacular characteristics. To attract tourists to sightsee and experience and form rural cultural and creative industries, it is suggested that the efforts of villagers be pulled together to transform idle buildings into cafés, bars, drawing studios, game booths, teahouses, handicraft stores, folk art production workshops, creative experience stores and boutique stores while leading them to build thematic cultural and creative commercial streets. To attract foreign tourists to experience and consume and lead the villagers to achieve common prosperity, it is suggested that various rural resources be made good use of to transform idle rural buildings into a field architecture complex integrating scientific popularization, parent-child activities, the theme park as well as health & wellness & culture experience while hosting a music festival, drama festival, straw festival and folklore festival in different periods to create a new form of field culture and tourism architecture hub integrating entertainment experience and leisure consumption.

5.2 *Sustainable transformation design of lingnan vernacular landscape*

5.2.1 *Approaches to improvement of landscape design ecosystem*

Approaches include taking a comprehensive consideration of the climate and light conditions, geological environment, plant resources, hydrological characteristics and other elements of the vernacular regions and exploring the laws of natural changes therein so that the various natural conditions and resources of the ecosystem and the sustainable landscape transformation design are effectively combined and reasonably compatible. Also, they include creating a more ideal landscape space with vernacular characteristics on the basis of the balance of the ecosystem, forming an optimal ecological sustainable landscape from the aspects of health, ecology, nature, human physiological and psychological needs in a scientific way, enhancing the comfort of the landscape design, improving the structural form of the vernacular space, putting in place a consummate landscape design ecosystem, and materializing the concept of sustainable development of the landscape ecosystem.

5.2.2 *Development and utilization of renewable resources of lingnan vernacular landscape*

The development and utilization of renewable resources of Lingnan vernacular landscape should follow the original traffic flow and spatial organization of the Lingnan countryside, reasonably transform the courtyards into open or semi-open courtyard landscape spaces for local villagers to rest, converse and have fun, offer service functions to satisfy villagers' needs and improve the quality of the residents' living environment. It should also create a linear landscape pattern for Lingnan vernacular landscape design, further improve the landscape ecological network, integrate wetland resources in landscape design, apply landscape ecological restoration, landscape stormwater management and resilient landscape strategies, establish water ecological corridors or green ecological corridors, further improve the accessibility of sustainable landscape design and ecological vernacular construction, and improve the natural resilience of the ecological environment in landscape design, etc (Zhang 2021). Moreover, wetland parks, green space centers, tourist farms and experience farms should be built by taking into account the local natural environment, vegetation and landscape resources of Lingnan vernacular regions to create an integrated community of "agriculture + cultural tourism + community", so that the renewable resources of vernacular landscapes can be exploited continuously.

5.2.3 *Sustainable transformation design of lingnan vernacular landscape from the perspective of cultural contents*

From the perspective of cultural contents, sustainable transformation design of Lingnan vernacular landscape should tap into the potential of the local Lingnan culture and tap into the combination of public Lingnan cultural landscape art and theme culture of the featured town, implant Lingnan culture in the original countryside, highlight Lingnan cultural theme and Lingnan regional characteristics, create Lingnan vernacular cultural theme landscape and enable the vernacular transformation to achieve historical and cultural heritage.

It also should integrate vernacular idyllic scenery and vernacular culture, collect Lingnan vernacular local materials and recyclable resources for public art creation of Lingnan vernacular landscape, enable a vernacular culture-based public art, ensure that the vernacular courtyard retains Lingnan vernacular culture and regional characteristics, create exquisite courtyards of Lingnan style, and embody Lingnan vernacular cultural characteristics.

5.3 *Challenges addressable through sustainable transformation of lingnan vernacular architecture and landscape*

The following four major challenges will be addressed after researching the sustainable transformation of Lingnan vernacular architecture and landscape hereinabove. Firstly, the lack of systematic theory and methodology for sustainable transformation planning of beautiful vernacular architecture and landscape in Chinese beautiful vernacular construction will be addressed; secondly, the problems of disorder in Chinese beautiful vernacular construction, disorganized architectural styles, unexpected landscaping, and incompatibility of architecture and landscape with the original vernacular regions will be addressed, and a holistic planning and design method for Lingnan vernacular architecture and landscape will be formed; thirdly, such challenges as the squeezing of the local culture of vernacular architecture and landscape ecology by modern and foreign cultural elements in the construction of beautiful countryside, the absence of vernacular history and culture, and the disappearance of indigenous vernacular characteristics will be addressed; fourthly, the problem of unsatisfactory rural development and rural economic development as a result of the drain of rural workforce due to poor rural environment will be addressed.

Figure 2. Methodology of sustainable transformation design of Lingnan vernacular architecture and landscape.

6 CONCLUSION

By reasonably utilizing Lingnan vernacular natural landscape resources, employing recyclable resources, local renewable resources and vernacular artisan techniques, and through the above-mentioned approaches to sustainably transform and design Lingnan vernacular architecture and landscape, we will effectively protect Lingnan vernacular ecological environment, embellish Lingnan vernacular living environment, preserve and develop vernacular culture, preserve and revitalize

Lingnan traditional architectural relics and cultural heritage, reshape the vernacular natural appearance and ecological values, make the vernacular environment more charming, turn vernacular Lingnan into a beautiful ecologically livable place, and effectively improve people's living conditions and quality of life. Thanks to the sustainable transformation of Lingnan vernacular architecture and landscape, localities will form rural industrial chains with local characteristics, retain the rural workforce, enable villagers to live in harmony and work contentedly in their native land, promote agricultural development and rural prosperity and increase farmers' income, and boost rural revitalization and even the development of the whole society.

FUND PROGRAM

This paper is a phased outcome of "Research on Sustainable Transformation Design of Lingnan Vernacular Architecture and Landscape" (Project No. 2021ZDZX4084), a 2021 Provincial Key Area Research and Development Program under Key Research and Platform Program of Department of Education of Guangdong Province), and "Research on the Design of Rural Tourism Industry System in Conghua Under the Guidance of Circular Economy (Project No. GD20XYS13)", a 2020 Project of Philosophy and Social Sciences Planning of Guangdong Province Under the Thirteenth Five-Year Plan.

REFERENCES

Chen Dongfang. (2021). A study on how to foster farmers' awareness of ecological progress from the perspective of rural revitalization strategy [J]. *Agricultural Economy*, (04): 77–79.

Deng Bijuan. (2015). A study on cultural landscape development in foreign small towns and its implications [J]. *Flowers*, (17): 30–31.

He Yanhua, Wu Jianguo, Zhou Guohua, Zhou Bingbing. (2020). Discussion on rural sustainability and rural sustainability science [J]. *Acta Geographica Sinica*, 75(04): 736–752.

Wang Jinnan, Jiang Hongqiang, He Jun, Wang Xiahui. (2017). Strategic tasks for ecological civilization construction of the socialism with chinese characteristics for a new era [J]. *Chinese Journal of Environmental Management*, 9(06): 9–12.

Zhang Xianglu, Chen Zhuojian, Zeng Huixian, Tan Hao, Chen Ping. (2021). Research on landscape design strategies under the concept of sustainable development[J]. *Journal of Green Science and Technology*, 2021, 23(03): 36–38.

*Frontiers in Civil and Hydraulic Engineering – Mohamed A. Ismail and
Hazem Samih Mohamed (Eds)
© 2023 The Authors, ISBN 978-1-032-38247-0*

Simulation and verification of indirect modeling method for urban wind environment

Yun Wang
CGN New Energy Holdings Co., Ltd, China

Chao Wang*, Chong Zhang & Zhenqing Liu
Huazhong University of Science and Technology, Wuhan, Hubei, China

ABSTRACT: As the most densely populated area, the city has a series of problems such as ventilation and urban wind energy evaluation. However, the previous research on how to accurately simulate actual urban wind fields is limited, which does not consider the effect of atmospheric stratification. Meanwhile, the efficiency of urban building modeling is low. Therefore, the urban wind environment is explored in this study considering atmospheric stratification, in which LES (large eddy simulation) is carried out for a single building and multiple buildings. Besides, an indirect modeling method is proposed. Taking actual urban buildings as examples, CFD (computational fluid dynamics) simulation of an actual urban wind environment is carried out based on this method. The results indicate that this method is reliable for numerical simulation of actual urban wind environment, which only takes 5–8 min to model the urban buildings within an area of 1 km × 1 km.

1   INTRODUCTION

As one of the most widely distributed clean energy, wind energy can be used by large-scale wind power plants in most cases. However, in cities and suburbs with limited land resources, there are no conditions for the establishment of wind farms. Therefore, urban wind resources can be used based on research on the urban wind environment. Compared with large-scale wind farms, urban wind power generation has its unique advantages in the direct use of the power generated by wind energy without considering the transmission problem. Further, it is of great significance to carry out the method of urban wind resource evaluation and simulation calculation for the full utilization of wind energy.

There are many numerical methods to solve the urban wind field, among which CFD can simulate the impact, separation, surrounding, reattachment, and other phenomena of turbulence in the atmospheric boundary layer around the building. Consequently, with the development of computer science, research based on CFD has been gradually increasing in recent years. Cheng & Porté-Agel (2015) proposed a simulation based on the idealized urban building surface, while Hertwig et al. (2012) simulated the urban wind environment based on the semi-idealized building cornice. The results verified the accuracy of the two methods. Castro et al. (1983) took the triangular prism as a model and carried out stratified flow experiments in salt water with a density gradient. They studied the influence of stratified flow on the flow field around the obstacle by changing the Froude number. Subsequently, Dong et al. (2017) considered the influence caused by temperature stratification in LES of the flow around the cylinder when studying the characteristics of the flow field around the cylinder under different stratification numbers. The results showed that the gas

---

*Corresponding Author: wangchao_wc@hust.edu.cn

stratification effect transited near K=0. Moreover, with the development of CFD in the urban wind environment, research on urban building modeling is also advancing. Razak et al. (2013) studied the characteristics of wind fields around urban buildings with cubes based on the LES turbulence model, in which five different vertical and horizontal spacing ratios were set for the cubes array. The results indicated that the front area ratio of buildings was an important parameter for estimating the pedestrian wind environment between buildings. Ricci et al. (2017) carried out a numerical simulation study on the wind field between buildings in a block of Livorno, Italy. Xiao et al. (2019) conducted full geometric modeling of urban buildings near London South Bank University.

However, in current numerical simulation research on urban wind environments carried out around the world, it was mainly based on the assumption of a neutral atmospheric boundary layer, which was rare in actual projects. Instead, the temperature effect of the urban atmosphere will have a great impact on wind field characteristics and turbulent structure. Meanwhile, when simulating the urban wind environment, previous research needs to carry out full geometric modeling of complex urban buildings, and manually animate the mesh, to simulate the real urban building wind field. For urban areas with dense buildings and complex building geometry, the modeling workload is huge with low efficiency.

Therefore, this study proposed a self-made program based on the harmonic synthesis method to generate the fluctuating wind speed time history that meets the requirements of the target characteristics. First, LES is carried out for a single building and multiple buildings under different atmospheric stable stratification conditions. Meanwhile, the wind field characteristics of simplified urban models under different atmospheric stratification conditions (K = 0, K = 1, K = 2, K = 3) are studied to discuss the comprehensive effects of urban building density and atmospheric stratification on urban wind fields. Then, based on the generated urban wind field, an innovative indirect modeling method is proposed for urban wind environment simulation. This method adopts a C++ program and is based on a fluent fluid computing platform to realize the full automation of urban building modeling, CFD parameter setting, and numerical simulation. Based on the classic calculation example of flow around a square column, this indirect modeling method is used to compare simulation results with the wind tunnel test data and simulation results of direct modeling, which verifies the accuracy of this indirect modeling method. Finally, the method is used to simulate the wind field of some actual urban buildings under the condition of atmosphere stable stratification. The calculation results of the indirect modeling method can be used as the numerical basis of real engineering.

2 THEORY

2.1 *Atmospheric stratification*

In the atmospheric boundary layer, buoyancy related to atmospheric stratification plays a key role, which directly suppresses or enhances the vertical movement in stratified flow. The atmosphere stratification intensity can usually be described by two dimensionless parameters: Froude number *Fr* and stratification number *K*.

The physical meaning of the Froude number is the ratio of inertia force to gravity magnitude, which can be expressed by Equation 1:

$$Fr = \frac{U}{Nh} \tag{1}$$

where U is the incoming wind speed, h is the obstacle height, and N is Brunt-Väisälä frequency.

The stratification number K does not refer to the number of atmospheric layers, but a dimensionless parameter describing the strength of atmospheric stratification, which can be expressed by Equation 2:

$$K = \frac{NH}{\pi U} \tag{2}$$

where H represents the height of stable stratification, which is usually taken as the height of the computing domain in practical research.

2.2 *Indirect modeling method*

The main idea of the indirect modeling method is to simulate the obstruction of urban buildings to the wind field by adding a resistance source term to grid cells at the location of urban buildings in the calculation domain. The N-S equation of the filtered LES turbulence model is expressed as Equation 3:

$$\rho \frac{\partial \tilde{u}_i}{\partial t} + \rho \frac{\partial \tilde{u}_i \tilde{u}_j}{\partial x_j} = \frac{\partial}{\partial x_j}\left(\mu \frac{\partial \tilde{u}_i}{\partial x_j}\right) - \frac{\partial \tilde{p}}{\partial x_i} - \frac{\partial \tau_{ij}}{\partial x_j} + f_{\tilde{u},i} \tag{3}$$

where $f_{\tilde{u},i}$ represents the resistance source term, which is used to represent the obstruction of urban buildings to the wind field.

3 METHOD VERIFICATION

3.1 *Model*

Based on numerical simulation of the ideally shaped building under single and multiple conditions, this study proposes an indirect modeling method, which is verified by the classic calculation example of the flow around a square column. Two working conditions, including direct and indirect modeling conditions, are set respectively, to obtain numerical simulation results, which are compared with the test data. The specific example model is shown in Figure 1. The bottom surface center of the square column is the origin, and the side length of the square column is D, with a calculation domain: 30 D × 20 D × 4 D. The inlet of the calculation domain is 10 D away from the square column center, and the side boundary is 10 D away from the square column. The inlet adopts uniform wind, of which the wind speed is U_0, the outlet condition is outflow, and the Reynolds number is $Re = 2.2 \times 10^4$. Two working conditions are set respectively: working condition 1 is the direct modeling working condition which establishes a geometric model of the square column set, and grids are divided, as shown in Figure 2a; Condition 2 is an indirect modeling condition, in which all grids in the calculation domain are divided, and the resistance source term is added at the location of the square column through the program. The grid diagram is shown in Figure 2b.

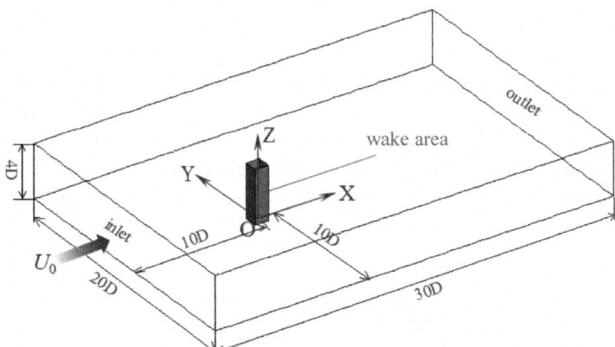

Figure 1.   Verification example model and size diagram.

3.2 *Verification results*

The comparison results of instantaneous velocity nephogram around the square column under two working conditions are shown in Figure 3. It can be clearly seen that under the indirect modeling condition, the flow around the square column can also be obtained by adding the resistance term at the position of the square column. The Karman vortex is formed in the wake area, which is almost consistent with the flow field under the direct modeling condition, with an obvious vortex shedding phenomenon.

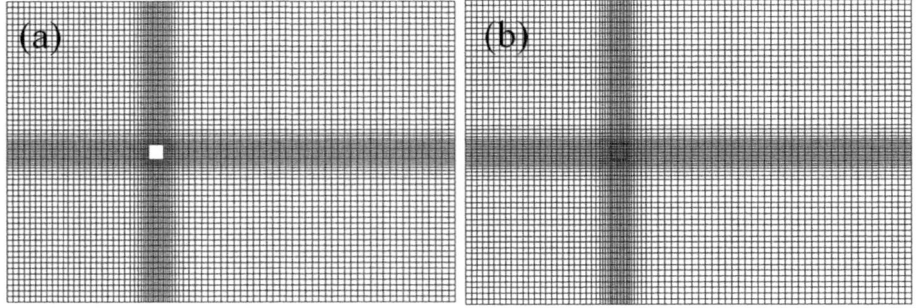

Figure 2.   Grid diagram of verification example model. (a) Direct Modeling (b) Indirect Modeling.

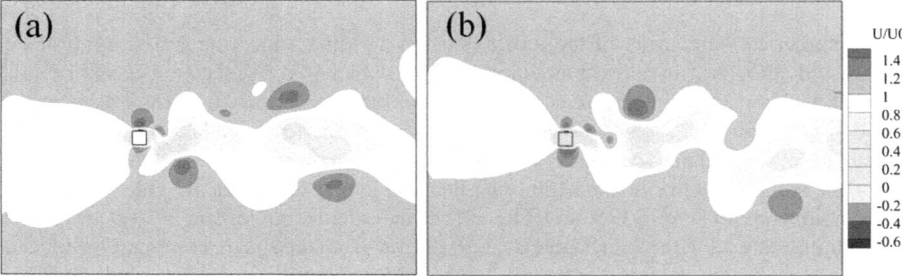

Figure 3.   Comparison of instantaneous velocity nephogram under two working conditions. (a) Direct Modeling (b) Indirect Modeling.

4   NUMERICAL EXAMPLE

4.1   *Building model*

Based on the indirect modeling method proposed above, it can accurately and efficiently model urban buildings, as shown in Figure 4. The left figure shows an actual urban building model, while the right figure shows the cloud diagram of adding resistance source terms to the location of buildings using the indirect modeling method. As seen in Figure 4, the characteristics of actual urban buildings can be completely captured by automatically identifying and adding urban models through the program.

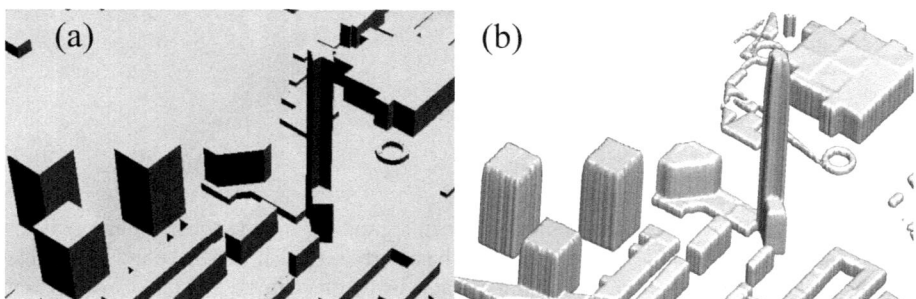

Figure 4.   Comparison diagram of urban models using indirect modeling method. (a) Actual Buildings (b) Model Buildings

Taking actual urban buildings in Beijing as an example, a cubic computing domain model is established, and the resistance source term is added to the grid cells with buildings through automatic judgment by the program. The computing domain model is shown in Figure 5. The computing domain size is 2 km × 2 km × 2 km, with the highest urban building as the center of the computing domain, of which the height is 160 m. All urban buildings are located in a circle with a diameter of 600 m in the computing domain. The grid in the circle is locally encrypted, and then gradually transited. The minimum size of the grid is 3 m in the horizontal direction, and 1 m in the vertical direction, which is densified near the ground. The grid growth ratio in both directions does not exceed 1.1. In this study, three working conditions with different wind angles are set up, as shown in Figure 5 (NW, N, and NE). The uniform wind speed is adopted at the entrance, in which the inlet speed is 10 m/s, and atmospheric stratification $K \approx 1$ is estimated according to the change in the local maximum vertical temperature, the height of the computing domain, and the wind speed. The outlet adopts outflow, and the other boundaries are set as walls. Three different observation points are set between buildings. Fluid computing is carried out based on the open-source fluid computing platform OpenFOAM 2.1.3, and the standard $k-\varepsilon$ turbulence model and PIMPLE algorithm are adopted to decouple the pressure and velocity.

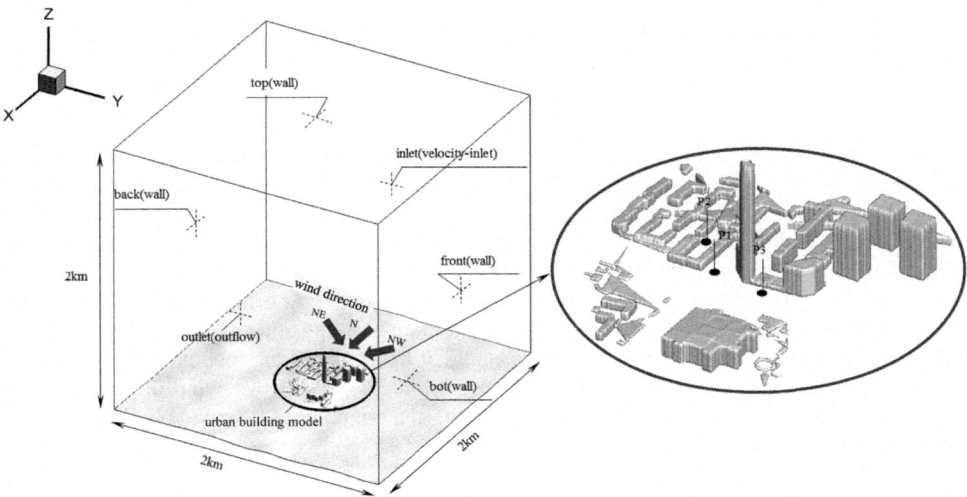

Figure 5.   Schematic diagram of computing domain and model of actual city.

4.2 *Numerical example results*

Figure 6 shows the downwind ground wind velocity cloud map under different wind angles. Due to the obstruction of urban buildings, airflow has separated at the corner of the building's windward side, and the wind speed is small at the back of the buildings and downwind of the city. In addition, as the wind direction angle is different, the distribution of the urban wind field varies greatly. When the wind direction is N and NW, the airflow forms an obvious vortex between the block buildings. The place with a higher wind speed appears at the block intersection, while the wind speed in the densely populated place is small. On the contrary, due to the obstruction of buildings, when the wind direction is NE, the airflow is brought downstream by the incoming flow before it has time to generate vortices at the block intersection, of which the wind speed in the block is greater than that under other working conditions.

Figure 7 shows the vertical wind profile at the observation points under different wind directions. In Figure 7, U represents the downwind wind speed, Uinlt represents the wind speed at the entrance, and H represents the height from the ground. Observation point P1 is located at the block intersection. The wind profile at P1 is also different when the wind direction is different. When the wind

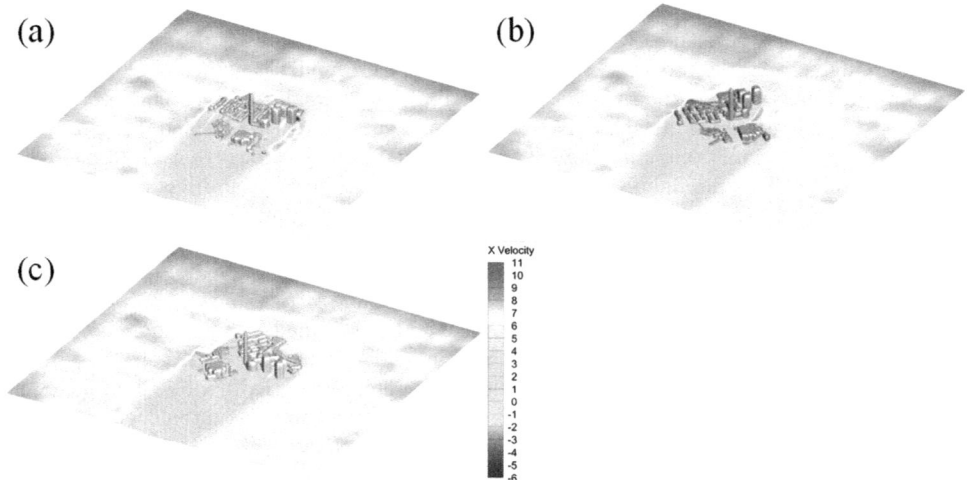

Figure 6. Nephogram of downwind velocity at the bottom under different wind angles. (a) Wind Direction N (b) Wind Direction NE (c) Wind Direction NW.

direction is NW, the incoming flow is blocked by high-rise buildings upstream, making the wind speed within 160 m above the ground of the monitoring point very small, which gradually equals the incoming flow wind speed after more than 160 m. As the incoming flow of wind direction N is not hindered by buildings, its wind speed is higher than that under other working conditions. The observation point P2 is surrounded by dense buildings, and the wind direction has little effect on its wind profile, indicating that the wind speed in the building height is small, and then increases. The wind profile at P3 is greatly affected by the front main building and annex building. When the wind direction is NE, the main building is located just upstream of the incoming flow, resulting in a small wind speed within the height of the main building. When the wind direction is NW, the annex building blocks the incoming flow, making the wind speed near the observation point smaller, which then gradually increases.

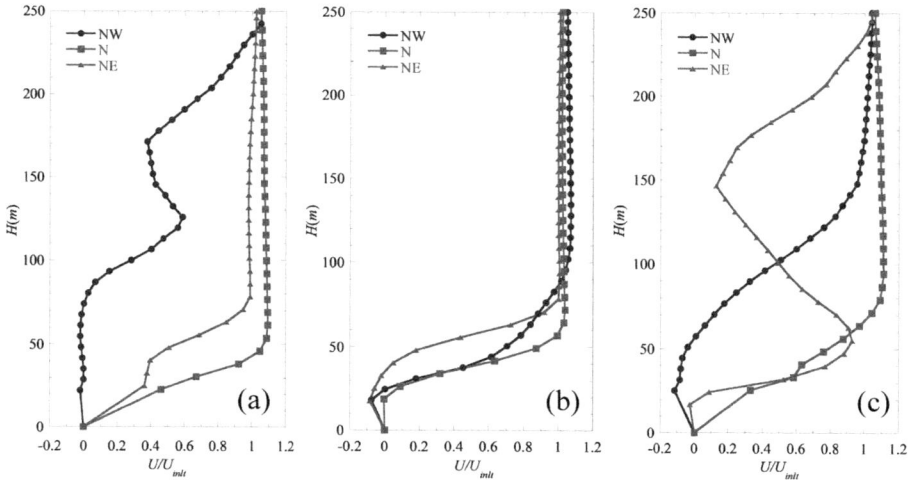

Figure 7. Downwind wind speed profile at different observation points (a) P1 (b) P2 (c) P3.

5 CONCLUSION

This study proposes an indirect modeling method for urban wind environment and verifies its feasibility. Subsequently, a program module for the actual urban wind environment simulation is put forward, which simulates some actual urban building wind fields. The verified results indicate that this indirect modeling method can accurately and effectively simulate the obstruction effect of obstacles on the flow of the field. Moreover, the indirect modeling program is used to simulate the wind environment under the effect of atmospheric stratification within 1 km of the urban area. The results show that this method is fast and reliable for numerical simulation of the actual urban wind environment, with a less time cost of 5–8 min to one-click "modeling", which improves the efficiency of urban building modeling.

REFERENCES

Castro, I.P., Snyder, W.H., Marsh G.L., et al. (1983). Stratified flow over three-dimensional ridges. *J. Fluid Mech.* 135(1), 261–261.

Cheng, W.C. & Porté-Agel, F. (2015). Adjustment of turbulent boundary-layer flow to idealized urban surfaces: A large-eddy simulation study. *Bound-Lay Meteorol.* 155(2), 249–270.

Dong, H., Cao, S., Ge Y. (2017). Large-eddy simulation of stably stratified flow past a rectangular cylinder in a channel of finite depth. *J. Wind Eng Ind Aerod.* 170, 214–225.

Hertwig, D., Efthimiou, G.C., Bartzis, J.G. et al. (2012). CFD-RANS model validation of turbulent flow in a semi-idealized urban canopy. *J. Wind Eng Ind Aerod.* 111, 61–72.

Razak, A.A., Hagishima, A., Ikegaya, N. et al. (2013). Analysis of airflow over building arrays for assessment of urban wind environment. *Build Environ.* 59 (Jan.), 56–65.

Ricci, A., Kalkman, I., Blocken, B. et al. (2017). Local-scale forcing effects on wind flow in an urban environment: Impact of geometrical simplifications. *J. Wind Eng Ind Aerod.* 170, 238–255.

Xiao, D., Heaney, C.E., Mottet, L. et al. (2019). A reduced order model for turbulent flows in the urban environment using machine learning. *Build Environ.* 148, 323–337.

Frontiers in Civil and Hydraulic Engineering – Mohamed A. Ismail and Hazem Samih Mohamed (Eds)
© *2023 The Authors, ISBN 978-1-032-38247-0*

Comparative research of fault tree, bayesian networks, and fuzzy bayesian networks for system reliability assessment in civil engineering

Ziyao Sun*

College of Civil Engineering, Huaqiao University, Xiamen, Fujian, China

ABSTRACT: Safety analysis and evaluation in civil engineering are necessary to prevent accidents that may lead to accidents. This paper aims to select appropriate methods for risk prediction by traditional fault tree (FT), Bayesian network (BN), and Fuzzy Bayesian network (FBN), and compare them in the same accident situation. The conclusion of this paper shows that the three methods have their own advantages, but FBN can contribute to reliable risk analysis and evaluation, especially in complex and uncertain construction.

1   INTRODUCTION

In the history of human development, civil engineering is undoubtedly the cornerstone of human progress. In recent years, with the innovation and breakthrough of civil engineering construction technology, it has become the trend of civil engineering development to better serve human beings through the development and utilization of larger underground space, followed by various potential risk factors in the complex engineering environment. These risks may bring incalculable economic losses and even threaten the life safety of workers. On March 23, 2005, due to the lack of reliability assessment, an explosion occurred at a refinery in Texas, which killed 15 persons (Zhang 2010). On July 6, 2010, a tunnel collapsed in Prague of the Czech Republic, resulting in a 15 meters wide pit (Khakzad 2011). On January 12, 2007, the subway station on Metro Line 4 of the aquarium in Sao Paulo of Brazil collapsed, which killed 7 persons (Khakzad 2011).

Facing the risk factors that may bring serious consequences to the project, many researchers have studied different methods in accident risk analysis. For example, LeBeau used FT when analyzing the accident cause of the Schoharie Creek bridge collapse (LeBeau 2007). Daniel used the BN model in the probability modeling of the tunnel excavation process (Straub 2013). Sun conducted the risk assessment of tunnel collapse based on FBN (Sun 2018). Because the three methods—FT, BN, and FBN—have better accuracy than other prediction methods in risk analysis and evaluation, and can reasonably and systematically predict the causes of accidents, there are more and more cases of accident analysis using these methods in the field of civil engineering, which effectively ensures the safety of civil engineering construction.

FT is mainly used in the field of reliability engineering. This method is not only applied to civil engineering risk analysis but also applied to risk identification and assessment in aerospace, nuclear power, chemical, and other high-risk industries. The FT method is used by Andrew in the aerospace field (Andrew 2013).

---

*Corresponding Author: 1796160490@qq.com

    DOI 10.1201/9781003344209-72

BN can be used to predict the probability of unknown variables. Bayesian can provide security analysis and risk assessment for systems.

FBN is another risk analysis method developed based on BN. It can have a more accurate performance in probability prediction. Comparing the results of FBN and BN, it shows that FBN has advantages in providing more detailed, transparent, and real results. (Zarei 2019)

In the existing articles, there are only pairwise comparative analyses of these three methods, and there are few comparative studies on these methods. Therefore, in order to further understand the specific application of these methods, FT, BN, and FBN will be introduced in this paper, and the accident risk analysis of high-risk highway sections will be taken as an example to compare the three in the structure, establishment, and prediction of the model.

2 METHODOLOGY

2.1 *Fault tree*

2.1.1 *Definition*
FT is a special inverted tree logical cause and effect diagram. It is one of the important tools for product reliability and safety analysis. In general, accidental or catastrophic failures of products are regarded as top events and tracked from top to bottom according to the causal logic of the failure. In the field of civil engineering, fault trees play a great role. For example, Johnson used fault tree analysis to show the interaction of complex processes of the pier and abutment erosion and determine the overall probability of bridge failures due to scouring and channel instability (Johnson 1999). McDaniel had shown that based on the case study of a bridge, in the risk assessment study of bridge structure, the analysis of fault through the fault tree model can help identify countermeasures to reduce risk failure (McDaniel 2013).

The structure function of the fault tree is defined as $Y = \phi(X1, X2, \cdots, Xn)$, It represents the mapping relationship between the top event state of the fault tree and each bottom event state, that is, whether the top event occurs when each bottom event occurs or does not occur. The structure function of fault tree can be expressed as $\sum_{i=1}^{r} MCSi = \sum_{i=1}^{r} \prod_{CMCSi} Xj$ (Zhang 2010).

2.1.2 *Model establishment and evaluation*
2.1.2.1 *Accident risk analysis of highway high risk section based on the fault tree*
Taking the high-risk road section accident as the research object, the fault tree is established, that is, the high-risk road section accident is considered to be a major event, which is described as a top event, as shown in Figure 1. Table 1 shows the specific meaning of each event in Figure 1.

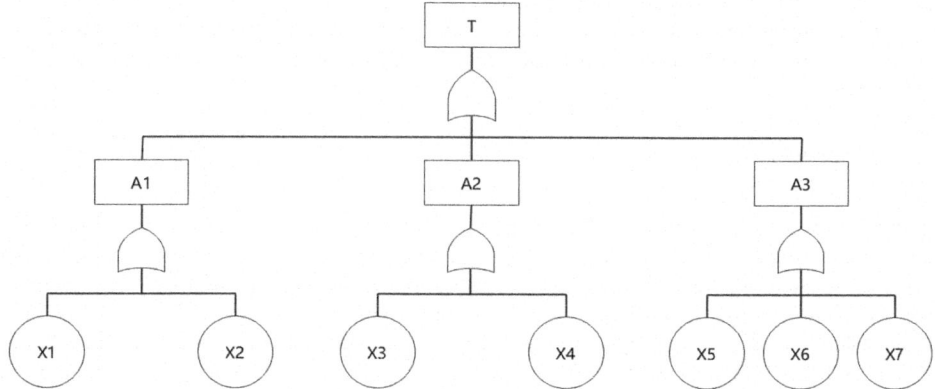

Figure 1. Fault tree of highway high risk section.

Table 1. Meaning of symbols in figure 1.

Code	Basic event	Code	Basic event
T	High risk road accidents	X3	Brake failure
A1	Road reasons	X4	ABS failure
A2	Vehicle reasons	X5	Drunk driving
A3	Human reasons	X6	Fatigue driving
X1	Long road curve	X7	Call while driving
X2	Long bumpy road section		

Cut set and minimum cut set are one of the important features of fault tree and play an important role in subsequent analysis. The top of the cutting event set is defined as the basic event set that occurs. The minimum cut set refers to the cut set that has the least number of basic events but can still ensure the occurrence of top-level events.

2.1.2.2 Qualitative analysis

According to the above definition of the minimum cut set, it can be deduced that the minimum cut set of the fault tree is 7. If the duality of the original fault tree is used to solve the minimum radius set, the unique minimum radius set of the fault tree can be obtained: $P = \{X1, X2, X3, X4, X5, X6, X7\}$.

2.1.2.3 Quantitative analysis

When investigating the causes of hotel fire, Hu used the minimum radius set and ranking method to calculate the importance of fire causes for quantitative analysis (Hu 2016). The direct method to solve the minimum radius set is to use the duality of the original fault tree, that is, replace the "and gate" of the fault tree with "or gate", calculate the minimum cut set of the success tree, that is, the minimum radius set of the fault tree, and then rank the importance of the bottom events according to the minimum radius set:

$$I[X5] = I[X6] = I[X7] > I[X1] = I[X2] > I[X3] = I[X4]$$

According to the arrangement and comparison, it can be seen that human factors are the most important causes of accidents in high-risk sections, with the highest importance.

2.1.3 Advantages, disadvantages and development prospect of FT

The fault tree can analyze accidents caused by many factors, and the causal relationship between events is very clear, but it has limitations in dealing with uncertainties and does not have the ability to update the probability. For example, through the comparative analysis of fault tree and Bayesian network in actual cases, Khakzad concluded that FT has weak inductive reasoning and uncertainty processing ability and no more flexible structure (Khakzad 2011). Due to the limitations and shortcomings of the fault tree, people continue to innovate and explore the establishment of reliability analysis models based on the fault tree. Bayesian network have come into being at the right moment.

2.2 Bayesian network

2.2.1 Definition

Bayesian network is also known as the causal model. Since it was proposed by pearl in 1988, it has become a research hotspot in recent years. Bayesian network is a directed acyclic graph model, which is composed of nodes representing variables and directed edges connecting these nodes.

BN represents the joint distribution probability of variables: p(U), $U = (x1, x2, \ldots, xn)$. Therefore, for a Bayesian network composed of N nodes, we have the representation of joint probability distribution: $p(U) = \prod_{i=1}^{n} p(xi|\theta)$, $p(xi|\theta)$ is the local conditional probability distribution associated with the node "i", recorded as $\theta = xPi$, and Pi is the indicator set of the parent node of marked node i (if node i has no parent node, Pi can be empty) (Khakzad 2011). This model has fault tolerance and can effectively deal with qualitative, uncertain and incomplete information. It has a high prospect for intelligent monitoring of civil engineering construction. For example, Tang used Bayesian network to evaluate basic engineering design information (Tang 2017). Huang carried out structural system identification and damage assessment through the Bayesian network (Huang 2019).

2.2.2 Model establishment and evaluation

There are three main ways to build the Bayesian network model: (1) based on expert knowledge; (2) based on database self-learning; (3) combined with the above two ways.

The first is to construct the Bayesian network by determining the selection of network nodes through expert knowledge, which can make full use of expert knowledge, but it is highly subjective and has uncertain factors. The second is entirely based on database data. The network structure established is objective and rigorous, but it is usually not easy to obtain due to the huge amount of data required. The third is not only the efficiency of self-learning, but also the experience and judgment of experts.

2.2.2.1 Accident risk analysis of highway high risk section based on Bayesian network

Aiming at the shortage of accident data in high-risk road sections, according to the characteristics of Bayesian network model, a comprehensive modeling method is adopted. Taking the length of the rough road ($S1$) as the evaluation unit, three groups of experts evaluate the road section. The directed acyclic relationship of Bayesian network is used, that is, the parent node index $S1$ points to the child node composed of the evaluation results $E1$, $E2$ and $E3$ of the three groups of experts, and a relationship diagram is established, as shown in Figure 2 (Hu 2014).

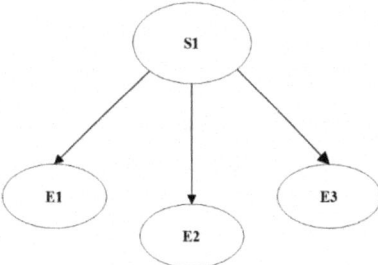

Figure 2.   Expert opinion.

When the expert group Ei ($i = 1, 2, 3$) conducts risk assessment on an evaluation unit, the index Sk ($k = 1, 2, \cdots\cdots$) is considered to be in a state Eij($j = 1, 2\cdots\cdots$ m) (m is the number of criteria given by experts), and the probability of traffic accident is P($Sk = $ Yes$|$Eij), according to Bayesian theorem:

$$P(Eij|Sk) = \frac{P(Sk|Eij)P(Eij)}{P(Sk)} \tag{1}$$

$$P(Sk = Yes|E1, E2, E3) = \frac{P(E1, E2, E3|Sk = Yes) \times P(Sk = Yes)}{P(E1, E2, E3)} \tag{2}$$

$$P(E1, E2, E3) = \sum_{Sk} P(E1, E2, E3, Sk) \tag{3}$$

P(Eij|Sk) - under Sk condition, the probability that the index falls into each state; P($Sk = $ Yes$|$E1, E2, E3)-Probability of occurrence of index SK under given results E1, E2, E3; P($Sk = $ Yes$|$Eij)-Experts believe that the accident probability of Sk index is in j state; P(Eij)-Assuming $1/m$, P(Sk|Eij)'s value can be given directly by experts; P ($Sk = $ Yes$|$E1, E2,E3) can be calculated by using Equations (2)–(4), so as to complete the prediction of accident probability. Three groups of experts evaluate the length of road turbulence (Si), and the criteria given are shown in Table 2:

Table 2. Evaluation criteria of road bumpy length by different experts.

Evaluating indicator	Expert group 1		Expert group 2		Expert group 3	
	Criteria	$P(Si = Yes \mid Ei1)$	Criteria	$P(Si = Yes \mid Ei2)$	Criteria	$P(Si = Yes \mid Ei3)$
Bump length (m)	>60	0.65	>60	0.6	>60	0.7
	[30,60]	0.3	[30,60]	0.3	[45,60]	0.3
	<30	0.1	[15,30)	0.15	[30,45)	0.2
			<15	0.1	[15,30)	0.15
					<15	0.1

According to the table, there are three criteria (m = 3) and three state values given by the first group of experts, namely $E11$, $E12$, $E13$. The conditional probability is calculated according to Formula (1): $P(E11 \mid S1 = Yes) = \frac{P(S1 \mid E11)P(E11)}{P(S1)} = \frac{0.65}{0.65+0.3+0.1} = 0.6190$, and similarly,

$$P(E12 \mid S1 = Yes) = 0.2587, P(E13 \mid S1 = Yes) = 0.0952$$

From the above table, the probability distribution of accidents given by the first group of experts for the road bumpy length within each range can be known. Similarly, the probability of each range of experts in groups 2 and 3 is:

$$P(E21 \mid S1 = Yes) = 0.5217, P(E22 \mid S1 = Yes) = 0.2609$$

$$P(E23 \mid S1 = Yes) = 0.1304, P(E24 \mid S1 = Yes) = 0.0970$$

$$P(E31 \mid S1 = Yes) = 0.4828, P(E32 \mid S1 = Yes) = 0.2069$$

$$P(E33 \mid S1 = Yes) = 0.1379, P(E34 \mid S1 = Yes) = 0.1034, P(E35 \mid S1 = Yes) = 0.0690$$

2.2.3 Advantages and disadvantages of Bayesian network

BN has high prediction efficiency, a simple algorithm, and good performance in practical application, but it usually needs to artificially assume the a priori probability. The assumed a priori model may lead to the difference in prediction data between different expert groups. For example, in the accident analysis of typical sections of rural roads, different expert groups have classified the curve radius differently (Hu 2014).

2.3 Fuzzy Bayesian network

2.3.1 Definition

A fuzzy Bayesian network is the combination of a Bayesian network and fuzzy set theory. FBN uses expert heuristics and fuzzy theory to determine the probability and uses the same reasoning and reasoning algorithm as the traditional BN for prediction analysis and probability update. The fuzzy set theory was first proposed by Jordan (Jordan 1999), trying to solve the uncertainty caused by imprecision and fuzziness. FBN provides a basis for generating widely applicable problem-solving technologies (Chen 2012).

When building the BN model, analysts are faced with the problem of insufficient root node probability data. Therefore, it is necessary to roughly estimate the probability (Hanss 1999). In this uncertain situation, it is not appropriate to use the traditional BN to calculate the system failure probability. FST provides an analysis framework that can deal with the blade root probability estimation of the input failure probability. Fuzzy marginalization rules and fuzzy Bayesian rules can be calculated by Equations (4) and (5):

$$P(T=tj)=\sum_i P(X=xi)\bigotimes P(T=tj|X=xi) \qquad (4)$$

$$P(X=xj|T=tj)=[P(X=xi)\bigotimes P(T=tj|X=xi)]\emptyset P(T=tj) \qquad (5)$$

(Eleye-Datubo 2008)

Here, T represents the leaf root and Xi represents the root node. According to the mathematical model of FBN, the FBN model can be established.

2.3.2 Model establishment and evaluation

2.3.2.1 Accident risk analysis of highway high risk section based on Fuzzy Bayesian network

The road bump length is still used as the evaluation unit for the road accident risk probability analysis based on FBN, and the fuzzy range predicted by experts is added on the basis of Table 2, as shown in Table 3:

Table 3. Increase the probability evaluation standard of expert fuzzy prediction.

Evaluating indicator	Expert group 1		Expert group 2		Expert group 3	
	Criteria	$P(Si=Yes|Ei1)$	Criteria	$P(Si=Yes|Ei2)$	Criteria	$P(Si=Yes|Ei3)$
Bump length (m)	>60	0.65–0.8	>60	0.6–0.7	>60	0.7–0.8
	[30,60]	0.3–0.45	[30,60]	0.3–0.5	[45,60]	0.3–0.4
	<30	0.1–0.2	[15,30)	0.15–0.3	[30,45)	0.2–0.3
			<15	0.1–0.25	[15,30)	0.15–0.2
					<15	0.1–0.15

Combining the first prediction probability and the second fuzziness probability of experts, the following is obtained:

$$P(E11|S1=Yes)=0.6190-0.5517, P(E12|S1=Yes)=0.2587-0.3103$$

$$P(E13|S1=Yes)=0.0952-0.1379, P(E21|S1=Yes)=0.5217-0.4$$

$$P(E22|S1=Yes)=0.2609-0.2857, P(E23|S1=Yes)=0.1304-0.1714$$

$$P(E24|S1=Yes)=0.0970-0.1429, P(E31|S1=Yes)=0.4828-0.4324$$

$$P(E32|S1=Yes)=0.2069-0.2162, P(E33|S1=Yes)=0.1379-0.1622$$

$$P(E34|S1=Yes)=0.1034-0.1081, P(E35|S1=Yes)=0.0690-0.0811$$

2.3.3 Advantages and disadvantages of FBN

According to the case, it can be found that when determining the a priori probability, the decision-maker can define the range under fuzzy conditions. Therefore, when there is not enough information to evaluate the probability, it will be easier for decision makers to allocate the probability value. However, the FBN based method also has some limitations. Many field experts participate in data collection, prediction and sorting, and the analysis cost is high (Zhang 2016).

3 COMPARISON

3.1 *Model structure*

In terms of structure, FT expresses the logical relationship through the symbolic relationship between the event and it, so as to carry out event analysis, find out all possible causes of the event, and predict the probability of accident cause through quantitative analysis. However, FT cannot deal with some aspects and modeling problems of the BN structure, such as multi state variables, real-time update and uncertainties. Generally speaking, the structure of BN is more flexible than FT and is suitable for various accident scenarios. FBN has the advantages of BN in structure, and FBN provides more powerful knowledge representation and reasoning tools under fuzziness and uncertainties, but its structure is more complex.

3.2 *Establishment aspect*

Firstly, the FT needs to determine the top event, logically connect the top event with all possible accident factors, and finally display it in the form of FT. BN needs to choose the corresponding establishment method according to the adequacy of the existing data of the analysis object, and then predict the accident through the risk probability calculated by the combination of Bayesian theorem and expert prediction probability. FBN is established through BN and fuzzy data, which makes the selection of data more extensive, reduces the dependence on data and makes modeling easier.

3.3 *Prediction aspect*

FT and BN methods have similar estimates of the probability of case accidents, but BN probability prediction is more specific and can update the a priori belief of accidents and generate a posteriori probability by considering new information, which can better give results for the studied events. For FBN, the fuzzy probability is determined on the basis of expert knowledge and fuzzy theory, which makes the prediction result more reliable without a large amount of data support.

4 CONCLUSION

At present, the three are mainly used in the field of civil engineering, such as construction environment prediction, safety and reliability evaluation and accident cause analysis. In application, the three have unique advantages in their respective fields. Fault tree analysis is easy to use in rough accident analysis and does not need professional probability prediction. BN can update data in real time and reasonably according to the experience of experts, which can be used for less data but more detailed analysis. When the data is fuzzy and it is difficult to determine the a priori probability, FBN can be used. The proposal of these methods is an important step towards the new era of safety and reliability evaluation, and opens up a new vision for the proposal of other reliability methods. These new methods will play an active role in civil engineering and other fields and better serve the mankind.

REFERENCES

Davis-McDaniel C., Chowdhury M., Pang W., and Dey K., (2013) Fault-tree model for risk assessment of bridge failure: Case study for segmental box girder bridges. *Journal of Infrastructure Systems*, vol. 19, no. 3, pp. 326–334.

Eleye-Datubo A.G., Wall A., Wang J., (2008) Marine and offshore safety assessment by incorporative risk modeling in a fuzzy Bayesian network of an induced mass assignment paradigm. *Risk Analysis*, 28(1): 95–112.

Hanss M, (1999) *On the implementation of fuzzy arithmetical operations for engineering problems. Fuzzy Information Processing Society*. NAFIPS. 18th International Conference of the North American IEEE, New York. pp. 462–466.

Hu S.T., Zhu Y.R, Xiang Q.J., (2014) accident risk assessment method for typical sections of rural roads. *Journal of Huaiyin Institute of Technology*. 23(05), pp. 30–35.

Hu Y. (2016) Research on the application of fault tree analysis for building fire safety of hotels. *Procedia Engineering*, vol. 135, pp. 524–530.

Huang Y., Shao C., Wu B., Beck J.L., and Li H. (2019) Recent progress of Bayesian reasoning in structural system identification and damage assessment. *Advances in Structural Engineering*, vol. 22, no. 6, pp. 1329–1351.

Johnson P.A. (1999) Fault tree analysis of bridge failure due to scour and channel instability. *Journal of Infrastructure Systems*, vol. 5, no. 1, pp. 35–41.

Jordan M.I. (1999) *Learning in Graphical Models*. Cambridge, MA: MIT Press.

Khakzad N., Khan F., and Amyotte P. (2011) Safety analysis in process facilities: Comparison of fault tree and Bayesian network approaches. *Reliability Engineering & System Safety*, vol. 96, no. 8, pp. 925–932.

Kornecki A.J. and Liu M. (2013) Fault tree analysis for safety/security verification in aviation software. *Electronics*, vol. 2, no. 1, Art. no. 1.

LeBeau K.H. and Wadia-Fascetti S.J. (2007) Fault tree analysis of schoharie creek bridge collapse. *J. Perform. Constr. Facil.*, vol. 21, no. 4, pp. 320–326.

Li P.-C., Chen G.-H., Dai L.-C., Zhang L. (2012) A fuzzy Bayesian network approach to improve the quantification of organizational influences in HRA frameworks. *Safety Science*; 50(7):1569–1583.

Špačková O. and Straub D. (2013) Dynamic Bayesian network for probabilistic modeling of tunnel excavation processes. *Computer-Aided Civil and Infrastructure Engineering*, vol. 28, no. 1, pp. 1–21.

Sun J., Liu B., Chu Z., Chen L., and Li X. (2018) Tunnel collapse risk assessment based on multistate fuzzy Bayesian networks. *Quality and Reliability Engineering International*, vol. 34, no. 8, pp. 1646–1662.

Tang W.H. (2017) *A bayesian evaluation of information for foundation engineering design*. pp. 214–214.

Zarei E., Khakzad N., Cozzani V., and Reniers G. (2019) Safety analysis of process systems using fuzzy bayesian network (FBN). *Journal of Loss Prevention in the Process Industries*, vol. 57, pp. 7–16.

Zhang L., Wu X., Skibniewski M.J., Zhong J., and Lu Y. (2010) Bayesian-network-based safety risk analysis in construction projects. *Reliability Engineering & System Safety*, vol. 131, pp. 29–39.

Zhang L., Wu X., Qin Y., Skibniewski M.J., and Liu W. (2016) Towards a fuzzy bayesian network based approach for safety risk analysis of tunnel-induced pipeline damage. *Risk Analysis*, vol. 36, no. 2, pp. 278–301.

*Frontiers in Civil and Hydraulic Engineering – Mohamed A. Ismail and
Hazem Samih Mohamed (Eds)
© 2023 The Authors, ISBN 978-1-032-38247-0*

Developing a smart city system for land use and sustainable urban renewal in Hong Kong

Boyu Huang*

The Hong Kong Polytechnic University, Hong Kong, China

ABSTRACT: Hong Kong, as one of the most densely populated cities in the world, has been suffering from land shortage, urban decay, and dilapidated buildings for a long time. Land is one of the most precious and scarce resources in Hong Kong. Sustainable land use planning has become a challenge for the Hong Kong government. In addition, due to the increasing population and limited land supply, Hong Kong has been committed to developing every piece of urban land to achieve its expected potential. Urban renewal is regarded as a strategy to meet these challenges because it aims to solve the problem of urban recession, optimize, and improve land use patterns, and enhance the image of the city. The concept of sustainable development has its advantages, but it is not always well applied in practice. The urban renewal plan has received considerable criticism, including the destruction of local culture, profit driven characteristics, the expulsion of low-income residents and the poor quality of urban renewal stock. Effective decision support is needed to reduce or eradicate these negative effects before starting the update plan. To better apply sustainability to urban renewal, an in-depth study should be carried out to determine what factors influence sustainability and how to promote sustainability. In addition, although the role of GIS in urban planning has been widely recognized, 2D systems are no longer enough, and there are few studies on the application of the 3D GIS technology in land development. To fill these research gaps, this paper sets two goals: (1) Determine influencing factors from social, economic, and environmental aspects and how these factors affect sustainable land use planning; (2) Establish a 3D modeling platform for the investigating land development in Hong Kong. To achieve these goals, the paper designs and implements two stages: the first stage reviews the advantages and disadvantages of the existing research on urban renewal and urban planning system from the perspective of practice and theory based on comprehensive literature review and literature analysis. Based on the in-depth understanding of the urban renewal mechanism of Hong Kong or other cities, this paper defines the research problems and objectives.

1 INTRODUCTION

How to carry out the urban renewal process in Hong Kong? Urban renewal has always been regarded as an effective way to improve land value, improve urban style and solve many social problems (Adams & Hastings 2001). Although Hong Kong has only a history of more than 150 years, the extremely sharp contradiction between people and land, as well as the high-density development strategy relying on the metropolitan area, make the old city reconstruction of Hong Kong very arduous. There are a large number of dilapidated or inadequate private buildings in the urban area. At the current rate of redevelopment, the proportion of such buildings over 30 years old in private buildings will double from 20% to 40% in the next 10 years (Ou & Wu 2014). Undoubtedly, it is of great strategic significance for the development of Hong Kong in the 21st century to properly handle various problems in the process of urban renewal.

The research method is to clarify the construction of smart city 3D system through the case study of Qingdao West Coast Free Trade Zone. After determining the influencing factors of land

---

*Corresponding Author: benny.huang@connect.polyu.hk

DOI 10.1201/9781003344209-73

use, a 2D urban information database is constructed according to the existing urban land data and GIS data, and then imported into the 3D smart city model. This study proposes a 3D smart city system to promote urban renewal and urban sustainable development in Hong Kong. The 3D smart city model proposed in this study can make it easier for the government and even the people to understand and plan Hong Kong's future urban sustainable development and encourage more social forces to participate in Hong Kong's urban renewal process.

2 LITERATURE REVIEW

2.1 Urban renewal

Urban renewal is defined as a process of "actual changes or changes in use or use intensity of land and buildings" (Coach 1990). It could also be seen as a process of old city cleaning, urban area reconstruction, heritage protection and restoration (Coach 2011). Akkar (2011) also regards it as a comprehensive vision and action to solve multifaceted urban problems and improve the physical, economic, social, and environmental conditions in backward areas. Urban renewal is widely regarded as the construction of new buildings and the retention, reconstruction and demolition of existing buildings. To sum up, urban renewal focuses on social, economic and environmental improvement of developed or backward areas in the city through a series of methods, including redevelopment, improvement and protection. Reconstruction and repair are important aspects of renewal (Zheng 2014). According to Akkar (2011), urban renewal improves the appearance of the city in many ways. It can not only improve the urban ecological environment, but also improve the way residents know and feel, and strengthen the emotional relationship between residents and the city. The diversity and sustainability of cities should be preserved in the process of urban renewal, which are two important characteristics of the city. Therefore, to achieve social stability, sustainable economy, and environmental protection, it is necessary to implement sustainable urban renewal.

2.2 Sustainability

According to textual research (Bromley et al. 2005), the term of sustainable development could be traced back to the 1970s, but it did not appear in the field of urban planning until the 1990s and was combined with the policy of urban renewal. At present, the concept of sustainable city is relatively vague and complex. To better study this concept, many studies have interpreted it from different angles. Lorr (2012) proposed three highly recognized sustainability perspectives, including intergenerational equity and justice perspectives, comprehensive environment, economy, fair change perspectives and free market sustainable development perspectives. The author applies these views to some big cities in North America and puts forward the definition of urban sustainability. Although the concept of sustainability is diverse, there is a widely accepted view that the three cores of sustainable development include social, economic, and environmental sustainability. Urban development is closely related to these elements. If it is not combined with sustainability, urban renewal may be difficult to implement.

2.3 Urban renewal in Hong Kong and other cities

Hong Kong is famous for its population density and development. With the increasingly serious problem of urban decline, it has become an urgent task to implement urban renewal projects, improve citizens' living conditions and improve urban environmental quality. Like other historic cities, urban renewal is now a hot topic in Hong Kong. Many old buildings and sites in developed areas need to be rebuilt or activated to meet changing needs.

The large-scale urban renewal of Hong Kong can be traced back to more than 40 years ago. It was not until 1988 that the Land Development Corporation (LDC) was formally established to implement, promote, promote and promote the old city reconstruction project in Hong Kong (Adams & Hastings 2001). The company is a self-financing business organization whose main goal is to promote urban renewal by implementing urban renewal projects in cooperation with private developers.

Although LDCs are more flexible than the public sector in the use of private resources, they have a fundamental weakness in coordination between governments and private developers, which requires a long government approval process (Adams & Hastings 2001). As a result, the condition of old buildings in Hong Kong remains unsatisfactory and poses a threat to public security. In recent years, the government of the Hong Kong Special Administrative Region has become more and more interested in incorporating the concept of sustainable development into urban development policies and has put forward similar ideas on local urban renewal. However, many reconstruction projects cannot solve the problem of urban decay. To solve the problem of urban aging and accelerate the pace of urban reconstruction, the urban reconstruction office was established in 2001 to replace the rural development company and become a new management organization responsible for improving the implementation of urban reconstruction plans. UCB's role in reconstruction projects is more flexible because UCB can rebuild itself or develop privately through a joint venture or sell the project area on the open market after the acquisition. Since its establishment, UCB has adopted the 4R strategy, namely reconstruction, restoration, activation, and protection, to update the urban area. In particular, the city is being rebuilt to be better for reconstruction. Reconstruction helps to slow down the decline of cities and protect urban networks. The Municipal Construction Bureau will also protect cultural relics and repair old urban areas. Over the past decade, the authorities have announced 64 urban renewal plans, of which more than 50 ones have been passed.

For example, the local nickname "wedding card street" is better known as Lidong street. Its printer sells custom wedding cards in bright red paint to celebrate good luck. In recent decades, thousands of residents have bought marriage cards. This region is also the birthplace of Hong Kong's publishing industry. In 2003, the Municipal Construction Bureau announced the reconstruction of 8900 square meters of Leidong road and Macley road. It is planned to demolish 54 buildings, a total of 930 households, and build 4 residential buildings and 4 shopping centers. Similarly, leisure Street used to be a very busy bird market with poor environment. After reconstruction, Langya square, a new commercial skyscraper integrating office, hotel and shopping, has become a landmark of Mongkok.

Although the Urban Construction Bureau adheres to the policy of "people-oriented, regional-ization and public participation", it has brought many benefits to the society. The public generally believes that municipal construction management is profit-oriented and market-oriented, and there are many problems.

In fact, in the past decade, UCB has launched some controversial reconstruction projects. For example, the company was accused of buying an old house in Tugua bay at a record price of HK $15900 per square foot in 2017. Critics suspect that the city's goal is to build another luxury residential complex, and the supply of "astronomical prices" will inevitably affect rents in the area. Another example is the large-scale redevelopment project in Kwun Tong. It is reported that the Housing Commission has changed its original design, increased many free space and public facilities, and launched large brand and expensive housing, which has aroused the fear and dissatisfaction of residents. Although the agency apologized under public pressure and announced design improvements, some questioned whether it was "too friendly" to large developers rather than paying more attention to local people, concerns and needs.

Despite strong criticism, a former chief surveyor of the land development agency said it was unfair to blame the City Council. "Land reconstruction is a complex process involving multiple departments," he said. The planning shall comply with relevant regulations, including land use planning and building planning prepared by the Department. Considering the difficulties encoun-tered in the practice of urban renewal, such as: benefit sharing, land mining, inter departmental coordination, it is not easy to complete the reconstruction task, accompanied by social inequality and residents' protests and resettlement of affected residents. However, most of these sites were occupied by very small but very high buildings (Adams & Hastings 2001), which led to a long negotiation process aimed at centralizing different plots for greater reconstruction.

In addition, many stakeholders are involved in the planning of urban renewal practices, such as the government, developers, contractors, tenants and buyers. Each party has the right to speak and act for its own interests. Most people do not always know under what special circumstances reconstruction is possible. Although some individuals or groups insist on maintaining the status, others support reconstruction and emphasize the progress of project implementation.

In this context, urban renewal projects often go through a long and unpredictable development cycle, which will inevitably have an impact on the surrounding residents. Therefore, the government must strengthen communication among various stakeholders in order to shorten the reconstruction cycle.

In short, urban renewal plays an important role in improving urban environment, urban image and landscapes. Although the roles and relationships in the practice of urban renewal are somewhat complex, the relationship between them provides a possible direction for sustainable urban renewal.

2.4 Urban renewal in Singapore

The history of urban renewal in Singapore can be traced back to the Second World War, when it was still dependent on Britain. Before the war, Singapore was already a problem. Due to the war and relatively slow economic development, the previous housing conditions deteriorated from the 1940s to the 1950s. Since Singapore's independence in 1965, the government has made many efforts to clean up slums and destroy collapsed buildings in the city center (Western et al. 1973). These obstacles come from the resistance of people and residents who once lived and hesitated in slums and will be removed from office.

In response to the call for more efforts, the City Tax Bureau was established in 1974 as an independent body to take initiatives for renewal and reconstruction. Unlike Hong Kong, the URA of Singapore has accepted more tasks in Singapore, such as urban planning, rural policy, land procurement, education and sales. In addition, it has also implemented urban renewal projects (Yeh 1975). Public participation and transparency of decision-making process (URA 2016) have attracted more and more attention.

Compared with other cities, the state's intervention in Singapore's urban renewal can be regarded as a relatively majority of directives and "controversial". Since the 1960s, the government has launched a large-scale public housing development programme through the Housing and Development Commission. The "people's ownership" program has also been introduced to provide residents with real estates. Residential housing is considered the primary responsibility of the Singapore government (Statistics Singapore 2016).

2.5 Urban renewal in Shanghai

Throughout the history of modern cities in the world, they can be divided into two categories. The first is gradually formed cities. In a city, the urban structure is renovated through the social and economic reform, which gradually leads to the transformation of urban function. The second category is explosive cities. External forces have changed the operation mode of cities in a short time. Shanghai belongs to the second category of cities. Different from most traditional Chinese cities, Shanghai did not develop naturally and gradually, but was promoted by foreign forces. In less than a century, Shanghai developed rapidly from a small fishing city in the early 1940s to the largest modern city in China. The rapid development of Shanghai makes it the second highest skyline in China after Hong Kong (Denison et al. 2006; Sun 1999).

However, when the urban land market appeared in the 1990s, a large number of downtown residents were forced to move to remote suburbs without adequate compensation. The decoration of Shanghai has aroused public controversy and protest. It was not until 2001 that the State Council approved the master plan for the further development of Pudong New District in Shanghai (1999–2020). The plan also plans to build new towns and central urban areas, making Chongming a strategic area for sustainable development in Shanghai for decades. The master plan is based on different opinions. In order to collect comprehensive opinions and suggestions, social groups can directly negotiate or stage performances on the key themes of the new master plan. At the same time, the public can obtain the information on the formulation of the master plan through Wechat, Microblog and website and give timely feedbacks (Shanghai Planning and Land Resource Administration Bureau, Shanghai Urban Planning and Design Research Institute 2012).

2.6 *Smart city*

The smart city system is also known as City Information Modeling (CIM) system, which is a combination of Building Information Modeling (BIM) and Geography Information System (GIS). The BIM Technology has been widely used in urban development planning in the past. It is a process involving the generation and management of digital representation of physical and functional features of buildings. BIM Technology stores building information in the model in the form of electronic data for display and application. However, with the increasing complexity of urban information and the introduction of the concept of urban sustainability, professionals responsible for urban infrastructure systems need new tools to meet their needs in planning, projects, construction, urban equipment management and restoration. In this context, CIM is proposed.

CIM is considered not only the combination of all BIM models, but also represents the advanced level of infrastructure network, management and human activities. The model helps to analyze and monitor the urban environment to support city-wide projects and planning. So, it is an integration of all kinds of spatial data models. In order to explore how CIM is used for city development and how relevant authorities and government carry out multi-dimensional urban planning based on the CIM model, it is necessary to define the concept of sustainable city. One is based on the International City Standard, ISO37120, proposed during the Global Cities Summit, which established city quality indicators to provide development direction and evaluation indicators for sustainable cities.

CIM follows the IOS 37120 index principle to build an intelligent city system. In order to compare or measure different things, we need to reach a consensus. In recent decades, ISO has provided more than 20000 dynamic processes and frameworks to support comparison. ISO 37120:2018 describes a complete set of data points that cities or municipalities can use to determine performance. Its purpose may be to make progress in achieving its own goals or to compare with other cities or municipalities. It combines urban social, economic, transportation and other indicators to provide the authorities with a comprehensive and multi-dimensional real-time urban information monitoring platform. Users can also expand urban development and future urban planning based on the existing CIM platform to make it a real-time urban information platform for sustainable development.

With the rapid development of CIM technology, more and more researchers have applied 3D modeling and visualization technology to various decision-making processes supporting urban development. Ranzinger, Gleixner (1997), Pular and Tidey (2001) took the lead in studying the possibility of realizing 3D visualization/virtual reality in participatory planning. They studied the feasibility of using three-dimensional visualization technology and structured assessment method (Delphi) to support visual impact assessment and determination. The experimental results show that 3D visualization can support the process of urban and landscape planning. However, the research results are far from the purpose of inter group data exchange and visualization. Hudson Smith et al. (2005) developed a series of "virtual city" concepts based on conveying program information and supporting participation in the program. Petti et al. (2006) continued to study the effectiveness of the 3D visualization and virtual reality technology, pointing out that professional designers are very satisfied with visual specificity. However, people with basic computer skills are satisfied with the average level of visual specificity.

In the past decade, more and more researchers are committed to using three-dimensional spatial analysis to provide better solutions for urban environment. For example, Zhang et al. (2004) described some possible three-dimensional spatial analyses, namely visibility, flood, energy, solar panels and air pollution, to investigate urban development based on a three-dimensional urban model. Mak et al. (2005) measured the height and visibility of buildings on the ridge and skyline of Hong Kong using the CIM technology. The results show that CIM can effectively implement the recommendations of the "Hong Kong urban planning guidelines". Some researchers focus on other topics, such as urban roads, temperature and ventilation (Li et al. 2004). Air pollution, solar panels, flood risk, landscape analysis and urban management were recently discussed in Plock, Poland. At the same time, it draws attention to the development trend of three-dimensional spatial analysis technology and the data and policies that need to be improved in urban development. The problem is that few people are concerned about the environmental impact of controlling development, that is, relaxing the maximum PR limit.

3 RESEARCH METHODOLOGY

3.1 *Document analysis*

Document analysis is a kind of archival research for resolving research problems and questions by investigating various recorded information and published documents. Being a qualitative method, document analysis can be classified into two approaches in terms of the data source: content analysis and existing data analysis. The content analysis reviews documents from a theoretical perspective, and any communication mediums like written materials, picture and audiovisual records are involved (Dane 1990). As one of the major forms of content analysis, literature review is often used in social studies, providing a systematical understanding of existing knowledge, theoretical contributions and practical applications based on academic publications and other paper- or web-based resources (Rowley & Slack 2004).

In this study, a comprehensive literature review of land use planning and urban renewal from both theoretical and practical levels is conducted. Official publications and regulations issued by Hong Kong and other high-density cities, including policies on land use and urban renewal, urban planning standards and guidelines, are also reviewed to find the existing achievements in the current urban renewal practice. Research gaps and limitations are identified and summarized, serving as a solid reference foundation for the following analyses.

3.2 *Interviews*

Interviews can be defined as structured conversations that conduct data collection surveys (Bingham & Moore 1924). Interviews can be conducted in different ways, such as face-to-face, telephone or e-mail. The interview type can be structured, semi-structured or unstructured (Longhurst 2003). Structured interviews follow a predetermined list of questions, while unstructured interviews are based on social interaction between researchers and informants (Dunn 2000). When answering questions about the research topic, if the respondents have an accurate understanding of the questions, well-structured interviews are an effective way to collect the latest information. "Experts" refer to practitioners with experience or authority in certain research fields. Expert interviews are considered to be an effective means of collecting first-hand and in-depth information.

In addition to the literature review of the main factors affecting land use decision-making, this paper also interviewed two urban planning experts and experts from Hong Kong urban renewal Bureau. The administration is now making a number of recommendations on land use factors and decision-making for urban renewal. Two professors dedicated to land use planning from China and the United States were also invited to comment on the effectiveness of the proposed land use plan.

3.3 *Case study*

Case studies can be defined as "empirical research on contemporary phenomena in real life; the boundary between phenomenon and context is not obvious; and the use of various sources of evidence" (Yin 1984). In addition to conceptual or theoretical studies of different assumptions, case studies also focus on a detailed contextual analysis of some existing events and their relationships (Soy 1997). A major limitation of the case study is that the implementation of this scientific tool depends largely on the availability of specific data. Therefore, it is often used in combination with research methods such as expert interview, literature review, observation, and field research. After Yin (1994), the case study is generally divided into four stages: (1) design stage and implementation; 3) evidence analysis; 4) giving conclusions, recommendations and implications.

To implement the proposed intelligent method to support land use decision-making, let's take Qingdao as an example. The whole research process in the actual case is introduced, including the latest data and practical problems reflecting the local situation.

3.4 *Tools: CATIA*

CATIA is the largest CAD / CAE integration software in the world, which is widely used in many fields. CATIA's unique hybrid modeling technology and complex surface design module make CATIA a leading company in the field of design analysis and manufacturing integration. CATIA used in this paper has a rich design module warehouse (about 140 building modules), which can create a project development and research environment for the whole life cycle of the company.

4 CASE STUDY

4.1 *Case study on Qingdao*

This study is carried on in the smart city project during the internship in Dassault Systemes. This project is developing a Qingdao smart city 3D information modeling system, with the objective to help Qingdao, the west coast area in especial, authorities and government build a more inclusive, resilient and sustainable city.

The smart city system comprehensively and deeply integrates various social, economic and environmental indicators of Qingdao, including but not limited to land use type, energy structure, urban spatial planning and carbon neutralization. The smart city project was reported during the Qingdao Innovation Night on Oct. 21st, 2021. Equally importantly, Dassault Systemes showcased the value of 3DExperience platform playing a role in the smart city construction and development.

4.2 *2D GIS data collection*

In order to collect the GIS geographic data of Qingdao, especially the West Coast Economic Development Zone, this study detected the existing geographic data of Qingdao, and collected a large number of spatial and non-spatial data. Spatial data refers to road networks and points of interest (POIs) (such as CBD, schools, hospitals, subways/bus stations, parks, etc.). Non-spatial data refer to population distribution, land sales/housing price records, land use regulations, etc. It should be noted that most of this information is not readily available and has to be derived from the statistical data or reports from various sources such as the government website, Property Agency, Google Earth, etc. Layers of these data need to be established separately according to the exact positions.

4.3 *Coding*

API of Gaode map search POI: The current idea is to crawl all POI data in the target area (the author's small county), store it in the database as the original data, and then call it for other systems.

First, a developer account of Gaode map is registered, a key bound is applied from the web service, and then the newly registered developer account is authenticated to the Gaode map development platform. The purpose of account authentication is to improve the number limit and concurrent requests for daily access to Gaode map API interface. To collect information from Gaode map, we need to obtain its administrator permission. So, we need a key that can be geocoded. We enter Gaode map open platform (http://lbsaamap.com/) and register an account. A geocode will be finally created.

According to the introduction in the API address, it is divided into three searches, namely keyword search: conditional search by using POI keywords, such as KFC, Chaoyang Park, etc.; POI type search, such as bank perimeter search, search by keyword or POI type within the set range near the longitude and latitude coordinate points entered by the user; polygon search: search ID query in the polygon area and query the details of a POI through POI ID. This search is recommended to be used in conjunction with the input prompt API. My goal lies in all POIs in a region, so I choose the third kind: polygon search. The most important parameter of polygon search is polygon-longitude and latitude coordinate pair. I picked up the longitude and latitude coordinate pair of my target area in Baidu map coordinate pickup system.

Before officially starting the code, a test is performed and the browser is used to directly access the interface to see the returned data (of course, Gaode's API interface has a description of the returned data).

According to the API, the count is the number of search schemes (the maximum value is 1000), so each request will return the number of POIs currently searched, and there is no way to obtain POIs greater than 1000. Then, if I want to query all the data of a region, I can divide the region into a collection of smaller regions (obviously a recursive operation), and then combine all the POI data of these regions where all POIs can be found, which is the data finally needed.

Coding is started. Because the whole process of calling API is inseparable from the longitude and latitude, a longitude and latitude description class is firstly defined.

Then, a method is needed to call the API to obtain the returned data. The method parameter is a rectangular block. Of course, the page number is also needed, that is, the current method obtains the data on the page of a rectangular area (25 POIs on each page, 20 POIs by default).

Of course, as mentioned above, if the rectangular block returns data count = 1000, it means that the current rectangular block needs to be divided. My idea is relatively simple. The rectangular block in the horizontal center and vertical distraction is divided according to the above sketch, and one rectangular block will be divided into four small rectangular blocks. The method is as follows:

At present, you can get the set of longitude and latitude pairs in the rectangular area, and there are also methods to obtain API data. Then, you can traverse the number of pages to obtain data and customize the operation data. When the number of POIs returned by a paging request is less than the maximum number per page, it is considered that the current region POI has been fully requested.

Time is a problem. At present, API interfaces are requested once every 50 ms. It takes about ten minutes to run the data (tens of thousands of pieces) of the small county. It takes one day to run the main data of the whole urban area intermittently, and finally nearly 27 W data is run.

4.4 *Application on construction of smart city system*

Then, the prepared POI data can be compatible with CATIA in the format of ShapeFile file to complete the establishment of 3D model of Qingdao, shown in Figures 1 to 6.

Figure 1.   Importing POI data into 3D city model.

Figure 2. Intelligent construction of city facilities.

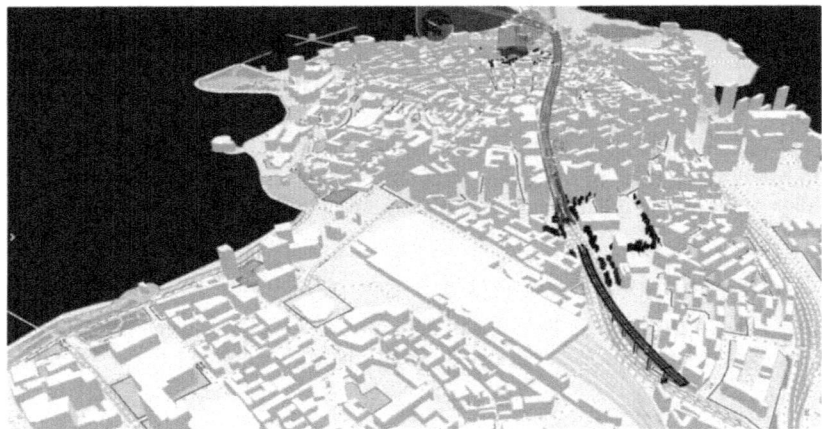

Figure 3. City sand table construction.

Figure 4. City information measurement & control.

Figure 5. Traffic accident tracking and detection.

Figure 6. Land use type planning.

5 FINDINGS & CONCLUSION

Taking Qingdao as an example, this study briefly introduces the establishment of the urban smart city system. The establishment of the smart city system based on CIM technology provides an effective and convenient platform for urban land use and urban renewal. Urban government departments, relevant departments and citizens can directly or indirectly participate in urban project decision-making and management through this open and transparent urban monitoring and management system.

However, due to COVID-19, field visits and data sampling in Hong Kong encountered difficulties. This study does not directly provide the establishment process of Hong Kong's smart city system. The establishment process is applicable to most large cities with a large population. However, due to the differences in specific problems between cities, such as the openness of government information and the details of map records, the construction mode of smart city in Hong Kong may be different, which will be the main challenge in the next stage of research.

REFERENCES

Adams, D., & Hastings, E. (2001). Urban renewal in Hong Kong: transition from development corporation to renewal authority. *Land Use Policy*, 18(3), 245–258.

Akkar Ercan, M. (2011). Challenges and conflicts in achieving sustainable communities in historic neighbourhoods of Istanbul. *Habitat International*, 35(2), 295–306.

Bingham, W.V.D. and Moore, B.V. (1924). *How to interview*. New York: Harper & Row.

Bromley, R.D., Tallon, A.R., & Thomas, C.J. (2005). City centre regeneration through residential development: Contributing to sustainability. *Urban studies*, 42(13), 2407–2429.

Couch, C. (1990). Urban renewal: theory and practice. Macmillan.

Couch, C., Sykes, O., & Börstinghaus, W. (2011). *Thirty years of urban regeneration in Britain, Germany and France: The importance of context and path dependency. Progress in Planning*, 75(1), 1–52.

Dane, F.C. (1990). *Research Methods*. CA, USA: Brooks/Cole

Denison, Edward, and Guan Yu Ren. (2006). *Building Shanghai—The story of China's gateway*. Chichester: Wiley-Academy, USA.

Li, W., Putra, S.Y. and Yang, P.P.-J. (2004). GIS analysis for the climatic evaluation of 3D urban geometry – The development of GIS analytical tools for sky view factor. Proceedings of GISDECO, May 10–12, 2004, Malaysia.

Longhurst, R. (2003). Semi-structured interviews and focus groups. *Key methods in geography*, 117–132.

Lorr, M.J. (2012). Defining urban sustainability in the context of north american cities. *Nature and Culture*, 7(1), 16–30.

Mak, A.S.-H., Yip, E.K.-M. and Lai, P.C. (2005). Developing a city skyline for Hong Kong using GIS and urban design guidelines. *URISA Journal*, 17(1), 33–42.

Ou, G., & Wu, G. (2014). *Theory and Practice of Urban Renewal: A Case Study of Hong Kong and Shenzhen*. In Proceedings of the 18th International Symposium on Advancement of Construction Management and Real Estate (pp. 147–153). Springer, Berlin, Heidelberg.

Pullar, D.V. and Tidey, M.E. (2001). Coupling 3D visualization to qualitative assessment of built environment designs. *Landscape and Urban Planning*, 55, 29–40.

Ranzinger, M. and Gleixner, G. (1997). *GIS Datasets for 3D urban planning*. Comput., Environ. And Urban Systems, 21(2), 159–173.

Rowley, J. and Slack, F., (2004). *Conducting a literature review*. Management Research News 27, 31–39.

Shanghai Planning and Land Resource Administration Bureau, *Shanghai Urban Planning and Design Research Institute*. (2012). Shanghai in transition: Urban planning strategy. Shanghai: Tongji University Press.

Soy, S.K. (1997). *The case study as a research method. Unpublished paper*, University of Texas at Austin.

Sun, Ping (ed.). (1999). *The history of Shanghai urban planning*. Shanghai: Shanghai Academy 150 of Social Science Press.

Western, Weldon, P.D., & Haung, T.T. (1973). Poverty, Urban Renewal, and Public Housing in Singapore. Environment and Planning. A, 5(5), 589–600. https://doi.org/10.1068/a050589.

Yin, R. (1994). *Case study research: Design and methods* (2nd ed.). Sage Publications, Thousand Oaks, CA.

Yin, R.K. (1984). *Case Study Research: Design and Methods*: Beverly Hills, CA. Sage Publications.

Zheng, H.W., Shen, G.Q. & Wang, H. (2014). *A review of recent studies on sustainable urban renewal*. Habitat International, 41, 272–279.

Frontiers in Civil and Hydraulic Engineering – Mohamed A. Ismail and
Hazem Samih Mohamed (Eds)
© 2023 The Authors, ISBN 978-1-032-38247-0

Sustainability study of a reinforced concrete building based on the LCA-Emergy method in China

Junxue Zhang* & Shanshui Wang*
*School of Civil engineering and Architecture, Jiangsu University of Science and Technology,
Zhen Jiang, China*

Li Huang
School of Design Art, Lanzhou University of Technology, Lanzhou, China

Qifang Kong
School of Architecture, Sanjiang University, Nanjing, China

ABSTRACT: At present, the sustainable analysis of building systems is a hot topic. In order to achieve the ecological state, the emergy method has been selected to execute the assessment with a series of emergy indicators. This paper applies the emergy approach to analyze the ecological sustainability of the building system, especially for the full life cycle of the building. For the whole building, building material emergy and operational stage emergy play a critical role and account for 92.4% of the entire emergy, which is the primary contributor. As the vital indicator, emergy sustainability index (ESI) has been displayed based on EYR and ELR, which is 0.669 (unsustainable).

1 INTRODUCTION

As a necessary condition for human existence, the architectural system plays an important role. With the growth of population and environmental degradation, building systems are becoming increasingly unsustainable. In this context, it is increasingly critical to assess the sustainability of ecological buildings (Donald et al. 2021).

The emergy method is a kind of important sustainable assessment method, which has been utilized in building systems. For instance, the emergy theory and building information modeling are combined into a building system to evaluate sustainability (Suman et al. 2021). For the building refurbishment, several strategies have been conducted based on the emergy-LCA method (Wenjing et al. 2021). As the basic components of building systems, the sustainability of building materials has been also concerned by scholars (Junxue et al. 2020; Tilba 2021). By using emergy evaluation, major highway building has been assessed for decision-making in Italy (Silvio & Francesco 2019). Taking a zero-energy building, for example, American scholars have redefined zero-energy buildings based on the emergy approach (Hwang et al. 2017). Lin and William studied the high-density and high-rise buildings by using emergy analysis to calculate density parameters (Jae & William 2017). In order to measure the renovation effect, the emergy method has been selected to assess environmental performance (Ivan et al. 2017). This paper aims to assess the ecological sustainability in a building system through the LCA-emergy method.

---

*Corresponding Authors: zjx2021@just.edu.cn, wangshanshui123@163.com, weeds215@sina.com and 837577110@qq.com

DOI 10.1201/9781003344209-74

2 INTRODUCTION OF EMERGY METHOD

The emergy methodology was proposed by H. T. Odum firstly to evaluate the sustainability of ecological systems (Odum 1996). Emergy's spelling has the letter "m" and it is different from energy. Unlike energy, the emergy theory contains embodied energy and can connect the mass flow and energy flow in order to have a more accurate assessment result for system sustainability. Odum defines the emergy as the amount of available energy, which is directly or indirectly put into use in the system. The concept of emergy allows us to compare different types of energy contributions in a system with the same energy standard.

3 EMERGY CALCULATION PROCESSES

3.1 Case study

The Building of Southeast University is a typical office building, which is in Nanjing city, along the Yangtze River in China. The specific parameters involve a height of 71.5 m, a total land area of 4610 m^2, and a total construction area of 16873 m^2 (including a ground area of 15419 m^2 and an underground area of 1454 m^2), as well as 15 floors above ground in the main building, 1 underground floor, and 3-floor podiums.

3.2 Building materials emergy

In Table 1, there are 14 types of building materials, including the usage of building materials and the corresponding amount of emergy. The result can also be clearly obtained from Figures 1 and 2. Figure 2 shows the emergy ratio of primary materials in Yifu Building. It can be seen that the 14 types of building materials can be sorted by emergy usage: concrete, cement, steel, paint, tile, stone, gravel, glass, lime, organic materials, bricks, wood, alloys, and copper. The three largest emergy usages are concrete, cement and steel, accounting for 60.7%, 20.3%, and 8.65%, respectively. The total amount of emergy is 3.72 E+19 sej. In terms of quantity, the descending order of usage is concrete, cement and steel. From the perspective of emergy, the descending order of usage is concrete (2.2 6 E+19 sej), cement (7.69 E+18 sej), and steel (3.67 E+18 sej). This means the main contributors contain three types, which should be analyzed regarding the sensitivity changes.

Table 1. Emergy calculation of main building.

Item	Materials	Data	Unit	Transformity (Sej kg^{-1})	Emergy (Sej)
1	Cement	2971870	Kg	2.59 E+12	7.69 E+18
2	Steel	527122	Kg	6.97 E+12	3.67 E+18
3	Aluminum alloy	1168.6	Kg	1.27 E+13	1.48 E+16
4	Copper	204	Kg	6.77 E+13	1.38 E+16
5	Concrete	12467950	Kg	1.81 E+12	2.26 E+19
6	Brick	23609.1	Kg	2.52 E+12	5.95 E+16
7	Gravel	384440	Kg	1.00 E+12	3.84 E+17
8	Stone	440265	Kg	1.00 E+12	4.4 E+17
9	Limestone	121799.8	Kg	1.69 E+12	2.06 E+17
10	Ceramic tile	236723	Kg	2.52 E+12	5.97 E+17
11	Painting	64375.5	Kg	1.52 E+13	9.79 E+17
12	Glass	48525	Kg	7.87 E+12	3.82 E+17
13	Lumber	29379.61	Kg	8.79 E+11	2.58 E+16
14	Organic material	17249.1	Kg	6.88 E+12	1.19 E+17

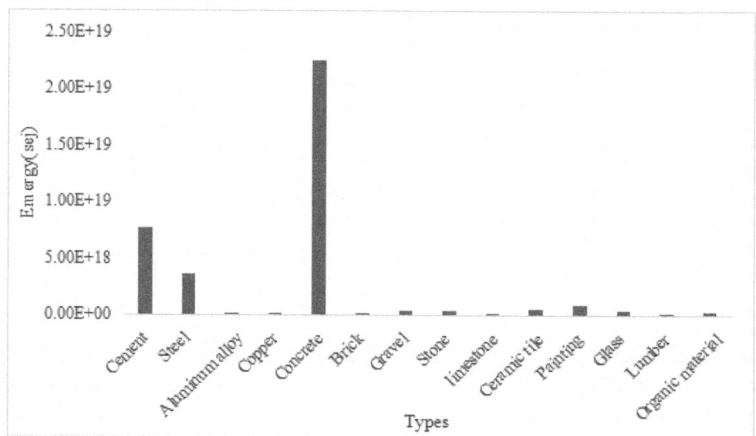

Figure 1.   Emergy of main building materials.

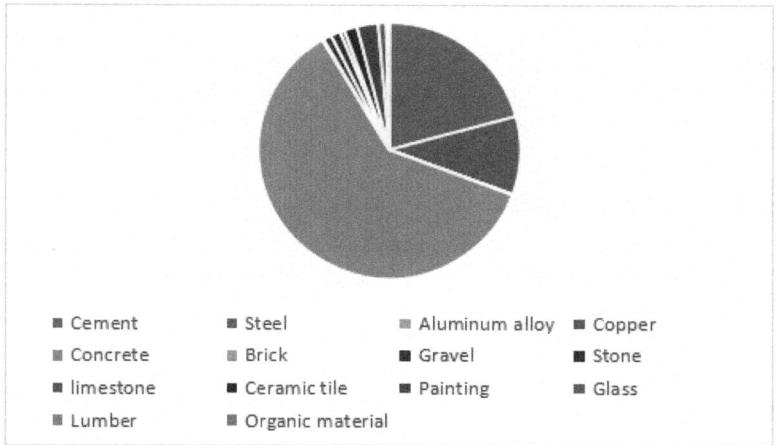

Figure 2.   Emergy proportion of main materials.

3.3 *Emergy of building materials in transport*

According to the principle of the standard of GB30510-2018, 5% of the total building material emergy is used as the transport emergy. The result is 1.86 E + 18 sej.

3.4 *Emergy in the construction stage*

The order of main equipment emergy is pressure welding machine, followed by butt welding machine and tower crane, respectively accounting for 2.7 E+18 sej, 2 E+18 sej and 9 E+17 sej. The total emergy amount of construction equipment is 80.1 E+17 sej.

The main power consumption can be calculated on the basis of Equation (1).

$$P = 1.1 \times \left(\frac{K_1 \cdot \sum P_1}{\cos \theta} + K_2 \sum P_2 + K_3 \sum P_3 + K_4 \sum P_4 \right) \tag{1}$$

The meanings of the equation are as follows:

$$\sum P_1 \text{— Total motor power, } kW;$$

$$\sum P_2 \text{— Welding equipment total power, } kW;$$

$$\sum P_3 \text{— Indoor lighting total power, } kW;$$

$$\sum P_4 \text{— Outdoor lighting total power, } kW;$$

The power electrical calculation result is 335.9 kW. Lighting power is also a major aspect of building energy consumption. According to electricity statistics, it is estimated based on 10% of total electricity consumption. Therefore, the total electricity consumption of lighting power is 370 kW. Taking transformity of electricity as 2.1 E+05 sej/j, the total emergy consumption of the equipment is 2.82 E+18 sej.

3.5 *Emergy in the operation stage*

In accordance with the electric power standard of buildings in China, Yifu Building, which is a public building, has a service life of 50 years. In view of the hot summer and cold winter district in Nanjing, the emergy consumption amount of a single area of the Yifu Building is 117.963 kWh/(m^2· a). The amount of emergy in the construction operation stage is calculated as 7.52 E + 19 sej.

3.6 *Emergy in the building demolition stage*

According to the official standard of JGJ 147-2016(Chenyang, 2016), the power consumption during the demolition phase accounts for 10% of the construction stage emergy and it is 4.52E + 18 sej.

4 CONCLUSIONS

The primary conclusions can be summarized as follows:

(1) Renewable resource emergy aspects: The emergy of renewable resources is calculated as 3.39 E+19 sej, consisting of solar energy, geopotential energy of rain, the chemical potential energy of rain, wind energy, and geothermal energy, respectively.
(2) The emergy of non-renewable resources involves five stages on the basis of LCA-emergy assessment. The emergy of building materials and that in the operational stage plays a critical role and account for 92.4% of the entire emergy, which is the primary contributor.
(3) Compared with the emergy of the whole life cycle of buildings, the proportion of the emergy in the operational stage is the largest (61.64%), followed by that in the building material stage (30.49%), the demolition stage (3.7%), the construction stage (2.31%), and the material transportation stage (1.52%).

ACKNOWLEDGEMENT

The work described in this paper was supported by the Open Fund of the State Key Laboratory of Silicate Materials for Architectures (Wuhan University of Technology) (SYSJJ2022-16).

REFERENCES

Cui W.J., Hong J.K., Liu G.W., Li K.J., Yuanyuan Huang and Lin Zhang. Co-benefits analysis of buildings based on different renewal strategies: The emergy-Lca approach. *Int. J. Environ. Res. Public Health* 2021, 18, 592.

Donald L. DeAngelis, Daniel Franco, Alan Hastings, Frank M. Hilker, Suzanne Lenhart, Frithjof Lutscher, Natalia Petrovskaya, Sergei Petrovskii, Rebecca C. Tyson. Towards building a sustainable future: positioning ecological modelling for impact in ecosystems management. *Bulletin of Mathematical Biology* (2021) 83:107.

Hwang Yi, Ravi S. Srinivasan, William W. Braham, David R. Tilley. An ecological understanding of net-zero energy building: Evaluation of sustainability based on emergy theory. *Journal of Cleaner Production* 143 (2017) 654–671.

Ivan Andri, Andre Pina, Paulo Ferrao, Bruno Lacarriere, Olivier Le Corre. The impact of renovation measures on building environmental performance: An emergy approach. *Journal of Cleaner Production* 162 (2017) 776–790.

Jae Min Lee, William W. Braham. Building emergy analysis of Manhattan: Density parameters for high-density and high-rise developments. *Ecological Modelling* 363 (2017) 157–171.

Odum. H.T. Environmental Accounting, 1996. *Emergy and Environmental Decision Making* [M] NewYork: John Wiley, 1996. 32–34.

Silvio Cristiano, Francesco Gonella. To build or not to build? Megaprojects, resources, and environment: An emergy synthesis for a systemic evaluation of a major highway expansion. *Journal of Cleaner Production* 223 (2019) 772–789.

Suman Paneru, Forough Foroutan Jahromi, Mohsen Hatami, Wilfred Roudebush and Idris Jeelani. Integration of emergy analysis with building information modeling. *Sustainability* 2021, 13, 7990.

Tilba Thomas, A. Praveen. Emergy parameters for ensuring sustainable use of building materials. *Journal of Cleaner Production* 276 (2020) 122382.

Wang C.Y. *Study on Carbon Emissions of Office Buildings in the Yangtze River Delta Region* [D]. 2016 (in Chinese).

Zhang J.X. and Ma L. Environmental Sustainability assessment of a new sewage treatment plant in china based on infrastructure construction and operation phases emergy analysis. *Water* 2020, 12, 484.

Zhang J.X., Ravi S. Srinivasan and Peng C.H. A Systematic approach to calculate unit emergy values of cement manufacturing in china using consumption quota of dry and wet raw materials. *Buildings* 2020, 10, 128.

Frontiers in Civil and Hydraulic Engineering – Mohamed A. Ismail and
Hazem Samih Mohamed (Eds)
© 2023 The Authors, ISBN 978-1-032-38247-0

Temperature effects of complex roof structures with extra-length span effect—Take Guangzhou Baiyun station as an example

Wancai Zhong & Jun Liu
China Railway Construction Group Corporation, Guangzhou, Guangdong, China

Shiwei Zhang
China State Railway Group Corporation, Engineering Quality Supervision Administration, Guangzhou, Guangdong, China

Ren Wang & Yuzao Tang
China Railway Construction Group Corporation, Guangzhou, Guangdong, China

Jiaoyan Zhang*
School of Civil Engineering, Lanzhou Jiaotong University, Lanzhou, Gansu, China

ABSTRACT: The influence of the temperature effect on the complex roof structure with an extra-long span is mainly reflected in both the stress and deformation. This paper combines the project of the complex roof structure of Guangzhou Baiyun Station with an overgrown span, based on the use of SAP2000 and ANSYS finite element software in combination with experimental research, expecting to study the change law of the stress and deformation of the truss + mesh roof under the non-uniform temperature field and uniform temperature field. Due to the complex structure form, the deformation process is much more complicated than other structures. The numerical simulation results are compared with the test results to verify the correctness of the finite element simulation method for the temperature stress. The results show that the effect of temperature on the steel structure cannot be ignored. Under the same radiation angle and strength, the cross-sectional size of the round steel pipe has a more obvious effect on the nonuniform temperature field, and the larger the outer diameter size of the cross section, the larger the maximum temperature stress, the maximum of the joist can reach 0.816×10^8 Pa, the maximum of the net frame can reach 0.780×10^8 Pa, the temperature change law is the same as the temperature stress law, the maximum temperature difference of the truss can reach $21.6°C$, the test can reach $21.4°C$, the error is within the safety range, the maximum temperature difference of the net frame can reach $18°C$, the stress of the lower structure of the roof is affected by the temperature change, the upper structure is less affected by the temperature change, but the overall structural stress state is within the safety range. The research results in this paper are intended to provide theoretical and technical reference for the construction and operation of the same type of steel structures.

1 INTRODUCTION

Extra-long-span complex steel structures with a novel body shape and long construction time are mostly complex high sub-super-stationary space structures, and also these structures have high requirements for the construction technology (Dong 2010). However, in the structural design process, the temperature effect is often ignored by designers, while the temperature stress has an important influence on the overall stability of the building structure. Under the temperature effect, the temperature stress will have a certain influence in the length direction and height direction

---

*Corresponding Author: 1712721964@qq.com

   DOI 10.1201/9781003344209-75

of the building, and with numerous space structure bars and more complex structural forms, the temperature effect will have a negative impact on the structure. Therefore, a more refined numerical simulation is of important engineering significance for large-span complex steel structures.

The temperature will impose both deformation and temperature stresses on very long-span steel structures, and many scholars at home and abroad have made preliminary studies on the temperature effect and steel structures. Chen D K (2017) took three types of steel members as the research object, designed the daylight nonuniform temperature field and the experimental research on its effect, and combined with engineering examples for numerical simulation and experimental research. Zhao Z W et al. (2015) studied the change law of the maximum stress of the bottom ring beam and the upper mesh shell with temperature difference based on the steel structure of Yujiabao transportation hub station house, and used ANSYS simulation to study the change trend of each support reaction force with temperature differences. Tian Limin et al. (2012) combined a steel structure project of the main stadium of the World University Games, carried out the theoretical calculation, numerical simulation and actual monitoring of the temperature action, and compared the results to check the accuracy of the numerical simulation. Liu Z (2018) established a temperature field model for the construction of complex steel structures, analyzed the ambient temperature effect during the construction process, and analyzed the influence of butt welds on the static performance of complex steel structures. Liu J et al. (2020) proposed that the temperature effect should consider the annual temperature effect, the sunlight temperature effect and the temperature difference of members at different locations, and compared the temperature effect of transverse single-bay trusses and longitudinal trusses of roof covers by using SAP2000 for multiple calculations, and finally concluded three methods that could reduce the temperature stress. Fan J S et al. (2021) conducted a parameter-controlled indoor baking lamp radiation test for reinforced concrete bridges to study the temperature field of composite materials. Zhou M et al. (2020) proposed a new method to simulate the non-uniform temperature field of large-span steel structures based on the calculation of solar radiation of tubular members considering the shadowing effect, and obtained a significant non-uniformity of the temperature field of large-span structures. Hélder D. Craveiro et al. (2016) studied the evolution of the yield strength and elastic modulus of steel under the effect of temperature and evaluated the thermal properties of steel by tensile tests.

In summary, the temperature effect on the super-stationary steel structure is not negligible. Therefore, this paper takes the super-long span complex roof structure of Baiyun Station in Guangzhou as the engineering background, and studies the effect of the nonuniform temperature field on the stress and displacement of the key members of the complex steel structure through the refined finite element numerical simulation analysis.

2 PROJECT OVERVIEW

Baiyun Station is located in the south of Baiyun District, Guangzhou City, about 5 km south of Guangzhou Liuhua Railway Station and 2 km east of Baiyun New Town. The Baiyun Station building and related works are divided into two parts: the railway station building and related works as well as integrated and connected works. Baiyun Station has a construction area of 144,366.4 m², and the main body of the station building contains three storeys, with five storeys at local parts. The total height of the building is 27.0 m (from the platform level elevation to the eaves), and the maximum height from the platform level to the roof is 37.0 m. The upper roof of the station building is a three-dimensional curved shape, mainly using the steel truss + mesh structure, the projected area of the station building roof is 95224 m², the maximum size of the plane projection is 252.5 m×412 m, the steel roof consists of the central waiting room roof, the north-south wavy "ribbon", and the east-west light valley "petal". The central waiting room roof and the wavy "ribbon" on the north and south sides adopt a combination of the space tube truss + mesh structure. The steel roof is composed of the central waiting room roof, the north-south wavy "ribbon" and the east-west light valley "petals", with the central waiting room roof and the north-south wavy "ribbon" adopting the combination form of the space tube truss + mesh frame structure, of which the side effect of the main structure axis on the roof of Guangzhou Baiyun Station is shown in Figure 1.

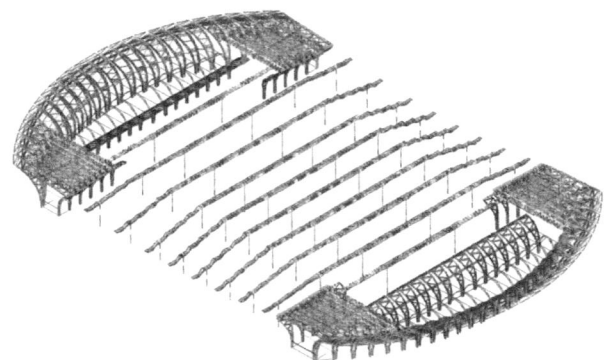

Figure 1. Effect drawing of main structure axis on the roof of Guangzhou Baiyun Station.

3 ANALYSIS OF WORKING CONDITIONS

3.1 *Temperature load values*

According to the construction organization design of the new Guangzhou Baiyun Station and related works as well as the total schedule network diagram, the following information is obtained: a total of 145 days for the construction of the first phase roof (February 5, 2022–June 30, 2022), and a total of 80 days for the construction of the second phase roof for the main structure (June 10, 2023–August 30, 2023), querying the atmospheric temperature at the construction site in Baiyun District of Guangzhou using the Chinese weather station (23.392°N, 113.299°E), as shown in Figure 1; for the 2023 atmospheric temperature with reference to the 2022 atmospheric temperature, we get the basic temperature of Baiyun District, Guangzhou City, where the minimum temperature is 6°C, the highest temperature is 36°C, and the average annual temperature of the region is 21°C.

Figure 2. Temperature query process of Baiyun Station.

3.2 *Analysis of temperature differences and temperature stresses for steel pipes*

Based on the ANSYS program, the temperature field analysis program of steel members under the action of solar radiation is prepared using the APDL language to study the temperature field distribution law of all typical members on the complex roof under the action of solar radiation, the steel pipe is simulated by the SOLID unit, the steel pipe primer is the water-based inorganic zinc-rich primer, the intermediate paint is the intermediate epoxy paint for iron, and the top coat is the fluorocarbon top coat. Among them, Tables 1 and 2 show the parameters of typical members of the complex roof structure, and Table 3 shows the basic parameters used in the numerical simulation using ANSYS. Figures 3 to 5 give the non-uniform temperature field distribution of the upper chord, lower chord and web members of the truss, respectively, where the lowest temperature is taken as 6°C and the highest as 36°C. Figures 6 to 8 give the non-uniform temperature field distribution of the upper chord, lower chord and web members of the grid, respectively.

Table 1. Parameters of Q355B seamless steel pipe materials for the truss.

No.	Group 1			Group 2			Group 3		
Cross-section type	Round steel pipe			Round steel pipe			Round steel pipe		
Location	Top string			Lower string			Ventral rod		
Material	Q355B			Q355B			Q355B		
Section size/mm	450×20	450×25	450×30	168×7.5	180×7.5	194×7.5	500×30	600×30	600×40
Length/mm	500			500			500		

Table 2. Parameters of Q355B seamless steel pipe materials for the grid.

No.	Group 1			Group 2			Group 3		
Cross-section type	Round steel pipe			Round steel pipe			Round steel pipe		
Location	Top string			Lower string			Ventral rod		
Material	Q355B			Q355B			Q355B		
Section size/mm	450×20	450×30	500×40	140×5	140×5.5	140×6	60×3.5	63.5×3.5	68×3.5
Length/mm	500			500			500		

Table 3. Parameter values of numerical temperature field simulation model.

Material	Density/ (kg·m^{-3})	Modulus of elasticity/MPa	Poisson's ratio	Linear expansion Coefficient/°C	Yield Strength /MPa	Thermal conductivity /[W/(m·°C)]	Specific heat capacity/(kg·°C)
Q335B	7.850×10^3	2.0×10^{11}	0.3	1.2×10^{-5}	100×10^6	4510	465

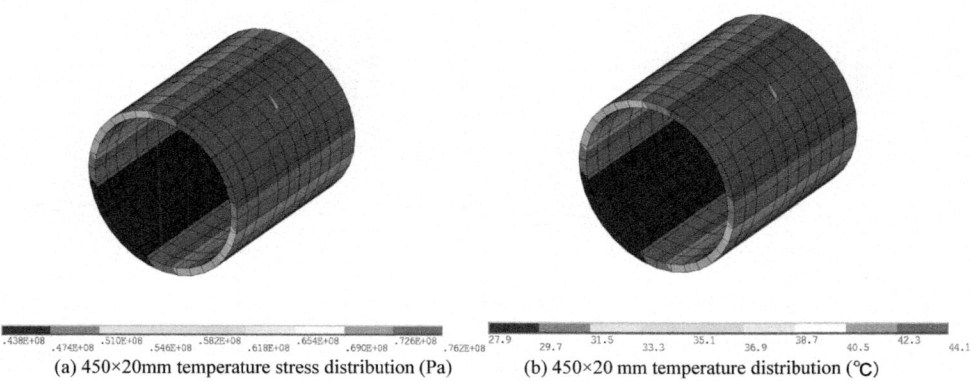

(a) 450×20mm temperature stress distribution (Pa)	(b) 450×20 mm temperature distribution (°C)

Figure 3. Temperature stress distribution of top chord of the truss.

According to Figure 3 displaying the top chord of the truss, under the temperature effect, the maximum temperature stress of the 450×20×500 (mm) steel pipe is $0.762×10^8$ pa, the minimum temperature stress is $0.438×10^8$ Pa, and the section temperature difference is 16.2°C; similarly, under the temperature effect, the maximum temperature stress of the 450×25×500 (mm) steel pipe is $0.762×10^8$ Pa, the minimum temperature stress is $0.438×10^8$ Pa, and the section temperature difference is 16.2°C. Under the temperature effect, the maximum temperature stress of the 450×30×500 (mm) steel pipe is $0.762×10^8$ Pa, the minimum temperature stress is $0.438×10^8$ Pa, and the cross-section temperature difference is 16.2°C.

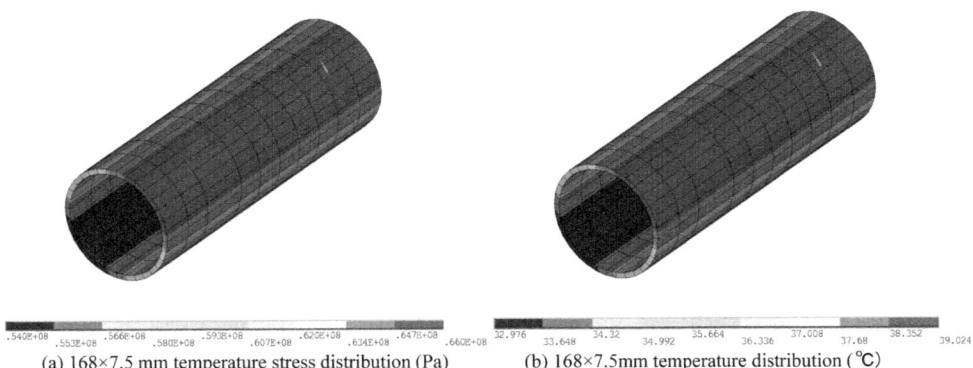

(a) 168×7.5 mm temperature stress distribution (Pa)   (b) 168×7.5mm temperature distribution (°C)

Figure 4.   Temperature stress distribution of bottom chord of the truss.

According to Figure 4 showing the top chord of the truss, under the temperature effect, the maximum temperature stress of the 168×7.5×500 (mm) steel pipe is $0.660×10^8$ Pa, the minimum temperature stress is $0.540×10^8$ Pa, and the section temperature difference is 6.048°C; similarly, under the temperature effect, the maximum temperature stress of the 180×7.5×500 (mm) steel pipe is $0.665×10^8$ Pa, the minimum temperature stress is $0.535×10^8$ Pa, and the section temperature difference is 6.48°C; under the temperature effect, the maximum temperature stress of the 194×7.5×500 (mm) steel pipe is $0.670×10^8$ Pa, the minimum temperature stress is $0.530×10^8$ Pa, and the section temperature difference is 6.984°C.

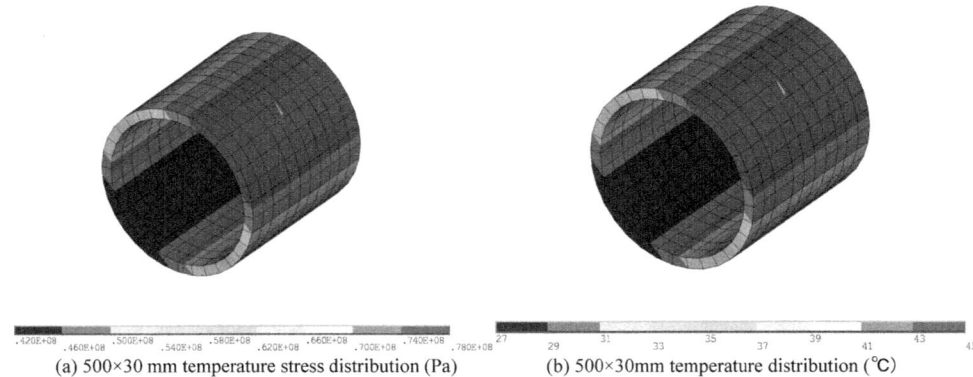

(a) 500×30 mm temperature stress distribution (Pa)   (b) 500×30mm temperature distribution (°C)

Figure 5.   Temperature stress distribution of web member of the truss.

According to Figure 5 exhibiting the web member of the truss, the maximum temperature stress of 500×30×500 (mm) steel pipe is $0.780×10^8$ Pa, the minimum temperature stress is $0.420×10^8$ Pa, and section temperature difference is 18°C; under the temperature effect, the maximum temperature stress of the 600×30×500 (mm) steel pipe is $0.816×10^8$ Pa, the minimum temperature stress is $0.384×10^8$ Pa, and the cross-section temperature difference is 21.6°C; under the temperature effect, the maximum temperature stress of the 600×40×500 (mm) steel pipe is $0.816×10^8$ Pa, the minimum temperature stress is $0.384×10^8$ Pa, and the cross-section temperature difference is 21.6°C.

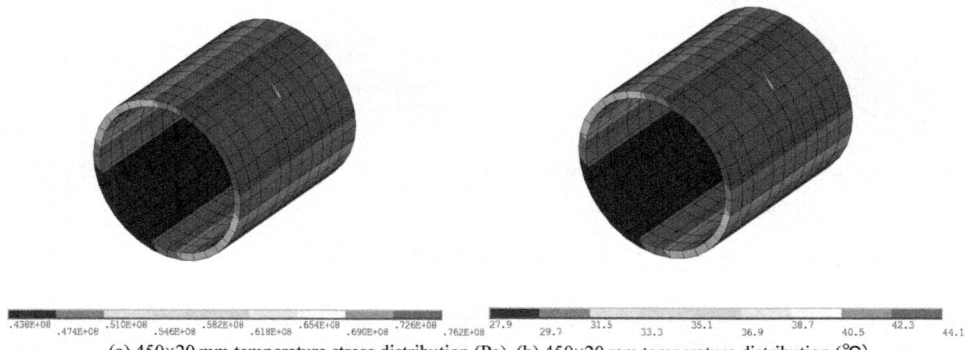

(a) 450×20 mm temperature stress distribution (Pa) (b) 450×20 mm temperature distribution (℃)

Figure 6. Temperature stress distribution of upper chord of the grid web.

According to Figure 6 showing the upper chord of the grid web, under the temperature effect, the maximum temperature stress of the 450×20×500 (mm) steel pipe is $0.762×10^8$ Pa, the minimum temperature stress is $0.438×10^8$ Pa, and the section temperature difference is 16.2°C; similarly, under the temperature effect, the maximum temperature stress of the 450×30×500 (mm) steel pipe is $0.762×10^8$ Pa, the minimum temperature stress is $0.438×10^8$ Pa, and the section temperature difference is 16.2°C; under the temperature effect, the maximum temperature stress of the 500×40×500 (mm) steel pipe is $0.780×10^8$ Pa, the minimum temperature stress is $0.420×10^8$ Pa, and the section temperature difference is 18°C.

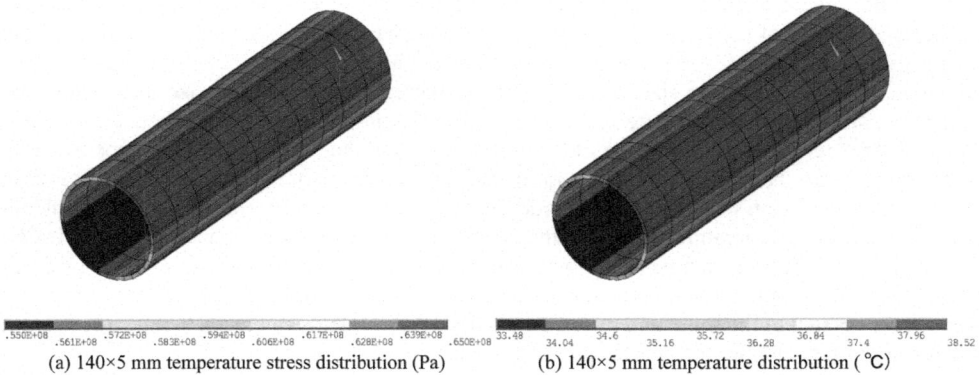

(a) 140×5 mm temperature stress distribution (Pa) (b) 140×5 mm temperature distribution (℃)

Figure 7. Temperature stress distribution of lower chord of the grid structure.

According to Figure 7 displaying the lower chord of the grid structure, under the temperature effect, the maximum temperature stress of the 140×5×500 (mm) steel pipe is $0.650×10^8$ Pa, the minimum temperature stress is $0.550×10^8$Pa, and the section temperature difference is 5.04°C; similarly, under the temperature effect, the maximum temperature stress of the 140×5.5×500 (mm) steel pipe is $0.650×10^8$ Pa, the minimum temperature stress is $0.550×10^8$ Pa, and the section temperature difference is 5.04°C; under the temperature effect, the maximum temperature stress of the 140×6×500 (mm) steel pipe is $0.650×10^8$ Pa, the minimum temperature stress is $0.550×10^8$ Pa, and the section temperature difference is 5.04°C.

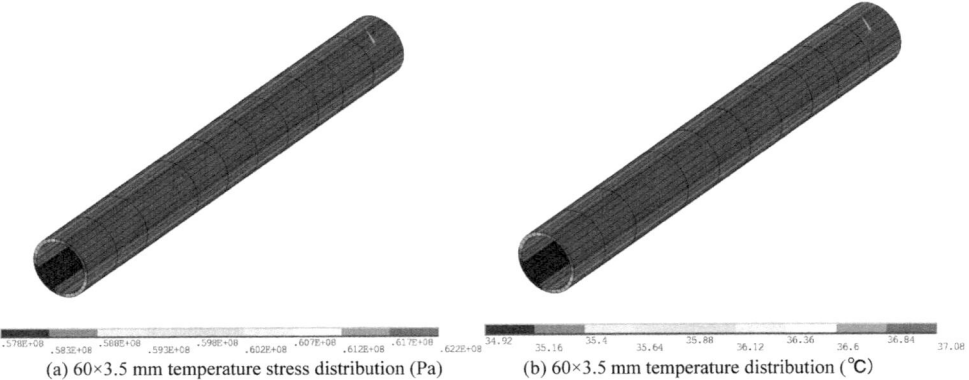

(a) 60×3.5 mm temperature stress distribution (Pa)   (b) 60×3.5 mm temperature distribution (°C)

Figure 8.   Temperature stress distribution of web member of the grid structure.

According to Figure 8 exhibiting the web member of the grid structure, under the temperature effect, the maximum temperature stress of the 60×3.5×500 (mm) steel pipe is 0.622×10^8 Pa, the minimum temperature stress is 0.578×10^8 Pa, and the section temperature difference of 2.16°C; similarly, under the temperature effect, the maximum temperature stress of the 63.5×3.5×500 (mm) steel pipe is 0.623×10^8 Pa, the minimum temperature stress is 0.577×10^8 Pa, and the section temperature difference is 2.286°C; under the temperature effect, the maximum temperature stress of the 68×3.5×500 (mm) steel pipe is 0.624×10^8 Pa, the minimum temperature stress is 0.576×10^8 Pa, and the section temperature difference is 2.448°C.

In the temperature stress and temperature distribution diagram of the truss in Figures 3 to 5, under the same radiation angle and intensity, the section size of the round steel pipe has an obvious influence on the non-uniform temperature field. The larger outer diameter of the section, the higher the maximum value of the temperature stress (Liu, 2018), being up to 0.816×10^8 Pa; when the pipe diameter is unchanged and the wall thickness changes, the maximum value of the temperature stress remains unchanged; when the wall thickness remains unchanged and the outer diameter changes, the minimum value of the temperature stress remains unchanged, and the maximum value of the temperature stress increases with the increasing outer diameter. When the outer diameter and wall thickness, that is, the section size, change, with the increase of the section size, the greater the maximum value of the temperature stress, the smaller the minimum value of temperature stress, the greater the temperature difference of the component, and the more easily the component will be destroyed under the non-uniform temperature field, but the maximum value and minimum value of the temperature stress are in the same order of magnitude. The temperature change law is the same as that of the temperature stress, and the maximum temperature difference can reach 21.6°C.

From Figures 6 to 8, the temperature stress and temperature distribution diagram of the grid can be obtained. The temperature stress distribution law and temperature distribution law of the grid are the same as that of the truss, and its maximum temperature stress can reach 0.780×10^8 Pa, and the maximum temperature difference can reach 18°C.

3.3   *Temperature stress analysis of the grid model*

Based on ANSYS finite element software modeling (Chen), in the middle area of the complex roof cover with overgrown spans, there are many identical net frame models, the upper chord has 5 spans in the X and Y directions, and the plane size is 10.0×10.0 (m); the lower chord has 4 spans in the X and Y directions, the plane size is 8.0 m×8.0 m, the length of both the upper and lower chord planes is 2.0 m, and the height of the net frame (the vertical distance between the upper and lower chord planes) is 2.0 m. The nodes around the upper chord plane are hinged, and the rest of the nodes and the lower chord plane are free, considering the self-weight of the structure, to study the internal force and deformation of the mesh structure when the temperature changes. The mesh frame bars are simulated using the LINK180 unit, and the steel tube size is mostly 152×6 (mm) to simplify the calculation. According to the data in Section 3.1, the temperature difference is 5.5°C, and the temperature change is applied to the whole structure with the lowest temperature of 6°C and the highest temperature of (36 + 5.5) 41.5°C, in order to obtain the deformation and internal force distribution of the structure. Among them, the finite element model of the mesh frame after applying the constraints is shown in Figure 9, and the temperature stress cloud and the local details of the cloud are shown in Figures 10 and 11.

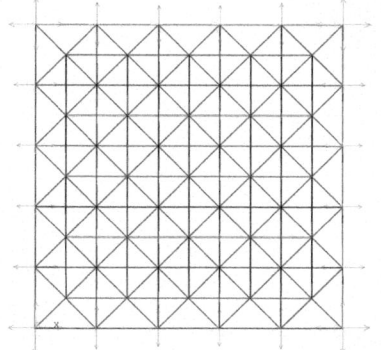

Figure 9. Finite element model of the space Truss with Constraints.

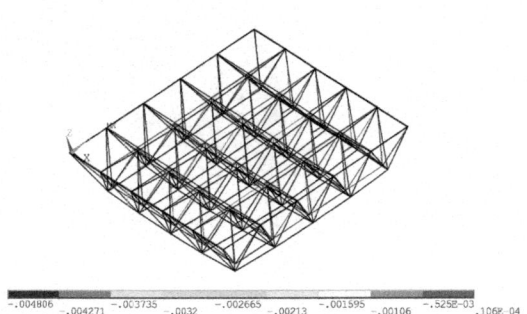

Figure 10. Temperature stress cloud diagram of the grid.

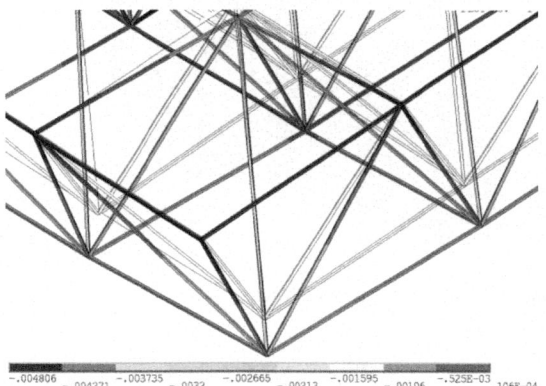

Figure 11. Local details of the grid temperature stress cloud map.

As can be seen from Figure 10, the overall maximum temperature stress of the grid frame is 0.106×10^{-4} Pa and the minimum temperature strain is -0.004806 m; as can be seen from Figure 11, the stress in the lower structure of the roof is affected more by temperature changes and the upper structure is less affected by temperature changes, but the overall structural stress state is within the safety range (Cheng 2019).

3.4 *Overall temperature analysis of roof cover*

To simplify the calculation, the most used steel pipe size is 152×6 (mm). According to the data in Section 3.1, the temperature difference is 5.5°C, the lowest temperature is 6°C, the highest temperature is (36+5.5) 41.5°C, and the unit in SAP is KN, mm. SAP2000 is used to establish the overall model of the overgrown-span roof structure (You 2020) as shown in Figure 14.

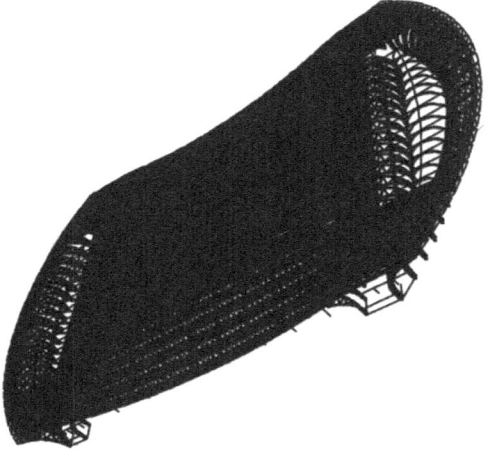

Figure 12.　3D Drawing of the roof structure.

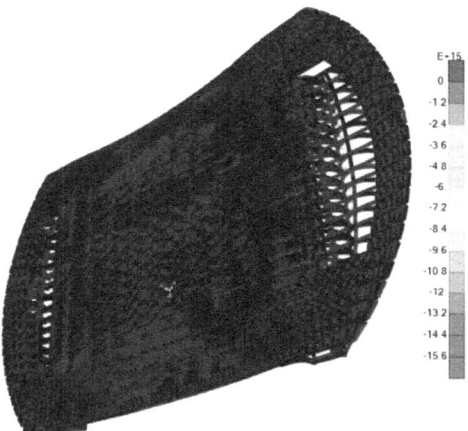

Figure 13.　Cloud diagram of temperature stress of roof.

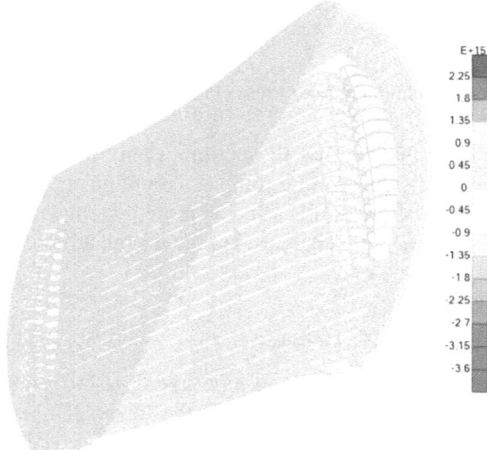

Figure 14.　Cloud diagram of temperature displacement of the roof.

In Figure 13, when the minimum ambient temperature is 6°C and the maximum temperature is 41.5°C, the overall maximum temperature stress of the roof is 0 and the minimum temperature stress is -15.6×10^{15} Pa. From Figure 14, the total maximum upper displacement is 2.25×10^{15} mm and the minimum displacement is -3.6×10^{15} mm, from which it can be concluded that the effect of the upper displacement is greater than that of the lower displacement (Guan 2020).

4 EXPERIMENT

4.1 Experimental scheme

In the test, the middle section of the 1:10 steel joist sample is selected to conduct the indoor non-uniform temperature field test, and a short-wave infrared baking lamp instead of a solar radiation light source is designed to rotate the temperature to simulate the sun rising and setting (Fang 2016).

The baking lamp selected for the test is LD01-6A, which can be rotated up and down arbitrarily, and the lamp head can be rotated 360° up and down, and can be freely adjusted from 0 to 100°C, so as to realize the loading of different temperatures. The lamp source consists of 6 short-wave infrared lamps with a total radiation power of 6600 W, as shown in Figure 15. According to the actual measurement, when the lamp base is 1 m away from the specimen, the surface radiation of the specimen can reach about 100°C. The distance between the baking lamp and the main illuminated surface of the steel joist is controlled within 1 m, the height and angle are adjusted by lifting and rotating the lamp head, and the uneven temperature field of the joist is measured by the temperature sensor.

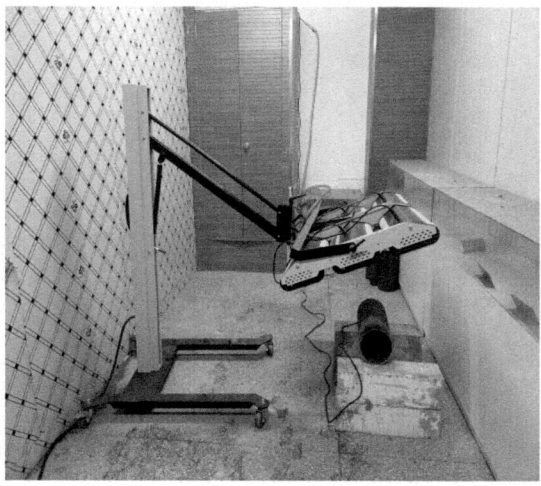

Figure 15.   Process of baking steel pipe with short-wave infrared baking lamp.

4.2 Test results

At 12:25 p.m. on August 6, 2022, the temperature data was obtained using infrared thermography at the location of the truss web rod with an outside diameter of 600×40 (mm) and ambient temperature of 36°C. The measured steel pipe surface as well as the infrared thermograms is shown in Figures 16 and 17 below.

The maximum temperature on the steel pipe measured by the portable pocket thermal imaging camera is 45.1°C, the minimum temperature is 3.77°C, and the temperature difference is 21.4°C. According to the numerical simulation results in Section 2.1, the maximum temperature difference is 21.6°C, and the error rate is 0.9%, which is within the error range, thus verifying the reliability of the numerical simulation.

Figure 16. Surface drawing of 600×40 (mm) steel pipe.

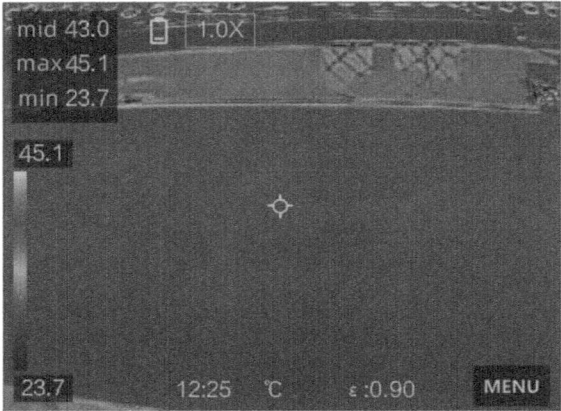

Figure 17. Infrared thermal image of 600×40 (mm) steel pipe.

5 CONCLUSIONS

In this paper, the response law of complex structures to the temperature load is studied systematically in the context of actual engineering. The temperature is a key factor affecting the stress and deformation of the structure, and the temperature effect on the steel structure cannot be neglected.

1) Under the same radiation angle and strength, the cross-sectional size of the circular steel pipe has a more obvious effect on the non-uniform temperature field, and the larger the outer diameter size of the cross-section, the greater the maximum temperature stress, being up to 0.816×10^8 Pa, and the maximum mesh can reach 0.780×10^8 Pa, the temperature change law is the same as the temperature stress law, the maximum temperature difference of the truss can reach 21.6°C, and the maximum temperature difference of the mesh can reach 18. The maximum temperature difference can be 21.6°C for trusses and 18°C for nets.
2) The stress of the lower structure of the roof is influenced by the temperature change, and the upper structure is less influenced by the temperature change, but the overall structural stress state is within the safety range, and the overall upper displacement of the roof is greater than the lower displacement.

3) The temperature difference in the steel pipe measured by the temperature sensor is 21.4°C at 12:25 noon on August 6, 2022 when the outer diameter of truss is 600×40 (mm), and the maximum temperature difference is 21.6°C. According to the numerical simulation result in Section 2.1, the error rate is 0.9%, which is within the error range, thus verifying the reliability of the numerical simulation.

REFERENCES

Chen D.S. *Study on the non-uniform temperature field of space steel structure with sunlight and its effect* [D]. Harbin: School of Civil Engineering, Harbin Institute of Technology, 2017.

Chen C.-H., Liu H.-B., Zhou T., et al. *APDL Parametric Calculation and Analysis of Space Steel Structures* [M/OL].

Cheng K. *Temperature Field Monitoring and Analysis of Dome Structure Stadium* [D]. Zhejiang University of Technology, 2019.

Craveiro H.D., Rodrigues J.P.C., Santiago A., et al. Review of the high temperature mechanical and thermal properties of the steels used in cold formed steel structures—the case of the S280 Gd+Z steel[J]. *Thin-Walled Structures*, 2016, 98: 154–168.

Dong S.L. Development and prospects of space structures in China [J]. *Journal of Building Structures*, 2010,31(06): 38–51.

Fan J., Liu Y., Liu C. Experiment study and refined modeling of temperature field of steel-concrete composite beam bridges[J]. *Engineering Structures*, 2021,240: 112350.

Fang Y. Qg. *Computational analysis of temperature effect of large span steel truss structure*[D]. Guangzhou University, 2016.

Guan P.W., Yang G.Y., Zhang D.L., et al. Analysis of the effect of temperature on complex steel structures[J]. *Structural Engineer*, 2020,36(02): 80–87.

Liu H.-B., Chen Z.-H., Zhang Y.-L., et al. Non-uniform temperature effect on steel structure of Erdos Airport Terminal: The 12th National Symposium on Modern Structural Engineering [C], 2012.

Liu J., Liu Y. Sy, Li G., et al. Temperature stress analysis and countermeasures for large-span steel roof of Shanghai National Convention and Exhibition Center[J]. *Building Structure*, 2020, 50(12): 40–45.

Liu Z. *Study on the temperature effect of complex steel structures during construction and operation*[D]. Tianjin University, 2018.

Tian L.M., Hao J.P., Wang Y., et al. Analysis of temperature effects on large-span space structures and study of closure temperature [J]. *Journal of Civil Engineering*, 2012, 45(05): 1–7.

You Y., Zhang Z.T., Wang J., et al. Stress and deformation variation law of space steel structures considering the effect of temperature [J]. *Space Structures*, 2020, 26(04): 58–63.

Zhao C.-W., Chen Z.-H., Wang S.-Dun, et al. Study on temperature effect and closing temperature of complex steel structures with large span [J]. *Space Structures*, 2015, 21(02): 40–45.

Zhou M., Fan J., Liu Y., et al. Non-uniform temperature field and effect on construction of large-span steel structures[J]. *Automation in Construction*, 2020,119: 103339.

Frontiers in Civil and Hydraulic Engineering – Mohamed A. Ismail and
Hazem Samih Mohamed (Eds)
© 2023 The Authors, ISBN 978-1-032-38247-0

The application of 3D design of hydraulic engineering based on dynamo

Ning Wang*, Bo Luo* & Donghai Chen*
Changjiang Survey Planning Design and Research Co., Ltd, Wuhan, China

ABSTRACT: At present, due to the limitations of mainstream platforms, the 3D design of hydraulic engineering is generally completed by multi-platforms, which may cause the loss of data conversion information and an increase in application and promotion costs. How to complete the main design work on the same platform has become a research hotspot. Whereas the limitations of the original function of the Revit platform for the modeling of the complex surface shape of hydraulic engineering, according to the technical characteristics of Revit and Dynamo, this paper discusses Dynamo parametric modeling technology and the data exchange/interaction mode with Revit. This technology has been applied in the construction drawing design of a pumping project, which provides an important reference value for the 3D design of a hydraulic project with the Revit platform in the future.

1 INTRODUCTION

In recent years, BIM (Building Information Modeling) technology has gradually been widely promoted and applied in the water conservancy and hydropower industry. Compared with construction projects, water conservancy projects generally have the characteristics of large volume, complex structure, many disciplines involved, long design and construction cycle, and are greatly affected by environmental factors. In particular, the three-dimensional fine modeling of various complex special-shaped surface components (such as volute, draft tube, arch dam) in hydraulic engineering is the main difficulty of BIM three-dimensional design of hydraulic engineering, and it is also a technical problem that often plagues designers.

At present, there are feasible solutions on the mainstream software platform, such as CATIA with strong surface modeling and parametric modeling functions, which can realize the three-dimensional modeling function of special-shaped surface components of hydraulic structures (Huang 2009). However, the data conversion of models between platforms hinders the collaborative design between platforms, e.g., the integration of the hydraulic 3D model created in CATIA and the auxiliary building model created by Revit. Although the data of CATIA and Revit software can be exported to each other through plug-ins or interfaces, there are some problems in this format conversion, such as one-way irreversibility, loss of accessory information of components, or possible loss of components. Generally, there are two solutions in the industry: One is to develop a data interface or a general exchange data format to convert the data of one platform to another software platform for data integration under the condition of ensuring the completeness of information to the greatest extent, so as to realize the collaborative design between multiple platforms; the second one is trying to complete all three-dimensional modeling, calculation, and mapping on the same platform, because of immature design data conversion between platforms.

---

*Corresponding Authors: wangning@cjwsjy.com.cn; luobo@cjwsjy.com.cn and chendonghai@cjwsjy.com.cn

DOI 10.1201/9781003344209-76

The main research route of this paper is based on the second idea above, that is, to explore the main design modeling work of designing water conservancy projects on a single software platform, so as to avoid the loss of data information caused by data mutual guidance between multiple platforms. In view of the wide application of Revit software in the construction industry and its strong building modeling and drawing functions, due to the limited modeling ability of special-shaped surfaces, there are obstacles to the full application of Revit software in large and medium-sized complex water conservancy projects. Whether the parametric surface shape can be created quickly and conveniently in Revit has become a key technical problem in the whole process of three-dimensional design of BIM of hydraulic engineering on the Revit platform. In recent years, with the emergence of dynamo plug-in in the Revit platform, the ability of parametric surface modeling of the Revit platform has been greatly improved, so it also provides a feasible technical path for realizing the three-dimensional design of large and medium-sized hydraulic projects on a single Revit platform.

After comparing and selecting the existing mainstream platforms, this paper analyzes the technical characteristics of the Revit software platform and dynamo plug-in, combs the technology of dynamo parametric modeling and the data exchange/interaction with Revit, and introduces the technical process of design modeling in detail through the examples of typical special-shaped components of hydraulic engineering (spiral case, double-curvature arch dam). This technology also provides a new idea for the 3D design of hydraulic projects based on the Revit platform in the future.

2   3D DESIGN PLATFORM OF HYDRAULIC ENGINEERING

2.1   *Comparison and selection of three-dimensional design platforms for hydraulic engineering*

According to the current market share, there are three mainstream BIM 3D design software platforms in the water conservancy and hydropower design industry: Revit series software of Autodesk, MicroStation series software of Bentley, and CATIA software of Dassault. These three design platforms have their own characteristics. Revit is the most popular in the industrial and civil construction industry, and has a high degree of integration with BIM Technology-related applications; the Bentley platform has a unified format for all professional data, without format conversion problems, and has advantages in the field of industrial architectural design and infrastructure; CATIA software has powerful surface modeling and parametric modeling capabilities, which can easily complete the parametric modeling of complex special-shaped components.

Table 1.   Comparison of 3D design platform characteristics of mainstream hydraulic engineering.

Platform name	Provider	Advantages and characteristics	Application in water sector	Irregular surface modeling
Revit	Autodesk	The industrial and civil construction bank has high popularity, convenient three-dimensional modeling, mapping capabilities, and other functions, low technical threshold, and low investment cost.	Universal	General
MicroStation	Bentley	All professional data are in a unified format and have advantages in the field of industrial architectural design and infrastructure.	Universal	Strong
CATIA/3DE	Dassault	With strong surface modeling and parametric modeling capabilities, it basically monopolizes the high-end manufacturing industry.	Universal	Very Strong

2.2 Features and limitations of Revit

Revit software is the BIM three-dimensional design and modeling software for the construction industry developed by Autodesk. It provides a working mode of multi-disciplinary collaborative design including architecture, structure, water supply and drainage, HVAC, electrical and other disciplines. Its main functions include the creation of design models of various disciplines, pipeline integration, collision inspection, production drawings, sunlight, daylighting analysis, etc. In view of the market share, software maturity, ease of use, and cost performance of Revit, many water conservancy design units have adopted Revit for three-dimensional design in the implementation of water conservancy and hydropower projects and achieved certain results. However, the surface modeling function of the Revit software system itself is weak. For example, in architectural design, the surface modeling is generally carried out through third-party software (such as Rhino, and CATIA), and is imported into Revit. After that, the corresponding surface is picked up in Revit by using the mass mode to generate volume in Revit. Finally, the volume is further divided into surfaces, the generation of wall floors, and other deepening design operations in Revit.

A prominent problem with this mode is that once the irregular mass is generated, it cannot be modified after being imported into Revit. This scheme should be modified, if necessary, in the third-party software and re-imported into Revit, and the detailed design work in Revit basically needs to be done again, which greatly increases the repeated workload. In order to solve the technical weakness of Revit's complex surface modeling, Autodesk introduces a dynamo plug-in, which provides a feasible technical route for Revit to realize the three-dimensional design of large and medium-sized hydraulic projects on a single platform and avoid data conversion between multiple platforms.

2.3 Dynamo plug-in

Dynamo is a parametric modeling plug-in for Revit. It is a design tool developed based on the concept of visual programming for design. At first, the dynamo was open-source software, and later it was directly integrated into subsequent versions of Revit 2017 by Autodesk. Dynamo is similar to the Grasshopper plug-in on Rhino. It completes complex design operations through visual module parametric driving, so as to realize the user's design intention. Dynamo can realize parametric shape creation, interact with Revit models (such as element adjustment, statistical calculation, and color matching.), and analyze external data to create shapes, such as generating terrains according to contour lines or point sets, creating shapes according to parameter tables, etc. Dynamo software's powerful surface modeling creation function remedies the deficiency of Revit's native surface modeling function and can simplify some cumbersome and repetitive operations so that designers can get rid of constraints and focus on the design itself.

2.4 Overview of dynamo design

Dynamo has been applied in the construction industry to a certain extent (Xu 2017). For example, Wang S. (2015) realized the batch automatic layout of building components and the creation of special-shaped building shapes by using dynamo. Li Y. (2016) directly used dynamo to adjust the family parameters of the building components (columns, walls, slabs, slopes) of the prefabricated parking building.

In recent years, some units in the water conservancy industry began to explore the application of dynamo plug-ins in the parametric design modeling of water conservancy projects. For example, Song D. (2020) and Zhao J. M. (2020) respectively studied the three-dimensional modeling technology of dynamo plug-in to realize double-curvature arch dam and buttressed retaining wall in Civil3d software, and Cao Y. & Xu H. L. (2018) discussed dynamo's modeling technology for hydraulic machinery design in Revit, and both conducted in-depth research on dynamo's modeling of special-shaped surface design in hydraulic engineering, which also gave some technical inspiration to this research.

3 KEY TECHNOLOGIES OF DYNAMO PARAMETRIC MODELING

3.1 *Typical design process*

The process of parametric modeling of curved surface components of hydraulic engineering by dynamo software is shown in Figure 1.

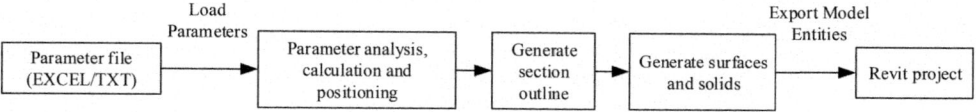

Figure 1. Flowchart of complicated surface modeling by dynamo.

3.2 *Parameter reading and analysis*

In Dynamo, hydraulic structures can be driven by reading table data. First, the standard data related to the dimensions of building components should be stored in the corresponding files in the form of tables, and the relationship between the data in the tables and the characteristic parameters of the three-dimensional model should be established. The designer achieves the goal of changing the geometric dimensions and obtaining the model of the required hydraulic facilities by selecting different records in the table. In Dynamo parametric design, the designer can read the relevant parameters in the form of Excel files into Dynamo's list container through the node module for further processing.

3.3 *Geometric shape generation*

Various geometry types are provided in Dynamo, such as Curve, Surface, and Solid. Designers can generate curves, surfaces, entities, and other geometries by lofting or sweeping through type node functions. For example, curves are generated by common point fitting curves, surfaces are generated by lofting and enclosed areas, entities are generated by lofting, sweeping, intersection, and so on, and surface modeling of structures is achieved by creating curves, surfaces, and entities.

3.4 *Interaction with revit*

Dynamo interacts with Revit (data exchange and interactive control) in four ways: '*Import Instance*', '*DirectShape*', '*ExportToSAT*', and '*Family Instance*'. Each of these four methods has its own characteristics and obvious advantages and disadvantages. In general, '*Import Instance*' and '*DirectShape*' are commonly used when Dynamo shape modeling is imported into Revit; '*Export-ToSAT*' model transformation is required through SAT format; '*Family Instance*' is often used in the process of parametric modeling of bridge tunnels [12, 13]. Therefore, the designer can choose the appropriate import mode according to the specific situation. In summary, the advantages and disadvantages of the four methods are shown in the following table.

Table 2. Comparison of four interactive methods between dynamo and revit.

	Advantage	Disadvantage
ImportInstance	Dynamo native node function, Suitable for surface parametric driving modeling	Represented as "import symbol" in Revit; components cannot be edited
DirectShape	Dynamo native node function, Suitable for surface parametric driving modeling	Represented as a "normal model" in Revit; components cannot be edited; the shape is converted into a triangle, which reduces the accuracy of the model.

(continued)

Table 2. Continued.

	Advantage	Disadvantage
ExportToSAT	Dynamo native node function. Third-party software can be imported via intermediate SAT format	Represented as a "normal model" in Revit; components cannot be edited; SAT is required as an intermediate format
FamilyInstance	Generated components can be edited and attributes can be modified, which is suitable for linear engineering parametric layouts such as bridges, roads, tunnels, etc.	Belongs to the third-party node package; modeling needs to be achieved by calling a pre-built family of parameterized components

4 APPLICATION EXAMPLE OF DYNAMO MODELING

4.1 *Example of metal spiral case modeling*

A Spiral case is a common component with complex surface structure in hydraulic and Hydropower projects, and it is also a typical hydraulic component that cannot be quickly modeled based on the native function of Revit software. As the design phase of engineering projects continues to advance, equipment manufacturers will provide the detailed size of spiral cases, so designers need to build three-dimensional models based on the above data.

The main technical idea of parameterized three-dimensional modeling of drive volute by Dynamo is to establish each section by reading the design section parameters in Excel format and to generate the shell entity by lofting and surface thickening operations. The seat ring consists of three main components, i.e., upper seat ring, lower seat ring, and fixed guide vane. The contour of the fixed guide vane is created by the curve tool, and then several fixed guide vane entities are generated by lofting, translation, and rotation operations. The contour of the seat ring is created by the curve tool, and then the seat ring is generated by sweeping the contour curve along the axis. Finally, the generated spiral case is integrated with the seat ring part and the spiral case is imported into Revit by *Import Instance* or *DirectShape*. The main process of spiral case modeling is shown in Figures 2a–2d.

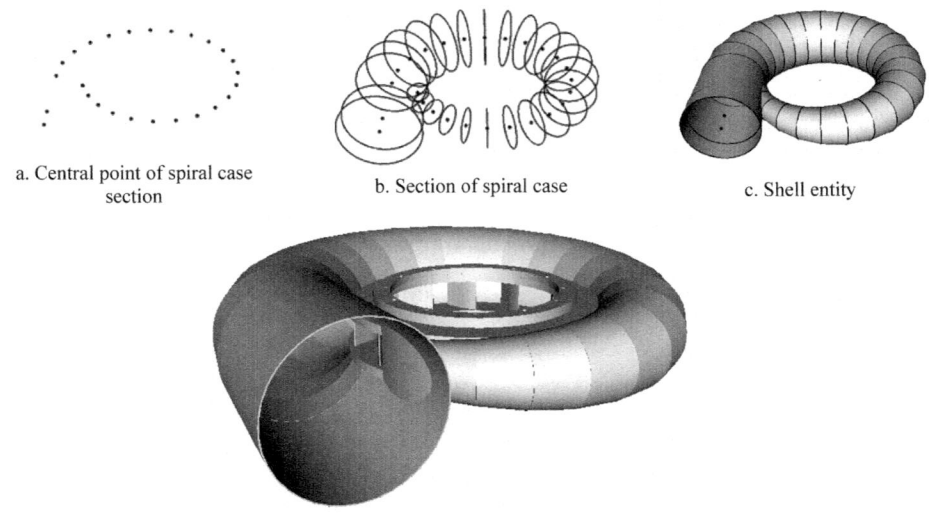

a. Central point of spiral case section

b. Section of spiral case

c. Shell entity

d. Integrated model

Figure 2. Process diagram of spiral case modeling.

4.2 Modeling example of double curvature arch dam

The hyperbolic arch dam is a bi-directional (horizontal and vertical) bending arch dam. It is the most representative dam type of arch dam and is also a common hydraulic structure with a curved body in hydraulic engineering. Because it is difficult for Revit to create a hyperbolic arch dam as such a curved surface by using the original modeling function. Similar to CATIA, the three-dimensional modeling idea of the hyperbolic arch dam body in Dynamo is: to determine the curve of the arch ring through the center line of the arch ring at different elevations, plot the corresponding elevation of the arch ring, and then set out the surface of the arch ring at each elevation to generate the basic shape of the hyperbolic arch dam.

Three-dimensional modeling of hyperbolic arch dam firstly needs to determine each elevation arch ring, which needs to be calculated according to relevant parameters of each elevation of the dam body. The steps to obtain the arc curve are as follows: firstly, the imported design parameters are generated into the arc centerline; then the curve of the arch ring is drawn according to the calculation formula of the centerline of the arch ring, and the thickness of the arch ring. The centerline of the arch ring of the hyperbolic arch dam is determined based on the parameter equation. In Dynamo, the center line equation of the arch ring can be achieved by block code. The curve in Dynamo is fitted by sampling points. In this example, the center angle of the arch end on the left and right bank is taken as the center angle. With ϕ_l and ϕ_r as the sampling parameters, the sampling interval is $0.5°$.

It is important to note that the sampling interval of curve fitting in Dynamo is related to the accuracy of the model. Setting a high accuracy will result in a large amount of calculation and modeling time, and directly affect the efficiency of surface fitting and entity generation after dynamo. When the centerline of the arch ring is determined, the thickness of the arch ring at the specific point of the center line of the arch ring is calculated, and the sampling points of the outer and inner curves of the arch ring are determined by combining the vector calculation module. Finally, the curve of the arch ring is obtained by fitting.

When all elevation arch rings (Figure 3b) are drawn, the dam surface (Figure 3c) can be generated by lofting the joint against the closing curve of the arch ring in Dynamo and the three-dimensional entity of the dam body (Figure 3d) can be finally generated.

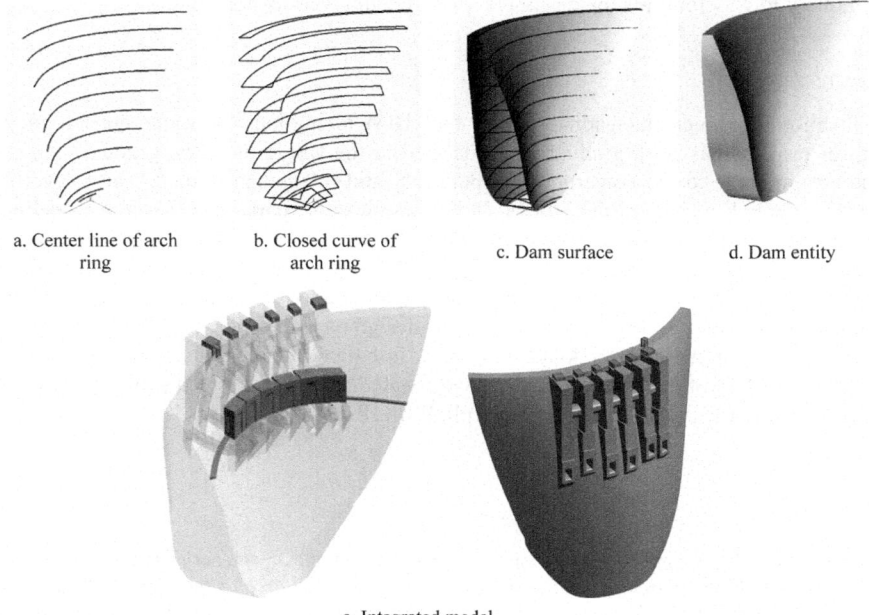

| a. Center line of arch ring | b. Closed curve of arch ring | c. Dam surface | d. Dam entity |

e. Integrated model

Figure 3.   Arch curve modeling.

Through a similar modeling process to the arch dam shape, surface-hole and middle-hole water release structures are created, and the arch dam shape is imported into Revit software by using the Dynamo interface, which is integrated with the powerhouse and computer room built by Revit software. Eventually, an integrated model of the arch dam is shown in Figure 3e.

5 SUMMARY AND EXPECTATION

5.1 *Summary*

Due to the characteristics of hydraulic and hydropower projects, 3D parametric modeling of complex shapes is the main technical difficulty in the design stage. Although some platforms have been better solved (e.g., CATIA), at present, all of the main design platforms have certain technical shortcomings. Therefore, we choose Revit, the main software of the industrial and civil construction industry, as the design platform, but in the case of insufficient Revit surface modeling function, the parametric modeling of surface components is realized by means of the Dynamo plug-in, and the project design and drawing are completed in cooperation with Revit. This approach has the following characteristics:

(1) *Integrity of information*. The architect can complete the entire work by using a single platform (Revit with Dynamo) to avoid information loss caused by data format conversion between software platforms.
(2) *Complementarity*. The Dynamo plug-in has been built into Revit 2017 (and later versions) and becomes the Revit standard plug-in. Dynamo is closely integrated with Revit, which makes up for the limitation of Revit's native function of modeling.
(3) *Flexibility and scalability*. Dynamo can be quickly modeled by presetting parameters and can be parameterized by adjusting design parameters. Dynamo has its own rich functional nodes and supports Python to model complex surfaces.
(4) *Technical threshold*. Similar to Rhino's grasshopper plug-in, Revit's dynamo plug-in has high requirements for designers' spatial thinking and programming technology and has a high technical threshold. Due to the lack of reference materials at present, the improvement of dynamo technology mostly requires the designers' research and work experience.

5.2 *Expectation*

The application of three-dimensional design and BIM technology in water conservancy and hydropower projects has been gradually popularized in the whole industry, gradually extending from the design stage to the construction, operation, and maintenance stages, and covering all stages of the whole life cycle of the project. In the design stage, parametric surface modeling and multidisciplinary collaborative design are the main technical basis for the full implementation of 3D design. At present, several mainstream design platforms have obvious advantages and disadvantages. Taking the Revit platform as an example, the advantages of the parametric design of dynamo software are introduced into the three-dimensional design of hydraulic engineering. Through two typical examples of the volute and double curvature arch dam, the technological process of three-dimensional modeling of the dynamo surface is illustrated. It provides a technical reference for the realization of hydraulic 3D design on a single platform (Revit).

REFERENCES

Cao Y. and Xu H.L. (2018). Application of parametric design in hydraulic machinery design based on dynamo for revi. *Journal of Information Technology in Civil Engineering and Architecture*. 10 (2), 29–34.
Huang Y.F. and Li X.S. (2009). Application of CATIA software in the design of double curvature arch dam. *Yangtze River*. 40 (21), 26–28.

Li Y., Wang Q.M., et al. (2016). Parametric generative design of prefabricated parking structure. *Journal of Information Technology in Civil Engineering and Architecture*. 8 (5), 51–57.

Liu H.Q., Zhang K.C., Li Z.P. (2017). On rapid BIM modeling of measured pipeline data hunan hydro & power, *Municipal Engineering Technology*, 35 (6), 137–138, 193.

Song D. (2020). *Modeling of double-curved arch dam using BIM technology*. Henan Science and Technology. (10), 79–83.

Wang G.X., Cai D.M., et al. (2018). *Revit Rapid Modeling Method of Municipal Box Section Structure*. Municipal Engineering Technology, 36 (1), 68–70, 74.

Wang S. (2015). Computational algorithm design plug-in in Visual programming language—initial research of Dynamo. *Fujian Architecture & Construction*. (11), 105–110.

Xu Z., Bai X.H., Ba J. (2017). *A Study of Componentization Based on Building Information Modeling* (BIM). (4), 19–22.

Zhao J.M. (2020). Design of buttress retaining wall by applying civil 3D. *Journal of Information Technology in Civil Engineering and Architecture*. 12 (6), 177–182.

Frontiers in Civil and Hydraulic Engineering – Mohamed A. Ismail and
Hazem Samih Mohamed (Eds)
© 2023 The Authors, ISBN 978-1-032-38247-0

Research on risk management of a green building project based on computer-aided technology

Sen Li*
Management School, Shenyang Jianzhu University, Shenyang Liaoning, China
Institute of Applied Ecology, Chinese Academy of Sciences, Shenyang Liaoning, China

Hongyuan Liang*
Management School, Shenyang Jianzhu University, Shenyang, China

ABSTRACT: Green buildings have the advantages of green, energy saving, and environmental protection, so they have become the main development direction of the future construction industry. Compared with traditional building forms, green buildings have higher requirements for design, construction, operation, and other links, so they may face more risk factors. The application of computer-aided technology, represented by BIM, can greatly improve the risk management ability of construction enterprises, which has been widely concerned by the community. Under this background, according to the whole life cycle theory, the whole process risk evaluation indicator system of green building projects is constructed, and the weight of each risk factor is determined by the entropy weight method. Taking a project as an example, this paper discusses the performance of three types of construction enterprises in green building risk management: BIM Technology, traditional computer-aided technology, and no computer-aided technology. Finally, on the basis of the empirical analysis, this paper puts forward countermeasures and suggestions to improve the risk management ability of green building projects, so as to accelerate the healthy development of China's green building industry.

1 INTRODUCTION

In recent years, China's construction industry has made great progress, the output value of the construction industry is increasing year by year, and its importance in the national economy is also increasing. Under the dual constraints of resources and environment, people have higher requirements for the construction industry, and the concept of green building emerges as the times require. Green buildings aim to save resources, protect the environment, reduce pollution and provide people with healthy, applicable, and efficient use space in the whole life cycle. Therefore, green buildings have become the future development trend of the construction industry. Compared with traditional building forms, green building projects are more difficult in design, construction, and operation, and face higher risks. In this context, more and more enterprises begin to apply computer-aided technology to improve the working ability of construction enterprises. Therefore, it is necessary to study the influence of computer-aided technology on the risk management ability of green building projects.

2 APPLICATION OF COMPUTER-AIDED TECHNOLOGY IN THE CONSTRUCTION INDUSTRY

In the field of construction, computer-aided technology usually refers to the use of various software technologies in the design and construction of construction projects, so as to improve the work

---

*Corresponding Authors: lisen@sjzu.edu.cn and ls0123@163.com

DOI 10.1201/9781003344209-77

efficiency and the level of construction enterprises and achieve the goal of fine management in the construction process. Traditional computer-aided technology includes CAD, ERP, and so on. These software systems play a great role in the construction industry. In recent years, BIM software technology (building information modeling) has been used by some construction enterprises. In the BIM system, the geometric information, professional properties, and state information of building components are included, and the state information of non-component objects (such as space and motion behavior) is also included. With the help of this three-dimensional model containing construction engineering information, the degree of information integration of construction engineering is greatly improved, so as to reduce the risk probability of construction projects in each link.

3 RISK FACTORS OF GREEN BUILDING PROJECTS

Green buildings have experienced many links. If we can't effectively analyze the risks in different links, we can't build a scientific green building project risk management indicator system. Based on the whole life cycle theory, this paper divides green buildings into four stages, including the decision-making stage, design stage, construction stage, and operation stage. In BIM software, the corresponding modules can also be found. In different stages, combined with the research status at home and abroad, as well as the practice of green project risk management, the green building risk management indicator system is constructed.

The decision-making stage: The decision-making stage is the beginning of the whole process. In the decision-making stage, the budget ability of construction investments and the demand forecasting ability of the green building market are mainly considered. Compared with traditional buildings, the budget of green buildings is about 10% higher within the same building area, so their price will be higher. Enterprises need to conduct full research on the market demand to reduce the risk of enterprises.

The design stage: In the design stage, the function design ability and the green building design standard are considered. As a highly complex architectural form, green buildings have higher requirements for designers. At the same time, the standards of the green building industry are still improving. Design enterprises should adjust the architectural design according to the relevant standards in time to avoid risks.

The construction stage: The construction stage is the most time-consuming and cost-effective stage in the whole process. In the construction stage, we mainly consider the ability of green building construction technology and the construction cost control. Green buildings have higher requirements for construction technology, and need to use some new materials and technologies, which is difficult for some construction enterprises. At the same time, cost control is also a key risk point.

The operation stage: The operation stage mainly refers to the use and maintenance of green buildings, focusing on the ability of green buildings to achieve the expected effect and the ability

Table 1. Risk factors of green building project.

Stage	Risk factor
Decision-making stage	Construction investment budget capacity X1; market demand forecasting capacity X2
Design stage	Function design capability X3; design standard X4
Construction stage	Construction technology capacity X5; Construction cost control capacity X6
Operation stage	Achieve the expected capacity X7; maintenance capacity X8

of green building maintenance. Compared with traditional buildings, green buildings have specific requirements for energy saving and environmental protection. At the same time, the follow-up maintenance ability of enterprises also determines the use effect of green buildings.

To sum up, this paper constructs a green building project risk factor indicator system (see Table 1). In this study, five experts in the field of construction were invited to score the risk factors of green building projects, which were divided into 9 to 1 points according to the importance. The scoring results are shown in Table 2.

Table 2. Descriptive statistical analysis of risk factors of green building projects.

Risk indicator	Maximum	Minimum	Average	Risk indicator	Maximum	Minimum	Average
X1	7	4	5.6	X5	9	7	7.8
X2	9	6	7.8	X6	6	4	5.6
X3	8	6	6.8	X7	7	3	5.2
X4	7	4	5.6	X8	7	5	6

4 WEIGHT ANALYSIS OF RISK FACTORS IN GREEN BUILDING PROJECTS

After constructing the indicator system of the green building project's risk factors, the entropy weight method is further used to determine the weight of each risk factor. The entropy weight method is an effective method to determine the impact of indicators on the comprehensive evaluation. In the weight calculation process, the following steps are mainly included:

First, the original data is standardized. There are many methods for data standardization, and the max-min method is chosen in this paper.

$$y = \frac{x_i - \min_{i=1..n}\{x_i\}}{\max_{i=1..n}\{x_i\} - \min_{i=1..n}\{x_i\}} \tag{1}$$

Secondly, the entropy and effective value are calculated. In this study, the sample scale is 5, so in the following formula, the value of M is 5.

$$e_j = -\frac{1}{\ln m}\sum_{i=1}^{m} y_{ij} \ln_{ij} (i = 1..n; j = 1..m) \tag{2}$$

$$d_j = 1 - e_j \tag{3}$$

Among them, e_j is the entropy of the j (th) indicator, M is the number of samples, y_{ij} is the proportion of the i (th) year under the j (th) indicator, and d_j is the redundancy of the entropy. After obtaining the entropy and redundancy of the indicator, the weight of risk factors of the green building project is obtained.

$$w_j = \frac{d_j}{\sum_{j=1}^{m} d_j} \tag{4}$$

Finally, the comprehensive score of alternatives is calculated. After determining the weight of the indicator, the final comprehensive score can be obtained by combining the specific data of the indicator. The calculation formula is as follows:

$$s_i = \sum_{j=1}^{m} w_j \cdot p_{ij} \tag{5}$$

After calculation, we determine the weight of green building risk factors, as shown in Table 3.

Table 3. Weights of green building risk factors.

Risk factors	X1	X2	X3	X4
Weight	0.1059	0.0883	0.2090	0.1059
Risk factors	X5	X6	X7	X8
Weight	0.2090	0.0819	0.0981	0.1018

From Table 3, we can see that the weights of risk factors are different, among which the weights of functional design capacity and building construction capacity are the highest, both of which are 0.2090, indicating that the two factors have the highest impact on the risk management of green buildings. At the same time, the weight of construction cost control ability is 0.0819, ranking last among all factors, which has relatively a small impact on the risk of green buildings.

5 CASE ANALYSIS OF THE GREEN BUILDING PROJECT'S RISKS

After the risk factors of green building projects are determined, a green building project is taken as an example. This paper analyzes the performance of three types of construction enterprises, which are BIM Technology, traditional computer-aided technology, and non-computer-aided technology, in the risk management of green buildings. Project H is a comprehensive playground located in the south of Shenyang City, Liaoning Province. Its green building is rated as star 2.

For the green building project H, we choose enterprises A, B, and C as the candidate construction enterprises, which respectively represent BIM Technology, traditional computer-aided technology, and non-computer-aided technology. Similarly, five experts and scholars mentioned above are selected to score the ability of three enterprises in each risk factor. According to the enterprise's ability level from high to low, the score is divided into 1–9 points, that is, the highest ability level is 9 points, and the lowest ability level is 1 point. In this way, when calculating the comprehensive score, the enterprise with the highest risk score is the enterprise with the strongest risk management ability for the green building project.

After scoring, we obtained the average scores of the three enterprises on eight risk factors, as shown in the table below:

Table 4. Scores of three enterprises in each indicator.

Enterprise	X1	X2	X3	X4	X5	X6	X7	X8
A	8.5	8.2	8.3	8.1	8.3	8.6	7.9	7.8
B	7.8	8.1	8.3	7.9	7.6	7.8	7.6	7.5
C	7.6	7.9	7.9	7.7	7.8	7.9	7.4	7.2

It can be seen from Table 4 that among all the risk factors, enterprise A adopting BIM Technology has the best performance, showing strong risk management ability of the green building project. Enterprise B, which adopts traditional computer-aided technology, ranks second among multiple indicators. Enterprise C, which does not use any computer-aided technology, ranks last in many indicators. Further considering the weight of each factor, the comprehensive scores of the three enterprises are determined.

In Table 5, we get the comprehensive score of the three enterprises regarding the green building project's risk management. The comprehensive score of enterprise A is 8.2256, ranking first. The comprehensive score of enterprise B is 7.8496, ranking second. The comprehensive score

Table 5. Comprehensive scores of 3 enterprises.

Enterprise	X1	X2	X3	X4	X5	X6	X7	X8	CS
A	0.9003	0.7239	1.7350	0.8579	1.7350	0.7046	0.7752	0.7937	8.2256
B	0.8262	0.7151	1.7350	0.8367	1.5887	0.6390	0.7457	0.7632	7.8496
C	0.8050	0.6974	1.6514	0.8156	1.6305	0.6472	0.7261	0.7327	7.7058

of enterprise C is 7.7058, ranking third. It can be seen that computer-aided technology plays an important role in the risk management of green building projects, which can greatly improve the risk management ability of construction enterprises.

6 SUGGESTIONS

Therefore, in order to improve the risk management ability of green construction enterprises in China and accelerate the development of the green construction industry in China, the following countermeasures and suggestions are put forward:

First of all, green building enterprises should apply BIM and other advanced computer-aided design software to improve their risk management ability in the field of green buildings. From the perspective of input-output, although enterprises may increase certain costs in the early stage, they have outstanding advantages in cost control and risk management of green buildings, with considerable comprehensive incomes. Secondly, enterprises should increase their investment in innovation, accelerate the integration of CAD and green buildings, and help enterprises seize the development opportunities of the green building industry and quickly occupy the market. Finally, the state should introduce relevant policies to encourage and guide green building enterprises to apply BIM and other advanced computer-aided design systems, which can be used as a candidate enterprise standard in the process of project bidding, so as to promote more enterprises to apply this technology. At the same time, the use of computer-aided technology can also improve the risk response capacity of construction projects, which has a positive impact on the quality and delivery time of construction projects and achieve a win-win situation for both sides of green building projects.

7 CONCLUSIONS

This paper compares the performance of different enterprises in the risk management of green building projects from the perspective of computer-aided design. The risk factor model of green building projects is constructed, and the weight of each risk factor is determined by the entropy method. It can be seen from the analysis results that the weight of functional design capacity and construction capacity is the highest, which has the greatest impact on the risk management of green buildings. At the same time, the weight of the construction cost control ability ranks the last among all factors, which has relatively a small impact on the risk of green buildings. On this basis, taking green building project H as an example, three enterprises are selected for a comparative study. The research finds that BIM Technology is more capable of enterprise risk management, which has obvious advantages compared with traditional technologies and machine-aided technologies. Finally, the paper puts forward the countermeasures and suggestions regarding the risk management ability for green buildings.

ACKNOWLEDGMENTS

This work was financially supported by 1. Liaoning Federation of Social Sciences *Study on the coupling of coordinated development of real estate and the urban economy in Liaoning Province*

(NO.2022lslwtkt-049) 2. Department of Education of Liaoning Province *Research on coupling development of strategic emerging industries and traditional industries in Liaoning Province* (NO.lnqn 202031).

REFERENCES

Ahlam B.Q.A., Rahim Z. Abdul. 2021. A review of risks for BIM adoption in malaysia construction industries: multi case study. *IOP Conference Series: Materials Science and Engineering*, 1051(1).

Albert Yau, Samuel K.M. Ho. 2014. Fire risk analysis and optimization of fire prevention management for green building design and high-rise buildings: Hong Kong experience. *Nang Yan Business Journal*, 3(1).

Hwang B.G., Zhao X.B., See Y.L. & Zhong Y. 2015. Addressing risks in green retrofit projects: The case of Singapore. *Project Management Journal*, 46(3).

Li Z.D. 2021. Research on risk management of green building development. *IOP Conference Series: Earth and Environmental Science*, 647(1).

Li M., Li G.B., Huang Y.Q., Deng L.Y. 2017. Research on investment risk management of chinese prefabricated construction projects based on a system dynamics model. *Buildings*, 7(2).

Nida Javed, Muhammad Jamaluddin Thaheem, Beenish Bakhtawar, Abdur Rehman Nasir, Khurram Iqbal Ahmad Khan, Hamza Farooq Gabriel, 2019. Managing risk in green building projects: Toward a dedicated framework. *Smart and Sustainable Built Environment*, 9(2).

Zhao X.B., Hwang B.G., Gao Y. 2016. A fuzzy synthetic evaluation approach for risk assessment: A case of Singapore's green projects. *Journal of Cleaner Production*, 115.

Yang Y.R. 2021. Analysis on risk control of civil engineering cost based on BIM technology. *IOP Conference Series: Earth and Environmental Science*, 643(1).

Yaser Farajiasl, 2015. An analytic network process model to analyze and identify the risks associated with green building projects[J]. *Researcher*, 7(2).

Frontiers in Civil and Hydraulic Engineering – Mohamed A. Ismail and
Hazem Samih Mohamed (Eds)
© 2023 The Authors, ISBN 978-1-032-38247-0

Analysis of sustainable water resource utilization in Qinghai Province based on ecological footprint theory

Ru Xu, Haifeng Zhang* & Luqing Yan
School of Geography Science, Qinghai Normal University, China

ABSTRACT: The sustainable use of water resources is the basis of regional sustainable development, and the ecological footprint model provides a new idea for the quantitative evaluation of the sustainable use of regional water resources. Qinghai Province, located on the Qinghai-Tibet Plateau, is an important part of the national ecological barrier and an important water source for the Chinese Water Tower and the Asian Water Tower and has a great responsibility for national ecological safety and sustainable development. Using the data on water resources, economy, and population of Qinghai Province from 2010 to 2020, the ecological footprint of water resources and the ecological carrying capacity model of water resources were constructed to analyze and study the sustainable utilization status of water resources in Qinghai Province and each city and prefecture. The results show that: (1) The ecological footprint of water resources per capita in Qinghai Province from 2010 to 2020 showed a decreasing trend, the ecological carrying capacity of water resources per capita showed an increasing trend, and water resources were in ecological surplus. (2) The efficiency of water resources utilization in Qinghai Province has been improving year by year, the value of the ecological pressure index for water resources is less than 0.50, the value of the sustainability index is more than 0.70, and water resources are in a sustainable state, and the degree of sustainable development is high. (3) The spatial variability in the ecological surplus of water resources and the sustainable development level in Qinghai province from 2010 to 2020 was significant, manifesting that it is closely related to natural conditions such as total regional water resources, and human conditions such as the economic development level and population scale.

1 INTRODUCTION

Water is the source of life, the key to production, the foundation of ecology, and the material basis for sustainable development of society, and plays an important role in people's life and stable social development (Du 2015; Liu 2003). As China's urbanization process continues to accelerate, the population scale continues to expand, the national economy develops rapidly, and the demand for water resources is also increasing, how to use water resources scientifically and reasonably is an important factor in achieving the sustainable use of water resources and the stable social development (Wang 2015). From the perspective of the ecological footprint, existing research believes that the spatial and temporal evolution of the ecological footprint of water resources is an important element to evaluate the sustainable development of water resources. (Wackernagel 1999; Zhu 2020). Therefore, the analysis of the dynamic evolution of the ecological footprint of regional water resources and the current situation of ecological profits and losses can better determine whether the water resources in the region are in a sustainable development situation.

The ecological footprint of water resources is an ecological statistical method proposed by Hoekstra et al. to describe the ecosystem services of watersheds based on the ecological footprint theory,

---

*Corresponding Author: haifzhang@126.com

DOI 10.1201/9781003344209-78

i.e., the productive land area required to sustain the population and economic development of a region or to carry the waste discharged by humans (Hoekstra 2005; Ma 2013; Zhang 2021). Since then, many international scholars have studied the ecological footprint of water resources. Venetoulis (2001), Luck (2001), Jenerette (2006), and others have introduced the ecological footprint of water resources into different areas such as universities, cities, and urban corridors in the United States to study the development and utilization of water resources and their sustainability. In recent years, scholars in China have systematically studied the dynamic evolution of the ecological footprint of regional water resources from different perspectives based on the principles and methods of water resources' ecological footprints. From the research scales, Lu Y L (2021), Huang J (2019), Zhao M S (2020), Hou J C (2020), Xiong N N (2019), Yue C (2021), Xu D Z (2019), and Zhang M Y (2019) have respectively studied the spatial and temporal evolution characteristics of water resources' ecological footprints and ecological carrying capacity of nine provinces (regions) in the Yellow River Basin, Shandong Province, northeast Sichuan urban agglomeration, small and medium-sized cities, Chengdu City, Beijing City, Shanxia Reservoir Area, Yanbian Prefecture, etc.; from the research contents, Lan J Q (2020), Jing P R (2021), Wang N (2020), Liu Y B (2020), and Jiao S X (2020) have respectively studied the sustainable use of water resources, the coordinated the development of water resources use and economy, and the distribution characteristics and driving factors of water resources' ecological footprints based on the ecological footprint theory of water resources. From the perspective of research methods and based on the improved water ecological footprint model, Fan Y H (2021), Hao S (2021), and An H (2021) established the spatial-temporal data analysis framework and BP neural network, enriching the research methods of water resources' ecological footprint.

At present, scholars in China have conducted extensive and in-depth studies on the assessment of ecological footprints of water resources, but few studies have been conducted in Qinghai-Tibet region. Qinghai Province is located in the northeastern part of Qinghai-Tibet Plateau, which is an important ecological function area and water-supporting area in China, but there is a lack of relevant studies; at the same time, due to the great differences in natural and human geographic factors within Qinghai Province, the supply and demand of water resources are different, so it is necessary to evaluate the ecological footprint of water resources in different regions of Qinghai Province. Based on this, this paper takes Qinghai Province and each city and prefecture as the research area, uses the ecological footprint theory of water resources as the theoretical basis to comprehensively analyze the spatial and temporal evolution characteristics of ecological footprints of water resources in Qinghai Province and each city and prefecture from 2010 to 2020 and reveals the influencing factors; at the same time, it introduces the evaluation index for the sustainable use of water resources to comprehensively evaluate the sustainable development level of water resources in each city and prefecture of Qinghai Province, so as to provide a theoretical basis for water resources planning and management in Qinghai Province.

2 MATERIALS AND METHODS

2.1 *Overview of the study area*

Qinghai Province is located in the northeast of the Tibetan Plateau, the source of the Yangtze, Yellow, Lancang, and Black rivers, an important water source in China, and the largest and most concentrated area in Asia to breed large rivers, known as the "the source of three rivers" and "Chinese water tower". Its geographical location is between $89°35' - 103°04'$E and $31°9' - 39°19'$N. It is bordered by Gansu in the north and east, Xinjiang in the northwest, Tibet in the south and southwest, and Sichuan in the southeast, and is located in the Qinghai-Tibet region of the four geographic divisions. Qinghai Province covers a total area of 722,300 km^2, with 2 prefecture-level cities and 6 autonomous prefectures under its jurisdiction. The terrain is generally high in the west and low in the east, with a trapezoidal descent. The average altitude of the province is above 3000 m, and the climate is continental on the plateau. By the end of 2020, the province's average precipitation was 367.1 mm, the total volume of water resources was 101.191 billion m^3, the resident population was 5,927,900, and the total GDP was 300.592 billion yuan.

2.2 Research methods

2.2.1 Ecological footprints of water resources

The ecological footprint of water resources is the number of water resources consumed and converted into the corresponding account of water resources' land use area. The ecological footprint of water resources is divided into four types: agricultural water use, industrial water use, domestic water use, and ecological environment water use. Its calculation formula is as follows.

$$EF_W = N \times ef_w = Y_w \times (W/P_W) \tag{1}$$

$$EF_A = Y_w \times (W_a/P_W) \tag{2}$$

$$EF_B = Y_w \times (W_b/P_W) \tag{3}$$

$$EF_C = Y_w \times (W_c/P_W) \tag{4}$$

$$EF_D = Y_w \times (W_d/P_W) \tag{5}$$

In the formula: EF_W is the total ecological footprint of water resources (hm^2); N is the year-end resident population; ef_w is the per capita ecological footprint of water resources (hm^2/person); Y_w is the global equilibrium factor of water resources, using the WWF 2002 accounting results (Fan 2005), and its value is 5.19; W is the total water resources consumed (m^3); P_W is the global average production capacity of water resources (m^3/hm^2), and it is characterized by the global multi-year average water production modulus, which is 3140 m^3/hm^2; EF_A is the ecological footprint of water used for agriculture (hm^2); W_a is the amount of water used for agriculture (m^3); EF_B is the ecological footprint of water used for the industry (hm^2); W_b is the amount of water used for the industry (m^3); EF_C is the ecological footprint of water used for domestic purposes (hm^2); W_c is the amount of water used for domestic purposes; EF_D is the ecological footprint of water used for ecological purposes (hm^2); W_d is the amount of water used for ecological purposes (m^3).

2.2.2 Ecological carrying capacity of water resources

The ecological carrying capacity of water resources refers to the support capacity of regional water resources for sustainable social development. According to the existing research (Huang 2008), 60% of the water resources developed and utilized in a region need to be deducted to maintain the ecological cycle, so in the calculation, the coefficient of 0.4 should be multiplied, and its calculation formula is as follows.

$$EC_w = N \times ec_w = 0.4 \times \varphi \times Y_w \times Q/P_W \tag{6}$$

In the formula: EC_w is the ecological carrying capacity of water resources (hm^2); N is the year-end resident population; ec_w is the per capita ecological carrying capacity of water resources (hm^2/person); φ is the yield factor of regional water resources; Y_w is the global equilibrium factor of water resources; Q is the total amount of water resources (m^3); P_W is the global average production capacity of water resources (m^3/hm^2); where the yield factor of regional water resources (φ) is calculated as follows:

$$\varphi = P/P_W \tag{7}$$

$$P = V/S \tag{8}$$

In the formula: φ is the yield factor of regional water resources; P is the average production capacity of regional water resources (m^3/hm^2); P_W is the average production capacity of global water resources (m^3/hm^2); V is the total amount of water resources (m^3); S is the calculated area (hm^2); According to the calculation of the research data, the water resource yield factors of Qinghai Province and cities and prefectures from 2010 to 2020 are shown in Table 1.

Table 1. Water Resource Yield Factors in Qinghai Province and Various Cities and Prefectures.

Region	Xining City	Haidong City	Haibei Prefecture	Hainan Prefecture	Huangnan Prefecture
Yield factor	0.63	0.42	0.60	0.29	0.61

Region	Guoluo Prefecture	Yushu Prefecture	Haixi Prefecture	Qinghai Province	
Yield factor	0.64	0.47	0.16	0.35	

2.2.3 *Ecological surplus/deficit of water resources*
The ecological surplus/deficit of water resources is the difference between the ecological carrying capacity of water resources and their ecological footprint, which is used to measure the state of water resources' ecological safety. Its calculation formula is as follows.

$$\text{Ecological surplus/deficit of water resources} = EC_w - EF_W \tag{9}$$

In the formula: When the difference value is >0, water resources present ecological surplus and are in an ecological safety state. If the difference value is 0, water resources are in equilibrium. When the difference value is <0, water resources present ecological deficits and are in an unsafe state.

2.2.4 *Water resources' use efficiency index*
The water resource use efficiency index is the ratio of water resources' ecological footprint to GDP. The smaller the ratio, the higher the water use efficiency, and vice versa. Its calculation formula is as follows.

$$\text{Ecological footprint of } 10,000\text{GDP} = EF_W/GDP \tag{10}$$

2.2.5 *Water resources' ecological stress index*
The water resources' ecological stress index is the ratio of the ecological footprint of water resources to their ecological carrying capacity. The ecological surplus/deficit is used to determine whether the regional water resources are in a safe state, and the water resources' ecological stress index is used to indicate the degree of ecological safety of water resources (Ren 2005). Its calculation formula is as follows.

$$EPI_W = EF_W/EC_w \tag{11}$$

In the formula: EPI_W is the ecological stress index of water resources. EPI_W classifies the state of water safety into four levels: safe ($EPI_W < 0.5$), relatively safe ($0.5 \leq EPI_W < 0.8$), critical ($0.8 \leq EPI_W < 1.0$), and unsafe ($EPI_W \geq 1.0$).

2.2.6 *Water resources' sustainability index*
The water resources' sustainability index reflects the sustainable development level and utilization efficiency of water resources (Fang 2014). Its calculation formula is as follows:

$$ESI_W = EC_w/(EC_w + EF_W) \tag{12}$$

In the formula: ESI_W is water resources; sustainability index; $0 < ESI_W < 1$, the higher the ESI_W value is, the higher the sustainable development level and utilization efficiency. The value of 0.5 is the demarcation point between sustainable and unsustainable utilization of water resources. ESI_W divides the sustainable development level and utilization efficiency of water resources into six grades: highly sustainable ($ESI_W \geq 0.80$), moderately sustainable ($0.65 \leq ESI_W < 0.80$), weakly sustainable ($0.50 \leq ESI \leq 0.64$), weakly unsustainable ($0.35 \leq ESI_W < 0.49$), moderately unsustainable ($0.20 \leq ESI_W < 0.34$), and highly unsustainable ($ESI_W < 0.20$).

2.3 Data sources

The data related to this study include the resident population and GDP data of Qinghai Province and each city and prefecture from the Statistical Yearbook of Qinghai Province (2010–2020), and the data related to water resources from the Water Resources Bulletin of Qinghai Province (2010–2020).

3 RESULTS AND ANALYSIS

3.1 Ecological footprint analysis of water resources

3.1.1 Time scale

From 2010 to 2020, the ecological footprint of water resources per capita in Qinghai Province showed a general decreasing trend (Figure 1). The range of its change during the period was 1.063–0.677 hm²/person, with the highest value in 2011 being 1.065 hm²/person and the lowest value in 2020 being 0.677 hm²/person, with a difference of 0.388 hm²/person. In terms of the ecological footprint of water resources in each secondary account, agriculture used the most water, accounting for 75% of the ecological footprint of water resources per capita on average; the ecological footprint of industrial water consumption per capita accounted for 14%; the ecological footprint of domestic water consumption per capita accounted for 9%, and the ecological footprint of ecological environment water consumption per capita accounted for the smallest proportion of only 2%. As shown in Figure 2, the ecological footprint of agricultural water consumption per capita in Qinghai Province decreased significantly, indicating that water-saving agriculture had developed better and

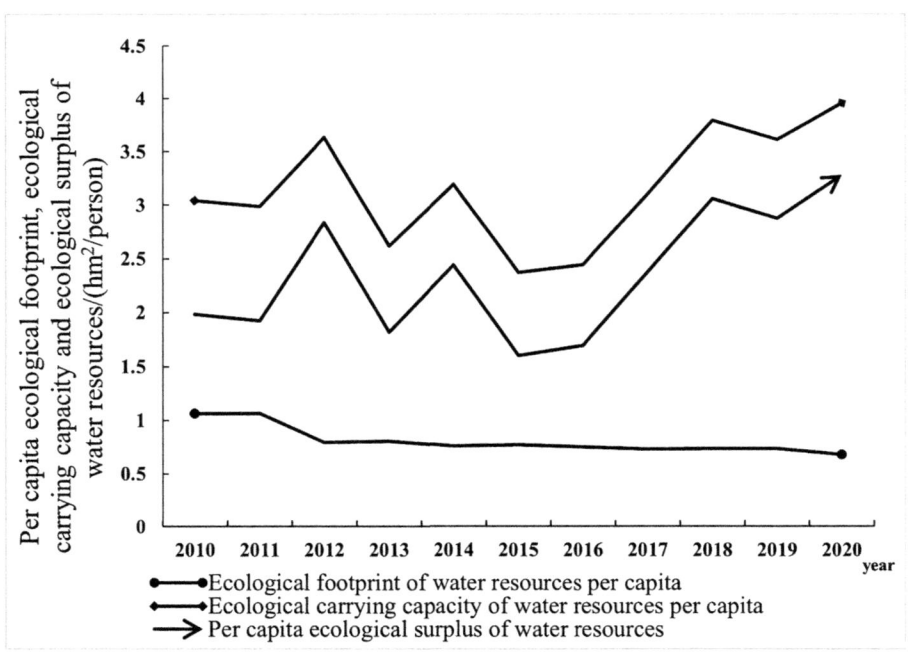

Figure 1. Ecological footprint, ecological carrying capacity and ecological surplus of water resources per capita in Qinghai Province from 2010 to 2020.

594

the agricultural water efficiency had improved significantly during the period; the ecological footprint of industrial water consumption per capita was larger in 2010 and 2011, and had fluctuated steadily since 2012 with a decreasing trend in general, indicating that Qinghai Province has paid attention to the protection of water ecological environment, reduced industrial water consumption, and vigorously developed industrial water conservation in recent years. The ecological footprint of domestic water consumption per capita shows an overall increasing trend, indicating that with the increase in the population and living standard, the amount of living water consumption is also increasing; the ecological footprint of ecological environmental water consumption per capita is increasing year by year. It can be seen that the relationship between water resources, ecological environment, and stable social and economic development in Qinghai Province is improved.

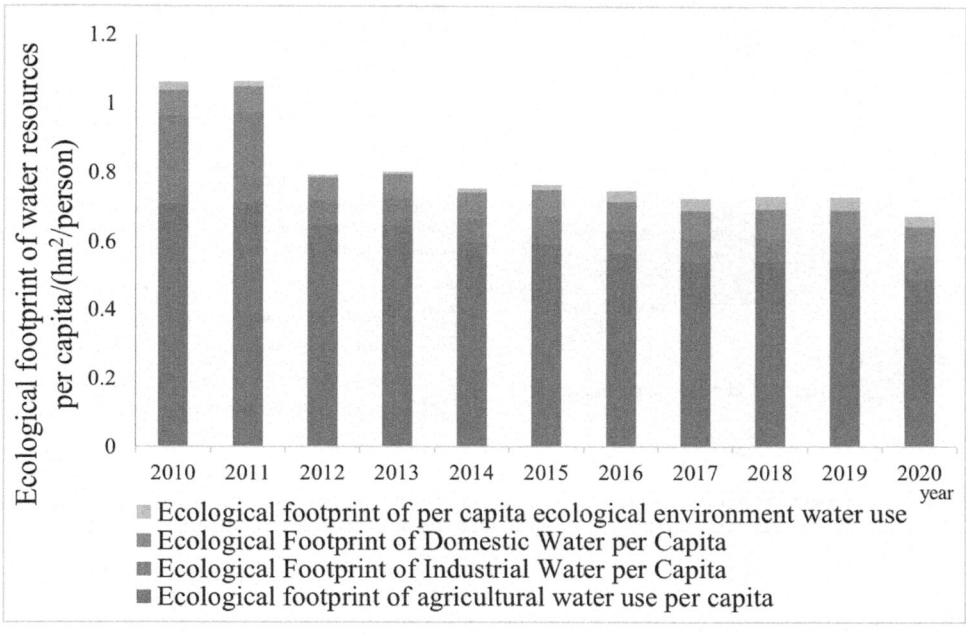

Figure 2. Ecological footprint of water resources per capita in Qinghai Province from 2010 to 2020.

3.1.2 *Spatial scale*

Taking 2010, 2015, and 2020 as the time point, the ecological footprint of water resources per capita in the eight cities (prefectures) of Qinghai Province generally showed the pattern characteristics of high in the north and low in the south (Figure 3). The natural environment and human conditions of the eight cities (prefectures) in Qinghai Province are different, which makes the water resources utilization of each city and prefecture different. Haixi Prefecture is a key development zone in the Qaidam Basin, with a high demand for water resources; Haibei Prefecture and Hainan Prefecture enjoy the better development of agriculture and animal husbandry, and more water is used for agriculture, so the ecological footprint of water resources per capita is higher; Yushu Prefecture and Guoluo Prefecture have a small population and slow economic development, so the demand for water resources for social development is relatively small, and the ecological footprint of water resources per capita is smaller; Xining City is a key economic development zone in Qinghai Province, and the proportion of tertiary industry was 65.3% in 2020. Since the water consumption of tertiary industry is low, its per capita ecological footprint of water resources is lower than that of Haibei and Hainan prefectures. From the spatial and temporal evolution of the ecological footprint of water resources per capita in each city and prefecture (Figure 3), the ecological footprint of water resources per capita in Hainan Prefecture and Xining city decreased significantly from 2010 to 2020; for the water consumption of each secondary account, the ecological footprint of ecological water consumption per capita in each city and prefecture showed a stable increasing trend from 2010 to 2020.

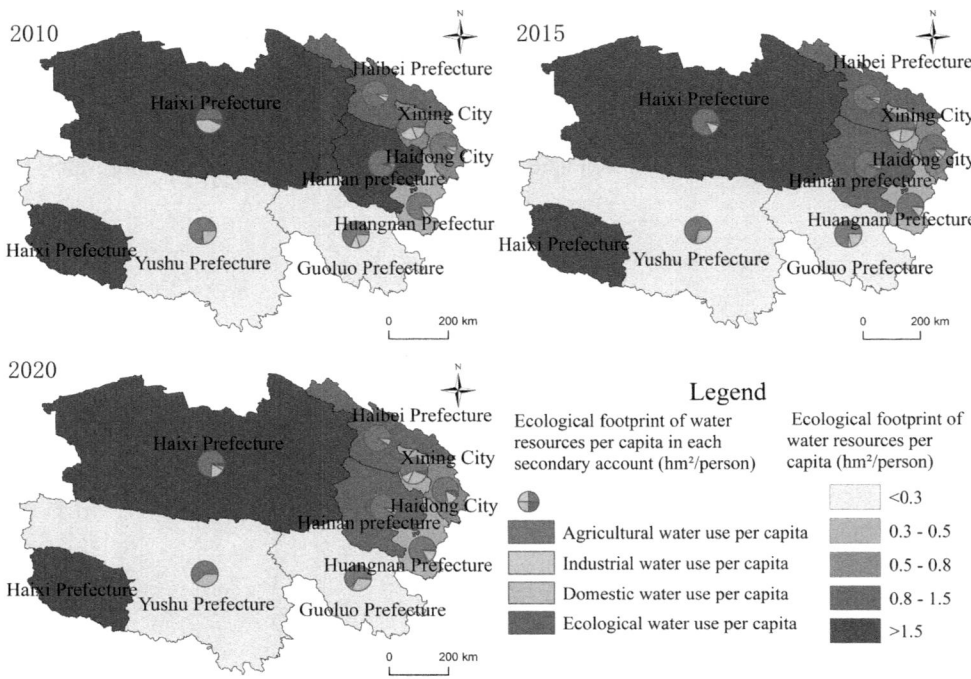

Figure 3. Ecological footprint of per capita water resources in qinghai province from 2010 to 2020 and spatial distribution of per capita water resources' ecological footprint of various water use types.

3.2 *Ecological carrying capacity analysis of water resources*

The per capita ecological carrying capacity of water resources in Qinghai Province from 2010 to 2020 ranged from 3.043-3.950 hm^2/person, with large fluctuations and an overall upward trend (Figure 1). The highest value is 3.950 hm^2/person in 2020 and the lowest value is 2.364 hm^2/person in 2015, with a difference of 1.586 hm^2/person between the two. As shown in Figure 4, there is a significant positive correlation between the ecological carrying capacity of water resources and the precipitation in Qinghai Province from 2010 to 2020 (positive correlation coefficient = 0.869), i.e., the ecological carrying capacity of water resources is also at a higher level in years with more precipitation. From different regions, the per capita water resources' ecological carrying capacity of each city and prefecture in Qinghai province varies significantly (Figure 5). In 2020, the precipitation in Guoluo and Xining is larger, being 663 mm, and 626.1 mm, respectively; the precipitation in Yushu and Haixi is smaller, being 456.1 mm, and 184.4 mm, respectively. As shown in figure, the per capita water resources' ecological carrying capacity of Guoluo and Yushu is larger, while that of Xining and Haidong City is smaller, indicating that the significant variability of the per capita ecological carrying capacity of water resources in each city and prefecture is not only ascribed to the precipitation. Xining City, Haidong City, and Haixi Prefecture are economic development zones with population concentration in Qinghai Province, and their social development puts certain pressure on the water environment accordingly; while Guoluo Prefecture and Yushu Prefecture have a small population size and relatively backward economic technology, so the ecological carrying capacity of water resources is much larger than that of Xining City and Haidong City.

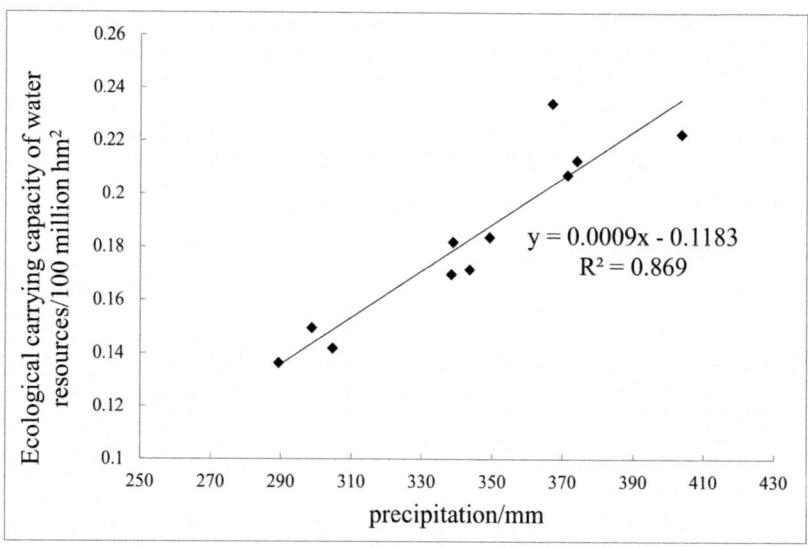

Figure 4. Correlation between precipitation and water resources' ecological carrying capacity in Qinghai Province.

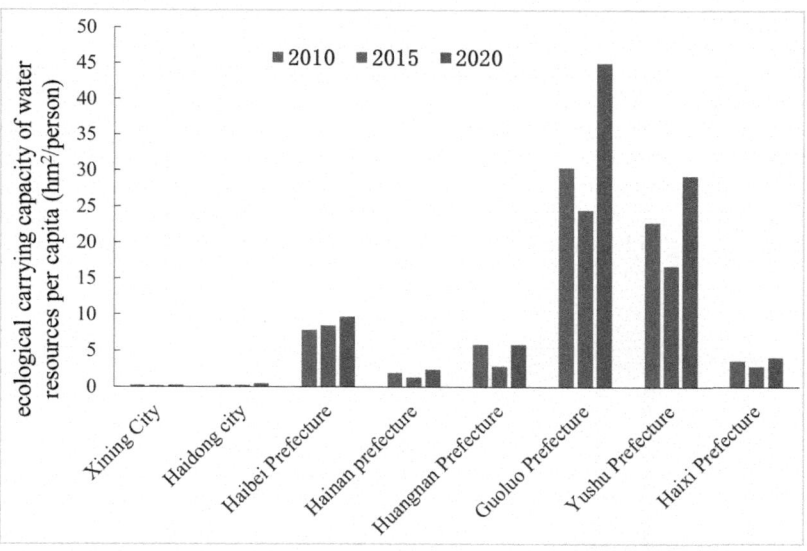

Figure 5. Ecological carrying capacity of water resources per capita in cities and prefectures in Qinghai Province from 2010 to 2020.

3.3 Ecological surplus/deficit analysis of water resources

From 2010 to 2020, water resources in Qinghai Province are all in a safe state with ecological surpluses, while its interannual variation is consistent with the interannual variation of ecological carrying capacity of water resources (Figure 1). The spatial variability of the per capita water resources' ecological surplus in Qinghai province is significant, and the overall trend is increasing (Figure 6). The areas with larger per capita water resources' ecological surplus during the period are

Yushu and Guoluo; the areas with smaller per capita water resources' ecological surplus are Xining, Haidong, and Haixi. Yushu Prefecture and Guoluo Prefecture have lower ecological footprints and higher ecological carrying capacity, and larger ecological surpluses, while Xining City, Haidong City, and Haixi Prefecture have larger ecological footprints and smaller ecological carrying capacity, and relatively smaller ecological surpluses. Among them, the ecological surplus of water resources in Xining City and Haidong City is consistently less than 0 during the period, and the water resources are in an insecure state.

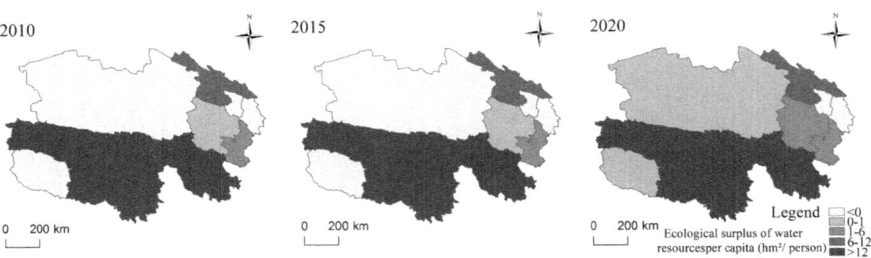

Figure 6. Spatial distribution of ecological surplus of water resources per capita in Qinghai Province from 2010 to 2020.

3.4 Analysis of water use efficiency index and ecological pressure index

As shown in Figure 7: From 2010 to 2020, the ecological footprint of water resources per 10,000 yuan of GDP in Qinghai Province showed a significant decreasing trend, the ecological stress index of water resources was in a safe state (less than 0.5), and the water resources were enjoying sustainable development. The ecological footprint of water resources per 10,000 yuan of GDP decreased from 0.443 hm^2/10,000 yuan in 2010 to 0.134 hm^2/10,000 yuan in 2020, with a decrease rate of 70%. This indicates that Qinghai Province has not depleted its water resources excessively in the process of economic development, fully practicing the ecological concept of "green water and green mountains are golden mountains" proposed by General Secretary Jinping Xi. For the ecological pressure index of Qinghai Province, it fluctuated from 2010 to 2014, and showed a decreasing trend

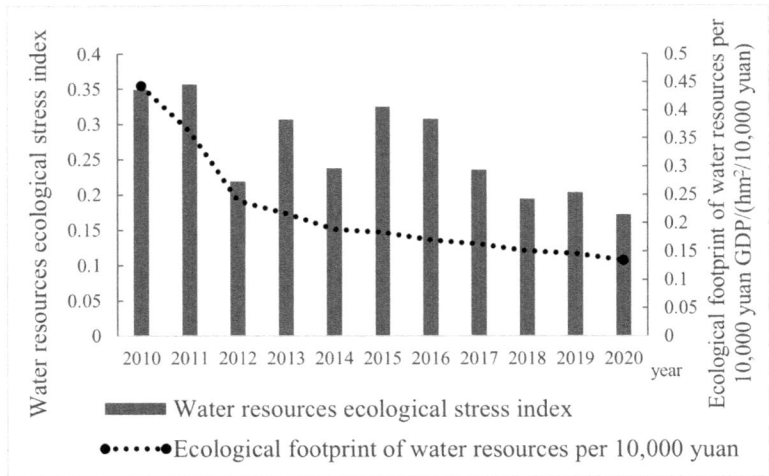

Figure 7. Ecological footprint and ecological pressure index of water resources per 10,000 yuan of gdp in Qinghai Province from 2010 to 2020.

in general from 2015 to 2020. Combined with the variation pattern of per capita water resources; ecological surplus in Figure 1, the highest per capita water resources' ecological surplus in 2020 corresponds to the lowest ecological pressure index; the lowest per capita water resources' ecological surplus in 2015 corresponds to a higher ecological pressure index. It shows that the variation trend of water resources' ecological pressure index and the ecological surplus are correlated.

As shown in Figure 8, the ecological footprint of water resources per 10,000 yuan of GDP in each city and prefecture from 2010 to 2020 showed a decreasing trend in general, with Hainan, Haibei, and Haixi Prefectures showing a larger decrease. This shows that Qinghai Province has continuously attached importance to the utilization efficiency of water resources in the process of socio-economic development, which makes full use of water resources on the one hand and reduces the waste of water resources on the other hand. Moreover, this is conducive to the sustainable development of water resources and promotes the benign development of ecological environment. As shown in Figure 9, the regional differences in water resources' ecological pressure index in each city and prefecture are significant. During the period, Haibei, Huangnan, Guoluo, and Yushu are in the ecological safety of water resources; Hainan is in the critical-relatively safe-safe level; Haixi is in the unsafe-critical level; while water resources in Xining City and Haidong City are always in an unsafe state.

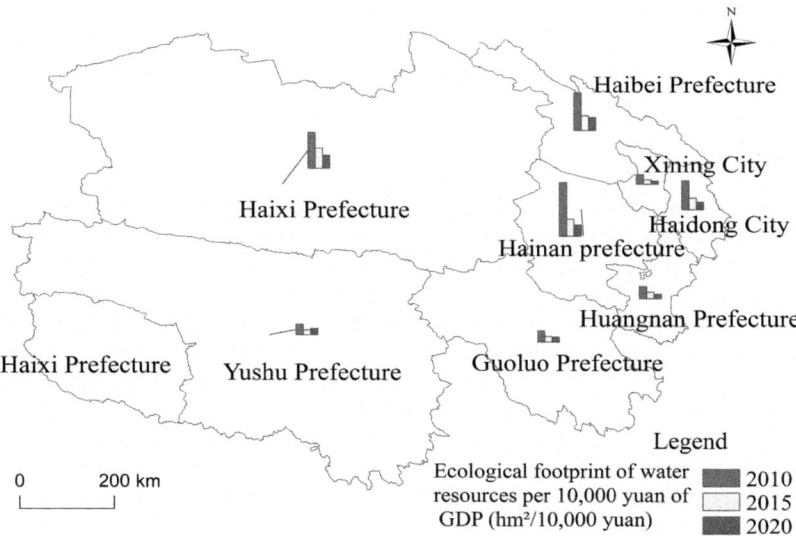

Figure 8. Ecological footprint of water resources per 10,000 yuan of GDP in Qinghai Province.

3.5 *Water resources' sustainability index analysis*

As calculated according to Equation (12), the sustainable development index of Qinghai Province from 2010 to 2020 is greater than 0.70, and the degree of sustainable development and utilization of water resources is high. From different regions, the degree of sustainable development of water resources in each city and prefecture is significantly different. Among them, Haibei, Huangnan, Guoluo and Yushu Prefectures all have sustainability indexes greater than 0.80, and their water resources are at a "strong" level of sustainability; Hainan and Haixi Prefectures have a relatively low level of sustainability, but their sustainability indexes are greater than the threshold of 0.50; while Xining and Haidong cities have sustainability indexes below 0.50, and their water resources are at an unsustainable level.

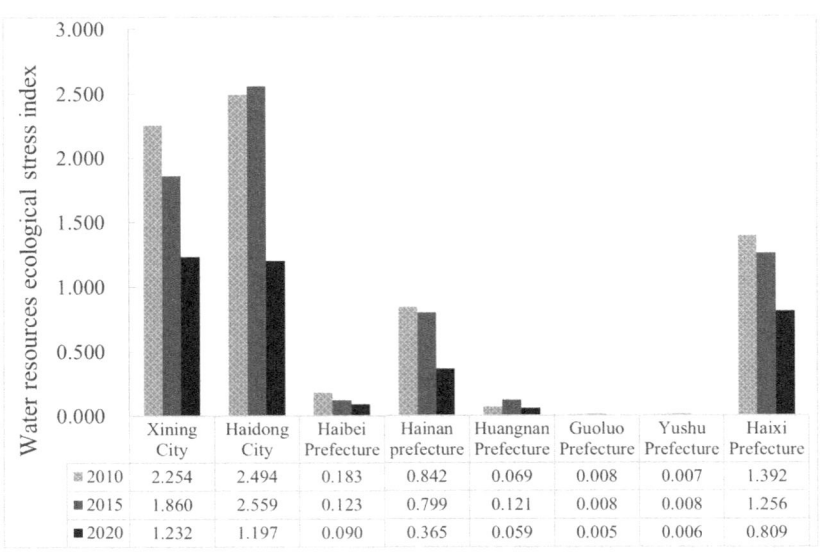

	Xining City	Haidong City	Haibei Prefecture	Hainan prefecture	Huangnan Prefecture	Guoluo Prefecture	Yushu Prefecture	Haixi Prefecture
▨ 2010	2.254	2.494	0.183	0.842	0.069	0.008	0.007	1.392
▪ 2015	1.860	2.559	0.123	0.799	0.121	0.008	0.008	1.256
▪ 2020	1.232	1.197	0.090	0.365	0.059	0.005	0.006	0.809

Figure 9.   Ecological pressure index of water resources in various prefectures in Qinghai Province.

4   DISCUSSION

Qinghai Province, located on the Qinghai-Tibet Plateau, serves as an important part of the "Chinese Water Tower", and even the "Asian Water Tower", and the sustainable use of its water resources is related to the sustainable development of the middle and lower reaches of the basin, but there are few studies in this region. The ecological footprint theory of water resources evaluates the ecological profit and loss of regional water resources from the perspective of ecological occupation; at the same time, it can be combined with the ecological footprint model to comprehensively evaluate the sustainable development of regional water resources. This study provides a reference basis for judging the sustainable development status and ecological health level of Qinghai Province from the ecological footprint of water resources. The research perspective and methodology are different from those of scholars Shang Q K (2020) and Wang X Q (2005), but the conclusions are basically consistent with their research conclusions and have certain reliability. However, this paper has the following shortcomings.

(1) The ecological footprint model, established by Willam Rees and Wackernagel, can better evaluate the degree of human utilization of natural resources and the ability of nature to provide life services for human society, providing a new perspective and method for studying regional sustainable development. However, the model only evaluates the fishery function of surface waters and ignores the other functions of groundwater and surface water as water resources. The ecological footprint model of water resources and ecological carrying capacity makes up for the deficiency of the ecological footprint model in water resources to a certain extent and provides a new idea for the quantitative evaluation of the sustainable utilization of water resources. Three basic parameters are involved in its calculation: average production capacity of water resources, yield factor of water resources account, and equilibrium factor of water resources account. Due to the huge difference in water resources endowment in different regions and considering the spatial scale of the study, the three basic parameters mentioned above are determined as global hectare method, national hectare method, and regional hectare method. In order to be consistent with Wackernagel's ecological footprint calculation model, the calculation of water resources' ecological footprint in Qinghai is based on the global water balance factor and the global average water resource production capacity determined by WWF (2002), and the yield factor

is based on the regional calculation of Qinghai province. The calculation based on the national hectare method was not carried out in this study. In fact, the global water resource equilibrium factor and the global average water resource production capacity are not fixed and will change with the global climate change, so the comparative selection of parameters in the future needs to be further studied.

(2) Due to the data availability, when calculating the ecological footprint of water resources in Qinghai Province, only the ecological footprint of water quantity is considered without considering the ecological footprint of water quality, such as the pollution of water bodies by industrial and mining industries and agricultural production activities; at the same time, the study does not consider the changes in the water resources carrying capacity and ecological footprint under the background of unexpected events and extreme weather. In summary, how to incorporate pollution factors into the model for the scientific analysis needs to be further explored.

5 CONCLUSION

(1) The overall sustainable development level of water resources in Qinghai Province is good. During 2010–2020, the per capita water resources continue to be in an ecological surplus, the overall trend of water resources utilization efficiency index is decreasing, the water resources' ecological pressure index values are less than 0.5, and the sustainable development index values are greater than 0.70.

(2) The sustainable development level of water resources in each city and prefecture is significantly different. During 2010–2020, water resources in Xining and Haidong are continuously in ecological deficits, with water resources' ecological pressure index being greater than 1, and water resources sustainable development index being less than 0.5; the degree of sustainable development in Haixi prefecture is weaker; this is the most densely populated and industrial area, which needs to further enhance efforts into water conservation and environmental protection, improve water conservancy engineering facilities, and ease the pressure caused by the economy and society on the natural environment. This is also a key and critical area to enhance the level of sustainable development in the province. Haibei, Huangnan, Guoluo, and Yushu prefectures all have sustainability index values greater than 0.80, and their water resources are at a "strong" level of sustainability, which, as an ecological balance area, should gradually improve the ecological footprint of ecological water use.

REFERENCES

An H., Fan L.J., Wu H.L., et al. (2021). Water ecological footprint analysis and prediction in the Huaihe River basin based on BP neural network[J]. *Yangtze River Basin Resources and Environment*, 30(5): 1076–1087.

Du X.H., Zhang T. (2015). Simulation of coupling development of water resources environment and socio-economic system—A case study of Dongting Lake Ecological Economic Zone[J]. *Scientia Geographica Sinica*, 35(9): 1109–1115.

Fan Y.H., Chen L., Tang W. Wg (2021). Analysis of sustainable water resources use in Shenzhen based on improved water ecological footprint model[J]. *Hydroelectric Energy Science*, 9(4): 36–39+35.

Fang W.C., Sun C.F. (2014). Research on water sustainability in Dongguan City based on ecological footprint model of water resources[J]. *Hydropower Energy Science*, 32(1): 25–28.

Fan X.Q. (2005). *Research and Application of Ecological Footprint of Water Resources*[D]. Nanjing: Hohai University, 2005:25.

Hao S., Sun C.Z., Song Q.M. (2021). Evaluation of water ecological footprint and water ecological stress in China based on ESTDA model[J]. *Journal of Ecology*, 41(12): 4651–4662.

Hou X.C., Xie S.Y. (2020). Spatial and temporal analysis of ecological footprint and ecological carrying capacity of water resources in small and medium-sized cities: Suqian City as an example[J]. *Journal of Southwestern University* (Natural Science Edition), 42(12): 134–141.

Hoekstra A.Y. & Hung P.Q. (2005). Globalization of water resources: international virtual water flows in relation to crop trade[J]. *Global Environmental Change*, 15(1): 45–56.

Huang J., Xu C.G., Man Z. (2019). Study on the carrying capacity of water resources in Shandong Province based on ecological footprint[J]. *People's Yangtze River*, 50(2): 115–121.

Huang L.N., Zhang W.X., Jiang C.L., et al. (2008). Ecological footprint calculation method for water resources[J]. *Acta Ecologica Sinica*, 28(3): 1279–1286.

Jenerette G.D., Wu W.L., Goldsmith S., et al. (2006). Contrasting water footprints of cities in China and the United States[J]. *Ecological Economics*, 57(3): 346–358.

Jing P.R., Guo L.D. (2021). Study on the coordinated development of water resources utilization and economy in Zhejiang Province based on ecological footprint[J]. *Water Conservancy and Hydropower Technology*, 52(6): 42–51.

Jiao S.X., Wang A.Z., Chen L.F., et al. (2020). Water ecological footprint measurement and its driving effect analysis in Henan Province[J]. *Hydrology*, 40(1): 91–96.

Liu C.M., Wang H.R. (2003). An analysis of the relationship between water resources and population, economy and social environment[J]. *Journal of Natural Resources*, 18(5): 635–644.

Lan J.Q., Xie S.Y. (2020). Sustainable analysis of water resources in Qingdao based on ecological footprint theory[J]. *Journal of Southwest Normal University* (Natural Science Edition), 45(10): 55–62.

Luck M.A., Jenerette G.D., Wu J.G., et al. (2001). The urban funnel model and the spatially heterogeneous ecological footprint[J]. *Ecosystems*, 4(8): 782–796.

Lu Y.L., Xu S.S., Si B.J., et al. (2021). Study on the dynamic evolution of the environmental carrying capacity of water resources in nine provinces (districts) of the Yellow River basin[J]. *People's Yellow River*, 43(11): 103–108.

Liu Y.B., Yan Y.N. (2020). Temporal distribution characteristics of water ecological footprint in Chengdu and its influencing factors[J]. *South-North Water Transfer and Water Conservancy Science and Technology* (in English), 18(2): 93–98.

Ma J., Peng J. (2013). Advances in water footprint research[J]. *Acta ecologica sinica*, 33 (18): 5458–5466.

Ren Z.Y., Huang Q, Li J. (2005). Quantitative analysis of ecological safety and spatial differences in Shaanxi Province[J]. *Journal of Geography*, 60(4): 597–606.

Shang Q.K., Yin K.X., Mi W.B. (2020). Evaluation of water resources utilization in Qinghai Province based on water footprint theory[J]. *Arid Area Resources and Environment*, 34(5): 70–77.

Venetoulis J. (2001). Assessing the ecological impact of a university: the ecological footprint for the University of Redlands[J]. *International Journal of Sustainability in Higher Education*, 2(2): 180–197.

Wang N., Chun X., Zhou H.J., et al. (2020). Study on the relationship between water resources utilization and economic development in arid areas: the case of Erdos City[J]. *Water Conservation and Irrigation*, (6): 108–113.

Wang T., SimayiZibibula·Zibibula, Chen S., et al. (2015). Dynamic analysis of water footprint and water resources carrying capacity of Urumqi city[J]. *China Rural Water Conservancy and Hydropower*, (2): 42–46.

Wackernagel M., Onisto L., Bello P., et al. (1999). National natural capital accounting with the ecological footprint concept[J]. *Ecological Economics*, 29(3): 375–390.

Wang X.Q., Lu Q., Li B.G. (2005). Study on evaluation of water resources carrying in Qinghai Province by fuzzy comprehensive evaluation method. [J]. *China Desert*, (6): 152–157.

Xiong N.N., Xie S.Y. (2019). Study on the spatial and temporal evolution of ecological footprint and carrying capacity of water resources in Chengdu[J]. *Journal of Southwestern University* (Natural Science Edition), 41(6): 118–126.

Xu D.Z., Zhang W.S., Peng H. (2019). Study on the spatial and temporal evolution of ecological footprint and carrying capacity of water resources in the Three Gorges reservoir area[J]. *People's Yangtze River*, (5): 99–106.

Yue C., Liu F., Yang, et al. (2021). Ecological footprint and ecological carrying capacity of water resources in Beijing from 2010 to 2019[J]. *Soil and Water Conservation Bulletin*, 41(3): 291–295+304.

Zhao M.Q., Xu Y., Li W.P., et al. (2020). Spatial and temporal variation of the ecological footprint of water resources in northeast Sichuan urban agglomeration[J]. *People's Yangtze River*, 51(6): 73–78+106.

Zhang M.Y., Zhao C.Z. (2019). Research on the ecological environment of water resources in Yanbian Prefecture based on the ecological footprint method[J]. *Journal of Northeast Normal University* (Natural Science Edition), 51(3): 135–142.

Zhang S.L., Zhang H.J., Zhang H.J., et al. (2021). Ecological footprint and sustainable use of water resources in Shaanxi Province[J]. *People's Yangtze River*, 52(4): 130–136.

Zhu G.L., Zhao C.Z., Zhu W.H., et al. (2020). Evaluation of sustainable water resources utilization in jilin province based on ecological footprint model [J]. *Journal of China Agricultural University*, 25(9): 131–143.

*Frontiers in Civil and Hydraulic Engineering – Mohamed A. Ismail and
Hazem Samih Mohamed (Eds)*
© 2023 The Authors, ISBN 978-1-032-38247-0

Water hammer protection design of complex water conveyance system in sponge cities

Zichun Song* & Tianchi Zhou*
Suzhou Changsong Engineering Consulting Co., Ltd, Suzhou, China

ABSTRACT: The water conveyance project of Linping supplies four waterworks. The diameter of the pipeline is 2.6 –1.4 m, the total length is 30.9 km, and the prospective water supply scale is $43 \times 10^4 \mathrm{m}^3/\mathrm{d}$. The project adopts a gravity flow water supply mode, with many water users and complex lines, so the calculation and analysis of the hydraulic transient process are complex, and the design of water hammer protection is difficult. In the paper, the hydraulic transient process simulation software is used for modeling and calculation, and the pressure change of the water supply system in the hydraulic transient process is obtained. According to the hydraulic characteristics of the system, the water hammer protection design of the whole water supply system is carried out. After a lot of trial calculations and analyses, when the regulating valve in front of the waterworks is closed in 120 s, the water hammer problem is effectively solved by setting pressure relief valves with a diameter of 1m in front of each waterworks. The paper provides a reference for the water hammer protection design of similar water supply projects.

1   INTRODUCTION

Water transfer projects are the most direct and effective way to solve the uneven distribution problem of water resources in time and space and are the main way for countries around the world to supply water to water-scarce cities and regions. The water supply pipeline will change the flow rate in it during the opening and closing of the valve so that its pressure will change, causing the water hammer, and the pressure in the pipeline will change with the change in the flow rate, which is called the hydraulic transient process (Chaudry 2013; Yu et al. 2010).

The water hammer is a common physical phenomenon in water transmission devices, which is very harmful to the safe and steady operation of the pipe network, and due to the improper setting of technical measures for water hammers, pipe burst accidents occur from time to time (Zheng et al. 2008). In December 2008, the Fengjiashan water transmission project in Shaanxi Province experienced three consecutive pipe bursts, and the project was abandoned. In June 2013, a water hammer occurred during the opening of the pump in the pump room of the Weifang waterworks, and in less than 5 minutes, the pump room was completely submerged. Therefore, the calculation and analysis of the hydraulic transient process should be carried out in the engineering design to prevent the destruction of the pressure pipeline by the maximum water hammer and the splitting of the water column, or the negative pressure in the pressure pipeline, so as to ensure the safe operation of the project (Chaudry 2013; Yu et al. 2010).

With the development of urbanization, water supply projects have also shown a trend of increasing complexity and enlargement, and the problem of the hydraulic transient process has become increasingly prominent. Based on the typical multi-user complex gravity flow water supply system—the water supply project of Linping waterworks, this paper calculates and analyzes the hydraulic transient process under typical control conditions and selects the pressure relief valve as the water hammer protection facility, and optimizes its parameters, which can provide a reference for the

---

*Corresponding Authors: songzichun@foxmail.com and 9821416334@qq.com

DOI 10.1201/9781003344209-79   603

hydraulic transient process analysis and water hammer protection design of similar water supply projects.

2 PROJECT OVERVIEW AND BASIC INFORMATION

2.1 *Project overview*

The first part of the project is located in the Yuhang watershed point of the Qiandao Lake water diversion project, and the water supply objects are Linping waterworks, Renhe waterworks, Hongpan waterworks, and Tangqi waterworks. The recent scale is 200,000 m^3/d, and the long-term scale is 430,000 m^3/d.

The whole project adopts a gravity flow pressure water supply mode, with the characteristics of a long water supply pipeline, a large scale of water supply, complex operating conditions, etc. The opening and closing process of the valve in front of the waterworks is prone to cause greater water hammer pressure, threatening the safe operation of the project, so it is necessary to calculate and analyze the hydraulic transient process and design water hammer protection.

2.2 *Basic information*

2.2.1 *Water level of the reservoir and waterworks*
The Qiandao Lake water supply project draws water from the Xianlin reservoir, and the designed water level of the Xianlin reservoir is 52 m. There are Xianlin waterworks, Jiuxi waterworks, and Xiangfu waterworks between the Xianlin reservoir and the Yuhang watershed point, and after the Yuhang watershed point is the Jiaxing branch line. After the calculation and analysis of constant flow, the node head of the Yuhang watershed point is 22.55 m under the normal operation, which is selected as the upstream control head of the water supply project of Linping waterworks.

According to the requirements for the surface elevation and water pressure of the waterworks, the water head of Renhe waterworks is 11.6 m, that of Linping waterworks is 9.2 m, that of Hongpan waterworks is 11.7 m, and that of Tangqi waterworks is 5.5 m.

2.2.2 *Pipeline parameters*
The main pipeline parameters of the four waterworks involved in the water supply project from the Xianlin reservoir to the Yuhang watershed point, Yuhang Watershed, Jiaxing branch line, and Linping waterworks are shown in Table 1.

Table 1. Pipe parameters.

Pipe	L/(km)	D/(m)	A/(m^2)	n
Yuhang-Linping	1.40	2.6	5.31	0.013
Linping-Linping waterworks	1.43	2.6	5.31	0.013
Linping-Renhe	5.00	2.6	5.31	0.013
Renhe-Hongpan	4.04	1.4	1.54	0.013
Hongpan-Hongpan waterworks	0.62	1.4	1.54	0.013
Hongpan-Tangqi waterworks	5.05	1.4	1.54	0.013

2.2.3 *Water level of reservoirs and waterworks*
The relevant parameters of the valve initially selected for each outlet are:
Linping waterworks: flow control valve with a diameter of 2.2 m;
The inlet valve of Renhe waterworks: flow control valve with a diameter of 2.2 m;
The inlet valve of Hongpan waterworks: flow control valve with a diameter of 1.2 m;
The inlet valve of Tangqi waterworks: flow control valve with a diameter of 1.2 m.

2.3 Control conditions

The maximum pressure of water hammers along the general water transmission system is controlled to be <1.3 times of the hydrostatic pressure. The internal water pressure during the normal operation of the waterworks of this project is small at about 20 m of water head. If the hydrostatic pressure is 1.3 times of that, the maximum allowable pressure of the pipeline in the transient process is only about 26 m of water head. Besides, the working pressure of pipe is designed as 50 m of water head, and the pipe is selected as 100 m of water head, which is much greater than 26 m. In summary, the maximum pressure of the water in the pipeline in the transient process state is considered as 80 m of water head, which can not only ensure the operation safety of the water transmission system but also make better use of the pipeline's carrying capacity.

During the normal operation, the water pressure is not less than 2 m of water head; during the transient process, there should be no water column fracture in any part of the water transmission system, and the minimum momentary pressure of the steel pipe section is not less than 0 m of water head.

2.4 Calculation of working conditions

Linping waterworks is a new waterworks, and Renhe waterworks, Hongpan waterworks, and Tangqi waterworks are the current waterworks; the pipeline from the Yuhang watershed point of Qiandao Lake to the Renhe waterworks has been completed, and other water pipes need to be built. The pipeline flow rate under each typical operating condition is shown in Table 2.

Table 2. Pipeline flow under each condition.

Pipe	C1	C2	C3	C4	C5	C6
Yuhang-Jiaxing	4.98	4.98	4.98	4.98	4.98	4.98
Yuhang-Linping	0.00	0.00	2.32	2.66	2.32	4.98
Linping-Linping waterworks	4.98	4.98	2.66	2.32	2.66	0.00
Linping-Renhe	4.98	2.31	0.00	2.32	1.50	0.00
Renhe-Hongpan	0.00	1.74	1.74	0.00	1.16	0.00
Hongpan-Hongpan waterworks	0.00	0.58	0.58	0.00	0.64	0.00
Hongpan-Tangqi waterworks	0.00	1.16	1.16	0.00	0.52	0.00

According to the above 6 combinations, for Renhe waterworks, the pipeline flow is the largest under Condition 1, so this working condition is the most dangerous working condition; for Linping waterworks, the pipeline flow is the largest under Conditions 6, so this working condition is the most dangerous working condition; for Hongpan waterworks, the pipeline flow rate is the largest under the Condition 5, so this working condition is the most dangerous working condition; for Tangqi waterworks, the pipeline flow is the largest under the Condition 2, so this working condition is the most dangerous working condition. Therefore, working conditions 1, 2, 5, and 6 are selected as the control conditions for calculating and analyzing water hammer protection.

3 CALCULATION MODEL

For a complex waterway system, a matrix equation expression can be used as a numerical calculation model for calculating the hydraulic transient process of the system (Brekke et al. 1988):

$$[E]\vec{H} = \vec{Q} + \vec{C} \tag{1}$$

where, $\vec{H}$ is the node head vector; $\vec{Q}$ is the node input flow vector, with the inflow as the positive and the outflow as the negative; $\vec{C}$ is a supplementary vector related to the nonlinearity of the system; $[E]$ is the system structure matrix.

The system structure matrix is built by various elements in the system according to certain laws. The premise of building the system matrix is to establish an element matrix. The element matrixes of three typical hydraulic elements of pressurized pipe elements, impedance elements, and reservoir elements are introduced briefly as follows.

3.1 Matrix of pressurized pipe elements

For the pressured pipe, i is assumed to be its upstream end and j is its downstream end, and the matrix equation of momentary flow for the elements of the pressurized pipe is:

$$\begin{bmatrix} \frac{-1}{Z_C} & 0 \\ 0 & \frac{-1}{Z_C} \end{bmatrix} \begin{bmatrix} H_i \\ H_j \end{bmatrix} = \begin{bmatrix} Q_i \\ Q_j \end{bmatrix} + \begin{bmatrix} \frac{C_n}{Z_C} \\ \frac{C_m}{Z_C} \end{bmatrix} \tag{2}$$

$$Z_C = \frac{a}{gA} \tag{3}$$

where, Z_c is the characteristic impedance of the pressurized pipe element; H_i, H_j, Q_i, and Q_j are the head and flow rate of the pipeline terminals i and j at that moment, respectively; C_n and C_m can be obtained from the pipeline's characteristic parameters and the head flow of the pipeline ends i and j at the previous moment; a is the water hammer wave velocity; g is the gravitational acceleration; A is the cross-sectional area of the pipeline.

3.2 Element matrix of impedance elements

The impedance elements include throttles, partially open valves, and gates, local head loss points, etc. i is assumed to be the inlet end of the impedance element and j is the outlet end of the impedance element, and the momentary flow matrix equation of the impedance element is:

$$\begin{bmatrix} \frac{-1}{Z} & \frac{1}{Z} \\ \frac{1}{Z} & \frac{-1}{Z} \end{bmatrix} \begin{bmatrix} H_i \\ H_j \end{bmatrix} = \begin{bmatrix} Q_i \\ Q_j \end{bmatrix} + \begin{bmatrix} -Q_{i0} - \frac{h_{ij}}{Z} \\ -Q_{j0} + \frac{h_{ij}}{Z} \end{bmatrix} \tag{4}$$

$$Z = 2k|Q_{0i}| = 2k|Q_{0j}| \tag{5}$$

$$h_{ij} = k|Q_{0i}|Q_{0i} = k|Q_{0j}|Q_{0j} \tag{6}$$

where, Z is the characteristic impedance of the impedance element; H_i, H_j, Q_i, and Q_j are the head and flow rate of the pipeline ends i and j at that moment, respectively; h_{ij} is the difference between the water heads at both ends of the impedance element; k is the overflowing head loss coefficient of the impedance element. Q_{0i} and Q_{0j} are the flow of the impedance element's ends i and j at the previous moment, respectively.

4 CALCULATION ANALYSIS AND PROTECTION DESIGN OF WATER HAMMERS

4.1 Water hammer calculation of closing valve

The water hammer of the closing valve is mainly caused by the closure of the valve due to the maximum positive pressure and minimum negative pressure of the water pipeline. According to the characteristics of the project, due to the length of the water transmission system, it's quite possible that the maximum positive pressure and the minimum negative pressure along the water transmission system in the process of closing the valve cannot meet the requirement of the control situation. Therefore, the calculation should be carried out according to the water hammer of the closing valve by viewing the calculation results of the closing time of each valve, and whether the system needs water hammer protection can be judged. After calculation, under the 4 control conditions, the change law of the internal water pressure in the pipeline is basically the same when the valve is closed. As follows, the working condition 1 is taken as example to discuss in detail.

Condition 1 is the condition of the branch line flow of Renhe waterworks with the largest flow, so the valve closing time of 120 s, 300 s, and 600 s is used to calculate the transient process of Renhe waterworks when closing the valve, and the calculation result is as shown in Figure 1.

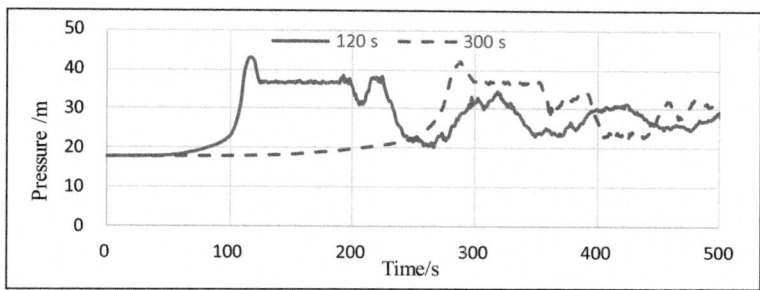

Figure 1. Pipeline pressure from linping distribution point to renhe distribution point (at 2 + 200 m).

It can be seen that when the valve in the waterworks is closed without water hammer protection, the maximum pressure in front of the valve and the maximum pressure of the pipeline are more than 80 m, and the negative pressure appears under most working conditions within the closing time of 120 s, 300 s, and 600 s. Under a few working conditions using the closing time of 600 s, the maximum pressure in front of the valve and the maximum pressure of the pipeline are reduced to less than 80 m without negative pressure, which are also close to the situation of the control value. But the safety margin is small, and the closing time of 600 s is not conducive to the flexible operation of the waterworks. Therefore, it is necessary to set up water hammer protection measures throughout the system.

4.2 *The selection of water hammer protection measures*

The water hammer protection measures that are often used in the design of water supply engineering are: (1) increasing the pressure level of the pipeline; (2) optimizing the opening and closing time of the outlet valve; (3) setting up pressure relief valves at high-pressure points at outlets and along water pipelines; (4) setting up a two-way pressure regulating tower at the outlet. This project is a long-distance gravity flow project, so from the perspective of the water hammer characteristics of the gravity flow, (2) and (3) usually can better control the water hammer.

The pressure relief valve is a valve that automatically discharges part of the medium to adjust and stabilize the pipeline pressure when the pressure is higher than the set value. The valve opening pressure can be adjusted within a certain range, and when the pressure returns below the set value, the valve is automatically closed to prevent the medium from continuing to discharge.

After a large number of trial calculations, the relevant parameters of the selected pressure relief valve are as follows: A pressure relief valve is set respectively with a diameter of 1.0 m in front of Linping waterworks, Renhe waterworks, Hongpan waterworks, and Tangqi waterworks. The opening pressure of the valve is set as 40 m for Linping waterworks, Renhe waterworks, and Hongpan waterworks, and that for Tangqi waterworks is 30 m. At the same time, different closing time is used to calculate the four control conditions. The calculation results are shown in Figures 2–5.

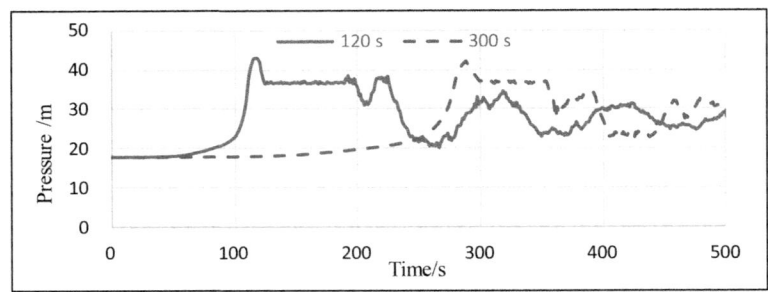

Figure 2.   Pressure in front of the valve in renhe water plant (C1).

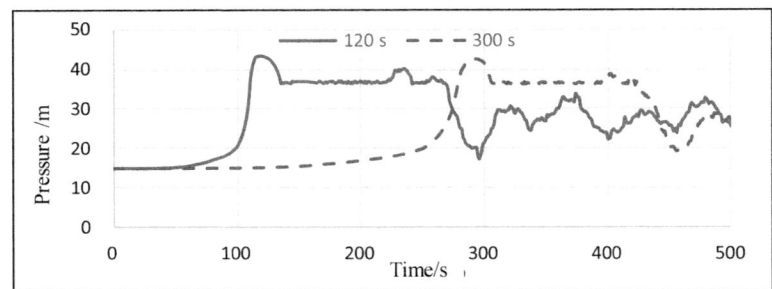

Figure 3.   Pressure in front of the valve in linping water plant (C2).

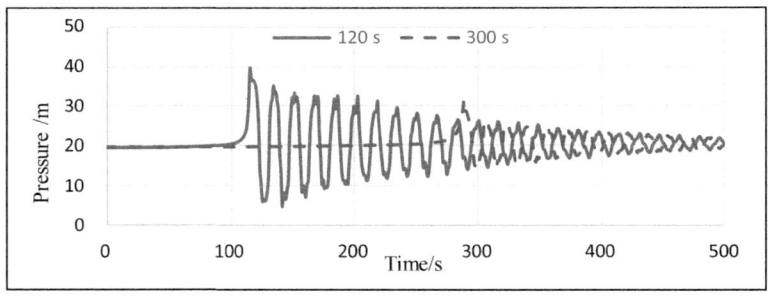

Figure 4.   Pressure in front of the valve in hongpan water plant (C5).

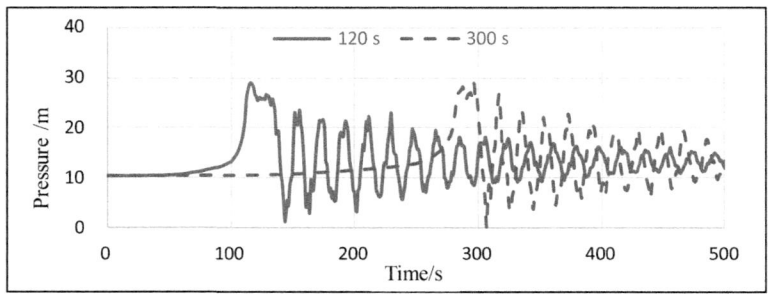

Figure 5.   Pressure in front of the valve in tangxi water plant (C6).

It can be seen that when the valve in the waterworks is closed, the outlet valve of the waterworks adopts 120 s as the closing time, and the pressure relief valve is set up simultaneously in front of the four waterworks, namely, Linping Waterworks, Renhe Waterworks, Hongpan Waterworks, and Tangqi Waterworks for protection. The maximum pressure in front of the valve and the pipeline is effectively reduced without negative pressure. So, it is recommended that the overpressure relief valve can serve as a water hammer protection measure.

5 CONCLUSIONS

In this paper, the calculation and analysis of the hydraulic transient process are carried out for the typical complex gravity flow water supply system—Linping waterworks project. The water hammer protection design is carried out according to the analysis results, and the safety of the water supply system is reviewed according to the selected water hammer protection measures. After the calculation and analysis, the following valve opening and closing time and the pressure relief valve's parameters are determined: The opening time of each waterworks can be selected within 60 s to 300 s. In order to reserve a certain safety margin and convenient operation, 120 s of valve opening time can be adopted, and the simultaneous opening of valves should be avoided. 120 s of closing tim is applied to the inlet valve of each waterworks, and at the same time, the protective means of the pressure relief valve are taken. The diameter of the pressure relief valve in front of Linping waterworks, Renhe waterworks, Hongpan waterworks, and Tangqi waterworks is 1.0 m, the set opening pressure value of the pressure relief valve of Linping waterworks, Renhe waterworks, and Hongpan waterworks is 40m, and the set opening pressure value of the pressure relief valve of Tangqi Waterworks is 30 m.

REFERENCES

Brekke, H. & Li, X.X. (1988). *A New Approach to the Mathematical Modelling of Hydropower Systems*. Conference of CONTROL 88, London, 152–160.
Chaudry, M.H. (2013). *Applied hydraulic transients*. New York: Springer, 255–257.
Cheng, J.Y. (2007). *Mathematical simulation and control of hydraulic transient process*. Chengdu: Sichuan University Press, 254–255.
Yang, K.L. (2000). *Hydraulic transient and regulation in power station and pumping station*. Beijing: China Water Power Press, 75–81.
YU, J.Y. (2010). *Study on the hydraulic transient process of safe operation of long-distance water transmission pipeline*. Harbin Industrial University, 2010, 124–128.
Zheng, Y. & Zhang, J. (2008). *Transition process of hydraulic unit*. Beijing: Peking University Press, 52–69.

Frontiers in Civil and Hydraulic Engineering – Mohamed A. Ismail and
Hazem Samih Mohamed (Eds)
© 2023 The Authors, ISBN 978-1-032-38247-0

GIS+BIM-based source control platform of construction wastes

Chengyu Zhu
School of Civil Engineering, Jiaying University, Meizhou, Guangdong, China

Shuqin He
Guangdong Research Institute of Water Resources and Hydropower, Guangdong Provincial Science and Technology Collaborative Innovation Center for Water Safety, Guangdong Guangzhou, China

Caiying Huang & Hualong Cai*
School of Civil Engineering, Jiaying University, Meizhou, Guangdong, China

ABSTRACT: The source management & control of construction wastes is an important element in the comprehensive management and industrial development of urban construction wastes. This paper analyses the current situation and problems of China's construction waste source control, namely the lack of construction waste source control and management, the lack of unified construction waste source reduction and sorting management, and the lack of a complete industrial chain. With the objective of improving the utilization rate of construction wastes and facilitating industrial development, Geographic Information System (GIS) and Building Information Modeling (BIM) technology are introduced into the field of construction waste supervision, based on which the construction waste source control platform is designed, and the corresponding functional modules are designed to achieve real-time monitoring of construction waste source control in response to the existing requirements for intelligent control at the source and waste reduction at the end.

1 INTRODUCTION

In order to comply with the requirements of sustainable social and economic development, the pace of construction planning in towns and cities has been accelerating, and the demolition and renovation of old houses have produced a lot of construction wastes, which will occupy a large number of land resources and cause great damage to the ecological environment. At present, China treats construction wastes with the traditional direct landfill technology, which cannot separate its harmful heavy metals from usable materials, resulting in serious land pollution and waste of resources (Wang 2020). The source control of construction wastes aims to make full use of new technologies and techniques to solve the current management problems of waste generation, transportation, disposal, and utilization in the construction waste disposal industry and to improve the resource utilization of construction wastes and the intelligent control of the industry. Based on this, this paper designs and develops a control platform for the resource utilization of construction wastes based on GIS+BIM technology and gives full play to the characteristics of efficient information integration in computer technology.

The integration of GIS+BIM technology into the construction waste source control platform is mainly the combination of three-dimensional parametric design and engineering design, which makes use of its visualization, virtualization, and sharing characteristics to form a visual digital model of the building components, which is positioned and displayed isometrically on the GIS platform. By giving full play to the characteristics of GIS technology in the collection, calculation, analysis, and storage of geospatial elements, a large amount of engineering information in the BIM model is analyzed and displayed on the same electronic map, providing accurate and complete

---

*Corresponding Author: whucad@whu.edu.cn

DOI 10.1201/9781003344209-80

information for construction project designers and construction waste handlers to visualize and control construction wastes.

2 SYSTEM COMPONENTS AND FUNCTIONS

The GIS+BIM-based construction waste source control platform includes four functional modules: a corrosion detection module, a digital metering module, an intelligent transport intelligent control module, and a consumption supervision module. The modules are based on computer application technology, provide accurate information collection and processing analysis of construction projects and construction wastes, and are designed to provide highly targeted and easy-to-use functional services for each stage of construction waste generation and disposal, to achieve the effect of interconnecting and complementing information and functions.

2.1 *Corrosion detection module*

Users need to upload CAD drawings of the house components, and the system combines BIM and AutoCAD to analyze the CAD drawings and build several visual models of the building's structural components. Users use the models to understand whether there are cracks in the concrete components, to obtain accurate load-bearing parameters of the walls and the types of materials inside the walls, and to obtain corrosion points, impact types, and damage parameter values through corrosion detection of the building components. On this basis, it is easy for architects to adjust the structural design plans according to the corrosion parameters, which can effectively slow down the corrosion of materials (Ma 2020). In addition, the system can calculate the proportion of reinforced concrete materials used in the load-bearing bottom and the interior part of the building based on crash detection indicators. For local component models with abnormal parameters, the system reconstructs the model and performs corrosion detection on the local components based on the data obtained from the engineers' recalculation and analysis.

2.2 *Digital metering module*

This module proposes a quantitative control function for construction wastes (Lv 2021), meaning that engineers need to upload the estimated amount of construction works and the actual amount of construction process, and the system will calculate the consumption of materials based on the type of the construction project and the location where the materials are used and predict the quantity of construction waste generated on this basis. In addition, the system automatically analyzes and digitally measures the basic information on construction wastes, and the location and type of material losses uploaded by the user, and separates out and classifies information on hazardous metal materials in construction wastes. By analyzing the total amount of building demolition, demolition methods, and opinions over construction waste disposal, the system estimates the cost of building demolition projects based on the actual amount of work and through a variety of pricing methods, and separately prices construction wastes according to corporate waste disposal standards and market prices, thus estimating the total cost of building demolition budgets and reducing the generation of construction wastes during construction.

2.3 *Intelligent transport control module*

The said intelligent transport control module is designed based on GIS technology and the highly integrated RF identification system (Zhou 2021), with functions such as transport personnel verification, transport vehicle tracking, and video supervision and positioning.

Construction waste transporters are required to register their personnel identification and face recognition information in this module before entering or leaving the building construction site and the construction waste disposal site. Access to the site will be granted to the staff authorized by the system, otherwise, the system's alert system will be triggered and the relevant manager will be alerted to verify the non-authorized user. In the process of construction waste removal, the remote office administrator starts the reader of the integrated radio frequency identification system on the construction transport vehicle, and the system automatically collects the basic information of construction wastes on the transport vehicle as well as the transport information and uploads it to the management platform after analysis by the computer to facilitate waste information checking, and transport route planning and tracking. The system uses GIS technology to monitor and locate the transport vehicles in real time and transmits the video data to the processing end of the cloud platform through the network exchange system to monitor and manage the transport process in real time.

2.4 *Consumption regulation module*

Building construction units are required to upload information on the construction site, the type of construction, and the estimated total amount of construction waste before construction, fill in the module's waste disposal application form within the regular disposal period, and book the disposal unit in advance. In this way, the disposal enterprise can reasonably arrange staff and waste disposal points according to the information and submit the preliminary transportation plan and construction waste disposal method, which will eventually form the waste disposal charge. Both the builder and the abatement unit will need to register staff identification information for the work interface and staff supervision. Users can view the transport profile of the truck and the video information of the disposal site through the monitoring management function to achieve real-time supervision of the construction waste transport and disposal process.

3 SYSTEM ARCHITECTURE

3.1 *Software architecture system*

The software architecture system of the GIS+BIM-based construction waste resource control platform consists of four layers: the presentation layer, the business layer, the processing layer, and the data layer. The presentation layer includes government departments, business units, clients, and community units (Rong 2021). The operational layer consists of a corrosion detection module, a digital metering module, a transport intelligence control module, and a consumption regulation module, whose operations include services such as component modeling, cost estimation, and transport tracking. The processing layer contains six data acquisition systems, BIM modelling systems, GIS spatial analysis systems, and video equipment, which are capable of efficient data acquisition, analysis, and calculation. The data layer consists of two databases: the information database and the equipment database, which enable the storage and management of engineering data and waste data in categories. The system enables the accurate collection and calculation of data through computer technology, and the interactive sharing of data in a secure network system, as shown in Figure 1.

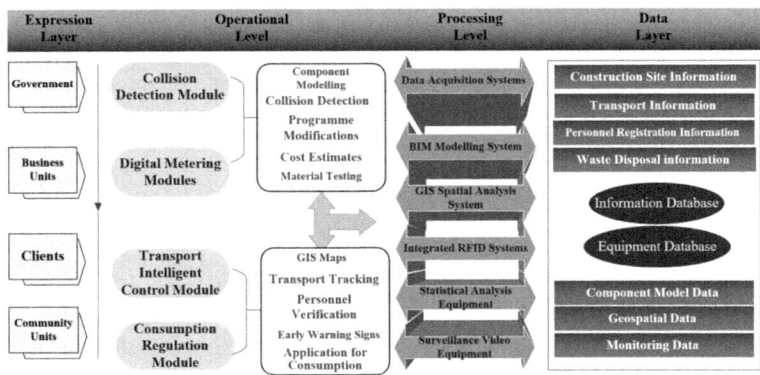

Figure 1.   Software architecture diagram.

3.2 *Hardware architecture system*

During the design phase, the building structure is modeled using the BIM model analysis system, and corrosion detection is carried out using a detector, and the structural corrosion model is automatically transferred to the designer's end of the platform (Jiang 2021). Video surveillance of building construction sites, demolition sites, and transit sites is realized using surveillance systems and data analysis through computer systems respectively. The radio frequency identification system can be effectively combined with the GIS system, detectors, and transmission system to achieve the purpose of unified collection and management of staff and transport vehicle information. The monitoring system for each of the city's disposal points is mobilized in the monitoring center, and the workload, access information, and system usage information of each disposal point is processed through the cloud-based processing system, and such information is transmitted to the government side, the enterprise side, and the official platform of the client and community units respectively, thus achieving the unified management of construction waste disposal information (Figure 2).

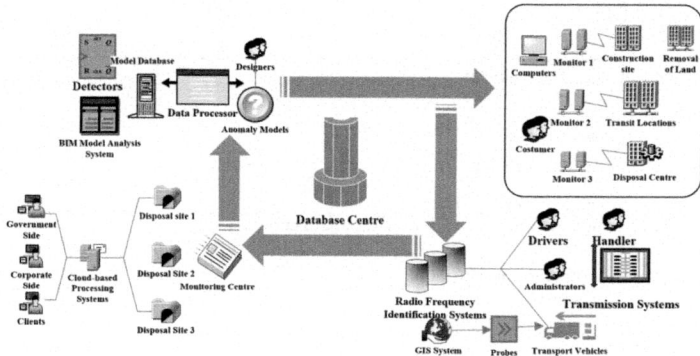

Figure 2.   Network topology diagram.

4   SYSTEM IMPLEMENTATION

4.1 *Main system interface*

The user logs in to the main interface of the construction waste resource management platform, clicks on the personal settings function to supplement the user's identity information, working rights, password protection, etc., and accesses the corresponding system services by clicking on each function module. This interface allows users to search for information on construction sites, construction workers, famous buildings, etc.; by entering the name of a company, information on related projects can be obtained, as shown in Figure 3.

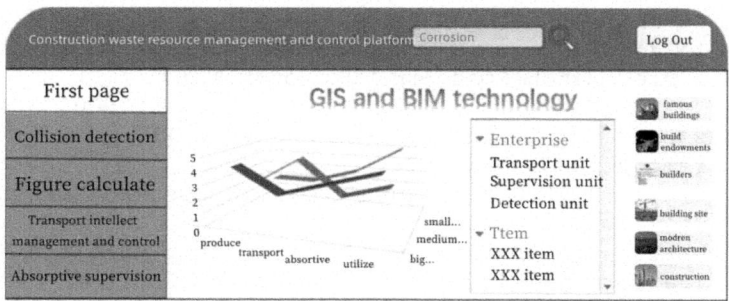

Figure 3.   Main interface of the construction waste resource management platform.

4.2 Related system interfaces

By clicking into the Dissipation Detection module and querying a project work in this module, the user will be able to obtain the safety of the construction site, the type of construction, and the predicted total amount of construction waste for the project. Staff checks the information on the construction unit and transportation unit on the abatement application form and shares the abatement application form through the WeChat and Weibo channels, as shown in Figure 4.

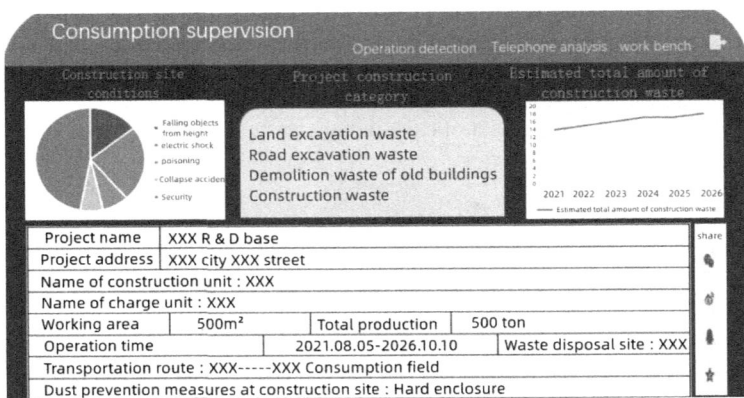

Figure 4. Consumption regulation interface.

4.3 System part code

Based on the PHP environment and combined with Html, the development of MySQL database technology for the construction waste resource control platform in the user login system requires strict legitimacy verification of the user login account and password entered, and the relevant code to verify the legitimacy of the user account and password is as follows:

```
// Login authentication class
public function login (){
  // Perform login authentication when the server is post
  if (Request::isPost()) {
      $account = trim(input('post.account'));
      // Verify the legitimacy of the account
      if (empty($account)) {
            $this->returnCode('90000001');}
      $pattern="/\^{}([0-9A-Za-z$\-\-_\\.] +)@([0-9a-z]+  \\.
          [a-z]{2,3}(\\. [a-z]{2})?) $/i";
      if (!preg_match($pattern, $account)) {
                    $this->returnCode('90000006');}
      $password = trim(input('post.password'));
      if (empty($password)) {
                $this->returnCode(1, 'password cannot be
                    empty');}
      // Verify the legitimacy of the account password
      if ($aUser['password'] ! = md5($password)){
            $this->returnCode('90000031');}
            $remember = input('post.remember');}
```

5 CONCLUSION AND OUTLOOK

Based on GIS + BIM technology, the construction waste resource control platform is built, with computer application technology to collect and analyze data in all aspects, and combined with data open ports to achieve data interconnection and sharing, forming an information-based construction waste control service system. The system improves the calculation efficiency of construction waste generation and the accuracy of emission estimation by analyzing BIM building model data, reduces errors in building construction, and saves human and material resources; overlays GIS maps with construction sites, demolition sites, and transportation routes to achieve real-time supervision of the whole process of construction waste disposal with accurate positioning and complete transportation information. The system aims to solve the problems of construction waste production forecasting, unreasonable measurement, and deployment, scattered information on waste source components, etc. Through a series of information technology function modules, the efficiency of engineers' access to information is improved, and comprehensive control of construction wastes is improved.

Due to regional development differences, there are differences in the functional needs of enterprises for the system, and the system platform has not yet reached a diverse control model. With the development of core equipment and the optimization and innovation of waste treatment processes, the author's team will delve into new modes of construction waste control, break the fixed regional layout and further improve the construction waste control platform.

REFERENCES

Jiang, J.J. & Li H. (2021). Let construction waste management go to standardized intelligence. *N. Nantong Daily*. 08–26.

Lv, S.R., Jin, Y.J. & Ren, F.M. (2021). Research on the quantification method of construction waste based on BIM. *J. Construction Technology Development*. 48, 55–57.

Ma, Z. & Chen, J.J. (2020). Research on the application of BIM corrosion detection technology in green construction of assembled buildings. *J. Value Engineering*. 39, 222–223.

Rong, Y.F., Sun, X.K. & Zhang, Y.Y. (2021). Resource utilization management of construction waste under the concept of "no waste". *J. Urban Architecture*. 18, 16–20.

Wang, N., Lou, D., Chen, D.Q. et al. (2020) A preliminary study on the information management platform for precise control of construction waste based on "BIM+GIS" technology. *J. Environmental Engineering*. 38, 46–50.

Zhou, Q.J., Liu, J.J. & Tian, R. (2021). Design of intelligent management system for urban construction waste transportation vehicles. *J. Heavy Duty Vehicles*. 25–26.

Frontiers in Civil and Hydraulic Engineering – Mohamed A. Ismail and Hazem Samih Mohamed (Eds)
© *2023 The Authors, ISBN 978-1-032-38247-0*

Research on the application of "Dracaena-Fractal" in architecture

Xiaosong Zhang & Mingyu Jin*
North China University of Technology, Beijing, China

ABSTRACT: The branch structure of the Dracaena is very characteristic and is a typical fractal. This article deeply understands the fractal in the Dracaena and the fractal pattern and uses the modeling software rhino and grasshopper to simulate the fractal in the Dracaena. The "truth" and "beauty" of the dracaena fractal are shown by analyzing the architectural cases of the dracaena fractal.

1 INTRODUCTION

There are many complex and changeable forms in nature, such as rugged and messy mountain outlines, fluffy cloud edges, jagged coastlines, and so on. However, there are often logical and clear mathematical principles behind these complex forms. The branch structure of Dracaena is also rugged and changeable, and it is difficult to explain this shape with traditional mathematical principles. Therefore, we need to introduce a new mathematical principle to explain this shape—fractal.

Fractal, a mathematical term, is an abstract object in mathematics used to describe things that exist in nature. Fractals have self-similar properties and are usually defined as "a rough or fragmented geometric shape that can be divided into several parts, each of which is a reduced shape of the whole" (as shown in Figure 1). Fractal geometry is a geometry that takes irregular geometric forms as its research object. Because there are many irregular geometric forms in nature, such as coastlines, mountains, mineral dendrite fractals, and even clouds, fractal geometry is also called to describe the geometry of nature.

After understanding the definition of fractal, we can better understand the branch structure of dracaena, so as to extract the geometric shape of dracaena fractal. The purpose of artificially controlling the shape of dracaena fractals is achieved by using modeling software to carry out computer modeling of dracaena fractals. By adjusting the parameters of the dracaena fractal, it can be applied to the building structure and building skin. This paper innovatively applies the dracaena fractal to the structure of the landscape pavilion, making full use of its natural form, and trying to create an architectural landscape that is better integrated with the natural environment. And by summarizing architectural cases, the application advantages of "dracaena fractal" in actual buildings are analyzed.

2 FRACTAL HISTORY

In the 17th century, the mathematician Gottfried Leibniz studied recursive self-similar graphs, and the mathematical basis of fractals has been developed since then. Two centuries later, in 1872, Karl Weierstrass gave the first definition of fractal at the Royal Prussian Academy of Sciences: a fractal is a functional graph with counter-intuitive properties such as being continuous everywhere, but not differentiable everywhere.

---

*Corresponding Author: jinmingyu@ncut.edu.cn

                                    DOI 10.1201/9781003344209-81

Figure 1. Mountains.

In 1904, Niels von Koch described a function-like hand-drawn figure that gave a more geometrical definition of a fractal, a hand-drawn figure known as a Koch snowflake. The curves that make up the Koch snowflakes are called Koch curves, which are infinite in length and are continuous and nowhere differentiable. This article uses Rhino3D and Grasshopper plug-in to simulate Koch snowflakes. The basic figure of the Koch snowflake is an equilateral triangle, and an equilateral triangle is randomly generated outward or inward with each side as the base. This process is called an iteration. Figure 2 shows a graph of a battery generating a Koch snowflake. Figure 3 shows the Koch snowflake after three iterations.

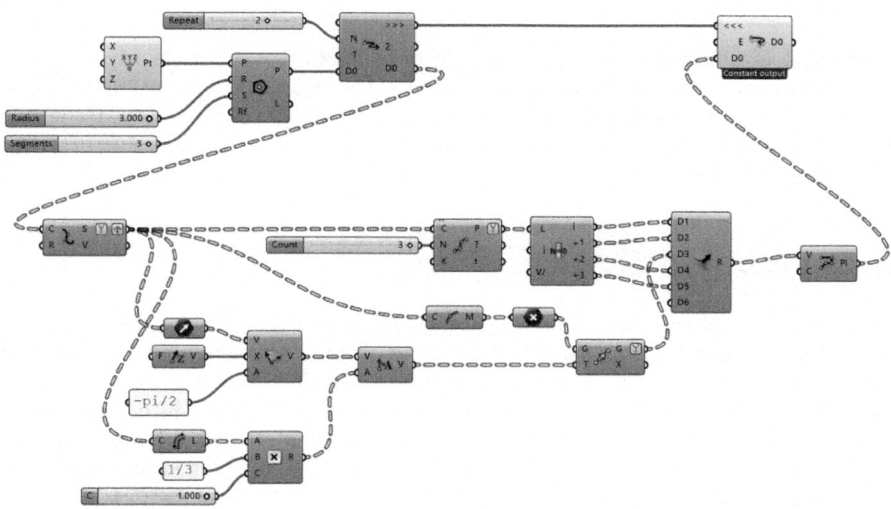

Figure 2. Koch snowflake battery pack.

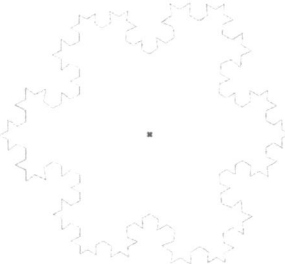

Figure 3. Results after three iterations of the koch snowflake.

In 1960, Benhua Mandelbrot published a very important paper in the field of fractals, "How long is the coastline of England? Statistical Self-Similarity and Fractional Dimensions". In this paper, Mandelbrot discusses self-similar curves between one and two dimensions. Although the word "fractal" is not used, the self-similar curves in question are fractal curves. After that, Mandelbrot coined the term "fractal" in 1975 and used computer-generated images to describe this mathematical definition. These intuitive fractal patterns are full of beauty and science fiction, giving the public a popular understanding of the concept of "fractal" (see Figure 4).

Figure 4. Mandelbrot graphics.

The morphological structure in plants has high self-similarity, which cannot be explained by ordinary mathematical models. Only the fractal theory can describe the morphological structure of plants well. From the geometrical point of view, the shape of leaves and flowers, the length and thickness of branches and roots, etc., all have fractal characteristics. The branching structure of dracaena is a typical fractal (Yang 2015).

3 FRACTAL IN DRACAENA

For Dracaena, which belongs to an arbor species, the trunk is short and thick, the surface is light brown, and the sap is dark red, so it is named dragon's blood tree. As shown in Figure 5, the branches of the dracaena are very distinctive, and each growth is a "divided into two".

This branching of the dracaena is called bifurcated branching in botany. The definition of bifurcated branching is that the apical meristem is divided into two halves during branching, and each forms an independent branch. After a period of time, each branch grows in the same way. The growth mode of the dracaena is consistent with the definition of fractal that "every part is the reduced shape of the whole". Therefore, the branching structure of the dracaena is a typical fractal.

Figure 5.   Dracaena.

4   SIMULATION OF FRACTALS IN DRACAENA

The fractal structure in the dracaena is simple and logical, and it is extremely applicable to other disciplines. However, due to the high level of digitalization in today's society, the natural form cannot be directly applied to other fields. The fractal in the dracaena needs to be parameterized. It can be combined with other disciplines to create greater social values.

This article will use the modeling software rhino3D and its own grasshopper plug-in to simulate the growth of the dracaena. L-system algorithm will also help in the analysis of fractals in dracaena in this paper. This algorithm is often used to simulate the growth process of plants, and it is also used to simulate the shape of organisms. Most importantly, the L-system algorithm can be used for Generate fractal patterns (Lin 2014).

The fractal simulation process of dracaena is shown in Figure 6. First, the rules of the L-system battery are rewritten, and the initial state F is input, and the rule is F=F[^\-"F][&\+"F]. If the number of iterations reaches 9, the obtained character set is input into the rabbit plug-in, and then the step size is determined as 5, the scaling ratio as 0.9, and the angle as 22. Finally, the shape is output to obtain the simulated shape of the dracaena fractal (Xu 2018). Figure 7 is the pattern generated by the program.

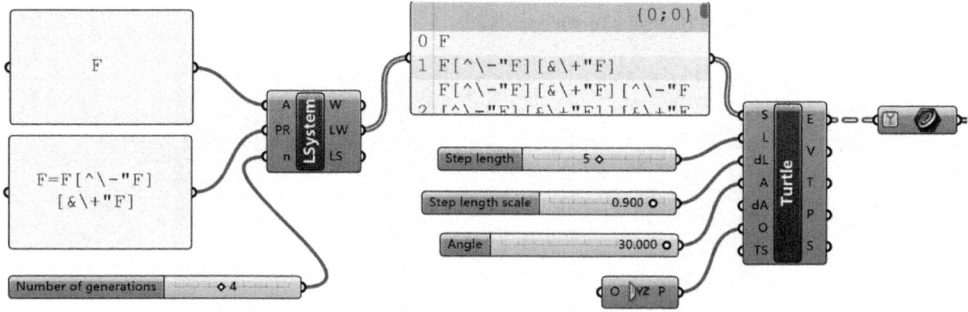

Figure 6.   Dracaena fractal battery pack.

Fractal software such as L-system algorithm and rabbit can make the dracaena fractal artificially controllable, and because the dracaena fractal is very regular, the generated form is simple and

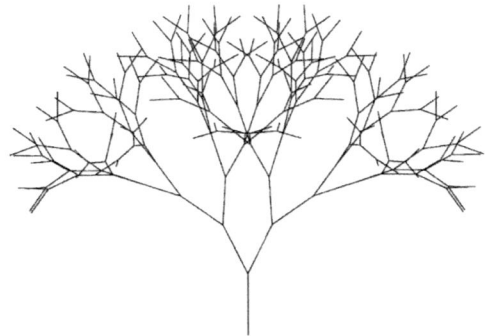

Figure 7.   Simulation results of dracaena fractal.

clear, making the dracaena fractal applicable in architecture. Architects can create forms or patterns that fit the architectural application by adjusting the simulation parameters.

This paper takes the fractal algorithm of the dracaena as an example and designs a landscape pavilion by adjusting the fractal simulation parameters of the dracaena. The generation logic is as follows: First, the dracaena fractal algorithm is used to generate the supporting trunk of the pavilion, then the end point of the curve generated by the last iteration is used to generate a surface, and the surface is processed to obtain the pavilion top. By generating this structure, it can be known that when the dracaena fractal is parameterized, it will be widely used in the field of architecture. Figure 8 is a perspective view of the bionic pavilion. Figure 9 is a top view of the bionic pavilion.

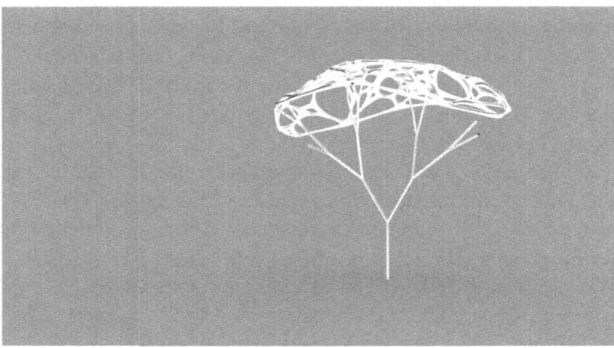

Figure 8.   Perspective view of dracaena bionic pavilion.

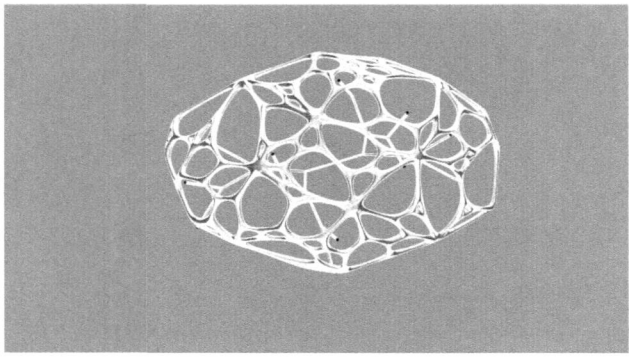

Figure 9.   Top view of dracaena bionic pavilion.

5 THE APPLICATION OF DRACAENA FRACTAL IN ARCHITECTURE

Whether it is the building structure or the building skin, or even the division of the building space, we can see that the architects are using the dracaena fractal, a form extracted from nature. From ancient times to the present, architects have never stopped imitating the shape of the dracaena fractal.

5.1 *Sagrada familia*

The columns of the Sagrada Familia in Barcelona imitate the fractal form of the dracaena, which is not only beautiful but also conforms to the mechanical properties, and is known as the pioneering work of modern bionic architecture (Zhao 2019) (see Figure 10). However, it can be observed that because the fractal of the dracaena has not been parameterized at that time, it only imitates its simple form, and its application in architecture is relatively rough.

Figure 10.   Sagrada familia pillars.

5.2 *Tote house*

In modern architecture, the application of dracaena in architecture becomes delicate and flexible by parameterizing the fractal of dracaena. The internal structure of the Tote House designed by Serie Architects in India has the biomimetic fractal form of dracaena. Figure 11 is a diagram of the interior space of the building.

Figure 11.   Interior space of tote house.

During the preliminary investigation, the architect felt that the natural environment around the site was well protected, and the entire building was wrapped in lush rainforest plants. Therefore, in order to unify the structure of the building with the surrounding environment, the architect adopted the fractal structure of the dracaena, creating a space type without clear boundaries but with coherence and functional differentiation. Figure 12 is the scheme generation diagram.

Figure 12.   Structural generation of tote house.

The fractal structure of the dracaena in the Tote House not only plays a supporting role in the building but also serves as a decorative component, creating a natural indoor environment. In order not to waste the top space because of the special-shaped structure, light troughs and fire sprinkler facilities are also installed in the gaps of the fractal structure of the dracaena. This shows that the dracaena fractal has been standardized for production and application, which is in line with modern aesthetics and meets the basic needs of architecture.

In this case, the dracaena fractal is used by architects due to its strong natural properties, and because of the parametric operation, the generated building components can be adjusted according to the needs of the site. This is the advantage of the dracaena fractal applied to building structures.

5.3   Stuttgart Airport

Stuttgart Airport, designed by GMP, also uses a structure similar to the dracaena fractal in its interior. The roof loads at Stuttgart Airport are transferred from a 4 m–5 m support grid to the tree branches, each of which is then transferred down to the trunk, thus forming the airport's structural system. The large canopy contrasts with the slender supporting structure, which gives the entire airport a sense of lightness and adds a natural touch to the modern industrial building due to its tree-like structure. These are the unique advantages of the dracaena fractal structure. Figure 13 shows the internal structural columns of the building.

Looking inward from the outside of the airport through the glass of the facade, the tree-like branches stand inside the airport to support the ceiling. Although being slender, they are full of power, just like the goddess column of Erechtheion Temple. Figure 14 is the night view of the building.

Figure 13.   Interior structural column in Stuttgart Airport.

Figure 14.   Night view of Stuttgart Airport.

6   CONCLUSIONS

The fractal in Dracaena finds a balance between order and disorder, and builds a bridge between human society and nature. When the beauty of the dracaena fractal, which is full of natural attributes, appeared on the building, we were all convinced by him. In the future, the development of the dracaena is inseparable from the participation of parameterization. In the future, architects will no longer be confused about how to apply the dracaena fractal. What troubles architects is how to standardize the dracaena fractal pattern, how to get it, and how to convert the 2D point and line into the 3D space required by the building.

Through the in-depth excavation of the dracaena fractal and the use of computer modeling, this paper innovatively applies the dracaena fractal to the design of landscape pavilions, creating a landscape architecture that interacts harmoniously with the natural environment. Through the analysis of architectural cases, dracaena fractal is widely used in architecture, especially in the field of building structure. Dracaena fractal expands the form of building structures, beautifies the interior space of buildings, and becomes a natural and architectural bridge.

ACKNOWLEDGMENT

General project of Beijing Natural Science Foundation: Research on the construction and design of non-standard inter-embedded modules based on industrial robots (ITEM NUMBER: 8212008).

REFERENCES

Lin Q.D. (2014). *Building form generation based on fractal theory*. D. Tsinghua University, 19–20.

Xu W.G., Li N. (2018). *Biomorphic architectural digital illustration*. China Construction Industry Press, 166–169.

Yang P. (2015). *Research on the modeling method of virtual plant morphology based on fractal features*. D. Shandong Normal University, 10–11.

Zhao P. (2019). *Research on the fractal mimic design of the skin of public buildings*. D. South China University of Technology, 50–53.

Frontiers in Civil and Hydraulic Engineering – Mohamed A. Ismail and Hazem Samih Mohamed (Eds)
© 2023 The Authors, ISBN 978-1-032-38247-0

A study of urban ecological landscape design based on low impact development (LID)

Jun Yan* & Xiaoqian Chen*
Jingdezhen Ceramic University, Jiangxi, China

ABSTRACT: Today, with the rapid growth of modern society and the economy, people's living standards are constantly improving, which makes people's thinking ideas change greatly, and gradually puts forward higher requirements for an urban ecological environment. To ensure that the urban ecological environment meets the requirements of modern people, the application of Low Impact Development (LID) should be strengthened. Based on this, this paper briefly introduces the low-impact development and then analyzes the design of a modern urban ecological landscape. Based on this, taking a wetland park as the research object, it discusses the application of low-impact development in detail and provides support for strengthening the transformation of the modern urban ecological environment.

1 INTRODUCTION

With the rapid growth of China's economy, the damage to the natural environment has intensified, and the problems such as water shortage, air pollution and soil erosion have become more and more serious, causing great interference to human existence and life. Against this background, the state and local governments have paid more attention to environmental protection, and have formulated many rules and regulations related to improving the urban ecological environment, such as the Technical Guide for Sponge City Construction. Under the guidance of these rules and regulations, strengthening the design of the urban ecological landscape has alleviated the destruction of the urban ecological environment to a certain extent. However, after in-depth study, we can find that there are still many defects, which will still have a certain impact on modern people's lives. Therefore, in the process of modern urban construction, the urban ecological landscape should be further constructed, and to achieve this goal, the concept of low-impact development should be effectively applied.

2 OVERVIEW OF LOW IMPACT DEVELOPMENT (LID)

In the early 1990s, with the continuous improvement of people's green environmental protection concept, after years of research, the Maryland Environmental Resources Department of George Province, USA, developed a brand-new rainwater management system, that is, Low Impact Development (LID) technology. The application of this technology can strengthen the management of water resources, reduce the damage and waste of water resources, and promote the sustainable development of modern society. Compared with the conventional rainwater management system, the LID system has many advantages, which can evolve into the rainfall of various scales and strengthen the supplementary effect of groundwater, and will not cause erosion to rivers. It can

---

*Corresponding Author: 003240@jcu.edu.cn and 3555817651@qq.com

DOI 10.1201/9781003344209-82

be effectively integrated with the landscape, with a high utilization rate of natural resources and low operation and maintenance cost, so it is widely used in the design and planning of the modern urban ecological landscape (Peng 2022).

3 URBAN ECOLOGICAL LANDSCAPE DESIGN BASED ON LOW IMPACT DEVELOPMENT (LID)

3.1 *Design principles*

3.1.1 *Principle of ecological protection*
LID-based urban landscape design mainly simulates the natural ecological environment, and all links of the design should follow the development incentives of the ecological system. For the urban ecological landscape, there are usually many functions. However, under the LID concept, the prevention of ecological functions is put at the core, and the design and planning of other functions serve to enhance the ecological functions of the ecological landscape. In urban ecological landscape planning, while keeping the natural ecosystem in its original state, we should strengthen the control and treatment of rainwater based on the natural development mechanism. The protection of ecologically sensitive areas should be strengthened to prevent the destruction of natural drainage systems. The impervious area should be reduced and the continuous water permeability of the ground should be improved.

3.1.2 *Principle of integrity*
The application of the LID concept is not only about the application of one or more landscape elements, but the integration of these elements for the joint planning of the urban landscape. Only in this way can the LID concept play its greatest role. When designing the ecological landscape, the whole ecological landscape should be designed from the perspectives of the water ecosystem, surrounding buildings and roads, so as to ensure that each subsystem can play its due role, and will not interfere with other subsystems. Rainwater can smoothly enter the ground in many ways, roads and buildings can be used normally, and the green area on the surface meets the requirements of modern residents' life.

3.1.3 *Principle of adapting to local conditions*
For different cities, there are certain differences in development history, cultural surrounding environment, etc., which makes each city have different requirements for ecological landscape. Therefore, when applying the LID concept, we should follow the principle of adapting to local conditions, and grasp the specific situation of the city by investigating the environment, climatic characteristics, and geological and hydrological conditions of the renovation site. On this basis, we can choose reasonable green vegetation and adopt a scientific rainwater management scheme to construct an urban ecological landscape that is in line with the local actual situation.

3.1.4 *Principle of ornamental*
The design and planning of the urban ecological landscape are not only to improve the ecological environment of a city, but also to enhance the beauty of the city, and bring a pleasant feeling to urban residents. So the urban residents can relax from their stressful work and life and improve their physical and mental health. Given this, when using LID to design urban ecological landscape, while paying attention to ecology and integrity, we should also consider the ornamental value of the ecological landscape, to ensure that the whole ecological landscape has a beautiful appearance and brings a stronger visual impact to the urban residents, and also to ensure that the ecological landscape plays an ornamental function (Xiong 2022).

3.2 *Design process and content*

3.2.1 *Site investigation*

To effectively design the urban ecological landscape, designers must have an accurate understanding of the transformation site. Therefore, before carrying out the design work, it is necessary to strengthen the investigation of the reconstruction site. For the investigation, there are two aspects: first, through communication with meteorology, geological survey, land and resources, we can obtain documents related to urban ecological landscape design, such as local meteorological data, hydrological data and geological data; second, site investigation includes the location analysis, that is, the analysis of the specific location of the reconstruction site and the configuration of various pipelines in this area. The terrain analysis is the analysis of the specific situation of elevation, slope and aspect of the reconstructed area. The hydrological analysis is to survey and transform the distribution and total area of the regional water ecosystem. The vegetation analysis refers to the analysis of the types and coverage levels of green vegetation in the region.

3.2.2 *Urban ecological landscape planning based on LID*

In urban ecological landscape planning, two aspects are mainly considered, namely:

(1) Layout planning

According to the requirements of the LID concept, we should ensure the integration of the urban ecological landscape and rain-flood ecosystem: find the core positions of the rain-flood ecosystem, and strengthen the planning of these core positions based on conditions to make the urban ecological landscape play the greatest role. Specifically, there are three core locations, namely: the source area, that is, the water collection area within the whole system is usually located in a high-lying area. The construction of urban parks in this area can improve the infiltration capacity of rainwater, thereby reducing the pressure of rain and flood. The runoff area, that is, the area through which water flows, is a relatively low-lying area in the city. Belt parks can be built to block rainwater and prevent waterlogging. The catchment area, that is, the end of the system, can form lakes and other systems due to the accumulation of a large number of water resources. In this area, lakes can be the core, and cities can be built far enough to purify rainwater.

(2) Urban ecological landscape-scale planning

According to the LID concept, in order to make the urban ecological landscape play the greatest role, it is necessary to determine a reasonable ecological landscape scale. In traditional ecological landscape design, the per capita use area is usually the main factor, while in the LID concept, more attention is paid to the balance of water quantity. Firstly, according to the total area of the catchment area, the total water quantity of the whole area is deduced, and the formula is:

$$P_a = \frac{1}{A} \int A \times P_{dA} \tag{1}$$

Where P_a represents rainfall; a represents the area of the catchment area; P represents the rainfall on the finite element dA.

$$W = \sum_{A-1}^{N} P_A F_A \tag{2}$$

Where W represents the total water volume and F_A represents the total area of the catchment area.

On this basis, the scale of the ecological landscape park can be deduced, and the calculation formula is:

$$A = \Psi \times H / W \tag{3}$$

Where A represents the total area of the ecological landscape park; H represents the average annual rainfall depth, and ψ represents the permeability coefficient of the surface (Zhang 2022).

3.2.3 *Urban ecological landscape design based on LID*

In the design of the urban ecological landscape, six aspects are mainly considered, namely:

(1) Vertical design, also called terrain design, mainly includes flat vertical design. It refers to the garden green space whose elevation is flush with the surrounding area. From the point of view of landscape design, if the site is relatively flat, it will be more billing visually, but it is easy to produce a boring feeling. From the point of view of rain and flood control, it can strengthen the absorption of rainwater and help prevent waterlogging. Therefore, to ensure that the topographic design plays a role in controlling rain and flood, and at the same time prevent people from feeling bored, it is necessary to adopt the way of convex or concave. Among them, proper artificial wetlands should be built around the convex surface to prevent the loss of rainwater.

(2) Vertical design. In the city, we can often see all kinds of slopes. The effective vertical design of these slopes can make the urban ecological landscape play its ecological function and be more ornamental. Under the concept of LID, different schemes need to be adopted in the vertical design of sloping land. If the slope is above 10%, a stepped enclosure can be constructed, and a grass planting ditch of appropriate size can be excavated at its top to avoid soil erosion and reduce the incidence of flooding. If the slope is in the range of 2.5% ~ 10%, the stepped grass planting ditch will be constructed directly, and the slope of each longitudinal slope should not be higher than 2.5%. At the periphery of the grass planting ditch, build a water retaining plate and lay a gravel layer. Water should be placed to wash the green belt. If the slope is lower than 2.5%, choose the plan of a stepped gravel wall, and build a grass planting ditch on each wall. For two adjacent gravel walls, the interval should be controlled at 5~6 m.

(3) Traffic design. When designing the road, it should be parallel to the contour line as far as possible to avoid damage to the natural drainage system. On both sides of the road, a certain space should be reserved for the construction of LID facilities. On the pavement, materials with good water permeability should be used to improve the permeability of rainwater and avoid rainwater staying on the pavement for a long time. In areas with few people and cars, a permeable paving scheme can be selected (Wang 2022).

(4) Architectural design. In the urban ecological landscape, architecture is an indispensable element, and it should be designed reasonably. On the one hand, we should pay attention to the management of rainwater. Some green vegetation can be planted on the roof of the building to absorb rainwater, and a good drainage system should be built around the roof to discharge rainwater into the ground in time. On the other hand, choose a reasonable building location, and it is forbidden to build in the catchment area, so as not to cause damage to the water ecosystem.

(5) Green vegetation design. To meet the demand for the LID concept, in the design of the urban physical landscape, reasonable green vegetation should be selected according to the principle of local conditions. On the one hand, ensure that vegetation can grow normally. On the other hand, it should also reflect the natural transition characteristics of "terrestrial-wet-aquatic". Among them, in terms of terrestrial vegetation, the broad-leaved forest is the main one, and some low trees and shrubs are arranged to block rainwater. In terms of wet vegetation, the vegetation with deep roots is the main one to purify rainwater. In the aspect of aquatic vegetation, vegetation with strong stress resistance is mainly used for the deep purification of rainwater (Pan 2022).

(6) Rainwater management system. Starting from the source, midway and end of the water ecosystem, a reasonable rainwater utilization system is constructed. After simple treatment of rainwater, it is applied to social development and human life.

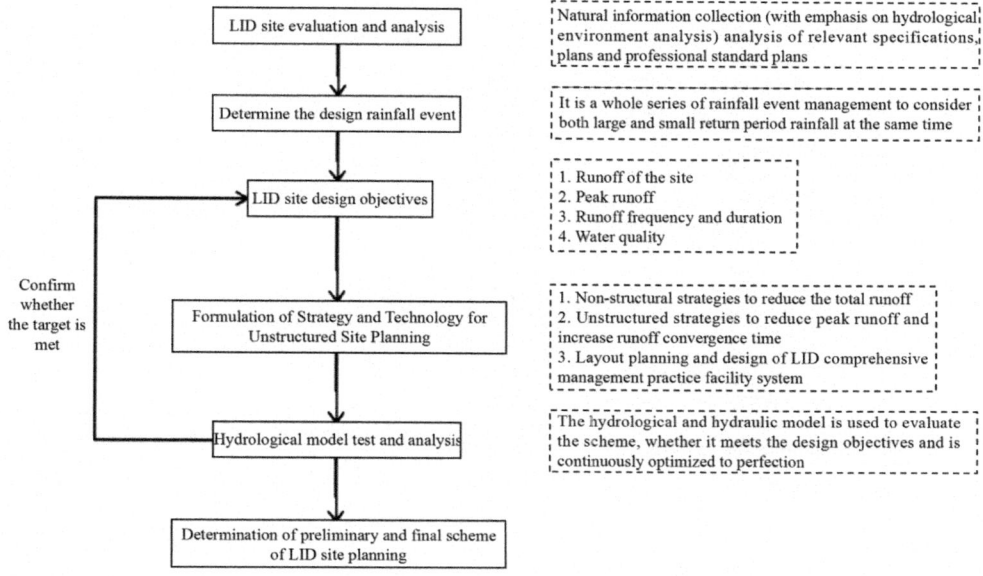

Figure 1. Low-impact development design process and content diagram.

4 DESIGN CASE

The LID concept was adopted in the construction of a city wetland park. The park has many functions such as flood control, purification and landscape. The elevation of the park site is in the range of 427~443 m, which is characterized by the northeast being higher than the southwest. The slopes are all less than 5%, and the whole park is relatively flat. In terms of hydrological characteristics, this area belongs to the wetland ecosystem. In important areas, there are many aquatic ecosystems, accounting for more than 60%. The whole aquatic ecosystem is divided into many small systems, which are scattered in various regions. There are a large number of green islands among the small systems. Through the treatment of rainwater by the islands, floods and other disasters can be prevented in the park. In terms of vegetation, there are 60 families of higher vegetation, totaling 168 species belonging to 140 genera, among which angiosperms are the most, with 51 families, 131 genera and 157 species. Sendai vegetation is the same as gymnosperms, with 2 families, 2 genera and 2 species, and ferns with 5 families, 5 genera and 7 species.

According to the above-mentioned site conditions, a park has been designed as follows:

(1) As for the vertical design, based on keeping the original characteristics of flat land, a concave scheme is adopted to deal with it.
(2) Traffic design. According to the demand for the LID concept and the specific situation of the park site, a traffic scheme as shown in Figure 2 is designed. It consists of two parts: the expressway, which is 6 m wide and used for riding and driving in sightseeing buses, and is located at the outermost side of the park. Slow lane, used for the connection of various areas, including the main sightseeing lane, with a width of 3 m; The second sightseeing road is 1.5 m wide.

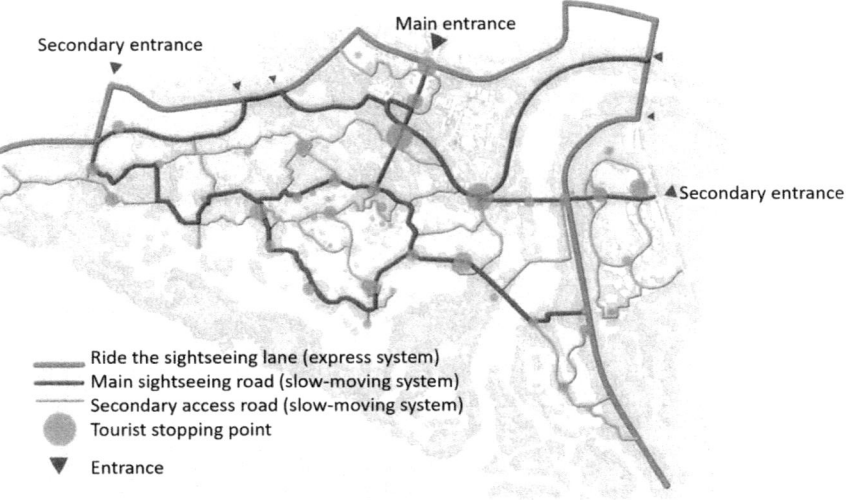

Figure 2.　Traffic design diagram of a certain park.

(3) Architectural design. There are museums, bird-watching pavilions, and other buildings in the park, which are used for tourists' viewing. These buildings cover an area of about 4,362 m². On the roof of the building, a small amount of green vegetation is planted, which is used to increase the green area of the building on the one hand and strengthen the management of rainwater on the other. As shown in Figure 3.

Vegetation can tolerate Drought and flooding

Growth medium

Drainage layer

Drainage blanket

Root barrier and waterproof film
Ceiling

Insulating layer

Vapor barrier

Structural support

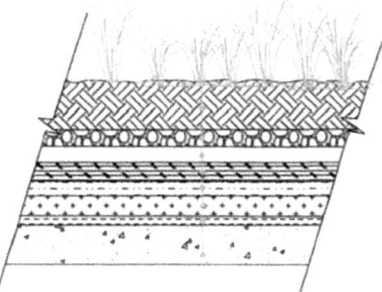

Figure 3.　Architectural design drawing of a certain park.

(4) Vegetation design. In terms of vegetation, vegetation suitable for local growth was selected, among which *Salix mandshurica* and *cosmos* were selected as terrestrial vegetation. The wet plants are *Typha angustifolia*, *Dianthus australis* and *Oenanthe cress*. *Hydrilla verticillata* and *Brasenia schreberi* were selected as aquatic plants.

5 CONCLUSION

To sum up, with the rapid development of modern society, we should strengthen the application of the LID concept. On this basis, combined with the principles of ecology, integrity and appreciation, the design of urban ecological landscapes should be strengthened from the aspects of the vertical surface, transportation, architecture, green vegetation and rainwater management, to promote the sustainable development of modern cities.

ACKNOWLEDGEMENT

Project fund: 2021 Science and Technology Research Project of Jiangxi Education Department (Project Name: Research on Ecological Approaches and Design Strategies of Urban Landscape Based on Low Impact Development (LID) Technology) (Project No. GJJ211337).

REFERENCES

Pan Chan, Guo Xiaoqin, Sima Cong (2022). A study on campus sponge renovation based on low-impact development [J]. *Industrial Safety and Environmental Protection*, 48(06): 79–81.
Peng Li, Lu Liping (2022). A practical study on low-impact development of rainwater system in urban green space [J]. *Municipal Engineering Technology*, 40(06): 132–136+142.
Wang Pei (2022). A study on low impact design strategy of wetland landscape in Western Sichuan Plateau—A case study of zoige hardjo wetland reserve [J]. *Decoration*, 01(02): 142–144.
Xiong Yuan, Lu Lijun (2022). Development design of low impact urban runoff pollution control based on parameter analysis [J]. *Guangdong Chemical Industry*, 49(09): 146–150+170.
Zhang Quan, Xue Shanshan, Zou Chengdong (2022). Rainwater and flood management of chaji ancient village in Jingxian county, anhui province based on low-impact development [J]. *Journal of Chinese Urban Forestry*, 20(02): 111–117.

Frontiers in Civil and Hydraulic Engineering – Mohamed A. Ismail and
Hazem Samih Mohamed (Eds)
© 2023 The Authors, ISBN 978-1-032-38247-0

Research on the application of intelligent detection robot system for airport pavement operation safety

Libin Han
Yunnan Airport Group Co., Ltd., Kunming, China

Kaidi Liu*, Lei Guo, Zonghe Li & Guoliang Zhai
Beijing Super-Creative Technology Co., Ltd., Beijing, China
China Airport Construction Group Co., Ltd., Beijing, China

ABSTRACT: The safe operation of airport pavement is the guaranteed basis for the overall safety of civil aviation. However, the current safety operation detection methods have problems such as low efficiency, poor accuracy and low coverage, which cannot solve the FOD detection of airport pavement and the surface of the pavement caused by tire rubber and snow cover. The problem of missing functions, which in turn jeopardizes flight safety. Therefore, it is of great significance to develop full-coverage, high-precision and high-efficiency pavement safety condition detection equipment to meet the needs of airport pavement safety operation, breaking through the theory and method of shallow strength detection. Aiming at the safety inspection requirements of civil aviation airports for runway surface topography characteristics, strength characteristics, foreign object obstacles, etc., tackle key problems in pavement surface texture acquisition and topography characterization, pavement skid resistance and durability testing, wide-view pavement FOD identification and disposal, Multi-robot path planning and coordination and intelligent navigation control, road surface status data analysis and safety prediction integration technology, forming an integrated technology and robot equipment for airport road surface safety detection, decision-making and disposal, and will lead the civil aviation road surface safety inspection industry. The development of industrialization supports the construction of safe and smart airports.

1 GENERAL INSTRUCTIONS

Air transportation is an important mode of transportation in addition to road transportation and rail transportation. In recent years, China's civil air transport industry has developed rapidly (see Figure 1). In 2021, the passenger throughput of China's civil aviation transport airports will be 907 million, an increase of 5.9% over the previous year, and the cargo and mail throughput will be 17.828 million tons, an increase of 10.9% over the previous year. In 2019, before the outbreak of the new crown epidemic, the passenger throughput of China's civil aviation transport airports reached an all-time high of 1.352 billion people.

The airport is an important infrastructure for the development of the air transportation industry. It plays a very important role in changing the transportation conditions, optimizing the allocation of urban resources, stimulating the local economy and promoting cultural exchanges (Luyin 2008). Through decades of development and construction, the number of transport airports in China has increased rapidly. As shown in Figure 2, by the end of 2021, the number of transport airports certified by the Civil Aviation Administration of China has reached 248, with a total of 275 runways, 7,133 parking spaces, and a road surface area of more than 300 million square meters.

---

*Corresponding Author: liukaidijc@163.com

 DOI 10.1201/9781003344209-83

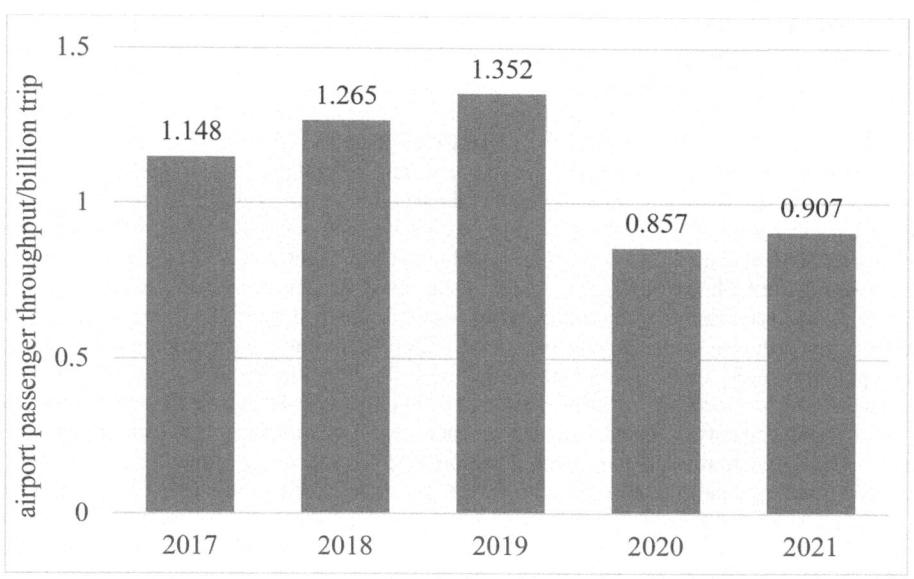

Figure 1.　China's civil aviation transport airport passenger throughput.

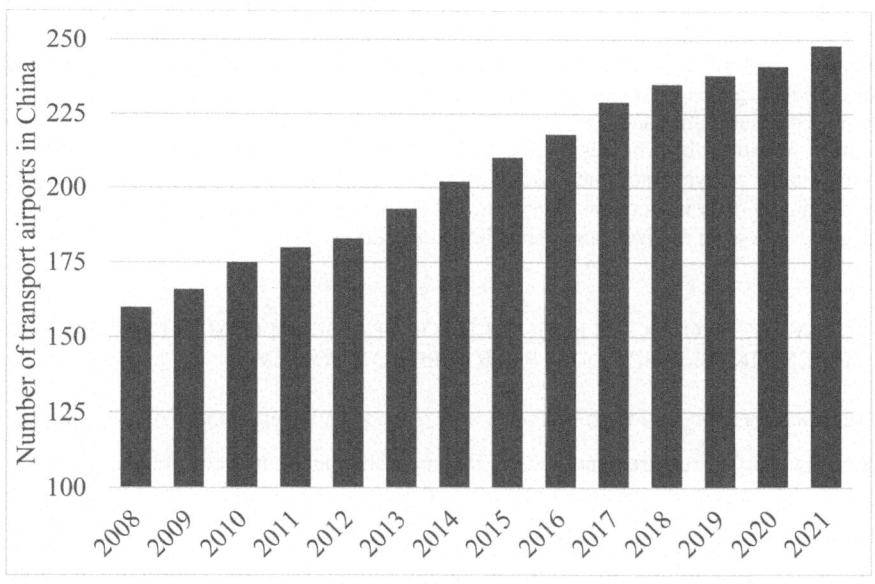

Figure 2.　Number of transport airports in China.

As one of the main infrastructures of the airport, the airport pavement is an important carrier for aircraft to taxi, take off and land. In the past decade or so, with the introduction of large aircraft, airport throughput and pavement utilization, the increasingly frequent takeoffs and landings have brought severe challenges to the airport pavement quality. The quality of airports directly affects the flight safety of aircraft (Gendreau & Soriano 1998). Rain and snow, rubber residue contamination and surface polishing will cause the lack of anti-skid ability of the pavement, which will easily lead to the difficulty of effective braking of the aircraft, and even cause lateral drift or even deviate or rush out of the runway, resulting in a serious loss of life and property (Chen et al. 2021). On the other hand, the FOD that often appears on the road surface will not only easily puncture the tires, but also cause the engine to inhale foreign objects and fail during the take-off and landing process. Great hidden danger (Chauhan et al. 2020). One of the most typical is the crash of Air France Flight 4590. On July 25, 2000, shortly after the flight Concorde took off, its left wing caught fire and crashed, killing all 109 passengers and crew on board. The official investigation report believes that on the day of the air crash, there was a long metal part on the runway that fell out of the engine of a McDonnell Douglas DC-10 airliner. During takeoff of Flight 4590, the wheel of the aircraft ran over the component, causing the tire to burst, and the debris penetrated the wing cylinders at high speed, severing the cables of the landing gear, causing sparks to ignite an oil spill and fire (Cramoisi 2010). It can be seen that the consequences of accidents caused by FOD are extremely serious. Therefore, given the use of the airport pavement, timely and effective detection and management of the operation safety of the pavement is a key link that the airport must pay attention to. The traditional detection methods of airport pavement mainly include geometric size, compaction, flatness, bearing capacity and anti-skid performance, etc. There are various detection methods (Yi & Sato 2018). These traditional pavement detection methods are simple in principle and can reflect the actual situation of pavement to a certain extent, but there are also some drawbacks. The growth rate of air traffic at large airports in China is higher than the design consideration rate, resulting in a significant increase in the number and frequency of road surface diseases. However, with the increase in flight volume, the operating time of the airport is prolonged, and the time for road surface maintenance is also reduced accordingly. The traditional inspection of pavement generally takes a long time. On the other hand, the traditional inspection techniques for the internal structure of the pavement, such as the heavy hammer deflectometer and the Beckman beam deflection test, are point-by-point non-destructive testing, with low work efficiency and poor representativeness, and the most common coring methods are also inaccurate. The original pavement structure will cause a certain degree of damage. At the same time, most domestic airport runway inspections are mainly completed manually by pavement inspectors, and the runway will be closed during pavement inspection, which makes the flight capacity not only inefficient and unreliable, but also takes up valuable runway use time (Zou et al. 2021). Finally, the industry standard requires the airport to conduct daily inspections every day, but due to technical conditions, only simple manual inspections can now be realized, and the technical requirements are difficult to implement.

2 THE KEY TECHNOLOGIES REQUIRED FOR THE DEVELOPMENT OF THE AIRPORT PAVEMENT INTELLIGENT DETECTION ROBOT SYSTEM

2.1 *Real-time task decision-making and control technology in multimodal robot collaboration*

Due to the diversity of requirements such as the operating range, time constraints, detection granularity, and effective area of the airport pavement detection task, as well as the obvious differences in the movement capabilities and characteristics of each modal robot body, it is necessary to construct collaborative task decision-making and collaborative control research. More complex models, taking into account more environmental constraints (Green et al. 2008). This has brought new challenges to the research on the theory and method of multi-robot cooperation, which is at the forefront of the field of robotics. It is necessary to design a comprehensive index function based on scene division and its distributed optimization algorithm to solve the optimal set of key poses and realize multiple robots. Goal Routing and Task Optimization Algorithms (Mahmud et al. 2008).

2.2 *Airport pavement safety assessment and danger early warning technology based on heterogeneous spatiotemporal big data analysis*

The existing indicators for evaluating the airworthiness of the pavement cannot be fully applied to the pavement safety inspection robot. It is necessary to construct a new evaluation index for the airworthiness of the pavement and integrate it into a set of indicators with complete internal logic. And based on this, design fusion criteria of data at different levels and dimensions, analyze the time and space characteristics of data in different dimensions of road surface airworthiness, and build a data-driven pavement that can intelligently adjust its structure, parameters, and algorithm characteristics. The safety evolution model realizes the scientific early warning of the airworthiness of the road surface.

2.3 Analysis Technology of "One Measurement and Multiple Uses" for the Safety Status of Airport Pavement Operation

Aiming at the one-sided problem of identification criteria caused by the traditional automatic detection method of pavement inspection, which is "specialized for special instruments", a set of detection data-multiple indicators output is proposed. Research the three-dimensional high-precision measurement technology of pavement surface appearance, and develop a high-speed acquisition device with full coverage of multi-source data of pavement surface based on this. A multi-source heterogeneous and multi-scale pavement surface appearance data fusion model based on a general-purpose structural operator is constructed to obtain standardized pavement multi-dimensional measurement data. Based on this, the pavement surface disease and FOD detection methods based on deep learning and feature engineering are studied. Finally, a set of "one-measurement-multiple-use" analysis technology systems that can accurately, quickly and universally monitor the running safety status of the pavement surface is formed.

3 RESEARCH AND DEVELOPMENT OF INTELLIGENT DETECTION ROBOT SYSTEM FOR AIRPORT PAVEMENT

3.1 The key technology of the detection robot for the safety status of the airport pavement operation

3.1.1 Long-term synchronization technology of pavement environment modeling and detection information for airport detection robots

Aiming at the problems of the wide airport environment, complex pavement structure and short time window, we use prior geometric knowledge such as airport runway plane drawings to guide the heuristic exploration of the robot environment and guide the robot to model the environment. Using RTK fusion edge lights and runway signs as road sign features, and based on the real-time positioning of the robot, a hierarchical map is generated. A one-to-many mapping relationship is established based on the feature points and the results of laser, visual and infrared road surface detection data, and a local update technology of the road surface map is established to maintain the consistency of the runway environment map. The synchronization mechanism maintains the continuity of road surface map and road surface detection data mapping and provides basic method support for robot planning and detection prediction.

3.1.2 Safe navigation and control method of airport pavement detection robot

Aiming at the real-time positioning requirements of robots for large-scale coverage detection of airport pavement and accurate fixed-point detection in local areas, a high-robust real-time positioning method for airport pavement environment based on multi-source information fusion is established to realize wide-local long-term positioning in the airport environment. High and reliable positioning. Facing the highly robust trajectory tracking control requirements of omnidirectional robots with distributed drive and independent steering, a method for analyzing and identifying dynamic obstacles on the road surface, dynamic collision risk assessment among multiple robots, and collision avoidance decision-making method is established. Considering the multi-objectives such as control error, energy saving, and collision avoidance, as well as complex constraints such as the complex dynamics of omnidirectional robots and changing adhesion characteristics, a highly robust trajectory tracking method based on a model prediction framework is established to achieve accurate and dynamic trajectory tracking. An omnidirectional robot control distribution and driving wheel anti-skid and stable motion control method based on optimal adhesion control is established to realize the safe navigation of the robot in all-weather detection and operation.

3.1.3 *Airport patrol robot coverage detection path planning and multi-robot collaborative strategy*

For airport detection tasks, establish human-machine interfaces and mechanisms, and establish task allocation and status feedback mechanisms with the control center. According to the rasterization of the known environment of the airport, based on the comprehensive analysis of the airport processing area, detection tasks, multi-type robot characteristics, full coverage optimization objectives, time windows, instruction constraints, etc., the knowledge of detection tasks and robot road surface coverage capabilities is established library. According to the evaluation of robot load configuration and road surface coverage capability under highly dynamic and multi-task conditions, a multi-robot trajectory planning optimization method based on the dynamic multi-traveling salesman problem (DMTSP) is proposed. On the premise of satisfying task safety, optimality, rapidity and robustness, this paper analyzes the optimal redundant team design, flexible deployment and low-cost operation of three types of robots covering detection, fixed-point sampling and FOD recovery. Allocate specific areas to robots with redundant coverage and use local coverage detection algorithms, relying on the road surface environment map to plan robot paths, fixed-point sampling paths and FOD recovery paths to form a complete set of robot collaborative detection and recovery technology solutions for specific airport environments.

3.2 *Development of a complete set of technologies for rapid detection of airport pavement operation safety status*

3.2.1 *Multi-source data comprehensive collection technology*

Aiming at the one-sided problem of pavement index detection criteria caused by the "special instrument-specific" mode of one type of detection equipment corresponding to one detection index in traditional automatic detection, a high-dynamic, high-precision, large-width data center with high-precision three-dimensional sensors as the core is established. Comprehensive collection technology. The high-precision 3D sensor uses high-energy line structured light as the light source. By analyzing the comprehensive influence mechanism of the light source energy, the light source line width, the CCD acquisition frequency, and the CCD acquisition width on the acquisition effect, a high-precision measurement speed of 60 km/h can be developed. The three-dimensional sensor, combined with the fully automatic high-precision calibration technology developed at the same time, realizes the high-speed and accurate acquisition of the three-dimensional data of the pavement. Based on the integration of the above acquisition-transmission-storage technologies, the system-level thermal design and EMC design concepts are adopted to form a comprehensive multi-source data acquisition system for road surfaces with high integration, good data acquisition accuracy and stable working conditions.

3.2.2 *Multi-source heterogeneous data fusion technology*

A fusion registration technology based on multi-source heterogeneous data of various sensor data and various data types and dimensions collected by the multi-source data comprehensive acquisition system of the pavement is established. Aiming at the problem of different acquisition frequency and acquisition range among various sensor data, considering the problem of less saliency structure of pavement surface, the multi-source data registration method based on NSCT method based on the overall texture characteristics is analyzed, so as to complete the same time and space. Point multi-source data fusion problem. According to the "comprehensive" requirement in road surface detection, the multi-dimensional data splicing technology based on detail perception can complete the multi-temporal splicing of homologous data. The change law of the simultaneous space point data in the heterogeneous data relative to the adjacent domain is introduced, and finally a high-precision and standardized multi-source heterogeneous data fusion structure is obtained.

3.2.3 *Pavement safety index detection technology based on fusion data*

Based on the standardized fusion data, an object-oriented pavement index detection method is established. Aiming at the anti-skid performance of the pavement, according to the contact stress distribution characteristics of the aircraft tire and the pavement, the elastic-solid contact mechanics theory is used to construct the statistical and spatial distribution map of the contact stress. Based on the principle of energy conservation, a tire/road friction mechanics model is established, and a prediction method of anti-skid performance under different pavement conditions is proposed by comprehensively using the information on pavement texture and pavement water layer thickness. Aiming at the problem of pavement deformation diseases, a theoretical pavement topography restoration technology based on prior knowledge and texture information is established and based on this, high-precision detection of pavement deformation diseases is completed. Aiming at the targets with obvious color information such as pavement FOD and tire rubber, a highly robust identification and classification method based on the standardized database and deep learning method is established, and combined with high-precision 3D data, it is quantitatively detected. Combining the above-mentioned technologies, a set of object-oriented pavement index detection systems based on standardized multi-source data is constructed to realize a comprehensive, comprehensive and modular pavement safety index detection method. The high-precision detection of pavement cracks based on fusion data is shown in Figure 3.

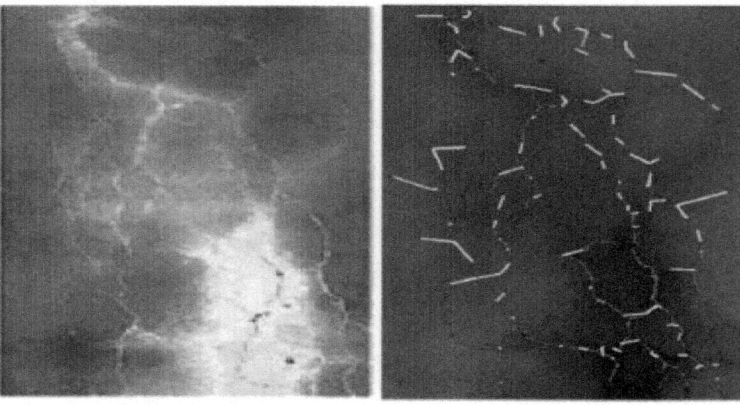

Figure 3. Number of transport airports in China.

3.3 *Integration and Demonstration of Robotic System for Airport Pavement Operation Safety Status Detection*

3.3.1 *Integration and development of pavement condition detection robot*

According to the requirements of coverage detection tasks, fixed-point sampling detection tasks and FOD recovery tasks, three types of robot systems are integrated and developed: coverage robots, fixed-point sampling robots and FOD recovery robots. According to the pavement detection requirements and conditions of civil airports, the integrated framework of the robot pavement detection system is designed, and subsystem modularization, standardized design and overall integration are carried out for units such as load, perception, drive and electrical, and the interconnection of each subsystem is realized based on the bus method. To ensure the reliability of the robot system under multi-climate conditions, various robot sub-systems are shown in Figure 4. To achieve full coverage

637

of airport pavement, multi-load, multi-task operation safety inspection, and multi-robot coordination is required to complete, for this purpose, an airport runway pavement inspection robot system platform is built. This platform consists of three modules: the robot control center, the robot team and the data calculation and management center. The above three modules can realize information exchange before. Among them, the robot control center realizes the information exchange between the road surface inspection robot system platform and the civil aviation airport control center, and realizes the analysis of inspection tasks and robot state load data; the robot team is composed of three types of robots to realize flexible team formation; the data calculation and management center is composed of The road surface communication base station and the road surface RTK system are composed to realize the communication and positioning functions.

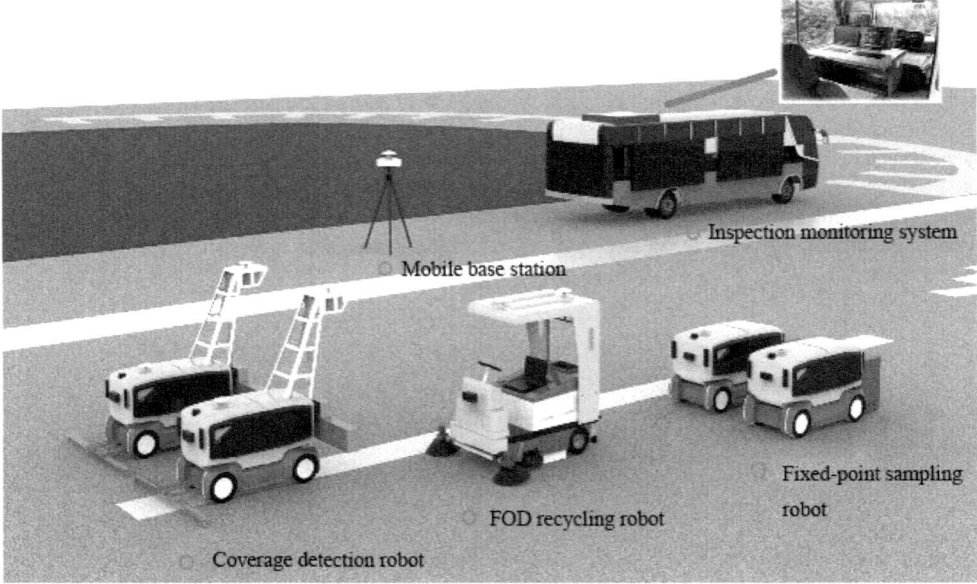

Figure 4. Robot system.

3.3.2 *Application demonstration of airport pavement detection robot system*

Aiming at the safety requirements of civil aviation airports for runway surface topography, strength characteristics, foreign object obstacle detection and recovery and other road surface operation safety requirements, and facing the needs of airport road surface safety operation coverage detection and fixed-point detection, the corresponding road surface status scenarios are constructed to verify the airport road surface. The coverage detection and fixed-point detection capabilities of the detection robot system for the surface state of the airport pavement are proposed. Based on the analysis results of the pavement state detection data, the construction method of the pavement state heat map fused with the airport pavement map is proposed, and the evaluation and verification are carried out. Facing the FOD recovery requirements of the airport pavement safety operation, a common FOD intrusion pavement scene on the airport pavement is constructed, and the target position and recovery path are given by the coverage detection robot to verify the task completion ability of the FOD recovery robot. Facing the complexity of the airport pavement detection task, three types of robot fusion application scenarios are constructed to verify the collaborative operation ability of the airport pavement detection robot system. Aiming at the uncertainty of the detection time window and the emergency of the airport, the rapid evacuation mechanism, breakpoint location and continuous inspection planning method of the robot system are designed, and the evaluation and verification are carried out. Based on the above achievements, combined with the actual operation status of China's civil airports, the application demonstration of the operation safety detection robot system under the complex environment, different airport scales, different road surface types and different service years is carried out. Based on the long-term detection data statistics of pavement conditions, the evolution law of pavement state is analyzed to verify the accuracy of early warning of pavement operation safety status.

Figure 5 shows the ISM values of the two survey lines of an airport runway using a detection robot. Impact stiffness modulus ISM value can directly reflect the comprehensive bearing capacity of pavement and foundation. It can be seen that the distribution trend of the ISM values of the two survey lines of the runway is basically the same, showing the distribution characteristics of "high at both ends and low in the middle", and the ISM value of the R-F-W2r survey line of the runway is slightly lower than that of the R-F-E1r survey line. It is because the main operating models of the airport are C-type and below models, and the main wheel track belts of the aircraft are mainly distributed in the first track panels on both sides of the runway centerline. The ISM value of the web is slightly higher than that of the second web. The intelligent road surface detection robot in the airport flight area can accurately detect the road surface condition.

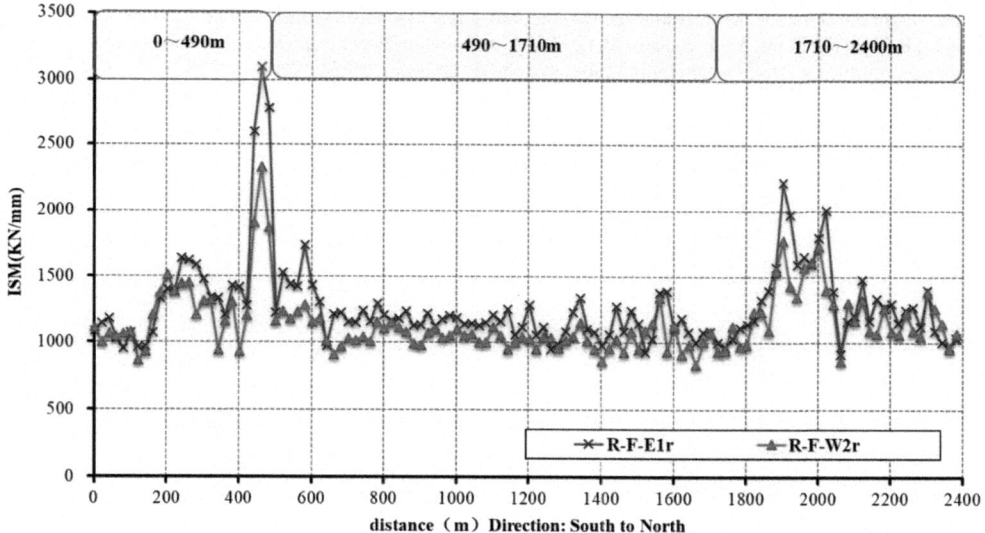

Figure 5.   ISM value distribution map of runway pavement deflection survey line.

4   CONCLUSION

Airport pavement inspection is the core work to ensure the safety of civil aviation. The traditional manual and semi-automatic detection methods are difficult to meet the requirements of ultra-high safety and precision, ultra-narrow operation time limit, and ultra-large operation range in terms of detection efficiency and accuracy. With the evolution of robotics and artificial intelligence technology, airport pavement detection technology shows a development trend from static detection to dynamic early warning, from manual to single robot to multi-mode robot systems. Robotics and artificial intelligence technology will be intelligent in airport pavement and play an important role in detection solutions. This paper focuses on the "accurate, efficient and intelligent" requirements of airport pavement safety inspection. Starting from the special operation scenarios and detection of airport pavement safety inspection, the overall scheme of the airport pavement safety inspection robot system is proposed, which verifies the airport pavement safety inspection. Application value and engineering benefits of robotic systems.

REFERENCES

Chauhan, T., Goyal, C., Kumari, D., & Thakur, A.K. (2020). A review on foreign object debris/damage (FOD) and its effects on the aviation industry. *Materials Today: Proceedings*, 33, 4336–4339.

Chen, M., Xu, F., Liang, X., & Liu, W. (2021). MSD-based NMPC aircraft anti-skid brake control method considering runway variation. *IEEE Access*, 9, 51793–51804.

Cramoisi, G. (2010). *Air Crash Investigations: The End of the Concorde Era, the Crash of Air France Flight*, 4590. Lulu. com.

Gendreau, M., & Soriano, P. (1998). Airport pavement management systems: An appraisal of existing methodologies. *Transportation Research Part A: Policy and Practice*, 32(3), 197–214.

Green, S.A., Billinghurst, M., Chen, X., & Chase, J.G. (2008). Human-robot collaboration: A literature review and augmented reality approach in design. *International Journal of Advanced Robotic Systems*, 5(1), 1.

Luyin, S. (2008). A study on the development patterns of world airport economy. *World Regional Studies*, 17(3), 19–25.

Mahmud, M.S.A., Abidin, M.S.Z., Buyamin, S., Emmanuel, A.A., & Hasan, H.S. (2021). Multi-objective route planning for underwater cleaning robot in water reservoir tank. *Journal of Intelligent & Robotic Systems*, 101(1), 1–16.

Yi, L., Zou, L., & Sato, M. (2018). Practical approach for high-resolution airport pavement inspection with the yakumo multistatic array ground-penetrating radar system. *Sensors*, 18(8), 2684.

Zou, L., Kikuta, K., Alani, A.M., & Sato, M. (2021). A study of wavelet entropy for airport pavement inspection using a multistatic ground-penetrating radar system. *Geophysics*, 86(3), WA69–WA78.

Frontiers in Civil and Hydraulic Engineering – Mohamed A. Ismail and
Hazem Samih Mohamed (Eds)
© 2023 The Authors, ISBN 978-1-032-38247-0

Analysis of the application of LID in cities of China

Qiqi Tan, Liangbing Hu, Zhenyu Yun*, Xiaolei Ma, Yao Zhang & Yutong Wang
China National Institute of Standardization, Beijing, China

ABSTRACT: With the rapid advancement of China's urbanization construction, urban built-up areas expands rapidly. In the process of urban development and expansion, the original natural ecological environment is also changed. The emergence of a large number of hard ground and the encroachment of river systems cause various water problems and contradictions are prominent. Facing all kinds of severe water problems in cities, the state has been carrying out urban water environment governance and rainwater management, control, and utilization in an orderly manner since 2013. This paper takes the practice of low-impact development in Beijing, Shanghai, and Shenzhen as examples to introduce the application of low-impact development in China.

1   GENERAL INSTRUCTIONS

The low-impact development model has been widely used in urban development and construction in the United States with innovative ideas, ecological technical measures, and good operation effects. It has been widely used for references in other countries and regions. Since 2007, more than 360 cities have suffered from waterlogging, with Beijing, Shanghai, Guangzhou, and Shenzhen all experiencing severe waterlogging disasters. In the face of various severe water problems in cities, China has been methodically carrying out urban water environment governance, rainwater management, control, and utilization since 2013. Released in October 2015, the State Council General Office on the Sponge City Construction Guidance put forward the general work target for urban construction through the Sponge Mix "permeability, hysteresis, storage, and net" with measures, such as minimizing urban development and construction of the ecological environment influence, and 70% of the rainfall on the given and use. By 2030, more than 80% of the urban built-up area will reach the target. Since 2015, the Ministry of Finance, the Ministry of Housing and Urban-Rural Development, and the Ministry of Water Resources have organized local governments to apply for pilot cities for sponge city construction. The pilot covers seven regions in China, including East China, North China, South China, Central China, Northeast China, Southwest China, and Northwest China.

As an important part of sponge city construction, the practical application of low-impact development mode in relevant cities in China has been carried out on a large scale. In the paper, the practice of low-impact development in Beijing, Shanghai, and Shenzhen is taken as an example to introduce the application of low-impact development in China.

2   APPLICATIONS OF LOW-IMPACT DEVELOPMENT IN BEIJING

2.1   *Rainwater collection and utilization system in Beijing Olympic Forest Park*

Beijing Olympic Forest Park is the first large-scale urban park in China to adopt a rainwater harvesting system, with a stormwater utilization rate of up to 95%. The park adopts diversified

---

*Corresponding Author: yunzy@cnis.ac.cn

DOI 10.1201/9781003344209-84

comprehensive ecological water conservancy planning and design, including rainwater collection, recycled water utilization, recycling filtration purification, wetland purification, and other engineering facilities. The external form of the park is the landscape pavement, leisure green space, sunken garden, dragon-shaped water system, forest park, etc. Park pavement involving granite pavement, brick pervious pavement, and aeolian sand pervious pavement has good water permeability, which can reduce the groundwater, improve the flood control ability of the area, and increase the supply of groundwater. On both sides of the park's central axis avenue, a permeable stormwater collecting ditch is set up, which can better collect rainwater. The green space in the park is 50–100mm lower than the surrounding road surface or square, and the rainwater on the road surface and square can be infiltrated or discharged through the green space. The park also has rainwater infiltration wells to remove the dirty water at the beginning of the rainfall process (Liu 2015).

2.2 Beijing sponge city pilot road project construction

The comprehensive utilization route of rainwater and flood resources in Beijing road engineering construction is shown in Figure 1 (Li Xiaoyan 2022). First of all, the pavement is paved with permeable materials. Most of the rainwater directly seeps down, and a small part of the rainwater flows into the green isolation zone. At the intersections and low points where water is easy to accumulate, a group of environment-friendly rainwater inlets is added near the sidewalk side of the nonmotorized driveway. At this point, most of the rainwater from the sidewalk directly seeps down, and a small part confluence into the environment-friendly rainwater inlets, and finally enters the downstream channel through the municipal pipe network. Second, the rainwater on the nonmotor road enters the green isolation zone through the open kerb stone. Because the rainwater on the pavement and nonmotor road surfaces contains fewer pollutants, it can rely on the purification effect of the green isolation zone without the need to build the initial rainwater purification facilities. Finally, the pollution control of the rainwater in the motorway should be carried out. Centralized interception is adopted to control the initial rainwater in the motorway, and the environment-friendly

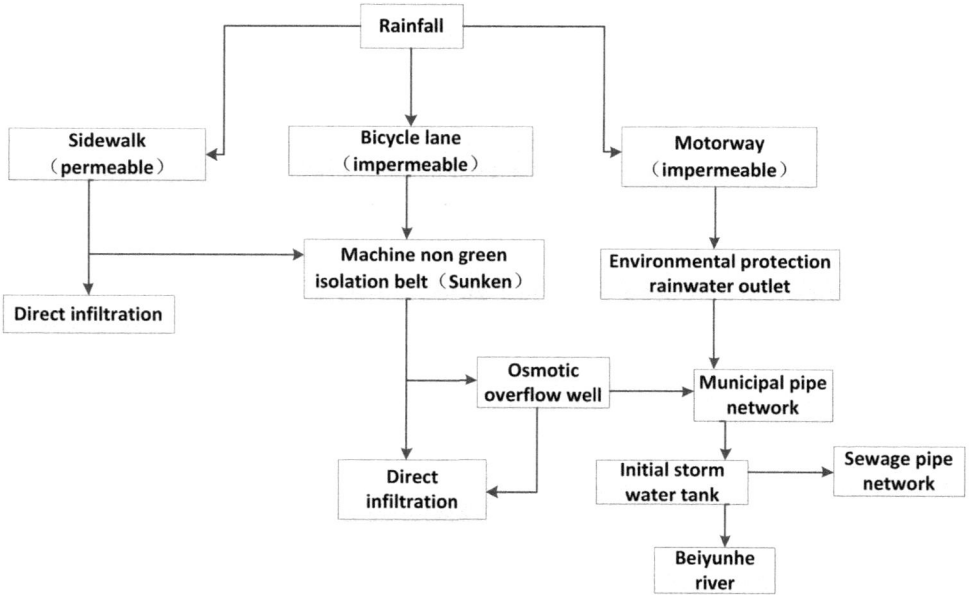

Figure 1.   Comprehensive utilization routes of rainwater and flood resources in Beijing road engineering construction.

rainwater inlet is adopted to purify the initial rainwater at the source. After the initial purification through the municipal pipe network into the initial stormwater pool, the rainwater went through the sewage network unified treatment later.

3 APPLICATIONS OF LOW-IMPACT DEVELOPMENT IN SHANGHAI

3.1 *Shanghai Gongkang Greenbelt low-impact development rainwater control and utilization demonstration project*

In August 2015, in the Gongkang forest belt on Changlin Road in Zhabei District, Shanghai completed the first demonstration project of low-impact development technology application in the city's public green space system, covering an area of about 1 hm^2. Gongkang Greenbelt was originally a high-pressure forest belt built in the 1990s, in which the plant planting density was too high, the canopy density was high, and the plant growth environment was poor. The height difference between the green space and the road is about 30 cm, which cannot meet the natural drainage demand of the green space. In the rainy season, the forest zone is chronically stagnant, which brings a lot of inconvenience to residents' leisure activities. To change the Gongkang Greenbelt into a public green space with rainwater storage functions, the underlying surface of the current green space was transformed, infiltration facilities were installed, bedding structures such as coarse gravel layer, non-woven fabric, and pebbles were laid, and excavation and backfilling were completed by manpower, so as to protect the original forest resources. In terms of plant configuration, ponds and rain gardens are dominated by a large number of water plants with strong moisture tolerance and resistance to pollution. The optimized green space is concave, about 25 cm below the ground level. The road is made of pervious pavement materials, and the ecological grass planting ditch on the side of the road will channel surface runoff into the pond or discharge into the municipal pipe network to reduce the pressure on the stormwater pipe network. In addition to storage and transport functions, ecological grass ditches and rain gardens also have the function of purifying water quality, and pollutants in the collected rainwater runoff can be absorbed by the roots of plants such as canna and yellow flag. After adjustment, the capacity of rainwater storage regulation, cooling and humidification effect, and landscape effect of green space were improved. The reference values of the runoff coefficient of the technical measures selected for the project are shown in Table 1 (Li Rong 2022).

Table 1. Value of runoff coefficient.

Type of underlying surface	Value of rainfall-runoff coefficient
Hard roof	0.85
Hard road	0.85
Permeable pavement	0.30
Ecological parking space	0.30
Transplanting grass ditch	1.00
Rain Garden (Ecological Retention Pond)	1.00
Sinking green space	1.00
Green space	0.15

3.2 *Rainwater collection system in Shanghai World Expo Park*

The rainwater collection system of Shanghai Expo Park is a relatively complete and comprehensive rainwater utilization system in China. Shanghai has abundant rainfall, with an annual rainfall of more than 1000 mm. The comprehensive utilization of rainwater in the Expo Park can reduce the discharge of rainwater, reduce the investment and operation cost of drainage facilities, and achieve the purpose of replenishing groundwater, controlling non-point source pollution, and improving urban ecology. Rainwater collection and utilization in the park are mainly reflected in three aspects: first, roof rainwater application. The core area of the Expo Park collects and utilizes rainwater from the roof. The design adopts a perfect rainwater utilization system to collect and treat the air conditioning condensate and roof rainwater for road washing and greening watering. The water-saving irrigation technology such as program-controlled green space sprinkler irrigation or drip irrigation is adopted to improve the utilization efficiency of water resources. Second, rainwater storage seepage application. In the Expo Park, the entire ecosystem has a drainage area of the application, permeable asphalt pavement in the pedestrian plaza, parking lot and other regions. At the same time, a large number of porous materials reduces waterproof performance of road surface pavement, effectively reducing the park's rainwater runoff and promoting rainwater infiltration. The existing drainage facilities can significantly improve the drainage area. Third, the initial rainwater storage pool application. In the rainwater drainage pump station of the Expo Park, there is an initial rainwater storage tank, and the initial rainwater storage tank of 3000 m^3 and 8000 m^3 has been built in Houtan and Puming rainwater pumping stations, respectively. During the rainstorm, the initial rainwater with high pollutants in the rainwater pipe network first enters the rainwater storage. After the rainstorm stops, the collected rainwater is slowly transported to the municipal sewage pipe network, and the rainwater in the middle and later stages is discharged into the water body of Huangpu River by the rainwater pumping station, which reduces the pollution effect of the initial rainwater on water quality (Ma 2013).

4 APPLICATIONS OF LOW-IMPACT DEVELOPMENT IN SHENZHEN

4.1 Guangming new district municipal road

For the light district municipal road demonstration projects, according to the low-impact development concept, rainwater measures were comprehensively adopted. The concave greenbelt and permeable pavement engineering measures were adopted, for example, the road red line within the scope of the rainwater collection priority access to the biological stranded on either side of the belt for percolation, processing, storage will be used to supplement the groundwater runoff, which plays a role in runoff pollution control, peak flow reduction, hydrological ecological restoration, and other aspects. A variety of new materials, such as warm mixing and recycled permeable building materials, are used to build a concave green belt in the center and the green belt on both sides of the road. After collecting the rainwater, it is filtered, retained, permeated, and purified by the concave green belt. The rainwater exceeding the designed capacity will enter the municipal rainwater pipe through the overflow mouth. The runoff coefficient of the demonstration road is controlled at 0.5, and the moderate and light rain does not produce confluence. When the rainfall occurs at 4–11 mm per hour, the total runoff and peak reduction rate can reach 95% and 84%, the flood peak delay is 12–34 minutes, the pollutant reduction rate is over 40%, and the drainage capacity of the stormwater pipeline is increased from the original design return period of once in two years to once in four years (Ding 2015).

4.2 Guangming new district park green space

Guangming New District Park green space demonstration project mainly adopts the construction of grass planting ditch, detention pond, underground module reservoir, and other engineering measures. For example, the new city park, covering an area of more than 50 hectares has grass-planting ditches and retention ponds for transportation and purification when stranded in the rain and flood peaks. At the same time, all PeiJian 900 m^3 of underground reservoirs collect rainwater. Rain pipes are no longer separated to collect and recycle water of 15000 m^3, covering 250000 m^3 of groundwater and reducing the municipal rainwater pipe drainage downstream pressure. In addition, the construction of Niushan Science and Technology Park in Guangming New District is also designed with the concept of low-impact development. The park almost does not use rainwater drainage pipes. The park is built along the road grass gully, road rainwater, and green rainwater runoff by the grass gully collection, infiltration, and discharge. Except for the driveways, the rest of the walkways in the park are paved with permeable bricks or grass grids to reduce the runoff of rainwater. To maximize the utilization of rainwater resources in the park, five underground reservoirs are constructed according to the park topography, including four 300 m^3 and one 250 m^3, which are used to collect rainwater in the rainy season and for greening irrigation in the dry season. The reference value of the pollutant removal rate provided by the technical measures selected for the project is shown in Table 2 (Ding 2012).

Table 2. Pollutant removal rate.

Type of facilities	Pollutant removal rate
Rain Garden (ecological retention pond)	70–95%
Permeable pavement	80–90%
Sinking green space	35–90%
Ecological parking space	80–90%
Transplanting grass ditch	35–90%
Rainwater collecting tank	80–90%

5 CONCLUSIONS

The construction of sponge cities in China is developed based on the theory of foreign low-impact development and the theory of green infrastructure and integrates the development ideas and technical measures. It aims at ecological protection and urban sustainable development, making people paying attention to the ecological benefits of site development. Currently, based on these two theories and related research results, it has been gradually applied to urban rainfall management in China.

REFERENCES

Ding Nian, Hu Aibing, Ren Xinxin (2012). Analysis of low-impact development municipal roads in Guangming New District, Shenzhen. *J. Shanghai Urban Planning*. (06).
Ding Shufang, Ren Xinxin, Yang Chen (2015). Research on incentive policy of comprehensive utilization of rainwater in low impact development in guangming new district. *J. China Water and Drainage*. 31(17).
Li Rong (2022). Research on the application of sponge city technology in landscape design: A case study of shanghai lingang hongxing tianbo residential project. *J. Home*. (20).
Li Xiaoyan (2022). Comprehensive utilization of rainwater and flood resources in sponge city construction: A case study of Beijing Sponge city pilot road project. *J. Soil and Water Conservation in China*. (07).
Liu Chaoyang, Wang Shudong(2015). Analysis on the application of low impact development mode in landscape design in Beijing. *J. Journal of Beijing Agricultural College*. 30(03).
Ma Yating, Yang Kai (2013). Implications of international low-impact development practices for urban rainwater management in Shanghai. *J. World Geographic Research*. 22(04).

Frontiers in Civil and Hydraulic Engineering – Mohamed A. Ismail and
Hazem Samih Mohamed (Eds)
© 2023 The Authors, ISBN 978-1-032-38247-0

Construction demonstration of Guangdong-Macau in-depth cooperation zone in Hengqin—research on the application of pre-fabrication technology in the Macau new neighborhood

HoiIan Tam
Faculty of Humanities and Arts, Macau University of Science and Technology, Avenida Wai Long, Taipa, Macau, China

Ionkei Chan
Faculty of Innovation and Design, City University of Macau, Avenida Padre Tomás Pereira Taipa, Macau, China

Yile Chen* & Liang Zheng
Faculty of Humanities and Arts, Macau University of Science and Technology, Avenida Wai Long, Taipa, Macau, China

ABSTRACT: The prefabricated building consists of factory-made components which are assembled on-site. It is characterized by standardized design, factory production, prefabricated construction, integrated decoration, and information management. It integrates various business fields from R&D design, production and manufacturing, and on-site assembly, to achieve a new sustainable building construction method that promotes energy saving, environmental protection, and maximization of full-cycle value of building products. This research takes Macau New Neighborhood, a cutting-edge cooperation project in the current construction of the Guangdong-Hong Kong-Macau Greater Bay Area, as an example, and focuses on its experience in the improvement and design of prefabricated parts and the optimization of transportation and logistics in its prefabricated buildings.

1   INSTRUCTIONS

1.1   *Construction background of double carbon in the construction industry*

The Outline Development Plan for the Guangdong-Hong Kong-Macau Greater Bay Area proposes to support Zhuhai and Macau in cooperating in Hengqin to build a comprehensive livelihood project integrating community nursing, residence, education, medical care, and other functions, and to explore the direct application and extended coverage of Macau's medical system and social insurance (Yu 2021).In September 2020, President Xi Jinping announced that "China will strive to peak carbon dioxide emissions by 2030 and strive to achieve carbon neutrality by 2060" (Caineng 2021). According to the 2020 China Building Energy Consumption Research Report, the total carbon emissions in the entire building process in 2018 was 4.93 billion tons, accounting for 51.3% of the national carbon emissions (Wang 2021). Under the goal of "Dual Carbon", the construction industry urgently needs to realize green transformation. The implementation of dual carbon in the construction industry generally requires reforms in technology, production mode, and operation mode in three stages including building materials production, building construction, and facility

---

*Corresponding Author: chenyile1996@163.com

DOI 10.1201/9781003344209-85

operation. In order to promote carbon emission reduction in the construction industry, all countries are actively exploring ideas and directions for carbon emission reduction.

Therefore, in April 2020, the Zhuhai People's Government signed a contract with Macau Urban Renewal Limited, a public institution wholly owned by the Macau Special Administrative Region government. The land for the "Macau New Neighborhood" project with a total area of about 190,000 square meters was sold (MURL 2020).

1.2 *Macau new neighborhood*

In terms of design, the Hengqin "Macau New Neighborhood" project will make full use of the natural advantages of the backing of the mountain and the river, and the building spacing is designed to create a housing environment with convection ventilation and natural lighting. There are 27 residential buildings in the project, with a height of 20 to 26 floors, which will be constructed by an eco-friendly construction method of prefabricated components, which can provide about 4,000 finished residential units. Among them, two-bedroom units account for about 80%, each with a construction area of about 90 square meters; three-bedroom units account for about 20%, each with a construction area of about 100 square meters to 120 square meters.

In terms of supporting facilities, the project has commercial shops, schools, health stations, elderly service centers, and family community service centers. The residential buildings will be completed simultaneously with the supporting facilities to meet the living needs of the residents. The entire construction project is subject to strict quality control by a third party, and the construction period is 3 years. Macau Urban Renewal Limited is currently carrying out the construction of the project and will kick off the purchase application process in the next stage. Since Macau New Neighborhood is an important demonstration project of cooperation between Guangdong and Macau, its progress has also attracted the attention of the public (Figure 1).

Figure 1.   Project renderings. (Image source: Provided by Macau Urban Renewal Limited).

1.3 *Types of prefabs used*

The Macau New Neighborhood Project consists of two plots with a total of 27 residential buildings. The tower consists of two layout planes (T7 and T8). The distribution of building serial numbers

647

is shown in Figures 2 to 3. The types of prefabricated components include ① Prefabricated bay windows; ② Prefabricated stairs; ③ Autoclaved Lightweight Concrete (ALC) Wall panels.

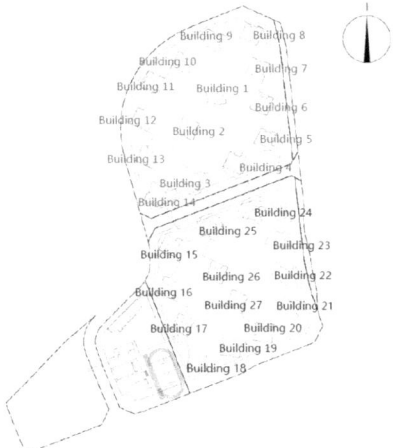

Figure 2. Building serial number.

Figure 3. Project renderings. (Image source: Provided by Macau Urban Renewal Limited).

2 PREFABRICATED SCHEME DESIGN

2.1 *Prefabricated bay window*

The design of the prefabricated bay window incorporates the layout habits of the early and present dwellings in Macau (Figure 4). The horizontal seam of prefabricated components adopts a combination of three waterproof design methods: the outermost part adopts building waterproof sealant, the middle part is a decompression space formed by a flat-mouthed physical cavity, and the inner part is a compressed rubber strip. The inner side is cast-in-place concrete, which has an excellent waterproof effect. In the vertical waterproof design, the vertical joints between prefabricated components and cast-in-place concrete are mainly self-waterproof through reinforced concrete. Set up 100mm wide grooves and make rough surfaces at the connection, and set up distribution ribs to achieve a waterproof effect after pouring (Xu 2019). The design of the prefabricated bay window incorporates the layout habits of the early and present dwellings in Macau (Figure 4).

Figure 4. Prefabricated bay window. (Image source: Provided by Macau Urban Renewal Limited).

The waterproofing problem of the connection seam will be considered in the prefabricated components, it is divided into horizontal seams and vertical seams, the design methods are as below:

Horizontal seam:
In the Horizontal waterproof design, the horizontal joint would adopt the silicone sealant at the outdoor part and the inner side part is a cast-in-place concrete structure, which has an excellent waterproof effect.

Vertical seam:
In the vertical waterproof design, the vertical joints between prefabricated components and cast-in-place concrete are mainly self-waterproof through reinforced concrete. The prefabricated bay window is Set up with 100mm wide grooves on both sides. In addition, the seam which will be connected with pouring concrete would be made to roughen the surface. This increases the contact area of concrete aggregate between the prefabricated bay window and the pouring concrete, achieving reduced connection cracks and improving the waterproof affection.

The installation steps for prefabricated bay windows are as follows:

- The concrete pouring and curing of the lower structure is completed;
- Place height-adjusting bolts to adjust the height to be flushed with the structural elevation;
- Install on the lower prefabricated bay window;
- Lift the component in place and remove the spreader;
- Install tilt braces and adjust the position of the components;
- Tying the steel rebar of the wall and column on this floor;
- Installing the propping and aluminum formwork for wall, column, and next layer beam and slab.;
- Tying the steel rebar of beam and slab for the next layer, and pour the concrete of this layer;
- After the construction is completed, apply a weather-resistant sealant to the seam position.

2.2 *Prefabricated stairs*

Prefabricated stairs are one of the commonly used components in prefabricated buildings. The prefabricated components of prefabricated stairs are mainly concrete prefabricated treads, platform slabs, and supporting structures. It has the advantages of high efficiency, fast construction speed, and high-quality controllability. At the same time, the prefabricated stairs (evacuation stairs in the core tube) of the Macau New Neighborhood Project are equipped with anti-skid strips, which do not require on-site wet plastering (Figure 5).

Figure 5. Prefabricated stairs. (Image source: Provided by Macau Urban Renewal Limited).

2.3 Prefabricated inner partition wall (Autoclaved lightweight concrete wall panel)

According to the characteristics of the project and the adaptability of material properties, autoclaved lightweight aerated concrete wall panels ALC are used. It is a porous concrete product made of cement, silica sand, various additives, etc. as the main raw materials, reinforced by anti-rust treated steel bars and cured by high temperature, high pressure, and steam. It has excellent sound insulation and sound absorption performance, and has good thermal insulation performance; the proportion of lightweight is 0.5, which is 1/4 of that of ordinary concrete, which greatly reduces the self-weight of the wall and reduces the basic cost of the building.

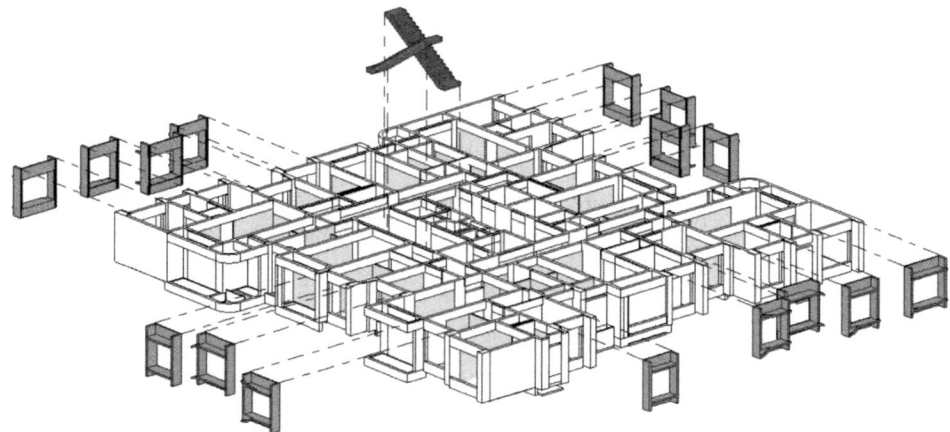

Figure 6. Schematic diagram of the location of prefabricated components. (Image source: Provided by Macau Urban Renewal Limited).

The above three components have been used maturely in the neighboring cities of Macau and have certain experience in quality control and craftsmanship. In addition, considering the in-depth cooperation between Guangdong and Macau, it is the first time that Macau Urban Renewal Limited has used prefabricated concrete components and the positioning of this project, The aim is to try out prefabricated component technology and to promote it to the construction industry in Macau. (Figure 6).

3 LOGISTICS AND TRANSPORTATION PLANNING

In order to reduce the greenhouse gas generated during the transportation of components and ensure the quality of components, priority will be given to prefabricated parts manufacturers in the vicinity or in the same city (provided that the quality fulfills the project requirements) to reduce transportation distance. Taking Macau New Neighborhood as an example, the primary consideration for transport is the location of the factory where the prefabricated components will be produced (Figure 7). In addition, the type of transport vehicle and its maximum load, the logistics route, and the location of the storage for access to the site.

At present, the manufacturer factories of prefabricated components can be reached in about 1 hour by vehicle at the construction site. During transportation, the prefabricated components are securely fixed with appropriate equipment. Finally, before entering the site, plan the storage location of the prefabricated parts to avoid affecting the logistics and construction of the site, and the storage location can be used for subsequent lifting and installation operations. In addition, the logistics routes after entering the site should be pre-planned to ensure that the relevant transportation

Figure 7. Transportation plan. (Image source: Callouts drawn by the author).

floors have sufficient bearing capacity to allow the relevant wheels to pass. Now, the prefabricated parts have been installed in Building No. 25 of Macau's New Neighborhood (the building on the left in the picture below). And Building 24 (the building on the right in the picture below) is entering the construction of the floor slab (Figure 8).

Figure 8. The precast installation has been carried out and concrete pouring is in progress. (Image source: Provided by Macau Urban Renewal Limited).

4 CONCLUSION

Europe, the United States, Japan, Singapore, and Hong Kong have successively implemented prefabricated buildings because of housing shortages and are eager to solve them. At the same time, the labor force is seriously insufficient, and its labor cost has accounted for 50% of the construction cost. Indeed, our country encounters the same problem. According to statistics, prefabricated projects can reduce the number of on-site construction workers by 30%. Under the circumstance that the demographic dividend has gradually dried up, a more cost-effective, more efficient, and environmentally friendly prefabricated construction method will become the future development direction. Therefore, the national implementation of prefabricated buildings is forward-looking. At present, the prefabricated concrete elements (Prefabricated bay window) and prefabricated staircases of Macau New Neighborhood have been produced and installed. After the improvement of construction components and transportation planning, Guangdong and Macau will definitely join hands to build more prefabricated building projects in the future, making contributions to the construction of the Guangdong-Hong Kong-Macau Greater Bay Area.

REFERENCES

Building the new Macau Neighborhood Project. Progress Report Macau Urban Renewal Limited (2020) and https://www.mur.com.mo/macau_new_neighborhood.

Caineng, Z. O. U., Dongbo, H. E., Chengye, J. I. A., Bo, X., Qun, Z., & Songqi, P. (2021). Connotation and Pathway of World Energy Transition and its Significance for Carbon Neutral. *Acta Petrolei Sinica*, 42(2), 233.

Wang, J., Yu, C. W., & Cao, S. J. (2021). Technology Pathway of Efficient and Climate-friendly Cooling in Buildings: *Towards carbon neutrality. Indoor and Built Environment*, 30(9), 1307–1311.

Xu, Q., & Sun, L. (2019). *Application Analysis of Prefabricated Buildings Under Green Construction*. In ICCREM 2019: Innovative Construction Project Management and Construction Industrialization (pp. 527–534). Reston, VA: American Society of Civil Engineers.

Yu, H. (2021). The Guangdong-Hong Kong-Macau Greater Bay Area in the Making: Development Plan and Challenges. *Cambridge Review of International Affairs*, 34(4), 481–509.

Frontiers in Civil and Hydraulic Engineering – Mohamed A. Ismail and
Hazem Samih Mohamed (Eds)
© *2023 The Editors and Contributors, ISBN 978-1-032-38247-0*

Author index